Tidal Wetlands Primer

Tidal Wetlands Primer

An Introduction to Their Ecology, Natural History, Status, and Conservation

RALPH W. TINER

University of Massachusetts Press
AMHERST AND BOSTON

Printed in the United States of America

ISBN 978-1-62534-022-1

Designed by Dennis Anderson
Set in Sabon
Printed and bound by Sheridan Books, Inc.

Library of Congress Cataloging-in-Publication Data

Tiner, Ralph W.
Tidal wetlands primer : an introduction to their ecology, natural history, status, and conservation / Ralph W. Tiner.
p. cm.
Includes bibliographical references and index.
ISBN 978-1-62534-022-1 (pbk. : alk. paper) 1. Wetlands--North America.
2. Wetland ecology--North America. 3. Wetland conservation--North America. I. Title.
QH87.3T57 2013
333.91'80973--dc23

2013018306

British Library Cataloguing-in-Publication Data
A catalogue record for this book is available from the British Library.

publication of this book is supported by a grant from

The Haven Fund

Dedicated to those who provided me with opportunities to explore and to write about wetlands over the course of four decades and to others who have broadened our knowledge of tidal wetlands or helped conserve and restore them

Contents

Foreword

Why do we care about tidal wetlands? What is the future of salt marshes? How many kinds of tidal wetlands are there? Ralph Tiner introduces us to their ecology and leads us through a fascinating history, including a time when most people thought of wetlands as nuisances and wastelands. Salt marshes blocked access to the shore. So what to do? Fill them in, dredge them out, or drain them. Boston's Back Bay was built on filled salt marshes. There are salt marshes along the Long Island and Jersey shores that have been dredged and filled and covered with houses. Marshes on Cape Cod were dredged to make marinas. At times, our attack on marshes was less direct: they were cut off from full tidal exchange by roads and railroads.

Despite society's willingness to alter salt marshes, some people valued them. Even now a few salt marshes are used to harvest salt hay as the first European colonists on the Atlantic Coast did. And a few people began to recognize that the seaside marshes provided foraging and nesting areas for birds, and nursery grounds and feeding areas for marine fishes. Some marsh mudflats were good clamming areas. Some people living on the landward edges of marshes valued them just for the view.

Now salt marshes and tidal wetlands are protected by federal, state, and local laws. Many people recognize the values of marshes and swamps. Children are taught these values both in schools and in community events. Artists draw and paint marshes for others to admire. People visit marshes, walk along their edges, listen to the birds living in them, and hear the rustling of the marsh grasses. What happened to change public attitudes about wetlands? Ralph Tiner provides an excellent review and history of tidal wetlands and people, describing eloquently how, in the past half century, society began to look at salt marshes in a new light based on the research done on their value both to nature and to society.

Tiner's book is a wonderfully complete history and explanation of tidal wetlands, their biology, geology, history, and relationship to people.

John M. Teal
Scientist Emeritus,
Woods Hole Oceanographic Institution

Preface

Today tidal wetlands are widely recognized as among the world's most valuable natural resources. This has not always been the case as many such wetlands have been filled for development of various kinds and degraded by pollution, hydrologic modification, and other human actions. My introduction to these habitats came during my graduate studies at the University of Connecticut in 1970 when I was hired as part of a survey crew that would conduct Connecticut's first on-the-ground inventory of tidal wetlands. The state had just passed a law protecting tidal wetlands from development and placed a moratorium on any construction work in these wetlands until the tidal wetland mapping effort could be completed. This inventory involved walking around the state's approximately 20,000 acres (8,100 ha) of tidal marshes, marking the boundaries with either wooden stakes or blazes on trees, recording those locations on large-scale maps, and taking notes on the vegetation and wildlife observations. Public hearings were then held in each town/municipality, adjustments were made by the hearing examiner based on other considerations, the boundaries were made official, and regulations took effect. From Connecticut my career led me to South Carolina, where I oversaw the mapping of more than 400,000 acres (162,000 ha) of tidal wetlands in addition to providing environmental reviews of projects proposing to fill, impound, or otherwise alter tidal wetlands. In the mid-1970s, I returned to the Northeast to direct a regional wetland mapping effort as part of the federal government's National Wetlands Inventory.

Over the past forty years, I've visited thousands of wetlands (mostly nontidal types) and have never lost my love of salt marshes. These fertile, verdant plains exude life, teeming with wildlife from herons, egrets, and sharp-tailed sparrows to mummichog, menhaden, and young bluefish in tidal creeks to fiddler crabs, ribbed mussels, mud snails, and marsh snails, plus the hordes of deerflies and salt marsh mosquitoes that make the marshes unbearable to most would-be visitors. The smell of brine and even the faint essence of hydrogen sulfide offer a sense of aquatic productivity and testimony to their linkage with the sea.

The purpose of this book is to introduce the world of tidal wetlands to nonspecialists, environmental professionals from other fields of study working in these wetlands, students in the environmental and natural sciences, and others with an interest in learning about these interesting and valuable resources. It is an introduction and a fairly comprehensive overview; it is not a synthesis of all the scientific literature to date, although there are plenty of references cited for interested scientists. The book covers many topics well enough to give readers a better understanding of the development, natural history, functions and values, current status, trends, conservation, restoration, and probable future of tidal wetlands. By design, it does not delve heavily into the biophysiochemical interactions and other more technical processes associated with these habitats or into ecological theories like a standard college textbook might because there are other books available that do this. Readers looking for more detailed information on particular topics can consult the numerous technical references listed for these details. Instead, this book will give readers a much better sense of what these wetlands are, how they differ, why they are valuable resources, the challenges they face, and what can be done to help improve their future.

The manuscript originally began as an expansion of the introductory material on coastal wetland ecology for an update of my tidal wetland plant field guide, *A Field Guide to Coastal Wetland Plants of the Northeastern United States* (University of Massachusetts Press, 1987). This section took a life unto itself as more and more material was added and eventually reached the point that it was worthy of publication in its own right. I met with Bruce Wilcox, Director of the University of Massachusetts Press, and he agreed that the sheer volume of the material warranted a separate publication as it would be too much to add to the revised field guide. At that point, I began to do a more exhaustive review of the available literature and added even more material to the book than I had originally intended. Today it serves as a companion book to the updated field guide as well as to my field guide for the tidal wetlands of the Southeast.

In preparing this primer, I tried to walk the line between a highly technical document and a natural history book for the nonspecialist seeking more information about these habitats, with a leaning toward the latter. In spite of my best intentions, it was difficult to avoid more technical discussions of some topics, especially when addressing specifics, so my apologies if some material is too detailed for the nonscientist. The book can therefore be considered a hybrid between a natural history book and a science textbook for advanced students. Also, although I covered topics that I felt were important to understanding and appreciating wetlands, some scientists have differing opinions on topics also worthy of coverage and emphasis. The book includes a wealth of references that can be reviewed for more detail on all aspects of tidal marsh science for individuals seeking more knowledge in specific subject areas. Believe me, anyone researching a book like this learns much through the process, and it is a humbling experience to see how much information is out there and how relatively little one individual actually knows. And, after working on this manuscript for too many years, I know that I've skimmed only the surface as more information continues to be published on a variety of topics involving tidal wetlands. When I started working in tidal wetlands in 1970, there were a mere handful of references that could be found on this subject, whereas today there are thousands of journal articles available. Clearly, our knowledge of tidal wetlands has taken a quantum leap over the past forty years as wetlands have become a research topic for many scientific disciplines.

The emphasis of the text is on vegetated wetlands—the tidal marshes and swamps—with some discussion of eel-grass meadows, rocky shores, beaches, and tidal flats. While there is an intentional Northeast focus in terms of describing tidal wetland vegetation and wildlife, much of the material is of a general nature and relevant to tidal marshes and swamps elsewhere in North America. To broaden the geographic scope of this primer, I have made numerous references to wetlands from other regions of North America, especially in discussing plant communities. Although mangroves are mentioned, they are not discussed in any detail since they have limited distribution in North America and several books have been written about them, such as

The Biology of Mangroves and Seagrasses (Hogarth 2007), *Mangrove Ecosystems* (Lacerda 2002), *Mangrove Ecology, Silviculture and Conservation* (Saenger 2002), and *The Botany of Mangroves* (Tomlinson 1986). Likewise tidal freshwater wetlands have received less treatment in this book because *Tidal Freshwater Wetlands* (Barendregt et al. 2009) provides a comprehensive examination of these types.

This book should be useful as an introductory textbook on tidal wetland ecology for college students as well as to general biologists, park naturalists, environmental professionals, natural resource managers and planners, and people wanting to know more about these wetlands than can be gained from a general natural history book. For use in the classroom, it can be supplemented with specific technical references to emphasize key points and concepts and to present local examples. It can also be used in a general ecology course to provide an overview of tidal wetland ecology. For nonstudents, the material presented provides a basic understanding of these wetlands, their environment, their history, and the challenges they face. Armed with this knowledge, it is my hope that readers will, like me, gain a lasting appreciation for tidal wetlands; support efforts to protect, conserve, and restore these wetlands; and pass this enthusiasm on to their families, friends, and coworkers.

Acknowledgments

Several people stimulated my interest in tidal wetlands and gave me opportunities to develop my skills as a wetland ecologist over the years. For their support, I'd like to thank John Rankin and Michael Lefor (University of Connecticut), Michael McKenzie and Rob Dunlap (South Carolina Marine Resources Division), Curt Laffin and Dave Riley (U.S. Fish and Wildlife Service), and Peter Veneman (University of Massachusetts).

This book would not have been possible without the efforts of many researchers who have spent years investigating tidal wetlands. Several of them have helped me produce a primer that will serve as a foundation for would-be scientists and others interested in learning more about tidal wetlands. I'd like to acknowledge and thank the many reviewers of the first draft of this manuscript and others who helped critique near-final chapters in preparation for publication. Reviewers of the draft manuscript include John Teal (Woods Hole Biological Laboratory), Andrew Baldwin (University of Maryland), Candy Bartoldus (George Mason University), Gail Chmura (McGill University), James Perry (Virginia Institute of Marine Science), Richard Stalter (St. John's University), Charles Roman (U.S. Geological Survey), Ron Rozsa (Connecticut Department of Environmental Protection), Sean Basquill (Atlantic Canada Conservation Data Centre), Alan Hanson (Canadian Wildlife Service), Martin Jean (St. Lawrence Center, Montreal), and Lee Swanson (New Brunswick Department of Natural Resources & Energy). Many of these scientists also provided material or referrals to other papers that helped improve and expand the scope of this manuscript, while some also reviewed later chapters. Reviewers of specific chapters related to their expertise or interests were Mallory Gilbert, William Sipple, Joshua Collins (San Francisco Estuary Institute), Jon Kusler (Association of State Wetland Managers), Jan Smith (Massachusetts Coastal Zone Management), Don Leopold (Environmental School of Forestry, State University of New York), Christopher Craft (Indiana University), David Burdick (University of New Hampshire), Matthew Hatvany (Université Laval), Paul Adamus (Oregon State University), Donald Cahoon (U.S. Geological Survey), Robert Buchsbaum (Massachusetts Audubon Society), Amy Deller Jacobs and Mark Biddle (Delaware Department of Natural Resources and Environmental Control), Laura Brophy (Institute of Applied Ecology, Oregon), Janet Morlan (Oregon Department of State Lands), John Hefner (Atkins North America), Kevin Hess (Pennsylvania Office of Coastal Zone Management), and three students from the Virginia Institute of Marine Science—Sean Charles, Wes Hudson, and Lori Sutter. Their reviews helped improve the depth of coverage of many topics. I am especially grateful to John Teal for providing the foreword to this book.

Numerous individuals provided information that was helpful in preparing various chapters: Michael Bartlett, Dan Belknap, Luc Brouillet, David Burke, Laura Cammon, Wendy Carey, Bruce Carlisle, David Chamberlain, Michele Dionne, John Dorney, Karen Duhring, Sherif Fahmy, John Gallegos, Stephen Gill, Glenn Guntenspergen, Jon Hall, Nate Herold, John Holman, John Jeglum, Michael Kearney, Matthew Kirwan, Eric Lamont, Curtis Larson, Claude Lavoie, Jan Mackinnon, Ken Metzler, Julie Michaelson, Randy Milton, Kim de Mutsert, Matthew Perry, John Potente, C. Edward Proffitt, Maryellen Sault, Dale Schweitzer, Richard Shaw, Joe Smith, Steve Sollod, Jerry Tande, Mark Tedesco, Bill Teschek, Irwin Ungar, Vinton Valentine, Kelly Vertucci, Ken Webb, and Matt Whitbeck. Others assisting in researching materials were Marc Pollett, Terri Winchcombe, and Leo Cheverie.

Aerial imagery comes from government sources except for copyrighted satellite images from MDA Information Systems, Inc. (Gaithersburg, MD). Thanks to Greg Koeln and Mike O'Brien for providing these images that show the broad-scale view of tidal wetlands in several estuarine ecosystems. Also I am grateful to Lindsey Lefebvre who prepared the interactive maps of tidal wetlands for the conterminous United States that are accessible through ScholarWorks@UMassAmherst and to Bob Lichvar for his support in this effort. While most of the wetland type photos are mine, wildlife and other photos were graciously provided by a host of photographers or were available from government websites. Photographers are credited in the respective caption. Illustrations were obtained from a variety of sources including government publications and journal articles. Special thanks to Richard Heard for allowing use of drawings from his book on estuarine invertebrates and to various organizations permitting use of copyrighted material. Josh Collins and Ruth Askevold (San Francisco Estuary Institute) provided figures showing tidal wetland losses in the Bay area. My son Dillon helped redraw some figures for this book.

I owe my wife, Barbara, an enormous thank you for her tolerance of the various piles of references, handwritten notes, and other paraphernalia that were scattered about our home during my decade-long effort to get this manuscript prepared and also for the weeks I spent in my own world as I researched, wrote, proofread, and prepared the book for publication. In the final stages, she helped by typing some draft tables, putting tables in print-ready format, and helping me check the citations in the text to the corresponding references.

Finally, I wish to acknowledge the support of the University of Massachusetts Press in producing this book and to offer my sincere thanks to Bruce Wilcox and Jack Harrison for their help and encouragement throughout this process.

Purpose and Organization of the Book

The significance of tidal wetlands as coastal landscapes, their vital contribution to estuarine productivity, and their utilization by fish and wildlife of recreational and commercial importance were largely responsible for generating interest in their conservation. After witnessing a decade or more of accelerating wetland destruction and degradation after World War II, many state governments passed special laws during the 1960s through the 1980s to protect these wetlands in the United States. These laws have worked well to save the remaining tidal wetlands from destruction by human actions. The federal government also began to put more emphasis on minimizing the environmental impact of proposed construction in these wetlands and associated waters. More recently, attention has focused on restoring tidal flow to wetlands with restricted connections to the sea and to former estuarine wetlands separated from the estuary. With worldwide recognition of climate change and global warming, significant concern now exists over the future of tidal wetlands given predicted sea level rise, armoring of shorelines, and continued development in the coastal zone.

Several books have been written about coastal wetlands including the best seller *Life and Death of the Salt Marsh* by John Teal and Mildred Teal (1969) and technical references such as *Salt Marshes and Salt Deserts of the World* (Chapman 1960), *Wet Coastal Ecosystems* (Chapman 1977), *The Ecology of a Salt Marsh* (Pomeroy and Wiegert 1981), *Conservation of Tidal Marshes* (Daiber 1986), *Coastal Marshes: Ecology and Wildlife Management* (Chabreck 1988), *Saltmarsh Ecology* (Adam 1990), *Ecology of Tidal Freshwater Forested Wetlands of the Southeastern United States* (Conner et al. 2007), *Coastal Wetlands: An Integrated Ecosystem Approach* (Perillo et al. 2009), *Tidal Freshwater Wetlands* (Barendregt et al. 2009), and *Tidal Marsh Restoration: A Synthesis of Science and Management* (Roman and Burdick 2012). Scientific community profiles on specific types of tidal wetlands have been published by the U.S. Fish and Wildlife Service (e.g., Nixon 1982; Whitlatch 1982; Seliskar and Gallagher 1983; Odum et al. 1984; Teal 1986; and Wiegert and Freeman 1990). The 1969 Teal book did a superb job at educating average citizens and students of ecology on the value and plight of salt marshes. Since then, there has been great improvement in the status of and advances in our knowledge of salt marshes and other wetlands. A book covering the full range of tidal wetland types, summarizing much of what we've learned, and describing current protection and management would complement both the Teal book and the more technical publications.

Previously I have written two comprehensive field guides for identifying plants representative of tidal wetlands: one for the northeastern United States (Tiner 1987) and another for the southeastern United States (Tiner 1993a). Each book contained a brief introduction to coastal wetland ecology. In expanding the former field guide to include plants of northern tidal wetlands in neighboring Canada and plants of tidal freshwater wetlands for a 2009 revision, I saw the need to include more background on coastal wetlands. I had recently written a very successful popular book on wetlands of the Northeast—*In Search of Swampland: A Wetland Sourcebook and Field Guide*—that included a wetland primer as the first part of the book (Tiner 2005). In updating *A Field Guide to Coastal Wetland Plants of the Northeastern United States,* I thought readers would benefit from a primer on tidal wetland ecology. As I worked on expanding this section, however, I found that the new material was worthy of a separate volume, especially since there was not a single reference that covered the ecology of tidal wetlands of the northeastern United States and adjacent Canada, or the natural history of tidal wetlands and recent efforts to restore these valuable habitats. This two-volume approach would also allow the field guide to focus on plant identification and not be encumbered by the bulk of material included in a primer.

This book is a primer to tidal wetlands using examples mostly from the eastern United States and Canada but also referencing similar wetlands from the Pacific Coast of North America. It provides a fairly broad but not exhaustive treatment of coastal wetlands, their formation, ecology, distribution, status, trends, protection, and restoration. The emphasis is on vegetated tidal wetlands, mainly marshes and swamps. Some topics are discussed only briefly (e.g., sea grasses, mangroves, beaches, tidal flats, rocky shores, tidal animals, and physiological adaptations of plants to the rigors of the tidal environment); nonetheless, readers should come away with a more than basic understanding of tidal wetlands after reading the primer, a strong foundation of knowledge to build upon, plus volumes of references for investigating topics in more detail.

This book contains 12 chapters. Chapter 1 is an introduction to tidal wetlands, estuaries, and their classification. Chapter 2 discusses the origin and formation of tidal wetlands. Chapter 3 describes the intertidal environment while Chapter 4 addresses plant response to these conditions. The variety of plant communities growing along North American coasts is summarized in Chapter 5, with emphasis on those of the North and Mid-Atlantic coasts. Chapter 6 focuses on the use of tidal wetlands by wildlife including rare species, while a general discussion of tidal wetland functions and values is presented in Chapter 7. The distribution and current status of tidal wetlands, threats, and human use of tidal wetlands are covered in Chapter 8. Chapter 9 reviews wetland conservation and management efforts, whereas Chapter 10 introduces wetland identification-delineation practices and functional assessment approaches. Chapter 11 describes and gives examples of wetland restoration, creation, and monitoring. Chapter 12 addresses the future of tidal wetlands emphasizing the impact of sea-level rise. Appendix A lists wetland areas in North America designated as wetlands of international importance. Appendix B provides profiles of 11 tidal wetland restoration projects to give readers a closer look at the success of these projects, along with additional references for more detailed information. Since there are hundreds of references to plants in this book and I have used common names for the ease of non-technical readers, I have included the scientific name the first time the plant is referenced in each chapter and in every table.

Readers should take note that many of the photographs presented in black and white in this book can be viewed in color online at ScholarWorks@UMass (http://scholarworks.umass.edu/); look under University of Massachusetts Press. This site also provides an interactive set of maps showing the distribution and general type of tidal wetlands in the conterminous United States.

Scientific plant names in this book follow the U.S. Department of Agriculture's national plants database; most can be identified using two field guides: *Field Guide to Tidal Wetland Plants of the Northeastern United States and Neighboring Canada* and *Field Guide to Coastal Wetland Plants of the Southeastern United States* (Tiner 2009, 1993a). Scientific names may change over time, so studies referenced in the book have sometimes used what are now obsolete scientific names. An effort was made to update these names to current taxonomy when referencing the results of these studies in the text but it is possible that some former names remain. Readers should also recognize that plant taxonomists continue to review species relationships and herbarium specimens, which means that scientific names of some species may have been changed by the time this book is read. Consult the national plants database (http://plants.usda.gov/) or similar references for the current name (e.g., Corps of Engineers national wetland plant list website at http://wetland_plants.usace.army.mil). Scientific names of animals are, for the most part, those given in the cited references, with some updated nomenclature where more recent articles have been reviewed.

Tidal Wetlands Primer

1 Definitions and Classification of Tidal Wetlands and Estuaries

Tidal wetlands are saline and freshwater marshes, swamps, banks, and shores subjected to flooding by tides. They are mainly low-lying, relatively flat plains formed by deposition of river-carried (fluvial) and/or marine sediments. They also include gently sloping beaches and other sloping landforms in areas with large tide ranges. Some are intertidal shorelines at the base of cliffs formed on material eroded from the cliffs. Tidal wetlands may cover vast areas that are miles wide or be limited to narrow, fringing bands only a few feet or meters wide. Regardless of their origin, their location beside oceans, estuaries, and tidal rivers promotes an exchange of water, nutrients, sediments, and biota between the land and the sea that has made tidal wetlands some of the world's most productive natural environments. Throughout this book, the terms "tidal wetlands" and "coastal wetlands" are used interchangeably. Readers should recognize, however, that wetlands along the Great Lakes have also been referred to as "coastal wetlands" because of the lengthy coastlines associated with these huge waters. Incidentally, these freshwater lakes are also tidally influenced due to their huge size, yet the tidal effect is minimal—only an inch or more. Water levels vary more due to seiches in which strong winds in one direction cause water levels to rise on one side of the lake and fall on the opposite side. Also note that a few researchers have applied the term "coastal wetlands" more broadly to include all wetlands in coastal watersheds and not strictly to tidal wetlands (Field et al. 1991; Stedman and Dahl 2008a; Stedman et al. 2010).

Vegetated tidal wetlands are typically found in deltas of major rivers, sheltered embayments behind barrier islands and spits and along the shores of bays, and on the floodplains of tidal rivers (Figure 1.1). They also occur in open water along low-energy, unprotected shores. Intertidal sites that are protected from the full force of ocean waves and strong tidal currents, and that are alternately flooded and exposed by tides, favor colonization by vascular hydrophytes (water-tolerant plants). More exposed locations are typically devoid of such plants (e.g., mudflats and beaches). Seaweeds (macroalgae) may colonize high-energy, surf-pounded rocky shores in northern climes where they find suitable substrates for attachment by their holdfast organs.

Hydrology, salinity, sedimentation and erosion rates, and other factors profoundly influence plant and animal life and make coastal wetlands interesting places for scientific research as well as for nature observation. Most tidal wetlands along the Atlantic Coast are marshes colonized by hydrophytic plants: salt-tolerant (halophytic) plants in salt and brackish regions and freshwater hydrophytes upstream along fresh tidal waters (Figure 1.2). Where tidal ranges are high or in exposed locations in which

a

b

Figure 1.1. Aerial view of tidal wetlands: (a) along the New Jersey coast showing tidal wetlands formed behind barrier islands and along the Mullica River and (b) tidal wetlands in a Connecticut embayment off Long Island Sound. (a: Copyright Geospatial Division, MDA Information Systems Inc., Gaithersburg, MD)

a

b

Figure 1.2. Marshes dominated by herbaceous species are the most common type in North America, while tidal flats predominate in regions with extremely high tidal ranges: (a) Nova Scotia salt marsh and (b) Maine tidal flat.

currents and wave action are strong, intertidal flats devoid of macrophytic plant life are particularly extensive due largely to longer periods of submergence. As we shall later see, human influences not withstanding, the fate of tidal marshes and swamps ultimately rests on their response to changes in sea level (i.e., their ability to maintain a certain elevation above mean sea level), the presence of low-lying lands to allow for their continued landward movement, and the ability of hydrophytes to colonize nonvegetated sites (e.g., tidal flats, deltas, and overwash sands on the leeside of barrier islands).

Since tides affect bodies of water ranging from the ocean (marine waters) to estuaries to freshwater tidal rivers (tidal fluvial), a diverse group of wetlands have formed along their shores. They include vegetated types such as salt marshes, brackish marshes, estuarine forests, oligohaline marshes, and tidal freshwater marshes and swamps as well as nonvegetated types such as coastal beaches, tidal flats, and rocky shores. Marshes are dominated by herbaceous (nonwoody) plants, whereas swamps are characterized by trees and shrubs. Other coastal wetland types may be found elsewhere along the world's coasts, for example, intertidal oyster reefs, saline flats (salt barrens), and mangrove swamps (mangals). Aquatic beds are often associated with tidal wetlands, either in marsh pools or ponds, adjacent shallow water, impoundments, or on intertidal flats. Specific types of tidal wetlands are described in detail later in this primer (Chapter 5).

Types of Estuaries

Since the majority of coastal wetlands form in estuaries, one might ask: What is an estuary? In classic terms, an estuary is a semi-closed body of water having a free exchange with the open ocean where seawater is measurably diluted by freshwater runoff from the land (Pritchard 1967). The degree of dilution depends on their location and linkage to freshwater rivers and streams. The exchange between marine and fluvial ecosystems is fundamental in the classic concept of an estuary—the mixing zone where rivers meet the sea. In arid and tropical regions, hypersaline lagoons (shallow bays) may form behind barrier beaches parallel to the coast where reduced or no freshwater runoff, high evapotranspiration, and weak or intermittent tidal connection cause salinities to exceed sea strength (e.g., Laguna Madre of Texas). The classic concept of estuary excludes such lagoons, while other definitions have referred to them as "inverse, reverse, or negative estuaries" due to their hypersalinity or simply as a type of estuary that although periodically closed, the aquatic environment is affected by both freshwater and marine water (Pritchard 1952; Edgar et al. 1999). The wetland classification system devised by the U.S. Fish and Wildlife Service includes these lagoons in the estuarine system (Cowardin et al. 1979). Some definitions restrict the estuary to the maximum penetration of ocean-derived saltwater (to 0.5 parts per thousand or practical salinity units; Cowardin et al.1979), while others may include the full reach of tidal influence (Caspers 1967; see Elliott and McLusky 2002 for an in-depth discussion of estuary definitions). The Great Lakes have been referred to as estuaries or freshwater estuaries by some scientists (e.g., Herdendorf 1990) and in the Estuaries and Clean Water Act of 2000 because they are similar in form and function to more traditional estuaries.

In summary, estuaries include coastal embayments, lagoons, coastal ponds (some with only periodic connection to the sea), tidal rivers, and their associated wetlands. Different types of estuaries may be classified by geomorphology, circulation patterns, tidal ranges, and other factors. Major characteristics of estuaries in the Northeast, South Atlantic, and west coasts of North America have been summarized (Roman et al. 2000; Dame et al. 2000; Emmett et al. 2000, respectively).

Geomorphic Types

From the geomorphologic perspective, three widely recognized types of estuaries occur in North America: 1) bar-built estuary; 2) drowned river valley estuary; and 3) fjord-type estuary, with the first two types being most common and the latter restricted to northern glaciated regions (Table 1.1). A fourth type—tectonic estuary—formed by tectonic processes (geologic faults, volcanic activity, and landslides) does not occur along the Atlantic Coast of North America, but does occur along the Pacific Coast where seismic activity is more pronounced. San Francisco Bay is the best U.S. example of a tectonic estuary. Some Alaskan estuaries have been affected by earthquakes. An uncommon type of estuary present in North America is the drowned basin estuary, with Long Island Sound being the best example. Small estuaries ("rocky headland bay wetlands," Tiner 2003a) may develop along rocky marine coasts where they are semi-enclosed by peninsulas and/or islands. Freshwater flow may enter these estuaries

Table 1.1. Examples of different estuary types in North America

Estuary Type Region	Example (State or Province)
Bar-built	
Atlantic Coast	Kouchibouguac Bay (NB), Malpeque Bay (PEI), Little River (ME), Plum Island Sound and Little Pleasant Bay (MA), Hammock River (CT), Great South Bay (NY), Barnegat Bay (NJ), Assawoman Bay (DE), Chincoteague Bay (MD/VA), Bogue and Pamlico Sounds (NC), North and Murrells Inlets (SC), Indian River Lagoon (FL)
Gulf Coast	Pensacola Bay (FL), Little Lagoon (AL), Graveline Bay (MS), Laguna Madre and Matagorda Bay (TX)
West Coast	Mugu Lagoon, Humboldt and Arcata Bays (CA), Tillamook and Netarts Bays (OR), Grays Harbor and Willapa Bay (WA), McIntyre Bay (BC)
Bar-built Coastal Pond	
Atlantic Coast	East Pond (Étang de l'Est; QC), Round Pond (PEI), Tisbury and Edgartown Great Ponds (MA), Winnapaug, Trustom, Green Hill, and Ninigret Ponds (RI), Mecox Bay and Georgica Pond (NY)
West Coast	Garrison Lake (OR)
Drowned River Valley	
Atlantic Coast	St. Lawrence River (QC), Chaleur Bay and Restigouche River (NB, QC), Shepody Bay and Petitcodiac River (NB), Chezzetcook River (NS), Kennebec River (ME), Narragansett Bay (RI), Connecticut River (CT), Hudson River (NY), Delaware River and Delaware Bay (NJ, DE, PA), Chesapeake Bay, Susquehanna and Potomac Rivers (MD, VA), Cape Fear River (NC), Winyah Bay and Pee Dee River (SC), Santee River (SC), Savannah River (SC/GA), St. Johns River (FL)
Gulf Coast	Apalachicola and Escambia Bays (FL), Mobile Bay (AL), Pascagoula Bay and Bay St. Louis (MS), Mississippi Delta (LA), Corpus Christi Bay (TX)
West Coast	San Francisco Bay, Sacramento and San Joaquin Rivers (CA), Coos Bay and Coquille River (OR), Columbia River (OR/WA), Willapa Bay and Grays Harbor (WA), Fraser River (BC)
Drowned Basin	Long Island Sound (CT, NY), Bras d'Or Lake (NS)
Fjord	Saguenay Fjord (QC), Strait of Juan de Fuca and Puget Sound (WA), Howe Sound (BC), Tracy Arm and College Fjords (AK)
Tectonic	San Francisco and Tomales Bays (CA)

Note: Some estuaries represent more than one type as many drowned river valleys have a bar-built estuary at their mouths.

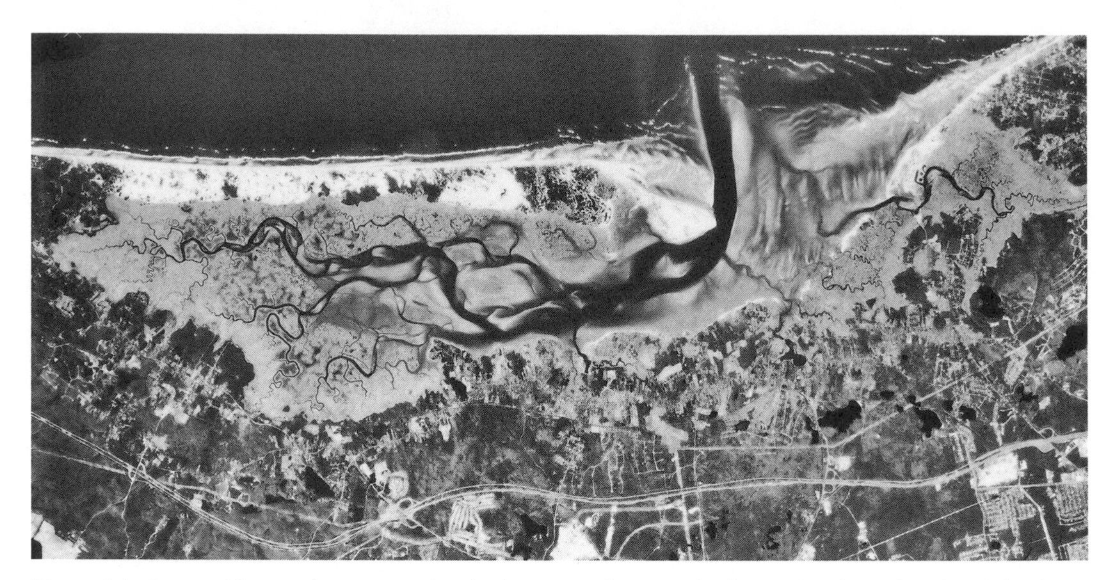

Figure 1.3. Barnstable Marsh (MA) is a bar-built estuary that was the focus of early studies that helped describe tidal wetland formation.

through one or more streams. While they may be viewed as part of the estuarine system, they also share characteristics with marine waters as their salinities are typically at or near sea strength.

Bar-built estuaries form behind barrier beaches (spits and islands) where sand deposits accumulate along tidal shores, more or less enclosing a water body (e.g., lagoon) behind the beach (Figure 1.3). Sometimes called "barred" or "lagoon" estuaries, they typically promote the development of salt marshes in the sheltered embayments. A variation of the bar-built estuary is one in which the inlet closes seasonally, usually in summer, due to reduced freshwater river discharge or low tidal flow and a buildup of beach deposits; these have commonly been called a "coastal pond" or "barachois pond" (pronounced *bara-schwa*) (e.g., Trenhaile 1990) in the northeastern United States and Canadian Maritimes, and referred to as "blind estuaries" elsewhere (Day 1981; Seliskar and Gallagher 1983). In their natural state, inlets to these ponds often remain closed for months or years and are opened by major storms (Figure 1.4). Today, inlets of many large estuaries are kept open by jetties or periodic dredging.

Drowned river valley estuaries become established when river valleys are inundated as sea level rises. These estuaries are essentially linear systems—permanently to periodically flooded valleys with salinities varying from the river mouth upstream. Their lower reaches may include broad embayments as illustrated by Delaware Bay (Delaware River) and Chesapeake Bay (Susquehanna River) (Figure 1.5). Drowned river valley estuaries are often associated with coastal plain land masses and have therefore been called "coastal plain estuaries" (Biggs 1978). They may include bar-built estuaries formed by spit development at the mouth of coastal rivers as exemplified by those behind sea islands in South Carolina and Georgia. Along the Louisiana coast, large estuarine lakes (e.g., Lake Pontchartrain, Calcasieu Lake, White Lake, Grand Lake, and Sabine Lake) have formed within its extensive marshlands.

Where river discharge is great, draining an extensive area of land and carrying enormous sediment loads, huge deltas form where the river meets the sea. This type of drowned river valley estuary may be described as "river-dominated." The Mississippi River is the best example of this type

Figure 1.4. Coastal ponds are common along Rhode Island's south shore (Trustom and Card Ponds). Note that temporary inlets are closed by sand deposits but are visible by their nonvegetated surfaces.

Figure 1.5. Chesapeake Bay and Delaware Bay are excellent examples of drowned river valley estuaries: Susquehanna and Delaware Rivers, respectively. (Copyright Geospatial Division, MDA Information Systems Inc., Gaithersburg, MD)

Figure 1.6. The Mississippi Delta receives sediments from nearly half of the United States and supports more than 20 percent of the nation's tidal wetlands. (Copyright Geospatial Division, MDA Information Systems Inc., Gaithersburg, MD)

in the United States (Figure 1.6). This river drains nearly 41 percent of the continental United States, bringing millions of tons of sediment to the Gulf of Mexico annually: 400 to 500 million tons before dam construction and 205 million tons today (Blum and Roberts 2009). Postglacial deposition of fluvial sediment at the river mouth created an extensive delta resembling the shape of a bird's foot from the air. River-dominated estuaries in macrotidal regions (tides >13.2 ft [4 m]) where marine influences are strong have deltas that contain deposits of both marine and fluvial origins. Examples include the following rivers: Ganges-Brahmaputra (Bangladesh; Figure 1.7), Amazon (Brazil), Yangtze-Kiang (China), Ord (Australia), Shatt-al-Arab (Iraq), and Klang (Malaysia) (Wright 1978). Other deltas may be dominated by wave action, with the Nile River Delta being an excellent example (Figure 1.8).

Fjord-type estuaries are found in northern areas (generally above 45° latitude) where glaciers have cut steep-walled valleys (Figure 1.9). These former glacial valleys are the deepest of estuaries; their basins can exceed 2,625 feet (800 m) in depth (Tomczak 1996). A shallow sill at their mouth is a typical characteristic of these estuaries. Linear, fringing wetlands tend to be the predominant wetland type because of the fjord's steep walls. Where a river empties into a fjord, a delta and more extensive wetlands may form (Figure 1.10). In the North Atlantic, fjord-type estuaries are common in Newfoundland and Labrador, while Alaska and British Columbia may have the best examples in North America (Oberrecht 1997). Saguenay River in Quebec and Somes Sound in Maine are the southernmost examples in eastern North America.

Hydrologic Circulation Types

Mixing of freshwater with saltwater in estuaries varies with river flow and tides. During heavy river discharge periods (e.g., early spring in the Northeast), tidal flow is suppressed upstream and virtually no mixing occurs in these reaches. During low

Figure 1.7. The Ganges-Brahmaputra Delta has a shape quite different from the Mississippi Delta due to a strong marine influence from extremely high tides. This delta contains the largest continuous stand of mangroves in the world. (Copyright Geospatial Division, MDA Information Systems Inc., Gaithersburg, MD)

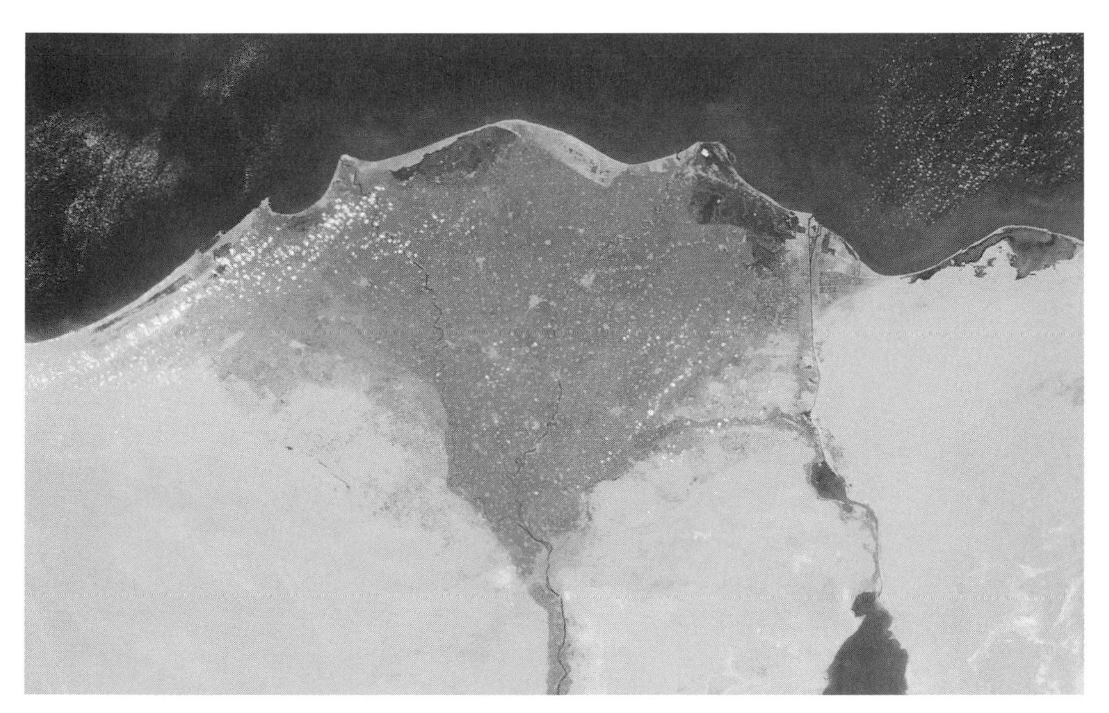

Figure 1.8. The Nile Delta is a wave-dominated delta where strong wave action from the Mediterranean Sea smooths the seaward edge of the formation. (NASA image)

Figure 1.9. Fjord-type estuaries are typical of the North Pacific coast from British Columbia north. This image shows the fjord at Port Valdez; note delta formation where the Valdez River meets the fjord (right side of image). (Copyright Geospatial Division, MDA Information Systems Inc., Gaithersburg, MD)

Figure 1.10. Close-up of mouth of Valdez River where it empties into Port Valdez Fjord showing tidal wetlands. (Dennis Cowals, U.S. National Archives)

flow periods, however, saltwater moves considerable distances upstream along the bottom creating a highly stratified condition: freshwater on top and the heavier saltwater at the bottom. At the river mouth where tidal currents are strong, freshwater is typically well mixed with saltwater. Although estuarine hydrodynamics are often quite complex, a simple classification recognizes four basic types: 1) well-mixed (homogeneous) estuary; 2) partially mixed (partially or slightly stratified) estuary; 3) highly stratified (salt-wedge) estuary; and 4) fjord-entrainment type (Figure 1.11; Officer 1976).

The *homogeneous type* has a well-mixed water column, with salinity more or less the same from top to bottom (no significant stratification or layering). The homogeneous type is usually associated with high-energy systems, open embayments, and low river discharge. Examples include Buzzards Bay (MA), lower Delaware Bay, San Diego Bay (CA), and Pamlico Sound (NC). The *partially mixed type* has different salinities from the top to the bottom of the water column, but it is not highly stratified. This type is associated with low-energy systems (e.g., Chesapeake Bay and Long Island Sound) or very deep estuaries such as San Francisco Bay. The *highly stratified estuary* has distinct stratification with a layer of freshwater at the surface and most saline waters at the bottom. A "salt-wedge" moves back and forth in the estuary with the tides, penetrating farthest inland during low freshwater runoff periods (late summer in the eastern United States). This type is associated with high river discharge and low tidal amplitude and occurs in tidal rivers (the drowned river valley estuary) such as the Hudson, Connecticut, Mississippi, Columbia, and Fraser rivers. In these estuaries, flood tides typically dominate flow in the lower parts of the estuary, while ebb tides predominate upstream, resulting in a point in the estuary where neither dominates—there is no net flow upstream or downstream. This point is called the "null point" and is important

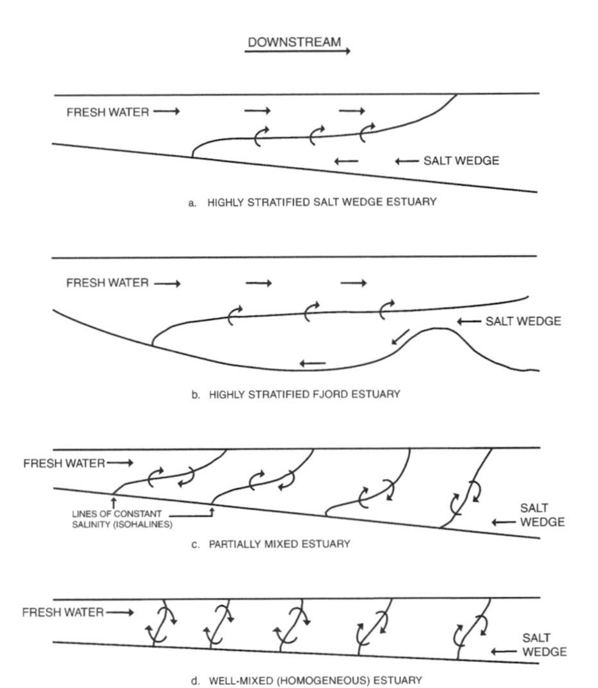

Figure 1.11. Four types of hydrologic circulation in estuaries: (a) highly stratified salt wedge; (b) highly stratified fjord estuary; (c) partially mixed; and (d) well-mixed type. Salinity is higher at the bottom of the estuary and fresher at the top because saltwater is denser than freshwater. (Redrawn from U.S. Army Corps of Engineers 1991)

because this is a region of major shoaling in estuaries (increased sedimentation). In highly stratified estuaries, this region is often bounded by the limits of saltwater intrusion at high tide and low tide (U.S. Army Corps of Engineers 1991). The *fjord-entrainment type* is associated with fjord estuaries where the presence of a sill restricts flow of bottom waters resulting in relatively stagnant, more saline deep water overlain by a thin layer of freshwater river flow. These estuaries are common along the coasts of British Columbia and Alaska. Estuarine circulation patterns are not static or uniform as they can change due to variations in river discharge, wind action, and other factors. For one part of a large estuary to exhibit one pattern while another section has a different pattern is not uncommon. For example, at low river discharge, the upper portion of Mobile Bay (AL) is highly stratified, while the lower bay waters are nearly well-mixed (homogeneous) by the tides (Schroeder 1978). The circulation pattern changes during periods of high river flow when the upper bay becomes homogeneous due to mixing of bay water with high volumes of river water and the lower bay becomes stratified. This mixing affects salinity levels throughout the estuary, which is important for estuarine plants and animals.

Tidal Range Types

Estuaries may also be categorized by the height of their tides. Three types are recognized: 1) microtidal (tides <6.6 ft [2 m]); 2) mesotidal (6.6–13.2 ft [2–4 m]); and 3) macrotidal (>13.2 ft [4 m]) (Davies 1973, 1974) (see Figure 1.12 for the general location of these tidal ranges in North America). Some classification systems extend the mesotidal to 19.2 feet (6 m), with the macrotidal range >19.2 feet (>6 m) (Digby et al. 1998). The microtidal range might benefit from further subdivision as estuaries at the lower end of the range have water levels that are greatly influenced by winds (i.e., wind tides). Microtidal and mesotidal estuaries are often associated with bar-built estuaries (Boothroyd 1978). Most of estuaries in Nova Scotia on the Atlantic and Gulf of St. Lawrence are microtidal, and the St. Lawrence Estuary contains both microtidal and mesotidal regions (Davis and Browne 1996; Gail Chmura, pers. comm., 2005). The Bay of Fundy is the best example of a macrotidal estuary in the North Atlantic—its highest tide is 50 feet (15.6 m) in the Minas Basin, which is among the greatest in the world. Further discussion of tides can be found in Chapter 3.

Tidal Wetland Classification

Tidal wetlands can be categorized in many ways to emphasize certain properties. Some distinguishing features include salinity, dominant vegetation, flooding frequency and duration, position in the estuary, and estuary type. At the most general level, tidal wetlands can be separated into estuarine (salt and brackish) and tidal freshwater wetlands, and vegetated and nonvegetated types. The U.S. and Canadian governments have developed national wetland classification systems to describe their wetland resources. These systems are summarized below, with emphasis on how tidal wetlands are classified.

United States Classifications

The U.S. Fish and Wildlife Service's *Classification of Wetlands and Deepwater Habitats of the United States* (Cowardin et al. 1979) serves as the U.S. national standard for wetland classification for mapping wetlands and reporting on national wetland trends (library.fws.gov/FWS-OBS/79_31.pdf). The terms "coastal wetlands" or "tidal wetlands" are not used in this system, but can easily be identified from its scientific nomenclature (e.g., systems, subsystems, and water regimes). The classification system is an open-ended hierarchical system represented by a number of levels providing increased information about a particular wetland: system (marine, estuarine, palustrine, lacustrine, and riverine), subsystem, class, subclass, and four types of modifiers (water regime, water chemistry, soil, and special) (Figure 1.13). Tidal wetlands fall mostly within two ecological systems influenced by ocean-derived saltwater—marine (open ocean and associated shorelines) and estuarine (the freshwater–saltwater mixing zone), but also occur in the other three systems: palustrine (freshwater and other inland wetlands), riverine (rivers and streams, from bank to bank) and lacustrine (lakes including those under tidal influence). The former two systems are saltwater ecosystems, while the latter three encompass tidal freshwater ecosystems. Wetlands may be classified further by their vegetation or substrate (classes and subclasses), hydrology, water chemistry, and other factors. Principal classes for coastal wetlands include nonvegetated types: unconsolidated shore (e.g., beaches and tidal flats) and rocky shore; and vegetated types: aquatic bed, emergent (herbaceous) wetland,

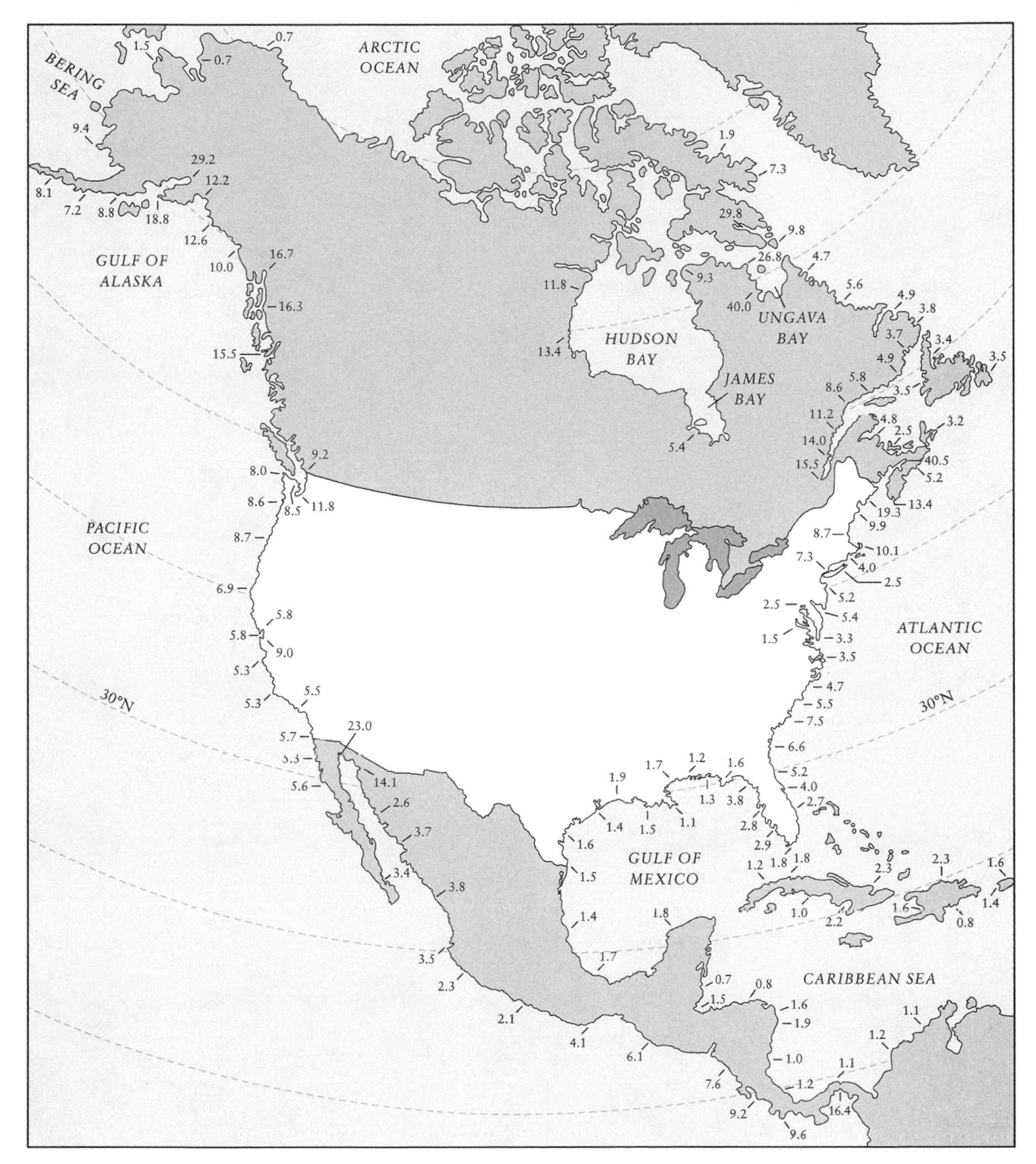

Figure 1.12. Spring tidal ranges across North America. Values are in feet. (Data from NOAA and Canadian Hydrographic Service)

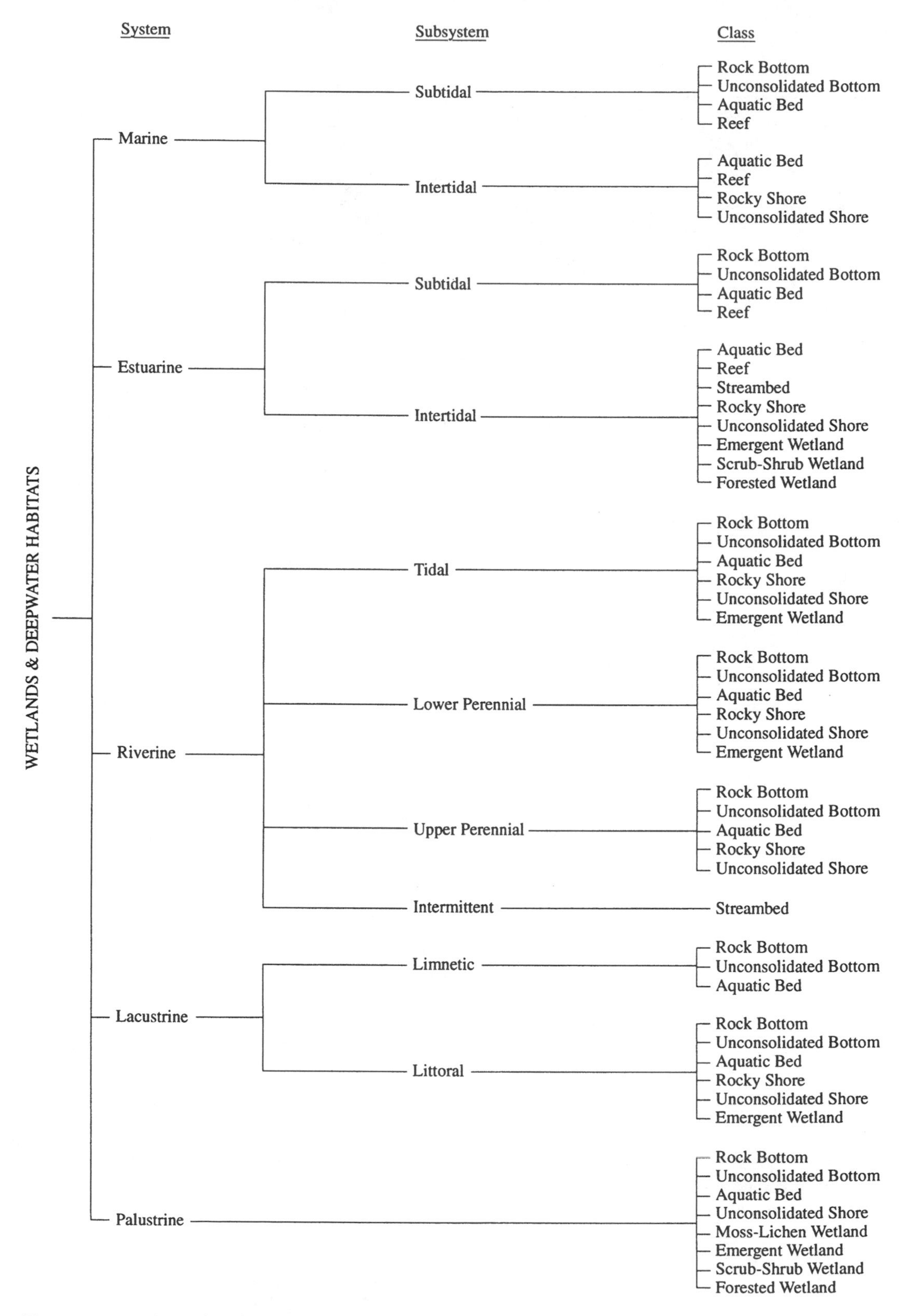

Figure 1.13. Habitat classification hierarchy according to the U.S. federal wetland classification system. (Cowardin et al. 1979)

scrub-shrub wetland, and forested wetland; subclasses further define substrate types for the former, and vegetation life form for the latter (Table 1.2). Aquatic beds beyond the intertidal zone are considered deepwater habitats (not wetlands) in the marine and estuarine systems, while those in freshwater (either tidal or nontidal) at depths less than 6.6 feet (2 m) are considered wetlands. The classification system includes modifiers to describe other wetland characteristics: 1) water regime or hydrology (e.g., regularly flooded, irregularly flooded, seasonally flooded-tidal, or subtidal; Table 1.3); 2) water chemistry (halinity [salinity from ocean-derived salts]—hyperhaline, euhaline,

Table 1.2. Classes applicable to tidal wetlands and waters

Class	Brief Description	Subclasses
Rock Bottom	Generally permanently flooded area with bottom substrates consisting of at least 75% stones and boulders and less than 30% vegetative cover	Bedrock, rubble
Unconsolidated Bottom	Generally permanently flooded area with bottom substrates consisting of at least 25% particles smaller than stones and less than 30% vegetative cover	Cobble-gravel, sand, mud, organic
Aquatic Bed	Generally permanently flooded area vegetated with plants growing principally on or below the water surface	Algal, aquatic moss, rooted vascular, floating vascular
Reef	Ridgelike or moundlike structure formed by the colonization and growth of sedentary invertebrates	Coral, mollusk, worm
Streambed	Channel whose bottom is completely dewatered at low water	Bedrock, rubble, sand, mud, cobble-gravel, organic, vegetated (by pioneer plants)
Rocky Shore	Wetland characterized by bedrock, stones, or boulders with areal coverage of 75% or more and with less than 30% areal coverage by vegetation	Bedrock, rubble
Unconsolidated Shore	Wetland characterized by unconsolidated substrates with less than 75% coverage by stones, boulders, and bedrock and less than 30% vegetative cover, except by pioneer plants	Cobble-gravel, sand, mud,organic, vegetated (by pioneer plants)
Emergent Wetland	Wetland dominated by erect, rooted, herbaceous hydrophytes	Persistent, nonpersistent
Scrub-shrub Wetland	Wetland dominated by woody plants less than 20 feet (6 m) tall	Broad-leaved deciduous, needle-leaved deciduous, broad-leaved evergreen, needle-leaved evergreen, dead
Forested Wetland	Wetland dominated by woody plants 20 feet (6 m) or taller	Broad-leaved deciduous, needle-leaved deciduous, broad-leaved evergreen, needle-leaved evergreen, dead

Source: Cowardin et al. 1979.
Note: Subclasses are also listed.

Table 1.3. Tidal water regimes

Type of Water	Water Regime	Degree of Tidal Flooding
Salt/Brackish	Subtidal	Permanently inundated
	Irregularly exposed	Mostly inundated, exposed less often than daily
	Regularly flooded	Inundated at least once daily
	Irregularly flooded	Inundated less often than daily; exposed to air for longer periods
Fresh	Permanently flooded-tidal	Permanently inundated
	Regularly flooded	Inundated at least once daily
	Semipermanently flooded-tidal	Inundated for most of the growing season and subject to tidal fluctuations
	Seasonally flooded-tidal	Inundated irregularly by the tides and seasonally for extended periods by river overflow
	Temporarily flooded-tidal	Inundated irregularly by tides and seasonally for brief periods during the growing season by river overflow

Source: Cowardin et al. 1979.

mixohaline, polyhaline, mesohaline, oligohaline, and fresh; Table 1.4); 3) soils (organic or mineral); and 4) human- or beaver-induced alterations—"special modifiers" (e.g., partially drained/ditched, diked/impounded, excavated, and spoil). The classification hierarchy also includes the option to classify "dominance type," which may be used to identify one or more dominant species representing the wetland.

Table 1.4. Halinity modifiers for coastal wetlands and waters

Halinity Modifier	Salinity (ppt)
Hyperhaline	>40.0
Euhaline	30.0–40.0
Mixohaline (Brackish)	30.0–0.5
Polyhaline	30.0–18.0
Mesohaline	18.0–5.0
Oligohaline	5.0–0.5
Fresh	<0.5

Source: Cowardin et al. 1979.
Note: The term "halinity" is used because tidal waters are dominated by sodium chloride. Salinity values given are in parts per thousand (ppt), which are virtually equivalent to the "practical salinity units" (psu) used by some oceanographers. Seawater averages 35 ppt (35 grams of salt per 1000 grams of water).

This level of the classification has not been used for the national inventory of wetlands and a list of dominance types has not been developed. The classification system is open-ended so that other significant properties can be added as needed.

Recently, the U.S. Fish and Wildlife Service's Northeast Region developed additional hydrogeomorphic-type descriptors to better characterize individual wetlands mapped by the National Wetlands Inventory and to aid in predicting wetland functions. A set of keys titled "Dichotomous Keys and Mapping Codes for Wetland Landscape Position, Landform, Water Flow Path, and Waterbody Type Descriptors" (Tiner 2003a, 2011b) make it possible to expand the characterization of both tidal and nontidal wetlands. Three basic types of descriptors address landscape position (proximity to a water body), landform (physical shape or form of a wetland), and water flow path (hydrodynamics). For tidal wetlands, landscape positions include marine, estuarine, and lotic (tidal river and tidal stream gradients), while landforms include fringe, island, and basin (tidal wetlands with altered hydrology due to road crossings, dikes, etc.). Coastal landforms may be defined further

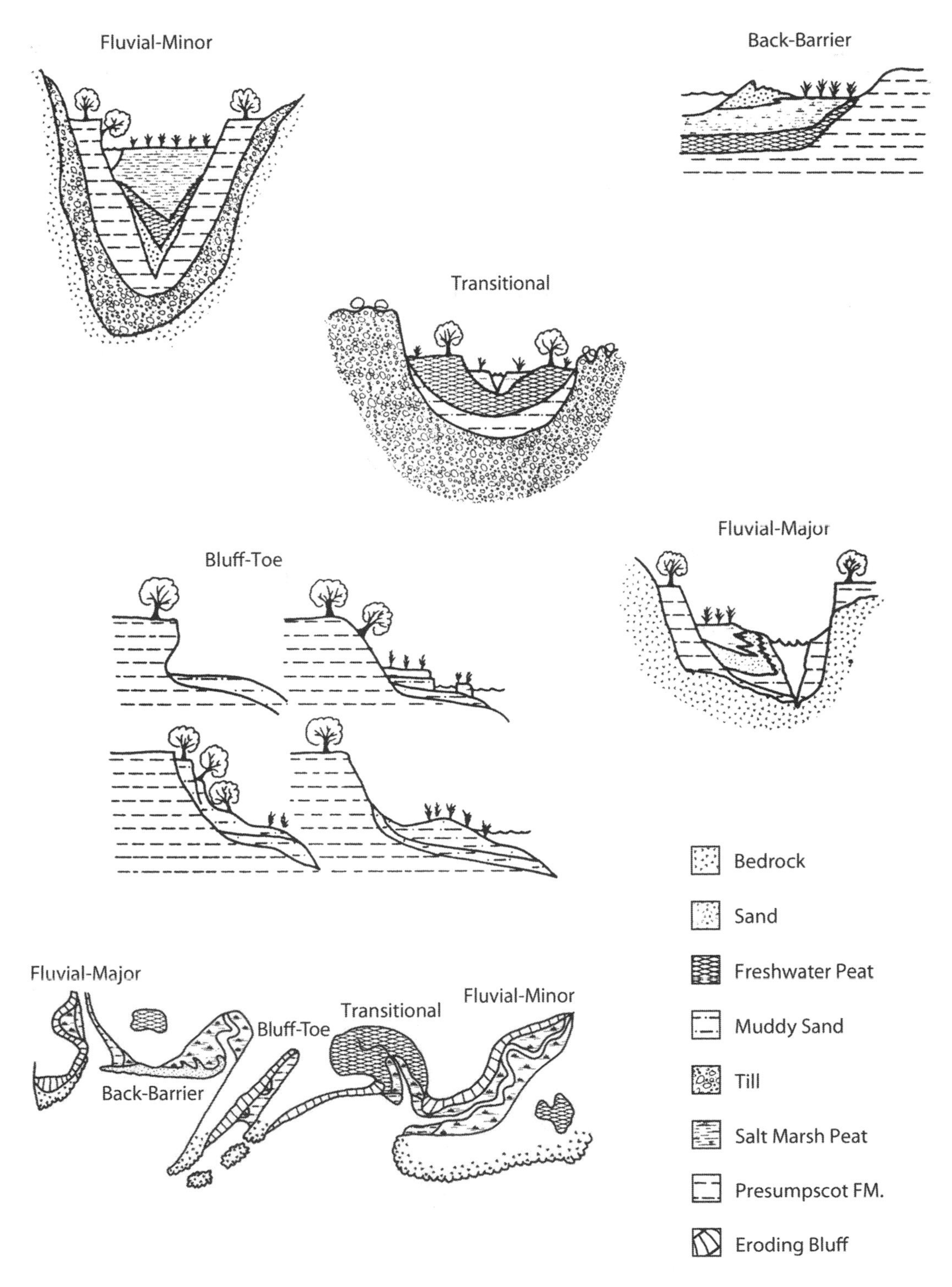

Figure 1.14. Tidal marsh types classified for Maine estuaries. (Recompiled Figures 3 and 8 from Kelley et al. 1988; *Journal of Coastal Research*; copyright and reprinted with permission from the Coastal Education & Research Foundation, Inc.)

Figure 1.15. Swale marsh between ridges on a South Carolina barrier island.

by their position in the estuary (e.g., barrier island, barrier beach, bay, coastal pond, river, headland, or ocean). Water flow path is bidirectional due to tidal forces, except where tidally restricted or controlled. Specific estuary types are also defined (e.g., drowned river valley, bar-built, tectonic, and fjord), and modifiers are available to describe tidal range (microtidal, mesotidal, or macrotidal) and hydrologic circulation patterns.

Other classifications have been developed for scientific research. A few are introduced below. A geomorphology-based classification system classifies Maine salt marshes (Kelley et al. 1988). Four types have been recognized: back-barrier, fluvial, bluff-toe, and transitional (Figure 1.14). Back-barrier salt marshes form behind barrier islands and spits. Fluvial salt marshes develop in drowned valleys along rivers and streams (fluvial-major and fluvial-minor). Where cliffs are eroding and depositing sediment into the estuary below, bluff-toe marshes become established. Transitional salt marshes are those that form where saltwater invades low-lying freshwater wetlands. Another geomorphic classification system was created for investigating marsh evolution and sedimentation rates in barrier island systems (Oertel and Woo 1994). Three landscape settings are recognized and several marsh types occur within each setting (listed in parentheses): 1) mainland fringe-marshes (valley, headland, interfluve, hammock, and tidal-channel); 2) mid-lagoon marshes (tidal-channel, marsh island, platform, and hammock); and 3) back-barrier fringe-marshes (washover-fan, storm-surge platform, swale, flood-delta, and platform; Figure 1.15). The National Oceanic and Atmospheric Administration (NOAA) is creating a hierarchical marine and estuarine ecosystem and habitat classification system to provide a national framework for their investigations of the marine environment (Standards Working Group 2010). While the focus of this classification is on deepwater habitats that are not well covered by the Cowardin et al. classification system, it does address intertidal environments to encompass all habitats of interest to the agency. Under this system, the estuarine system extends to the limits of tide (i.e., where the mean tide range becomes less than 0.2 ft [0.06 m]). Estuarine wetlands are within the shallow water intertidal and shallow water tidal riverine subsystems. Further subdivisions for intertidal habitats are based on composition of the substrate (i.e., surface geology component including classes such as unconsolidated substrate, rock substrate, coral reef, and faunal reef) and biota (benthic biotic component including classes such as faunal reef, coral reef, faunal bed, aquatic bed, emergent wetland, scrub-shrub wetland, and forested wetland). The classification also identifies different types of estuaries: riverine, lagoon, embayment, and fjord with additional subtypes. Since this classification is still in development, readers should refer to the Federal Geographic Data Committee's website (www.fgdc.gov/) for the latest information.

Canadian Classifications

The Canadian national wetland classification system identifies wetlands by three main features: wetland class, wetland form, and wetland type (National Wetlands Working Group 1997). *Class* relates to properties that reflect the wetland's origin and the nature of the environment, such as vegetation, hydrology, and water chemistry. *Form* is the shape of the wetland based on surface morphology, landform, and its physiographic position relative to uplands, waters, and other wetlands. *Type* separates wetlands based on plant morphology. Tidal wetlands fall mainly within the marsh class. Two marsh forms address these wetlands: estuarine marsh and tidal marsh, with the former differentiated from the latter by a major input of freshwater, thereby creating brackish to fresh conditions (Figure 1.16). Four subforms of estuarine marsh are listed: 1) estuarine bay marsh (fringes of tidal flats and channels along river embayments); 2) estuarine delta marsh (at mouths of rivers and streams); 3) estuarine lagoon marsh (in embayments, lagoons, and open embayments); and 4) estuarine shore marsh (along the shores of coastal rivers). The tidal marsh form receives insignificant freshwater inflow and typically develops behind beaches and along islands where there is shelter from wave action. Four tidal marsh subforms are recognized: 1) tidal basin marsh; 2) tidal bay marsh; 3) tidal channel marsh; and 4) tidal lagoon marsh. The tidal basin marsh form is a periodically inundated depression that does not drain at low tide; it occurs in the supratidal zone (above normal tides) where it derives salt from spray and infrequent flooding by storm surges. The tidal bay marsh is located along an open water body (ocean, bay, or inlet) where currents are not strong enough to erode the marsh. The tidal channel marsh forms in tidal channels that drain at low tide. The tidal lagoon marsh is the typical salt marsh forming along a lagoon (coastal or barachois pond) behind a coastal barrier (e.g., barrier island or spit) or reef. Tidal swamps are mentioned as one of the eight forms of swamp, with two subforms cited as "tidal saltwater swamp" and "tidal freshwater swamp." Various wetland types are further recognized by the general physiognomy of the vegetation: aquatic (floating and submerged), forb (non-graminoid herb), graminoid (grass, low rush, reed, sedge, and tall rush), shrub (low <0.5 m, tall >0.5 m, and mixed), and treed (coniferous, hardwood, and mixed). This classification tends to focus on vegetated wetlands, with nonvegetated tidal wetlands (beaches, tidal flats, and rocky shores) apparently classified as "tidal" and "estuarine" water forms.

General Classification for This Book

For this book, tidal wetlands are divided into 11 types: 1) rocky shore; 2) intertidal

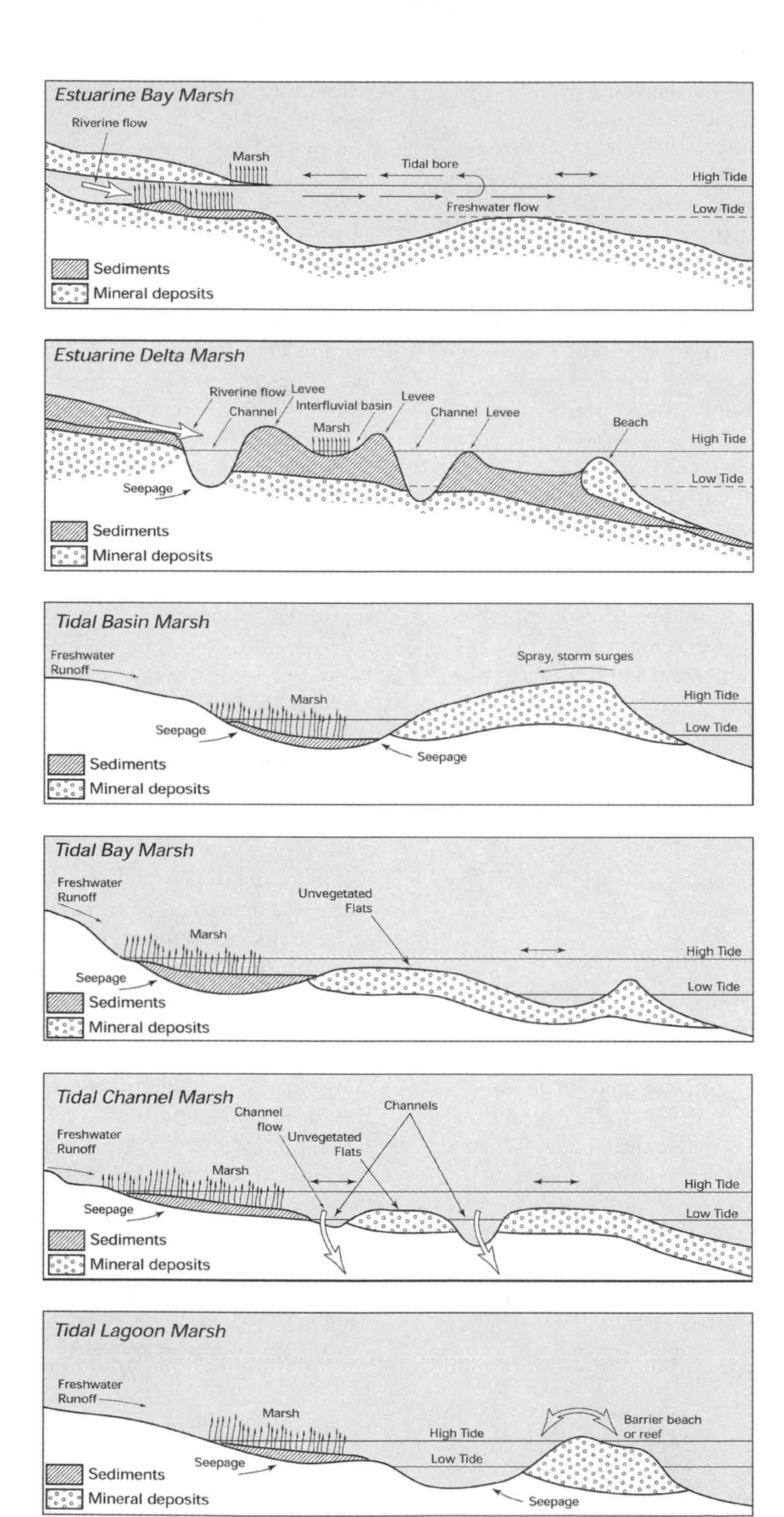

Figure 1.16. Estuarine wetland types according to the Canadian wetland classification system. (National Wetlands Working Group 1997)

beach; 3) tidal flat; 4) salt marsh; 5) salt barren/flat; 6) estuarine shrub swamp; 7) estuarine forest (including mangrove swamp); 8) brackish marsh; 9) tidal fresh marsh; 10) tidal freshwater swamp; and 11) aquatic bed (Figure 1.17). All types except the last are strictly intertidal habitats (alternately flooded and exposed by the tides). Aquatic beds are mostly subtidal (permanently inundated) habitats, but some may be intertidal. Each of these types is described in more detail in Chapter 5. Table 1.5 correlates these general types with the U.S. and Canadian national wetland classification systems.

Further Readings

Classification of Wetlands and Deepwater Habitats of the United States (Cowardin et al. 1979)

The Canadian Wetland Classification System (National Wetlands Working Group 1997)

"Estuaries of the South Atlantic Coast of North America: their geographical signatures" (Dame et al. 2000)

"Geographic signatures of North American west coast estuaries" (Emmett et al. 2000)

"Estuaries of the northeastern United States: habitat and land use signatures" (Roman et al. 2000)

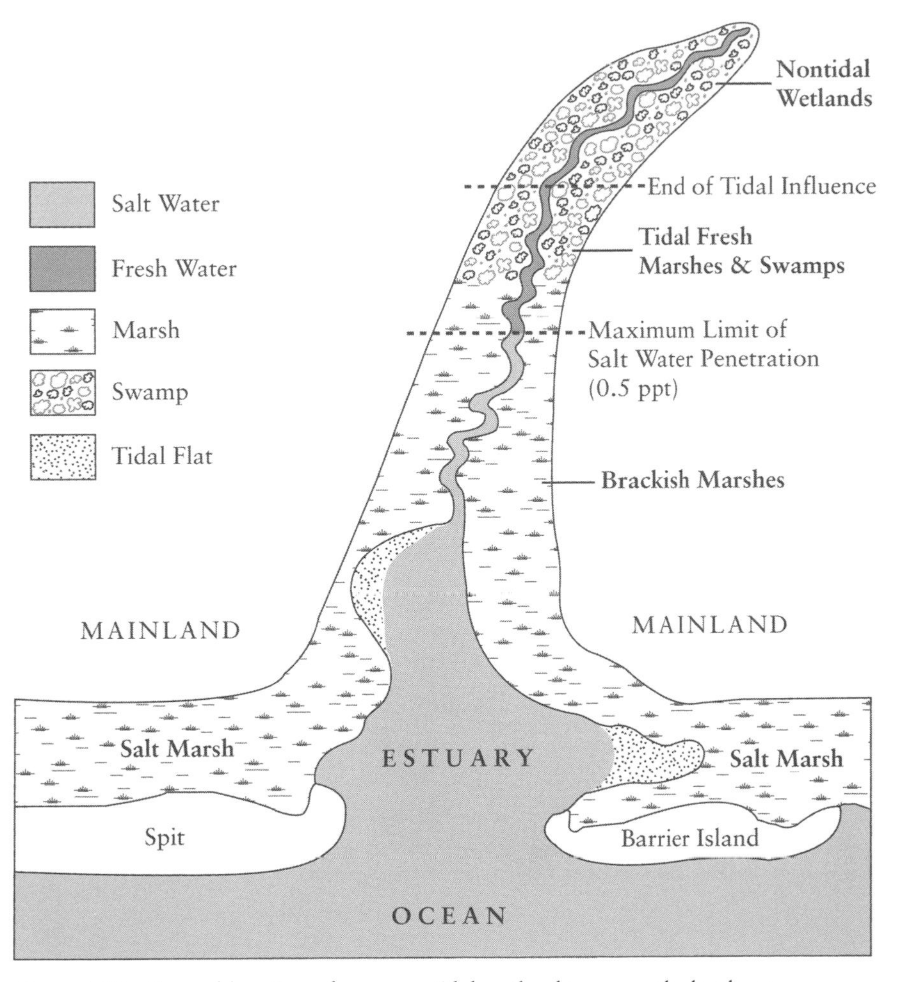

Figure 1.17. General location of common tidal wetland types on the landscape. (Tiner 2009)

Table 1.5. Correlations between wetland types referenced in book and types defined in U.S. and Canadian national systems

Generic Wetland Type	U.S. Wetland Type	Canadian Wetland Type
Rocky Shore	Marine Intertidal Rocky Shore	*
	Estuarine Intertidal Rocky Shore	*
Beach	Marine Intertidal Unconsolidated Shore	*
	Estuarine Intertidal Unconsolidated Shore	*
Tidal Flat	Marine Intertidal Unconsolidated Shore	*
	Estuarine Intertidal Unconsolidated Shore	*
	Riverine Tidal Unconsolidated Shore	*
Salt/Brackish Marsh	Estuarine Intertidal Emergent Wetland	Tidal Marsh
		Estuarine Marsh
Estuarine Shrub Swamp	Estuarine Intertidal Scrub-Shrub Wetland	Tidal Saltwater Swamp
Estuarine Forest	Estuarine Intertidal Forested Wetland	Tidal Saltwater Swamp
Tidal Fresh Marsh	Riverine Tidal Emergent Wetland nonpersistent	Estuarine Marsh
	Palustrine Emergent Wetland (tidal water regime)	Estuarine Marsh
Tidal Freshwater Swamp	Palustrine Scrub-Shrub Wetland (tidal water regime)	Tidal Freshwater Swamp
Coastal Aquatic Bed	Marine Subtidal Aquatic Bed (deepwater habitat)	Tidal Water
	Estuarine Subtidal Aquatic Bed (deepwater habitat)	Tidal Water, Estuarine Water
	Riverine Tidal Aquatic Bed (deepwater habitat)	Estuarine Water
	Palustrine Aquatic Bed (tidal water regime)	Estuarine Water
Mangrove Swamp	Estuarine Intertidal Scrub-Shrub Wetland	Not applicable (does not occur)
	Estuarine Intertidal Forested Wetland	
Salt Barren or Salt Flat	Estuarine Intertidal Unconsolidated Shore (irregularly flooded)	Not applicable (does not occur)

Note: *The Canadian system does not address rocky shores, beaches, and tidal flats as specific wetland types; they appear to be included within the "tidal marsh" or "tidal water" types.

2 Origin and Formation of Tidal Wetlands

Low-relief landscapes such as coastal plains favor the development of extensive coastal wetlands as such lands are more likely to be flooded than are steeper lands where mountains meet the sea. The latter regions may have narrow, steep continental shelves and may lack sufficient sediments for significant tidal wetland formation. Vast wetlands may also be found in macrotidal regions where huge tides expose extensive mudflats at low tide (e.g., the Bay of Fundy and Alaska).

Tidal wetlands commonly form in three basic locations: 1) along the ocean and its embayments; 2) in estuaries; and 3) along tidal reaches of freshwater rivers and streams (Figure 2.1). They may also develop along the shores of tidally influenced freshwater lakes, but this situation is uncommon. Near the ocean, vegetated tidal wetlands usually require some form of protection from wave action for establishment. Here they are typically found behind a protective barrier such as a spit or barrier island or along the shores of a sheltered cove or embayment. Yet in open water where there is little or no exposure to significant wave action, they may grow out to the water's edge as they do along tidal rivers and in sheltered coves. In the northeastern corner of the Gulf of Mexico (from Anclote Key to Apalachee Bay), wave action is virtually nonexistent allowing marshes to extend out into Gulf waters (Stout 1984).

One of the keys to understanding tidal wetlands is recognition that their world is in constant flux. Nature's forces—the tides, littoral currents, wave action, sea-level fluctuations, and changes in freshwater runoff and sedimentation rates—coupled with more than 300 years of development by modern societies—continue to shape coastlines and their wetlands. Sea level has risen and fallen many times over the Earth's history. Since the Atlantic Ocean opened about 200 million years ago with the formation of the Appalachian Mountains, many advances and retreats of the North American ice sheet, and even higher sea levels than current day levels have occurred. For example, during the last interglacial period—the Sangamon Stage (about 72,000–125,000 years ago)—sea level was 20 to 25 feet (~7 m) above present-day levels (Shafer et al. 2007). So, although subjected to profound changes over geological time, the events of prime significance that have shaped today's coastal wetlands chiefly result from the most recent glaciation and the effects of glacial retreat (melting ice) on sea level. From a geological perspective, tidal wetlands are ephemeral features, relatively short-lived, and dependent on sea levels and local topography. Today's coastal wetlands are geologically young features having formed sometime after the retreat of the continental glacier. Human actions have also had and continue to have a major impact on tidal wetlands, with the human role in global warming and shoreline development perhaps being the most significant

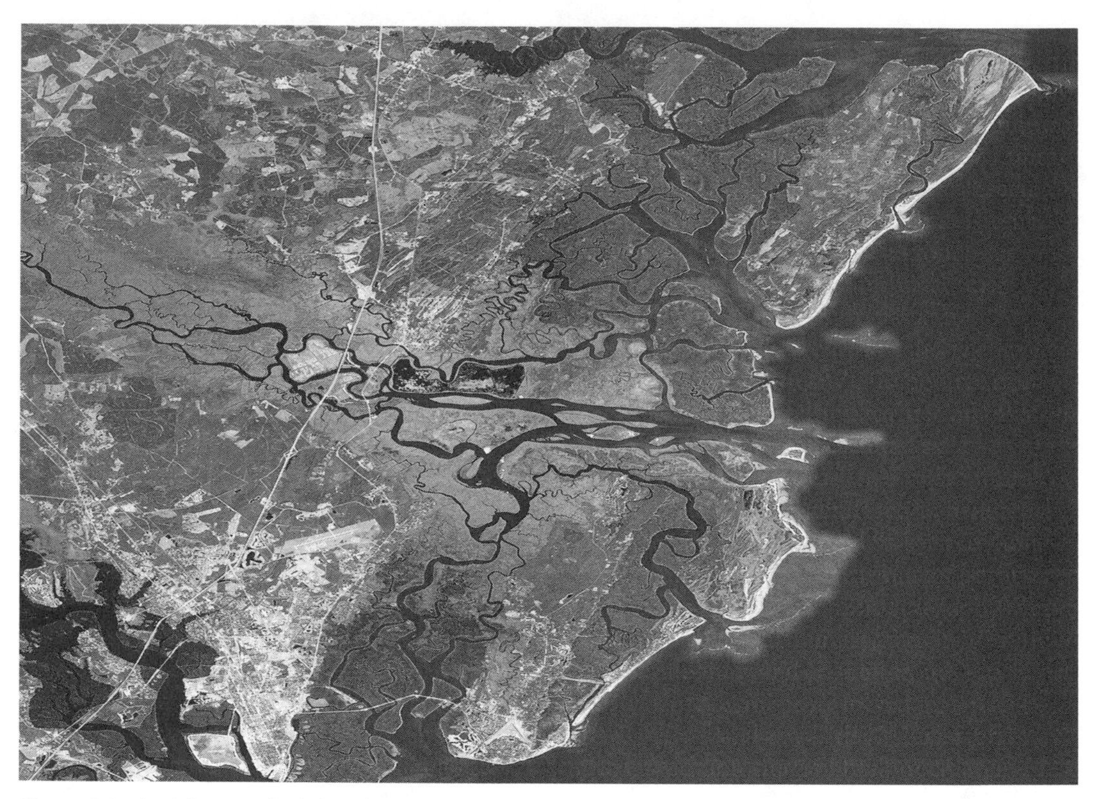

Figure 2.1. Aerial view of tidal wetlands behind some of Georgia's barrier islands and along the Altamaha River. (Copyright Geospatial Division, MDA Information Systems Inc., Gaithersburg, MD)

factors controlling the fate of the remaining tidal wetlands in the short term.

Glaciation

The Earth's climate changes from glacial (cold) periods to interglacial (warming) periods back to glacial periods roughly every 100,000 years. This cycle is believed to be related to the elliptical orbit of the Earth around the Sun, periodic changes in the angle of the tilt in the Earth's axis, and the wobble in its rotation—the Milankovitch cycle (Milankovitch 1941; Roe 2006). These changes affect the amount of solar radiation reaching the Earth, influencing climate and glaciation.

The first major cycle of Northern Hemisphere glaciations and deglaciations began in the late Pliocene epoch (approximately 2.7 million years ago) (Sewall 2008). The North American continental ice sheet (the Laurentide Ice Sheet) formed in the Arctic and started moving south. Since then repeated advances and retreats of the ice sheet have taken place every 80,000 to 120,000 years (Rutherford and D'Hondt 2000). The last advance—the Wisconsinan Stage of glaciation—started in Canada roughly 85,000 years ago and moved as far south as northern New Jersey and eastern Pennsylvania and to northern Washington on the West Coast. About 15,000 to 20,000 years ago, Canada and much of the northeastern United States were buried beneath glacial ice up to one mile thick, with few exceptions (Wolfe 1977). Glacial ice covered about one-third of the Earth's land mass. This ice sheet placed great weight on the land, depressing the landscape much like what happens to a sofa cushion when you sit down. Some depressed coastal areas from Massachusetts to Nova Scotia were flooded due to coastal subsidence, while other lands slumped beneath the weight of the glacier. Lands south of the glacier bulged, with the center of the bulge located near Cape Hatteras, North Carolina (Emery and Aubrey 1991). At its peak advance, the ice sheet covered about half of North America. The formation of the Laurentide Ice Sheet and other continental ice sheets caused sea level to drop 320 to 400 feet (100–125 m) below today's levels along the Mid-Atlantic coast because much of the Earth's freshwater was held in the glacial ice sheets (Wolfe 1977; Fairbanks 1989). This shoreline was at the edge of the continent shelf—about 100 miles (161 km) offshore from present-day beaches in New Jersey—and the coastal plain was more than twice its current size and covered by meadows, wetlands, and boreal, hemlock, and hardwood forests (Schubel 1986; Kraft 1988; Figure 2.2). Georges Bank (today's prime fishing grounds) off the coast of New England was dry land at this time. The South Atlantic and Gulf of Mexico coasts were about 62 miles (100 km) and 62 to 155 miles (100–250 km) offshore, respectively (Anderson and Lockaby 2007). Barrier islands formed along former shores during times of relatively stable sea levels. For example, an ancient barrier island system has been discovered 0.12 to 1.1 mile (0.2–1.7 km) off the New Jersey coast approximately 66 feet (20 m) below the water and buried beneath 66 feet (20 m) of sediment (Wellner et al. 1993).

Postglacial Conditions

About 21,000 years ago, the climate warmed, causing the glacier to melt. Gradually the great ice sheet retreated, leaving New York, New England, and Nova Scotia about 10,000 to 15,000 years ago, Quebec about 7,000 years ago, and returning to the Arctic (Roland 1982; Pielou 1991; Lewis 1995; Fensome and Williams 2001). Three major events happened with the melting of the ice sheet: 1) sea level rose, reflooding the continent shelf and some lands depressed by the glacier; 2) land formerly depressed under the weight of glacial ice rebounded (this process is called "glacial isostatic adjustment"—it is like what happens to a sponge when you take your finger off it or to a sofa cushion when you get up from the couch); and 3) land that had bulged (much of the Atlantic Coastal Plain) began to subside.

Glacial Rebound

Much of today's coast from Maine to the head of the Bay of Fundy was underwater during the last glaciation, and when the glacier receded these marine bottoms rebounded well above tidal influence (Belknap and Shipp 1991). Marine clays can be found beneath upland soils in these locations and in coastal New Hampshire and parts of eastern Massachusetts (Figure 2.3). Nova Scotia beaches were raised up to 130 feet (40 m) above current shoreline levels (Davis and Browne 1996). The Champlain Sea extended from Quebec into northern New York, northwestern Vermont, and southeastern Ontario covering what are now the St. Lawrence River and Lake Champlain valleys. Glacial rebound eliminated much of this sea as the submerged lands rose above the reach of the tides. About 10,000 years ago, the sea was greatly reduced in size, becoming what are now the Gulf of St. Lawrence and the St. Lawrence estuary. The shorelines of the former Champlain Sea are now 492 to 656 feet (150–200 m) above sea level (Lasalle and Rogerson 2007). Glacial rebound is still occurring in northern latitudes as glacial ice melts due to climate change, and shorelines and sea floors are raised when the weight of glacial ice is removed from the Earth's surface (see Chapter 12 for additional discussion).

Sea-Level Rise

As the glacial ice sheet melted, sea level may have first risen at very high rates, up to 16.5 feet (5 m) per century, with some shorelines retreating up to 50 feet (15 m) per year

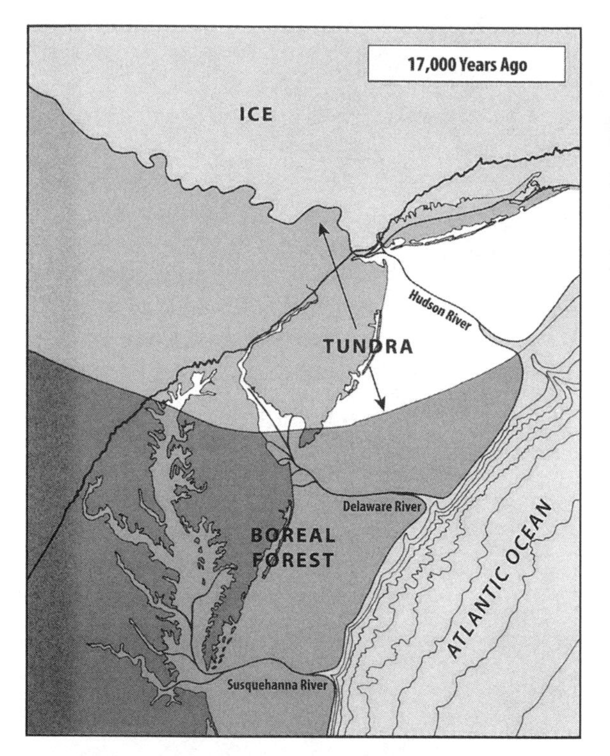

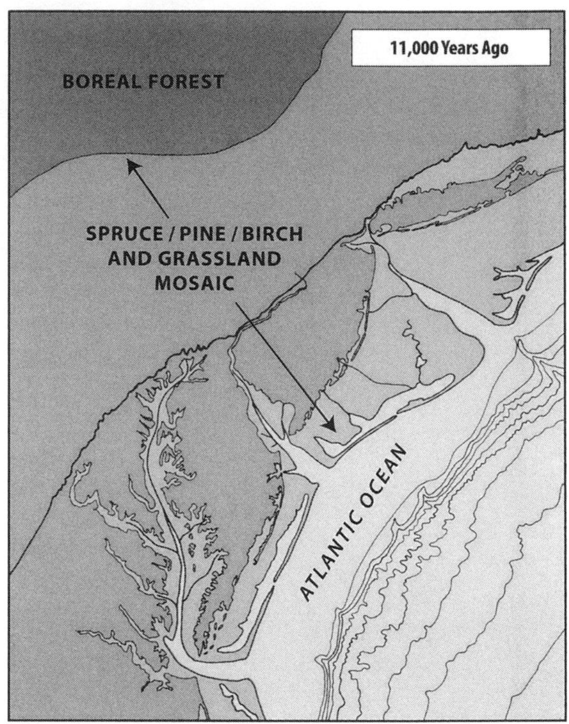

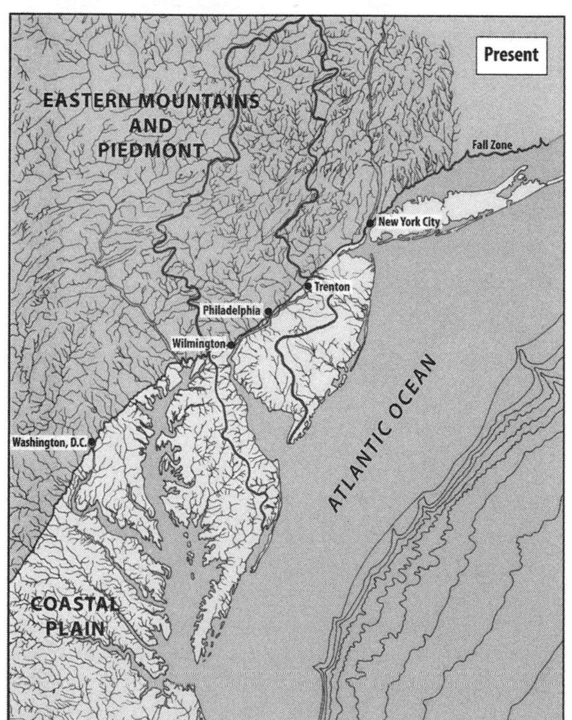

Figure 2.2. Dramatic changes in the Mid-Atlantic landscape occurred over the past 17,000 years. During the last glacial epoch, a thick sheet of ice covered much of North America to northern New Jersey along the Atlantic Coast. Sea level was about 426 feet (130 m) below current levels exposing much of the continent shelf. Tundra vegetation predominated to northern Delaware, and south of this area boreal forest (taiga) was present. As the continental ice sheet receded, sea level rose and ancestral estuaries migrated landward. Climatic shifts in vegetation took place with arctic/subarctic vegetation replaced by temperate forests and grasslands. As the rate of sea-level rise slowed, estuaries began to form in their present-day locations. (Kraft 1988; courtesy of University of Delaware, College of Earth, Ocean, and Environment and the Delaware Sea Grant College Program)

(Schubel 1986; Figure 2.4). This rise eventually drowned the former coastal plain (now part of the continental shelf). Other estimates project lower initial rates than this but with a couple of intervals of higher rates later (Gornitz 2007). From 15,000 to 5,000 years ago, the rate of rise eventually dropped to about 3.2 feet (1 m) per century, and about 10,000 years ago, the ocean reached the mouth of Chesapeake Bay. Even as sea level rose, during times of slow submergence rates tidal marshes formed on the continental shelf. Dating of continental shelf peat deposits from the northeastern United States suggests that marshes developed in various places from 8,600 to 4,700 years ago in response to changes in the rate of sea-level rise (Rampino and Sanders

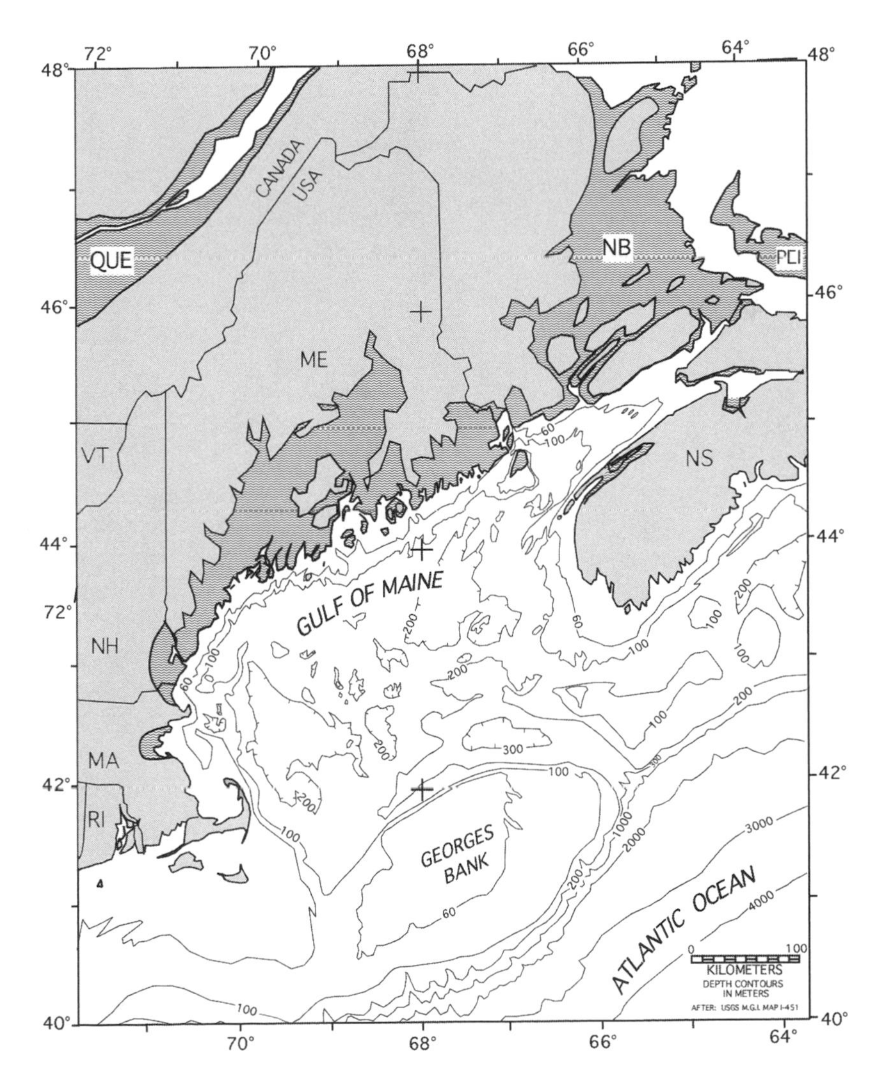

Figure 2.3. As the ice sheet melted and sea level rose quickly, some areas depressed by the weight of the ice were flooded until their surfaces rebounded. The wavy gray regions along the coast from Prince Edward Island to Massachusetts represent former marine bottoms. (Modified from Belknap and Shipp 1991; courtesy of Dr. Daniel Belknap, University of Maine)

1981). About 5,000 to 6,000 years ago, glacial ice melting virtually ceased, slowing sea-level rise to a rate between 0.5 and 1.0 foot (15–30 cm) per century (Craft 1988; Gornitz 2007). This allowed barrier beaches and islands to form near their present-day locations, and coastal wetlands started to form behind them and on floodplains along drowned river valleys. Between 2,000 and 3,500 years ago, sea-level rise in the Northeast dropped to an annual rate of 0.04 inches (1 mm) (Redfield 1965; Keene 1971; Van de Plassche et al. 1989; Warren 1995). This low rate facilitated more tidal wetland development. Chesapeake Bay and other large bays probably looked much like they do today, excluding human impacts, of course. With sea level still rising, there is a continued drowning of the coastline and a shoreward advance of barrier islands and salt marshes (see Chapter 12 for more discussion).

Tectonic Processes

Where the Earth's major plates meet, one plate often goes beneath another causing areas of volcanic and earthquake activity. Along the Pacific Coast of North America, the Pacific plate is sliding beneath the North American plate (subduction zone), building mountains along the edge of the continent and contributing to the "ring of fire" (active volcanic-earthquake zone) along the edges of the entire Pacific ocean basin. Earthquakes may create tidal wetlands by uplifting sea bottoms into the intertidal zone or by lowering uplands to this zone. These processes are particularly active in Alaska where, for example, the 1964 earthquake raised elevations in some areas by as much

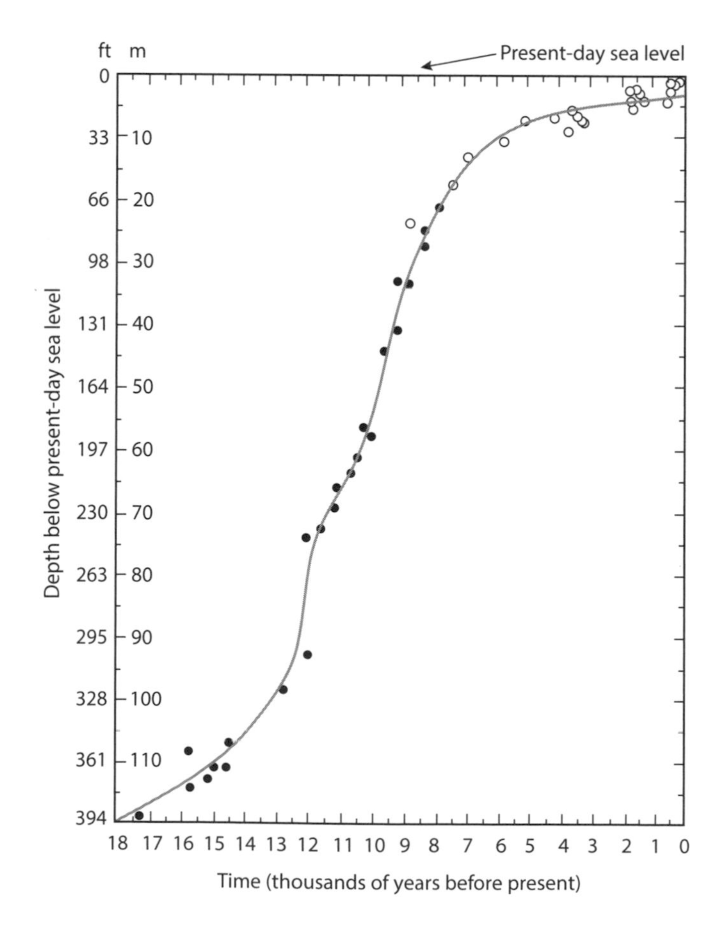

Figure 2.4. Global sea-level changes over the past 18,000 years. Around 6,000 years ago, sea-level rise began to slow and around 3,000 years ago began to level off. Recent changes are not reflected in this table due to scale. (Redrawn from Fairbanks 1989)

as 6 feet (1.8 m), while causing up to 7.5 feet (2.3 m) of subsidence in other places (Committee on the Alaska Earthquake 1972). Consequently, the Gulf of Alaska is one of the few areas where negative sea-level rise rates (declining sea level) can be found in North America. Volcanic eruptions can trigger massive landslides and mudflows that discharge sediment and debris into rivers and streams and eventually into tidal waters where they may contribute to accretion in the intertidal zone.

Tidal Wetland Formation

Wherever the tide flows and ebbs exposing substrates to the air for some time, tidal wetlands can be found. They include non-vegetated tidal flats and rocky shores as well as intertidal areas colonized by grasses and other herbs (marshes) or by trees and shrubs (mangroves and tidal swamps). These wetlands naturally form on floodplains along tidal rivers, in sheltered embayments and overwash deposits behind barrier islands and beaches, on river deltas, on former upland submerged by rising seas, and on former sea bottoms raised to intertidal levels by tectonic forces. Man-made wetlands can also be created, intentionally or accidentally, by mimicking natural processes that create intertidal substrates.

Since the mid-1800s scientists have proposed theories on how salt marshes form. Some scientists focused on the buildup of marsh peat as the marsh subsided (Mudge 1862; Davis 1910, 1911), while others emphasized the expansion of marshes onto tidal flats in a depositional environment (Shaler 1886). During these early times there was a general belief that the effect of deglaciation on sea level was over and that seas were not rising (Nixon 1982). Although neither theory fully explained marsh formation, later scientists found that each theory had some merit depending on the local conditions (e.g., Johnson 1925; Knight 1934). More recently, detailed studies of Barnstable Marsh on Cape Cod (Redfield and Ruben 1962; Redfield 1965, 1972) concluded that both of the early theories were correct and the differences were due partly to examining different locations of the marsh: high marsh development by B. F. Mudge and low marsh by N. S. Shaler. A sufficient supply of sediment is essential for the latter, while once established the former processes take over and marshes do advance over low-lying uplands in response to rising sea level. Since tidal wetlands include more than just salt marshes, deltaic, floodplain, and other processes are also important for tidal wetland formation.

Coastal Lagoon Infill

Once barrier islands and beaches stabilized, embayments formed behind them where waterborne sediments could drop out of suspension and build up bay bottoms. Mud, sand, and silt were deposited in the embayment creating a shallow water habitat. As sedimentation continued, eventually a portion of the shallow water area became a tidal flat exposed to air daily by the tides (Figure 2.5). A further increase in elevation (to mean sea level) allowed primary succession to commence—colonization of the substrate by pioneer species such as smooth cordgrass (*Spartina alterniflora*) and common glasswort (*Salicornia maritima*). The presence of these plants accelerated sedimentation by slowing the flow of water. Once plants are established, deposition rates can be five times that of adjacent tidal flats (Eisma and Dijkema 1997). Storms and spring tides continued to deposit sediments leading to a further buildup of marsh elevations. Levees may form along the edge of the marsh. A study of sedimentation of a Virginia salt marsh found that most of the sediments were deposited within 33 feet (10 m) of the shore and that the majority of deposition was brought in by spring tides (Christiansen 1998).

With vegetation firmly established, organic matter begins to accumulate in marsh substrates. Permanently saturated substrates typically experience anaerobic conditions due to lack of oxygen, except for a very thin oxygenated layer at the surface. Such conditions inhibit decomposition of dead vegetation causing peat or muck deposits to accumulate, further increasing marsh elevations. As the elevation rises above the daily tidal flooding level, environmental conditions change sufficiently to allow other plants to grow. Organic matter accumulation and sedimentation continue to raise marsh elevations until equilibrium is reached between erosion and deposition/accumulation. A deep soil core would show from top to bottom: high marsh, low marsh, and mudflat under the most simplistic circumstance. The development of a typical New England barrier spit marsh has been described for Barnstable Marsh on Cape Cod (Figure 2.6; Redfield 1965, 1972), while sequences of lagoon infill of Delaware's coastal bays have been detailed (Kraft 1971). More recently, establishment of smooth cordgrass marsh on a tidal sandbar has been documented, with vegetation, soil, and nutrient pool properties compared with those of mature salt marshes (Krull and Craft 2009).

On the seaward side, the advance of young *Spartina* marsh onto tidal flats depends on sedimentation (Figure 2.7). In time, pools may form within the marshes in depressions where water accumulates and peat rots, or in northern areas where rising tides lift out blocks of frozen peat. These processes continue today. Once the marsh is established, vegetation succession is determined by changes in sea level, salinity regimes, tidal influence, marsh accretion rates, human impacts, and other factors (see Chapter 3). The oldest salt marshes are only about 4,000 to 5,500 years old (e.g., Redfield and Rubin 1962; Kaye and

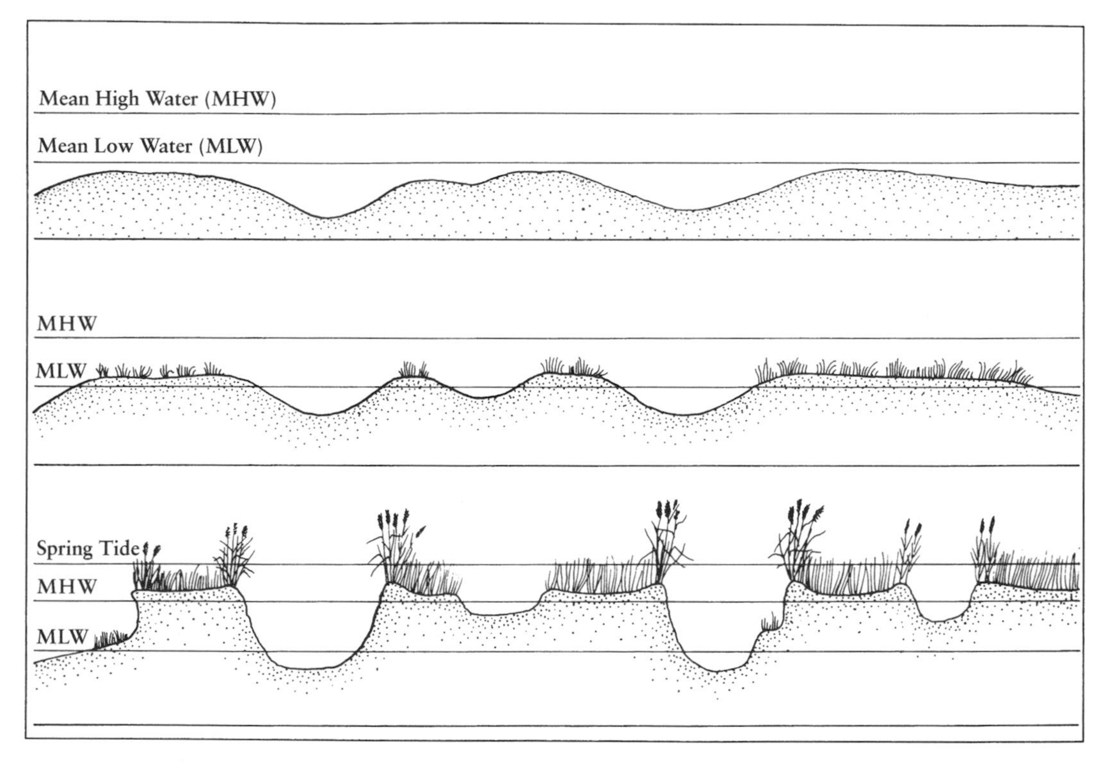

Figure 2.5. With increased sedimentation in shallow bays, substrates eventually reach the intertidal zone. Once elevations approach the mean high water level, smooth cordgrass (*Spartina alterniflora*) can colonize and the building of tidal marshland begins. (Courtesy of University of Delaware, College of Earth, Ocean, and Environment and the Delaware Sea Grant College Program)

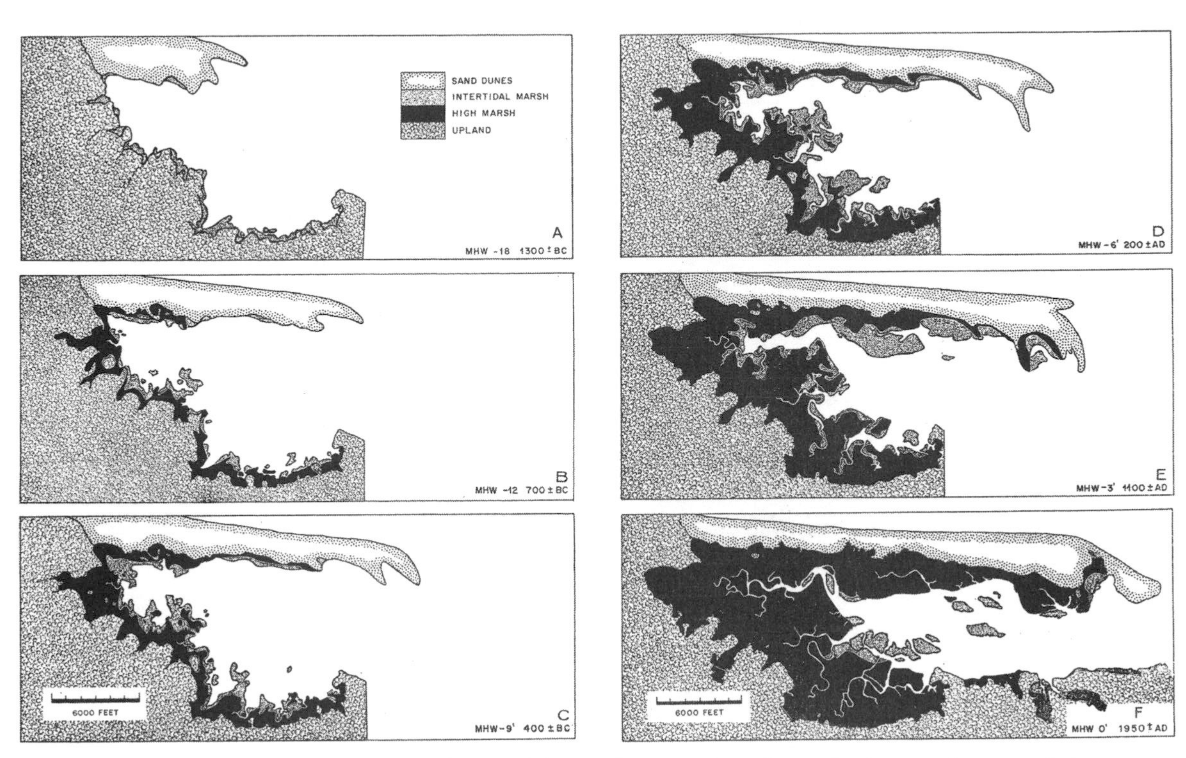

Figure 2.6. Postglacial development of Barnstable Marsh. Expansion of a sand spit created a more sheltered environment that facilitated formation of tidal flats and establishment and growth of the salt marsh. (Redfield 1972; reprinted by permission of the Ecological Society of America)

Barghoorn 1964; Gardner and Porter 2001). New England salt marshes typically have thick peat deposits up to several feet thick, while the youngest marshes have little or no peat. Soils of South Atlantic marshes are typically organic-rich deposits of sand, silt, and clay.

Marine Transgression

Rising sea level causes salt and brackish marsh migration in two directions: 1) landward into adjacent lowlands (uplands or freshwater wetlands) and 2) upriver into tidal fresh marshes and swamps. These changes are the result of a process called "marine transgression." It is one form of "paludification"—the swamping of land where marshes, swamps, and bogs advance landward or upslope as water tables rise due to climate change (wetter conditions) or, in the case of bogs, the amazing ability of peat moss to absorb water and expand its areal coverage. Marine transgression has been ongoing since deglaciation as sea levels rose and flooded former uplands and nontidal wetlands that now lay hundreds of feet below the ocean surface waters on the continental shelf. Peat deposits with smooth cordgrass rhizomes have been found at depths up to 194 feet (59 m) below today's waters on Georges Bank off the Massachusetts coast (Emery et al. 1965).

TRANSFORMATION OF COASTAL LOWLANDS

Salt marshes are known to advance landward as higher sea levels frequently flood low-lying uplands and adjacent freshwater wetlands, elevating local water tables and eventually creating saline soil conditions favoring halophytic (salt-tolerant) plants over freshwater or dryland species. Rising seas appear to initially raise water tables in lowlands bordering tidal marshes prior

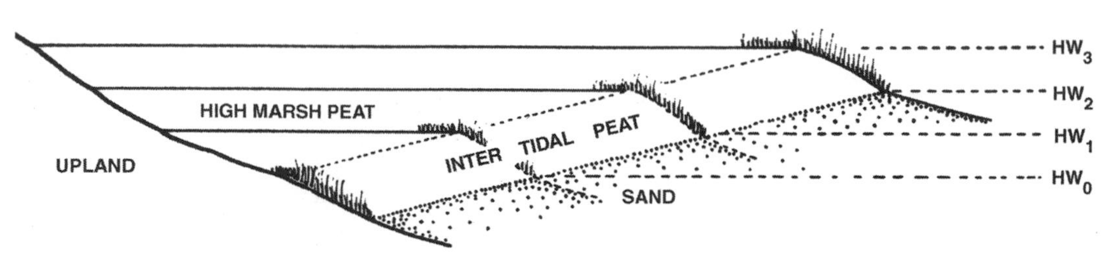

Figure 2.7. Salt marsh formation over sandflat on the seaward end and over upland on the landward end—Redfield's model. (Redfield 1972; reprinted by permission of the Ecological Society of America)

to tidal flooding. This may cause a shift in vegetation from woody plants to freshwater marsh plants in advance of tidal inundation. For example, analysis of estuarine sediments in the Severn estuary in Great Britain found estuarine clays above a layer of sedge peat that overlaid alder peat (Hewlett and Bimie 1996).

Since the mid-1800s, North American scientists have reported evidence of dead trees or stumps in the marshes or of bog deposits or former uplands beneath salt marshes (e.g., Dawson 1855; Mudge 1862; Ganong 1903; Penhallow 1907; Bartlett 1909, 1911; Harshberger and Burns 1919). The remains of Atlantic white cedar have been found at the bottom of a Cape Cod bay (Bartlett 1909). In Massachusetts, salt marsh peat from high marsh species was discovered above archeological remains (stone tools and hearths) providing evidence of high marsh overtaking upland as sea level rises, while other excavations found salt marsh peat over freshwater sedge peat (Raup 1959; Kaye and Barghoorn 1964). Former agricultural fields cultivated before the Civil War on Maryland's Eastern Shore are now covered by 10 inches (25 cm) or more of salt marsh peat (Darmody and Foss 1978). At the Magdalen Islands in the Gulf of St. Lawrence, the following signs of submergence have been found: 1) tree stocks and terrestrial peat submerged along sand dunes; 2) erosion of most of the salt marshes; and 3) thick deposits of terrestrial peat at water depths to about 21 feet (6.5 m) (Dubois and Grenier 1993). Carbon dating of peats has shown that this area has been undergoing submergence at a mean rate of ¹⁄₂₅th inch (1 mm) per year for at least 6,500 years. In other places, peat banks can be observed in front of existing beaches where they are now subject to the erosive forces of wave action—further evidence of marine transgression (Figure 2.8).

Soil coring studies elsewhere have found forest soils and tree roots below high marsh vegetation as well as marshes buried below estuarine sediments (e.g., Kraft 1971; Gardner and Porter 2001). Analysis of high marsh soils in Maryland and Virginia found that they possessed properties similar to adjacent uplands, leading researchers to conclude that the marshes were submerged uplands (Darmody and Foss 1978; Edmonds et al. 1985). Likewise in South Carolina, salt marsh soils have been found overlying relict Pleistocene beach-ridge sands and former forest soils (Thibodeau 1997). Analysis of deep soil cores determined that the establishment of black needlerush (*Juncus roemerianus*) marsh preceded colonization by smooth cordgrass in some places, while in other areas smooth cordgrass formed on tidal flats (Gardner and Porter 2001).

Direct evidence of contemporary marine transgression includes the presence of recently killed trees or dying salt-stressed trees in the marshes, or seemingly healthy trees with salt marsh vegetation growing beneath them (Figure 2.8). The latter trees eventually show signs of salt stress (e.g., chlorosis—yellowing of pine needles) and later succumb to the saline environment. In the northeastern United States, the area where the process of marine transgression

Figure 2.8. Signs of marine transgression: (a) an eroding peat bank in front of a sandy beach and dunes (NJ); (b) dead trees in a salt marsh (NB); and (c) salt marsh vegetation invading dying pine forest (MD).

is most widespread, is Maryland's Eastern Shore of Chesapeake Bay—part of the land mass that bulged during the last glaciation. Two soil types (Honga and Sunken series) described as "submerged upland soils" now support tidal marsh vegetation, with the latter being in a transitional stage with salt marsh vegetation growing beneath dying or dead pines (Brewer et al. 1998a). Atlantic white cedar swamps and red maple swamps along some creeks on the New Jersey shores of Delaware Bay are experiencing similar transgressions (e.g., Dennis Creek and Old Robins Branch), as are mainland forests along the barrier island lagoons of Virginia (Brinson et al. 1995) and lowland forests along the South Carolina coast (Gardner et al. 1992). Besides dead, dying, or salt-stressed trees, another sign of recent submergence in New England salt marshes is a change in marsh plant communities from salt hay grass (*Spartina patens*) to salt grass (*Distichlis spicata*), as the latter species appears to dominate waterlogged sites (Niering and Warren 1980; Roman et al. 1997), or from high marsh species to stunted smooth cordgrass (Warren and Niering 1993).

TRANSFORMATION OF UPSTREAM WETLANDS

As sea level rises, tides penetrate farther upriver, causing nontidal marshes and swamps to become tidal wetlands, while downstream tidal swamps may be converted to marshes and neighboring marshes may become tidal flats or open water. In the late 1700s, naturalist William Bartram wrote that Carolina rice planters seeking to drain these marshes found stumps of bald cypress and other trees 3 to 4 feet (0.9–1.2 m) beneath the marsh surface (Van Doren 1928). More recently, investigation of salt marsh formation in the lower reach of the Pataguanset River (East Lyme, Connecticut) provided a long-term view of the process of marine transgression in a New England river valley (Figure 2.9; Orson et al. 1987). With rising sea level about 4,000 years ago, brackish marshes dominated by salt marsh bulrush (*Schoenoplectus robustus*) and common three-square (*Schoenoplectus pungens*) replaced freshwater marshes and low-lying uplands in the river valley. Salinities increased over time and around 3,000 years ago, salt marsh vegetation began to colonize the marshes. Salt grass was the main pioneer species, eventually joined by other species. There was no evidence of black grass (*Juncus gerardii*) in peat samples dated 500 years and older, leading the investigators to suspect that it may have been a European transplant. Since this species was an important salt hay crop for early colonists, it may have been here before the arrival of Europeans or established quickly after they arrived. (*Note:* Roots and pollen of rushes do not persist in peat, so pollen analyses would be inconclusive; Gail Chmura, pers. comm. 2005.) A similar pattern of salt marsh transgression over freshwater wetlands has been described for Maine (Kelley et al. 1988).

Formation of some tidal wetlands along the Delaware River followed a similar progression but with humans playing a role in the process (Orson et al. 1992). About 2,500 years ago, the Woodbury Creek site was an oak-dominated forested wetland interspersed with sedge-dominated depressions. As sea level continued to rise, elevated groundwater levels created wetter conditions that eventually allowed alders to replace the oaks, and cattail and sedges to form marshland. By the time Europeans began to settle the area (about 350 years ago), the former forest had become freshwater marsh and wet meadow, and tidal fresh marsh occurred along the main channels of the creek. By the early 1800s, settlers had clear-cut lowland forests, dammed the creek, and diked marshes to create pastures and cropland. Abandoned agriculture in the late 1800s allowed second-growth forest to take over the diked open land. Later the dikes began to fail, allowing tides and waterborne sediments to enter the diked lands. In 1940, the dike collapsed and within a few years, tidal fresh marsh occupied the entire site. Today, the marsh is flooded twice daily by the tides and characterized by typical tidal fresh marsh herbs.

A recent study of vegetation change over a 10-year period in a Virginia freshwater tidal marsh found that salt-tolerant species, mainly big cordgrass (*Spartina cynosuroides*) and shoreline sedge (*Carex hyalinolepis*),

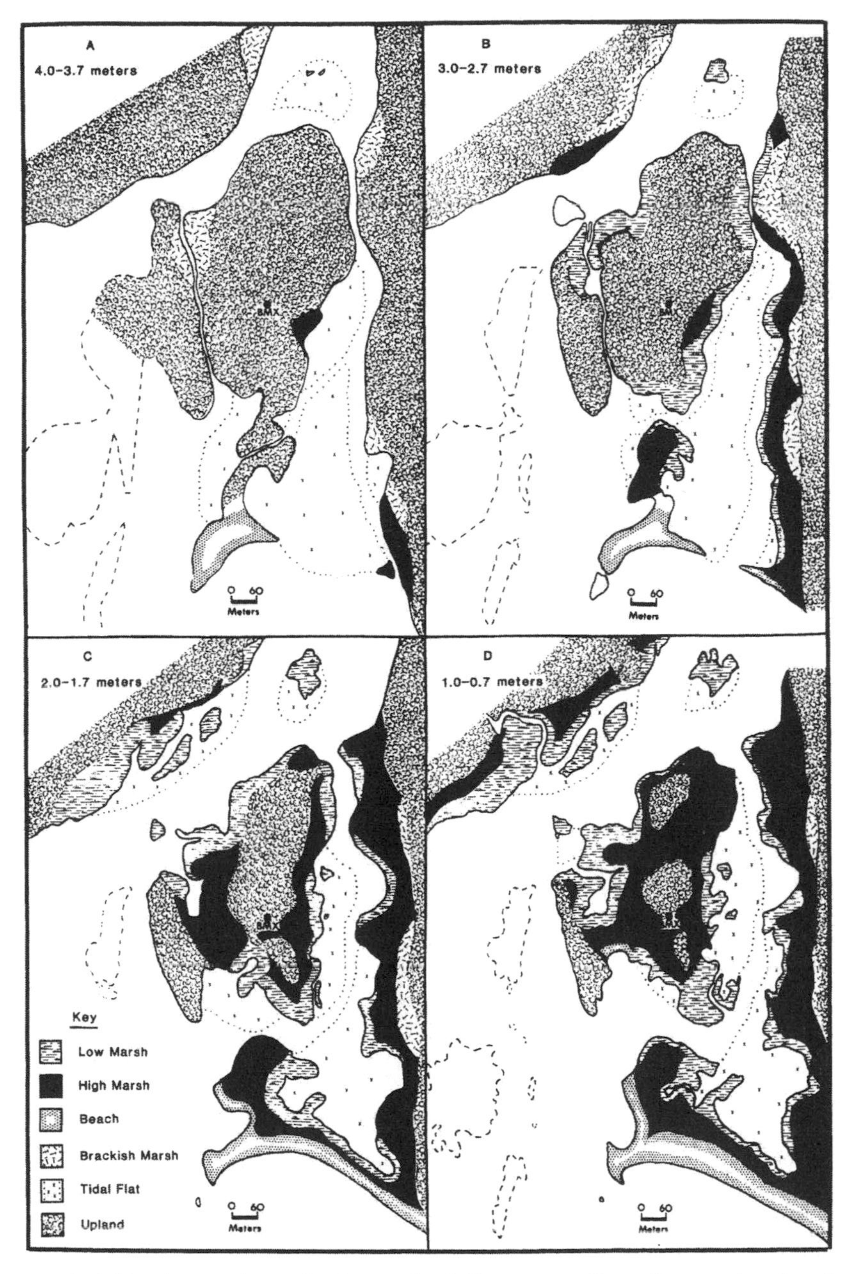

Figure 2.9. Postglacial landscape change in Connecticut's Pataguanset River Estuary showing the submergence of upland and expansion of tidal wetlands induced by rising sea level. This sequence is based on sea-level rise rates of about 3.3 feet (1 m) per 1,000 years, which are much lower than today's rate. (Figure 3 from Orson et al. 1987; *Estuaries*; copyright Estuarine Research Federation; reprinted by kind permission from Springer+Business Media B.V.)

replaced the dominant freshwater species—arrow arum (*Peltandra virginica*)—in some areas, suggesting active marine transgression (Perry and Hershner 1999). Some other examples of recent changes in coastal plant distribution attributed to sea-level rise and global warming are presented in Chapter 12 (in the section Observed Effect of Recent Sea-level Rise on Tidal Wetlands).

OVERWASH MARSH FORMATION

Relatively narrow barrier islands or spits with low dunes and no forests are often retreating coastal features—migrating landward as a unit with rising sea level (Figure 2.10). Wetlands are created, destroyed, and rebuilt immediately behind them over time. Storm-generated waves and water levels cause the island to "roll over" itself, moving the sand toward and often into the back-barrier lagoon or embayment creating "washes." This overwash process occurs periodically and usually deposits 4 to 6 inches (10–15 cm) of sand on the back-barrier marshes (NOAA 2010). While thicker deposits may eliminate fringing wetlands through deep burial, existing vegetation (e.g., salt hay grass or common three-square) buried beneath thinner deposits can push their stems through the material, re-establishing salt marsh. When washes spread onto tidal flats or shallow water, elevations may be raised sufficiently for new salt marsh to establish. Smooth cordgrass may colonize these sites. In just 14 years, this species built up a peat layer 4 inches (10 cm) thick on a North Carolina overwash deposit (Godfrey and Godfrey 1974). So as existing marshes are buried, new wetlands (tidal flats and marshes) may form in the shallow back-barrier lagoon. On the ocean side, beach migration may eventually expose old salt marsh deposits allowing marsh plants to re-establish on low-energy shorelines. This roll-over process can occur within 100 years (e.g., Core Banks in North Carolina; Godfrey 1976). A barrier beach at St. Catherines Island in Georgia moved at an average rate of 12.5 feet per year (3.8 m/yr) from 1945 to 1990 (Goodfriend and Rollins 1998).

The most severe storms, especially hurricanes, can break through the dunes of these narrow islands, creating inlets that separate the barrier island into smaller units. When the inlet forms, material that once formed part of the original barrier island is redeposited in the embayment inside the mouth of the new inlet. These sandy shoals may become tidal flats as they receive more sediment, and with continued sedimentation, elevations may become high enough to support marsh vegetation. Depending on local conditions, these "flood-tide deltas" may continue to receive sediments and the inlet may eventually fill in with sand, restoring the former barrier island as a single unit with fringe marshes behind (Figure 2.11). Inlet migration over time may create sizeable wetlands along the backside of barrier islands in otherwise sediment-starved estuarine lagoonal systems.

Deltaic Wetland Formation

Rivers with large watersheds often carry huge amounts of sediments and deposit them at their mouths where the river meets the sea. Here as river velocity decreases upon interacting with ocean currents and tides, the bed-load drops out of suspension forming a depositional plain called a "delta" (see Figure 1.6). Tidal wetlands form on these intertidal plains. When viewed from the air, deltaic sediments may give the appearance of having been spilled out of the river in different directions forming a broad plain comprising a number of distributary channels and deltaic lobes. These deltas continue to expand as long as the sediment supply is sufficient. Natural levees may form along these rivers and periodically high floods cause breaks (crevasses) in the levees (Figure 2.12). These breaks discharge fluvial sediments into open bays, thereby forming subdeltas upon which new wetlands become established. When this sediment supply is significantly diminished, coastal erosion may cause deterioration of wetlands,

Figure 2.10. Overwash deposits moving onto back-barrier marshes: (a) Cape Cod (MA); (b) Matagorda Island (TX); and (c) Metompkin Island (VA). (c: Bryan Watts)

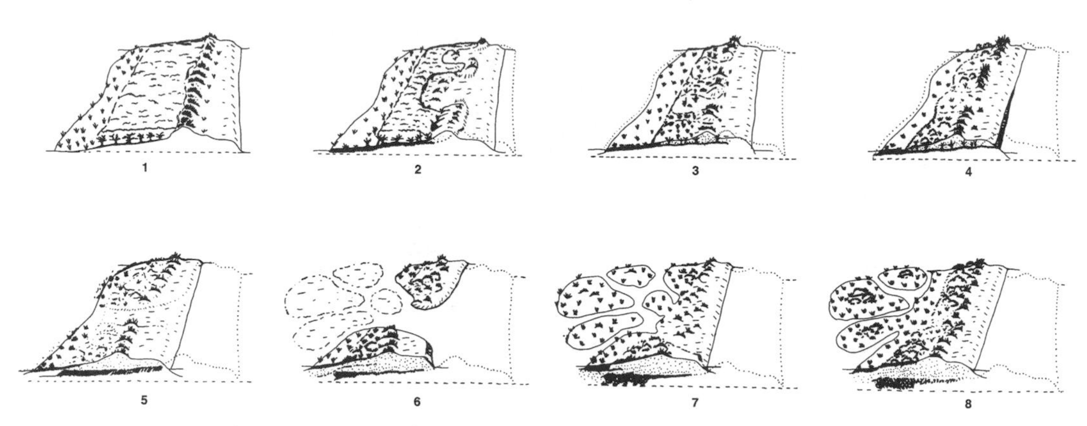

Figure 2.11. Sequence showing dynamics of barrier islands, inlet formation, and overwash. In each stage, overwash deposits are recolonized by plants depending on the site (salt marsh, fresh marsh, or dune species). Marshes also form on tidal deltas when inlets are open. Over time, the barrier island rolls over itself and former salt marshes (peat deposits shown as dark pattern beneath the island or exposed along the ocean shore). Barriers will continue to advance landward with rising sea level. (Redrawn from Godfrey and Godfrey 1974)

especially when combined with rising sea-level impacts. Natural changes in the river course ("channel avulsion" or "delta switching") therefore affect the distribution of suspended sediment causing a buildup of sediments and marsh formation in new discharge areas at the expense of maintaining wetlands along the old channel.

Since most tidal wetlands form in depositional environments, they are well represented on tidal deltas either as intertidal flats, salt marshes, or mangroves (in the tropics). The actual shape of a delta is the product of a number of factors including river flow, sedimentation, tectonic stability, shoreline currents, and climate (Boggs 2000). What most people think of when visualizing a delta—a conspicuous lobe or series of lobes protruding outward at the mouth of a river—is one where fluvial processes predominate. In tidal regions, this type usually forms where the tide range is low and the ocean currents weak. The best example of this is the Mississippi River Delta, which is commonly referred to as the "bird's foot" due to the pattern formed by four major distributaries (see Figure 1.6). The enormous size of the Mississippi Delta, and the fact that it has been transporting sediments to the Gulf for thousands of years forming new deltas over time with each change in the river's course, are the reasons why more than one-quarter of the estuarine wetlands in the coterminous United States are located in Louisiana (Figure 2.13). With changes in distributaries, new deltas form and abandoned deltas degrade as land subsides. Deltas forming along coasts with moderate to high tidal ranges and strong wave action and currents, tend to be truncated (see Figure 2.1). Here riverborne sediments are moved along the shore by longshore currents to be deposited on down-drift beaches. Deltas of many rivers along the South Atlantic Coast exhibit this shape, while others are associated with coastal plain rivers and smaller drainages. In macrotidal areas, linear ridges may form at the mouth and an intricate pattern of numerous creeks may characterize the intertidal zone (Wright 1978). India's Ganges-Brahmaputra Delta is a tide-dominated delta where channels and islands are created perpendicular to the coast (see Figure 1.7).

Floodplain Wetland Formation

Besides forming at the mouths of rivers on deltaic floodplains, tidal wetlands also

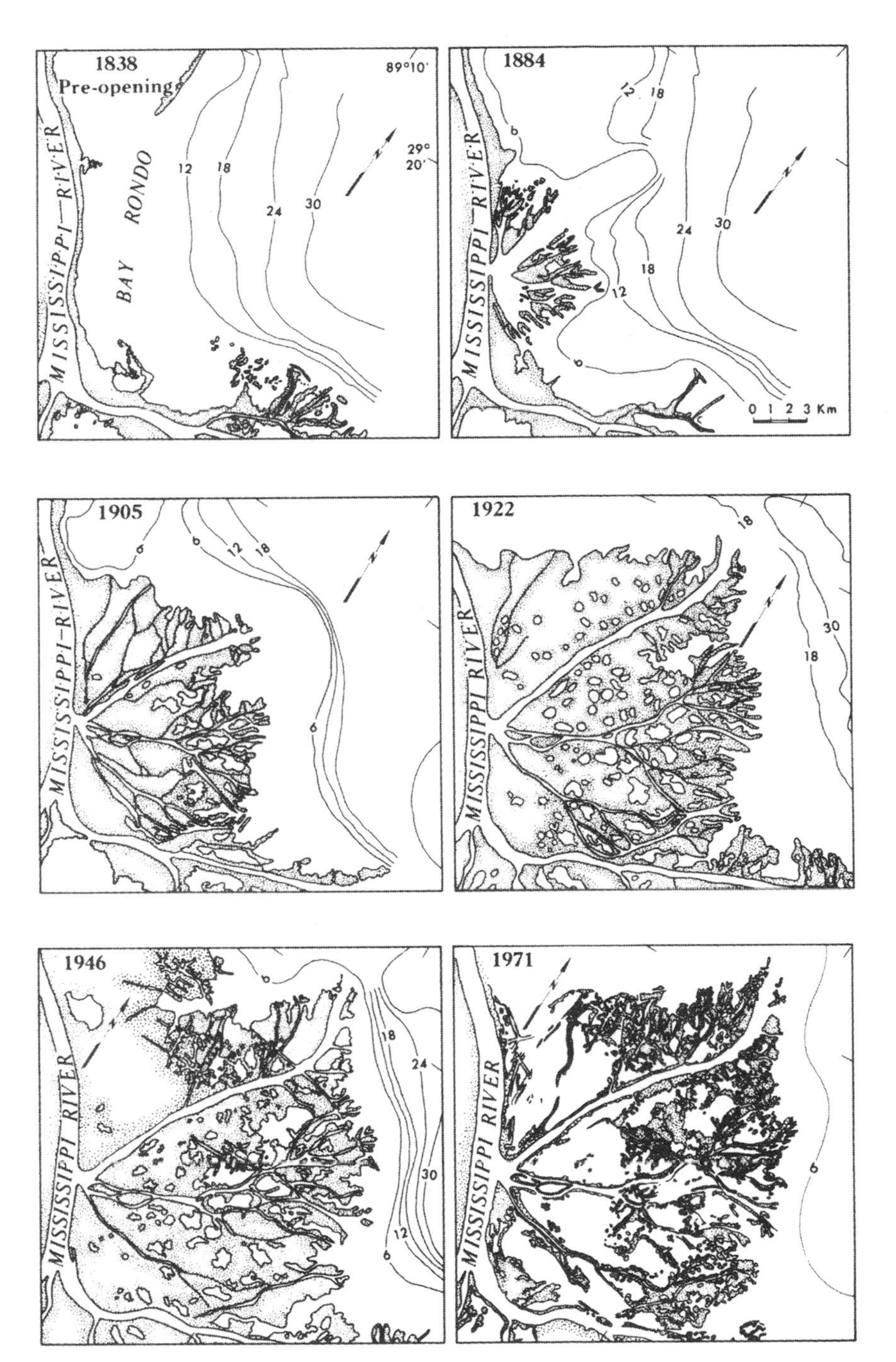

Figure 2.12. Development of a subdelta of the Mississippi River Delta at Cubits Gap. Prior to 1862, daughters of a fisherman named Cubit excavated a ditch in the natural levee near the seaward end of the Mississippi Delta for passage of shallow-draft boats. The flood of 1862 caused a crevasse break that widened the ditch and by 1868, it was about six times its original width. In time, more sediment was deposited in Bay Rondo creating a subdelta. Progradation continued until sometime before 1946 when evidence of marsh deterioration and sedimentation limited to the seaward ends of certain distributaries indicated the advance of marine waters (possible subsidence) and reduction of sediment supply. By 1971 significant erosion of the marshes had taken place. Note that land loss was initiated and greatest near the crevasse break as sediments are deposited at the end of the distributaries. The wetlands formed at Cubits Gap are now part of Delta National Wildlife Refuge. (Wells et al. 1982; Gosselink 1984)

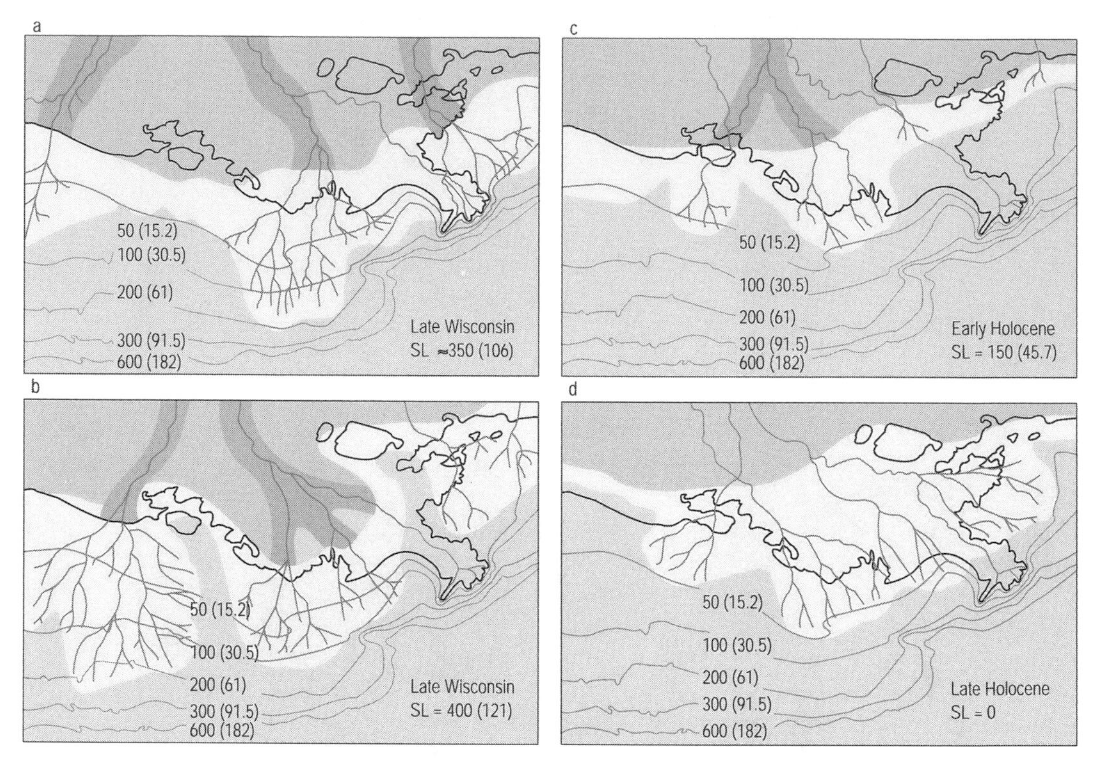

Figure 2.13. Major deltas of Louisiana since the last ice age: (a) 25,000 to 20,000 years before present; (b) 15,000 years before present; (c) 12,000 to 10,000 years before present; and (d) 5,000 to 1,000 years before present. Depth contours are in feet (meters in parentheses) below today's sea level. For example, in the Late Wisconsin epoch, sea level was 400 feet (121 m) below current levels, therefore much of the Gulf's continental shelf was marsh or dry land. (Gosselink et al. 1998)

become established along the shorelines where other floodplains have developed (see Figure 2.1). Fluvial sediments transported downstream by rivers are deposited along shores by lateral accretion and overbank deposition during high water (e.g., episodic) events in both nontidal and tidal reaches. When the tide shifts (ebb to flood tide and flood to ebb tide, "slack tide"), currents slow, causing waterborne sediments to begin settling out of suspension, especially in the brackish water zone and tidal fresh zone.

Tidal flats develop along the margins of coastal rivers. These flats are the first sign of floodplain development where marshes, forested wetlands, and non-wetland alluvial forests later may arise. The rate and extent of sedimentation and continued buildup of these floodplains depend on several factors including the original size and age of the estuary, present erosion rates upstream, deposition by the river and tides, and currents (Reid 1961).

Depending on the water's salinity, salt-tolerant (halophytic) or freshwater plants become established. In the northeastern United States, smooth cordgrass is the pioneer species in salt and brackish areas, while water hemp (*Amaranthus cannabinus*) joins this species along tidal creek banks of moderately brackish waters. Plants such as spatterdock (*Nuphar luteum*), arrow arum, common three-square, wild rice (*Zizania aquatica*), and a variety of less conspicuous, low-growing mudflat species are among the first species to colonize freshwater tidal flats in this region. The presence of these pioneer plants helps further slow the velocity of water, inducing more sedimentation. Marsh vegetation eventually will add significant amounts of organic matter to the substrate, leading to a further increase in wetland elevations and a transformation to tidal shrub swamps and eventually tidal forested wetlands in freshwater reaches. Where rivers carry significant sediment loads, the forested wetlands may eventually become alluvial wetlands (floodplains above tidal influence) that are only infrequently subjected to flooding during spring freshets or are flooded only by catastrophic events. The extent of alluvial coastal wetlands varies with the width of the "valley" and its elevation relative to flood tide ranges. They may be vast along rivers with broad valleys or limited to linear features in narrow stream valleys.

Along the Hackensack River in New Jersey, Hackensack Meadowlands became coastal wetlands under different circumstances (Heusser 1949; Sipple 1971). The Meadowlands had its origins as a glacial lake—Lake Hackensack, formed by waters melting from the Laurentide Ice Sheet and runoff from adjacent uplands. Following the retreat of the glacier, the depressed land rebounded to the point that the lake drained. This exposed the lake bed for plant colonization and freshwater wetlands became established. Rising sea level eventually entered the former lake basin through its southern end (lowest point). In the early 1800s, an Atlantic white cedar swamp occupied the Secaucus area, with boreal plants such as black spruce (*Picea marina*), larch (*Larix laricina*), and bunchberry or dwarf cornel (*Cornus canadensis*) also present. As sea level rose in the Hackensack River, trees subjected to tidal flooding and sediment deposition died and were replaced by brackish species, namely Olney three-square (*Schoenoplectus americanus,* formerly *S. olneyi*). Human activities quickened the vegetation change and demise of the cedar swamp. By the late 1800s, much of the wetland had been ditched, increasing tidal flow and saltwater intrusion. Common reed (*Phragmites australis*) invaded the bog and took over. Reduced freshwater inflows due to increased water use by the growing New York City–Newark population and channel deepening of rivers for navigation further accelerated the conversion of freshwater wetlands to tidal brackish wetlands. By the mid-1900s, the wetland vegetation was drastically different from what it was in the early 1800s. The Hackensack Meadowlands is now the largest estuarine wetland ecosystem in northern New Jersey. Evidence of these vegetation changes can be found in the substrate where peat deposits from swamp forests lie beneath today's estuarine deposits.

Other Natural Factors in Wetland Formation

While the main forces stimulating tidal wetland formation have been discussed above, other processes are also operating to create and reshape these wetlands. Erosion of cliffs will deposit material in tidal waters. Where sufficient material is deposited to reach an elevation at which the substrate is periodically exposed by the tides, a small wetland will become established. A nonvegetated one such as a rocky shore or tidal flat is the first wetland observed, and perhaps later if conditions are favorable, a vegetated wetland will become established, either an algae-covered rocky shore or a tidal marsh. Narrow fringing tidal marshes—bluff-toe marshes—have formed along the rocky coast of Maine where bluffs of muddy Pleistocene sediment have eroded (Kelley et al. 1988). Salt marsh development in northern Canada is limited by the availability of silt and clay and suitable substrates for vascular plant establishment (e.g., substrates are mostly boulders, pebbles, and sand) and by the harsh climate (ice scouring and frost action) (Glooschenko et al. 1988). For information on beach formation, consult other sources (e.g., Davis 1978).

The elevation of coasts in some regions of the world is rising due to glacial rebound or tectonic processes. When the continental glacier left New England approximately 10,000 years ago, the great weight of the ice sheet was removed from the land and the depressed land rebounded ("isostatic uplift"). Uplift exceeded sea-level rise in many areas, and today areas of former beaches and sea bottoms lie well above sea level (e.g., blue-gray marine clays form subsoil in much of Maine, see Figure 2.3). Glacial rebound or uplifting is still occurring in some regions where the continental ice sheet last departed. Along the St. Lawrence River in Quebec, the land is rebounding at less than ¼ inch (6 mm) per year (Delft Hydraulics 1992). Instead of being submerged by rising sea level, here coastal wetlands are uplifted. Tidal marsh plants move to newly emerged downslope areas where favorable water levels are found. The upper marshes gradually become freshwater wetlands as they are decoupled from marine influence (Dubois 1993). Soil profiles show a vegetation pattern opposite that of submerging coasts: freshwater marsh overlies high marsh with low marsh below. The Labrador coast, Quebec's northern coast (including the southwestern part of Hudson Bay), Russia's White Sea coast, and the northern coast of the Gulf of Bothnia (Sweden and Finland) are other examples of emerging coastlines (Delft Hydraulics 1992; Svensson and Jeglum 2000). These areas are experiencing uplift at a rate of up to 0.8 inches (20 mm) per year (Valentin 1954; Delft Hydraulics 1992). Glacial rebound along the Gulf of Bothnia has amounted to a rise of about 2,640 feet (800 m: 500 m during glacial melting and 300 m since then), and Norway spruce forests now occupy former tidal areas (Svensson and Jeglum 2000, 2001).

Tectonic processes such as mountain building ("orogenesis"), where continental and oceanic plates collide, and accompanying volcanic activity also can cause an increase in shoreline elevations, while earthquakes can have varied effects. The former process is responsible for an uplifting of the Oregon coast at rates of 0.04 to 0.14 inch per year (1.0–3.5 mm/yr) (Weldon et al. 2006). In 1964, an earthquake uplifted parts of the Copper River Delta in Alaska 5.9 to 11.2 feet (1.8–3.4 m) above their former elevation causing major shifts in coastal wetland plant communities (Thilenius 1990, 1995; Boggs 2000). This event raised slightly brackish marshes above the tidal zone and elevated pre-1964 subtidal areas into the intertidal zone. Former subtidal areas are now mudflats, former mudflats have become brackish marshes, and after 20 years, brackish species have been replaced by freshwater herbs, shrubs, and trees on the uplifted former brackish marshes. The 1964 Alaskan earthquake also caused land subsidence in other areas. For example, the ground in and around the town of Portage dropped at least 7.9 feet (2.4 m) converting about 7 square miles (18 km^2) of upland to tidal flats and fringing marshes (Ovenshine and Bartsch-Winkler 1978; Shennan and Hamilton 2006). A layer of sediment from 3.0 to 4.9 feet (0.9–1.5 m) thick was deposited mostly by tidal action on the newly flooded areas. Quicksand formed over the entire region as uncompacted silt became liquefied upon inundation. After 10 years, some revegetation of the intertidal zone was beginning to occur.

The Human Factor in Wetland Formation

For thousands of years, human activities have accidentally created tidal wetlands through poor land-use practices. The introduction of stone and later bronze axes made it possible to clear forests for early agriculture. In the Mediterranean region, this practice coupled with slash-and-burn agriculture may have been at least partly responsible for eroding sediments that silted in coastal embayments including important harbors—possibly as far back as 4000 B.P. (e.g., van Andel et al. 1990; Zangger 1991). Tidal wetlands undoubtedly formed in these new intertidal zones.

During settlement of North America by Europeans in the 18th to 19th centuries, deforestation and poor farming practices in coastal watersheds accelerated the erosion of topsoil. Enormous amounts of soil were eroded, carried by rivers and streams, and deposited in the coastal waterways. This process contributed significantly to tidal wetland formation in the Plum Island Estuary, Massachusetts (Kirwan et al. 2011), the Delaware River (Orson et al. 1992), tributaries emptying into Chesapeake Bay (Brush 1989; Orson et al. 1990; Hilgartner and Brush 2006), the Mississippi River, Chezzetcook and Petpeswick Inlets, Nova Scotia (Davis and Browne 1996), and East Machias River, Maine (Anderson et al. 1992). For the latter site, the reduction in logging and agriculture is believed to be partly responsible for the erosion of the marsh resulting from rising sea level since the sediment supply that created the wetland is no longer available in sufficient quantities to sustain it. This may also be true for other wetlands as well. An analysis of soil cores, pollen, and macrofossils (seeds, fruits, and rootlets) at Otter Creek Point in the upper Chesapeake Bay region provided evidence of infilling of tidal freshwater and a change from aquatic bed habitat to marsh, shrub swamp, and riparian forest following European settlement (Hilgartner and Brush 2006). Prior to this colonization, a subtidal aquatic bed maintained its presence in the creek's delta for 1,500 years while the watershed's forests remained intact. At that time, the estimated sedimentation rate was 0.02 inch per year (0.5 mm/yr). When colonists began clearing forests to create farmland, sedimentation rates increased; and when it reached 0.2 inch per year (6.0 mm/yr) the aquatic macrophytes disappeared. From 1840 to 1880 four major storms caused enormous soil erosion, raising sedimentation rates as high as 18.9 inches per year (480 mm/yr) and changing the remaining aquatic beds to marsh to shrub swamp and riparian forest. Areas have received more than 3.3 feet (1 m) of sediment since the 1700s. Vegetation changes were not gradual but were pulsed according to sediment input going from steady state to a pulsed state back to equilibrium (i.e., stasis-pulse-stasis model). Riverborne materials were believed to be the primary sediment source for the formation of salt marshes south of Cape Lookout, North Carolina (Meade 1982). In the 1800s, soil washed from logged sites and farmland filled shallow water and mudflats of Tomales Bay in California, creating extensive marshes that nearly doubled the extent of marshland (National Park Service and California State Lands Commission 2007).

Contemporary land-clearing operations have similar effects and at varying scales. Land clearing for silviculture in the mid-1960s produced sediments that dramatically increased the rate of delta and tidal wetland formation in a small estuary in North Carolina (Mattheus et al. 2009). In tropical countries (e.g., Brazil, Venezuela, Thailand, and New Zealand), widespread land clearing is contributing enough sediment to coastal waters that new mangrove swamps are forming (Schwartz 2003; Maia et al. 2006). (*Note:* The lack of sediment supply from a reduction of land clearing, improved soil conservation, and damming of rivers may ironically contribute to the erosion and submergence of tidal wetlands from rising sea level.)

Some marshes are unintended products of other human activities. For example, marshes may form behind causeways and similar structures that create sheltered tidal habitat (Figure 2.14). In 1839, construction of a solid pier in the Hudson River created a sheltered environment where sediments accumulated, allowing for the expansion of Piermont marsh (www.nerrs.noaa.gov/HudsonRiver/HistoryPiermont.html). Road construction in Chezzetcook Inlet in Nova Scotia since the 1950s provided sediments that increased mudflat elevations sufficient to support smooth cordgrass (Wells and Hirvonen 1988). Hydraulic gold mining in the Sierra Nevada Mountains of California from 1856 to 1887 washed huge volumes of sediments downstream into San Pablo Bay creating 16,000 acres (25 mi^2 or 64.74 km^2) of new mudflats (Jaffe et al. 1998). Marsh

Figure 2.14. Tidal marsh formed behind rip-rap pier.

vegetation colonized suitable sites and as much as one-third of the San Francisco estuary marshes may have been created in this way (Atwater et al. 1979). These marshes are about 120 years old (Brown and Pasternack 2005).

In some places, Colonial farmers created salt marshes from freshwater wetlands to grow salt hay—a valuable commodity. They breached natural barriers to bring tidal flow to nontidal wetlands. Fresh Pond and Flax Pond on Long Island, New York, are examples of this process (Heusser et al. 1975; Clark and Patterson 1985). For Fresh Pond, elevations were initially too low for salt hay grass establishment, so it was colonized by smooth cordgrass. By 1850, high marsh had replaced the low marsh. In the 1920s, common three-square colonized the marsh in likely response to a short-term decline in local sea level. The marsh was ditched in 1930 and by 1940, typical salt marsh species regained dominance.

After years of neglect and degradation, coastal wetlands have once again become recognized as valuable natural resources in the United States. Since the 1970s, this awareness and public concern about the future of tidal wetlands have prompted government agencies at both the federal and state levels to not only protect these wetlands but to restore damaged marshes and create coastal wetlands on fill sites that were formerly occupied by tidal wetlands (e.g., dredged material disposal sites) (Chapters 9 and 11). The U.S. Army Corps of Engineers has created numerous tidal marshes in conjunction with navigation projects (e.g., maintaining channels for deep-draft ships). Building diked marshes has been viewed by some as a beneficial use of dredged material, although such projects have often been completed at the expense of tidal flats and shallow bay bottoms. In some cases, however, projects are actually attempting to restore lost wetlands. For example, many islands in Chesapeake Bay are rapidly eroding and bordering marshes have been lost. On one such site, Poplar Island, the Corps has been rebuilding tidal marshes.

In Washington, DC, tidal freshwater marshes that were converted to mudflats by dredging have been restored by pumping in sediment and planting with native species (Neff and Baldwin 2005). Tidal freshwater marshes have also been created by excavating fill material to intertidal elevations (Leck 2003). To protect shorelines from erosion, fringing marshes have been built by contouring slopes and planting marsh grasses. These artificial marshes ("living shorelines") help reduce the erosive action of currents and waves and promote sedimentation necessary for bank stabilization. This process works best in low-energy environments and is not a remedy for controlling erosion in areas exposed to heavy wave action.

Despite these recent beneficial actions, man's overall effect on coastal wetlands has been negative. Most human impacts have involved conversion of wetlands to impoundments, cropland, or real estate, and degradation of remaining wetlands by ditching, levee construction, pollution, and other disturbances (Chapter 8). Also, society's role in contributing to global warming is having a significant impact on tidal wetlands as sea level is rising at rates high enough to put many tidal wetlands at risk (Chapter 12).

Further Readings

"The ontogeny of a salt marsh estuary" (Redfield 1965)
"Development of a New England salt marsh" (Redfield 1972)
"Coastal salt marshes" (Frey and Basan 1978)
"Development of a tidal marsh in a New England river valley" (Orson et al. 1987)
"The paleoecological development of a late Holocene, tidal freshwater marsh of the Upper Delaware River Estuary" (Orson et al. 1992)
"Coastal Louisiana" (Gosselink et al. 1998)
"Tidal salt marsh morphodynamics: a synthesis" (Friedrichs and Perry 2001)

3 The Dynamic Intertidal Environment

The intertidal zone is a unique environment where conditions fluctuate between aquatic habitat and semiterrestrial habitat as frequently as twice daily and for most tidal wetlands at least a few times a month. A variable pattern of often sinuous creeks typify the landscape of many tidal wetlands and serve as conduits for the tides to bring water, sediments, and nutrients into and out of coastal marshes. Such creeks provide transient marine and estuarine nekton (fishes and aquatic invertebrates) access to the marsh interior as well. Tidal wetlands fluctuate between aquatic habitat when inundated and semiterrestrial habitat when exposed. As such they are excellent examples of a pulsating ecosystem (Odum et al. 1995; Odum 2000). The alternating flooding and exposure also produce variations in salinity, soil saturation, aeration, temperature, and use by fish and wildlife. These fluctuations create a harsh environment for plants and sedentary animals attempting to colonize these habitats. Special adaptations are required for permanent residency. Salinity changes from one part of the estuary to another, within marshes, seasonally, and even daily in some places depending on rainfall and groundwater influences. The most upstream reaches in the tidal zone are strictly freshwater habitats. From a vegetation standpoint, these wetlands have more in common with nontidal wetlands than the salt or brackish wetlands downstream. Substrates differ among tidal wetlands and represent another significant component of the intertidal environment affecting plant and animal life. Coastal processes are constantly reshaping the intertidal zone. Episodic events, for example, hurricanes, can have devastating effects on tidal wetland vegetation but may actually help build marsh elevations to keep pace with rising sea level. Consequently, the intertidal environment is one shaped by tides and coastal processes and further influenced by salinity gradients, soil properties, and other factors. Superimposed on the intertidal environment are differences in climate including the frequency of wet (pluvial) periods and droughts that further add to the dynamics and rigors of the tidal wetland environment. The end result of all of these factors is a pulsating or dynamic ecosystem offering a wide range of intertidal habitats for plants and animals.

Wetland Hydrology

Flooding by tidal water is the unifying property or common denominator of all tidal wetlands. It is the driving force that creates, shapes, and maintains these habitats and makes them different from their nontidal counterparts—the inland wetlands. The intertidal zone is periodically flooded at regular or irregular intervals creating environmental conditions that appear to be in a constant state of flux (e.g., wetting, drying, and rewetting). Some marsh areas are continuously saturated or nearly so, while other places are naturally drained to some depth for varying periods.

Water Budget

Before addressing tides and their effect on the intertidal zone, an introduction to the water budget of tidal wetlands is worthwhile to demonstrate that other factors affect site wetness besides the tides. The "water budget" is an accounting of water inflows (gains or inputs) and outflows (losses or outputs). A simple model can show the dynamics of these flows (Figure 3.1). Site wetness at any particular time is related to its water budget. Inflows increase site wetness with water coming from four sources: 1) precipitation (P)—rain, snow, sleet, hail, or fog; 2) surface water inflow (Si)—rivers, streams, and surface water runoff from the land; 3) groundwater inflow (Gi)—water discharged to the wetland surface (seepage); and 4) flood or rising tides (Ti). Outflows reflect water losses through four circumstances: 1) evapotranspiration (ET)—the combination of evaporation and plant uptake of water and loss through transpiration; 2) surface water outflow (So)—water draining to rivers or streams; 3) groundwater outflow (Go)—water recharging underground aquifers; and 4) ebb tides (To)—falling tides that drain water off the surface and lower the water tables. The water budget is expressed by a simple formula that yields a net change in volume (ΔV):

$$\Delta V = [P + Si + Gi + Ti] - [ET + So + Go + To]$$

When the inputs are greater than the outputs, a positive value indicates water storage (saturated substrate and/or inundation) at the site, whereas a negative value reflects a net loss of wetness favoring drier conditions (lower water tables). From this equation, one can see that the tides are just one factor affecting site wetness, although the dominant one for tidal wetlands as illustrated by the following example. A 34-day water budget for a California tidal wetland island in Suisun Bay revealed mean flow inputs of 0.906 m^3 per second (0.89 m^3/s from flood tides, 0.016 m^3/s from precipitation) and outputs of 0.863 m^3 per second (0.62 m^3/s from ebb tides, 0.043 m^3/s from evapotranspiration, and 0.20 m^3/s from groundwater/unaccounted outflow) for a positive net storage of 0.043 m^3 per second (Ganju et al. 2005). As expected, tidal flows dominated the hydrology, accounting for 98 percent of the inflow and 72 percent of the outflow. Other water sources may make significant contributions to the water budget of individual wetlands. Along the upland edges of wetlands, groundwater inflow may cause local sites to remain wetter than the typical high salt marsh. For freshwater tidal river swamps, surface water inflow during

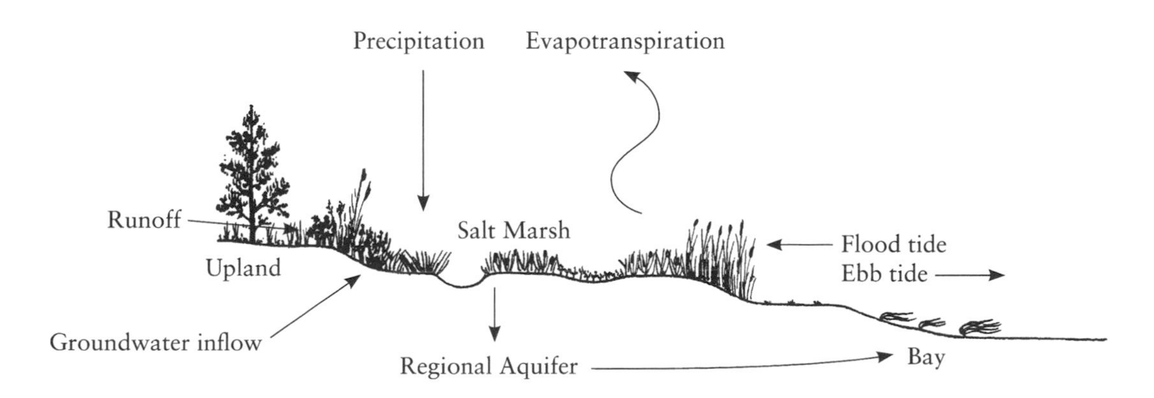

Figure 3.1. Elements of the water budget of a New England salt marsh. (Adapted from Nuttle 1988)

spring freshets is a major factor contributing to the annual water budget, especially for those infrequently flooded by tides.

The Tides

Observed tides are the result of the gravitational forces of the moon and the sun exerted upon the Earth and modified locally by weather conditions (e.g., wind and precipitation). Predicted tides are based on the gravitational forces and an analysis of measurements from tide gauges collected over at least an 18.6-year period that take into account most of the configurations between the Earth, moon, and sun.

Newton's law of gravity describes the attraction between two bodies—the force of attraction between two bodies is proportional to the product of their masses and inversely proportional to the square of the distance between their centers. So even though the moon is only 1/400th the size of the sun, the moon has a stronger influence on the tides because it is closer to Earth. More precisely, the moon's influence on the Earth is about 2.25 times greater than the sun's. Put another way, the sun's pull on the tides is only 46 percent of the moon's influence (Coker 1962; U.S. Army Corps of Engineers 1991; Bowditch 1995). The moon and sun both exert a pull on the Earth's water bodies that is detectable in large water bodies. Even the Great Lakes are subject to tidal influence, but the tide is just an inch or two and is masked by wind and other factors affecting water levels.

When the shoreline of a given water body is closest to the moon, a high tide occurs at that time, then 12 hours later when the moon is farthest away another high tide occurs. This paradox can be simply explained by differential gravitational forces. Since the moon's pull is weakest on the opposite side of the Earth and its pull is stronger on the Earth mass itself, the Earth is pulled away from that water creating a high tide on that side of the planet. Since it takes the moon 24.84 hours to circle the Earth (a lunar day), the tides arrive 50 minutes later each day. The moon's orbit is elliptical, so once a month the moon is closest to the Earth (at perigee) and the tides will be higher than usual; two weeks later it will be farthest away (at apogee) and tides will be lower. The Earth's orbit around the sun is also elliptical, being closest to the sun around January 2 (at perihelion) and farthest away around July 2 (at aphelion) with corresponding effects on the tides. When perigee and perihelion coincide, even higher tides should occur.

A tidal cycle consists of one high tide and one low tide. While recognizing that no two tide cycles are exactly alike, tides have been classified into three types: diurnal, semidiurnal, and mixed (Figure 3.2). Where only one tidal cycle occurs during a twenty-four-hour period, the tides are called "diurnal." *Semidiurnal tides* occur when two tidal cycles take place each day. *Mixed tides* are characterized by two daily tides of distinctly unequal height: typically, a "higher high tide" and "lower high tide" and a "higher low tide" and "lower low tide." Along the Atlantic Coast and eastern end of the Gulf of Mexico (Apalachicola Bay south), tides are typically semidiurnal and more or less equal in height. Tides in Miramichi Bay in Canada and generally in the Gulf of St. Lawrence tend to be mostly diurnal (http://cgca.mcan.gc.ca/coastweb). The best examples of diurnal tides in the United States are found along the northern coast of the Gulf of Mexico from Apalachicola Bay west into Texas. The Pacific Coast is characterized by mixed semidiurnal tides, yet diurnal tides occasionally occur (e.g., in parts of Alaska).

Depending on the position of the moon relative to the sun, other tides occur at two-week intervals (Figure 3.3). Spring tides are the highest astronomic tides and flood the largest expanse of wetland biweekly. They occur on new and full moons and produce higher high tides and lower low tides, resulting in a greater tidal range than normal (the tides appear to "spring" forth). At these times, the moon and sun are aligned together in a straight line with

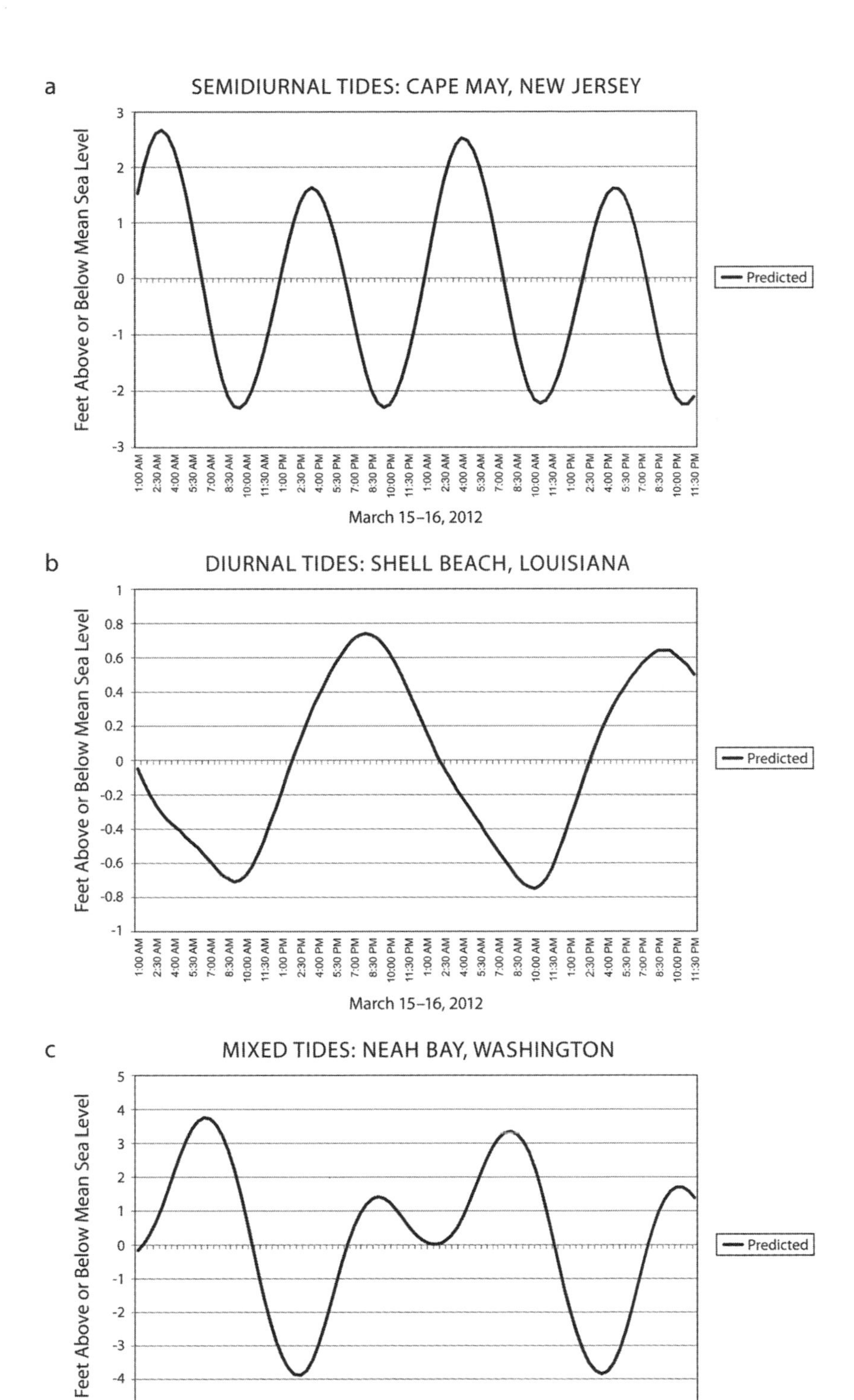

Figure 3.2. Three types of daily tides: (a) semidiurnal; (b) diurnal; and (c) mixed. (Prepared from NOAA data)

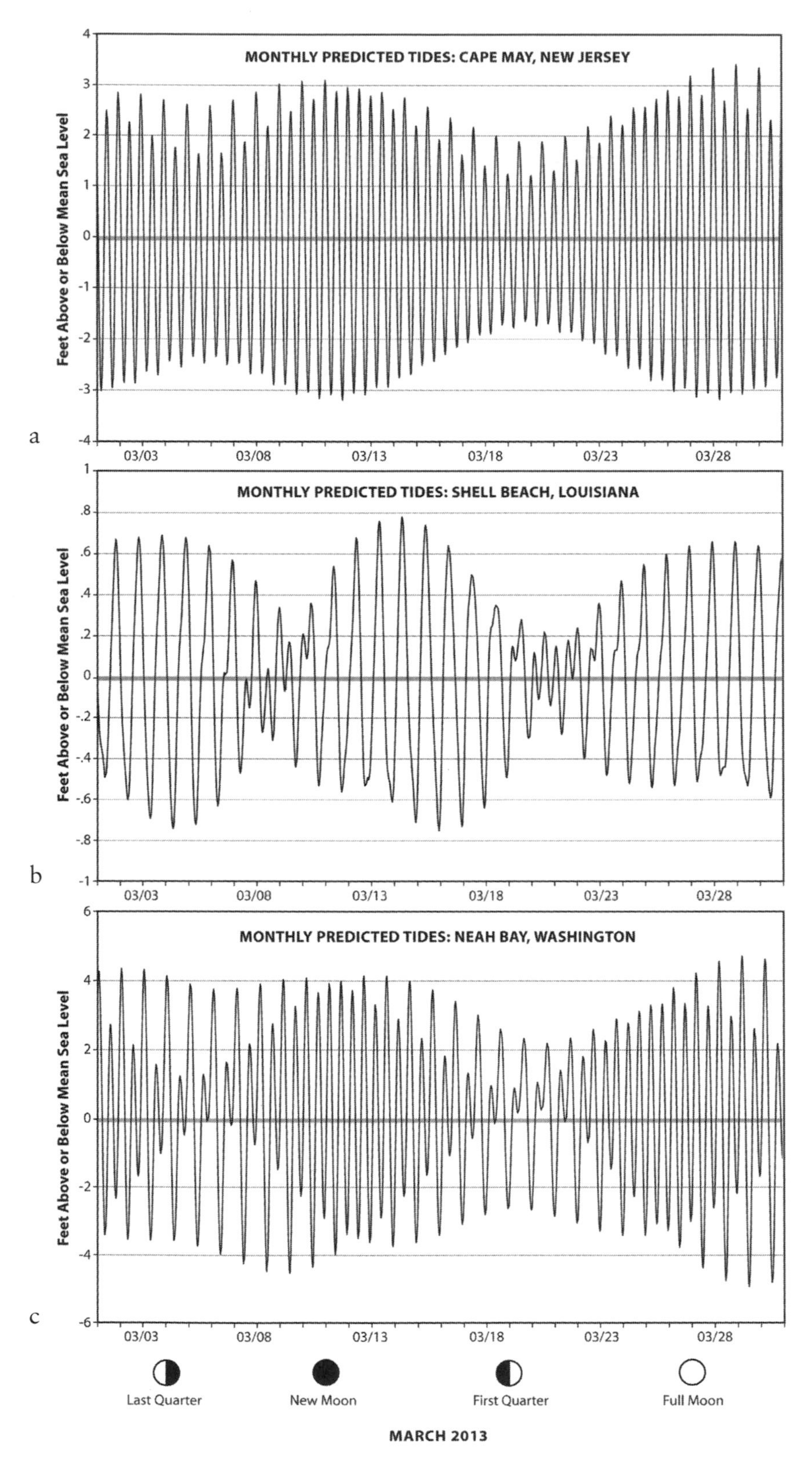

Figure 3.3. Changes in tides over a lunar month: (a) semidiurnal tide (Cape May, New Jersey); (b) diurnal (Shell Beach, LA); and (c) mixed tides (Neah Bay, WA). (Prepared from NOAA data)

the Earth ("syzygy") to exert a combined gravitational pull on the Earth's oceans (Figure 3.4). About a week after the spring tides, the sun and moon are 90° apart in relation to the Earth (i.e., moon's first and third quarters). Their gravitational forces work against each other, minimizing the pull on the Earth's oceans and resulting in moderated tides called "neap tides," which produce the lowest tidal range.

The angular distance of the sun and the moon north or south of the equator ("declination") can cause tides to change from one type to another during a given month. For example, tides in Los Angeles and Honolulu can change from diurnal at neap tide to semidiurnal at spring tide to mixed tide at other times (U.S. Army Corps of Engineers 1991). Areas typically experiencing diurnal tides may also change to mixed tides at neap tide (Figure 3.3b). The lunar cycle of the spring and neap tides repeats itself roughly every 30 days (see Figure 1.12 for general tidal ranges across the United States).

THE METONIC CYCLE

The moon takes 18.6 years to repeat its four phases at the same day of the solar year. This cycle is called the "metonic cycle," named after its discoverer Meton of Athens, a Greek astronomer from the 5th century BC. The significance of the cycle is that a 19-year period is used as the National Tidal Datum Epoch to calculate tidal data (International Marine 2008). The current tidal epoch is based on the period 1983 to 2001 (Hicks 2006). These data will be recalculated in 20 to 25 years nationally, while in areas of anomalous sea-level changes (Alaska and the Gulf of Mexico where elevations of coastal landforms change more frequently than other areas), data are recalculated every 5 years. Since lunar tides are not the same every day due to changing distances from the sun and the moon, orbit patterns, and axis rotations, at some point during the metonic cycle, the highest and lowest tides occur. Considering the variation in tides annually and over longer periods, even other tides have been defined. Some are simply the result of a numeric calculation such as "higher high water large tide" (average of the highest high water from each year of 19 years of prediction), "lower low water large tide" (average of the lowest low water from each year of 19 years of prediction), and "lowest normal tide, higher high water mean tide" (average from all the higher high waters from a 19-year period). The number of extreme high tides appears related to the declination of the moon (angle of the moon relative to the equator) over the 18.6-year period: about every 9.3 years the number of extreme high tides increases around the time of high-maximum lunar declination, whereas 9.3 years later or earlier, the lowest number of extreme high tides occur during the low-maximum phase (Wood 2001).

OTHER FACTORS AFFECTING TIDES AND WATER LEVELS

The amplitude of tides and the height of the tide at any given time are affected by other factors including the morphology of the estuary, river discharge, local weather conditions, and tsunamis (high waves created by earthquakes on the sea floor). The behavior of the tides in an estuary is largely governed by the size and shape of the estuary and offshore bottom characteristics. As a result, some estuaries have higher tides farther away from the sea (e.g., St. Lawrence River, Narragansett Bay, Long Island Sound, and Delaware River), whereas others do not (Table 3.1). Some estuaries have diurnal tides while others nearby have semidiurnal tides.

Local climatic conditions, especially winds and storms (e.g., hurricanes), influence the actual height, duration, and frequency of tidal flooding. Actual tides therefore may be different from the predicted tides (Figure 3.5). Storms generate the highest tides. In the North and Middle Atlantic region, "storm tides" are most frequent in winter (from November through February) (Dolan and Davis 1992; Parkes et al. 1999). Winter storms blowing in from the northeast ("nor'easters") cause much

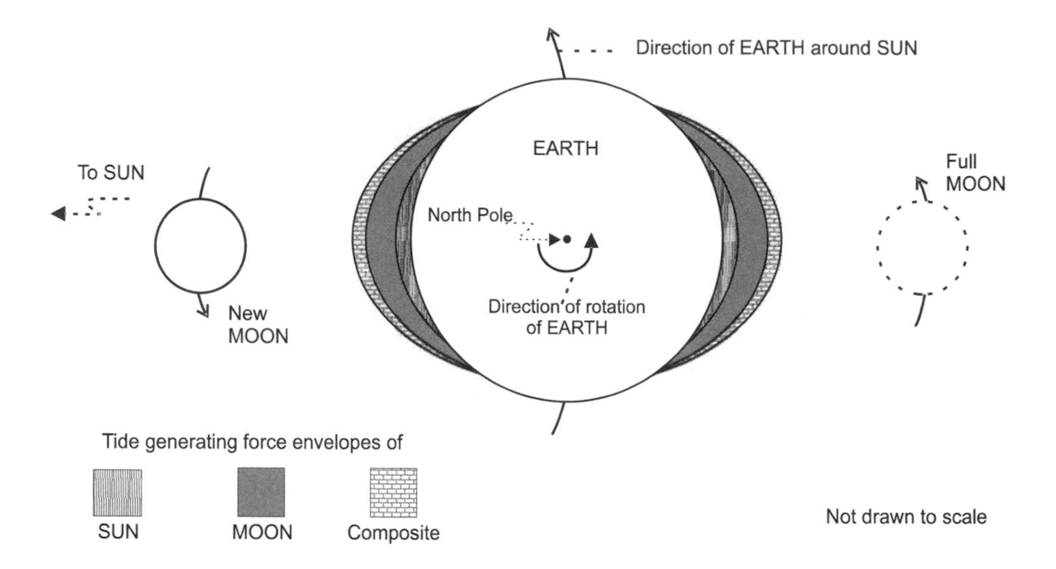

Spring tides

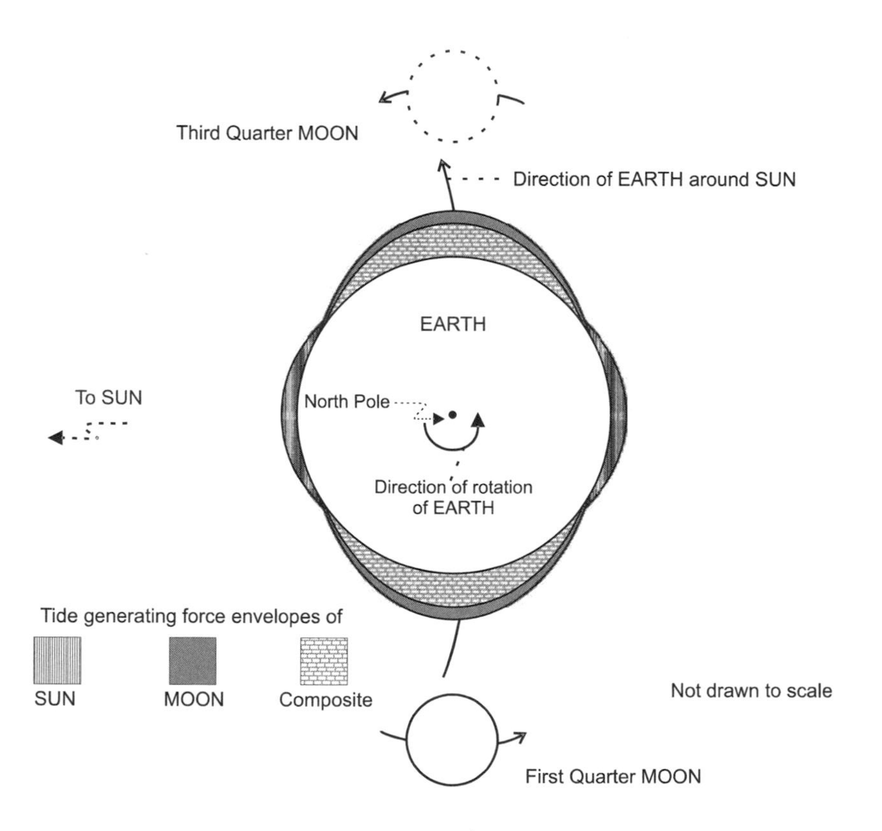

Neap tides

Figure 3.4. The gravitational pull of the moon and sun on the Earth affect tidal range and in some places the type of tides. Their effect is greatest when the moon and sun are aligned in a straight line (at full and new moons) producing what are called "spring tides" (top). When the moon and sun are positioned at right angles in respect to Earth, their forces work against each other to suppress tides, producing "neap tides" (bottom). (Hicks 2006)

Table 3.1. Examples of mean and spring tidal ranges in the selected estuaries in the United States and Canada

Location	Tidal Range in feet (m)	
	Mean	Spring
St. Lawrence River, QC		
Trois-Rivieres	0.7 (0.2)	1.0 (.3)
Quebec City	13.7 (4.2)	15.5 (4.7)
Cap Chat	7.0 (2.1)	9.0 (2.7)
Gulf of St. Lawrence		
Anticosti Island, QC	3.5 (1.1)	4.4 (1.3)
Dalhousie, NB	5.6 (1.7)	7.1 (2.2)
Miramichi River, NB	4.0* (1.2)	—
Malpeque Bay, PEI	2.5* (0.8)	—
Magdalen Islands, PEI	1.6 (0.5)	2.0 (0.6)
Bay of Islands, NF	3.3 (1.0)	4.2 (1.3)
Georgetown Harbor, PEI	2.8 (0.9)	3.5 (1.1)
Pictou, NS	3.2 (1.0)	3.9 (1.2)
Bay of Fundy		
Annapolis Basin, NS	21.2 (6.5)	24.6 (7.5)
Minas Basin, NS	38.4 (11.7)	43.5 (13.3)
Saint John, NB	20.8 (6.3)	23.7 (7.2)
Campobello Island, NB	17.0 (5.2)	19.4 (5.9)
Long Island Sound		
New London, CT	2.6 (0.8)	3.1 (0.9)
Saybrook Jetty, CT	3.5 (1.1)	4.2 (1.3)
Madison, CT	4.9 (1.5)	5.6 (1.7)
New Rochelle, NY	7.3 (2.2)	8.5 (2.6)
Connecticut River, CT		
Saybrook Point	3.2 (1.0)	3.8 (1.2)
Haddam	2.5 (0.8)	3.0 (0.9)
Hartford	1.9 (0.6)	2.3 (0.7)
Delaware Bay and River		
Cape May Point, NJ	4.8 (1.5)	5.7 (1.7)
Lewes, DE	4.1 (1.3)	4.9 (1.5)
New Castle, DE	5.2 (1.6)	5.7 (1.7)
Philadelphia, PA	6.2 (1.9)	6.6 (2.0)
Burlington, NJ	7.2 (2.2	7.6 (2.3
Trenton, NJ	8.2 (2.5)	8.5 (2.6)
Chesapeake Bay		
Fishermans Island, VA	3.0 (0.9)	3.6 (1.1)
Tangier Island, VA	1.4 (0.4)	1.7 (0.5)
Pocomoke Sound, MD	2.3 (0.7)	2.8 (0.9)
Baltimore, MD	1.1 (0.3)	1.3 (0.4)
Susquehanna River, MD	1.9 (0.6)	2.2 (0.7)
Winyah Bay, SC		
South Jetty	4.6 (1.4)	5.4 (1.6)
South Island Ferry	3.7 (1.1)	4.2 (1.3)
Waccamaw River	3.6 (1.1)	4.1 (1.3)
Bull Creek	2.5 (0.8)	2.9 (0.9)
Conway	1.2 (0.4)	1.4 (0.4)
Ossabaw Sound, GA		
Egg Islands	7.2 (2.2)	8.4 (2.6)
Fort McAllister	6.9 (2.1)	8.1 (2.5)
Florida Passage	7.6 (2.3)	8.8 (2.7)

Table 3.1. *(continued)*

Location	Tidal Range in feet (m)	
	Mean	Spring
St. Johns River, FL		
Mayport	4.6 (1.4)	5.3 (1.6)
Dame Point	3.2 (1.0)	3.4 (1.0)
Jacksonville	1.8 (0.5)	2.0 (0.6)
Welaka	0.4 (0.1)	0.5 (0.2)
Mobile Bay and Tensaw River, AL		
Mobile Point	—	1.2* (0.4)
Great Point Clear	—	1.4* (0.4)
Lower Hall Landing	—	1.3* (0.4)
Atchafalaya Bay, LA		
Eugene Island	—	1.9* (0.6)
Point Au Fer	—	2.0* (0.6)
Shell Island	—	1.5* (0.5)
San Francisco Bay Estuary and Sacramento River, CA		
Golden Gate	4.1 (1.3)	5.8 (1.8)
Oakland Airport	5.0 (1.5)	6.7 (2.0)
Palo Alto Yacht Club	6.2 (1.9)	7.6 (2.3)
Sausalito	4.0 (1.2)	5.7 (1.7)
Suisan Point	3.8 (1.2)	5.2 (1.6)
Antioch	2.8 (0.9)	3.9 (1.2)
Sacramento	2.3 (0.7)	2.9 (0.9)
Puget Sound, WA		
Seattle	7.7 (2.3)	11.4 (3.5)
Tacoma	8.1 (2.5)	11.8 (3.6)
Henderson Inlet	10.0 (3.0)	14.0 (4.3)
Burns Point	11.0 (3.4)	15.0 (4.6)

Sources: Tide Tables 1999 High and Low Water Predictions: East Coast of North and South America Including Greenland. International Marine, Division of The McGraw-Hill Companies, Camden, Maine; NOAA Tides and Currents 2010 Predictions.

Notes: Sites marked with an asterisk (*) have a diurnal tide (range is difference between mean higher high tide and mean lower low tide).

local flooding, especially when coupled with lunar spring tides. High-pressure air masses (fair weather) lower sea level and tides, whereas low pressures raise sea level and tides. Wind effects may be most pronounced in microtidal estuaries with winds blowing onshore producing higher water levels (setup), while offshore winds are generating the lowest water levels (setdown). For example, in New Jersey at Barnegat Bay, strong offshore winds can blow so much water out of the bay that hundreds of acres of tidal flats and bay bottoms (including normally submerged aquatic beds) are exposed to air. Strong onshore winds bring more ocean water into the bay and keep water levels higher for longer periods than normal. Wind tides along the northern coast of the Gulf of Mexico and Pamlico and Albemarle Sounds in North Carolina are the main source of water for their irregularly flooded marshes. In these areas, tide levels often do not exhibit a regular or predictable pattern (Figure 3.6). The effect on the tides is related to the orientation of the water body relative to the direction of the wind. In the Pamlico River Estuary, the highest tides are produced by winds blowing directly upstream from Pamlico Sound, while winds blowing in the opposite direction yield the

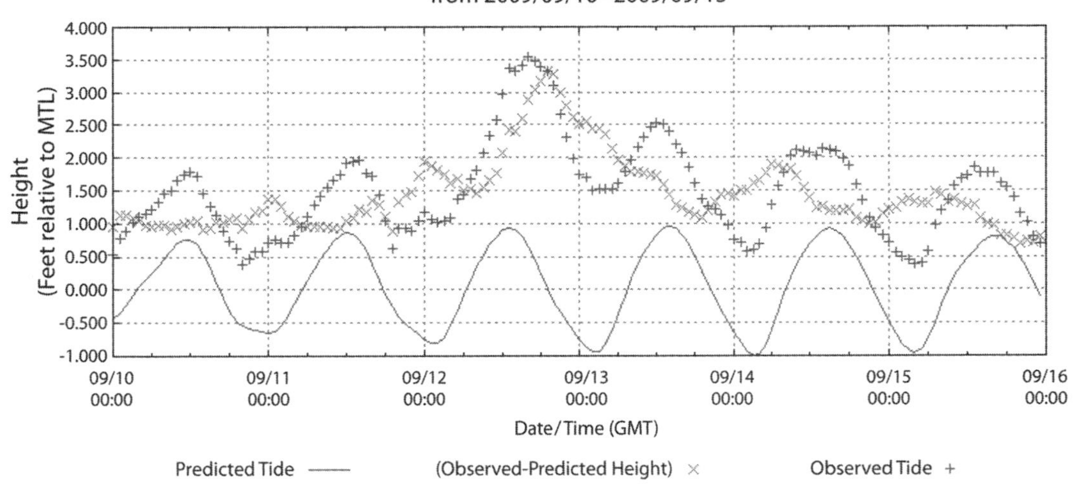

Figure 3.5. Observed tides often follow the pattern of the predicted tide but may vary due to local weather conditions. In this example, the tides were actually higher than predicted. NOAA website "Tides & Currents" allows you to display the differences between predicted and observed tides for many locations. (Redrawn from NOAA data)

lowest water levels (Copeland et al. 1984). Along the northeastern coast of the Gulf of Mexico, southerly winds increase marsh flooding in summer (Stout 1984). Northerly winds push water out of the tidal swamps and such an event could last for days. In Galveston Bay, strong southerly winds that often precede the arrival of a cold front produce water levels higher than the predicted tides, flooding the marshes for a few days (Rozas 1995). After the front passes, northerly winds and an increase in barometric pressure cause a quick drop in water levels and rapid drainage of the marshes. Of course, hurricanes with over 100 miles per hour (160.9 km/hr) winds generate the highest tides ("storm surges"), e.g., Katrina—up to 33 feet (10 m) along the Mississippi coast, and Andrew up to 17 feet (5.2 m) in Florida's Biscayne Bay and to 9 feet (2.7 m) in Louisiana's Terrebonne Bay (Lovelace and McPherson 1996; Fritz et al. 2007). Storm surges move inland for considerable distances—Katrina's surge penetrated at least 6.2 miles (10 km) along the coast and up to 12.4 miles (20 km) in Mississippi bays and rivers (Fritz et al. 2007). The longer the water remains upstream, the greater the ecological impact of saltwater intrusion and prolonged inundation.

Seasonal and annual shifts in sea level also play a role in tidal marsh flooding (Figure 3.7). The lowest levels occur in winter and highest levels in summer or early fall due to a combination of factors including thermal heating (decreases water density thereby increasing water volume), precipitation, and increased discharge of freshwater (spring to summer) (Nixon 1982; Rozas 1995). Along the Atlantic Coast, sea level is above the annual mean for about three months beginning in August. The situation is much different along the Gulf Coast. From June to December, the high marsh zone of northern Gulf coast marshes may be flooded for longer periods than high marsh in other regions because sea level is above the annual mean and prevailing winds blow onshore raising estuarine water levels (Provost 1973; Stout 1984). Late summer water levels may be twice that of winter levels (Patullo et al. 1955; Rozas 1995). Louisiana marshes experience the peak of flooding in early summer and early fall, with

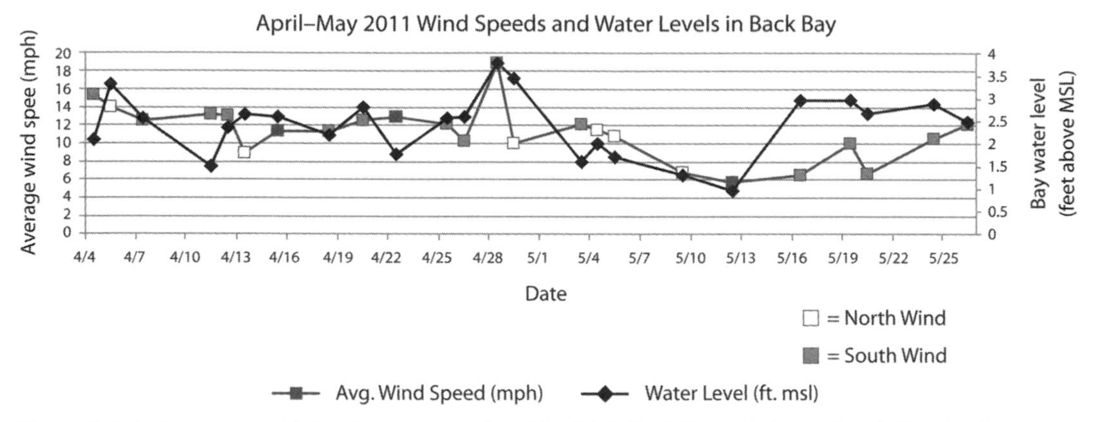

Figure 3.6. In large microtidal embayments such as Virginia's Back Bay, wind speed and direction have a great effect on water levels, creating "wind tides." When strong winds from the south blow across the bay, wind tides raise water levels, whereas when strong winds come from the north, water levels are lowered. (Prepared from U.S. Fish and Wildlife Service data)

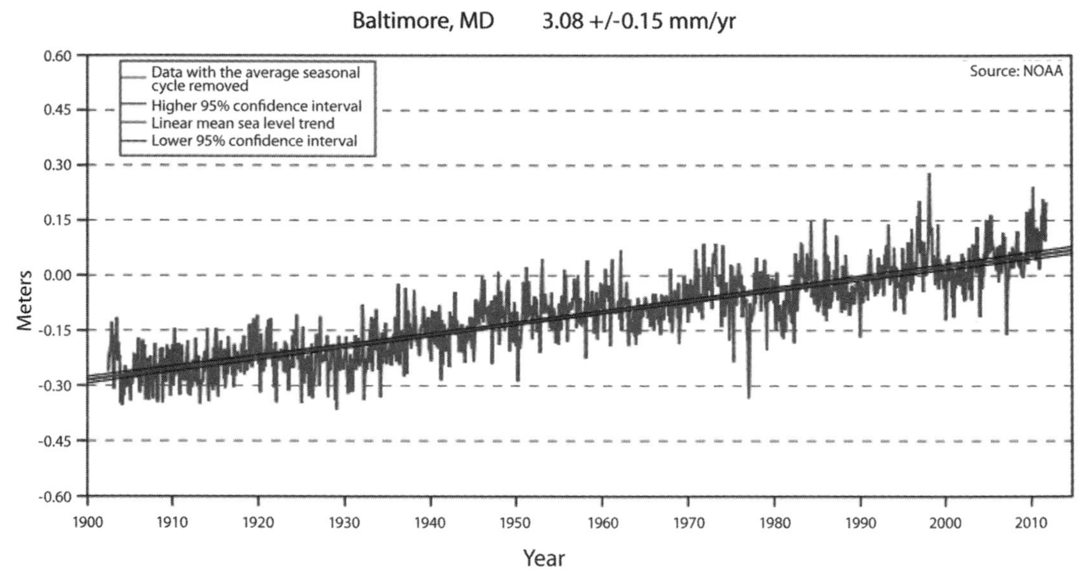

Figure 3.7. While sea level continues to steadily rise, water levels are quite variable seasonally and annually: Baltimore (MD) example. (Prepared from NOAA data)

minimum inundation in winter and a second low in mid-summer (Baumann 1987). Monthly sea levels also experience interannual variability. In any given year, sea level varied as much as 8 inches (20 cm) in a Massachusetts salt marsh (Teal and Howes 1996); however, the variation during the period of active plant growth (June–August) was much smaller—about 2 inches (5 cm).

Heavy runoff from contributing watersheds can raise water levels in estuaries as well as move the salt wedge seaward. In most of the United States this happens in spring during peak runoff. These discharges temporarily reduce the daily tidal range in local areas as their flows overwhelm the tides (Day et al. 2007b). Precipitation brought about by hurricanes can quickly raise water levels especially in tidal reaches of coastal rivers and in estuaries with low tidal ranges that receive freshwater inflow from major river basins. The combined effect of three hurricanes (Dennis, Floyd, and Irene) in September and October 1999 raised river levels significantly, causing almost two months of flooding in most of eastern North Carolina (Bales et al. 2000). The volume of freshwater flowing into Pamlico Sound during these two months was equivalent to 83 percent of the total volume of the sound, whereas under normal conditions inflow during this period amounts to roughly 13 percent of the volume. In California, winter precipitation associated with its Mediterranean climate may contribute to higher tidal levels from December to January (Josselyn 1983).

Constant discharge from major rivers whose watersheds cover enormous sections of continents, such as the Mississippi River, suppress the tide. The average daily discharge of 470,000 cubic feet per second from the Mississippi River has essentially restricted tidal influence to its mouth at low-flow periods (www.lacoast.gov/landchange/basins/mr/). Heavy discharge of the Mississippi River eliminates the tide at New Orleans, while at low flows diurnal tides average 0.8 foot (0.24 m). Discharge from all rivers dampens incoming tides with spring freshets having the greatest effect.

The Bay of Fundy has a rare geomorphology that creates extraordinary tides that are among the highest tides in the world, reaching heights of 50 feet or more (>16 m) (Parkes et al. 1999; Figure 3.8). The basin's funnel shape and gradual decrease in depth cause a great rise in tides from a 20-foot (6.0 m) tide at its mouth to a 53-foot (16 m) tide at its upper end. These huge tides are also the result of a so-called bathtub effect in the Bay of Fundy and the Gulf of Maine where water sloshes back and forth within these marine basins. This sloshing effect is nearly in sync with the lunar tides so these forces act in harmony to produce

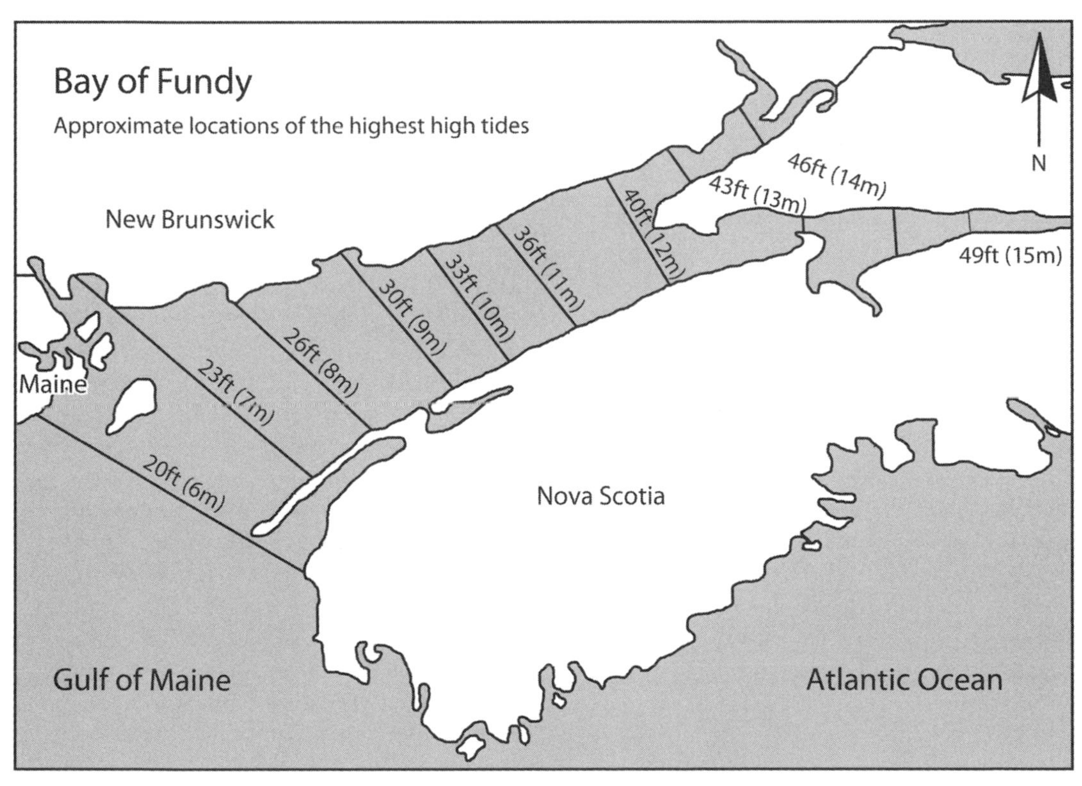

Figure 3.8. The unique funnel shape of the Bay of Fundy and the sloshing effect of tides in the Gulf of Maine produce what are claimed to be the highest tides in the world. (Redrawn from NOAA figure)

tides of enormous proportions. Storm tides can generate tides of 70 feet (21 m) or more. Peak tides occur at three cycles: 7 months, 4.53 years, and 18.03 years, with the last generating the highest tides (Desplanque and Mossman 1998). Ungava Bay along the northern coast of Quebec has tides of similar range to 52.5 feet (16 m).

Global climatological events related to warming and cooling of the tropical Pacific Ocean (El Niño/La Niña-Southern Oscillation, ENSO) also affect the hydrology of coastal wetlands in some places (e.g., Childers et al. 1990; Drexler and Ewel 2001). Pacific warming (El Niño) brings more precipitation to central-southern California, the Southwest, and the Southeast, thereby increasing river discharge and marsh flooding. The cooling event (La Niña) produces drier conditions in the Southwest and Southeast with less frequent and shorter duration flooding of their coastal wetlands (Childers et al. 1990). Interestingly, La Niña brings wetter conditions and higher water levels in the Pacific Northwest. These warming or cooling periods may last for 7 to 9 months or more and occur on average every 5 years. Increased frequency of ENSO events may pose significant challenges for coastal wetland plants and wildlife in affected regions.

Earthquakes are common along the Pacific Coast, and those of high magnitude can create tsunamis that generate waves reaching 100 feet (30.5 m) or more in height. For example, changes in land and sea floor surfaces from the March 28, 1964, Prince William Sound earthquake (8.3–8.6 magnitude on the Richter scale) produced tsunami wave heights of over 100 feet (30.5 m) in some communities along the Alaskan coast (e.g., Whittier and Valdez) and serious impacts as far south as California (Committee on the Alaska Earthquake 1972).

TIDAL ZONES WITHIN THE WETLANDS

Flooding frequency is related to elevation above mean low water—low elevations are flooded longer and more frequently than higher elevations. Permanently flooded areas are called "subtidal habitats"—they are open water habitats and may support submerged macroalgae and vascular plants. Tidal wetlands are intertidal habitats—alternately flooded and exposed to air. Several intertidal zones may be recognized based on the frequency of flooding. The lowest portions of tidal flats and rocky shores that are almost continuously under water and exposed to air only during extreme low tides are part of what may be called the "lower intertidal" or "irregularly exposed" zone. The majority of coastal wetlands, however, are not underwater that long but are subjected to alternate flooding and exposure for variable periods. The principal intertidal zones are the regularly flooded zone and the irregularly flooded zone. The former zone is flooded at least once daily by the tides, whereas the latter zone is flooded less often and is exposed to air for long periods (Figure 3.9). The regularly flooded zone includes tidal flats and low marsh, while the irregularly flooded zone is called the high marsh. Within these marsh zones, temporary or permanent pools or ponds may be found. A third zone—supratidal—occurs above the intertidal zone. These areas are typically not flooded often or long enough to be considered wetlands, but they are influenced by occasional flooding (e.g., winter or storm tides) or salt spray.

North Atlantic salt marshes are mostly irregularly flooded (high marsh) types as are those behind barrier islands in North Carolina (e.g., Pamlico and Albemarle Sounds) and along the northern coast of the Gulf of Mexico. While the low marsh is flooded at least once daily for a few hours, the high marsh zone in the northeastern United States and Atlantic Canada typically experiences flooding a few times a month on average (with spring and storm tides). From April to September the low marsh zone of a Rhode Island salt marsh—the tall form of smooth cordgrass zone (*Spartina alterniflora*)—experienced flooding 30.5

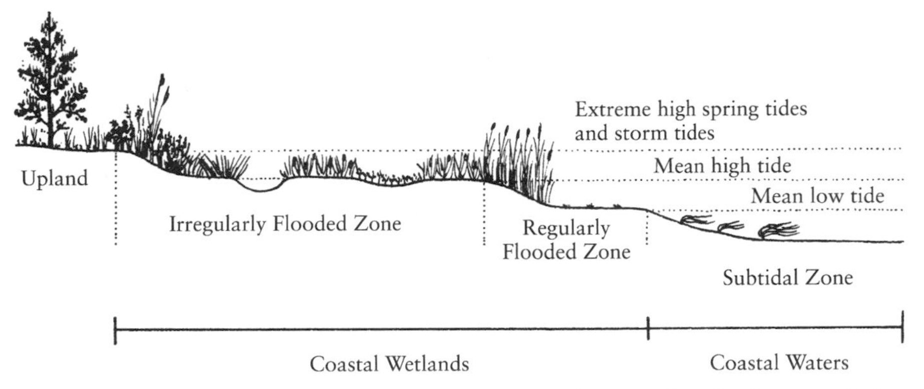

Figure 3.9. Two major intertidal zones can be described for coastal wetlands: 1) regularly flooded (inundated every day) and 2) irregularly flooded (inundated for variable periods depending on elevation). Coastal storms can produce tides that may inundate low-lying uplands. (Tiner 2009)

days/month, while the high marsh was inundated less frequently: short form of smooth cordgrass zone, 23.5 days per month; the salt hay grass zone (*Spartina patens*), 17.3; and black grass zone (*Juncus gerardii*), 8.7 days (Bertness and Ellison 1987). Bay of Fundy marshes in the Cumberland Basin are completely inundated once a week (Desplanque and Mossman 1998). The Fundy tides are also so powerful that they effectively "dam" outflowing river water during high tide. For example, spring freshets from the Saint John River bring volumes of water to the Bay of Fundy through a narrow channel (known as Reversing Falls). At high tide, river discharge is impeded causing water levels to rise upstream and to flood the lower Saint John River valley (Center for Science Advice 2009).

While some of their wetlands may be flooded daily, most of the salt marshes behind barrier beaches in microtidal estuaries such as Barnegat Bay (NJ), Currituck Sound/Back Bay (VA/NC), Albemarle and Pamlico Sounds (NC), and those along much of the Gulf Coast from northwest Florida to the Mexican border are flooded at irregular intervals with the highest tides produced by wind action (see Figure 3.6). Along the Louisiana coast where considerable subsidence is occurring, there is more low marsh than along the rest of the Gulf Coast. Barataria Bay (LA) salt marshes are flooded 15 to 25 times per month with the duration of flooding varying from around 20 to 70 percent of the time (Madden et al. 1988). Other Louisiana marshes are inundated an average of 25 to 40 percent of the time with peak water levels and flooding in early summer and early fall (up to 80% of the time) and low water levels in winter and mid-summer (Baumann 1987).

In contrast to those cited above, most of the salt marshes behind sea islands of South Carolina and Georgia are flooded twice daily. The low marsh dominated by the tall form of smooth cordgrass has a flood frequency of 80 to 100 percent, while the zone just above it dominated by the short form of this species has a flood frequency of 40 to 80 percent (Antlfinger and Dunn 1979). Irregularly flooded marshes also exist between the low marsh and neighboring upland (mainland or barrier islands). The rest of the high marsh has a flood frequency of 2 to 10 percent (salt flat at 5–10%, succulent zone at 4–8%, and the black needlerush–sea ox-eye zone [*Juncus roemerianus–Borrichia frutescens*] at 2–5%).

On the West Coast, the majority of salt marshes are irregularly flooded. For example, in the Tijuana salt marsh, the low marsh dominated by California cordgrass (*Spartina foliosa*) is submerged for about four hours every day with 5.1 to 6.7 inches (13–17 cm) of water, whereas the high marsh of perennial glasswort or pickleweed (*Salicornia depressa,* formerly *S. virginica*) is flooded only 5 to 6 days per month with less than 4 inches (10 cm) of water (Cahoon et al. 1996).

TIDAL INFLUENCE IN FRESHWATER RIVERS AND WETLANDS

Wind tides and river discharge play a more significant role influencing wetland hydrology of Mississippi Delta marshes than do astronomical tides. They are flooded at different frequencies than are typical salt marshes. Inland portions are inundated for an average of 16 to 27 hours as much as 263 times per year, while streamside levee marshes (2–3 inches [5.1–7.6 cm] higher in elevation) are flooded only 160 times for an average duration of 6.6 hours (Gosselink 1984). These marshes are flooded for 50 percent and 12 percent of the year, respectively, and for more than 80 percent of the time in September and October.

In coastal rivers with large contributing watersheds and high discharge, eventually a point is reached where the water is strictly fresh, with no trace of ocean salts, yet the water levels still fluctuate with the tides. This marks the beginning of the zone where tidal freshwater wetlands are found. Flood tides cause water levels to rise in these areas. A directional change in the water flow occurs in the lower freshwater tidal areas with water moving upstream during the flood tide and downstream during ebb tide. This is an area called the "point of tide reversal." Tidal influence, however, extends farther upstream, where downstream flow continues while water levels temporarily rise due to high tide conditions downstream. Upstream of this area where the water levels are not affected by the tides marks the beginning of the nontidal reach of the river. The National Ocean Service considers the point where the mean tide range is less than 2.4 inches (6 cm) to be the "head of tide" (Hicks et al. 2000).

Like their estuarine counterparts, tidal fresh marshes have regularly flooded zones (low marsh) and irregularly flooded (high marsh) zones. Along northeastern coastal rivers, frequently flooded low marshes may be more extensive than high marshes. Two zones of low marsh have been described for a Maryland freshwater tidal marsh (Jug Bay, Patuxent River): 1) spatterdock–pickerelweed–arrow arum marsh (*Nuphar lutea–Pontederia cordata–Peltandra virginica*)—flooded with 12 to 25.6 inches (30–65 cm) of water for 8 to 9 hours per tidal cycle; and 2) cattail zone (*Typha angustifolia* and *T. latifolia*)—inundated for 2 to 4 hours with 2 to 8 inches (5–20 cm) of water (Khan and Brush 1994).

Flooding of tidal swamps may be more complicated in the lower reaches (closer to the estuary) since their topography often consists of a mosaic of hummocks (mounds) and hollows (depressions). This contrasts with the tidal swamps near the upper limits of tidal influence that take on the typical relatively flat floodplain form. The hollows are frequently inundated as much as once or twice daily and when not flooded remain saturated near the surface, largely due to the short duration between inundations (Rheinhardt 2007). As expected, hummocks are flooded less often, by spring tides and other high-water events, but have water tables close to the surface. Inundation and soil saturation affect the vegetation with the hummocks supporting less-tolerant trees and the hollows colonized by more flood-tolerant herbaceous vegetation. The upstream floodplain tidal swamps are flooded by spring freshets, spring tides, and storm surges. Peak river discharges—spring freshets—may eliminate any tidal signal in the water levels and inundate the wetlands for extended periods. (*Note:* For a detailed description of the hydrology of southern tidal swamps see Day et al. 2007b.)

Soil Saturation

Water brought in by tides moves both as "sheet flow" on the wetland surface once the tide overtops creek or ditch banks and as "subsurface flow" through the soil-saturating, aerated portions of the soil. The height of the water table and duration of saturation vary due to several factors including soil type, topography, duration of flooding, groundwater interactions, evapotranspiration, precipitation patterns, and distance from tidal creeks. In a Long Island (NY) salt marsh, water table fluctuations occurred 2 to 16 inches (5–40 cm) below the surface, while below this the soil was continuously saturated (Taylor 1938). Water tables and soil saturation of nonflooded portions of the marsh are affected by the tides to some degree. In a Massachusetts salt marsh, the tidal signal from high tide was observed in the water table within 8.2 feet (2.5 m) of a tidal creek and from 8.2–49.2 feet (2.5–15.0 m) the water table fluctuated in response to tides, but beyond that point there was no horizontal movement of water (Nuttle 1988). In a brackish marsh along the Hudson River the high tide signal could still be detected in the water table up to 39.4 feet (12 m) from the creek (Montalto et al. 2006). For the latter marsh, the water table remained high in the interior marsh longer than in the creekside marsh—the water table was within 8 inches (20 cm) of the surface more than 90 percent of the time during a lunar month (four quarters of the moon) at a distance of 19.7 feet (6 m) or more from the creek, while the water table of the marsh closer to the creek had a water table at this depth for only half of the time. And beyond 78.7 feet (24 m) from the creek, the interior marsh was saturated to the surface for more than 90 percent of the time. Recognize that water table measurements do not reflect the upper limit of saturation as water is held in pores under tension above this zone ("capillary fringe"). For the Massachusetts salt marsh, desaturation of the marsh started to occur only when the water table fell below 4 inches (10 cm) (Nuttle 1988). Prolonged saturation creates anaerobic conditions in the soil. In mudflats and much of the salt marsh, oxygen may be present in only a very thin layer at the soil surface, while the aerated layer should be thicker in sandy soils given larger pore spaces and better internal drainage (Teal and Kanwisher 1961; Whitlatch 1982).

Salinity

Ocean-driven tides bring seawater into coastal rivers that otherwise would be strictly freshwater ecosystems. Seawater is saline because it contains minerals derived mostly from weathering of the Earth's surface: chlorine (55.1%), sodium (30.7%), sulfate (7.8%), magnesium (3.7%), calcium (1.2%), potassium (1.1%), and lesser amounts of other minerals (Millero 1974). The concentration of salts in seawater is 3.5 percent and salinity is usually expressed in parts per thousand units (ppt) or practical salinity units (psu). (Note: In 1978, the standard salinity scale was changed from parts per thousand [ppt or 0/00] to practical salinity units [psu]; the difference between the two values is very small. Some oceanographers use psu to report salinity, while others, including marine and estuarine scientists, typically use ppt [Libes 2009]. Because the sources used for this book report salinity in ppt, I have implemented this salinity unit as well.) Ocean water averages about 35 ppt but varies from 30 to 40 ppt worldwide (NASA Aquarius composite image, 15 August 2011–8 February 2012).

Tidal wetlands with little or no freshwater inflow have soil salinities near and above sea strength whereas those with significant freshwater contributions have reduced salinities. For tidal wetlands in drowned river valley estuaries, salinities follow a general pattern of decreasing salinity from the seaward end upstream into freshwater. Soil salinity in tidal wetlands varies with tides, seasons, elevation, evaporation, and contributions of freshwater from various sources (runoff, high river flows, groundwater discharge, precipitation, and in some places, irrigation). In general, the soil salinity of the zone adjacent to the water follows that of the water body, while parts of the interior marsh often have higher salinities due to evapotranspiration (especially in warmer regions). The upper marsh will have lower salinities due to infrequent saltwater flooding and to runoff from adjacent uplands and groundwater discharge. Salinities may vary continually with tides and precipitation.

Salinity Differences in the Salt Marshes

Changes in salt marsh vegetation often reflect differences in soil salinity. A Bay of Fundy salt marsh exhibited a general decrease in salinity (September) from low marsh to marsh upland border: 35 ppt in the smooth cordgrass zone, 34 ppt in seaside plantain (*Plantago maritima*), 32 in salt hay grass, and 24 in the more diverse marsh border, while pannes had an average salinity of 23 ppt and marsh pools 21 ppt (Magenheimer et al. 1996). In a Georgia marsh where the creek salinity was 20 ppt, the change in interstitial salinity also was reflected by a shift in species composition: 23 ppt in the tall-form smooth cordgrass growing on the natural levee, 33 ppt in the backmarsh dominated by the short form of this species, 127 ppt in the salt barren of the middle marsh, 41 ppt in the glasswort–saltwort community, 25 ppt in the black needlerush–sea ox-eye zone (*J. roemerianus–Borrichia frutescens*), and 15 to 20 ppt in the upper marsh that supported a more diverse community (Antlfinger and Dunn 1979; Pomeroy et al. 1981; Figure 3.10). Although salt barrens are rare in the Northeast, shallow depressions in the salt marsh interior (pannes) occur that collect saltwater that later evaporates, producing salinities in excess of 100 ppt (Sugihara et al. 1979). These are the most salt-stressed environments in the estuary. Hyperhaline conditions are characteristic of arid and semi-arid regions where tidal lagoons have a weak connection to the marine system and little freshwater inflow (Figure 3.11). For example, prior to dredging channels and passes, the Laguna Madre in Texas and Mexico's Tamaulipas were extremely hypersaline with salinities commonly over 100 ppt and sometimes reaching 295 ppt in Tamaulipas (Tunnell et al. 2002). Interior portions of California marshes from San Francisco Bay

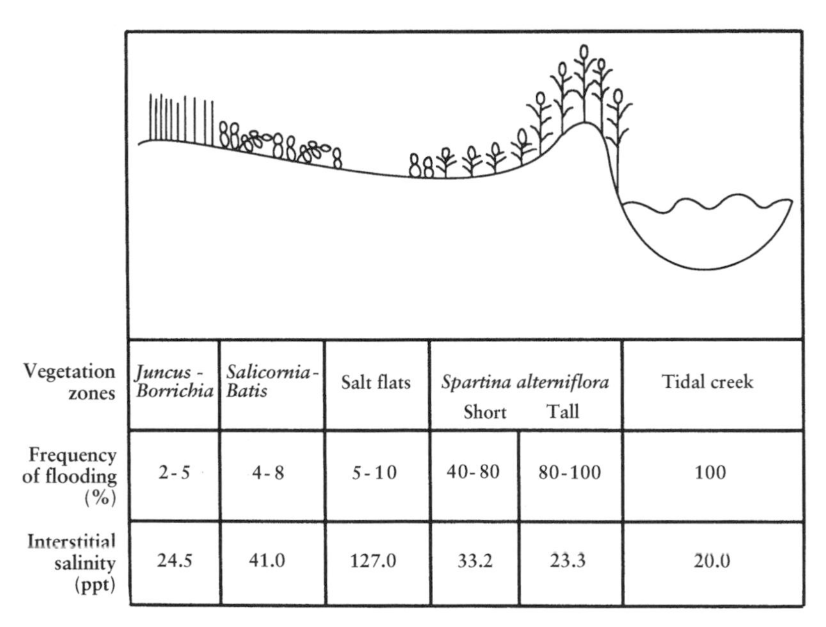

Vegetation zones	*Juncus - Borrichia*	*Salicornia - Batis*	Salt flats	*Spartina alterniflora* Short	*Spartina alterniflora* Tall	Tidal creek
Frequency of flooding (%)	2-5	4-8	5-10	40-80	80-100	100
Interstitial salinity (ppt)	24.5	41.0	127.0	33.2	23.3	20.0

Figure 3.10. Vegetation changes in a Georgia salt marsh due to a combination of salinity and frequency of flooding. (Wiegert and Freeman 1990; redrawn from Antlfinger and Dunn 1979)

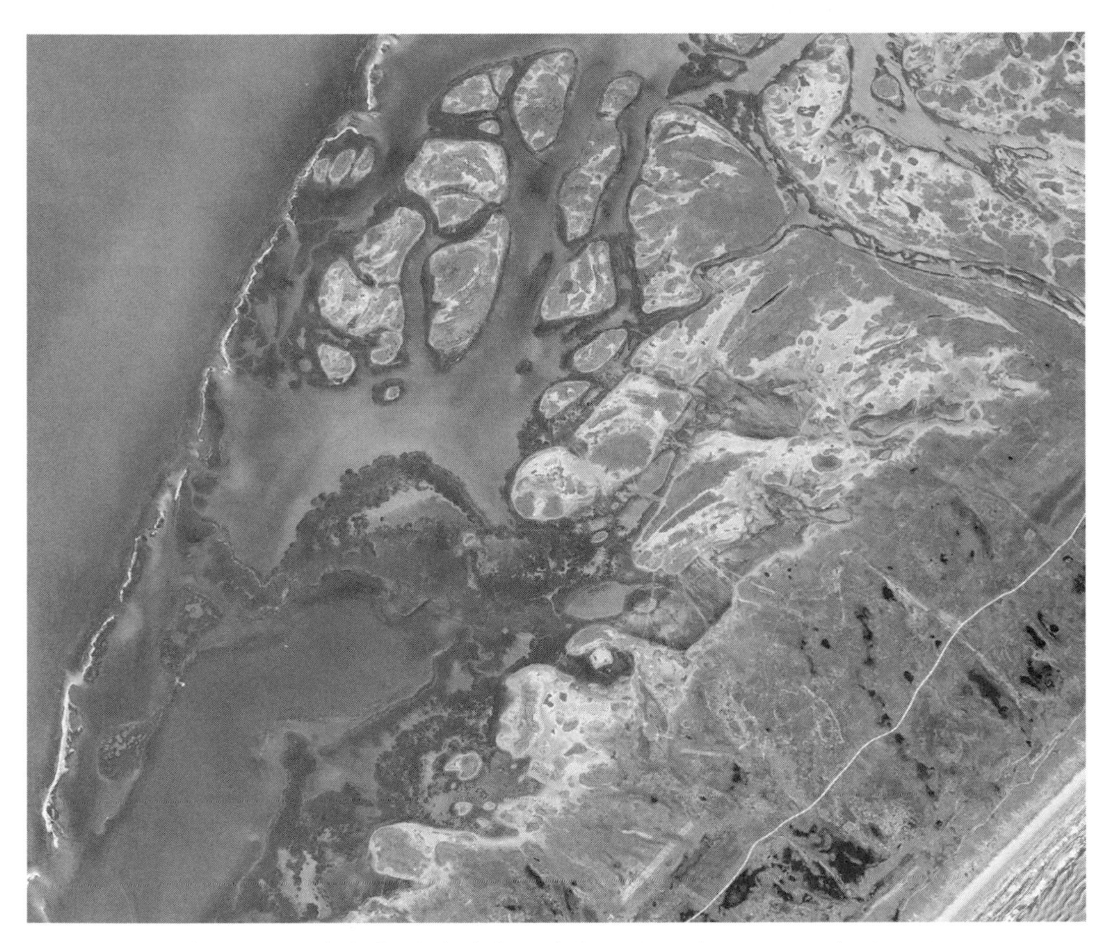

Figure 3.11. "Salinas" (irregularly flooded salt flats—light gray to white areas on this image) are common tidal wetlands along most of the Texas coast.

south are exposed to salinities exceeding 200 ppt during the summer months due to high evapotranspiration and lack of rainfall, while salinities elsewhere are reduced by seepage from neighboring irrigated lands (Purer 1942). Areas with salinities above 117 ppt are typically devoid of vegetation and often have a crust of white salt on the surface. Southern California experiences a Mediterranean climate (wet winters, dry summers), so soil salinities are typically highest from June to September.

Many salt marshes receive freshwater through groundwater seepage, especially along their upper borders. These areas may be readily detected by the presence of species more typical of brackish or freshwater conditions contiguous to salt marsh vegetation and by their constant wetness—saturated soils. Rose mallow (*Hibiscus moscheutos*) occurs at the upper edges of salt marshes with significant freshwater input, whereas farther upstream in brackish waters, they grow along creekbanks (Whigham et al. 1989). Other species such as narrow-leaved cattail (*Typha angustifolia*), three-squares (*Schoenoplectus americanus* and *S. pungens*), and common reed (*Phragmites australis*) are also found in these areas along the upper edge of salt marshes. These plant communities have been included in a zone called "brackish meadow" by some scientists (Nichols 1920). The brackish meadow represents the upper edge of the salt marsh where freshwater influence is significant, leading to the occurrence of species that are more common in less saline soils, for example, creeping bent grass (*Agrostis stolonifera* var. *palustris*), silverweeds (*Argentina* spp.), spike-rushes (*Eleocharis palustris* and *E. rostellata*), and marsh pink (*Sabatia stellaris*) along with typical salt marsh plants.

The interaction of groundwater, tides, and evapotranspiration on water tables in both the marsh and adjacent forest has received considerable study in South Carolina (Thibodeau 1997; Thibodeau et al. 1998; Gardner et al. 2002). As expected, the portions of the high marsh receiving considerable groundwater discharge are less saline (5–10 ppt) than other areas in which salinities (25–50 ppt) eventually became hypersaline favoring the growth of glassworts (*Salicornia* spp.). Salinities are most variable in the glasswort zone. During the year, salinities range from 13.7 to 85.3 ppt in the upper 3 inches (7.6 cm) of the soil to 104 ppt a little deeper to the 10-inch (25 cm) depth, with lowest soil salinities recorded in winter when rainfall is higher than average. Less saline conditions (typically less than 25 ppt) support black needlerush along the marsh-forest boundary. The tides also significantly affect water tables in the neighboring forest within 33 feet (10 m) of the marsh, and saltwater intrusion into the forest water table was noted during a prolonged drought (Gardner et al. 2002).

Salinity Differences Upstream in the Estuary

In coastal rivers, salinity is diluted by freshwater discharge from upstream watersheds. Across estuaries, salinities vary seasonally with river flows but also change with tides, after heavy rainfall events, and during floods and droughts. In the eastern United States, spring runoff creates the freshest condition throughout the estuary as heavy river discharge forces the salt wedge far downstream, while low flows in summer produce the highest salinities as the salt wedge moves farther upstream (Figure 3.12). The most extreme salinities in the uppermost tidal reaches are experienced during droughts. At these times, the salt–fresh water interface may be moved as much as 6 miles (10 km) upstream in the Merrimack River and more than 14 miles (23 km) in New York's Hudson River, for example (Caldwell and Crow 1992; Leck et al. 2009). Because salinities decrease upstream in coastal rivers, conditions for wetlands change from saltwater to brackish and eventually to freshwater. Plant species that were restricted to the edges of salt marshes due to salt stress increase their coverage in brackish marshes. For example, along the Altamaha River in Georgia, brackish marshes where soil salinities are usually less than 15 ppt are dominated by black needlerush, a species that is generally restricted to the upper margins of salt marshes (Wiegert et al. 1981).

Since the saltwater–freshwater interface is a dynamic boundary, plants growing in the middle and upper reaches of the estuary are exposed to considerable shifts in salinity regimes, yet even tidal freshwater wetlands can be subjected to saltwater influence. In the Mid-Atlantic region, tidal forested wetlands tend to dominate freshwater tidal regions. While their salinities are normally fresh (<0.5 ppt), during drier summers (extremely low river discharge) salinities can reach those of oligohaline and mesohaline (brackish) marshes (Darke and Megonigal 2003; Baldwin 2007). These infrequent "salt pulses" may be more important in affecting plant community structure than the long-term average (fresh) conditions. Concern about the effects of rising sea level has focused on tidal freshwater wetlands, especially in Louisiana, as saltwater intrusion will have a significant adverse impact on existing vegetation (e.g., McKee and Mendelssohn 1989; Pezeshki et al. 1989, 1990; Knighton et al. 1991; Allen et al. 1996, Shaffer et al. 2009; Weston 2011).

Hydrogen Sulfides

Prolonged anaerobic (anoxic), strongly reduced conditions, and readily available

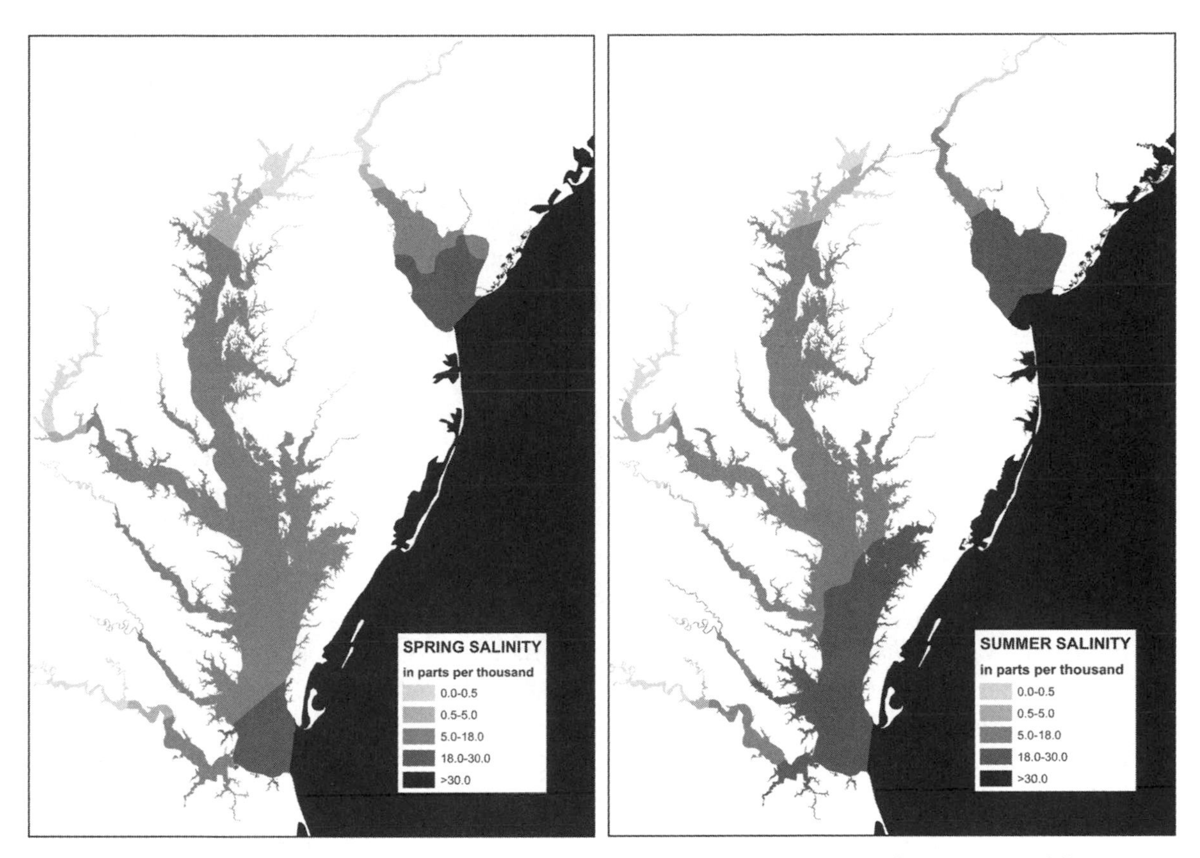

Figure 3.12. Surface water salinity differences between high flows (spring) and low flows (summer/fall) in Chesapeake and Delaware Bays. (Adapted from Chesapeake Bay Program maps and Bryant and Pennock 1988)

sulfide materials make salt marshes, mudflats, and mangrove swamps prime environments for sulfur-loving bacteria (e.g., *Desulfovibrio, Desulfotomaculum,* and *Desulfobacter*). Their activity reduces sulfates to sulfides (hydrogen sulfide; H_2S) and pyrite and is enhanced by the abundance of organic matter available in tidal wetlands (see Reddy and DeLaune [2008] for a detailed description of cycling of sulfur and other elements in wetlands). H_2S produces an unmistakable "rotten egg" odor that can be detected by a simple "whiff" test (Darmody and Foss 1978). These soils have been referred to as acid sulfate soils (Fanning 2002; Rabenhorst et al. 2002). While reduced sulfides are not harmful under natural tidal flooding conditions, when these soils are drained or excavated and left to dry, sulfides oxidize producing sulfuric acid and eventually jarosite, which lower soil pH (<4.0) and create strongly acidic conditions (Rabenhorst et al. 2002). Such conditions can kill plant life leaving the area barren as well as adversely affecting aquatic life and water quality (Anisfeld and Benoit 1997; Portnoy 1999; Melville and White 2002; Ohimain 2003).

Hydrogen Ion Concentration (pH)

The pH of tidal wetland soils typically varies between slightly acidic (e.g., pH = 5.6) and moderately alkaline (e.g., pH = 8.4). Clayey tidal soils in drier climates may be strongly alkaline (e.g., Barrada series along Texas coast and Novato series in California). Frequency of tidal flooding influences soil pH. Soil properties along a gradient from salt marsh into low-lying loblolly pine (*Pinus taeda*) forests show an increase in both pH and salinity with tidal flooding, yet the submerged soils remain more acidic than typical salt marsh soils (Hussein and Rabenhorst 2001). Prolonged periods without flooding can cause the pH of marsh soils to become quite acidic, falling to 3.5 to 4.0 (Howes et al. 1986). The soil pH at the salt marsh–upland border can be more acidic (e.g., pH <5.0) than the rest of the marsh, especially where the marsh grades into sand dunes, peat bogs, or upland forests (e.g., Griffin and Rabenhorst 1989). In the first two cases, vegetation more typical of acidic bogs may be found in the upper marsh including grass pink (*Calopogon tuberosus*), rose pogonia (*Pogonia ophioglossoides*), buckbean (*Menyanthes trifoliata*), and big cranberry (*Vaccinium macrocarpon*) among others (Wherry 1920).

Substrates and Soils

Depending on the type of coastal wetland and suspended sediment available for deposition, soil and substrate properties vary widely. Substrates may be consolidated materials (rock or stone), unconsolidated materials (e.g., sand, silt, clay, or gravel), organic soil (peat or muck), or mineral soils (sandy or nonsandy) (Table 3.2). Materials may form in place (i.e., peat and muck = autochthonous materials) or be brought in by tidal flood waters from eroding shores, marine sources, or river-borne alluvium (allochthonous materials) and deposited during low-energy and slack water periods or following storms (see Griffin and Rabenhorst [1989] for a detailed description of soil formation in tidal marshes, and Frey and Basan [1978] for discussion of processes affecting sedimentation in salt marshes). The nature of the soils and substrates affects plant and animal life as well as influences various functions such as denitrification, which should be higher in salt marsh sediments with high organic matter content (e.g., Davis et al. 2004).

Prior to 1999, the main differences between soils and substrates as classified by American soil scientists were that soils supported free-standing plants (such as emergent herbs, shrubs, and trees) and showed evidence of soil-forming processes ("pedogenesis"), while substrates did not (e.g., Soil Survey Staff 1998). Today, however, certain permanently flooded substrates are recognized as either submerged soils or subaqueous soils because they show evidence of pedogenesis through the presence of soil horizons affected by seagrasses, burrowing organisms, and biochemical reactions driven by microorganisms (Demas and Rabenhorst 2001;

Table 3.2. Particle sizes of unconsolidated materials and boulder

Material	Particle Size Diameter	
	U.S. Department of Agriculture	Federal Wetland Classification System
Boulder	—	>24 inches (>604 mm)
Stone	>9.8 inches (>250 mm)	10–24 inches (254–604 mm)
Cobble	3–9.8 inches (76–250 mm)	3–10 inches (76–254 mm)
Gravel	0.08–3 inches (2–76 mm)	0.08–3 inches (2–76 mm)
Sand	0.002–0.08 inches (0.05–2.0 mm)	0.003–0.08 inches (2–76 mm)
Mud	—	<0.003 inch (<0.74 mm)
Silt	0.00008–0.002 (0.002–0.05 mm)	—
Clay	<0.00008 inches (<0.002 mm)	—

Sources: Particle sizes defined by U.S. Department of Agriculture for soil classification (Soil Survey Division Staff 1993) and the U.S. Fish and Wildlife Service for wetland classification (Cowardin et al. 1979).
Note: There are also other classification systems for describing unconsolidated materials for engineering purposes.

http://nesoil.com/sas/sasinfo.htm). Submerged soils represent former uplands or wetlands that are now permanently inundated due to beaver dams, elevated water tables, or rising sea level, while subaqueous soils are those that have formed underwater, often in connection with submerged aquatic vegetation. To describe soils that are inundated for more than 21 hours of each day in all years, soil scientists created two new soil suborders in soil taxonomy—Wassents and Wassists for subaqueous entisols (e.g., sandy or young soils) and histosols (organic soils), respectively (Soil Survey Staff 2010). Knowledge of these soils is important for shellfish aquaculture and eel-grass restoration.

Tidal Flat Substrates

Tidal flats are substrates comprising variable amounts of sand, silt, clay, organic matter, and in some cases, cobble and gravel. While algae and diatoms are often abundant, mudflats and sandflats (rarely exposed peat beds) typically lack vascular plants, although the lowest elevations that are infrequently exposed to air may be colonized by submergent species, namely eel-grass (*Zostera marina*).

Mudflats have high proportions of silt and/or clay (mud has a >90% silt-clay fraction), while muddy sands are 5 to 50 percent silt-clay (Whitlatch 1982). Muddy substrates have more organic matter than sandy substrates and are anoxic almost to the surface. Higher energy environments tend to produce sand and cobble-gravel substrates. Sandy substrates are therefore found on beaches, in channels, and near the mouths of inlets, whereas mudflats occupy sheltered coves. The duration and frequency of flooding, anaerobic (anoxic) substrates, and littoral currents make life difficult for rooted vascular plants. Oxygen may penetrate 4 to 8 inches (10–20 cm) in sandy substrates of tidal flats, whereas nonsandy substrates have an anoxic zone within ⅕ inch (5 mm) or so of the surface (Whitlatch 1982). In mudflats, the aerobic zone (oxidized zone) may be only a millimeter thick on these generally blackish H_2S-laden substrates (Figure 3.13). For sandflats, three distinct layers of sand may be present, from top to bottom: a yellow layer representing the aerobic (oxic) zone, a gray sand zone where soil is becoming more reduced (i.e., redox potential is rapidly decreasing), and a black reduced zone with iron sulfide and ammonium present and little or no oxygen (Little 2000). Small changes in elevation (e.g., convex versus concave positions) and proximity to channels can reduce stress factors and allow for colonization of mudflats by different algae and associated invertebrates (Aníbal et al. 2007). Low-energy, sheltered environments make it possible for rooted aquatic plants to colonize these flats. The presence of such vegetation may change the nature of the substrate as the plants slow the current velocity allowing finer materials to settle out, contribute organic matter to

Figure 3.13. Black substrate of a mudflat. The lighter colored surface contains some oxygen, while the black layer is anoxic and contains high levels of hydrogen sulfide.

the substrate, and facilitate establishment of benthic invertebrates.

Some water-saturated sands are fluid-like and unstable, while others are more stable since their particles may be bound together with organic substances produced by diatoms and cyanobacteria and by plant roots (Little 2000; Trites et al. 2005). Quicksand is sand that appears solid, but when walked on, for example, the vibrations cause the particles to separate and form a sandy fluid (thixotropic substrate). Other sands may be very firm and become harder under pressure. When walking on these dilatant substrates, footprints leave a whitish mark in the sand. Such sands are inhospitable to both plants and burrowing animals.

Many northern shores in glaciated regions are represented by cobble-gravel beaches ("shingles" as they are called in England). Degree of exposure to wave action and the amount of sand, silt, clay, or organic matter associated with the stones influence plant growth (Packham and Willis 1997). These materials increase the water-holding capacity of the beaches and provide suitable substrates for establishment of vascular plants such as sea lungwort (*Mertensia maritima*) and beach pea (*Lathyrus japonicus*).

Hydric Soils

Wetland soils—known as hydric soils—are saturated or inundated (flooded or ponded) long enough during the growing season to develop anaerobic conditions in the upper part of the soil (Cowardin et al. 1979; Hurt et al. 2006; Vasilas et al. 2010). "Long enough" in this circumstance is usually one week of flooding or two weeks of saturation within 1 foot (30 cm) of the surface. No mention of the frequency or duration of tidal flooding is given in the definition of hydric soil. Hydric soils include soils typically considered to be poorly or very poorly drained, which are characteristic of marshes, swamps, bogs, and other waterlogged lands including tidal marshes and mangroves.

Since tidal wetlands are located along coastal waterways and are frequently flooded, their soils often have high amounts of inorganic sediments, with colonizing plants adding organic matter to the upper soil, mainly through decomposition of roots and rhizomes. The presence of vegetation slows the velocity of tidal waters allowing sediments to drop out of suspension, while periodic to near permanent waterlogging creates anaerobic (anoxic) conditions that retard decomposition of plant remains, leading to the buildup of organic materials at or near the soil surface and lower bulk density. Consequently, soils tend to have higher mineral content closer to bays, sounds, creeks, and sloughs, and more organic matter in the marsh interior farther away from these waters. In Chesapeake Bay, bayside and channel-margin marshes tend to possess the coarsest sediments and least organic matter, while the interior and submerged upland marshes have the finest, organic-rich sediments (Kastler and Wiberg 1996; Ward et al. 1998). The latter areas may be flooded for longer periods than are the streamside levee marshes, however, the low-velocity waters flooding them have typically deposited most of their sediments before reaching the marsh interior (Neubauer et al. 2002). Consequently, these interior marshes have been referred to as "sediment-starved" wetlands. Higher sedimentation in Louisiana streamside marshes creates denser soils, contributes to marsh accretion, provides more nutrients, and results in greater productivity when compared with interior marshes where plant growth is more important for accretion (DeLaune et al. 1979; Nyman et al. 2006). In the Suisun Marsh in California, soils along the sloughs (Reyes series—silty clay, sulfidic fluvaquents) have less than 15 percent organic matter, while soils farther away from the slough have increasingly more organic matter: 15 to 30 percent (Tamba series—mucky clay, fluvaquentic endoaquepts), 30 to 50 percent (Joice series—clayey muck, typic haplosaprists), and >50 percent (Suisun series—muck, typic haplohemists) (Environmental Services Office 2000). Organic matter content does increase in the middle of Georgia salt marshes, but silt and clay remain the major soil components throughout the marshes (Teal and Kanwisher 1961; Figure 3.14). Organic matter content was highest in the short-form smooth cordgrass marsh, whereas the uppermost marsh occupied by salt grass (*Distichlis spicata*), stunted smooth cordgrass, and coastal dropseed (*Sporobolus virginicus*) had the highest sand content, presumably due to their proximity to sandy uplands. The coarsest materials are therefore not creekside, but adjacent to the upland—this may be typical of back-barrier salt marshes.

Many salt marsh soils have developed from sandbars where elevations became suitable for plant colonization; and once primary succession started, substrate characteristics changed: 1) more organic matter was added by decomposition of above- and belowground biomass; and 2) sedimentation was enhanced. A study of marsh development on a sandbar near the mouth of the Altamaha River in Georgia found that after three years, the developing marsh contained less sand than the sandbar and its soil bulk density had decreased, while organic carbon, nitrogen, and percentage of silt increased (Krull and Craft 2009). Despite marked gains in the last three properties, levels were below those of the region's mature marshes. This is not unexpected since it takes time to develop organic or organic-rich soils. Neighboring salt marshes had roughly five times the amount of carbon and nitrogen in the upper 12 inches (30 cm) of soil as the new marsh. These findings generally agreed

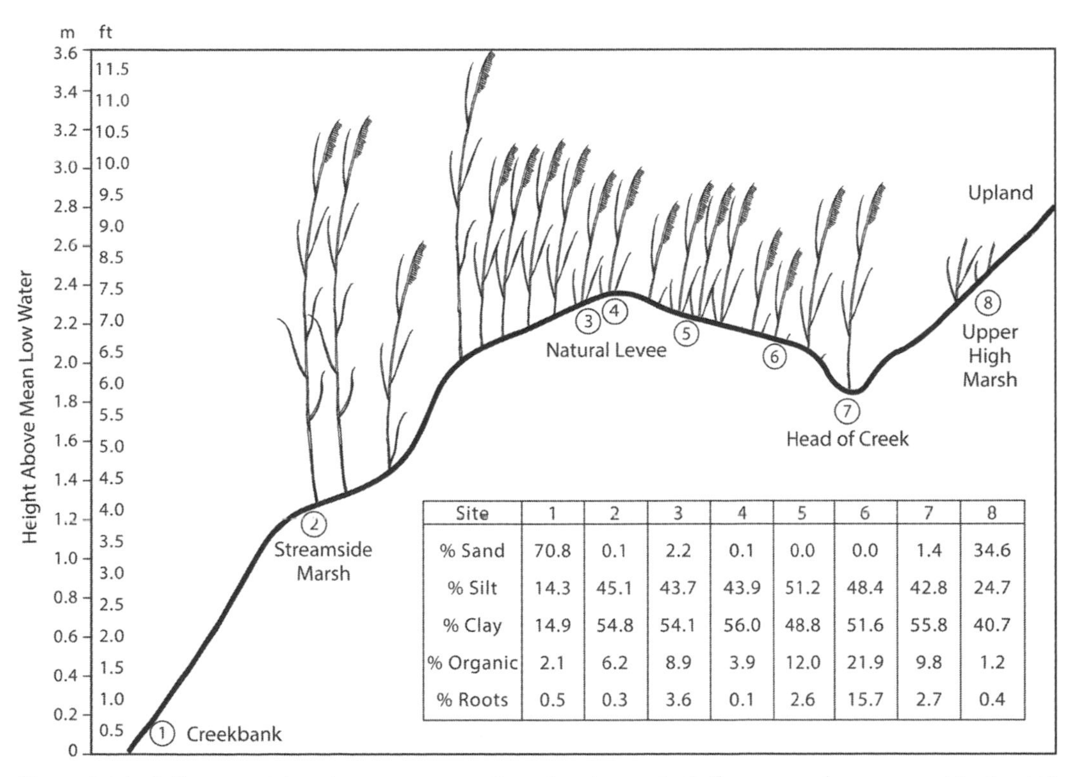

Site	1	2	3	4	5	6	7	8
% Sand	70.8	0.1	2.2	0.1	0.0	0.0	1.4	34.6
% Silt	14.3	45.1	43.7	43.9	51.2	48.4	42.8	24.7
% Clay	14.9	54.8	54.1	56.0	48.8	51.6	55.8	40.7
% Organic	2.1	6.2	8.9	3.9	12.0	21.9	9.8	1.2
% Roots	0.5	0.3	3.6	0.1	2.6	15.7	2.7	0.4

Figure 3.14. Soil composition changes across a Georgia salt marsh. Soil cores are from upper 4 inches (10 cm). Circles indicate different locations: creekbank (1), tall-form smooth cordgrass (*Spartina alterniflora*; 2, 7), short-form smooth cordgrass (5, 6), and higher elevations of the natural levee and upland border (3, 4, 8). Organic content of substrate was lowest near the low-water level and near the upland. It was highest in the short-form smooth cordgrass, which formed a peaty layer to a depth of 1 foot (30 cm). Creek stations had the blackest mud and strongest hydrogen sulfide odor. High clay content limited internal drainage at all but the creekside sites. Horizontal scale is distorted unevenly in this figure. (Adapted from Teal and Kanwisher 1961)

Table 3.3. Identification of different types of organic soils by proportion of fibers visible after rubbing

Organic Layer Type	Soil Texture	Percentage of Fibers Still Visible (%) (live roots are excluded)
Sapric	Muck	>33
Hemic	Mucky peat	33–67
Fibric	Peat	<67

Source: U.S. Army Corps of Engineers 2010.
Note: Soil should be rubbed gently 10 times between thumb and forefinger, then examined with a 10x hand lens.

with those reported for newly constructed marshes in North Carolina and naturally forming marshes in Virginia, except that the latter marsh (a sediment-starved back-barrier marsh) did not show any change in soil particle size (Osgood et al. 1995; Craft et al. 2003).

Given regional differences in climate, landforms, and soil parent materials coupled with local variations in suspended sediments, hydrology, and salinity, a wide variety of soils have formed in tidal wetlands across North America. While soil scientists recognize many soil types (orders, suborders, great groups, subgroups, and series) from a taxonomic standpoint (Soil Survey Staff 2010), hydric soils can be separated into two basic groups: organic hydric soils and mineral hydric soils. The former soils are composed of the remains of plants, whereas varying combinations of sand, silt, or clay dominate the latter. Mineral soils may also have surface layers of organic matter less than 16 inches (40 cm) thick.

Organic soils are defined further by the degree to which plant remains are decomposed, with peats (fibrists) having the least decomposed material, mucks (saprists) the most, and mucky peats (hemists) intermediate levels (Table 3.3). They can also be classified by the thickness of the organic matter (terric—shallow, less than 51 inches [130 cm], or typic—thicker) or shallowness attributed to bedrock (lithic). Since seawater contains about 4 percent sulfate, under prolonged anaerobic conditions typical of salt marshes, sulfate is reduced by sulphur-loving microbes to produce hydrogen sulfide. Because of the presence of this material within 40 inches (100 cm) of the soil surface and the level of organic matter decomposition, most organic salt marsh soils are classified as "sulfihemists" (Figure 3.15a), while the more decomposed ones are "sulfisaprists" (Soil Survey Staff 2010). Salt marsh soils may accumulate pyrite sulfur at rates up to 7 $g/m^2/year$ (Rabenhorst et al. 2002). Hemists are the predominant organic soils in tidal wetlands of the Northeast and the Pacific Northwest, whereas saprists are the dominant organic soil type in tidal wetlands along the South Atlantic and Gulf Coasts, and fibrists characterize northernmost tidal marshes (Rabenhorst 2001). Freshwater tidal wetlands are well represented by organic soils (typically mucks) that are not differentiated from their neighboring nontidal counterparts located just upstream.

Along the Atlantic Coast, mineral salt marsh soils are mostly "sulfaquents"—characterized by the presence of sulfidic materials within 20 inches (50 cm) of the surface (Soil Survey Staff 2010; Figure 3.15b). These soils include loamy and clay soils (e.g., mucky sandy loam, silt loam, silty clay loam, loam, silty clay, and clay). They are the predominant salt marsh soil in many southeastern salt marshes. Sandy hydric soils ("psammaquents") have been mapped in salt marshes along the South Atlantic and Gulf Coasts (Table 3.4).

"Submerged upland" soils that are now salt marshes are often characterized by a histic epipedon—an organic surface layer 8 to 16 inches (20–40 cm) thick over silty former mineral hydric soils. The thickness

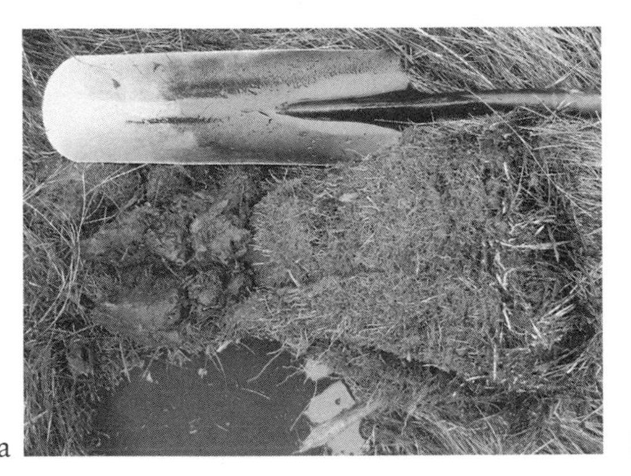

Figure 3.15. Examples of organic and mineral hydric soil of tidal marshes: (a) peat (sulfihemist; NJ) and (b) sandy soil (sulfaquent; MA).

of this layer is related to the period of submergence (i.e., how long the area has been "submerged"). At annual accretion rates of 0.06 to 0.13 inch (1.4–3.2 mm), it would take 125 to 286 years for a submerged Maryland lowland forest soil to accumulate enough organic matter to be classified as an organic soil (Hussein and Rabenhorst 2001). Two "submerged upland soils" have been described in Maryland: the Sunken and the Honga series, with the latter having thicker organic deposits resulting from longer submergence (Brewer et al. 1998a). Peat deposits of salt hay grass and salt grass now overlie former upland soils near the mouth of the Pataguanset River in Connecticut (Orson et al. 1987). In Virginia, submerged upland soils did not possess a histic epipedon, but were loamy soils with a low chroma matrix and prominent mottling (Edmonds et al. 1985). These high marsh soils have been classified as "typic natraqualfs" (e.g., Magotha series). Submerged spodosols in South Carolina have largely retained their spodic horizon but their A- (topsoil) and E-horizons (top soil and leached layer) have disappeared due to silt and clay deposition by the tides and the mixing of these materials by fiddler crabs to produce a gray-colored A-horizon (Gardner et al. 1992). Although the impact of flood frequency is the initial driver of soil development, once the area is permanently saturated, water chemistry, soil properties, and biochemical processes become more important (Hussein and Rabenhorst 2001).

The soils of northeastern U.S. salt marshes are often dominated by organic matter (remains of roots, stems, and leaves), especially in the irregularly flooded (high marsh) zone, whereas southeastern salt marshes are muddy with higher mineral content (e.g., silt and clay) (Table 3.4). This difference is related to many factors including climate (higher as compared with lower temperature effects on plant matter decomposition, bioturbation, and sediment resuspension), sediment source (finer soils from Piedmont and Coastal Plain or coarser materials from New England Highlands), higher tidal amplitudes (more effective at trapping sediment), higher frequency of flood-dominated tidal channels in the north, and a higher percentage of northeasters (storms that can bring marine sediments shoreward) (Stevenson et al. 1986). Local differences occur within regions. While most mineral salt marsh soils are grayish to black in color, tidal wetlands in the upper Bay of Fundy have silt and clay soils derived from red parent material—Triassic redbeds (Frey and Basan 1978). The typical salt marsh soil of Bay of Fundy—the Acadia association—is mostly diked today. It exhibits a range of characteristics with textures from silt loam to silty clay loam to silty clay and topsoil colors varying from reds of 10R5/4 to reddish browns 2.5YR 5/4 and

Table 3.4. Descriptions of several tidal wetland soils (taxonomic classification and series) representative of different coastal regions

Horizon	Depth	Description
Northeast Coast		
Typic Sulfihemists		Ipswich series (salt marshes; New Hampshire to New Jersey)
Oe1	0–18" (0–46 cm)	Dark grayish brown (2.5 Y 4.2) mucky peat
Oe2	18–42" (46–107 cm)	Very dark brown (2.5Y 3/2) mucky peat
Oa	42–62" (107–158 cm)	Very dark gray (2.5Y 3/1) muck
Terric Sulfihemists		Westbrook series (salt marshes; New Hampshire to Maryland)
Oe12	0–40" (0–102 cm)	Very dark gray (10YR 3/1) mucky peat
Oe3	40–48" (102–122 cm)	Dark olive gray (5Y 3/2) mucky peat
Cg1	48–64" (122–163 cm)	Very dark gray (5Y 3/1) silt loam
Cg2	64–99" (163–252 cm)	Dark gray (N4/) silt loam
Typic Sulfaquents		Broadkill series (salt marshes; New Jersey to Maryland)
Oe	0–6" (0–15 cm)	Very dark grayish brown (2.5Y 3/2) mucky peat
Ag	6–13" (15–33 cm)	Dark gray (5Y 4/1) and very dark gray (5Y 3/1) silty clay loam
Cg1	13–32" (33–81 cm)	Dark gray (5Y 4/1) and dark olive gray (5Y 3/2) silty clay loam
Cg2	32–38" (81–97 cm)	Dark olive gray (5Y 3/2) silt loam
Cg3	38–80" (97–203 cm)	Very dark gray (5Y 3/1) silt loam
Southeast and Gulf Coasts		
Typic Sulfaquents		Chincoteague series (salt marshes; Virginia, possibly to Florida)
A	0–6" (0–15 cm)	Dark gray (5Y 4/1) silt loam
Cg13	6–40" (15–102 cm)	Dark gray (5Y 4/1) silty clay loam
Cg4	40–65" (102–165 cm)	Dark gray (5Y 4/1) silt loam
Typic Sulfaquents		Capers series (salt marshes; North Carolina to Georgia, possibly Virginia, Florida, and to Louisiana)
A	0–16" (0–41 cm)	Very dark grayish brown (2.5Y 3/2) silty clay
Cg1	16–46" (41–117 cm)	Black (N 2/0) clay
Cg2	46–60" (117–152 cm)	Very dark gray (N 3/0) and dark gray (N 4/0) clay
Sodic Hydraquents		Scatlake series (salt marshes; Louisiana and Texas, possibly Mississippi)
Oe	0–6" (0–15 cm)	Very dark gray (10YR 3/1) mucky peat
A	6–12" (15–30 cm)	Very dark gray (5Y 3/1) mucky clay
Cg1	12–16" (30–41 cm)	Dark gray (5Y 4/1) clay
Cg2	16–18" (41–46 cm)	Black (N 2/0) muck and gray (5Y 5/1) clay
Cg3	18–21" (46–53 cm)	Gray (5Y 5/1) clay
Cg4	21–80" (53–203 cm)	Greenish gray (5GY 6/1) clay
(*Note:* Mineral horizons turn black when air-dried.)		
Typic Haplosaprists		Lafitte series (brackish marshes; Louisiana, possibly Mississippi and Alabama)
Oa1	0–6" (0–15 cm)	Very dark brown (10YR 2/2) muck
Oa2-5	6–52" (15–132 cm)	Black (10YR 2/1) muck
Oa6	52–75" (132–191 cm)	Very dark brown (10YR 2/2) muck
Ag	75–90" (191–229 cm)	Dark grayish brown (10YR 4/2) clay
Cg	90–100" (229–254 cm)	Gray (N6/0) clay
Typic Psammaquents		Carteret series (salt and brackish marshes; Virginia and North Carolina)
A	0–4" (0–10 cm)	Dark grayish brown (2.5Y 4/2) loamy sand
Ag	4–10" (10–25 cm)	Dark gray (N 4/0) loamy sand
C1g	10–34" (25–86 cm)	Gray (N 5/0) loamy sand
C2g	34–80" (86–203 cm)	Greenish gray (5GY 5/1) sand

Table 3.4. *(continued)*

Southeast and Gulf Coasts *(continued)*		
Sodic Psammaquents		Tatton series (salt barrens; Texas)
Anz	0–4" (0–10 cm)	Light yellowish brown (2.5Y 6/3) fine sand
Anz	4–12" (10–31 cm)	Light grayish brown (2.5Y 6/2) fine sand with light yellowish brown (2.5Y 6/3) strata
Cnzg1	12–18" (31–46 cm)	Grayish brown (10Y 6/1) loamy sand with 2% brownish yellow (10YR 6/6) redox concentrations
Cnzg2	18–26" (46–66 cm)	Light brownish gray (2.5Y 6.1) fine sand with 4% dark grayish brown (2.5Y 4/2) redox depletions
Cnzg3	26–40" (66–102 cm)	Dark greenish gray (5G 4/1) fine sand with few grayish green (5GY 6/1) strata
Cnzg4	40–80" (102–203 cm)	Dark bluish gray (5B 4/1) fine sand
Pacific Coast		
Typic Sulfaquents		Novato series (salt marshes; California)
A11g	0–2" (0–5 cm)	Light gray (5Y 7/1) clay with common dark brown (7.5YR 4/4) mottles
A12g	2–6" (5–15 cm)	Light gray (5Y 7/1) clay with common dark reddish brown (5YR 3/4 mottles)
A13g	6–15" (15–38 cm)	Gray (5Y 6/1) clay with common reddish brown (5YR 4/4) mottles
C1g	15–27" (38–69 cm)	Gray (5Y 6/1) clay with common reddish brown (5YR 4/4) mottles and pale brown (2.5YR 7/4) jarosite mottles
C2g	27–40" (69–102 cm)	Gray and light gray (5Y 6/1, 7/1) clay with common pale yellow (2.5Y 8/4) jarosite mottles
C3g	40–60" (102–152 cm)	Light gray (N 6/0) clay
Histic Humaquepts		Clatsop series (tidal fresh marshes; Oregon and Washington)
Oa	0–6" (0–15 cm)	Very dark grayish brown (2.5Y 3/2) muck
A	6–13" (15–33 cm	Very dark grayish brown (2.5Y 3/2) mucky silt loam
Cg	13–24" (33–61 cm)	Dark gray (5Y 4/1) silt loam
2Cg	24–60" (61–152 cm)	Very dark gray (5Y 3/1) silt loam

Sources: USDA Natural Resources Conservation Service, official soil descriptions. Soil Survey Division Staff 1993.
Notes: These are examples of soil profile descriptions; for mineral layers, textures typically vary. Horizons with same colors and textures may be separated into other horizons based on other features such as percentage of fiber content for mucky peats or chemical properties for mineral soils. O = organic horizon, e = hemic, A = A-horizon (topsoil), C = C-horizon, g = gleyed, n = accumulation of sodium, z = accumulation of salts more soluble than gypsum.

5YR 4.5/3 to very dark brown 10YR 2/2 (Wicklund and Langmaid 1953; Rees et al. 1996; Fahmy et al. 2010). In the salt marsh, these soils usually have a peat layer of variable thickness above these materials, while the reddish colors are most evident along unvegetated creekbanks. Glacial erratics (e.g., boulders) can be found in northern wetlands. Southern estuarine marshes, with exceptions in Florida and Louisiana, lack the thick peat deposits that accumulate in northern marshes, although the mineral soils are rich in organic matter, particularly in the more frequently flooded areas. The high marsh of southeastern back-barrier salt marshes is often characterized by sandy soils, especially where overwash processes are active. Many Louisiana marshes are floating on a nearly liquid substrate (organic ooze that grades into clay)—flotant marshes. Other Gulf Coast salt and brackish marshes often have 2 to 8 inches (5–20 cm) of organic matter over clay. On the Pacific Coast, thin layers of peat may be present from Oregon north, but most salt marsh soils are characterized by clays and silts (Macdonald 1977).

Coastal Processes

Ocean shorelines are ever-changing features, and one of the most dramatic examples and one important to tidal wetland establishment is "spit formation" (see Figure 2.6). Waves and littoral currents keep the beach in a relative state of flux. Storms play a major role in this process, for their powerful waves cause the greatest movement of beach particles (sand, gravel, or cobbles). Winter storms tend to erode beaches, while summer conditions promote deposition, so beaches narrow in winter and widen during the summer. In the Bay of Fundy, marshes receive seasonal deposits of clay and silt in summer, with erosion taking place in winter (Desplanque and Mossman 1998).

As sea level continues to rise, barrier beaches and islands migrate in a landward direction—a phenomenon that has occurred for thousands of years. It is an immutable property of the coast. Direct evidence of this can be seen in places where dunes have advanced over the once-protected marsh, thereby exposing beds of marsh peat to the full force of the ocean's waves and currents, or where the dead snags of former coastal forests can be found along the oceanfront beach zone of barrier islands. Chapter 12 contains a discussion of sea-level rise effects on tidal marshes.

Coastal processes cause changes in tidal wetlands in other ways. Certain bar-built estuaries such as coastal salt ponds and lagoonal estuaries may be subjected to periodic or seasonal closure of their mouths by sand deposited from littoral drift (see Figure 1.4). Some coastal ponds remain closed for years. Depending on the duration of the closure, salinity changes could be significant enough to cause plant mortality of intolerant species. What these estuaries have in common is a general lack of significant freshwater discharge or tidal flows to keep the natural breachways open year-round. In other cases, spit growth may eventually enclose a small estuary, separating it from tidal influence and allowing freshwater conditions to establish, or in the case of southern California wetlands, more saline conditions to form. Inlets of some larger coastal ponds have been stabilized and are now permanently open via jetties (breachways) and maintenance dredging (e.g., Ninigret Pond, RI).

Analysis of pollen, rhizomes, and macrofossils in marsh peat allows scientists to reconstruct the development of wetlands. About AD 700, Long Island's Fresh Pond was a low salt marsh dominated by smooth cordgrass that was later replaced by salt hay grass (Clark and Patterson 1985). From AD 1000 to 1680, the mouth of the pond was closed by a spit, cutting off the area from tidal flow and saltwater. Fresh Pond was truly a freshwater marsh at the time of settlement since farmers grazed and watered cattle there. Later it was opened by farmers to encourage salt hay growth given its high forage value, and it is still a salt marsh.

Storms

Excessive rainfall in upriver watersheds increases freshwater runoff to coastal rivers, moving the salt wedge farther seaward. Plants growing in estuaries, especially the middle reaches, must be highly tolerant of variable salinities as wide fluctuations are the norm. In the past century, the frequency of extratropical storms has increased along the Atlantic Coast and is expected to continue increasing through the 21st century (Emanuel 1987; Knutson et al. 1998; Hayden and Hayden 2003). With rising sea level, the impacts of coastal storms may worsen (Zhang et al. 1997). Also recent predictions call for increased activity of hurricanes along the Atlantic Coast as well as more intense hurricanes—likely due to climate change (Goldenberg et al. 2001; Webster et al. 2005; Intergovernmental Panel on Climate Change 2007).

Increased storm activity from 1954 to 1964 may have been responsible for a higher accretion rate of 0.16 inch per year (0.40 cm/yr) in a Connecticut salt marsh (Orson et al. 1998). Accretion during this decade allowed the marsh to compensate for decades of accretion deficits. Such episodic depositions may help sustain marshes in the future. In a Virginia salt marsh, storms contributed 27 percent of the sediment deposited on the marsh, with the rest deposited by normal spring tides (Christiansen 1998). In an Arctic salt marsh, storms deposited up to 7.9 inches (20 cm) of homogenized peat or sand, eliminating vegetation and essentially restarting vegetation colonization and marsh development (Funk et al. 2004). Storms also generate waves that erode lower marshes, especially those along exposed bay shores and mudflats. A 1975 storm in the Minas Basin of the Bay of Fundy removed a layer of mud up to 3.2 inches (8 cm) thick (Percy 1996a).

When storm-induced waves break through low breaches in the dunes of barrier islands, they may wash over into adjacent salt marshes, depositing a layer of sand on the marsh vegetation (Figure 3.16). One of the more notable marsh overwash areas in the northeastern United States can be found on the bayside of Assateague Island in Maryland. They are sites of former inlets that are still periodically flooded by the tides (Higgins et al. 1971). In New England, nor'easters can raise tide levels to the point at which they cause local flooding and move sand from low beach-dune complexes into back-barrier marshes. The New England Blizzard of 1978 overwashed Cape Cod's Nauset Spit depositing about 5.5 feet (1.65 m) of sand on the living salt marsh (Leatherman and Zaremba 1987). Over half of this material was subsequently blown back to the beach by offshore winds creating dunes. (*Note:* Offshore winds are those blowing from land to sea—the opposite of onshore winds.) Overwash processes can destroy some sections of marsh with too much overburden, while in areas of less deposition marsh surfaces are slightly elevated, which may actually benefit their long-term survival in the face of rising sea levels (see Chapter 12). A Virginia marsh lost 7.2 percent of its area in eight years due to overwash (Kastler and Wiberg 1996).

Figure 3.16. This soil sample shows a layer of sand deposited over peat, evidence of overwash processes.

Hurricanes push saltwater considerable distances inland while depositing and eroding sediments in coastal wetlands. The greatest effect of hurricanes is on coasts with shallow offshore waters, like the Gulf of Mexico. Storm surges from hurricanes may elevate water levels to 10 feet (3.1 m) or more in the northeastern United States (Pore and Barrientos 1976) and to higher levels on the southeast coast (e.g., Hugo, 20 ft [6.1 m] in South Carolina; Camille, 24.6 ft [7.5 m] in Mississippi; and Katrina, 28 feet [8.5 m] in Mississippi and 20 feet (6.1 m) in Louisiana; National Hurricane Center, www.nhc.noaa.gov/HAW2/english/history.shtml). Coastal wetlands may substantially benefit from these catastrophic events. Over the course of a few days, hurricanes can deposit four to eleven times the amount of sediment that accretes annually under normal circumstances (Nyman et al. 1995). Hurricane Agnes discharged more sediment into Chesapeake Bay in one week (June 1972) than had been deposited collectively over several previous decades (Schubel 1974). The 1938 hurricane, southern New England's most severe storm of recent record, deposited a layer of sand about 0.5 inch (1.25 cm) thick over some wetlands (Warren and Niering 1993). After Hurricane Isabel hit Virginia in September 2003, marsh elevations increased by 0.12 inch (0.3 cm), 0.16 inch (0.4 cm), and 0.6 inch (1.5 cm) in the low, middle, and high marsh, respectively (Willis et al. 2005). Along the Louisiana coast, in August 2005 Hurricane Katrina deposited 1.2 to 3.1 inches (3–8 cm) of sediment that compacted over time for a net increase of 0.28 inch (0.7 cm) and 0.67 inch (1.7 cm) at two locations by July 2007 (McKee and Cherry 2009). Elevations in oligohaline marshes along Lake Pontchartrain rose by 8.5 inches (215 mm) after hurricanes Katrina and Rita but declined in subsequent years (Reed et al. 2009). Much of this newly deposited material was highly organic matter transported from more seaward marshes, and over time this material decomposed since the elevated marsh was raised above normal tidal levels. The intensity of the storm surge and the location of the wetland relative to the storm track are important factors affecting the sediment deposition and compaction, as wetlands in the direct path of the hurricane should receive more sediments than others (Cahoon 2006). These low-frequency events may be vital to sustaining elevations of sediment-starved wetlands. An increase in the frequency of hurricanes and accompanying sediment deposition could help counter the adverse effects of rising sea level on coastal marshes. The long-term effect of these additions, however, will depend on other factors including plant growth (organic matter accumulation and root growth), sediment texture, resistance to erosion and compaction, soil swelling or shrinkage, and local subsidence.

The ever-changing tidal shorelines shaped by coastal processes, the range of climates in which tidal wetlands form, and the actions of people combined with the fluctuating environmental conditions (e.g., inundation, saturation, exposure, salinity, and temperature) present numerous challenges for plant and animal life. These forces have interacted to create a wide variety of habitats in the intertidal zone ranging from nonvegetated to vegetated sites, hypersaline to fresh waters, cobble-gravel substrates to sandy soils to organic soils, and nearly permanently flooded to infrequently flooded sites. Although the challenges are considerable, many plants and animals have found a way to utilize these habitats, and some have developed highly specialized features to cope with the most extreme conditions of a given zone.

Further Readings

Wet Coastal Ecosystems (Chapman 1977)
"Coastal salt marshes" (Frey and Basan 1978)
The Ecology of a Salt Marsh (Pomeroy and Wiegert 1981)
"Soils of tidal and fringing wetlands" (Rabenhorst 2001)
Understanding Tides (Hicks 2006)
Ecology of Tidal Freshwater Forested Wetlands of the Southeastern United States (Conner et al. 2007a)

4 Plant Response to the Tidal Environment

Tidal wetlands are subject to natural forces and human actions. From an organism's standpoint, the intertidal environment is rigorous, harsh, and characterized by fluctuating conditions. Tides ebb and flow exposing and then rewetting substrates; salinities vary with tides, precipitation, and river discharges; temperatures change seasonally, daily, and hourly; and the vagaries of weather all create unique conditions and challenges for both plant and animal life. Although many animals can move in and out of wetlands, occupying them at favorable times, plants do not have such freedom. Once established, plants must cope with the physical stresses posed by the intertidal environment, disturbances (e.g., storms, herbivory, and human activities), and competition. The interaction of physical, chemical, and biological factors affects the germination, growth, and reproduction of plants as well as productivity and species richness (e.g., McFalls et al. 2010). Species colonizing coastal wetlands must develop morphological and/or physiological adaptations and other strategies to cope with these challenging conditions. In this chapter I address plant response to various stressors—the physiochemical factors and biotic interactions that largely determine the presence or absence of vegetation and species composition of tidal wetlands. Each factor is discussed separately; however, it must be emphasized that multiple stressors characterize different environments and they operate both individually and collectively to affect plant life. Since the frequency and duration of tidal inundation are largely responsible for site wetness and salinity, separating out the influence of tidal hydrology from salinity on vegetation is understandably difficult in estuarine ecosystems. They, in large part, form the foundation of what is the intertidal estuarine environment and are strongly interrelated.

Hydrology

Plant distribution in the intertidal zone is in part attributed to differences in flooding and soil saturation. Salt marsh vegetation tends to grow above mean sea level, with aquatic vascular species and macroalgae colonizing the lower intertidal zone. Some plants appear to range widely within the marshes, while others may have predictable elevation ranges, so small changes in marsh topography are often reflected in plant distribution (e.g., Adams 1963; Bartoldus 1984; Chmura et al. 1997; Figure 4.1).

Flooding

Changes in the frequency and duration of flooding have significant effects on vegetation patterns and productivity in estuaries and tidal freshwater environments. Conditions range from permanently flooded (subtidal) to infrequent flooding by storm tides with periodic changes in the amplitude of all tides during the metonic cycle.

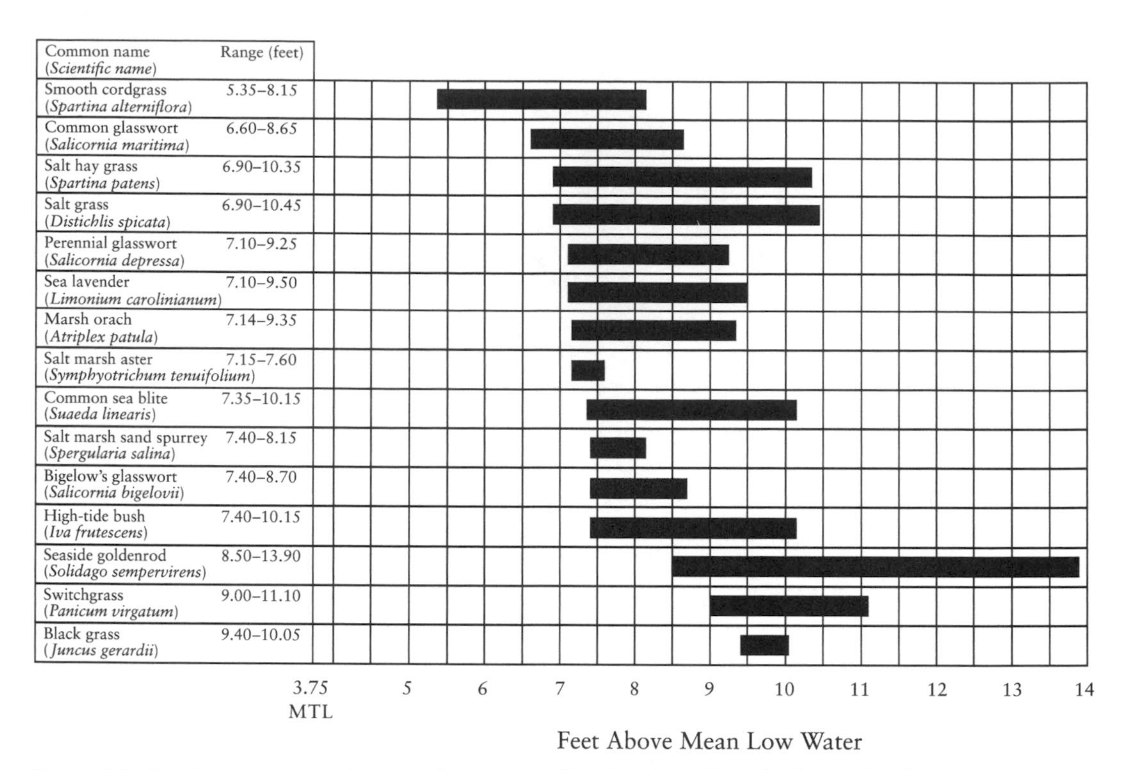

Figure 4.1. Species occurrence along an elevation gradient in a Long Island salt marsh relative to mean low water. Mean tide level (MTL) is 3.73 feet (1.14 m), while the mean tide range for this area is 7.01 feet (2.14 m) and the spring tide range is 7.60 feet (2.12 m). (Data from Bartoldus 1984)

In estuaries, differences in tidal hydrology also influence soil salinity. Since marsh herbs and mangroves grow to only a certain level below mean high tide, the effect of long-duration flooding on these vascular plants is evident. Their tolerance of submergence then sets the lower limit for these vascular plants, with lower areas being characterized by tidal flats and nonvascular plants. Vascular plants become abundant only when exposure to air is limited in subtidal waters, with eel-grass (*Zostera marina*) in temperate marine waters and a wider variety of species in the tropics (e.g., *Thalassia testudinum, Cymodocea filiformis, Halodule,* and *Halophila*). Smooth cordgrass (*Spartina alterniflora*), the dominant salt marsh emergent species along the water's edge in eastern North America, typically grows seaward to the midpoint of the mean tide range (the half-tide level), but can colonize lower levels about halfway between this mark and mean low water in some places (Figure 4.2; Lagna 1975; McKee and Patrick 1988). Its lower limit may be less in macrotidal regions as marshes in the Bay of Fundy are reportedly restricted to the upper quarter of the tidal range (Wells and Hirvonen 1988).

In the simplest terms, tidal wetlands can be separated into two zones: 1) regularly flooded zone (flooded at least once daily), and 2) the irregularly flooded zone (flooded less often). Within the latter, permanently flooded water bodies may be found (e.g., shallow ponds in northern marshes and even lakes in Louisiana's marshes). From April to September 1985, the average number of days that Rhode Island salt marsh species experienced tidal flooding varied by species: tall form of smooth cordgrass, 30.5 days per month; short form of smooth cordgrass, 23.5; salt hay grass (*Spartina patens*), 17.3; and black grass (*Juncus gerardii*), 8.7 days (Bertness and Ellison

1987). Although elevations were not provided, it is rather obvious that increasing elevation was responsible for lowering the flooding period. An earlier study of a Cape Cod marsh reported on differences in submergence relative to elevations above mean high water (Table 4.1). In a Mississippi tidal marsh, a tall form of smooth cordgrass grew in areas flooded during virtually every tidal cycle and inundated for as long as it was exposed—362 as compared with 361 hours during the year (Eleuterius and Eleuterius 1979). Interestingly, this zone was inundated continuously through several tidal cycles in April, May, and November of 1975, yet in December six tidal cycles failed to flood this low marsh. Smooth cordgrass dominated the intertidal marsh that was subject to more than 140 inundations per year, a region flooded from 10 to 87 percent of the time. With less flooding, smooth cordgrass assumed a shorter growth form. Above this zone, black needlerush dominated a region that was exposed most of the time—flooded only 16 to 99 times per annum and 0.8 to 5.0 percent of the time. On rocky shores, zonation of some macroalgae is mainly controlled by exposure (time out of water), which affects desiccation (Little and Kitching 1996). In tidal freshwater wetlands, the duration and timing of flooding significantly influences vegetation patterns as observed in studies of identical seedbanks subjected to different inundation regimes where 1 to 4 inches (3–10 cm) of flooding significantly reduced seedling recruitment and growth in many species (Baldwin et al. 2001). These changes that

Figure 4.2. Variations in the elevational range of smooth cordgrass (*Spartina alterniflora*) in North America. MTR = mean tide range; HTL = half tide level (the midpoint between mean high water and mean low water). Note: 1 m = 3.3 feet. (Figure 2 from McKee and Patrick 1988; *Estuaries*; copyright Estuarine Research Federation; reprinted by kind permission from Springer Science+Business Media B.V.))

Table 4.1. Flooding and submergence of a Cape Cod salt marsh

Feet above MHW (m)	Number of Tides/Year	Period of Submergence	
		Hours/Year	Percentage of the Year (%)
0 (0)	316	500	5.7
0.5 (0.15)	170	240	2.7
1.0 (0.30)	83	115	1.3
1.5 (0.46)	43	48	0.55
2.0 (0.61)	18	8	0.09
2.5 (0.76)	1	0	0

Source: Redfield 1972.
Notes: Below mean high water (MHW), the marsh is flooded every day. At one such location, flooding of the low marsh lasted for 6.8 hours or 55% of the time; tidal flats are exposed for about 4 hours after high tide and are flooded about 2 hours after low tide.

could result from urbanization of watershed, sea-level rise, or coastal subsidence would likely alter plant species composition.

Cyclical changes in tides may influence plant life by increasing or decreasing periods of submergence and soil saturation. Over an 18.6-year period (the metonic cycle), the amplitude of all tides varies by a few inches (several centimeters). Metonic highs create tides with amplitudes higher than "normal," while metonic lows produce tides of lower amplitude. These changes can affect vegetation patterns. For example, pollen analysis of soil cores from a Long Island salt marsh identified two or three shifts in dominant high marsh species from grasses to sedges from 1835 to 1930 during which time the metonic cycle completed four episodes (Clark 1986). Repeated surveys of another Long Island marsh documented 1) a seaward advance of smooth cordgrass from 1915 to 1927; 2) seaward migration of common three-square (*Schoenoplectus pungens*) into the high marsh; and 3) drift logs did not float as far inland as observed in the 1924 survey (Johnson and York 1915; Conard 1924; Conard and Galligar 1929). These and other studies support claims that sea-level rise was not a constant rise since 3000 B.P. but instead exhibited short-term fluctuations. Such changes may also facilitate common reed (*Phragmites australis*) expansion (Chambers et al. 2003). This invasive species may take advantage of the lower tides and better drainage conditions during metonic lows as common reed invasion of estuarine marshes from Connecticut to Maryland appeared to correspond with recent metonic lows (around 1951 and 1969) (Winogrond and Kiviat 1997; Windham and Lathrop 1999; Rice et al. 2000; Warren et al. 2001). Elevated sea levels during metonic highs can also affect plant productivity due to the increase in tidal range (Morris et al. 2002). A hydrologic study of the Piermont marsh in the Hudson River took place during the high point of the metonic cycle when water levels were 0.8 to 2.4 inches (2–6 cm) higher than average levels for this site (Montalto et al. 2006). These researchers pointed out the importance of knowing the stage of the metonic cycle, especially when planning water levels for tidal restoration projects.

PLANT ADAPTATIONS TO SUBMERGENCE

Salt or brackish areas that are permanently submerged (subtidal) are typically not considered wetlands, although in the United States, shallow, fresh tidal waters are included in the definition of wetland (Cowardin et al. 1979). Since these areas are true aquatic sites that are permanently underwater, factors such as light availability, turbidity, current patterns, substrate type, and salinity influence aquatic macrophyte composition and distribution. Some areas that are nearly permanently inundated are periodically exposed during extreme low spring tides or by winds blowing offshore ("irregularly exposed" according to Cowardin et al. 1979). The plants occupying aquatic habitats are mostly algae, yet a number of vascular (flowering) plants can be found there. In marine waters, the latter species are represented by seagrasses—a remarkable group of about 60 species with terrestrial origins that have successfully adapted to life in subtidal saltwater. Interestingly, they represent only 0.02 percent of the world's approximately 300,000 flowering species (Les et al. 1997).

Aquatic plants of tidal waters have assumed one of several plant life forms for living in water (Table 4.2), while wetland plants growing in the tidal marshes and swamps are mostly free-standing (emergent) herbs, shrubs, and trees, or vines. In addition to their obvious life forms, the aquatic plants have developed special adaptations for life underwater (Table 4.3). For example, the flat straplike or ribbonlike leaves or highly dissected leaves possessed by many aquatic plants flow freely in currents offering minimal resistance (Figure 4.3). They also aid in diffusion of gases and allow for deeper light penetration. Some species produce polymorphic leaves, with submerged leaves possessing these characteristics, and floating or emergent leaves being broad and flat. Aerenchyma tissue (gas-filled lacunae) in stems, leaves, and petioles provides aquatic plants with buoyancy as well as the ability to move gases internally (see further discussion below under Plant Adaptations to Anaerobic Conditions). A thin cuticle in aquatic plants facilitates nutrient and water uptake.

Emergent herbs and a few woody plants subject to extended periods of flooding also develop extensive aerenchyma tissue along with morphological and physiological adaptations. Adventitious roots (growing above the ground or in the water column) enhance oxygen transport and aid in absorbing water and nutrients (DeLaune et al. 1987b). Shallow root systems, hypertrophied lenticels, pneumatophores, and hypertrophied stems are other prominent adaptations of wetland plants (Figure 4.4; see Kozlowski 1984

Figure 4.3. A dense bed of submerged aquatics exhibiting typical adaptations, including flattened leaves and highly dissected leaves.

Table 4.2. Plant life forms growing in tidal waters

Life Form	Examples of Species
Rooted with submerged leaves	Eel-grass (*Zostera marina*), Widgeon-grass (*Ruppia maritima*), Horned pondweed (*Zannichellia palustris*), Turtle-grass (*Thalasia testudinum*), Shoal-grass (*Halodule wrightii*), Pondweeds (*Potamogeton* spp.), Sago pondweed (*Stuckenia pectinata*), Naiads (*Najas* spp.), Water-milfoils (*Myriophyllum* spp.), Waterweeds (*Elodea* spp.), Wild celery (*Vallisneria americana*), Awl-leaf arrowhead (*Sagittaria subulata*), Grass-leaved arrowhead (*Sagittaria graminea*), Quillworts (*Isoetes* spp.)
Free-floating plant	Duckweeds (*Lemna* spp., *Spirodela polyrhiza*), Coontail (*Cabomba demersum*), Mosquito-fern (*Azolla caroliniana*), Water chestnut (*Trapa natans*)
Free-floating with emergent plant parts	Common bladderwort (*Utricularia vulgaris*), Water hyacinth (*Eichhornia crassipes*), American frog-bit (*Limnobium spongia*), Maidencane (*Panicum hemitomon*)
Rooted with floating leaves	Spatterdock (*Nuphar lutea*), Water-shield (*Brasenia schreberi*), White water lily (*Nymphaea odorata*), Alligatorweed (*Alternanthera philoxeroides*), American frog-bit (*Limnobium spongia*), Water smartweed (*Polygonum amphibium*)
Rooted with submerged and floating leaves	Fanwort (*Cabomba caroliniana*), Ribbonleaf pondweed (*Potamogeton epihydrus*), Water chestnut, Marsh mermaid-weed (*Proserpinaca palustris*)
Rooted with submerged and emergent leaves on the same plant	Marsh mermaid-weed
Rooted with emergent leaves	Soft-stemmed bulrush (*Schoenoplectus tabernaemontani*), Spatterdock, Pickerelweed (*Pontederia cordata*), Stiff arrowhead (*Sagittaria rigida*), Grass-leaved arrowhead (*Sagittaria graminea*), Bur-reeds (*Sparganium* spp.), Pipeworts (*Eriocaulon* spp.), Water lotus (*Nelumbo lutea*), Water pimpernel (*Samolus valerandi* ssp. *parviflorus*), Water purslane (*Ludwigia palustris*), Mare's-tail (*Hippuris vulgaris*)
Shrub	Buttonbush (*Cephalanthus occidentalis*)
Tree	Bald cypress (*Taxodium distichum*), Water gum (*Nyssa aquatica*)

Note: Most of these plants are found in tidal freshwaters.

Table 4.3. Specialized adaptations for life in water

Adaptation	Plants Exhibiting Adaptation
Ribbonlike submerged leaves	Eel-grass, Wild celery
Threadlike submerged leaves	Widgeon-grass, Horned pondweed, Sago pondweed
Flat linear submerged leaves	Ribbonleaf pondweed, Flat-stem pondweed (*Potamogeton zosteriformis*), Robbins' pondweed (*P. robbinsii*), Baby pondweed (*P. pusillus*), Awl-leaf arrowhead
Highly dissected submerged leaves	Marsh mermaid-weed, Water chestnut, Water-milfoils, Fanwort, White water crowfoot (*Ranunculus trichophyllus*)
Floating leaves or plants	See Table 4.2
Chloroplasts concentrated in the epidermis	Coontail, Water-milfoils, Pondweeds
Aerenchyma	All aquatic plants

Sources: Sculthorpe 1967; Tiner 1999, 2009; Cronk and Fennessy 2001.

Figure 4.4. Morphological adaptations for growing in wetlands: (a) adventitious roots of common reed (*Phragmites australis*); (b) adventitious roots ("prop roots") of red mangrove (*Rhizophora mangle*); (c) shallow roots of swamp black gum (*Nyssa biflora*) and American holly (*Ilex opaca*); (d) buttressed and fluted trunk of bald cypress (*Taxodium distichum*) with pneumatophores ("knees"); (e) hypertrophied lenticels on aerial roots (pneumatophores) of black mangrove (*Avicennia germinans*); (f) hypertrophied lenticels on shallow roots of American holly; and (g) hypertrophied lenticels on buttressed trunk of water tupelo just above waterline (*Nyssa aquatica*). (e: Anna Armitage)

and 1999 for further discussion of these adaptations).

Effect of Prolonged Soil Saturation

While the mechanical action of tidal currents may be a factor in controlling plant establishment, the major impacts of the tides on plants (besides salinity) are caused by inundation and prolonged soil saturation (waterlogging). The effects of flooding and soil saturation on plant growth are well known (Kozlowski 1984; Pezeshki 2001). The frequency and duration of flooding and saturation during the growing season have the greatest impact.

Since tidal wetlands experience alternating periods of flooding and exposure, tide stage has an obvious effect on soil saturation. During spring tides when a Rhode Island high marsh was flooded every day, the black grass zone never drained and was saturated for 5 to 10 days (Bertness et al. 1992a). While the black grass zone was saturated at low tide, the high-tide bush (*Iva frutescens*) zone and the creekbanks drained readily to 4 inches (20 cm) below the surface. Saturation of the root zone of high-tide bush (2–4 inches [5–10 cm] below the surface) for more than a day appears to mark its lower limit. On days when the high marsh was not flooded during daily high tides, the water table at slack tides was below 8 inches (20 cm) from the surface across the entire high marsh.

Prolonged waterlogging creates anaerobic or oxygen-deficient (anoxic) conditions that greatly affect plant growth and survival. When soils are flooded, oxygen diffusion is reduced 10,000 times (Gambrell and Patrick 1978). When soils are saturated for extended periods, oxygen is depleted from the soil by microbes. Oxygen depletion causes reduction or the mobilization of chemical elements that may be toxic to most plants. Facultative and anaerobic microbes obtain energy (oxygen or electrons) by converting the most readily available compounds from oxidized forms to reduced forms in a specific order related to the ease at which these compounds accept electrons: nitrate to ammonium, manganic form to manganous (manganese), ferric to ferrous (iron), sulfate to sulfide, and finally carbon dioxide to methane. This process of anaerobic respiration produces what are called "reduced soils" containing reduced materials (e.g., hydrogen sulfide) that are readily taken up by plants and which may be toxic (Salinas et al. 1986). Since tidal wetlands have fluctuating water levels, they likely experience alternating periods of aerobic (oxic) and anaerobic (anoxic) conditions near the soil surface. The thickness of the aerobic and anaerobic zones undoubtedly varies with tides, seasons, groundwater influences, and other factors. In an Arctic salt marsh, the aerobic zone varied in thickness across the marsh ranging from 1.2 to 12.0 inches (3–30 cm) with the anaerobic zone typically within 4 inches (10 cm) of the surface (Funk et al. 2004).

Since seawater contains much sulfate, anaerobic conditions in salt marshes, the presence of sulfur-reducing bacteria, and an abundance of organic matter promote reduction of sulfate to hydrogen sulfide producing the characteristic "rotten egg" odor of their soils. Hydrogen sulfide is extremely toxic to higher plants. Sulfides reduce root uptake of nitrogen—negatively affecting plant growth—and wetland plants have exhibited differences in their tolerance (Havill et al. 1985; Koch et al. 1990). Salt marsh plants have developed adaptations to cope with hydrogen sulfide and other harmful by-products of anaerobiosis. Common glasswort (*Salicornia maritima*) appears to be unaffected by high sulfide concentrations present in salt marshes and colonizes places that more susceptible plants cannot (Ingold and Havill 1984). High concentrations of sulfides can be lethal for salt marsh plants; dieback of smooth cordgrass has been attributed to sulfide toxicity (Goodman and Williams 1961).

Productivity of smooth cordgrass in Louisiana marshes is mainly controlled by the accumulation of free sulfide, and oxidation of the soil directly influences this and microbiological processes (DeLaune et al. 1983a). The tall form of this species occurs on streambanks where flushing and the availability of sediment to precipitate free sulfides favors high productivity, while the short form grows in the marsh interior where opposite conditions prevail. Stronger anaerobic conditions (low soil redox potential) in the marsh interior are a major determinant of nitrogen uptake, which also affects plant growth and cordgrass productivity (Howes et al. 1986). Despite their hydrophytic nature, both smooth cordgrass and salt hay grass need periods of drainage (aerated soils) during the growing season as they are adversely affected by anaerobic soil conditions (Pezeshki and DeLaune 1996). Smooth cordgrass is more tolerant of prolonged waterlogging than is salt hay grass and therefore can survive under wetter conditions in salt marshes. This may explain why smooth cordgrass forms nearly monospecific stands in depressions just behind the natural creek levees, whereas salt hay grass dominates areas of relatively dry turf. Recent research has shown that root aeration condition near creeks is optimal and may make it possible for species less tolerant of anaerobic stress to become established, or for tolerant species to achieve higher productivity (Li et al. 2005). This helps explain the fact that creekbanks are better aerated habitats than the marsh interior and promote greater productivity (Schelske and Odum 1961; Howes et al. 1981, 1986).

In hypersaline tidal wetlands, soil saturation plays an important role in both germination and death of marsh annuals. For example, in southern California, soil saturation for two weeks or more is needed for germination by Parish's glasswort (*Arthrocnemum subterminale,* formerly *Salicornia subterminalis*), plus an additional two to three weeks of moist soil conditions thereafter to ensure seedling survival through the harsher spring and summer seasons (Kuhn and Zedler 1997). More than eight weeks of soil saturation, however, has adverse impacts on plant growth and survival.

PLANT ADAPTATIONS TO ANAEROBIC CONDITIONS

Since oxygen is critical for respiration in all plants, how do plants growing in anaerobic soils obtain it? Many plants develop shallow root systems to minimize the effects of oxygen depletion. The soil surface is in contact with air, so oxygen is available close to the surface. Roughly 96 percent of the root biomass of salt grass (*Distichlis spicata*) was observed within 4 inches (10 cm) of the soil surface (Seliskar 1983), while 65 percent and 44 percent of smooth cordgrass underground biomass was detected within 6 inches (15 cm) of the surface in high marsh and low marsh, respectively (Gallagher 1974). The differences in the latter species may be related to better soil aeration in the low marsh. Most salt marsh plants simply avoid marsh areas with prolonged anaerobic (anoxic) conditions by colonizing the marsh-upland border where drier conditions prevail. The best-adapted species have developed morphological properties or physiological mechanisms (e.g., ethanol diffusion, malate accumulation, and anaerobic metabolism in roots) for life in anaerobic substrates (Kozlowski 1984).

Many plants have specialized gas-filled tissues (aerenchyma) that facilitate air movement (oxygen diffusion) from leaves to roots (Figure 4.5). Aerenchyma tissue forms by the breakdown of cell walls ("lysigeny") or the separation of cells ("schizogeny") in response to anaerobic conditions induced by prolonged flooding and/or soil waterlogging. Wetland plants have up to 60 percent of their tissue by volume as aerenchyma, while upland plant tissues contain only 2 to 7 percent pore space (Armstrong 1982). Smooth cordgrass, salt grass, and black needlerush (*Juncus roemerianus*) have extensive aerenchyma (Anderson 1974). In greenhouse experiments, flooded plants of smooth cordgrass had more aerenchyma than did plants grown under drained conditions, and the amount of aerenchyma in roots increased with distance from the root tip: no aerenchyma at the tip to nearly 30 percent at 4.7 inches (12 cm) from the tip (Maricle and Lee 2002). Individuals of perennial glasswort or pickleweed (*Salicornia depressa,* formerly *S. virginica*) growing in wetter soils had more aerenchyma tissue than did other individuals growing under drier conditions in the same marsh (Seliskar 1985). The presence of iron oxides around its roots provides ample evidence of its ability to aerate the soil (Mahall and Park 1976). In experimental studies, salt hay grass appeared to develop the maximum amount of aerenchyma tissue within a month of flooding (Burdick and Mendelssohn 1990). The value of aerenchyma tissue is its ability to transport oxygen to plant roots, and it may reduce metabolic requirements and transport other gases or waste products (Teal and Kanwisher 1966; Morris and Dacey 1984; Maricle and Lee 2002).

Leakage of oxygen from roots creates an oxidized zone (oxidized rhizospheres) that may be 1⁄10 to 1⁄8 inch (2–3 mm) wide (Little 2000), covering the root with an orangish coating of iron precipitates (iron oxides). Rhizosphere oxidation and the ability of smooth cordgrass to take up dissolved sulfide and oxidize it enzymatically within its roots help counteract the toxic effect of hydrogen sulfide and successfully colonize intertidal soils (Carlson and Forrest 1982; Teal 1986). The tall form of this species with its increased leaf area allows for more oxygen diffusion to the roots and the formation of prominent oxidized rhizospheres, while the root surface of the short form contains less iron precipitate. More oxygen in the root zone allows the roots of the tall form to respire mostly aerobically and to exclude toxins, whereas high rates of alcoholic fermentation in the short form indicate mostly anaerobic respiration in these waterlogged soils (Mendelssohn and Postek 1982). The success of smooth cordgrass in the low marsh may also depend on the benefits derived from collective substrate oxidation by groups of individual plants (Bertness 1991b).

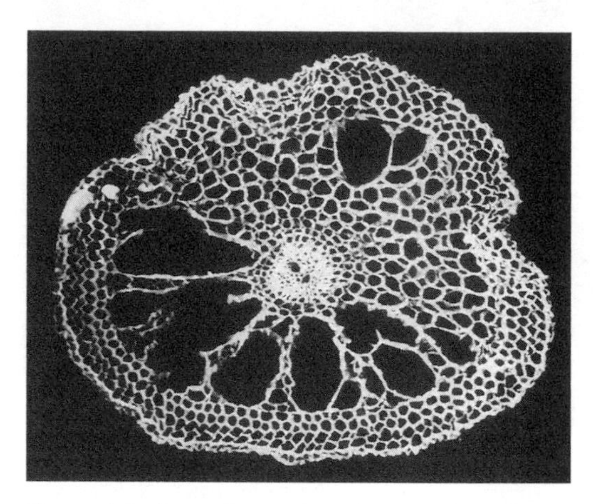

Figure 4.5. Aerenchyma tissue in smooth cordgrass (*Spartina alterniflora*). (Irv Mendelssohn, Louisiana State University)

Salt hay grass has also been shown to transport oxygen to its roots, and new roots develop significant aerenchyma within 25 days (Burdick and Mendelssohn 1987; Burdick 1989; DeLaune et al. 1990b). This ability appears to be limited in that this species does not colonize open bare ground in the low marsh adjacent to existing salt hay grass monocultures (Bertness 1991b). Cabbage palms (*Sabal palmetto*) may survive frequent tidal inundation because they oxidize their rhizospheres through well-developed root aerenchyma (Williams et al. 1999a). Mangroves (*Avicennia, Rhizophora*) may have aerial extensions of the roots (pneumatophores or prop roots) that contain aerenchyma tissue and possess enlarged pores (hypertrophied lenticels) that facilitate gas exchange (see Figure 4.4).

Burrowing animals aid plants by modifying soil-oxygen levels. For example, fiddler crab burrows provide pathways for oxygen to enter the soil thereby creating suboxic conditions as deep as 20 inches (50 cm) (Hoffman et al. 1984; Koretsky et al. 2005). In Georgia salt marshes, an estimated 20 percent of smooth cordgrass production has been attributed to the presence of fiddler crab burrows (Montague 1982). The relationship between smooth cordgrass and fiddler crabs has been described as one of facultative mutualism (Bertness 1985). The crabs can occupy soft substrates because cordgrass roots bind the soil and facilitate burrowing, while the burrows improve soil drainage, increase soil oxygen content, reduce belowground organic debris, and increase nutrient availability, resulting in higher productivity for the plants. Crabs benefit from the presence of the plants, which provide a source of food and protection from predators.

Other Tidal Effects on Plants

Besides periodic submergence and anaerobiosis, tides affect plants in other ways: 1) productivity; 2) desiccation effects when exposed at low tide; 3) changes in growth form; 4) sediment deposition; and 5) salt stress (discussed in next section). Photosynthesis of smooth cordgrass was reduced by 66 percent when inundated (Kathilankal et al. 2008). Tidal action may, however, stimulate increased plant productivity in regions of similar climate. For example, in Long Island Sound, salt marshes in the western end where tidal ranges are greater were more productive than were marshes on the eastern end of the sound (Steever et al. 1976). This effect is called "tidal subsidy" (Steever et al. 1976; Odum 1980).

Since tidal marsh plants live in an environment of alternating exposure and inundation, they must deal with desiccation (drying) as well as inundation, especially those species that are mostly submerged and exposed to air for only relatively short periods. Two growth forms common to intertidal flats that may reduce the effects of drying are low rosettes with short narrow leaves and low dense clumps of opposite-leaved plants rooting at the nodes (Ferren and Schuyler 1980). Macroalgae tolerance to drying plays a significant role in determining their position along rocky intertidal shores (Little and Kitching 1996). Significant changes in temperature may result from water level changes at a site—from exposed areas (warmer temperature in summer, but colder in winter) being flooded (cooler temperature in summer or warmer in winter) and from flooded areas draining and drying upon exposure to air (Adam 1993). The same species may exhibit different growth forms on mostly submerged sites versus mostly exposed sites. For example, grass-leaved arrowhead (*Sagittaria graminea*) develops sterile rosettes with linear to linear lance-shaped leaflike stems (phyllodia) on tidal flats, but in tidal fresh marshes, it grows as an erect emergent plant with broad leaves and bears flowers (Caldwell and Crow 1992).

Tides also contribute to various disturbances that affect vegetation. Strong tidal currents and wave action along shorelines may promote bank erosion. Water-carried soil material may temporarily coat leaf surfaces with a short-term effect on photosynthesis. Sands carried by strong wind or deposited by overwash processes may bury vegetation, forcing plants to either adapt to burial by sending up new shoots or perish. Tides may also bring water-carried debris (cans, wood, derelict boats) and dead vegetation (tidal wrack) onto the marshes. These and other disturbances are further discussed later in this chapter.

Salt Stress

Once a plant is adapted for life in wetlands (e.g., prolonged saturation and anaerobic conditions), salinity may be the most important natural factor affecting its growth in estuarine environments. Many studies have demonstrated the effect of salt stress on marsh plants (e.g., Good 1965; Nestler 1977; Parrondo et al. 1978; Smart and Barko 1978, 1980; Woodwell 1985; Portnoy and Valiela 1997; Mendelssohn and Morris 2000; Merino et al. 2010), while others have studied the effect on woody plants (e.g., Wainwright 1980; Pezeshki and Chambers 1986; Pezeshki et al. 1990; Allen et al.

1996; Perry and Williams 1996; Kozlowski 1997). Trees in temperate regions are sensitive to salinities of 2 ppt or more, and that is why tidal saline soils there are dominated by herbaceous vegetation and not by forests (Pezeshki and Chambers 1986; Pezeshki et al. 1989; Conner et al. 1997; Kozlowski 1997). In the tropics, of course, trees including mangroves have adapted to these saline conditions and dominate the intertidal zone that herbaceous species occupy in temperate and Arctic regions.

Most plants cannot tolerate prolonged exposure to seawater (32–37 ppt). Specially adapted plants called "halophytes" can tolerate high levels of salt and many such plants grow in soils where salinities are as high as seawater. Yet most halophytes cannot withstand salinities much higher than that. Salinities above 117 ppt were reported to create areas devoid of vegetation in southern California wetlands (Purer 1942). The two main negative effects of saltwater on plants are attributed to the toxicity of some of the ions and to osmotic impacts. While chlorine is an essential micronutrient for plants, and sodium may be essential to some plants, the problem with saltwater is too much salt. Sodium and chloride are toxic to many plants (e.g., destroying cell membranes). Salts may interfere with nutrient uptake and adversely affect water uptake. Non-specialized plants in saltwater lose freshwater due to osmosis—movement of water through a selectively permeable membrane from an area of lower solute (salt) concentration to one of higher solute concentration. This condition is called "physiological drought" or "salt-induced drought" and plants experiencing it lose water, their leaves wilt, and they eventually succumb if conditions persist. To prevent this from happening, plants must have adaptations that maintain a higher internal solute concentration by accumulating salt or organic solutes or otherwise preventing the outflow of water (Adam 1993). Maintaining salt and water balance is critical to plant survival in saline environments. Halophytes have developed mechanisms to prevent physiological drought from happening.

While the influence of salt stress on tidal plants is well established, other factors also operate simultaneously to influence plant composition. For example, in southern California marshes where the Mediterranean climate creates wet mild winters and dry hot summers, changes in both soil salinity and soil moisture largely affect germination of salt marsh annuals (Noe and Zedler 2000). In some species, salinity impacts are more pronounced when soil moisture is low, whereas the influence of soil moisture is greatest at high salinities. The combination of soil salinity and duration of soil saturation has been shown to affect germination of wetland plants (Baldwin et al. 1996; Kuhn and Zedler 1997). Experiments suggest that controlling water tables and salinity may be an effective means of combating the spread of invasives such as common reed (Hellings and Gallagher 1992).

Classification of Plants by Salt Tolerance

Plants may be classified into three groups based on their salt tolerance: 1) halophytes (most tolerant); 2) mesophytes (moderately tolerant); and 3) glycophytes (salt-sensitive). Halophytes will typically show increased growth, especially root growth, when subjected to salty conditions, while the glycophytes will usually cease growth (Loch et al. 2003). Mesophytes will continue to grow but will later decrease growth due to salt stress. The expansion of roots in halophytes may be an attempt to compensate for limitations on water uptake (low external water potential) due to soil-water salinity. The amount of salt in the soil or aquatic environment is important as non-halophytes do not appear to be affected by salinities around 3 ppt (Ranwell 1972). Perhaps that is why many freshwater species can be found in slightly brackish waters.

Because coastal wetlands exist between the open ocean and nontidal freshwater, they largely occur in a zone of transition or flux where seawater intermixes with freshwater—the estuary. Halophytes are adapted to these conditions, whereas strictly freshwater species (obligate glycophytes) are not. The term "halophyte" meaning "salt-loving" originated in the 1800s, and by the early 1900s it was in frequent use (Shaler 1886; Ganong 1903; Warming 1909).

Nearly all halophytic plants are more salt-tolerant than salt-loving and may be called "facultative halophytes." While the seeds of halophytes germinate in saltwater, in most species they are dormant at high salinities and wait until salinities are reduced to germinate (Ungar 1978). This is an evolutionary trait that allows these plants to colonize and survive in salt marshes where variable soil salinities are the norm. Halophytes thrive in salt marshes due to lack of competition from other wetland plants (Taylor 1939; Barbour 1978). Many grow better in freshwater but are not good competitors there and therefore are relegated to saline environments. For example, a transplant experiment relocating species within three types of estuarine marshes (salt, brackish, and oligohaline) found that the low-salinity species could not survive brackish or saline conditions, while the salt marsh species survived in all three types of marshes (Crain et al. 2004). The latter grew more luxuriantly in low-salinity wetlands when neighboring plants were removed, but were strongly suppressed when neighbors were present. Physiological stress (i.e., salt) appears to restrict the seaward distribution of the low-salinity species, whereas competition limited upstream distribution of salt marsh species. Experiments with Pacific cordgrass (*Spartina foliosa*) found that it grew better and had higher survival rates when grown in freshwater than in saltwater (Phleger 1971). In nature, it has found a niche containing little or no competition—the low intertidal zone in California salt marshes. Salinity affects plant distribution limiting most species to the upper edge of the salt marsh where freshwater runoff and infrequent flooding provide favorable conditions for growth and reproduction. These same species expand their range in brackish marshes since salt stress has been reduced by river and stream discharge.

Some species that dominate upland habitats have evolved salt-tolerant ecotypes that grow along the upper edges of salt marshes, with creeping bent grass (*Agrostis stolonifera* var. *palustris*) and red fescue (*Festuca rubra*) being common examples for the northeastern United States (Wainwright 1984). Intraspecific variation in salt tolerance has been observed in some species including smooth cordgrass, maidencane (*Panicum hemitomon*), and bald cypress (*Taxodium distichum*) (Hester et al. 1998; Conner and Inabinette 2005). Recognition of ecotypes and genotypes is important for marsh creation and restoration project planning. Note that some salt marsh halophytes also grow in inland alkaline or calcareous wetlands including arrow-grasses (*Triglochin* spp.), marsh orach (*Atriplex patula*), baltic rush (*Juncus arcticus*), seaside crowfoot (*Ranunculus cymbalaria*), sea milkwort (*Glaux maritima*), and salt marsh sand spurrey (*Spergularia salina*). Even within the saline environment, individual species have genetic differences that yield intraspecific variation in properties that may influence the plant's ability to deal with salt stress (e.g., Antlfinger 1981; Hester et al. 1996, 1998; Howard and Rafferty 2006; Howard 2010). That genetic diversity helps plants adapt to a wide range of environmental variables is widely recognized (e.g., Silander 1979; Linhart and Grant 1996).

Given the preceding discussion, one might ask: Are any plants obligate halophytes? Relatively few plants may actually require saltwater for growth and this has been a subject for debate. Laboratory experiments in the early 1900s demonstrated that some salt marsh species actually grew best in sand saturated with water at one-third sea strength: sea milkwort, common glasswort, and sea blite (*Suaeda maritima*), while common glasswort grew poorly in freshwater conditions (Barbour 1970). Other species showing optimal growth in saline soils included smooth cordgrass and groundsel-bush (*Baccharis halimifolia*). A classic study of zonation in North Carolina salt marshes concluded that smooth cordgrass and salt grass were obligate halophytes, while salt hay grass was listed as a facultative halophyte (Adams 1963). A test of the salt tolerance of 29 New Zealand salt marsh plants identified 17 species with a more than nominal salt requirement for maximum growth (Partridge and Wilson 1987). Only one species (*Suaeda novae-zelandiae*) grew best at salinity above 10 ppt (at 1.5% NaCl), while maximum growth for most species was achieved when salinity was less than 10 ppt; eight of the salt marsh species grew best in freshwater. The high number of halophytic species in the goosefoot family (Chenopodiaceae) and some other families suggest their lineage to ancient halophytic vascular plants (Wainwright 1984). Germination experiments have shown that most of the coastal plants examined could not germinate under full sea-strength salinities, but some salt marsh species did (Woodwell 1985). Those that did not germinate in seawater were stimulated by exposure to salinity and when placed in freshwater, they showed enhanced germination. Interestingly, many halophytes actually germinate best under freshwater or low salinity conditions (Ungar 1978). Rather than focusing on experimental studies of salt tolerance, from a practical standpoint plants growing exclusively in salt or strongly brackish marshes or on salt-sprayed shores could be considered obligate halophytes (Irwin Ungar, pers. comm. 2004). After all, these are the only places where they grow naturally.

Adaptations to Salt Stress

Plants must absorb water from the soil to grow. When placed in saltwater, glycophytes (non-halophytes) lose internal water as water moves from a region of higher water potential inside the plant to the area of lower water potential outside the plant—the saline environment, or from inside the plant where concentrations of solutes (salt) are lower to the saltwater (Cronk and Fennessy 2001). If a plant is unable to make adjustments, it dies. Since halophytes grow in saline soils, they have developed adaptations to effectively deal with salt stress (i.e., ion toxicity and osmotic effects of physiological drought). The first challenge is to be able to take up water in a saline environment. For this, these plants utilize a process called "osmotic adjustment" or "osmo-regulation." This process involves increasing internal solute concentrations with sodium chloride or other compounds referred to as "compatible solutes," such as glycinebetaine, praline, mannitol, and dimethylsulphonioproprionate (Cavalieri 1983; Cronk and Fennessy 2001). Once higher internal solute concentrations are attained, the roots can take up water from saline soils. Interestingly, all plants have the basic characteristics necessary for salt tolerance (e.g., acquire ions and compartmentalize them in vacuoles) but only about 0.25 percent of known flowering plants are halophytes (Flowers et al. 2010).

The next challenge is to exclude salt from the water being taken up. Plants utilize at least four approaches to this problem: 1) ion selection; 2) ion extrusion; 3) ion accumulation; and 4) ion dilution (Ranwell 1972). Most halophytes exclude salt uptake by their roots, yet this process is not perfect and some salt enters the plants (Little 2000). Smooth cordgrass excludes 91 to 97 percent of the salt from seawater (Bradley and Morris 1991), while some mangroves, including red mangrove (*Rhizophora mangle*), exclude 99 percent (Tomlinson 1986). Salt may also enter the plant through leaves and stems. To address this issue, some plants have developed adaptations to reduce their exposure to saltwater, for example, leaf surfaces covered with waxy flakes to make them saltwater-proof, and reduced leaf surfaces to minimize exposure to salt and evapotranspiration (including the ability to roll up leaves to reduce transpiration). Perhaps the final challenge for halophytes from the salt-stress standpoint is how to deal with any salt that enters the plant. Several mechanisms have evolved to handle this salt: 1) salt-secreting glands to excrete excess salt; 2) salt-concentrating organs (e.g., fleshy leaves and salt hairs) that are periodically shed; 3) succulence (fleshiness) to keep internal salt concentrations at acceptable levels; 4) internal organs (vacuoles) to isolate salt; and 5) cellular mechanisms to reduce internal salt concentrations (Table 4.4; Figure 4.6).

Plants may follow more than one of the above as well as other approaches (e.g., reduced transpiration) to minimize salt stress (for reviews of physiological mechanisms see Waisel 1972; Poljakoff-Mayber and Gale 1975; Cronk and Fennessy 2001; Flowers et al. 2010). Evidence of salt glands may be witnessed indirectly as salt crystals on the leaves of cordgrasses (Figure 4.7) or directly as a pair of raised organs on the leaf stalk of white mangrove (*Laguncularia racemosa;* Figure 4.5d). Some of the most salt-tolerant species concentrate more sodium (Na^+) in their leaves than chloride (Cl^-), including seaside plantain (*Plantago maritima*), sea blite (*Suaeda linearis*), seashore alkali grass (*Puccinellia americana,* formerly *P. maritima*), lamb's-quarters (*Chenopodium album*), and northeastern saltbush (*Atriplex glabriuscula*), whereas many salt-susceptible plants do the opposite. This may relate to ion toxicity tolerances (Wainwright 1984). Succulence is a water conservation mechanism and for plants growing in saline soils it reduces salt stress by dilution. The most succulent marsh plants such as perennial glasswort and saltwort (*Batis maritima*) have relative water contents of 93 to 98 percent, while plants with succulent leaves have lower values, e.g., 73 percent for sea ox-eye (*Borrichia frutescens*) (Antlfinger 1976). The leaves of the salt-tolerant ecotype of creeping bent grass have a waxy coating that prevents

Table 4.4. Examples of tidal wetland plants with morphological adaptations to salt stress

Adaptations	Species
Salt glands	Smooth cordgrass (*Spartina alterniflora*), Pacific cordgrass (*S. foliosa*), Salt hay grass (*Spartina patens*), Salt grass (*Distichlis spicata*), Sea milkwort (*Glaux maritima*), Sea lavenders (*Limonium* spp.), Creeping alkali grass (*Puccinellia phryganodes*), Key grass (*Monanthochloe littoralis*), Alkali aheath (*Frankenia salina*), Marsh jaumea (*Jaumea carnosa*), White mangrove (*Laguncularia racemosa*), Black mangrove (*Avicennia germinans*)
Salt hairs	*Chenopodiaceae* (goosefoot family including Coastal goosefoot, *Chenopodium rubrum* and Marsh orach [*Atriplex patula*])
Succulent stems	Glassworts (*Salicornia* spp.), Saltwort (*Batis maritima*), Slender sea purslane (*Sesuvium maritimum*), Sea milkwort (*Glaux maritima*), Slenderleaf iceplant (*Mesembryanthemum nodiflorum*), Sea fig (*Carpobrotus chilensis*)
Succulent leaves	Seaside plantain (*Plantago maritima*), Arrow-grasses (*Triglochin* spp.), Saltwort (*Batis maritima*), Sea blites (*Suaeda* spp.), Salt marsh sand spurrey (*Spergularia salina*), High-tide bush (*Iva frutescens*), Salt marsh asters (*Symphyotrichum tenuifolium, S. subulatum*), Seaside goldenrod (*Solidago sempervirens*), Sea rocket (*Cakile edentula*), Marsh orach, Sea ox-eye (*Borrichia frutescens*), Christmas-berry (*Lycium carolinianum*), Slender sea purslane, Sea milkwort, Seaside gerardia (*Agalinis maritima*), Slenderleaf iceplant, Sea fig
Waxy coat on leaves	Creeping bent grass (*Agrostis stolonifera* var. *palustris*)
Leaf roll up	Salt hay grass, Salt grass, Key grass, Seashore alkali grass (*Puccinellia americana*), Red fescue (*Festuca rubra*)

Sources: Anderson 1974; Hansen et al. 1976; Wainwright 1984; Naidoo et al. 1992; Tiner 1993a; Buck 2001.
Note: In-rolled leaves help reduce water loss through transpiration.

salt absorption through the leaf surfaces. Smooth cordgrass and other halophytes have cellular mechanisms such as the production of compatible solutes (e.g., proline and glycinebetaine) and selective uptake of potassium to exclude sodium (Cavalieri 1983; Wu et al. 1998). Retention of leaves by seedlings of woody plants may aid their colonization of environments exposed to frequent saltwater flooding (Williams et al. 1998). Cabbage palm and southern red cedar (*Juniperus virginiana* var. *silicicola*), whose seedlings retain their leaves when inundated with saltwater, grow in frequently flooded areas along the salt marsh fringe in northern Florida. Farther inland or on higher ground, live oak (*Quercus virginiana*) and sugarberry (*Celtis laevigata*) can be found; their seedlings lose their leaves when subjected to infrequent tidal flooding and resprout them later when favorable conditions resume. The retention of leaves by

Figure 4.6. Examples of morphological adaptations to salt stress: (a) succulent stems of perennial glasswort or pickleweed (*Salicornia depressa*); (b) succulent leaves of sea ox-eye (*Borrichia frutescens*); (c) succulent leaves in saltwort (*Batis maritima*); and (d) raised salt glands on leaf stalk of white mangrove (*Laguncularia racemosa*).

Figure 4.7. Salt crystals on leaf of Pacific cordgrass (*Spartina foliosa*), which are evidence of salt glands.

seedlings and resprouting of lost leaves may be important mechanisms influencing zonation in coastal forests (Williams et al. 1998).

Influence of Salinity on Plant Community Structure

The most obvious sign of salt stress occurs under extreme hyperhaline conditions, which prevent vascular plant establishment and create salt barrens or "salinas" in irregularly flooded marshes. These conditions can be found along the Atlantic and Gulf Coasts mainly from South Carolina to Texas, where the largest ones are found, and on the Pacific Coast in California from San Francisco south. In these regions, the interior portions of salt marshes often contain barren zones (salt barrens or salinas) varying in size with fluctuating environmental conditions. Interannual variations in precipitation have a dramatic effect on vegetation patterns, especially along the Texas and southern California coasts. Periods of high rainfall or freshwater inflows will lower soil salinities and moisten substrates sufficiently to allow temporary colonization of bare spots via seed germination by annuals such as glassworts and by clonal expansion from existing colonies of saltwort, sea ox-eye, and key grass (*Monanthochloe littoralis*) (Dunton et al. 2001; Alexander and Dunton 2002). Plant productivity in these marshes also increases during these times, and less salt-tolerant species like sea ox-eye may replace hyperhaline species like saltwort or may invade bare spots through expansion from neighboring areas. Marsh elevation appears to play a key role in the formation of these bare spots along the Texas coast where they occupy less than 5 percent of the marsh at elevations 7.1 inches (18 cm) above mean sea level (MSL) in contrast to approaching 100 percent of the marsh at elevations less than 3.9 inches (10 cm) above MSL (Dunton et al. 2001). In southern California, Parish's glasswort dominated the higher more saline parts of the marsh, whereas pickleweed characterized the low marsh where salinities were lower but flooding and waterlogging were more frequent (Pennings and Callaway 1992).

Since most halophytes avoid sites with the highest salinities (i.e., those above sea strength), there is a noticeable increase in plant diversity with decreasing salinity in coastal wetlands. This is evident as one moves from the low salt marsh to the upper marsh border (excluding any salinas, of course; Table 4.5) and as one travels upstream from the mouth of a tidal river into tidal freshwaters (Chapter 5). Within the salt marsh, soil salinity approaches sea strength at the seaward edge (around 30 ppt) and decreases to 10 ppt or less near the upland border (Wells and Hirvonen 1988). The highest salinities form in shallow depressions in the high marsh ("salt pannes") where salinities over 100 ppt can be attained. In one of the earliest studies of salt marshes, changes in soil salinity from the low marsh to the upper high marsh in a Long Island, New York, salt marsh were determined: 22 ppt at smooth cordgrass–salt hay grass marsh interface, 19 ppt in middle of salt hay grass marsh, 14 ppt in black grass zone, and 2.5 ppt where salt marsh bulrush (*Schoenoplectus robustus*) occurred (Transeau 1913). (Note that these salinity estimates were recalculated using a salinity of 25 ppt for the Cold Spring Harbor since percentages were given and harbor salinity value were not given in the original publication; the current salinity value for the harbor comes from Friends of the Bay [2006].) The upper salt marsh contains the most diverse assemblage of species because salt stress is lowest here and more plants can cope with this reduced level of salt (stress avoidance). Experimental evidence suggests that high marsh dominants—black grass and high-tide bush—are relatively intolerant of high salinities as their growth and photosynthetic rates were substantially reduced under such conditions (e.g., experienced significant mortality at moderate salinities), while salt panne species were not as affected (Bertness et al. 1992b). Freshwater plants such as marsh fern (*Thelypteris palustris*), sweet gale (*Myrica gale*), prairie cordgrass (*Spartina pectinata*), and poison ivy (*Toxicodendron radicans*) may be found along the upper edges of salt marshes where freshwater inflow is substantial (e.g., seepage areas). Mosses found in the upper high marsh zone with rushes (*Juncus arcticus, J. gerardii*) and cordgrasses (*Spartina patens* and *S. pectinata*) demonstrated considerable

Table 4.5. Species composition in zones of a northeastern salt marsh

Vegetative Zone	Plant Community	
	Dominant Plants	Common Associates
Low marsh	Smooth cordgrass (tall form) (*Spartina alterniflora*)	Smooth cordgrass (intermediate form), Rockweeds (locally), other algae
High marsh		
Lower high marsh	Smooth cordgrass (short form)	Glassworts, Salt hay grass (*Spartina patens*), Sea lavender (*Limonium carolinianum*), Filamentous green algae
Middle high marsh	Salt hay grass	Salt grass (*Distichlis spicata*; often codominant), Sea lavender, Black grass (*Juncus gerardii*), Marsh orach (*Atriplex patula*), Sea blites (*Suaeda* spp.)
Panne	Glassworts (*Salicornia* spp.), Smooth cordgrass (short form), Seaside plantain (*Plantago maritima*), Cyanobacteria (blue-green algae)	Seaside gerardia (*Agalinis maritima*), Sea blites, Seaside arrow-grass (*Triglochin maritimum*), Sea lavender, Salt grass, Salt hay grass, Sea milkwort (*Glaux maritima*; New England north)
Pool	Widgeon-grass (*Ruppia maritima*)	
Upper high marsh	Black grass	Salt grass, Salt hay grass, Perennial saltmarsh aster (*Symphyotrichum tenuifolium*), Sea lavender, High-tide bush (*Iva frutescens*), Salt marsh bulrush (*Schoenoplectus robustus, S. maritimus*), Seaside arrow-grass, Seaside goldenrod (*Solidago sempervirens*), Seashore alkali grass (*Puccinellia americana;* New England)
Marsh border	Switchgrass (*Panicum virgatum*), Prairie cordgrass (*Spartina pectinata*), Common reed (*Phragmites australis*), High-tide bush, Groundsel-bush (*Baccharis halimifolia*)	Seaside goldenrod, Grass-leaved goldenrod (*Euthamia graminifolia*), Salt hay grass, Annual marsh pink (*Sabatia stellaris;* New Jersey south), Creeping bent grass (*Agrostis stolonifera*), American germander (*Teucrium canadense*), Hedge bindweed (*Calystegia sepium*), Poison ivy (*Toxicodendron radicans*), Marsh fern (*Thelypteris palustris*), Baltic rush (*Juncus arcticus;* New England north), Sweet gale (*Myrica gale;* New England north), Northern bayberry (*Morella pensylvanica*), Salt marsh fimbristylis (*Fimbristylis castanea;* Long Island south), Wax myrtle (*Morella cerifera;* Delaware south)

salt tolerance in laboratory experiments (Garbary et al. 2008). In an Arctic salt marsh, only one vascular plant, creeping alkali grass (*Puccinellia phryganodes*), was found in the more saline portions of the marsh while 16 species were found in the uppermost marsh that was only rarely and infrequently flooded by saltwater (Funk et al. 2004).

Upstream in estuaries, freshwater plants such as spatterdock (*Nuphar lutea*), sweet flag (*Acorus calamus*), river bulrush (*Schoenoplectus fluviatilis*), wool grass (*Scirpus cyperinus*), pickerelweed (*Pontederia cordata*), and arrow arum (*Peltandra virginica*) may colonize slightly brackish (oligohaline) marshes, but they become dominant species in tidal fresh wetlands. They tolerate low salinities, particularly in mid- to late summer. Consequently, with decreased salt stress, plant diversity increases along tidal rivers. Since most trees, with the exception of mangrove species, do not tolerate saltwater, they are excluded from salt and brackish environments, but thrive in freshwater wetlands. In temperate wetlands, woody species tend to be found where salinities are less than 2 ppt, although during droughts, tidal freshwater swamps may have soil salinities higher than this (Baldwin 2007). These swamps have been referred to as "salt-pulsed tidal swamps."

Many examples of increasing diversity from the salt marshes to freshwater tidal wetlands can be found. Along the Connecticut River, 17 species were found in salt marshes, 36 in brackish marshes, and more than 150 species in tidal fresh marshes (Barrett 1989). Plant diversity for the Patuxent River in Maryland went from 14 species in a strongly brackish marsh to 56 species in the tidal fresh marshes (Anderson et al. 1968). Along the York and Pamunkey Rivers in Virginia, marsh plant diversity rose from 6 to 11 species in salt marshes to 56 species in tidal fresh marshes, although the oligohaline marsh in between had only 10 species (Perry and Atkinson 1997). Species richness increased over five times from salt to tidal fresh marshes along the Altamaha, Satilla, and Ogeechee Rivers in Georgia (Wieski et al. 2010). Similar increases were reported in Louisiana marshes where the number of species went from 17 in salt marshes to 40 in brackish marshes to 54 in "intermediate" marshes to 93 in tidal fresh marshes (Chabreck 1972). Some Gulf Coast estuaries do not exhibit this great rise in species richness from salt marshes to tidal fresh marshes. Salt marshes in the Rio Grande Delta contained 7 to 26 species, while brackish marshes had 7 to 24 species and tidal fresh marshes 15 to 31 (Judd and Lonard 2004). Plant diversity in Texas salt marshes is quite high compared with that of Atlantic Coast marshes (Judd and Lonard 2002, 2004; Kunza and Pennings 2008).

Salinity also affects the zonal distribution of individual species within estuaries. The ability of common three-square, a facultative halophyte, to tolerate submergence appears to be significantly affected by saltwater. Along the St. Lawrence River, it grows in freshwater areas submerged an average of 75 percent of the time, while in brackish waters (15–20 ppt) it may tolerate only 30 percent submergence (Deschenes and Serodes 1985). It does not grow in saline waters, but is restricted to the upper edges of salt marshes where freshwater influence is significant. Consequently, this plant is dominant in the middle and upper portions of estuaries. Many species restricted to the upper edges of salt marshes, such as salt marsh bulrush, seaside goldenrod, common reed, big cordgrass (*Spartina cynosuroides*), narrow-leaved cattail (*Typha angustifolia*), and black needlerush become dominant species in brackish marshes.

Saltwater intrusion into tidal fresh wetlands is causing changes in plant communities. Tree diversity in tidal swamps decreases with a salinity change from fresh to 1.2 ppt as plants such as water tupelo (*Nyssa aquatica*), sweet gum (*Liquidambar styraciflua*), and green ash (*Fraxinus pennsylvanica*) are eliminated (Krauss et al. 2007). This is a major concern given

Table 4.6. Comparative ratings of swamp tree (seedling) tolerance to flooding and salinity

Plant	Tolerance		
	Flooding	Flooding with 2 ppt Water	Salinity from Storm Surge*
Bald cypress (*Taxodium distichum*)	Most	Tolerant	Moderate
Water gum (*Nyssa aquatic*)	Most	Weak	Moderate
Buttonbush (*Cephalanthus occidentalis*)	Most	Weak	Moderate
Swamp black gum (*Nyssa biflora*)	Most	No	Moderate
Chinese tallow (*Triadica sebifera*)	Tolerant	No	Moderate
Red maple (*Acer rubrum*)	Tolerant	No	No
Red bay (*Persea borbonia*)	Not assessed	No	No
Overcup oak (*Quercus lyrata*)	Tolerant	No	No
Loblolly pine (*Pinus taeda*)	Moderate	Not assessed	No
Green ash (*Fraxinus pennsylvanica*)	Moderate	Weak	Moderate
Nuttall oak (*Quercus texana)*	Moderate	No	No
Water oak (*Quercus nigra*)	Weak-Moderate	No	No
Swamp chestnut oak (*Quercus michauxii*)	Weak	No	No

Source: Conner et al. 2007b based on several studies; this reference provides a good review of this literature.
Note: *Two days of exposure to 32 ppt is typical for hurricanes.

rising sea levels as increased salinity and hydroperiod are converting tidal swamps to marshes and open water along the Louisiana coast (Shaffer et al. 2009). Researchers are investigating salt tolerance in many species and in some cases searching for salt-tolerant ecotypes (e.g., Allen 1994; Allen et al. 1994; Conner et al. 2007b; Krauss et al. 2007). Experiments have demonstrated the salt-sensitivity of several species. Maidencane was found to be the most sensitive of four marsh species examined, followed by bull-tongue (*Sagittaria lancifolia*), then spike-rush (*Eleocharis palustris*), with Olney's three-square (*Schoenoplectus americanus*) being the most salt-tolerant (Howard and Mendelssohn 1999a, b). Seedlings of bald cypress (*Taxodium distichum*) and Chinese tallow (*Triadica sebifera*) tolerate 2 ppt but were killed by 10 ppt after 2 weeks and 6 weeks of flooding, respectively (Conner 1994). Later studies further documented the salt sensitivity of these two species as well as green ash and water tupelo (Conner et al. 1997). Seedlings of all four species survived 12 weeks of freshwater flooding with no ill effects—bald cypress and water tupelo seedlings were the most flood-tolerant. All seedlings except Chinese tallow were killed within 2 weeks of flooding with 10 ppt water; Chinese tallow seedlings lasted just 6 weeks under those conditions. Some wetland trees have been rated for their tolerance to flooding, salinity, and flooding with low-salinity water (Table 4.6).

Bald cypress may be the most tolerant freshwater tree in tidal swamps. Once established, it appears to be able to tolerate salinities of 5 ppt, although growth is substantially reduced (Yanosky et al. 1995; Krauss et al. 2007). It has survived short-term (acute) salinity pulses up to 18.5 ppt (Conner and Inabinette 2005). On the one hand, although drought conditions and corresponding acute salt pulses continue to pose risks, years of above average rainfall and sustained periods of high river discharge may help refresh swamp soils (lower salt concentrations) and promote tree survival even where subjected to chronic low salinities (Doyle et al. 2007). On the other hand, rising sea levels do not bode well for the future of these swamps.

Salinity Effect on Plant Height

Plant growth and reproduction can be stressed by increased soil salinity. Perhaps the most obvious sign of salt stress in plants is reflected in their height (Figure 4.8). Two of the dominant plants of Atlantic Coast salt marshes—smooth cordgrass and black needlerush—exhibit different growth forms in response to varied environmental conditions within the marshes. Taller forms grow in places of less stress, and stunted forms grow in the most stressful locations. Height differences may be related to flooding frequency, substrate, salinity, and nutrient availability. The tall form of smooth cordgrass (3.0–9.8 ft [0.9–3.0 m] tall or more)

a

b

Figure 4.8. Salinity differences in tidal marshes affect plant height: (a) the difference between the tall and short forms of smooth cordgrass (*Spartina alterniflora*) is evident in this Connecticut salt marsh: the tall form along the ditch and the short form on the high marsh (at left); and (b) stunted common reed (*Phragmites australis*) in the upper high marsh of a Delaware salt marsh—in freshwater wetlands it grows to 14 feet (4.3 m) or more.

grows along creekbanks flooded daily by tides, whereas the short form (generally less than 1.5 ft [0.5 m] tall) occupies shallow depressions where salts accumulate (above sea strength) and severe anaerobic conditions persist (Anderson and Treshow 1980). An intermediate form (about 1.5–3.0 ft [0.5–0.9 m] tall) can be found in the middle high marsh intermixed with other high marsh plants, especially salt hay grass. Various studies have offered mixed opinions on the causes of these growth forms. Some researchers found a correlation with physical factors (e.g., Mooring et al. 1971; Shea et al. 1975; Howes et al. 1986; Ornes and Kaplan 1989), while others suspect that genetics play at least some role (e.g., Gallagher et al. 1988; Seliskar et al. 2002). Variations in stem height and density of short-form smooth cordgrass transplants from different geographic locations remained five years after planting in a Delaware created marsh (Seliskar et al. 2002). An earlier experiment transplanting the tall form and the short form to irrigated garden plots showed no change in growth form after nine years in the same environment (Gallagher et al. 1988). Another study examining smooth cordgrass response to salinity found that this species achieved its greatest biomass in freshwater sediments (Smart and Barko 1980). Undoubtedly, both environmental factors and genetics are involved in producing different growth forms of smooth cordgrass and other species (Anderson and Treshow 1980; Proffitt et al. 2003).

Not surprisingly, salinity has similar effects on most species that can tolerate high levels of salt. The growth forms of black needlerush reflect differences in both salinity and soil properties. Tall plants (5–6 ft [1.5–1.8 m] in height) occur on more frequently flooded peaty soils with low salinity (0–6 ppt). Dwarf plants (about 1 ft [0.3 m] tall) grow on sandy soils underlain by clay with high salinity (60–300 ppt), while intermediate forms (about 3 ft [0.9 m] tall) occupy sandy clays of moderate salinity (5–20 ppt) (Eleuterius and Caldwell 1981). A reciprocal transplant study found that soil–water salinity was responsible for creating genetically different forms (Eleuterius 1989). Common reed grows to heights of 14 feet (4.3 m) in freshwater marshes but in salt marshes where it is restricted to the upper border, it may grow to only 6 feet (1.8 m) or less and its density is also reduced. Seaside plantain grows much taller at lower salinity sites as do arrow-grasses (*Triglochin* spp.) and sea lavender (*Limonium* spp.) (marsh border vs. salt pannes) (Chapman 1960; Tiner, pers. obs.). Sea ox-eye grows taller in marshes with salinities lower than that of typical salt marshes (Antlfinger and Dunn 1983). Even within salt marshes, a noticeable difference in the height of sea ox-eye occurs, with a stunted form occurring on salt flats and a taller form in the upper marsh where salt stress is less (Antlfinger 1981). Similarly, height differences were noted in southern California halophytes—pickleweed and Parish's glasswort, with individuals of shorter stature in the most salt-stressed parts of the salt marsh. Elsewhere, rose mallow (*Hibiscus mosheutos*) had much higher shoot heights (7.8–8.3 ft [2.39–2.53 m]) in freshwater marshes than in brackish sites (4.5–6.2 ft [1.37–1.89 m]), while tidewater hemp (*Amaranthus cannabinus*) had a similar response to salt stress reduction, growing to a height of 8.2 feet (2.5 m) in tidal fresh marshes, but reaching only 4.9 feet (1.5 m) in salt marshes (Kudoh and Whigham 1997; Bram and Quinn 2000).

Salinity Effect on Flowering

Flowering of some plants may be delayed by salt stress from episodic events (e.g., tropical storms). Delayed flowering (23–54 days) and suppressed growth (i.e., reduced height) due to increased salinity from irrigation has been observed in some ornamental plants (Zapryanova and Atanassova 2009). Soil pore water salinity was found to regulate flowering periodicity and intensity in three mangrove species (*Rhizophora mangle,*

Laguncularia racemosa, and *Avicennia germinans*) in Columbian mangroves (Sánchez-Núñez and Mancera-Pineda 2011). Reproduction of neotropical mangroves depended on seasonally contrasting water conditions. Dixie iris (*Iris hexagona*) in Louisiana delayed blooming in salt-stressed conditions (Van Zandt and Mopper 2002). Interstitial soil salinity differs within marshes and may cause individual plants to flower at different times during the spring. This may prevent cross-breeding among neighboring plants, which may promote genetic heterogeneity within the iris population. This mechanism may be a sign of adaptive plasticity necessary to grow in salt-stressed environments. Genetic variation within species (e.g., ability to tolerate salt or withstand prolonged flooding) undoubtedly plays an important part in the distribution of plants in the estuary and around the globe. With the threat of rising sea level, identification of salt-tolerant genotypes is worthy of further study.

Drought Effect on Salinity and Plants

Extended droughts reduce river discharge and allow saltwater to reach greater distances upstream in coastal rivers. This may cause a shift in plant species. The southeastern United States experienced a major drought from 1998 to 2002. One study along the Altamaha River in Georgia revealed that smooth cordgrass moved upstream along the riverbank during the drought, while big cordgrass declined where average high tide salinities were more than 14 ppt (White and Alber 2009). This study also showed that once established upstream, smooth cordgrass persists and can co-exist with big cordgrass in fresher waters, and that big cordgrass is more responsive to changing environmental conditions—it recolonized downstream riverbanks after normal salinity conditions returned. Salt and brackish marsh species replaced slightly brackish and tidal fresh marshes during the 2000 drought in Louisiana (Visser et al. 2002). These changes were facilitated by the occurrence of these species in low numbers in the former habitats. In more saline waters, smooth cordgrass experienced tremendous mortality (see discussion under Salt Marsh Dieback below). In California, reports are mixed on the effect of drought on Pacific cordgrass. In some places, it has migrated upstream colonizing new sites, while in other areas it has suffered heavy losses (Zedler et al. 1986; Collins and Foin 1992). Along the Texas coast, Olney's three-square, a dominant high marsh plant under normal conditions, temporarily disappears after prolonged droughts and "salt pulses" that are inflows of higher salinity water during storms (Miller et al. 2005).

Nutrient Availability

The availability of nutrients affects plant growth, community composition, and diversity. Numerous factors determine nutrient availability in wetlands including soil fertility, water quality, anaerobic conditions, and salinity (e.g., Ungar 1991; Cronk and Fennessy 2001; Pennings et al. 2002). In salt marshes, nitrogen is typically severely limited and may play a role in influencing plant distribution and plant vigor, while freshwater wetlands are either phosphorus-limited or co-limited by both nutrients (Valiela and Teal 1974; Bedford et al. 1999). Nitrogen and phosphorus are limiting in oligohaline marshes (Crain 2007). Some tidal freshwater wetlands appear to be nitrogen-limited as giant cutgrass (*Zizaniopsis miliacea*) increased aboveground productivity when nitrogen was added, and exhibited no response when fertilized with phosphorus (Frost et al. 2009).

Nutrient enrichment stimulates aboveground productivity but not necessarily in all species. For example, the short form of smooth cordgrass did not show an increase in production when an entire marsh area was fertilized during two growing seasons, while the tall form of this species and salt hay grass both grew taller (Deegan et al.

2007). Other in-field fertilization studies have demonstrated that so-called poor competitors outcompete their rivals when enriched. Smooth cordgrass is often viewed as a poor competitor that is relegated to live in the most stressful habitats; however, when field plots containing even mixtures of both smooth cordgrass and salt hay grass were fertilized, smooth cordgrass dominated after two growing seasons (Levine et al. 1998). Interspecific plant relationships were also reversed for other paired species in this experiment: salt hay grass outcompeted black grass, and salt grass outcompeted both salt hay grass and black grass. Fertilization experiments in estuarine marshes showed a change in species composition of low-salinity marshes in just three years, with dominance by plants with high aboveground biomass (Crain 2007).

If nutrient availability changes significantly, a shift in species composition may occur as experimentally demonstrated by the above examples. Shifts in salt marsh vegetation are already taking place due to nitrogen loading by coastal development (runoff from lawns, golf courses, and farms). In southern New England, these changes are allowing smooth cordgrass to replace salt hay, and common reed to eliminate black grass and numerous wetland forbs in the upper high marsh (Bertness et al. 2002). Such nutrient enrichment has made it possible for common reed to occupy full-strength seawater sites where such salinities would naturally limit its expansion. Shoreline development that removed natural vegetation borders caused an increase in nitrogen availability and decreased soil salinity in adjacent marshes (Silliman and Bertness 2004). These conditions facilitated common reed invasion of native salt marshes. Researchers suggested that maintaining natural borders may prevent significant alterations of the marsh environment and thereby preserve native plant communities. In northern New England, nitrate levels in marshes bordering development areas were

Figure 4.9. Dense colony of ribbed mussels (*Geukensia demissa*) with smooth cordgrass (*Spartina alterniflora*) in the regularly flooded zone of a Long Island salt marsh.

higher than for marshes in undeveloped areas in the same estuary (Fitch et al. 2009). Seaside arrow-grass (*Triglochin maritima*) was more abundant in the former locations, while smooth cordgrass occupied marshes at the latter sites. An increase in the former species in northern marshes may be an early indicator of excessive nutrient runoff from development.

Experimental studies have recently shown that while nutrient enrichment may increase aboveground biomass, it has negative effects on root and rhizome biomass and carbon accumulation (Turner et al. 2009). This will cause soil subsidence, increased decomposition of belowground organic matter, and make marshes more susceptible to erosion. Moreover, the results raise serious questions about whether marshes in coastal waters subject to heavy nutrient loading will be able to produce enough organic matter to raise their elevations in synch with rising sea level.

Another interesting observation is that ribbed mussels growing on creekbanks with smooth cordgrass add nitrogen to the soil (Figure 4.9; Bertness 1984b). This should stimulate more plant growth that may account, in part, for the tall height of smooth cordgrass in this zone. Creekbanks without these mussels also support the tall form of this species, however, so the influence may be minor compared with other factors. The mussels do stimulate root growth, and their byssal threads join together the roots of neighboring plants, thereby enhancing the ability of cordgrasses to withstand wave action and stabilizing creekbanks (Bertness 1992).

Substrate Type

The substrates of tidal wetlands are quite variable from bedrock or rocks to unconsolidated materials (cobble, gravel, sand, silt, clay, and organic matter). These substrates provide varying challenges for plant life with rocks, cobbles, gravel, and sands presenting the greatest challenges. Organic substrates and fine-textured mineral substrates (silt, loam, and clay) tend to hold moisture better than the other substrates and are more suitable sites for plant growth in wetlands and non-wetlands alike.

Several conditions make rocky coastlines difficult places for plant establishment: 1) the rock substrate (impenetrable by plant roots); 2) the variable flooding and exposure to air; 3) extreme temperature shifts; and 4) heavy wave action in the surf zone. Arguably, the first on a list of needed plant adaptations to colonize these sites is a mechanism for attaching to the rocks to keep the plant from washing away. Rockweeds (*Fucus* spp.), knotted wrack (*Ascophyllum nodosum*), Irish moss (*Chondrus crispus*), and dulse (*Palmaria palmata*) are among many algae that have developed holdfast organs to bind them to the rocks (Figure 4.10). Vascular plants typically need places where unconsolidated material (soil) is available for root growth. On rocky shores, plants such as seaside plantain, beach-head iris (*Iris setosa*), and alkali grasses (*Puccinellia* spp.) find such areas in cracks and crevices above the active wave zone. Surfgrasses (*Phyllospadix* spp.), a marine vascular plant related to eel-grass, utilize a system of thickened rhizomes and dense root hairs to colonize on rocks in the lower intertidal zone as well as subtidally on the Pacific Coast (Cooper and McRoy 1988). These types of plants are considered "foundation species" because once they colonize rocky shores they create environmental conditions that support an entire community of other plants and animals (e.g., Shelton 2010).

Many intertidal habitats that are made up of unconsolidated materials include beaches (sloping landforms that fringe islands and mainland) and flats (relatively level landforms in general). These habitats are generally devoid of macrophytic plants. Sandy beaches are inhospitable to plants due to shifting sands, high summer temperatures, extremely variable substrate temperatures, and low nutrients. An extremely low

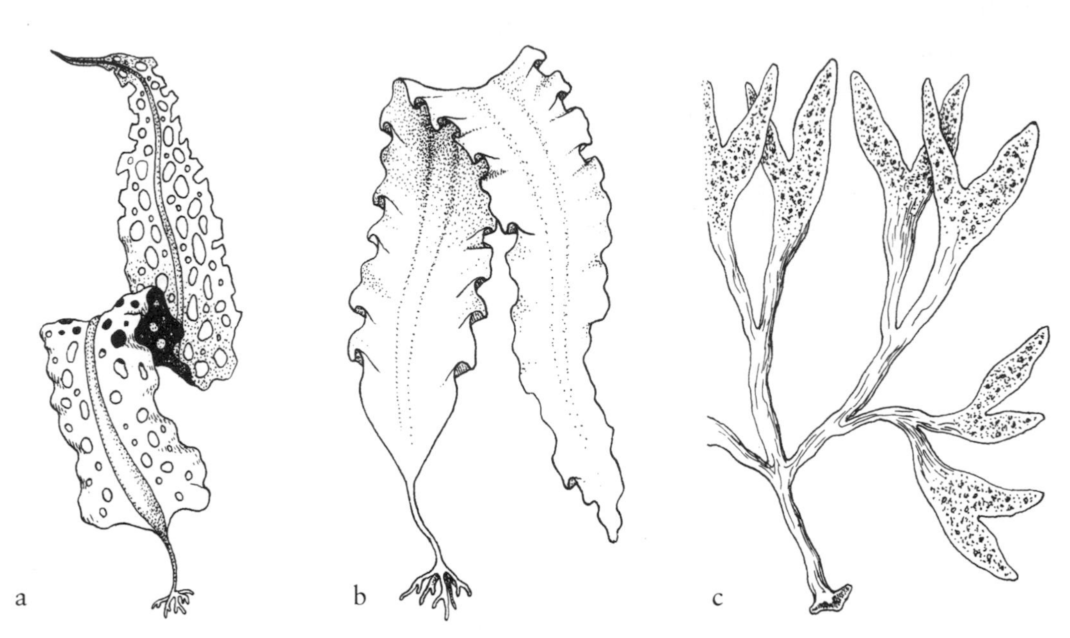

Figure 4.10. Many marine algae have holdfast organs at the base of their stems to secure them to boulders, bedrock, and other solid substrates: (a) sea colander (*Agarum clathratum*); (b) sugar kelp (*Laminaria saccharina*); and (c) sea wrack (*Fucus edentatus*). (Tiner 2009)

number of species have adapted to these situations (see Table 5.1). Plants seek out the upper levels of sandy beaches where conditions are more favorable. Accumulation of organic matter (tidal wrack) near the dune line increases soil productivity and moisture to allow for plant growth. Many northern shores in glaciated regions are represented by cobble-gravel beaches ("shingles"). These substrates pose similar difficulties for plant growth. Tidal flats may appear devoid of plants, yet diatoms and algae such as sea lettuce (*Ulva lactuca*) and hollow green seaweed (*Enteromorpha intestinalis*) are the more frequent colonizers of tidal flats in saline waters. In some regions, tidal flats are littered with rocks and boulders providing substrates suitable for marine macroalgae.

Plant growth varies with sediment type (van der Brink et al. 1995; Willis and Hester 2004) and is enhanced by the presence of nutrients. In salt marshes, organic soils have more nutrients than sandy soils have (Padgett and Brown 1999; Howard 2010), and yet, some species tend to be more abundant on sandy sites than on other mineral or organic soils (see Table 5.3). Intraspecific variation in salt marsh species can result in different growth responses in fine-textured soils versus sandy soils (Eleuterius and Caldwell 1981; Christian et al. 1983). Smooth cordgrass colonizing experimental plots of soft sediments were taller and more productive than the plants on peat plots or growing in uncontrolled plots, leading the researcher to suspect that peat may inhibit this species (Bertness 1988).

Climate

For all plants, climate exerts a major influence on their distribution both latitudinally (north to south) and altitudinally (valley to mountain top). In temperate and Arctic regions, estuarine wetlands are dominated by salt marshes from 25°N latitude to the Arctic Circle in the Northern Hemisphere and from 38°S to the tip of South America in the Southern Hemisphere (Chapman 1977). In between these regions, tropical coastal wetlands are characterized by woody species, mainly mangroves. On the

Atlantic Coast, smooth cordgrass ranges from the tropics to Newfoundland where it is replaced by Arctic species such as alkali grasses. In the United States, mangroves dominate from the tip of Florida to 25°N latitude, with one species, the black mangrove (*Avicennia germinans*) growing to about 30°N (St. Augustine–Jacksonville Beach, Florida) and to 29°N in Louisiana (Tiner 1993a; McKee 2004). Worldwide mangal distribution is limited by frost to locations generally between 32°N and 38°S (Walter 1977).

Species diversity of salt marshes tends to decrease with increasing latitude and decreasing temperature (Walter 1977). The harsher climate imposes a significant added stress on plant life. Interestingly, the species adapted to these climates appear to occupy a wider range of conditions in salt marshes than they do at lower latitudes (Pielou and Routledge 1976). In Nova Scotia, four vegetative zones were evident in the salt marsh: 1) smooth cordgrass; 2) salt hay grass; 3) chaffy sedge (*Carex paleacea*); and 4) baltic rush (*Juncus arcticus,* formerly *J. balticus*). In James Bay, Ontario, however, chaffy sedge dominated all but the upper and lower fringes of the salt marsh. Populations of northernmost species may, therefore, exhibit greater within-species genetic variability than is found in more southern populations of the same species, or perhaps reduced competition allows for expansion of its distribution within the more northern marshes.

Plants can be grouped into categories associated with different regions and climates (see Tables 5.5–5.7). Arctic and subarctic species include several that have colonized New England tidal marshes while some southern coastal species extend into the Mid-Atlantic region and southern New England. Creeping alkali grass dominates Canada's northernmost salt marshes but does not occur in New England or Atlantic Canada (Sean Basquill, pers. comm. 2004). Northern New England marshes experience more severe winters, cooler temperatures, and a shorter growing season than do southern New England salt marshes, with ice damage limiting low marsh vegetation and scouring portions of the high marsh to create depressions and waterlogged spots for subsequent colonization by marsh forbs (Evanchuk and Bertness 2004). Overwash deposition of fine clay may facilitate the poor drainage of these "forb pannes." The seaside plantain zone characteristic of northernmost marshes appears to be climate-related due to freezing and ice effects (Chmura et al. 1997).

The occurrence of salt barrens in the middle of southeastern United States and in southern California salt marshes, and the virtual absence of such zones in northern marshes, is a product of climate. Southeastern marshes are subjected to many months of high heat and solar radiation creating salinity extremes in this zone that prevent plant colonization, leaving the middle marsh with large nonvegetated areas (Bertness and Leonard 1997). A Mediterranean climate (wet winters and dry summers) in southern California creates hypersaline soils in salt marshes, which reduces vascular plant cover while increasing opportunities for algal growth (Zedler 1980). New England marshes are not exposed to severe temperatures for long periods, so such barrens do not form. Here small bare spots are usually colonized by more salt-tolerant species that reduce soil salinity and facilitate re-establishment by other high marsh species in two or three years (Bertness and Shumway 1993) or by mats of cyanobacteria (formerly blue-green algae).

Changes in annual rainfall in dry regions (e.g., Texas and southern California) may influence vegetation patterns in tidal marshes. For example, drought conditions increased soil salinity in Texas salt marshes and promoted the spread of perennial glasswort, whereas a series of wet years led to a decline of this species with a corresponding increase in sea ox-eye (Forbes et al. 2008). Human-induced changes in river flow regimes induce similar responses in downstream salt marshes in this region. Total annual rainfall creates differences in soil moisture and salinity that influence plant distribution of annuals in southern California salt marshes (Callaway et al. 1990; Callaway and Sabraw 1994). Wetter years provide opportunities for annual plants with lower salt tolerances that normally occur in the uppermost marsh to grow at lower levels.

Within their respective geographic ranges, individual species may have different tolerances based on climate. For example, smooth cordgrass tends to grow higher in elevation in marshes of the same tidal amplitude in warmer regions than in colder regions (McKee and Patrick 1988). This difference may be due to growth conditions including the length of the growing season and biological competition. Salt hay grass can recover from periodic overwash (sand deposition from high tides) in the South, but apparently not in the North (Leatherman 1979). The southern populations evolved behind broad, flat barriers subject to frequent overwash whereas northern populations developed along more stable shores behind tall dunes. Differences in the length of the growing season may be another explanation: northern plants may simply not have enough time with favorable growth conditions to recover.

Microclimate

Site-specific temperature changes also create environmental conditions that plants and animals must contend with. For example, when the tide changes in summer, the temperature of exposed beaches and rocky shores can fluctuate greatly. When covered by the tides, substrate temperatures are lower, but when the tide recedes surface temperatures rise. Anyone who has walked barefoot on beach sand has experienced this difference. Extreme temperatures subject plants and animals to the effects of drying (desiccation). In northern latitudes, winter may create opposite situations. Exposure in winter produces colder temperatures, while flooding likely promotes a slight decrease in temperature. Fluctuations in temperature are moderated by wet soils and water (Sculthorpe 1967). Along the Pacific Coast, fog created by cold ocean temperatures provides critical moisture to support tidal wetland vegetation during dry summer months (Leck et al. 2009).

Physical Disturbances

Coastal wetlands are among the most dynamic wetlands in the world. Natural forces, such as storms (e.g., hurricanes and nor'easters), wave action, tidal currents, and ice, reshape portions of these wetlands on a daily, seasonal, and intermittent basis. Some patterns are seasonal such as erosion of beaches in winter and deposition in summer, whereas other forces such as hurricanes are extreme events producing both short- and long-term impacts. Heavy rainfall and high river discharge may introduce more sediment into estuaries and elevate mudflats to levels suitable for plant colonization (e.g., Ward et al. 2003). Catastrophic seismic events such as major earthquakes uplift intertidal surfaces or cause coastal subsidence that impose rapid changes in tidal wetlands in some regions (e.g., Pacific Northwest to Alaska; see Chapter 2). Fire plays an important role in plant composition in many ecosystems, but the impact of fire from lightning strikes does not appear to be significant for tidal wetlands, although lightning-induced fires are a common occurrence in Gulf and South Atlantic marshes (Nyman and Chabreck 1995; Mitchell et al. 2006). Fire, however, is used as a wildlife management tool to promote wildlife food plants (Chapter 9).

Significant year-to-year variation in species composition and cover of annual plants is the norm for many tidal fresh marshes due to variable erosion and sedimentation patterns, while the abundance of perennials remains largely unaffected (Leck and Simpson 1995). Shifts in vegetation patterns may be caused by weather changes such as droughts or salt pulses as noted elsewhere in this chapter. The effect of some coastal processes appears one-directional, but if evaluated on a geologic time scale, they too show shifting patterns (e.g., sea-level change and barrier island migration). Other disturbances include salt spray, deposition of debris, brown marsh syndrome, and herbivory. The effects of sea-level rise on tidal wetlands are addressed in Chapter 12.

Coastal Processes

High-energy environments characterized by heavy wave action and fast-flowing currents cause scouring and conditions sufficient to uproot any vascular plant attempting to colonize such sites. The seedling stage of vascular plants is particularly vulnerable to these forces. Consequently, rooted vascular wetland plants are not found in such areas; they instead seek out sheltered locations. Algae are well adapted for life on high-energy shorelines. Cyanobacteria (blue-green algae) form the slippery "black" zone of the wave-washed rocky shores, the uppermost vegetated zone. Macroalgae colonize areas below where there are places for them to gain a firm footing and more tolerable wet conditions. In the northeastern United States and eastern Canada, the most common intertidal macroalgae are knotted wrack and various rockweeds. Holdfast organs attach them to the rocks and air bladders allow the plants to float during high water and buffer them against heavy wave action. Plant morphology may change in response to degree of exposure. For example, the fronds (leaves) of knotted wrack grow long (up to 3 feet [1 m] long) in sheltered areas of Nova Scotia, while in exposed locations, the fronds are less than 16 inches (40 cm) long (Little and Kitching 1996). Rockweeds (*Fucus vesiculosus*) grow smaller, take on a highly branched bushy habit, and lack air bladders in exposed areas, whereas individuals in more sheltered areas have longer, more open-branched fronds bearing air bladders. Growth rates of seaweeds growing on the lower shores may be higher than for those growing higher on the shore where stress to drying and temperature fluctuations are greater.

Hurricanes may alter wetland plant communities and even change vegetated wetlands to nonvegetated wetlands. Damage to wetland vegetation may include compression (mats of marsh vegetation pushed together like an accordion), mounding (creating large piles of vegetation), sedimentation (deposition may sink floating marshes), scouring (removing plants from their roots or removing plants entirely from the underlying substrate), or saltwater intrusion (change in salinity that kills freshwater plants or alters the composition of brackish marshes). Storm surges from hurricanes may elevate water levels many feet above normal causing structural damage to woody plants. The immediate impact of hurricanes Katrina and Rita in 2005 was conversion of an estimated 217 square miles (562 km^2) of wetland (mostly tidal) to open water as vegetation was sheared or ripped to the root mat or to the underlying clay substrates (Barras 2006). Hurricanes may also accelerate coastal erosion and wetland loss through heavy winds, wave action, and plant mortality (Guntenspergen et al. 1995; Stone et al. 1997; Cahoon et al. 1999, 2003; Whelan et al. 2009). Tidal surge from Hurricane Bret removed large areas of emergent vegetation from a Texas salt marsh, yet the change was temporary as subsequent freshwater flooding created conditions that later increased plant cover (Alexander and Dunton 2002). Portions of Florida's mangrove swamps have been converted to tidal mudflats by hurricanes (Smith et al. 2009).

Flooding of slightly brackish marshes and freshwater tidal swamps by more saline waters during storm events may have a significant effect on plant composition and structure (Howard and Mendelssohn 1999a,b; Conner and Inabinette 2003). Removal of aboveground growth may make some plants more susceptible to damage from increased salinities or prolonged flooding. For example, aboveground biomass of bull-tongue was significantly reduced by increased salinity (6 ppt) but unaffected by flooding, whereas salt hay grass was eliminated by year-round flooding but had no response to the higher salinity (Baldwin and Mendelssohn 1998). Clipped specimens of both species were almost eradicated when flooded with saline water, whereas unclipped individuals showed no ill effects. The adverse effects of the temporary increase in salinity may not be long lasting, at least for some species. In Louisiana, marsh plant cover reduced from 81 percent to 57 percent by a hurricane recovered to 75 percent in one year (Chabreck and Palmisano 1973). Hurricanes can also have more lasting effects, however, as Hurricane Hugo killed 77 percent of the bald cypress in a South Carolina forested wetland near the salt marsh, and saltwater intrusion curtailed tree regeneration for two to three years afterward creating a more open area for grasses and common reed to dominate (Conner and Inabinette 2003). Recognizing intraspecific differences in salt tolerance, most cypress grow in areas where salinity is less than 2 ppt (Allen et al. 1997), but some variants can tolerate quite brackish conditions (Conner and Inabinette 2005). In Louisiana, hurricanes Katrina and Rita increased salinities in brackish and freshwater wetlands with the effect lasting for months (Steyer et al. 2007). Maximum salinities recorded just a couple of months after the hurricanes were 8 ppt for freshwater swamp (normally less than 0.5 ppt), 26 ppt for fresh marsh (normally 0–3 ppt), 26 ppt for intermediate marsh (normally 2–8 ppt), and 34 ppt for the brackish marsh (normally 4–10 ppt). These changes have negatively impacted vegetation, with cattail showing only minimal recovery over six months later. On Chenier Plain in Louisiana, robust floating mat communities of maidencane were wiped out after two years of exposure to salinities above 6 ppt following Hurricane Rita (Sasser et al. 2009). Only salt-tolerant species have been able to recolonize the exposed peat.

Salt Spray and Other Wind Effects

While storm-driven winds have produced the highest tides on record and strong onshore winds may keep wetlands flooded for days, wind action affects some plants more frequently through salt spray and moving sand. Salt spray can give a pruned appearance to shrub thickets on neighboring dunes and other uplands. Species in the spray zone must have mechanisms to deal with salt stress. Grasses in this zone may have waxy surfaces, while the maritime ecotype of curly dock (*Rumex crispus*) has very dense panicles to withstand heavy salt spray and produces seeds with corky tubercules that float longer in water than do those of the inland ecotype (Adam 1993). Sea beach knotweed (*Polygonum glaucum*) and beach spurge (*Chamaesyce polygonifolia*) have adopted a prostrate habit (i.e., lying flat on the sand) to avoid both the effects of salt spray and wind-borne sand. Dune grass (*Ammophila breviligulata*) possesses tough leaves hardened with silicon to prevent the entry of salt, while the thick fuzzy leaf coating of dusty miller (*Artemisia stelleriana*) may reduce the abrasive effect of wind-blown sands and evaporative water loss (Johnson 1985).

Burial by sand affects all beach and dune plants and most possess rhizomes that grow upward when buried (e.g., southern populations of salt hay grass; Leatherman 1979). Interdunal swale species such as large cranberry (*Vaccinium macrocarpon*) and common three-square also have this adaptation. (*Note:* Sanding of commercial cranberry bogs is a management practice done to stimulate vine growth and berry production.) Once adapted, however, the key to survival appears to be the depth of burial. Other plants survive these conditions by their annual habit, relying on seed production to maintain their presence in this dynamic environment (e.g., sea rocket [*Cakile edentula*] and Russian thistle [*Salsola kali*]).

The most severe storms generate the highest tides, which may break through low dunes on barrier islands and transport sand into marshes on the bayside. These overwash deposits may completely bury marsh vegetation and convert the areas to upland dunes, or deposits may be thin enough so marsh vegetation can recover. Such areas, called "washes," often have an interesting collection of species including flowering herbs such as spring ladies'-tresses (*Spiranthes vernalis*), stiff yellow flax (*Linum medium*), Virginia meadow-beauty (*Rhexia virginica*), water-hyssop (*Bacopa monnieri*), purple gerardia (*Agalinis purpurea*), seaside gerardia (*Agalinis maritima*), salt marsh asters (*Symphyotrichum tenuifolium, S. subulatum*), and marsh fleabane (*Pluchea foetida*) (Higgins et al. 1971). Canada's threatened St. Lawrence or Laurentian Aster (*Symphyotrichum laurentianum*) is one of many plants adapted to periodic burial by sand (Alan Hanson, pers. comm. 2004).

Wind from hurricanes may break off stems and branches of woody plants, and uproot herbaceous and woody vegetation including rolling up marsh root masses into mats and depositing large sections of other root masses onto existing marsh surfaces (e.g., Cahoon 2006). In tidal swamps, the buttressed trunks of bald cypress and water tupelo seem to make these plants more resistant to blowdowns (uprooting), while other trees and shrubs in these forests may suffer more damage from wind action (Middleton 2009). This may affect forested wetland composition in the long term, as observed in Florida mangrove swamps. Winds of more than 143 miles per hour (230 km/hr) from Hurricane Andrew caused 60 to 85 percent mortality of mangrove trees in Biscayne Bay in 1992, which caused shifts in the relative abundance of mangrove species that persist today (Baldwin et al. 2001; A. Baldwin, pers. comm. 2007). This hurricane also defoliated trees in Louisiana coastal freshwater swamps (less than 10% of the trees eventually died) and produced a pulse of organic matter that adversely affected water quality (Rybczyk et al. 1995).

Ice Action

In northern regions, scouring by ice can remove patches of marsh vegetation, undercut creekbanks, and gouge out huge chunks of tidal mudflats. Regularly flooded smooth cordgrass marshes are reduced to stubbles in winter along with any natural levees created by deposition (Hatcher and Patriquin 1981). Harsh winters freeze soil to greater depths impacting the distribution of roots and rhizomes. In Atlantic Canada, maximum underground biomass in smooth cordgrass occurred from 1.6 to 4.8 inches (4–12 cm) below the surface and in salt hay grass from 1.6 to 3.2 inches (4–8 cm), while in southern New England, the belowground biomass for both species was greatest at depths of 0.8 to 1.0 inch (2–5 cm) (Valiela et al. 1976; Connor and Chmura 2000). The deeper depth at the former site may be attributed to northern freezing conditions. Ice shearing may be responsible for the development of a seaside plantain zone between the low marsh and high marsh in northern latitudes (Chmura et al. 1997). This zone appears to be absent south of Little River (Wells, Maine) where ice effects are not as intense. The predominance of three-square marshes along the St. Lawrence River has been attributed in part to its ability to withstand the mechanical action of ice (Marie-Victorin 1964).

High tides may lift out blocks of frozen marsh peat and transport them elsewhere in the estuary. Such action has been identified as a force creating marsh ponds (Dionne 1969; Reed and Moisan 1971; Richard 1978; Bleakney and Meyer 1979; Gauthier and Goudreau 1983; Gordon and Desplanque 1983; Bélanger and Bédard 1994; Gordon and Cranford 1994; Van Proosdij et al. 2000). This may be why marsh ponds are far more abundant in northern marshes than in southeastern marshes. When the peat blocks are relocated to a tidal flat, they slowly degrade or start the process of marsh formation on the flat. Ice rafts containing mineral sediments deposited on the marshes aid in marsh accretion (Argow 2004). In boreal and Arctic marshes, ice carries large boulders across the marsh creating depressions and interesting microrelief (Chmura 2003) in marked contrast to the relatively level terrain of more southerly marshes. Deposition of these boulders seaward of the marsh contributes to providing a more sheltered environment.

Deposition of Debris

Debris of all kinds deposited by storm tides can be found along the marsh-upland border in some marshes. This includes organic debris (dead plants) that floated in with the tides as well as derelict boats, wooden boards, steel drums, Styrofoam, and other human-derived flotsam that are brought in by storm tides. When deposited on marsh plants, the plants are smothered and die.

While the man-made materials will persist until removed by man or relocated by other storms, organic debris will decompose over time allowing for the recolonization by halophytes. Organic debris can be found in the form of the remains of dead leaves and stems or large woody material (dead snags). Tidal litter (wrack) is deposited in various locations in salt and brackish marshes, especially in the upper marsh at the edge of tall grass stands (e.g., common reed) and along causeways (Figure 4.11). Heavy mats of tidal wrack shade out and suffocate underlying vegetation, creating nonvegetated flats where pioneer species such as common glasswort can begin to grow (Reidenbaugh and Banta 1980; Hartman et al. 1983; Bertness and Ellison 1987; Brewer et al. 1998b). In New England, salt grass and common glasswort are more tolerant of wrack burial than are other salt marsh species (Bertness and Ellison 1987). Salt grass colonizes bare ground vegetatively by runners while common glasswort and smooth cordgrass establish by seeds. Over time these species are replaced by salt hay

Figure 4.11. Tidal wrack deposited in a Massachusetts salt marsh.

grass and black grass, which grow more slowly. Deposition of wrack may have more impact in the northern regions where salt marsh plants die back in the fall, producing huge amounts of organic litter for dispersal by tides in spring and early summer; whereas in southern regions (from North Carolina south), marsh plants exhibit some growth year-round, shed some of their aboveground parts throughout the year, and therefore produce smaller mats and less disturbance (Pennings and Bertness 2001).

In the Pacific Northwest, logs originating from downed timber in upstream watersheds are washed downstream and deposited on tidal marshes and coastal shores. The presence of large woody debris (LWD) influences vegetation patterns in estuarine marshes. LWD provides a substrate for both halophytic and upland species in salt marshes (Eilers 1975; MacLennan 2005). Upland species colonize higher logs that provided an elevated surface above normal tides, while lower logs create suitable habitat for halophytic species. In oligohaline (slightly brackish) marshes, the occurrence of sweet gale (*Myrica gale*) was strongly correlated with LWD (Hood 2007). Since sweet gale is a nitrogen-fixing plant, the presence of LWD may be influencing nitrogen dynamics in these marshes.

Salt Marsh Dieback

During a severe drought coupled with extremely high summer temperatures from 2000 to 2003, the vegetation in 158,000 hectares of Louisiana's salt marsh suddenly died causing the normally green marshes to turn brown (e.g., Schneider and Useman 2005). This dieback was coined "brown marsh syndrome," and later simply called "brown marsh" or "salt marsh dieback." Similar diebacks have been observed elsewhere along the Gulf Coast and also along the Atlantic Coast as far north as southern Maine (Figure 4.12; Alber et al. 2008). Smooth cordgrass was the main species impacted, but other affected plants include salt hay grass, black needlerush, salt grass, and black grass. While there has been no single cause identified for these lethal outbreaks, drought is believed to be the major cause in dieback in southern marshes. Drought conditions raise soil salinities and allow soils to oxidize, converting metal sulfides in saline soils to sulfuric acid which is lethal to plants. Plants stressed by droughts may be more susceptible to fungal infections, and a normally weak pathogen fungi (e.g., *Fusarium* spp.) may become more harmful at these times (Schneider and Useman 2005). After an intensive investigation, researchers found no evidence supporting fungal pathogens or snail grazing as the primary cause of the dieback in Louisiana. Instead, they concluded that drought during a time of low sea level and low river discharge caused soil acidification that produced elevated levels of potentially toxic metals (e.g., pyrite and acid-extractable iron and aluminum) in affected plants (McKee et al. 2006). On the positive side, brown marsh generally does not persist and most affected sites recover. For example, a review of Massachusetts brown marsh sites that were first detected in 2002 found that 21 of the 25 sites had no signs of dying vegetation by 2006 (Smith 2006; Smith and Carullo 2007). In some cases where bare areas persist, substrates have subsided or soils eroded to the point that the marsh surfaces (the marsh platform) are now at lower elevations, perhaps too low to support vascular plant growth. These sites will likely remain as mudflats. In Louisiana, rapid recovery occurred where smooth cordgrass roots survived, while areas that experienced more than 90 percent root mortality had low rhizome survival and experienced slow recovery (McKee et al. 2006). Rapid recovery may have been facilitated by increased nutrient and light availability. Keep in mind that all bare marsh areas are not attributed to brown marsh syndrome. Some denuded marshes may appear to be due to dieback but are actually the result of heavy grazing by snails, crabs, or other herbivores (see Herbivory below), marsh submergence by

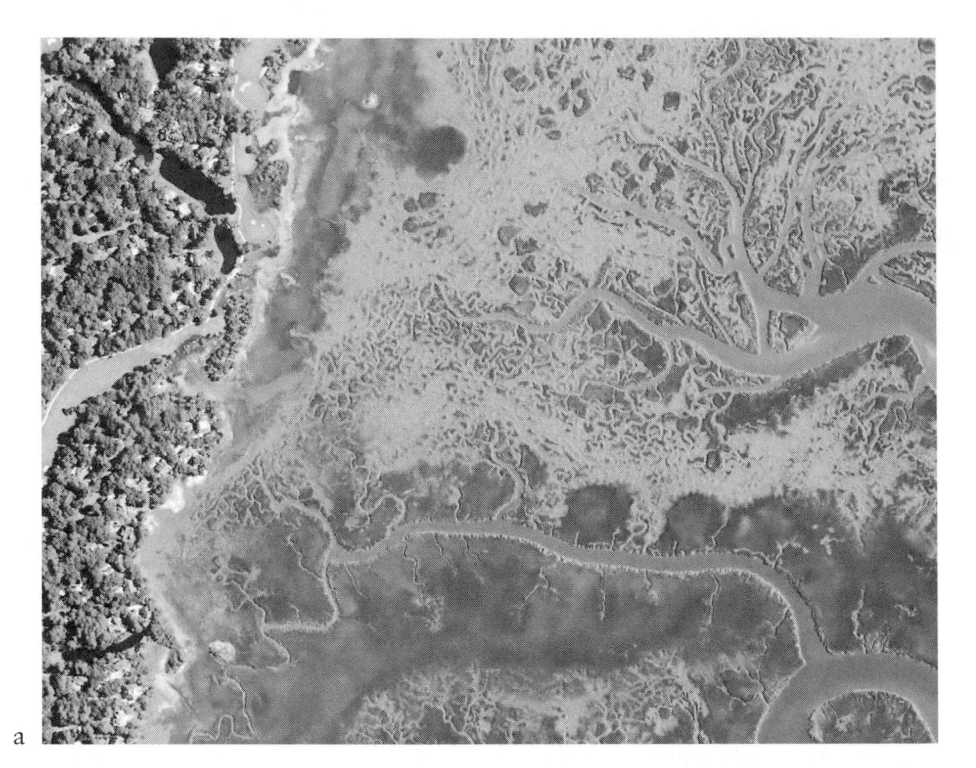

a

b

Figure 4.12. In the spring of 2002, marsh dieback was observed in several areas along the Georgia coast: (a) aerial view of affected marshes showing patchy distribution of surviving clumps of smooth cordgrass (*Spartina alterniflora*) and unaffected marshes; and (b) expansive mudflat was a former low marsh that was affected by brown marsh syndrome; this marsh has since recovered. (b: Jan Mackinnon)

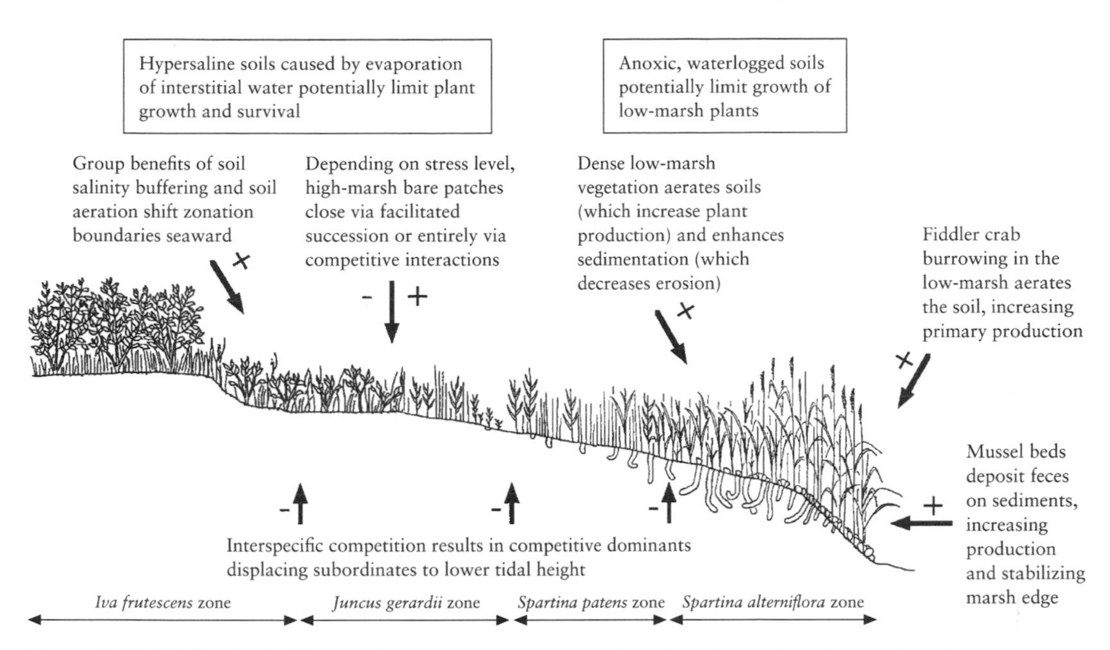

Figure 4.13. While physical factors impose great stress on plants in salt marshes, interactions between plants and animals help moderate these effects and influence plant distribution. (Bertness and Leonard 1997; reprinted by permission of the Ecological Society of America)

rising sea level (see Chapter 12), deposition of tidal wrack (see Deposition of Debris above), or altered hydrology.

Precipitation

While it may seem unorthodox to consider precipitation a natural disturbance, scientists studying tidal wetlands in semi-arid regions (humid, hot summer and mild, dry winter) have suggested this possibility since precipitation varies yearly with no predicted pattern (Dunton et al. 2001). This situation is quite unlike what is experienced in other tidal wetlands across temperate North America. Salt marshes along the Texas and southern California coasts are hypersaline with salinities frequently exceeding 100 ppt for extended periods. As mentioned earlier, under these conditions, extensive salt barrens or salinas devoid of vegetation have formed. Periods of heavy rainfall or freshwater flooding allow vegetation to temporarily colonize these areas and may cause shifts in pre-existing marsh vegetation. For example, fall flooding favors germination of an annual species—perennial glasswort—which colonizes bare spots and may even replace key grass (Alexander and Dunton 2002). Plant communities in arid and semi-arid tidal marshes are dynamic and relatively unpredictable as a result of the interannual variation in precipitation. In fact, native species in the interior of the Nueces River salt marshes in Texas are dependent on episodic precipitation events to reduce the hypersalinity of the system (Dunton et al. 2001).

Biological Factors

Animals and competition among plants also affect the composition of tidal wetland communities (Figure 4.13). Grazing may selectively eliminate or reduce the abundance of preferred plants, while predator–prey relationships offer some interesting possibilities that affect vegetation patterns. The ability to tolerate stresses imposed by the intertidal environment plays out when plants compete for resources. Invasive species are foreign competitors that seriously threaten the natural integrity of tidal wetlands, while human actions often give these species an edge over native species.

Herbivory

Grazing by animals has a direct effect on plants and tidal wetland communities. Early studies of this effect on salt marshes indicated only minor impacts by insects due to the unpalatability of most grasses, concluding that grazing affected only seven percent of smooth cordgrass production (Smalley 1960; Marples 1966; Kraeuter and Wolf 1974). The main grazers were grasshoppers (*Orchelimum fidicinum*) and planthoppers (*Prokelisia marginata*). Estimates of the carbon budget of salt marshes suggest that, on average, 31 percent of the net primary productivity goes to herbivory (Duarte and Cebrián 1996). Coastal development and estuarine eutrophication may be altering grazing impacts as well as salt marsh community structure. A Rhode Island study of insect herbivory and primary productivity of salt marshes in both natural and developed settings produced some interesting results (Bertness et al. 2008). Grazing was found to reduce smooth cordgrass production in the eutrophic marshes by 50 to 75 percent of what it would be in the absence of herbivory, while little impact from insect herbivory was noted in salt marshes surrounded by natural habitat. One conclusion of this work is that shoreline development and local nutrient enrichment are causing an increase in insect herbivory in local marshes. This perhaps is a more subtle influence of human society on coastal ecology. Intense grazing of Arctic salt marshes by lesser snow geese may alter soil properties, making the soil unsuitable for re-establishment of creeping alkali grass (McLaren and Jefferies 2004). As patch size increases, soil temperatures and salinity increase, soil moisture decreases, more organic matter decomposes, and nutrients are lost, which create adverse conditions for plant growth. Patch sizes 8 inches (20 cm) in diameter restrict re-vegetation.

The impact of salt marsh grazing by snails and crabs has received considerable attention given the interest in understanding predator control of grazer populations. Grazing of smooth cordgrass by the marsh periwinkle (*Littoraria irrorata,* formerly *Littorina irrorata*) is widespread from Maryland into Georgia (Silliman and Zieman 2001), yet there appears to be no obvious effect on the plant. One theory on this lack of any observed effect suggests that snail populations are regulated by predators, mainly blue crab (*Callinectes sapidus*), and that it is not due to the unpalatability of the grass (Silliman and Bertness 2002). When predators were excluded from study marsh plots, periwinkles converted the marsh plots to a mudflat in just eight months! In the Georgia salt marsh examined, predators appear to be controlling snail populations to the point that their grazing effect on marshes is minimal. If the snail predators are keeping the snail population in check, they may be responsible for sustaining highly productive southern salt marshes. The researchers speculate that overharvesting blue crabs may be linked to recent marsh diebacks along the Gulf Coast. Another experiment removed periwinkles from marsh plots and found that smooth cordgrass increased its density and biomass by 50 percent (Gustafson et al. 2006). In Massachusetts, populations of the marsh crab (*Armases* [*Sesarma*] *reticulatum*) may have increased due to reduced predation, which is why the crabs were able to denude creekside vegetation along nearly half of Cape Cod's salt marshes (Holdredge et al. 2008). This is yet another example that altering predator–prey interactions may be causing the collapse of marsh ecosystems previously thought to be exclusively under bottom-up control by physical factors (e.g., light, nutrients, salinity, anaerobiosis, and hydrology affecting plant growth and productivity). Marsh grazers of other halophytes appear more limited. For example, the major grazers of black needlerush marshes in the Southeast are adult grasshoppers (e.g., *Orchelimum concinnum, Conocephalus hygrophilus*) (Parsons and de la Cruz 1980). Other grazers include detritus-feeders (crabs, amphipods, isopods, and snails) that consume algae growing on the marsh surface or on plants.

While invertebrate herbivory in salt marshes may be minimal or on the rise, reports of heavy grazing by vertebrates have been recorded for tidal marshes. Heavy grazing creates large barren areas within the marsh called "eat-outs." Snow geese in fall and winter denude hundreds of acres of smooth cordgrass and common three-square marshes in the northeastern United States and Quebec (e.g., Smith and Odum 1981; Giroux and Bédard 1987). One study found that a flock of 5,000 snow geese removed smooth cordgrass from 300 acres (121 ha) of marsh in just six weeks (Griffith 1940). Goose grazing can have contrasting effects on plant diversity. In some cases, grazing creates openings that provide habitat for pioneer species and increases local plant diversity, while in subarctic salt marshes it can reduce plant diversity. For example, eat-outs of common three-square in St. Lawrence River marshes facilitated colonization by wild rice (*Zizania aquatica*), arrowheads (*Sagittaria* spp.), and other plants (Giroux and Bédard 1987; Bélanger and Bédard 1994). Yet in subarctic salt marshes, goose grazing was found to reduce plant diversity as marsh plots within geese exclosures had higher diversity than did grazed plots (Bazely and Jefferies 1986; Ungar 1998). In some situations, intensive herbivory may promote the formation of marsh ponds (Erwin et al. 2004). Intense grubbing by geese may lead to the degradation and loss of Arctic salt marshes as patches greater than 8 inches (20 cm) in diameter may be too wide for recolonization by creeping alkali grass (McLaren and Jefferies 2004). In a Texas salt marsh, open patches created by snow geese had higher salinities (due to exposure and increased evaporation) than vegetated areas and had a greater effect on plant distribution than other factors, favoring the growth of salt grass (Miller et al. 2005). Canada geese are known nuisances at salt marsh restoration planting sites, and protective netting may be installed to deter their feeding. Disturbance and herbivory of seedlings by carp and Canada geese have led to declines of wild rice and other annual low marsh species in tidal freshwater marshes (Baldwin and Pendleton 2003). Muskrats (*Ondatra zibethicus*) and nutria (*Myocastor coypus*) are important marsh herbivores in both slightly brackish and tidal fresh marshes. They also create substantial eat-out areas, with Olney's three-square being a preferred food. Other salt marsh herbivores include red-backed voles (*Clethrionomys gapperis*), meadow voles (*Microtus pennsylvanicus*), and in southern marshes and mangroves, marsh rabbits (*Sylvilagus palustris*). In Mid-Atlantic salt marshes, selective grazing by wild horses on smooth cordgrass has allowed salt grass to increase in abundance (Furbish and Albano 1994). In southern marshes, nutria and wild boar are other vertebrate herbivores that negatively impact coastal marshes by reducing underground biomass and expansion of root systems (Ford and Grace 1998).

For tidal fresh marshes, there is more evidence of grazing. As much as 10 percent of the live plant parts may be eaten by herbivores, mainly insects (Pfeiffer and Wiegert 1981; Figure 4.14). Grasshoppers, planthoppers, and aphids may be the chief grazers of living material, with the last two species sucking juices from the plants. Grasshoppers may be more important grazers of grasses in southern marshes where herbivore pressure is greater (Pennings et al. 2001; Pennings and Silliman 2005). Muskrats are abundant in oligohaline and tidal fresh marshes, and heavy grazing creates eat-outs. Exclosure experiments testing the effect of grazing on Louisiana marshes by the introduced nutria found substantially reduced plant biomass (up to 75%) and changes in plant community composition (Taylor et al. 1997; Evers et al. 1998). Small mammals including meadow voles, white-footed mice (*Peromyscus leucopus*), and meadow jumping mice

Figure 4.14. A mass of caterpillars (possibly the convict caterpillar, *Xanthopastis timais*) feeding on tidal freshwater marsh plants in South Carolina.

(*Zapus hudsonius*) may be important grazers in slightly brackish and tidal fresh marshes and may influence local vegetation patterns (Leck and Crain 2009). Other herbivores of these marshes include white-tailed deer (*Odocoileus virginianus*) and woodchucks (*Marmota monax*), while harvest of trees and shrubs by beaver (*Castor canadensis*) alters plant composition and hydrology of some tidal swamps.

Besides herbivory by marsh animals, introduction of livestock to tidal wetlands for grazing significantly affects vegetation structure and productivity (see Beeftink 1977 for review). Salt marshes formerly and actively grazed by livestock had lower aboveground productivity than did ungrazed marshes (Reimold et al. 1975). Moderate sheep grazing in French salt marshes was found to increase plant species richness and diversity, while overgrazing or no grazing decreased diversity and species richness (Bouchard et al. 2003). Cattle grazing in salt marshes of northern Germany had similar effects, but also showed evidence of reduced sedimentation, vegetation shifts from high marsh to low marsh species, and reduced litter production (Andresen et al. 1990). Cessation of grazing increased sedimentation and colonization by upper marsh and adjacent grassland species.

The distribution of macroalgae on rocky shores, especially in tide pools, can be controlled by grazing, and grazing can be regulated by predators with interesting consequences for tide pool vegetation (Little and Kitching 1996). One experiment showed that when periwinkles (*Littorina* spp.) and limpets were removed and excluded from portions of rocky shores, those areas eventually became colonized and heavily vegetated with algae, whereas control areas remained barren (e.g., Bertness 1984a). Interestingly, algae in tide pools may determine the fate of periwinkles entering the pools as larvae. One study found that pools with Irish moss had plenty of periwinkles, whereas pools with hollow seaweed (*Enteromorpha* spp.) did not (Lubchenco 1978). Irish moss is too tough to eat, while hollow seaweed and other green algae are preferred foods for periwinkles. Green crabs (*Carcinus maenus*) feed on young periwinkles and gulls feed on the crabs. Since Irish moss does not provide crabs with adequate cover from gulls, these pools lacked crabs but supported many periwinkles. Hollow seaweed provides cover for crabs, so green crabs were found in these pools and their presence protected this algae by removing a major grazer (i.e., crabs eat young snails as they settle from the plankton). Moreover, when snails were removed from the Irish moss pools, hollow seaweed colonized and later dominated the pool, and when adult snails were added to the hollow seaweed pools, this algae was virtually eliminated from the pools. These are other examples of predator–prey relationships affecting coastal plant distributions.

Biological Competition

Once a plant has adapted to the physiochemical rigors of the intertidal environment, it faces competition from other similarly adapted species that compete for nutrients, light, and space. Competitors may be native (endemic) species or species of foreign origin (exotics). Interspecific competition plays a significant role in determining plant distribution in tidal wetlands. Poor competitors cannot successfully colonize sites with mild stressors as they are outcompeted by other species. They are therefore relegated to more stressed habitats and thrive only if they develop the necessary adaptations to cope with the stressor. When stress is limited, competition between species becomes more intense. With salt stress removed, freshwater marshes are therefore places where competition among species is keen, resulting in more diverse plant communities than those of salt marshes.

That salt marsh plants have freshwater origins is widely recognized, but they do not grow in freshwater areas because they are outcompeted by other species (Barbour 1970, 1978; Ungar 1998; Crain et al. 2004). Salt marsh plants transplanted to salt, brackish, and tidal freshwater marshes grow best in freshwater wetlands when neighbors are removed, which aptly demonstrates that competition keeps these halophytic species out of freshwater habitats (Crain et al. 2004). Armed with adaptations to tolerate salt stress, they thrive in saline soils where they have less competition. Interestingly, plant nurseries grow smooth cordgrass in freshwater due to higher survival and growth rates when compared to cultivation under saline conditions (Candy Bartoldus, pers. comm. 2005).

The frequency of flooding in the low marsh produces physiological constraints that are limiting for most salt marsh species thereby providing a place for smooth cordgrass to flourish (Bertness 1991b). The upper limit of smooth cordgrass was found to be limited mainly by competition from high marsh species including salt hay grass (Reed 1947; Bertness 1991b). Parasitic species can influence competitiveness. For example, heavy infestation by salt marsh dodder (*Cuscuta salina*) weakens pickleweed, the dominant low marsh species, allowing Parish's glasswort (which dominates the area above) to occupy lower levels of southern California salt marshes (Callaway and Pennings 1998).

When plant distribution in most eastern salt marshes is examined, a single species—smooth cordgrass—dominates the low marsh and only a few species occupy the interior marsh, while diversity increases along the upland border where salt stress is less and soils are better aerated. Biological competition in marsh border determines species composition (Figure 4.15). Since marsh border species become the dominant species in brackish marshes and limit smooth cordgrass to the creekbanks, they are better competitors than plants restricted to salt marshes. Since flooding and soil salinity are major stressors in salt marshes, only the most tolerant species colonize the regularly flooded low marsh and the high-salinity pannes and barrens. The reduction of salt stress plays a major role in impacting species distributions throughout the estuary.

In general, the cordgrasses dominating low marshes across North America are poor competitors in the salt marsh environment. In southern salt marshes, black needlerush eliminates smooth cordgrass from the high marsh (Pennings et al. 2005).

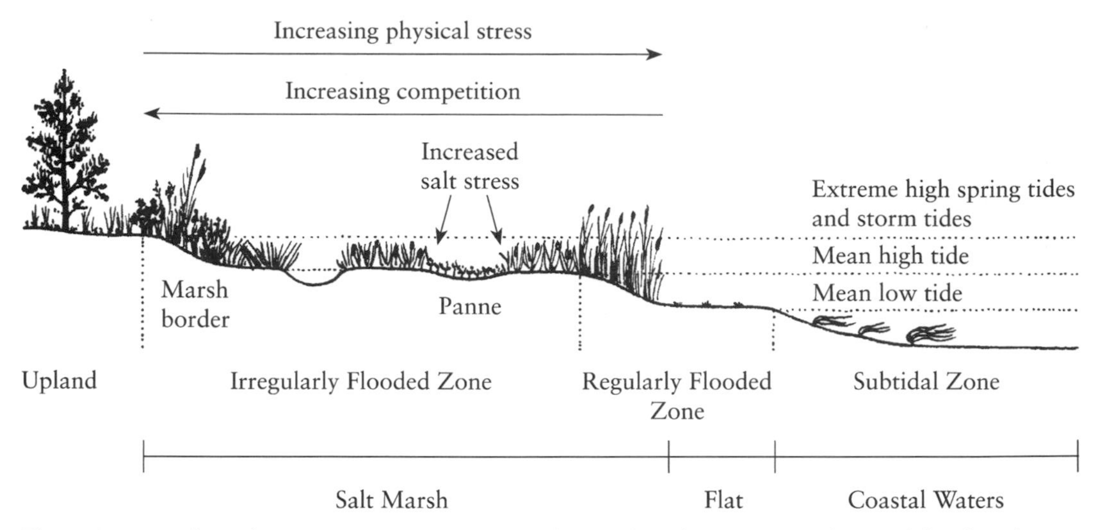

Figure 4.15. Biological competition increases across the marsh and is greatest at the marsh border where physical stress from salinity, inundation, soil saturation, and accompanying anaerobic conditions is less. Increased salt stress in pannes and salinas stunts plant growth. (Adapted from Emery et al. 2001)

The lower limit of black needlerush here is dictated by its tolerance to flooding and salinity as transplants of needlerush did poorly when planted in the smooth cordgrass zone, even when the cordgrass was removed. On the West Coast, salt marsh bulrush (*Schoenoplectus robustus*) is a better competitor in low salinity habitats and limits the distribution of Pacific cordgrass in these marshes (Pearcy et al. 1981). In salt marshes, this cordgrass grows best in areas with good tidal flushing. At other sites where soil salinity is higher, it is outcompeted by pickleweed, which can tolerate a wider range of salinities and soil moisture conditions (Zedler 1992). Studies have shown that when grown together, pickleweed reduces the density of Pacific cordgrass (Griswold 1988). Interestingly, the result of competition is different for their Gulf Coast relatives with smooth cordgrass suppressing the growth of Bigelow's glasswort (*Salicornia bigelovii*) in a Louisiana marsh (Proffitt et al. 2005). Common glasswort dominates the lower intertidal marshes of Europe (Adam 1993) but is nearly excluded from this zone in the eastern United States by smooth cordgrass. Glassworts, however, appear more common in this zone in northern New England and eastern Canada. Could it be that warmer climates favor smooth cordgrass, or are other factors involved? Does ice scour have an impact on the distribution of glasswort in northern waters by favoring an annual species? Seaside plantain appears to be a beneficiary of ice action in northern salt marshes (Chmura et al. 1997).

Black grass is not as salt tolerant as salt hay grass or salt grass, but it gets a competitive advantage by initiating growth earlier (in early February in southern New England) as compared with early May for the others (Bertness 1991a). This allows black grass to shade out the competition. Transplant experiments show that salt hay grass and salt grass placed in black grass vegetation were both suppressed unless black grass was removed, whereas when black grass was transplanted into salt hay grass and salt grass, black grass thrived (Bertness 1991a). This demonstrates that black grass is competitively dominant and able to suppress the other two marsh dominants. Consequently, salt hay grass and salt grass are relegated to the interior marsh, positioned between the smooth cordgrass–dominated low marsh and the upper marsh dominated by the more competitive black grass (Bertness and Leonard 1997). The thick turf formed by salt hay grass excludes many species. Wildlife managers in the southeastern United States burn this grass to provide opportunities for growth of desirable waterfowl food plants such as Olney's three-square and annual grasses (Gosselink 1984). Salt grass has the ability to produce runners that can quickly cover a bare surface, making it one of the pioneer species in the high marsh.

The occurrence of other species may enhance another species' ability to tolerate salt stress ("facilitation"). In particular, the presence of salt hay grass, black grass, marsh orach (*Atriplex patula*), seaside goldenrod (*Solidago sempervirens*), and high-tide bush may help more salt-sensitive species grow in salt marshes (Pennings et al. 2003). Shade provided by the canopy of salt-tolerant species reduces soil salinities and facilitates colonization by more salt-sensitive species (Bertness 1991a; Callaway 1994; Evanchuck and Bertness 2004). Bare ground exposes the marsh soil to increased evaporation and elevated salinities. While the seaward extent of high-tide bush in southern New England depends on the presence of the clonal tufts of black grass to grow in highly saline soils, along the upland border where salt stress is minimal, this shrub outcompetes and displaces the black grass (Bertness and Hacker 1994).

Competition is affected by nutrient and light availability as in other habitats. Fertilization experiments in New England found that stress-tolerators gained advantages over more salt-sensitive species when fertilized (Levine et al. 1998; Emery et al. 2001). Where black grass dominated mixed plots with salt hay grass in natural sites, when fertilized, salt hay grass flourished while black grass declined. In other mixed plots, the less abundant species (the competitive subordinate) became dominant after nutrient enrichment (salt grass over black grass, and smooth cordgrass over salt hay grass); however, this was the opposite of what was observed in the field. Competition for light may favor taller species. For example, when salt hay grass and black grass were transplanted in narrow-leaved cattail stands, they persisted only where cattails were removed (Leck and Crain 2009). Common reed invading English cordgrass (*Spartina anglica*) marshes eventually replaced most of this species, but when reed stems were cut out in patches, the remaining cordgrasses began sprouting shoots ("tillering") and flowering (Ranwell 1972).

INVASIVE SPECIES

Plants introduced into new regions have three likely outcomes: they perish, they achieve limited success, or they flourish. Invasive species, typically nonnatives, represent introduced forms of biological competition for native species. Given modern transportation systems and people's horticultural interests, the natural flora of most regions is under a biological siege of sorts. Invasive species are posing an increasing risk to native biota of wetlands and aquatic habitats around the globe, and coastal wetlands are no exception (see review by Zedler and Kercher 2004). Even though many tidal wetlands are salt-stressed habitats that are inhospitable to most plants, they are not without their foreign competitors (Table 4.7). Some of these plants have apparently come from people disposing of aquarium or aquatic garden plants. Rabbitfoot grass (*Polypogon monspeliensis*), an annual species, increases its abundance at the expense of native pickleweed during wet years and at sites subjected to increased freshwater flows from agricultural and urban runoff (Callaway and Zedler 1998). Efforts are under way to control all or most of these invasive species, but by the time most species are recognized as invasives, they usually have become well established in some locales and can be managed but not eradicated.

Of these, common reed is the most widespread and rampant in the Northeast, especially where tidal flow to salt marshes has

Table 4.7. Examples of invasive species in North American tidal wetlands

Common Name (*Scientific name*)	Wetland Type	Problem Area
Common reed (*Phragmites australis*)	Salt to fresh tidal marshes	Northeast
Purple loosestrife (*Lythrum salicaria*)	Oligohaline to fresh tidal marshes	Northeast, Pacific NW
Perennial pepperweed (*Lepidium latifolium*)	Salt marshes	New England, California
Japanese stilt-grass (*Eulalia vimineus*)	Tidal forested swamps	Northeast
Water chestnut (*Trapa natans*)	Tidal fresh waters	Northeast
Hydrilla (*Hydrilla verticillata*)	Tidal fresh waters	Northeast
Eurasian water milfoil (*Myriohyllum spicatum*)	Tidal	Northeast
Asiatic dayflower (*Murdannia keisak*)	Tidal fresh marshes	Mid-Atlantic, Southeast
Water lettuce (*Pistia stratiotes*)	Tidal fresh waters	Southeast
Water hyacinth (*Eichhornia crassipes*)	Tidal fresh waters	Southeast
Chinese tallow (*Triadica sebiferum*)	Brackish to fresh tidal marshes	Southeast
Rabbitfoot grass (*Polypogon monspeliensis*)	Salt and brackish marshes	California
Hybrid cattail (*Typha glauca*)	Brackish and tidal fresh marshes	Northeast and California
Redtop (*Agrostis stolonifera*)	Salt marshes	Oregon
Reed canary grass (*Phalaris arundinacea*)	Tidal fresh marshes	Northeast, Oregon
Yellow flag (*Iris pseudacorus*)	Tidal fresh marshes	Pacific Northwest
False indigo (*Amorpha fruticosa*)	Tidal fresh marshes	Pacific Northwest
Curly dock (*Rumex crispus*)	Salt marshes	Oregon
Brass-buttons (*Cotula coronopifolia*)	Salt marshes	Oregon
Salt marsh spurrey (*Spergularia salina*)	Salt marshes	Oregon
Smooth cordgrass (*Spartina alterniflora*)	Salt marshes	Puget Sound, California
Dense flower cordgrass (*Spartina densiflora*)	Salt marshes	Puget Sound, California
Common or English cordgrass (*Spartina anglica*)	Salt marshes	Puget Sound, California
Salt meadow cordgrass (*Spartina patens*)	Salt marshes	Puget Sound, California
Japanese eelgrass (*Zostera japonica*)	Marine waters	Puget Sound
Tamarisks (*Tamarix* spp.)	Salt marshes	California
Ripgut brome (*Bromus diandrus*)	Salt marshes	California
Oppositeleaf Russian thistle (*Salsola soda*)	Salt marshes	California
Slenderleaf iceplant (*Mesembryanthemum nodiflorum*)	Salt marshes	California
Hottentot fig (*Carpobrotus edulis*)	Salt marshes	California
Australian saltbush (*Atriplex semibaccata*)	Salt marshes	California

been reduced by undersized culverts or tide gates. Humans have played a major role in spreading this species (Bart et al. 2006). In just 20 years, common reed replaced 83 percent of the area occupied by native species on an undisturbed brackish marsh island in New Jersey (Windham and Lathrop 1999). Nutrient enrichment due to nitrogen loading from shoreline development is a major cause for its expansion in New England (Bertness et al. 2002). Also, common reed produces runners that have undoubtedly enhanced its invasive potential (Figure 4.16). Besides converting salt marshes to tall reed marshes, invasion by common reed changes soil properties: reducing microtopography (leveling of the land surface), surface soil salinity, and water levels as well as increasing soil redox potential. Although now recognized as an invasive species, this plant has been part of the American flora for a long time (Table 4.8). Evidence of common reed has been found in peat core sections estimated at 3,500 years old (Orson et al. 1987). Within the past decade, 11 varieties (haplotypes) of common reed native to North America (*Phragmites australis* ssp. *americanus*) have been identified (Saltonstall 2002). Why would a plant that has been part of the native flora suddenly become an aggressive

Figure 4.16. Common reed (*Phragmites australis*) is the most problematic of invasive species in northeastern salt marshes: (a) in flower and (b) runners help this species rapidly expand its presence in newly colonized sites.

weed and a threat to other plant communities? There has been much development of tidal wetlands since the 1950s, and the disturbed soils adjacent to filled wetlands appear to be particularly well-suited for this grass. Did the plant genetically mutate into a more aggressive form? Current research suggests that the invasive form (haplotype M) is an introduction related to Eurasian types to which it has genetic similarities (Saltonstall 2002). Moreover, this form does not have any genetic links to the 11 North American haplotypes. Type M is believed to have arrived in eastern U.S. ports during the early 1800s, existing in small populations that later expanded through human dispersal aided by railroad and road construction in the late 1800s and early 1900s. Since it is still actively expanding, researchers predict that it will continue to spread westward and northward on the continent. Besides introduction of the type M reed, altered hydrology, soil disturbance, pollution, and coastal development are factors involved in the spread of common reed (Meyerson et al. 2000).

Smooth cordgrass, a native on the Atlantic and Gulf Coasts, is among four cordgrasses introduced to five estuaries on the Pacific Coast (Daehler and Strong 1996). On the latter coast, it is viewed as an aggressive species that needs to be controlled because it is colonizing and expanding over intertidal mudflats at an alarming rate, reducing this already declining important shorebird habitat. In England, smooth cordgrass first appeared in the early 1800s, the likely result of seeds contained in shipping ballast (Thompson 1991). Initially it hybridized with the native cordgrass (*S. maritima*) to produce a sterile hybrid (*S.* x *townsendii*). Around 1890, the hybrid produced a new and fertile species, English cordgrass (*S. anglica*), by chromosome doubling. While smooth cordgrass and the sterile hybrid were not invasive, English

Table 4.8. Distinguishing features for separating native from introduced forms of common reed (*Phragmites australis*)

Feature	Differences
Persistence of leaf sheaths in winter	Native: they are non-persistent as most are shed by late fall Introduced: they persist and remain around stem
Stem surface from late season to summer (possibly longer)	Native: glossy above lower nodes from August to following June Introduced: dull, often with ridges, but can be glossy on rhizomes and on lower internodes
Stem color above lowest three or four internodes	Native: red from mid-July into October Introduced: never red
Stem spots	Native: spots typical Introduced: occasional spots observed

Source: B. Blossey, personal communication 2011.
Note: Persistence of the leaf sheaths is the best indicator for separating the two forms.

cordgrass could spread rapidly by seed dispersal. Today it is characteristic of British salt marshes. Like the Pacific Coast experience with smooth cordgrass, English cordgrass found an unoccupied niche in which to flourish. Its ability to promote sediment accretion is the foundation for its success in the low intertidal zone.

Human Actions

Human-induced impacts on wetland vegetation have been both direct and indirect, and mostly destructive. Changes in wetland hydrology, salinity, nutrient availability, the integrity of the natural environment, and coastal processes alter vegetation patterns along the coast. Some of these impacts affecting plant growth and survival are briefly discussed below, while the more destructive ones that eliminate or significantly degrade wetlands (e.g., filling, ditching, impounding, and hydrologic alteration) are addressed in Chapter 8.

Wetland plants are impacted by

- point and nonpoint source pollution (nutrient enrichment and chemical contamination from direct discharges and runoff),
- damming of rivers (reduces freshwater flows and sediment delivery and alters salinity patterns in estuaries),
- road and railroad crossings (modifies local hydrology),
- encroachment on wetland buffers by development, and
- climate change.

Multiple factors are usually responsible for vegetation changes. For example, several factors including dredged spoil disposal, landfills, bulkheading, damming of tributaries, and diking contributed to the extirpation and range reduction of brackish to tidal fresh intertidal plants in the Delaware River (Ferren and Schuyler 1980). Increased freshwater inflows from urban development allowed brackish and riparian species to colonize a California salt marsh (Greer and Snow 2003). Sedimentation from upland development and agriculture into estuaries has increased elevations of mudflats allowing for establishment of marsh vegetation (e.g., Ward et al. 2003). Increased elevation and changes in soil texture from silty clay to sands have lowered soil salinities and allowed colonization of a California salt marsh by brackish and freshwater species (Byrd and Kelly 2006). Interestingly, increased sedimentation from regional deforestation and poor agricultural practices led to the formation and expansion of tidal wetlands in many North American estuaries (see Chapter 2).

The effects of oil spills on wetlands

are often localized but can be persistent (Webb et al. 1981). More than 20 years after a relatively small oil spill in Falmouth, Massachusetts, high concentrations of oil still remain in subsoil (below 2.4 in [6 cm]) and erosion losses continue (Reddy et al. 2002). Plant sensitivity to fouling by oil varies among species and among populations within a species, by age of the plant, pre-existing condition of the plant (stressed or not), season of the spill, and soil organic matter content (Lin and Mendelssohn 1996; Pezeshki et al. 2000). For example, ecotypes of salt hay grass and smooth cordgrass from various locations along the Gulf of Mexico respond differently to oil fouling. Soil organic matter plays a key role in facilitating the penetration of oil into marsh substrates. High concentrations of oil in sediment can kill marsh plants and limit plant growth (Krebs and Tanner 1981; Baker 1989). Plants surviving initial oiling may die later if plant gas-exchange is impaired. Oil spill cleanup operations may pose additional harm to plants (Hoff 1995), especially the spraying of oiled fucoid algal beds with hot water and the burning of oiled marsh vegetation (Fukuyama et al. 1998; Lin et al. 2005). A study of oil effects on smooth cordgrass in Louisiana suggests that this species could tolerate oiling and that cleanup was not warranted (DeLaune et al. 1984). The best benefits achieved from cleanup would be removing as much oil as possible from the soil. Plant recovery may eventually take place without any restoration: vegetation returned to an oiled but uncleaned marsh in Brittany, England, within five years (Baca et al. 1987).

Spoil deposition on marshes was a common practice associated with dredging operations in harbors and along portions of the Atlantic Intercoastal Waterway. Depending on the change in marsh elevation, salt marsh vegetation may recover from such disturbance or be replaced by upland species. A study of the effects of spoil deposition on a New Jersey salt marsh found that if the deposit was less than 2 inches (5 cm) thick, the marsh recovered within two years, but thicker deposits prolonged the bare substrate condition and eventually led to colonization by shrubs (Burger and Shisler 1983).

Nitrogen and phosphorus inputs from human development (e.g., runoff from urban, residential, and agricultural lands, and discharge of treated wastewater) are causing eutrophication of U.S. coastal waters (Bricker et al. 1999). Tidal marshes tend to be nitrogen- and possibly phosphorus-limited ecosystems (Valiela and Teal 1974; Teal 1986; Burdick et al. 1989; Whigham and Nusser 1990; Kiehl et al. 1997; Frost et al. 2009). As previously noted, additions of nitrogen to tidal wetlands from urban development and other sources may give certain species advantages over others, leading to changes in plant community structure and some functions over time (Silliman and Bertness 2002). For example, the extent and density of salt hay grass were significantly reduced in Rhode Island salt marshes in watersheds with heavy residential development and corresponding high nitrogen inputs (Wigand et al. 2003). High nitrogen loads also allowed smooth cordgrass to replace salt hay, and common reed to eliminate black grass and numerous wetland forbs in the upper high marsh (Bertness et al. 2002). Direct application of wastewater to salt marshes will likely alter vegetation patterns especially in semiarid regions. When wastewater was added to a restored salt marsh in Texas, saltwort increased at the expense of perennial glasswort (Forbes et al. 2008).

Constant foot traffic across marshes creates minor impacts to vegetation, typically leaving a worn trail across the marsh. Riding recreational vehicles (e.g., all-terrain vehicles, trail bikes, and mountain bikes) through tidal marshes can create enormous damage. For example, some Australian salt marshes have lost half of their vegetative cover due to heavy use by recreational vehicles (Adam 2002). Only one or two passes of the vehicle may kill more sensitive species such as perennial succulents, while extensive use creates deep ruts that alter drainage patterns. Damage may be permanent or take many years for plants to recover, especially in northern climates and in sediment-starved wetlands.

Picking flowers may impact some species since biomass and flowers (vital for sexual reproduction and seed production) are removed. Sea lavender is collected by people in various locales for floral arrangements and wreath-making. Concern about overharvesting has led some towns to outlaw the picking of sea lavender. In Nova Scotia where sea lavender harvest is a local industry, a study has shown that the sustainable harvest rate would be 16 percent of the inflorescences (Baltzer et al. 2002). Since harvests in salt marshes tend to be more than twice this rate, current harvest rates may have a long-term effect on sea lavender populations but are not likely to eliminate the species. Mature plant survivorship is high, and harvesting does not affect plant growth. Nonetheless, sustainable use suggests harvesting at lower levels than are currently practiced in the province.

Further Readings

The Biology of Aquatic Vascular Plants (Sculthorpe 1967)

Ecology of Salt Marshes and Sand Dunes (Ranwell 1972)

The Biology of Halophytes (Waisel 1972)

The Ecology of Halophytes (Reimold and Queen 1974)

A Treatise on Limnology. Volume III: *Limnological Botany* (Hutchinson 1975)

Saltmarsh Ecology (Long and Mason 1983)

"Adaptations of plants to flooding with salt water" (Wainwright 1984)

Salinity Tolerance of Plants of Estuarine Wetlands and Associated Uplands (Hutchinson 1988)

Saltmarsh Ecology (Adam 1993)

Concepts and Controversies in Tidal Marsh Ecology (Weinstein and Kreeger 2000)

Wetland Plants: Biology and Ecology (Cronk and Fennessy 2001)

Ecology of Tidal Freshwater Forested Wetlands of the Southeastern United States (Conner et al. 2007a)

5 Tidal Wetland Types and Their Vegetation

Many types of tidal wetlands occur across North America from the edge of the ocean to a point where tidal action ceases in coastal rivers (Figure 5.1). Tidal wetlands include both vegetated and nonvegetated types. The latter occur either at lower elevations in the intertidal zone or above mean high tide in regions where salts accumulate and raise soil salinities well above sea strength to levels that prevent colonization by macrophytic plants. Tidal wetlands can be classified in numerous ways based on a wide range of properties (see Chapter 1). For this chapter, I have divided them into types based on differences in the presence or absence of vascular plants, the nature of the substrate, the life form of the dominant vegetation, and the salinity of their associated tidal waters. Ten general types are described: beaches, tidal flats, rocky shores, salt marshes, estuarine shrub swamps, estuarine forests (including mangrove swamps), brackish marshes, tidal fresh marshes, tidal freshwater swamps, and coastal aquatic beds. While the descriptions emphasize tidal wetlands in the northeastern United States and eastern Canada, similar types elsewhere in the United States are addressed under the subheading Regional Differences and through the use of tables. Macroalgae are referenced where they dominate the habitat (e.g., rocky shores). Animals frequenting tidal wetlands are covered in Chapter 6. *When reading the descriptions of coastal wetland types, readers must keep in mind the range limits of plant species, as some will not occur in certain estuaries of the northeastern United States or eastern Canada.* (*Note:* Two companion books—*Field Guide to Tidal Wetland Plants of the Northeastern United States and Neighboring Canada* [Tiner 2009] and *Field Guide to Coastal Wetland Plants of the Southeastern United States* [Tiner 1993a]—provide illustrations and descriptions including geographic ranges for most of the plants mentioned in this chapter.)

Beaches

In general, beaches are gently sloping sandy to cobbly shorelines found along the ocean and in coastal embayments (Figure 5.2). They may be miles long or restricted to narrow fringes along rocky headlands. The most familiar beaches are the sandy beaches of spits and barrier islands, but beaches may also be composed of cobble or gravel. These latter types are typical of rocky coastlines. Some cobble beaches have been called "musical beaches" for the sound made when the surf strikes the cobbles or when the swash recedes and the cobble stones strike one another, producing a rattling sound.

Sandy beaches are most common south of Portland, Maine, and form a nearly continuous band along the Atlantic shoreline from Cape Cod south. From Long Island south, they occupy the oceanfront intertidal zone of a series of barrier islands. In northern

Figure 5.1. Aerial view of wetlands along Connecticut's Long Island Sound shore.

New England and eastern Canada, many beaches are cobble-gravel shorelines typified by rather large rounded cobbles. Sandy beaches, however, comprise much of the southern shore of the Gulf of St. Lawrence including Prince Edward Island, the Magdalen Islands, Miscou Island, and the Northumberland Strait as well as sections of the southern shore of the St. Lawrence River.

Two zones are frequently described for beaches: 1) the foreshore (flooded frequently by the tides), and 2) the backshore (directly in front of the primary dunes and flooded by storm tides). Some beaches have only foreshore areas. The foreshore includes both sloping and nearly level landforms, while the backshore is more or less flat. The foreshore slope encompasses the swash zone where waves wash up and down the shore. A terrace or berm forms above this zone where sediments build up. Some beaches lack the berm, while in others the shape and number of berms may change from summer to winter due to seasonal beach erosion and accretion patterns (Leatherman 1979). In the Northeast, winter beaches tend to be narrower and steeper than summer beaches because winter storms generate higher tides, causing more erosion. Summer wave action is gentler, returning the sand to rebuild the beaches. Hurricanes and tropical storms from late summer to fall can have dramatic impacts on beach morphology.

Beaches are largely devoid of plant growth due to frequent wetting and drying,

a

b

c

Figure 5.2. All beaches are not alike: (a) gravel beach (Bay of Fundy, NS); (b) coarse, sandy beach (Cape Cod Bay, MA, with broad sand flats); and (c) broad, fine, sandy beach (GA).

wave action, lack of fine sediments, low nutrient availability, salt stress, temperature extremes, and drainage properties (of the upper beach). A few plants grow on the backshore at the toe of primary dunes where tidal litter ("wrack") of seaweeds and other plants often collect (Table 5.1). This wrack helps improve soil moisture, organic matter, and nutrient availability and moderates substrate temperatures, creating favorable conditions for seed germination, plant growth, and reproduction. American beach-grass (*Ammophila breviligulata*), the major dune-stabilizing species in the region, and seaside goldenrod (*Solidago sempervirens*) may also occur in the zone. In Canada, American dune grass (*Leymus mollis*) and sea lyme-grass (*L. arenarius*) may grow down to the upper edges of the beach (Couillard and Grondin 1986). On northern beaches with cobbles, pebbles, and sands, oysterleaf or seashore bluebells (*Mertensia maritima*), marsh orach (*Atriplex patula*), salt marsh sand spurrey (*Spergularia salina*), seaside plantain (*Plantago maritima*), ragweed (*Ambrosia artemisiifolia*), and scotch lovage (*Ligusticum scoticum*) may also be found.

Regional Differences. About one-third of the North American shoreline is beach with 23 percent associated with barrier islands, 2 percent with rocky headlands, and 8 percent being pocket beaches in coves (Dolan et al. 1972). Most beaches are largely composed of quartz sand, with feldspar being next in abundance. The best development of sandy beaches is in areas where huge quantities of sediments are available, and this occurs mainly along coastal plains and northern areas with glacial drift deposits (Davis 1978). Northern beaches tend to be composed of coarse sands or cobble-gravel substrates, while southern beaches are made up of fine sands with ground-up shells.

Table 5.1. Some plants growing on sandy beaches

Common Name (*Scientific name*)	Range
Redroot (*Amaranthus retroflexus*)	Newfoundland to Florida and Texas (tropical America native)
Beach wormwood (*Artemisia stellariana*)	Quebec to Virginia; also Florida and Louisiana (native of Japan)
Seabeach orach (*Atriplex cristata*)	New Hampshire to Florida and Texas
Marsh orach (*Atriplex patula*)	Prince Edward Island and Nova Scotia to Florida
Sea rocket (*Cakile edentula*)	Labrador to Florida and Louisiana
Sandbur (*Cenchrus tribuloides*)	Maine to Florida and Louisiana
Seashore spurge (*Chamaesyce polygonifolia*)	Quebec and Prince Edward Island to Florida and Alabama
Lamb's-quarters (*Chenopodium album*)	Newfoundland to Florida and Louisiana
Fireweed or Pilewort (*Erechtites hieracifolia*)	Newfoundland to Florida and Texas
Seabeach sandwort (*Honckenya peploides*)	Arctic to Virginia
Beach pea (*Lathyrus japonicus*)	Greenland to New Jersey
Carpetweed (*Mollugo verticillata*)	Quebec and Nova Scotia to Florida and Texas (tropical America native)
Bitter panicum (*Panicum amarum*)	Rhode Island and Connecticut to Florida and Texas
Seabeach dock (*Rumex pallidus*)	Newfoundland and Quebec to New York
Russian thistle (*Salsola kali*)	Newfoundland to Florida and Texas (Eurasian native)
Seabeach groundsel (*Senecio pseudo-arnica*)	Labrador and Newfoundland to Maine
Woodland groundsel (*Senecio sylvaticus*)	Newfoundland to Maine; also Massachusetts and New Jersey (European native)
Slender sea purslane (*Sesuvium maritimum*)	New York to Florida and Texas
Purple sand-grass (*Triplasis purpurea*)	Maine to Florida and Texas
Common cocklebur (*Xanthium strumarium*)	Maine to Florida and Texas

Source: Range along the Atlantic and Gulf Coasts according to Tiner (2009).
Note: Some of these plants are more characteristic of sand dunes, but occur in the uppermost tidal zone of beaches.

From Cape Hatteras south, beach sand has a high shell fragment, typically 10 percent or more, whereas northern beaches contain about 1 percent shell (Neal et al. 2007).

Tidal Flats

Tidal flats are barren areas in the intertidal zone. As their name suggests, tidal flats are nearly level landforms, in contrast to the sloping beaches, although some may slope noticeably. In northern regions, they generally form in somewhat sheltered embayments and along tidal rivers at elevations below tidal marshes, swamps, or rocky headlands (Figure 5.3). In southern and tropical regions, tidal flats may also become established above the reach of daily tides within irregularly flooded salt marshes where high salinities restrict plant growth. Where they occur below tidal marshes and swamps, tidal flats are usually inundated (regularly flooded) twice daily. Substrates of tidal flats are variable, including mud, clay, sand, cobble, gravel, and various mixtures. Like all tidal wetlands, intertidal flats range in size from narrow fringes to vast expanses of exposed substrates at low tide. The latter often exhibit a dendritic network of tidal drainage-ways when situated in sheltered embayments, or a series of low ridges parallel to the shoreline in exposed areas in front of beaches where exposed to wave action.

Tidal flats are the most extensive coastal wetland type in macrotidal waters. From Maine north, vast mudflats occupy the lower intertidal zone. Along the Bay of Fundy where tides of 50 feet or more (>16 m) expose huge areas at low tide, mudflats up to 2.5 miles (4 km) wide can be found. Like beaches, most tidal flats are largely devoid of macrophytic vegetation, yet macrophytes may dominate some flats. Eel-grass (*Zostera marina*), typically a submerged aquatic throughout the eastern United States, colonizes and even forms extensive beds on some tidal flats in northern waters (e.g., Nova Scotia according to Sharp and Semple 2004). Aquatic bed plants may be exposed by the lowest annual tides or by wind-driven blowouts (offshore winds). Regularly flooded flats along marine waters may support macroalgae such as rockweeds (*Fucus* spp.), knotted wrack (*Ascophyllum nodosum*), sea lettuce (*Ulva lactuca*), and hollow green seaweeds (*Enteromorpha* spp.), or microscopic diatoms. The first two algae predominate in colder waters of northern New England and Canada where they attach to rocks scattered on the muddy substrate. Several species of kelp occupy the lowermost shores. Diatoms sometimes give a golden sheen to the mudflat surface at low tide, and the common mud snail (*Ilyanassa obsoleta*) may be the most conspicuous organism of flats where hundreds, if not thousands, of individuals may be observed. Mussel reefs (blue mussel [*Mytilus edulis*]) may form on flats where rocks, stones, and other debris are abundant (Figure 5.4a). In areas of increased sedimentation, isolated clumps of smooth cordgrass may occasionally be found on the flats; they may represent the beginning of salt marsh formation or expansion of an existing marsh. Tidal flats along brackish and tidal freshwaters also appear as broad mudflats in most places in winter, but they may support a number of vascular plants (Table 5.2).

Regional Differences. Regularly flooded tidal flats occur along all coasts wherever the daily tides expose nonvegetated substrates. Along the Pacific Coast, at least two mussels form intertidal beds: the blue mussel and the California mussel (*M. californicus*), which attach to exposed rocks (Fukuyama et al. 1998). In the southeastern United States, intertidal reefs composed of the shells of eastern oyster (*Crassostrea virginica*) may occupy tidal flats; the reefs may be 2 feet or more in height (Figure 5.4b). In tropical waters, red mangrove seedlings often can be observed growing out in the flats, eventually extending the mangrove swamps seaward. From North Carolina into the tropics, nonvegetated tidal flats called "salt barrens," "salt flats," or "salinas" develop within

a

b

c

Figure 5.3. Examples of tidal flats: (a) gravel flat (Bay of Fundy, NB); (b) cobble-dominated flat in front of sandy beach (ME); and (c) mudflat (MA).

Table 5.2. Examples of vascular plants occupying tidal flats along brackish and tidal freshwaters in the northeastern United States

Common Name (*Scientific name*)	Range
Long's bitter-cress (*Cardamine longii*)	Maine to North Carolina
Pygmyweed (*Crassula aquatica*)	Quebec and Newfoundland to Maryland
Slender flatsedge (*Cyperus bipartitus*)	New Brunswick and Maine to Georgia
American waterwort (*Elatine americana*)	Quebec and New Brunswick to Georgia
Dwarf spike-rush (*Eleocharis parvula*)	Newfoundland to Florida and Texas
Parker's pipewort (*Eriocaulon parkeri*)	Quebec and Maine to North Carolina
Virginia hedge hyssop (*Gratiola virginiana*)	Quebec to Georgia and Florida
Water star-grass (*Heteranthera dubia*)	Quebec and New Brunswick to Florida and Texas
Kidney-leaf mud plantain (*Heteranthera reniformis*)	Connecticut to Florida and Texas
Riverbank quillwort (*Isoetes riparia*)	Quebec and Maine to South Carolina
Eastern lilaeopsis (*Lilaeopsis chinensis*)	Nova Scotia to Florida and Texas
Mudwort (*Limosella subulata*)	Quebec and Newfoundland to North Carolina
False pimpernel (*Lindernia dubia*)	Quebec and Nova Scotia to Florida and Texas
Water purslane (*Ludwigia palustris*)	Nova Scotia to Florida and Texas
Nuttall's mudflower (*Micranthemum micranthemoides*)	New York to Virginia
Golden-club (*Orontium aquaticum*)	Massachusetts to Florida and Texas
Heart-leaf plantain (*Plantago cordata*)	New York and Virginia
Seaside crowfoot (*Ranunculus cymbalaria*)	Labrador to New Jersey
Tidal sagittaria (*Sagittaria calycina* ssp. *spongiosa*)	Quebec and New Brunswick to North Carolina
Grass-leaved arrowhead (*Sagittaria graminea*)	Newfoundland and Labrador to Florida
Stiff arrowhead (*Sagittaria rigida*)	Quebec and Maine to Virginia
Awl-leaf arrowhead (*Sagittaria subulata*)	Massachusetts to Florida and Mississippi
Bluntscale bulrush (*Schoenoplectus smithii*)	Quebec and New Brunswick to Georgia

Source: Range along the Atlantic and Gulf Coasts according to Tiner (2009).

the irregularly flooded zone of salt marshes or mangrove swamps. They are discussed below under Salt Marshes.

Rocky Shores

Rocky shores dominate the coast of northern New England and adjacent Canada, especially along the Bay of Fundy, Cape Breton, the Atlantic Coast of Nova Scotia, the northern shore of the Gulf of St. Lawrence, Newfoundland, and Labrador (Figure 5.5). They can also be found in southern New England (e.g., Narragansett Bay, Rhode Island, and Norwalk Islands, Connecticut) and along the north shore of Long Island. Artificial rocky shores, namely "jetties" and "groins," have been constructed in many seaside communities along the entire Atlantic Coast to keep navigable channels (inlets) open and in an attempt to maintain private and public beaches, respectively. Other man-made rocky shores (e.g., rip-rap) have been built to protect eroding shorelines and developed lands.

Rocky shores exhibit a zonation pattern related to the frequency and duration of tidal flooding and the degree of exposure to air (desiccation) (Figure 5.6). The uppermost zone of rocky shores (the splash zone) is only occasionally flooded by the highest tides—storm tides. It is largely barren except for a few scattered vascular plants rising up between cracks in the rocks. On the North Atlantic Coast, these plants include seaside plantain (*Plantago maritima*), beachhead iris (*Iris setosa* var. *canadensis*), and seaside goldenrod. With decreasing elevation, there is increased submergence and less exposure to air. The uppermost level of note is a cyanobacteria (*Calothrix* spp. ["blue-green algae"]) zone, resembling a

a

b

Figure 5.4. Two common intertidal reefs in the eastern United States: (a) mussel reef (MA) and (b) oyster reef (SC).

painted black band across the rocks. Immediately below is the barnacle zone where blue mussels may also occur. The thick macroalgal zone begins below that with the occurrence of rockweeds. A rockweed lacking air bladders—spiral rockweed (*Fucus spiralis*)—occupies the uppermost portion of the rockweed zone, while rockweeds with air bladders—knotted wrack (*Ascophyllum nodosum*) and bladder wrack (*Fucus vesiculosus*)—dominate the middle region. Other common macroalgae in this zone are two red algae (*Rhodophyta*)—Irish moss (*Chondrus crispus*) and laver (*Porphyra* spp.)—and green algae (*Spongomorpha* sp. and *Cladophora* spp.). Sea wrack

(*Fucus edentatus*) is found at the lowest intertidal levels and shallow subtidal waters. Irish moss extends from the lower intertidal zone into deep water. Calcareous algae (*Lithothamnium* spp. and *Corallina officinalis*) may also occur in the lower tidal zone (Davis and Browne 1996). Knotted and bladder wracks and Irish moss are the predominate algae of rocky shores south of Cape Cod (Gosner 1971). Kelps are more characteristic of subtidal areas, but some species such as horsetail kelp (*Laminaria*

a

b

Figure 5.5. Examples of rocky shores: (a) typical fucoid-covered rocky shore (ME) and (b) bedrock shore along the St. Lawrence River (QC).

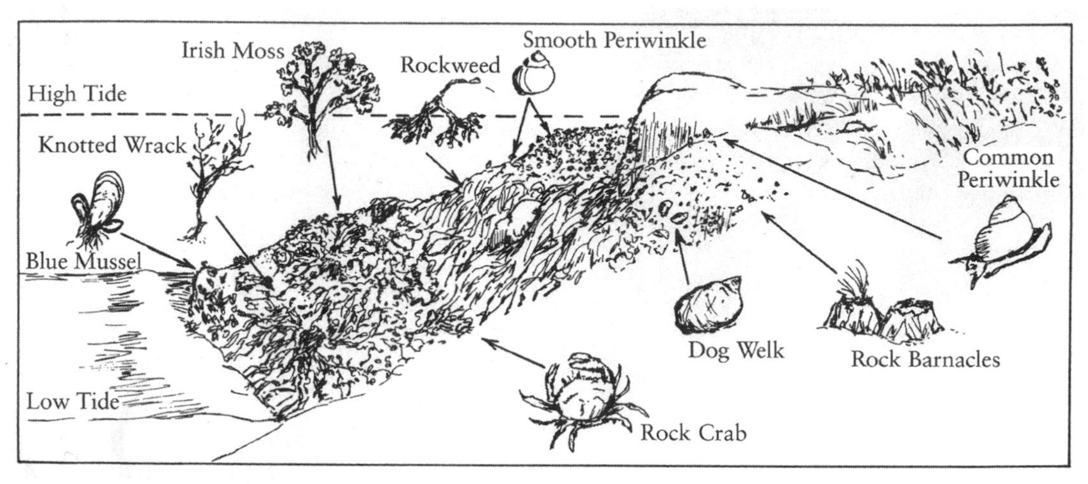

Figure 5.6. Sketch showing general zonation of the intertidal rocky shore. (Division of Wetlands 1978)

digitata) and sea colander (*Agarum clathratum*) may occupy the lower intertidal zone (Gregory et al. 1998).

Tidal pools of variable size and shape (water-filled depressions in stone that remain inundated at low tide) can be found on many rocky shores. They are among the most interesting places for nature observation for children of all ages, since they harbor many types of marine invertebrates (e.g., barnacles, snails, limpets, and crabs plus starfish and sea urchins on northern shores) and algae. (*Note:* When I was mapping tidal wetlands in Connecticut in the early 1970s, I saw hundreds of one of the smallest macroinvertebrates, a bluish-gray to somewhat blackish springtail [*Anurida maritima;* ⅛ inch long] in small tidal pools on a rocky outcrop. This was an impressive sight as the animals appeared to be walking on the water surface because the stiff, thick hairs covering their bodies facilitated this movement.)

Regional Differences. Rocky shores are also common along the Pacific Coast where they exhibit a similar zonation of plants from the upper intertidal to the lower intertidal level, although with different species. Interestingly, vascular plants called surfgrasses (*Phyllospadix torreyi, P. scouleri, P. serrulatus*) can be found in the lower intertidal zone of rocky shores along the Pacific Coast from California into Alaska. Limestone outcrops along the northeastern Gulf Coast create a unique type of intertidal rocky shore. Unlike the sloping rocky shores of northern waters, they are raised platforms marked by pitted and broken surfaces. Intertidal coral reefs may form hard shores in the tropics.

Salt Marshes

Salt marshes are intertidal saline environments colonized by grasses and other salt-tolerant plants ("halophytes"). While herbaceous plants predominate, shrubs are often present along the margins, and algae are abundant on the soil surface, on the lower parts of some marsh plants, and in the shallow depressions known as pannes. Along the Atlantic Coast, these marshes become the predominant coastal wetland type from York County, Maine, south, occurring as a broad band behind barrier beaches and islands, along the shores of protected embayments, and at the mouths of coastal rivers in high-salinity waters.

The occurrence of plant species in zones is striking and has been reported for salt marshes across North America since the early days of plant ecology (e.g., *eastern Canada,* Ganong 1903; *eastern U.S.,* Shaler 1886; Transeau 1909; *Connecticut,* Nichols

1920; Miller and Egler 1950; *Long Island,* Transeau 1913; Johnson and York 1915; Conard 1924; Conard and Galligar 1929; *New Jersey,* Harshberger 1909; *North Carolina,* Kearney 1900; Wells 1928; *Florida,* Laessle 1942; *Louisiana,* Penfound and Hathaway 1938; *Pacific Coast,* Purer 1942). The frequency and duration of tidal flooding, salinity gradients, and related factors strongly influence salt marsh vegetation patterns. Two vegetation zones are apparent from the water's edge to the upland: low marsh (or regularly flooded zone) and high marsh (or irregularly flooded zone). The low marsh occurs just below the mean high water level where it is flooded at least once a day. The high marsh begins above this level and is flooded weekly, monthly, or less frequently depending on elevation, tides, and other drivers. Plant composition varies within these zones and with latitude, whereas vegetation patterns shift over time due to changes in sea level, sedimentation rates, drought, and other factors (Chapter 4). In the Bay of Fundy region, the low marsh is submerged from 13 to 15 percent of the year and maximum flood duration is 4 hours per tide, while the high marsh is flooded only 1 to 2 percent of the time with a maximum duration of 2 hours per tide (Gordon and Cranford 1994). In Rhode Island, the low marsh is flooded every day for a few hours, while the lower high marsh is flooded for some time on most days (24 days per month), the middle high marsh for a little more than half the days each month (17 days), and the upper high marsh may be inundated on roughly 9 days during the month (Bertness and Ellison 1987).

Low Marsh

The low marsh represents the lowest zone where emergent (free-standing) herbaceous plants grow along the shore. This zone is normally flooded every day by the tides ("regularly flooded"). Along most of the Atlantic Coast, the low marsh is characterized by a virtual monoculture of smooth cordgrass (*Spartina alterniflora*). Here, it grows in a "tall form" to a height of 6 feet (2 m) or taller, although in the more northern waters of Canada's St. Lawrence estuary, smooth cordgrass may grow to only 3 to 4 feet (~1 m) tall in this zone. The low marsh typically forms a narrow band along tidal creeks in the New England and Mid-Atlantic states but may be wider in some places (Figure 5.7a). This is in marked contrast to the marshes behind the sea islands of South Carolina and Georgia where the low marsh is the predominant marsh type covering miles of estuarine habitat between the mainland and offshore barrier islands (Figure 5.7b). In eastern Canada, there appears to be more low marsh than in the northeastern United States; 40 to 61 percent of the marshes may be this type (Wells and Hirvonen 1988). (*Note:* They may be including the lower high marsh dominated by the short form of smooth cordgrass in this zone.) The low marsh is among the most productive wetlands in the world, producing more organic matter than a Midwest cornfield (see Chapter 7).

While the low marsh appears to be represented by a single species in most of the Northeast and Canadian Maritimes, a few other vascular plants may also be found here including sea lavender (*Limonium carolinianum*), salt hay grass (*Spartina patens*), common glasswort (*Salicornia maritima*), sea blites (*Suaeda* spp.), salt marsh sand spurrey (*Spergularia salina*), and dwarf spike-rush (*Eleocharis parvula*). In the Canadian Maritimes, alkali grasses (*Puccinellia* spp.) and marsh orach (*Atriplex patula*) have also been reported in this zone (Gordon et al. 1985; Couillard and Grondin 1986; Hanson and Calkins 1996). In Newfoundland and Labrador, smooth cordgrass and common glasswort are the main species of the low marsh, with the latter forming stands below the smooth cordgrass (Wells and Hirvonen 1988). Other low marsh associates in these northern waters include alkali grasses (*Puccinellia paupercula, P. phryganodes*), salt marsh sedge (*Carex recta*), seaside plantain

(*Plantago maritima*), seaside arrow-grass (*Triglochin maritima*), and salt marsh spike-rush (*Eleocharis halophila*). Pools in the low marsh may support eel-grass.

From Long Island north, morphological variants ("ecads") of brown algae species—knotted wrack and/or rockweeds—may occur at the seaward edge of the low marsh tangled around the stems of smooth cord-grass. Unlike the typical forms, these ecads are stunted (dwarf) and many-branched, lack a holdfast, and are rooted in the soil at the base of the cordgrasses (Brinkhuis 1976; Wallace et al. 2004). Both plants

a

b

Figure 5.7. In the Northeast, the low marsh is typically a narrow band of tall-form smooth cordgrass between the high marsh and tidal waters, while in the South Atlantic, the regularly flooded low marsh is the dominant salt marsh type: (a) low marsh and high marsh (NY) and (b) low marsh (SC).

benefit from their relationship as the thick algal carpet provides nutrients to enrich soils for cordgrass growth, while the smooth cordgrass canopy protects the algae from desiccation, photoinhibition, and biomass loss during storms (Gerard 1999). Ribbed mussels (*Geukensia demissa*) may also form colonies along low marsh banks.

High Marsh

The high marsh is represented by several habitats with varying environmental stresses: lower high marsh, middle high marsh, upper high marsh, and marsh border (Figure 5.8). These habitats make the high marsh more floristically diverse than the low marsh. From a distance, the marsh may appear as a broad vegetated flat plain—a low grassland, yet closer inspection will reveal slight topographic changes. Since small changes in elevation can have enormous impacts on plants, the distribution of plants in the high marsh may appear as more of a mosaic pattern represented by patches of characteristic species rather than a simple zonal change from creekbank to upland as it is often illustrated.

Throughout the northeastern United States the most prominent high marsh species are the short growth form of smooth cordgrass (less than 1½ feet tall), salt hay grass, salt grass (*Distichlis spicata*), and black grass (*Juncus gerardii*) (Figure 5.9). From Maine north, forbs also become dominant members of the high marsh community including seaside plantain (goose-tongue), arrow-grasses (typically *Trigochin maritima;* rarely, *T. palustris* and *T. gaspensis*), sea milkwort (*Glaux maritima*), silverweeds (*Argentina anserina, A. egedii*), salt marsh sand spurrey, sea lavender, marsh orach, sea-blite, and seaside crowfoot (*Ranunculus cymbalaria*) (Reed and Moisan 1971; Jacobson and Jacobson 1989; Beecher and Chmura 2004; Tiner, pers. obs.). Chaffy sedge (*Carex paleacea*) is a common dominant species of Atlantic salt marshes in Nova Scotia (Patriquin 1981).

In southern New England and the Mid-Atlantic region, the lower high marsh (just above the mean high water mark) is often characterized by the stunted or short form of smooth cordgrass. In some cases, it occurs in slight depressions where water

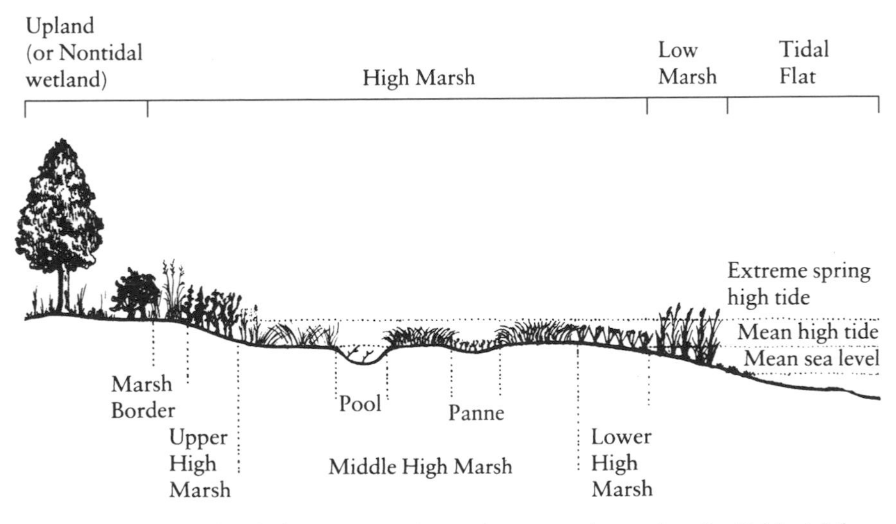

Figure 5.8. Generalized plant zonation in northeastern salt marshes. See Table 4.5 for species composition of zones. Plant distribution is largely related to elevation, which affects the degree of stress from flooding, salinity, exposure, and other factors. Small changes in elevation within individual marshes create more of a mosaic pattern rather than a strict zonation of species. (Tiner 2009)

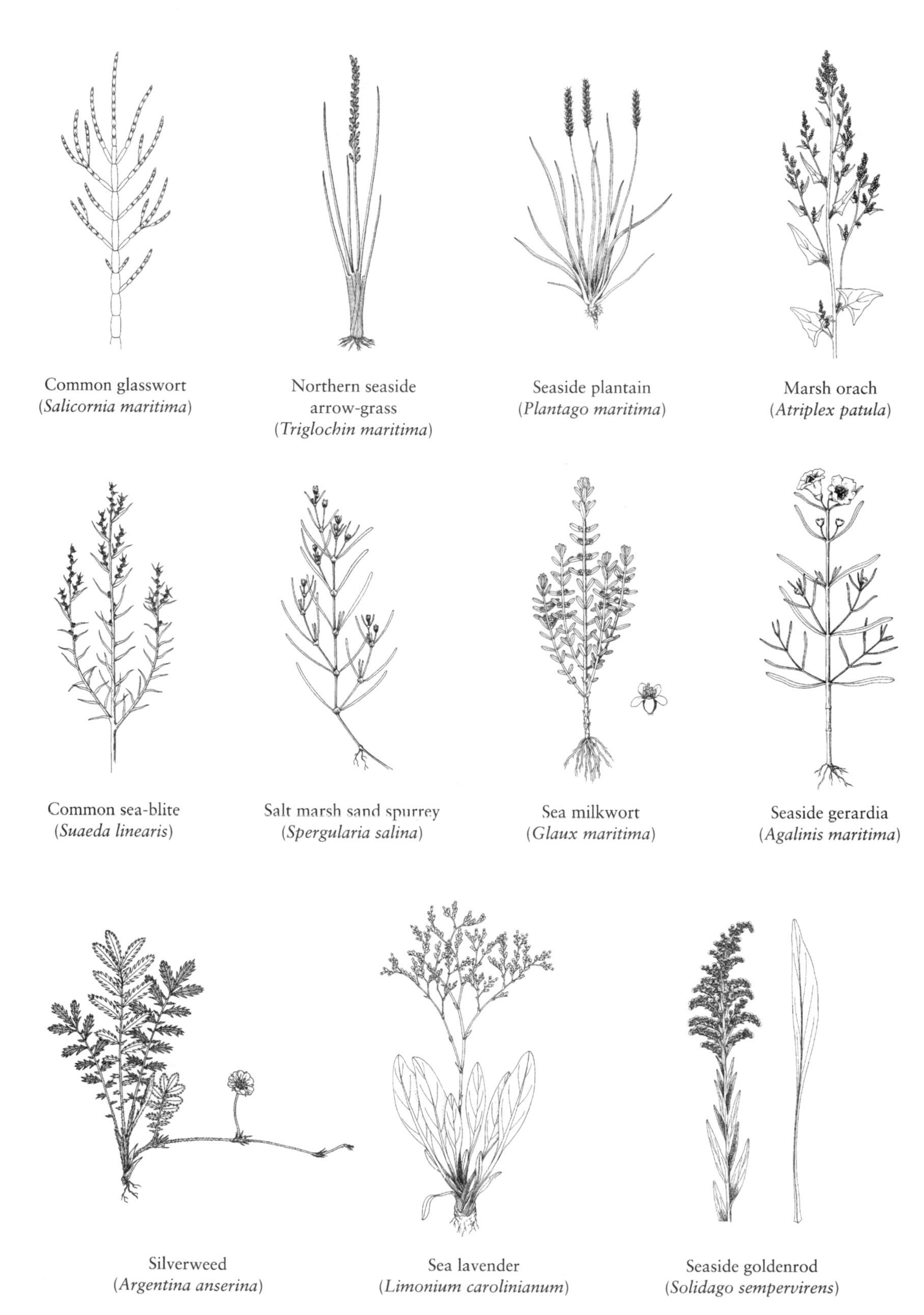

Figure 5.9. Examples of common wetland plants in North Atlantic salt marshes. (Tiner 2009)

American alkali grass
(*Puccinellia americana*)

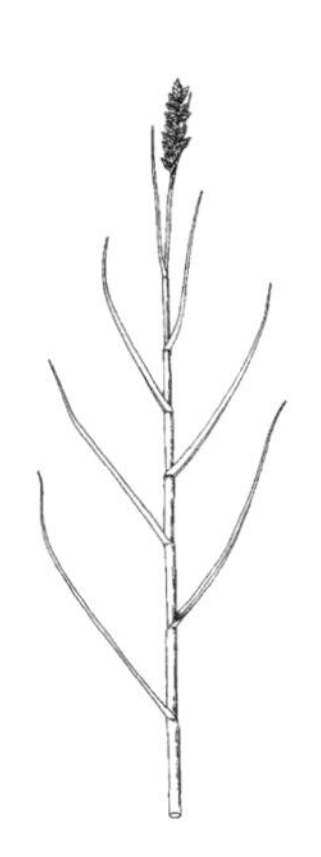

Salt grass
(*Distichlis spicata*)

Salt hay grass
(*Spartina patens*)

Smooth cordgrass
(*Spartina alterniflora*)

Prairie cordgrass
(*Spartina pectinata*)

Switchgrass
(*Panicum virgatum*)

Chaffy sedge
(*Carex paleacea*)

Salt marsh bulrush
(*Schoenoplectus robustus/ maritimus*)

High-tide bush
(*Iva frutescens*)

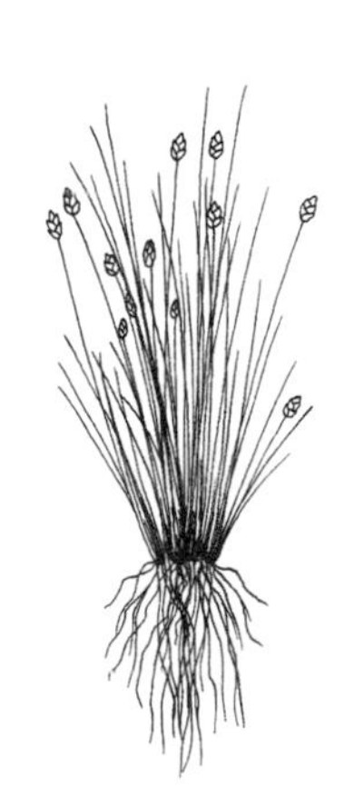

Dwarf spike-rush
(*Eleocharis parvula*)

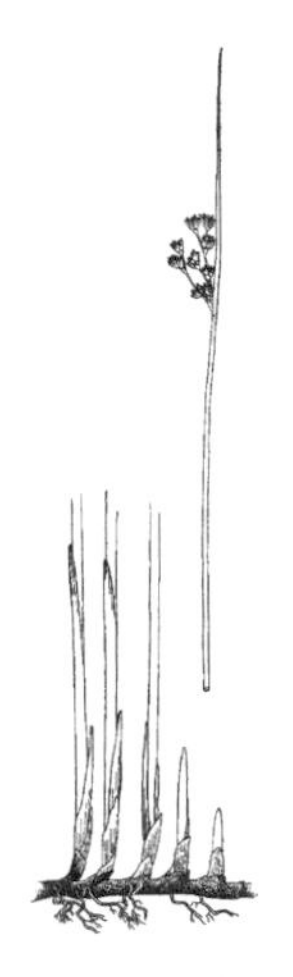

Baltic rush
(*Juncus arcticus*)

Black grass
(*Juncus gerardii*)

Groundsel-bush
(*Baccharis halimifolia*)

persists for significant periods. This may be the most anaerobic zone of the salt marsh creating great physiological stress on plants. Masses of the green algae (e.g., *Cladophora*) are often observed in these shallow basins.

In some northern marshes, an intermediate zone between the low marsh and high marsh occurs where ice removal of salt hay litter creates a suitable substrate for seaside plantain growth (Chmura et al. 1997). Smooth cordgrass may co-dominate this zone. Some of the more common associated species include sea milkwort, common glasswort, salt hay grass, northern sea lavender, and seaside arrow-grass.

Above this zone is the middle high marsh—a broad flat turf dominated by salt hay grass and/or salt grass. The former grass often appears as cow-licked or wave-swept patches that can be seen from some distance, while the latter may form monotypic stands in waterlogged areas in upper portions of northeastern U.S. salt marshes (Tiner 1987; Roman et al. 1997). The establishment of black grass (a rush) in the upper middle high marsh creates favorable soil conditions for other species (Hacker and Bertness 1999). This species is a "facilitator species" promoting floral diversity. Other high marsh plants that may be locally common include sea blites, salt marsh sand spurrey, common glasswort, silverweed, salt marsh bulrushes (*Schoenoplectus robustus, S. maritimus*), perennial salt marsh aster (*Symphyotrichum tenuifolium*), and seaside gerardia (*Agalinis maritima*). In northern marshes, Baltic rush (*Juncus arcticus*), alkali grasses (*Puccinellia americana, P. maritima, P. fasciculata*), red fescue (*Festuca rubra*), foxtail barley (*Hordeum jubatum*), and chaffy sedge may occur in this zone. Arctic alkali-grass (*Puccinellia tenella*) may co-dominate with salt hay grass in the Maritimes and the Gulf of St. Lawrence (Couillard and Grondin 1986; Hanson and Calkins 1996), but it is more typically found occasionally in salt pannes and in the lower marsh (Sean Basquill, pers. comm. 2004).

Shallow saline depressions, known as salt pannes, may occur in variable numbers within the middle high marsh (Figure 5.10). They retain water for extended periods and are subject to frequent wetting and drying. Evaporation concentrates salts over time

Figure 5.10. Salt panne dominated by common glasswort (*Salicornia maritima*) in a New Jersey salt marsh.

resulting in hypersaline conditions that may exceed 100 ppt—about three times that of seawater—during the summer (Sugihara et al. 1979). Few plants can tolerate these hypersaline conditions, but among them are the short form of smooth cordgrass, glassworts, salt grass, plantains, arrow-grasses, seaside gerardia, sea blites, perennial salt marsh aster, salt marsh sand spurrey, and sea lavender. Some depressions are virtually dominated by a single species such as stunted smooth cordgrass or common glasswort, while others contain a mixture of species. Pannes in northern New England may have higher plant diversity than those to the south (Evanchuck and Bertness 2004). Thick black mats of cyanobacteria (formerly identified as blue-green algae—*Cyanophyta*) may cover the substrate.

Open water pools form in salt marshes in several ways: 1) by blocking of tidal creeks; 2) behind natural levees where sedimentation is reduced and tidal water remains for extended periods; 3) in "rotten spots" where vegetation dies back due to excessive saturation causing the spots to form depressions as rhizomes decompose; and 4) from intertidal pannes (Figure 5.11; Teal and Teal 1969; Redfield 1972). In northern areas, depressions can form where frozen marsh turf and ice blocks are lifted out of the marsh by winter tides (Dionne 1969; Reed and Moisan 1971; Bleakney and Meyer 1979; Gauthier and Goudreau 1983; Gordon and Desplanque 1983; Gordon and Cranford 1994; Van Proosdij et al. 2000). High marsh pools may be colonized by widgeon-grass (*Ruppia maritima*), horned pondweed (*Zannichellia palustris*), sea lettuce, and other algae. Eel-grass may occur in salt marsh pools in the Maritimes. Northern salt marshes (e.g., along the St. Lawrence and some rivers in Maine) appear to have an extraordinary number of such pools when compared with similar marshes in the Mid-Atlantic region and farther south. Alkali bulrush (*Schoenoplectus maritimus*) can be found along the edges of some northern pools. Unditched marshes from

Figure 5.11. Aerial view of Maine salt marsh with numerous pools. (William French, U.S. Fish and Wildlife Service)

Connecticut to Maine have been found to contain three times the number of ponds (5 ponds/acre [13/ha]) and four times the surface water area per unit area of ditched marshes (Adamowicz and Roman 2005). Average size of these ponds is about 2,150 square feet (200 m^2) and their average depth is about 1 foot (29 cm).

Sandy sites in the high marsh may be colonized by several species depending on the geographic location (Table 5.3). Salt hay grass is common in these sites along much of the Atlantic Coast. Marshes located behind ocean beaches may be periodically covered by water-carried sands from coastal storms (overwash deposits). These "washes" support many of the species listed in the middle high marsh, plus others such as sea-beach knotweed (*Polygonum glaucum*), salt marsh asters (*Symphyotrichum* spp.), hairy

Table 5.3. High marsh plants occupying sandy site salt marshes of the Northeast

Common Name (*Scientific name*)	Range
Sea lavender (*Limonium carolinianum*)	Labrador and Quebec to Florida and northeastern Mexico
Seaside heliotrope (*Heliotropium curassavicum*)	Southern Maine and Massachusetts to Florida and Texas
Marsh pennywort (*Hydrocotyle umbellata*)	Nova Scotia to Florida and Texas
Annual salt marsh pink (*Sabatia stellaris*)	Southeastern Massachusetts to Florida and Louisiana
Slender sea purslane (*Sesuvium maritimum*)	New York to Florida and Texas
Salt marsh sand spurrey (*Spergularia salina*)	Quebec to Florida
Common sea blite (*Suaeda linearis*)	Maine to Florida and Texas
Gulf of St. Lawrence aster (*Symphyotrichum laurentianum*)	Prince Edward Island, Magdalen Islands, and New Brunswick
Fragrant galingale or flatsedge (*Cyperus odoratus*)	Massachusetts to Florida and Texas
Salt marsh fimbry (*Fimbristylis castanea*)	New York to Florida and Texas
Canada rush (*Juncus canadensis*)	Quebec and Nova Scotia to Florida

Source: Range along the Atlantic and Gulf Coasts from Tiner (2009).

smotherweed (*Bassia hirsuta*), and purple gerardia (*Agalinis purpurea*) as reported for Assateague Island in Maryland (Higgins et al. 1971).

The upper marsh edge typically exhibits the greatest plant diversity within the salt marsh since salt stress is reduced and inundation is less frequent. Many species are restricted to the marsh border including a few bulrushes and other sedges and several prominent grasses (Table 5.4). In New England, black grass (a member of the Rush Family, *Juncaceae*) is often dominant here as well as in elevated portions of the middle high marsh. From Maine north, Baltic rush, prairie cordgrass (*Spartina pectinata*), foxtail barley, and chaffy sedge are among the more abundant species in this zone. Seaside arrow-grass and sea milkwort may be common associates. Along the St. Lawrence estuary, the marsh border may be represented by one or more species including prairie cordgrass, bluejoint (*Calamagrostis canadensis*), Canadian burnet (*Sanguisorba canadensis*), timothy (*Phleum pratense*), red fescue, beachhead iris, seaside goldenrod, sow thistle (*Sonchus arvensis*), and hedge bindweed (*Calystegia sepium*) (Couillard and Grondin 1986; Tiner, pers. obs.). Interestingly, five mosses (*Campylium stellatum, Bryum capillare, Didymodon rigidulus, Mnium hornum, Amblystegium serpens*) have been found in the Baltic rush zone of Nova Scotian salt marshes (Garbary et al. 2008). Prairie cordgrass is a common border species in northern marshes, while switchgrass (*Panicum virgatum*) often forms a distinct border for many wetlands from New England south. Common reed (*Phragmites australis*) also occurs prominently in this zone, especially in developed areas and roadside marshes.

Shrubs tend to be more common on the marsh edges than in the interior marsh. An exception is high-tide bush (*Iva frutescens*), which colonizes the raised banks of "mosquito" ditches, thereby growing farther out into the high marsh than other shrubs. South of Maine, it is sometimes a dominant high marsh species where tidal flooding is somewhat restricted (e.g., on the marsh upstream of a causeway or road-crossing). Groundsel-bush (*Baccharis halimifolia*) is the second most abundant salt marsh shrub from Massachusetts south. It is usually restricted to the marsh edge where it sometimes forms part of a hedgelike border. It is most evident from late summer into fall when its numerous whitish flowers and cottony seedheads are highly visible. Somewhat higher up on the edges, northern bayberry (*Morella pensylvanica*), wax myrtle (*Morella cerifera,* from New Jersey south), and eastern red cedar (*Juniperus virginiana*) may be found. Poison ivy (*Toxicodendron radicans*) is common along the edges of many marshes. In New England, sweet gale (*Myrica gale*) is frequently observed along the marsh-upland border. An escaped exotic and now naturalized species, seaside rose (*Rosa rugosa*) and other roses (e.g., *R. palustris, R. virginiana*) may also be found here but they are usually outside the marsh.

Where freshwater runoff from the upland is heavy (e.g., road runoff or natural drainage-ways) or at groundwater seepage sites, brackish and freshwater plants may be found (Figure 5.12). Although part of the salt marsh complex, these areas may perhaps best be considered "brackish marshes" given their salinities and characteristic vegetation. Some of the more prevalent species in these situations are Olney's three-square (*Schoenoplectus americanus*), common three-square (*Schoenoplectus pungens*), common reed, marsh fern (*Thelypteris palustris*), narrow-leaved cattail (*Typha angustifolia*), salt marsh fleabane, rose mallow (*Hibiscus moscheutos*), seashore mallow (*Kosteletzkya virginica*), salt marsh loosestrife (*Lythrum lineare*), spike-rushes, and creeping bent grass (*Agrostis stolonifera* var. *palustris*). Soft rush (*Juncus effusus*), soft-stem bulrush (*Schoenoplectus tabernaemontani*), and giant bur-reed (*Sparganium eurycarpum*) may also occur in these places.

Along northern coasts, salt marshes may transition rather abruptly into bogs or fens. Brackish species such as twig rush (*Cladium*

Table 5.4. Some of the many plants associated with the upper edge of salt marshes

Common Name (*Scientific name*)	Range
Baltic rush (*Juncus arcticus*)	Labrador and Newfoundland to New Jersey
Seaside bulrush (*Blysmus rufus*)	Newfoundland to New Brunswick and Nova Scotia
Salt marsh sedge (*Carex recta*)	Labrador to Massachusetts
Chaffy sedge (*Carex paleacea*)	Greenland, Labrador and Quebec to southeastern Massachusetts
Salt marsh fimbry (*Fimbristylis castanea*)	New York to Florida and Texas
Alkali bulrush (*Schoenoplectus maritimus*)	Eastern Canada to Virginia
Salt marsh bulrush (*Schoenoplectus robustus*)	Nova Scotia to Florida and Texas
Lowland broom-sedge (*Andropogon glomeratus*)	Massachusetts to Florida and Texas
Red fescue (*Festuca rubra*)	Labrador and Quebec to North Carolina
Sweet grass (*Hierochloe odorata*)	Greenland to New Jersey and Maryland
Squirrel-tail or Foxtail barley (*Hordeum jubatum*)	Newfoundland to South Carolina
Switchgrass (*Panicum virgatum*)	Nova Scotia and Quebec to Florida and Texas
Big cordgrass (*Spartina cynosuroides*)	Massachusetts to Florida and Texas
Prairie cordgrass (*Spartina pectinata*)	Newfoundland and Quebec to North Carolina
Stiff-leaf quackgrass (*Thinopyrum pycnanthum*)	Nova Scotia to Rhode Island
Common reed (*Phragmites australis*)	Quebec and Nova Scotia to Florida and Texas
Creeping bent grass (*Agrostis stolonifera* var. *palustris*)	Newfoundland and Labrador to Virginia
Grass-leaved goldenrod (*Euthamia graminifolia*)	Newfoundland and Quebec to North Carolina
Scotch lovage (*Ligusticum scoticum*)	Greenland and Labrador to New York
Seaside goldenrod (*Solidago sempervirens*)	Newfoundland and Quebec to Florida and Texas
Silverweeds (*Argentina anserina, A. egedii*)	Greenland to Maryland
American germander (*Teucrium canadense*)	Nova Scotia and New Brunswick to Florida and Texas
Canadian burnet (*Sanguisorba candensis*)	Quebec, Labrador and Newfoundland to Maryland
Seaside angelica (*Angelica lucida*)	Greenland to New York
New York aster (*Symphyotrichum novi-belgii*)	Newfoundland and Nova Scotia to Maryland
Northern blazing star (*Liatris scariosa* var. *novae-angliae*)	Southern Maine to New Jersey

Source: Ranges along the Atlantic and Gulf Coasts from Tiner (2009), with minor modifications.

Figure 5.12. Occurrence of a broad swath of Olney's three-square (*Schoenoplectus americanus*) between the high marsh and lowland forest indicates significant groundwater influence.

mariscoides) and little green sedge (*Carex viridula*) may be observed in this tension zone, along with horned bladderwort (*Utricularia cornuta*) in shallow depressions. Sweet gale and cranberries (*Vaccinium macrocarpon, V. oxycoccus*) may grow along the upper edges in colder regions.

From Connecticut to Virginia, a rare wetland type called "sea-level fen" has been recently described for several locations (Bowman 2000; Swain and Kearsley 2001; Virginia Natural Heritage Program 2003). It occurs at the interface of salt marshes and upland seepage areas (including freshwater swamps). Diagnostic species include salt marsh sedge (*Carex hormathodes*), beaked spike-rush (*Eleocharis rostellata*), ten-angled pipewort (*Eriocaulon decangulare*), and Olney's three-square. Associated species are New York aster (*Symphyotrichum novi-belgii*), twig-rush, intermediate sundew (*Drosera intermedia*), Canada rush (*Juncus canadensis*), bog rush (*Juncus pelocarpus*), common reed, beakrushes (*Rhynchospora* spp.), swamp rose (*Rosa palustris*), common three-square, poison ivy, marsh St. John's-wort (*Triadenum virginicum*), slender blue flag (*Iris prismatica*), yellow-eyed grasses (*Xyris* spp.), and bladderworts (*Utricularia* spp.). Peat mosses (*Sphagnum* spp.) also occur in these wetlands (Ken Metzler, pers. comm. 2007). Sea-level fens possess several species that are rare in the affected states.

Regional Differences. From a biogeographic perspective, the northeastern United States appears to be somewhat of a transition zone or geographic ecotone where northern halophytes reach their southern limit and salt marsh flora with southern affinities establish their northernmost colonies (Table 5.5). Many salt marsh plants found in the North Atlantic region have wide distributions along the coastal plain with most of the common ones ranging into Florida and often into eastern Texas: smooth cordgrass, salt grass, salt hay grass (called "wire grass" in the South), black needlerush (*Juncus roemerianus*), and salt marsh bulrush. Some species are even abundant on both Atlantic and Pacific coasts, for example, seaside arrow-grass, sea milkwort, salt grass, silverweed, and perennial glasswort or pickleweed (*Salicornia depressa,* formerly

S. virginica), while other eastern species have virtual look-alike relatives common in Pacific Coast marshes, for example, sea lavender and New York aster on the East Coast and California sea lavender (*Limonium californicum*) and Douglas aster (*Symphyotrichum subspicatum*) on the West Coast.

Along the northeastern Gulf Coast, salt marshes are dominated by high marshes, whereas those of the South Atlantic Coast are predominantly low marsh, with few exceptions. Salt marshes in all regions tend to exhibit a zonation pattern related to elevation, hydrology, salinity, and other factors (Figure 5.13). Vast expanses of black needlerush characterize the northeastern Gulf Coast marshes (Figure 5.14), while smooth cordgrass typifies the South Atlantic marshes (see Figure 5.7b). On the latter

Table 5.5. Some halophytic species along the Atlantic Coast with range limits in the northeastern United States and eastern Canada

Species with Northern Affinities	Southern Limit
Gaspé Peninsula arrow-grass (*Triglochin gaspensis*)	New Brunswick and northeastern Maine
Salt marsh stitchwort (*Stellaria humifusa*)	Mid-coast Maine
MacKenzie's sedge (*Carex mackenziei*)	Southern Maine
Salt marsh sedge (*Carex recta*)	Northeastern Massachusetts
Chaffy sedge (*Carex paleacea*)	Southeastern Massachusetts
American alkali grass (*Puccinellia americana*)	Rhode Island, possibly Delaware
Salt marsh toad rush (*Juncus ambiguus*)	New York
Scotch lovage (*Ligusticum scoticum*)	New York
Canada sand spurrey (*Spergularia canadensis*)	New York
Arctic alkali grass (*Puccinellia tenella*)	New York
Seaside plantain (*Plantago maritima*)	New Jersey
Seaside crowfoot (*Ranunculus cymbalaria*)	New Jersey
Baltic rush (*Juncus arcticus*)	New Jersey
Sea milkwort (*Glaux maritima*)	New Jersey, possibly Virginia
Silverweed (*Argentina anserina*)	Maryland
Seaside goldenrod (*Solidago sempervirens* var. *sempervirens*)	Virginia
Species with Southern Affinities	**Northern Limit**
High-tide bush (*Iva frutescens*)	Nova Scotia
Sea blite (*Suaeda linearis*)	Maine
Annual salt marsh fleabane (*Pluchea odorata*)	Southern Maine
Perennial salt marsh aster (*Symphyotrichum tenuifolium*)	New Hampshire
Bearded sprangletop (*Leptochloa fusca*)	New Hampshire
Rose mallow (*Hibiscus moscheutos*)	New Hampshire
Groundsel-bush (*Baccharis halimifolia*)	Massachusetts
Foxtail grass (*Setaria parviflora*)	Massachusetts
Annual salt marsh pink (*Sabatia stellaris*)	Southeastern Massachusetts
Seaside goldenrod (*Solidago sempervirens* var. *mexicana*)	Southeastern Massachusetts
Perennial salt marsh pink (*Sabatia dodecandra*)	Connecticut
Salt marsh loosestrife (*Lythrum lineare*)	New York
Salt marsh fimbry (*Fimbristylis castanea*)	New York
Seashore mallow (*Kosteletzkya virginica*)	New York
Giant foxtail grass (*Setaria magna*)	New Jersey
Black needlerush (*Juncus roemerianus*)	Southern Delaware, possibly southern New Jersey
Southern seaside arrow grass (*Triglochin striata*)	Delaware and Maryland

Source: Ranges along the Atlantic and Gulf Coasts from Tiner (2009).

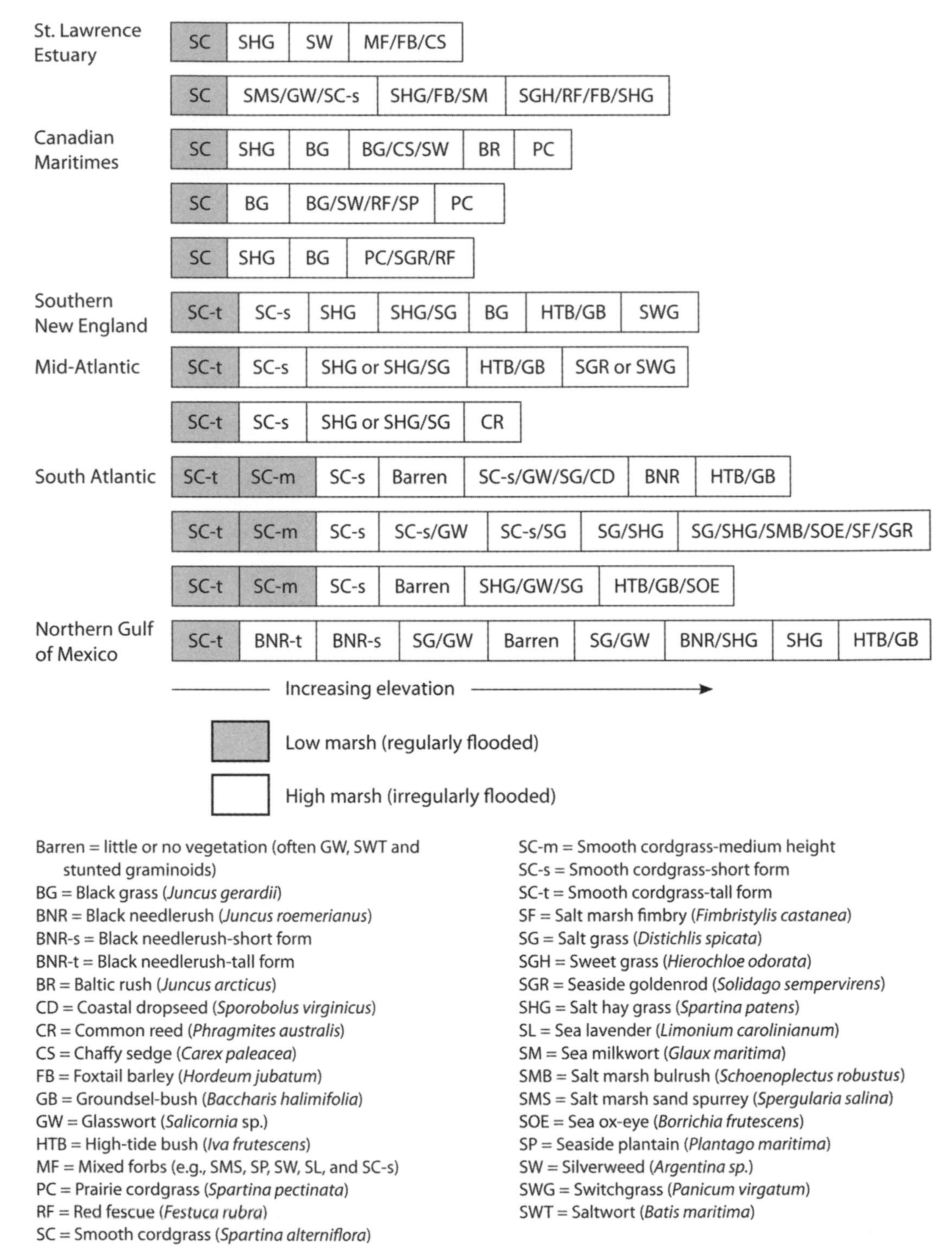

Figure 5.13. Bar graphs showing examples of changes in dominant species with increasing elevation in some eastern North American salt marshes. While vegetation zones are characteristic of salt marshes, local conditions dictate the actual distribution of species and the width or presence of a particular vegetation zone. (See also Figure 4.5 for more detailed representation for northeastern salt marshes.)

coast, black needlerush is often confined to a band of varying widths near the upper edges of the salt marsh and around the edges of salt barrens, but it becomes more abundant in the brackish marshes from Maryland south. Numerous herbs found in southeastern salt marshes do not occur in northern marshes (Table 5.6). A listing of typical salt and brackish marsh species found in other regions is given in Table 5.7.

Figure 5.14. Black needlerush–dominated salt marsh along the Gulf Coast (FL).

Table 5.6. Halophytic species unique to southeastern salt marshes

Common Name (*Scientific name*)	Range
Coastal water-hyssop (*Bacopa monnieri*)	Virginia to Florida and Texas
Sea ox-eye (*Borrichia frutescens*)	Virginia (possibly Maryland) to Florida and Texas
Common frog-fruit (*Phyla nodiflora*)	Virginia to Florida and Texas
Coastal dropseed (*Sporobolus virginicus*)	Virginia to Florida and Texas
Sandpaper vervain (*Verbena scabra*)	Virginia to Florida and Texas
Saltwater false willow (*Baccharis angustifolia*)	North Carolina to Florida and Texas
Climbing milkweed (*Cynanchum angustifolium*)	North Carolina to Florida and Texas
Marsh finger grass (*Eustachys glauca*)	North Carolina to Florida and Alabama
Salt marsh morning glory (*Ipomoea sagittata*)	North Carolina to Florida and Texas
Cabbage palm (*Sabal palmetto*)	North Carolina to Florida and Alabama
Sea purslane (*Sesuvium portulacastrum*)	North Carolina to Florida and Texas
Deer pea (*Vigna luteola*)	North Carolina to Florida and Texas
Saltwort (*Batis maritima*)	South Carolina to Florida and Texas
Christmas-berry (*Lycium carolinianum*)	South Carolina to Florida and Texas
Sand cordgrass (*Spartina bakerii*)	South Carolina to Florida; also Texas
Coastal leather fern (*Acrostichum aureum*)	Florida
Red mangrove (*Rhizophora mangle*)	Florida
Black mangrove (*Avicennia germinans*)	Florida to Texas
Key grass (*Monanthochloe littoralis*)	Florida to Texas
Silverhead (*Philoxeris vermicularis*)	Florida to Texas
Gulf cordgrass (*Spartina spartinae*)	Florida to Texas

Sources: Ranges along the Atlantic and Gulf Coasts are from Tiner (1993a, 2009).
Note: Many of these species also range into the subtropics.

Table 5.7. Representative salt and brackish marsh species in other regions of North America

Region (Climate)	Typical Species
South Atlantic (Warm Temperate–Humid Subtropical)	Graminoids: Smooth cordgrass (*S. alterniflora*), Black needlerush (*J. roemerianus*), Salt grass (*D. spicata*), Salt hay grass (*S. patens*), Big cordgrass (*S. cynosuroides*), Coastal dropseed (*Sporobolus virginicus*), Salt marsh fimbry (*F. castanea*), Lowland broomsedge (*Andropogon glomeratus*), Common reed (*P. australis*), Salt marsh bulrush (*S. robustus*), Olney three-square (*S. americanus*), Common three-square (*S. pungens*), Spikerushes (*Eleocharis* spp.) Flowering Herbs: Seaside goldenrod (*S. sempervirens*), Salt marsh aster (*S. tenuifolium*), Salt marsh pink (*Sabatia stellaris*), glassworts (*Salicornia maritima, S. bigelovii, S. depressa*), Sea-blite (*Suaeda linearis*) Shrubs: Sea ox-eye (*Borrichia frutescens*), Saltwort (*Batis maritima*), High-tide bush (*Iva frutescens*), Groundsel-bush (*B. halimifolia*), Saltwater false willow (*Baccharis angustifolia*)
Gulf Coast excluding Mississippi Delta (Warm Temperate–Humid Subtropical)	Graminoids: Smooth cordgrass, Black needlerush, Salt grass, Salt hay grass, Big cordgrass, Salt marsh bulrush, Olney three-square, Coastal dropseed, Gulf cordgrass (*S. spartinae*), Key grass (*Monanthochloe littoralis*) Flowering Herbs: Seaside goldenrod, Salt marsh aster, Salt marsh pink, glassworts, Sea-blite, Climbing milkweed (*Cynanchum angustifolium*) Shrubs: Sea ox-eye, Saltwort, Groundsel-bush, Saltwater false willow, Black mangrove (*Avicennia germinans*), Wax myrtle (*M. cerifera*), Christmas-berry (*Lycium carolinianum*)
Mississippi Delta (Warm Temperate–Humid Subtropical)	Graminoids: Smooth cordgrass, Black needlerush, Salt grass, Salt hay grass, Big cordgrass, Salt marsh bulrush, Olney three-square, Dwarf spikerush (*Eleocharis parvula*), Seashore paspalum (*Paspalum vaginatum*), Common reed Flowering Herbs: Coastal water-hyssop (*Bacopa monnieri*), Camphorweed (*Pluchea camphorata*) Shrubs: Saltwort
Southern California (Warm Temperate–Mediterranean)	Graminoids: Pacific cordgrass (*Spartina foliosa*), Salt grass, Key grass Flowering Herbs: Marsh jaumea (*Jaumea carnosa*), glassworts (*S. depressa, S. bigelovii, Arthrocnemum subterminale*), California sea lavender (*L. californicum*), Sea-blites (*S. esteroa, S. californica*), Spreading alkali-weed (*Cressa truxillensis*), Slender arrow-grass (*Triglochin concinna*), Salt marsh dodder (*Cuscuta salina*), Salt marsh bird's-beak (*Cordylanthus maritimus* ssp. *maritimus*) Shrubs: Alkali seaheath (*Frankenia salina,* formerly *F. grandifolia*), Saltwort, Box-thorn (*Lycium californicum*)
San Francisco Bay (Warm Temperate–Mediterranean)	Graminoids: Pacific cordgrass, Smooth cordgrass, Dense-flower cordgrass (*S. densiflora*), Common cordgrass (*S. anglica*), Salt grass, California bulrush (*Schoenoplectus californicus*), Pacific hair-grass (*Deschampsia holciformis*), Olney three-square, Alkali bulrush (*S. maritimus*), Baltic rush (*J. arcticus*) Flowering Herbs: Pickleweed (*Salicornia depressa,* formerly *S. virginica*), Marsh jaumea, Salt marsh dodder, Seaside arrow-grass (*T. maritima*), Oregon gumweed (*Grindelia stricta*), California sea lavender, Sea-blite (*Suaeda* sp.), Salt marsh sand spurrey (*S. salina*), Brass-buttons (*Cotula coronopifolia*), Marsh orach (*Atriplex patula*), Soft bird's-beak (*Cordylanthus mollis* ssp. *mollis*), Cattails (*Typha latifolia, T. angustifolia*) Shrubs: Alkali seaheath

Table 5.7. (*continued*)

Region (Climate)	Typical Species
Pacific Northwest and British Columbia (Warm Temperate–Marine)	Graminoids: Salt grass, Dwarf alkali grass (*Puccinellia pumila*), Lyngbye's sedge (*Carex lyngbyei*), Tufted hair-grass (*Deschampsia caespitosa*), Lime-American dune grass (*Leymus mollis*), Baltic rush, Common three-square, Salt marsh bulrush, Slough sedge (*C. obnupta*), Soft-stemmed bulrush (*Schoenoplectus tabernaemontani*), Spike-rushes, Spike bentgrass (*Agrostis exarata*), Creeping bent grass (*Agrostis stolonifera*), Squirrel-tail (*Hordeum jubatum*), Meadow barley (*Hordeum brachyantherum*) Flowering Herbs: Pickleweed, Seaside arrow-grass, Salt marsh dodder, Marsh jaumea, Brass-buttons, Seaside plantain (*P. maritima*), Sea milkwort (*G. maritima*), Pacific silverweed (*Argentina egedii*), Oregon gumweed, Marsh orach, Yarrow (*Achillea millefolium*), Douglas aster (*Symphyotrichum subspicatus*), Western lilaeopsis (*Lilaeopsis occidentalis*), Canada sand spurrey (*Spergularia canadensis*), Cattails (*Typha angustifolia, T. latifolia*), Salt marsh or cows clover (*Trifolium wormskioldii*)
Alaska (Cold Temperate–Tundra)	Graminoids: Alkali grasses (*Puccinellia nutkaensis, P. grandis, P. borealis, P. phryganodes, P. andersonii*), Fisher's tundragrass (*Dupontia fisheri*), Lyngbye's sedge, Ramensk's sedge (*Carex ramenskii*), Salt marsh sedge (*C. subspathacea*), sedges (*Carex ursina, C. mackenziei, C. pluriflora, C. glareosa, C. rariflora*), Marsh spike-rush (*Eleocharis palustris*), Slimstem reedgrass (*Calamagrostis stricta,* formerly *C. neglecta*), Squirrel-tail, Meadow barley, Red fescue (*Festuca rubra*), Marsh spike-rush (*Eleocharis palustris*), Largeflower speargrass (*Poa eminens*) Flowering Herbs: Seabeach sandwort (*Honkenya peploides*), Common glasswort (*Salicornia maritima*), Seaside arrow-grass, Marsh arrow-grass (*T. palustre*), Seaside plantain, Canada sand spurrey (*Spergularia canadensis*), Sea milkwort, Sea-blite (*S. depressa*), Pacific silverweed, Danish scurvygrass (*Cochlearia officinalis*), Salt marsh stitchwort (*Stellaria humifusa*), Seaside crowfoot (*Ranunculus cymbalaria*), Mare's-tail (*Hippuris tetraphylla*), Gmelin's saltbush (*Atriplex gmelinii*), Scotch lovage (*Ligusticum scoticum*), Marsh pea (*Lathyrus palustris*)
Hudson Bay (Cold Temperate–Subarctic and Polar–Tundra)	Graminoids: Alkali grasses (*P. phyraganodes, P. lucida*), Alkali bulrush, Baltic rush, Salt marsh sedge (*Carex recta*), Chaffy sedge (*C. paleacea*), Squirrel-tail Flowering Herbs: Common glasswort, Seaside arrow-grass, Seaside plantain, Pacific silverweed, Sea milkwort, Water hemlock, and Marsh orach

Note: Species are arranged by life-form within regions.

From North Carolina into the tropics, essentially nonvegetated areas, called "salt barrens," "salt flats," or "salinas," are common features in the high marsh just above the mean high water level (see Figure 3.11). Their soils have extremely high salinities—in excess of 100 ppt in places—that prevent colonization by most halophytes. These barrens are influenced by a combination of factors including tidal elevation, flooding, salinity, and climatic conditions. The size and location of these barrens may shift from year to year depending on these factors. Barrens may increase in size during periods of lower rainfall and less frequent flooding but may shrink in size with higher rainfall and more frequent flooding as neighboring salt marsh vegetation takes advantage of more favorable conditions. These changes can be detected on aerial photographs or satellite imagery and by examining soil profiles for differences in organic matter content (Hsieh 2001). On the Gulf Coast, salt barrens seem to develop above a critical point of frequent flooding and below the point where effective leaching by rainfall occurs (Stout 1984). A few plant species tolerate these conditions and can be found scattered in patches on the barrens including glassworts, saltwort (*Batis maritima*), salt grass, sea blites (*Suaeda* spp.), coastal dropseed (*Sporobolus virginicus*), and key grass (*Monanthochloe littoralis*) (Figure 5.15). California's Mediterranean climate (wet winters and dry summers) creates conditions at the end of the dry season in which salinities can be 200 ppt in tidal marshes (Zedler and Nordby 1986). Hypersalinities result from a combination of factors including infrequent tidal flooding, high evaporation,

a

b

Figure 5.15. Two examples of salinas: (a) South Carolina (dominated by Bigelow's glasswort, *Salicornia bigelovii*) and (b) Florida.

Figure 5.16. California salt marsh at San Francisco Bay National Wildlife Refuge. Salina (barren area) and dikes are readily observed.

and in some areas, low seasonal rainfall. For the most part, these areas are devoid of plants, covered by thin mats of cyanobacteria (blue-green algae) and/or sparsely vegetated by the most salt-tolerant of salt marsh plants.

With the exception of San Francisco Bay estuary wetlands and tidal wetlands along large rivers like the Fraser and Columbia, most of the coastal wetlands along the Pacific Coast are relatively small features and are scattered along the coast (Gallagher 1998). This is in marked contrast to the tidal marshes along the East Coast that virtually form a wide marshy belt behind barrier islands from Long Island to Florida. The steep topography with mountains meeting the sea and the absence of a coastal plain prevent the formation of extensive barrier islands. Instead estuaries form where rivers empty into the Pacific Ocean. Since rainfall is more seasonal here with winter rains, the mouths of some of these estuaries close periodically in summer creating "blind estuaries" (Seliskar and Gallagher 1983). Marshes in estuaries with rather small watersheds are subjected to large fluctuations in salinity. The interior portion of many California marshes is extremely salty due to high evapotranspiration. Pickleweed is a dominant there (Figure 5.16). Pacific cordgrass (*Spartina foliosa*) characterizes the low marsh, while other common high marsh species include salt grass, key grass, marsh jaumea (*Jaumea carnosa*), alkali seaheath (*Frankenia salina*), Parish's glasswort (*Arthrocnemum subterminale*), and salt marsh dodder (*Cuscuta salina*) (Peinado et al. 1994). The most extensive tidal wetlands in British Columbia are along the Fraser River Delta and Boundary Bay, with smaller marshes found in deltas at the head of fjords (e.g., Squamish River Delta at the head of Howe Sound; Glooschenko et al. 1988). Pacific Northwest salt marshes also exhibit vegetative zones from low marsh to the upland border. Characteristic species of the lower marsh may include perennial glasswort or pickleweed, marsh jaumea, salt grass, seaside arrow-grass, and Lyngbye's sedge (*Carex lyngbyei*), while high marsh dominants include Baltic rush, marsh jaumea, tufted hairgrass (*Deschampsia caespitosa*), Pacific silverweed, meadow barley (*Hordeum brachyantherum*), red-top (*Agrostis stolonifera*), and Douglas aster (Christy and Brophy 2007). While some plants are similar to those found on the East Coast, many species are unique to the Pacific region including several from western genera such as marsh jaumea, Oregon gumweed (*Grindelia stricta*), soft bird's-beak (*Cordylanthus mollis* ssp. *mollis*), spreading alkali-weed (*Cressa truxillensis*), and alkali seaheath (*Frankenia salina*) (Table 5.7; see Macdonald 1977 for summary of Pacific tidal wetland plant communities).

Creeping alkali grass (*Puccinellia phryganodes*), two sedges (*Carex subspathacea, C. ursina*), and Fisher's tundra grass (*Dupontia fisheri*) are common Arctic species creating a mosaic pattern of intertidal vegetation along Beaufort Bay (Funk et al. 2004). The first species is the dominant low marsh species in the Arctic region while numerous other species occupy the high marsh (Table 5.7; see Vierick et al. 1992 for descriptions of Alaskan types). Canada's northernmost wetlands occur along the Yukon and Northwest Territories, Hudson Bay, James Bay, and northern Quebec and Labrador. The Arctic climate creates harsh conditions for wetland plants especially due to ice scour. Many wetlands form a thin band along the shore, yet extensive wetlands can occur along the southwestern coasts of James Bay, Hudson Bay, and Foxe Basin. Salt marshes dominate the Hudson Bay lowland where they occupy 85 to 90 percent of the coast (Glooschenko et al. 1988).

Estuarine Shrub Swamps

A narrow belt of woody thickets often forms along the salt marsh-upland border from Massachusetts south (Figure 5.17). Low shrub swamps may replace former salt marshes where tidal flow has been moderately restricted. High-tide bush and groundsel-bush are the major species. They may also occur within salt marshes along mosquito ditch levees, on spoil piles, or along the edges of upland islands surrounded by marshland. Associated plants include typical high marsh plants such as black grass, salt grass, salt hay grass, sweet grass (*Hierochloe odorata*), prairie cordgrass (slough grass), switchgrass, American germander (*Teucrium canadense*), and seaside goldenrod. Big cordgrass (*Spartina cynosuroides*) is a co-dominant with high-tide bush forming narrow stands along levees in brackish waters of Maryland (Harrison and Stango 2003). Herb cover may vary from dense to sparse depending on shrub density. Rose mallow, sea lavender, marsh orach, smooth cordgrass, salt marsh loosestrife, and annual marsh pink (*Sabatia stellaris*) are other herbs that may occur. Woody associates include red cedar, common bayberry, wax myrtle, sweet gale, and poison ivy.

From Delaware Bay south, stands of wax myrtle may be found in slightly brackish (oligohaline) conditions, yet this species is more abundant upstream along fresh tidal waters and in nontidal wetlands. Common woody associates are chiefly groundsel-bush and swamp rose, with herbs represented by marsh fern, cinnamon fern (*Osmunda cinnamomea*), spike-rush (*Eleocharis fallax*), rose mallow, salt hay grass, wool grass (*Scirpus cyperinus*), climbing hempweed (*Mikania scandens*), switchgrass, and others (Tiner and Burke 1995; Harrison and Stango 2003).

Regional Differences. High-tide bush and groundsel-bush are dominants in southeastern salt swamps, along with saltwater false willow (*Baccharis angustifolia*), cabbage palm (*Sabal palmetto*), Christmas-berry (*Lycium carolinianum*), and others. Along the northern Gulf Coast, black mangrove (*Avicennia germinans*) forms dense shrub thickets in salt marshes. Associated plants include smooth cordgrass, saltwort, and glassworts, among other typical salt marsh species. Mangroves dominate south Florida's tidal wetlands (see discussion below under Estuarine Forests). In the Pacific Northwest, twinberry honeysuckle or black twinberry

Figure 5.17. High-tide bush (*Iva frutescens*) is the most abundant halophytic shrub in northeastern wetlands. Here it colonized a former Massachusetts cranberry bog that is now subject to tidal flooding.

(*Lonicera involucrata*), Pacific crabapple (*Malus fusca*), and dune willow (*Salix hookeriana*) form shrub thickets with an understory of brackish and freshwater herbs in the slightly brackish (oligohaline) zone of estuaries (Fox et al. 1984; Christy and Brophy 2007; Brophy 2009).

Estuarine Forests

Where low-lying forests occur along submerging coasts, direct evidence of the landward movement of salt marshes into these forests can be seen. During the initial stages of this marine transgression, trees still predominate. The first sign of this transgression is found in the herb layer with the occurrence of salt or brackish marsh herbs. In time, the canopy begins to open up as trees succumb to the salt stress, and eventually the swamp forest becomes an open marsh with a few dead snags and numerous stumps. Consequently, these "estuarine forests" are ephemeral or transitional communities destined to become salt or brackish marshes in the near future. In the northeastern United States, this condition is perhaps most widespread on the Delmarva Peninsula, especially in Somerset and Dorchester County, Maryland. This is not a recent phenomenon as similar conditions were reported in the early 1900s (Shreve et al. 1910). Interestingly, many of these estuarine forests are in designated wildlife management areas where controlled marsh burns are common practices. Such activities may have hastened the effects of sea-level rise as burning removes organic matter that would otherwise be incorporated into the soil and would help it maintain an elevation sufficient to support resident plants.

On the Delmarva Peninsula, these forests are dominated by loblolly pine (*Pinus taeda*) (Figure 5.18; Tiner and Burke 1995). Pines at the waterward edge are dying or severely stressed, exhibiting a distinctive yellowing (chlorosis) of their needles. Halophytes in the understory include salt hay grass, salt

Figure 5.18. Estuarine forests of loblolly pines (*Pinus taeda*) are a short-lived community as saltwater eventually converts them to salt marshes. This one is already well on its way to becoming a salt marsh.

grass, switchgrass, common reed, or black needlerush plus others that are usually less common (e.g., perennial salt marsh aster, salt marsh bulrush, rose mallow, and spike-rushes). The shrub stratum may be represented by wax myrtle, groundsel-bush, and high-tide bush, with the latter species common in more open areas. The presence of dead trees in salt or brackish marshes elsewhere in the region is conclusive evidence that these types of estuarine forests have occurred in other places in recent history. The longevity of these forests is unknown, but they are expected to be short-lived because significant turnover has been observed in less than a decade (Foulis et al. 1995).

Regional Differences. Marine transgression is also occurring along other coastlines, so these transitional estuarine forests undoubtedly exist elsewhere. A truly estuarine forested wetland type characterized by halophytic trees—the mangrove swamp—makes its appearance in Florida before dominating tropic tidal shores (Figure 5.19). Mangroves dominate about three-quarters of the world's coastline between 25°N and 25°S latitude and range somewhat farther north and south to about 30° in certain areas (Tomlinson 1986). Their range appears to be limited by freezing temperatures (Lugo and Snedaker 1974). In the tropics, they are typically forested wetlands reaching heights of 80 feet or more (25 m), but in Florida, they grow in a variety of height forms from low shrubs to tree size (Tiner 1993a). Mangrove swamps

Figure 5.19. Florida mangrove swamp dominated by red mangrove (*Rhizophora mangle*).

are common south of Cape Canaveral on the Atlantic Coast and south of Tarpon Springs on the Gulf Coast. They are best developed along the southwestern coast from Cape Sable to Everglades City where they form the Ten Thousand Islands. Three mangrove species characterize Florida's mangrove swamps: red mangrove (*Rhizophora mangle*), black mangrove, and white mangrove (*Laguncularia racemosa*) (Figure 5.20). Black mangrove is the most northerly ranging species, growing to about 30°N latitude to St. Augustine–Jacksonville Beach on the Atlantic Coast and along the northern Gulf Coast, with northernmost colonies reportedly in the Mississippi River Delta of Louisiana (Visser et al. 1998). The other mangroves occur only in peninsular Florida from Daytona Beach (Ponce de Leon Inlet) on the Atlantic to Cedar Key on the Gulf. The same three mangrove species grow on the Pacific Coast from Baja, California (28–30°N latitude) south (Macdonald 1977).

Red mangrove with its conspicuous arching prop roots typically occupies the regularly flooded zone along bay shores and tidal rivers. It can grow to a height of 80 feet (~25 m) in the tropics, but it is usually less than 35 feet (~10 m) tall in Florida.

Buttonwood (*Conocarpus erectus*)

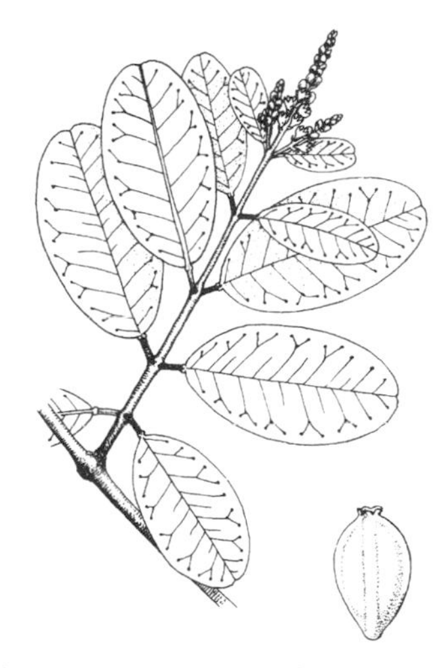

White mangrove (*Laguncularia racemosa*)

Red mangrove (*Rhizophora mangle*)

Black mangrove (*Avicennia germinans*)

Figure 5.20. Four tree species dominant in Florida's mangrove swamps. (Tiner 1993a)

Black mangrove with its cable root system of dense, fingerlike pneumatophores covering the ground (somewhat resembling a bed of nails) usually dominates the irregularly flooded swamp interior where the white mangrove also occurs. Both white and black mangroves can be found in overwash islands and beaches where red mangrove usually prevails. At the upper edges of the swamps another woody species, buttonwood (*Conocarpus erectus*), forms the border. Invasive woody species, Brazilian pepper (*Schinus terebinthifolius*) and Australian pine (*Casuarina equisetifolia*), may also take over the border areas or disturbed sites. Characteristic salt marsh plants such as perennial glasswort, saltwort, sea blite, and high-tide bush plus subtropical species including leather ferns (*Acrostichum* spp.), spider lily (*Hymenocallis latifolia*), cabbage palm, gray nicker (*Caesalpinia bonduc*), coin-vine (*Dalbergia ecastophyllum*), and rubber vine (*Rhabdadenia biflora*) are mangrove associates. Salt marshes dominated by black needlerush or saltwort and salinas are often intermixed with the mangroves and are typically found landward of these swamps.

Scientists have described six different types of Florida mangrove swamps based mainly on landscape position, which creates different hydrologic and other environmental conditions: 1) overwash forests (on islands frequently overwashed by tides); 2) fringing forests (narrow bands along waterways); 3) riverine swamps (regularly flooded floodplains); 4) basin forests (depressional wetlands); 5) hammock forests (slightly elevated sites); and 6) dwarf forests (on limestone marl) (Lugo and Snedaker 1974). Red mangroves averaging 25 to 30 feet (7.6–9.1 m) tall-form overwash forests and narrow bands along waterways and waterbodies, usually above mean high water. On river floodplains they grow to 65 feet (19.8 m) tall. Plant composition of basin mangrove forests depends on elevation, which affects the frequency and duration of flooding: red mangrove occupies regularly flooded basins, whereas the black and white mangroves dominate the irregularly flooded ones. Basin mangroves grow to about 50 feet (15.2 m) in height. All three mangrove species may be found in hammock forests where they are typically shrubs less than 15 feet (5.6 m) tall. Growth of mangroves on limestone marl is stunted, attaining a height of less than 5 feet (1.5 m) tall and forming dwarf forests.

In the Pacific Northwest, estuarine forested swamps occur in brackish regions (mesohaline to oligohaline). These swamps are characterized by Sitka spruce (*Picea sitchensis*), black twinberry, and Pacific crabapple with a diverse herbaceous layer similar to low mesohaline to oligohaline tidal marsh (Christy and Brophy 2007; Brophy 2009).

Brackish Marshes

Brackish marshes develop where freshwater significantly dilutes seawater creating moderately to slightly salty environments. Average salinities in brackish marshes range from moderately high (18.0–5.0 ppt; mesohaline) to essentially fresh (0.5–5.0 ppt; oligohaline) with drastic seasonal differences due to high runoff periods and droughts. These conditions present opportunities for halophytes and plants more typical of freshwater wetlands to co-exist and form rather complex plant communities, especially in lower salinity (oligohaline) waters where plant diversity appears to be the highest of estuarine marshes, sometimes rivaling that of neighboring tidal fresh marshes (e.g., Ferren et al. 1981; Sharpe and Baldwin 2009).

Brackish marshes are found in several places in the estuary: 1) along coastal rivers upstream of the salt marshes; 2) near the mouths of coastal rivers with heavy freshwater discharge that empty into bays and sounds with low tidal ranges; 3) between salt marshes and nearby uplands where freshwater streams enter the marsh or where groundwater seepage is significant; 4) around some coastal (barachois) ponds; 5) across the road or railroad embankment

Figure 5.21. Some examples of brackish marshes: (a) hummocky brackish marsh between a freshwater swamp and a salt marsh (MA); (b) narrow-leaved cattail (*Typha angustifolia*) marsh along a barachois (PEI); (c) black needlerush (*Juncus roemerianus*) marsh (MD); and (d) mixed community oligohaline marsh (MD).

from salt marshes where tidal flow has been restricted by undersized culverts or other structures; and 6) within marsh impoundments (Figure 5.21). In the northeastern United States, brackish marshes are most abundant in the Chesapeake Bay estuary, the largest estuary in the conterminous United States, while in eastern Canada, these wetlands are most extensive along the St. Lawrence River estuary from Pointe des Monts to Ile d'Orleans (Dubois 1993).

The vegetation of brackish marshes is highly varied, due largely to salinity differences. While many typical salt marsh plants remain abundant at their more seaward locations, brackish marshes can be recognized by the abundance of salt marsh border halophytes throughout the marsh or by more robust, taller forms of salt marsh plants attributed to reduced salt stress. Farther upstream in coastal rivers, the marshes become progressively fresher (or less salty) with increased freshwater inputs. The salinity of the uppermost brackish marshes varies from fresh in the spring at peak river discharge to slightly brackish (oligohaline) in summer during low river flows. Freshwater plants predominate in these marshes with relatively few brackish species present. Consequently, brackish marshes represent a vegetative continuum marked by a gradual intermixing of tidal fresh marsh species with salt marsh halophytes. With less salt stress, plant diversity of brackish marshes is often much higher than that of salt marshes. Finding three or four times as many species in these slightly brackish marshes as are found in salt marshes is not uncommon.

Smooth cordgrass remains a dominant low marsh plant in these wetlands. Three species may form a carpet beneath this tall grass: eastern lilaeopsis (*Lilaeopsis chinensis,* a low-growing herb with flattened linear basal "leaves"), dwarf spike-rush, and mudwort (*Limosella subulata*). Dwarf spike-rush has been reported forming beds along Delaware Bay (John Teal, pers. comm. 2003) and in the lower tidal zone on muddy shores in Atlantic Canada (Basquill 2003), with salt marsh sand spurrey, Gaspé peninsula arrow grass (*Triglochin gaspensis*), and other salt marsh species as common associates. Common three-square, water hemp (*Amaranthus cannabinus*), and wild rice (*Zizania aquatica*) may dominate the regularly flooded zone in moderate to weakly brackish areas.

As mentioned earlier, many of the dominant brackish plants are facultative halophytes that occupy higher elevations in the salt marsh. Reduced salinity allows them to cover large portions of the brackish marsh interior. These species include chaffy sedge and prairie cordgrass (i.e., Maine north), big cordgrass (especially from New Jersey south), common reed, switchgrass, creeping bent grass (e.g., New England), salt marsh bulrushes (*S. robustus, S. maritima*), three-squares (*S. americanus, S. pungens*), rose mallow, narrow-leaved cattail, seaside goldenrod, and high-tide bush. Cattail and rose mallow may co-dominate some marshes, while common reed thrives in tidally restricted situations and other low-salinity marshes. A southern species, black needlerush begins to form extensive virtually monotypic stands in brackish marshes from the Eastern Shore of Maryland south.

Other salt marsh plants also remain significant members of brackish marsh communities, especially in the more strongly brackish (mesohaline) ones. They include smooth cordgrass, salt hay grass, spike grass, perennial salt marsh aster, silverweed, Baltic rush, and alkali grasses. Salt hay grass may take on a more clumpy (tufted) habit giving the marshes a hummocky appearance in marked contrast to the turf-like lawns it forms in salt marshes (Figure 5.21a). Other halophytes of common occurrence, depending on their range, are salt marsh loosestrife, seashore mallow, salt marsh fleabane, and annual salt marsh pink. Groundsel-bush, a halophytic shrub, may occupy higher elevations in and around brackish marshes.

Some species making their first appearance in brackish marshes are water hemp, giant foxtail (*Setaria magna*), Nuttall's cyperus (*Cyperus filicinus*), twig rush, red milkweed (*Asclepias lanceolata*), late-flowering thoroughwort (*Eupatorium serotinum*), lance-leaved frog-fruit (*Phyla lanceolata*), water pimpernel (*Samolus parviflorus*), mock bishop-weed (*Ptilimnium capillaceum*), seaside crowfoot (*Ranunculus cymbalaria*), water parsnip (*Sium suave*), soft-stemmed bulrush, pickerelweed (*Pontederia cordata*), arrow arum (*Peltandra virginica*), tidal sagittaria (*Sagittaria calycina* ssp. *spongiosus*), blue flag (*Iris versicolor*), wild rice, eastern lilaeopsis, some species of spike-rushes (e.g., *Eleocharis rostellata, E. fallax*), and an herbaceous vine, climbing hempweed. Although more frequent in tidal fresh marshes, river bulrush (*Schoenoplectus fluviatilis*), purple loosestrife (*Lythrum salicari*a), and shoreline sedge (*Carex hyalinolepis*) may occur in slightly brackish marshes and be locally dominant. The latter, a wide-bladed sedge, may form dense colonies in tidal cypress swamps from Maryland south. The slightly brackish (oligohaline) marshes may be a species-rich community and, in many cases, are indistinguishable from neighboring tidal fresh marshes. The high species diversity of one such marsh was recorded in Massachusetts, with the overwhelming abundance of freshwater species being particularly noteworthy (Table 5.8).

In the Canadian Maritimes, two brackish marsh communities have been recognized—one dominated by common three-square and the other a mixture of prairie cordgrass, broad-leaved cattail (*Typha latifolia*), and soft-stemmed bulrush (Sean Basquill, pers. comm. 2011). The three-square community

Table 5.8. Tidal marsh plants observed along the Merrimack River, Massachusetts, and their relative abundance

Common Name (*Scientific name*): Abundance Code
Aquatic species
Tape-grass (*Vallisneria americana*): C
Fanwort (*Cabomba caroliniana*): O
Coontail (*Ceratophyllum demersum*): O
Curly-leaved pondweed (*Potamogeton crispus*): O
Long-leaf pondweed (*Potamogeton nodosus*): O
Clasping-leaved pondweed (*Potamogeton perfoliatus*): O
Robbins' pondweed (*Potamogeton robbinsii*): U
Water-meal (*Wolffia columbiana*): U
Mudflat
Water-wort (*Elatine americana*): A
Waterweed (*Elodea nuttallii*): A
Marsh herbs
Sweet flag (*Acorus calamus*): A
Water hemp (*Amaranthus cannabinus*): A
New York aster (*Symphyotrichum novi-belgii*): A
Nodding beggar-ticks (*Bidens cernua*): A
Boneset (*Eupatorium perfoliatum*): A
Jewelweed (*Impatiens capensis*): A
Swamp candles (*Lysimachia terrestris*): A
Pickerelweed (*Pontederia cordata*): A
Halberd-leaved tearthumb (*Polygonum arifolium*): A
Dotted smartweed (*Polygonum punctatum*): A
Grass-leaved arrowhead (*Sagittaria graminea*): A
Big-leaved arrowhead (*Sagittaria latifolia*): A
Water parsnip (*Sium suave*): A
Swamp beggar-ticks (*Bidens connata*): C
Eaton's beggar-ticks (*Bidens eatonii*): C
Marsh marigold (*Caltha palustris*): C
Turtlehead (*Chelone glabra*): C
Water hemlock (*Cicuta maculata*): C
Water horsetail (*Equisetum fluviatile*): C
Eastern Joe-Pye-weed (*Eupatoriadelphus dubius*): C
False pimpernel (*Lindernia dubia*): C
Water purslane (*Ludwigia palustris*): C
Water horehound (*Lycopus americanus*): C
Water mint (*Mentha arvensis*): C
Square-stemmed monkeyflower (*Mimulus ringens*): C
Arrow arum (*Peltandra virginica*): C
Clearweed (*Pilea pumila*): C
Japanese knotweed (*Polygonum cuspidatum*): C
Creeping buttercup (*Ranunculus repens*): C
Curly dock (*Rumex crispus*): C
Seaside goldenrod (*Solidago sempervirens*): C
Tall meadow-rue (*Thalictrum pubescens*): C
Indian hemp (*Apocynum cannabinum*): O
Devil's beggar-ticks (*Bidens frondosa*): O
Vernal water starwort (*Callitriche palustris*): O
Marsh herbs (*continued*)
Pennsylvania bitter-cress (*Cardamine pensylvanica*): O
Bulb-bearing water hemlock (*Cicuta bulbifera*): O
Asiatic dayflower (*Commelina communis*): O
Common horsetail (*Equisetum arvense*): O
Scouring rush (*Equisetum hyemale*): O
Parker's pipewort (*Eriocaulon parkeri*): O
Marsh bedstraw (*Galium palustre*): O
Yellow flag (*Iris pseudacorus*): O
Blue flag (*Iris versicolor*): O
Quillwort (*Isoetes echinospora*): O
Fringed loosestrife (*Lysimachia ciliata*): O
Lance-leaf loosestrife (*Lysimachia lanceolata*): O
Purple loosestrife (*Lythrum salicaria*): O
Slender-leaf dragonhead (*Physostegia leptophylla*): O
Pearlwort (*Sagina procumbens*): O
Mad-dog skullcap (*Scutellaria lateriflora*): O
Giant bur-reed (*Sparganium eurycarpum*): O
Narrow-leaved cattail (*Typha angustifolia*): O
Southern water plantain (*Alisma subcordatum*): U
Northern water plantain (*Alisma triviale*): U
Norwegian cinquefoil (*Potentilla norvegica*): U
Canadian burnet (*Sanguisorba canadensis*): U
Coltsfoot (*Tussilago farfara*): U
Marsh graminoids
*Common three-square (*Schoenoplectus pungens*): A
Soft-stemmed bulrush (*Schoenoplectus tabernaemontanii*): A
Wild rice (*Zizania aquatica*): A
Bluejoint (*Calamagrostis canadensis*): F
Three-way sedge (*Dulichium arundinaceum*): F
Small's spike-rush (*Eleocharis smallii*): F
River bulrush (*Schoenoplectus fluviatilis*): F
*Smooth cordgrass (*Spartina alterniflora*): F
*Prairie cordgrass (*Spartina pectinata*): F
Redtop (*Agrostis stolonifera* var. *stolonifera*): O
*Chaffy sedge (*Carex paleacea*): O
Shining flatsedge (*Cyperus bipartitus*): O
Yellow nut-grass (*Cyperus esculentus*): O
Taper-tip rush (*Juncus acuminatus*): O
Reed canary grass (*Phalaris arundinacea*): O
*Marsh straw sedge (*Carex hormathodes*): U
Stalk-grain sedge (*Carex stipata*): U
Orchard grass (*Dactylis glomerata*): U
Fall panic grass (*Panicum dichotomiflorum*): U
Smith's bulrush (*Schoenoplectus smithii*): U
Marsh shrubs
False indigo (*Amorpha fruticosa*): O

Source: Caldwell and Crow 1992.

Notes: Codes for abundance: A = abundant, C = common, F = frequent, O = occasional, U = uncommon. Plants common in more saline habitats are marked by an asterisk (*).

The authors considered this marsh a tidal fresh marsh using a 5 ppt break between tidal fresh and brackish marsh, but since its salinity was >0.5 ppt at low flow in September and smooth cordgrass was recorded as frequent, it is an estuarine oligohaline (slightly brackish) emergent wetland according to the U.S. Fish and Wildlife Service's wetland classification system (Cowardin et al. 1979). These wetlands are virtually indistinguishable from upstream tidal fresh marshes, and 0.5 ppt may not be a good dividing line given the predominance of typical freshwater species above this level.

may include bulrushes (*Schoenoplectus acutus, S. maritima*), sedges (*Carex paleacea, C. recta, C. viridula*), spike-rushes (*Eleocharis halophila, E. palustris, E. parvula*), rushes (*Juncus arcticus, J. pelocarpus*), grasses (*Spartina alterniflora, S. pectinata, Hierochloe odorata*), silverweed, seaside arrow-grass, marsh arrow-grass (*Triglochin palustris*), seaside plantain, marsh orach, Canadian sand spurrey (*Spergularia canadensis*), sea-milkwort, common glasswort, New York aster, and seaside goldenrod (Sean Basquill, pers. comm. 2011). The mixed brackish marsh occurs in areas where freshwater inputs are high. Several of the species listed above occur here along with swamp candles (*Lysimachia terrestris*), marsh fern, common skullcap (*Scutellaria galericulata*), blue flag, marsh willow-herb (*Epilobium palustre*), brass-buttons (*Cotula coronopifolia*), marsh bellflower (*Campanula aparinoides*), various grasses (*Agrostis, Calamagrostis, Festuca, Phragmites*), cow-vetch (*Vicia cracca*), and two shrubs—sweet gale and northern bayberry.

Regional Differences. Black needlerush is the dominant brackish marsh plant along the South Atlantic and Gulf Coasts where it forms nearly monotypic stands, with smooth cordgrass typically occupying the lowest marsh (Figure 5.22). Other typical salt marsh species may form green patches within the olive-green swards of needlerush. Southern species that colonize the more saline brackish marshes include Gulf cordgrass (*Spartina spartinae*) and climbing milkweed (*Cynanchum angustifolium*). Late-flowering thoroughwort or boneset (*Eupatorium serotinum*), a medium-height, white-flowered plant, is conspicuous among the rushes, although it is not particularly abundant. Halberd-leaved morning glory (*Ipomoea sagittata*) with its pinkish-purple funnel-shaped flowers often can be seen climbing the shrubs along the marsh border. Saw grass (*Cladium jamaicense*), the predominant graminoid of the Everglades, may dominate the border grading into lowland pine forests where freshwater runoff or seepage is significant. Big cordgrass is a dominant species regionwide. Wire grass (salt hay grass) is the dominant plant of Louisiana's brackish delta marshes, with salt grass co-dominating the more saline stands and common reed (locally called "roseau cane"), coastal water-hyssop, and bull-tongue (*Sagittaria lancifolia*, formerly *S. falcata*) being the main associates in more diverse "intermediate" marshes (Chabreck 1972; Gosselink 1984). Common three-square, salt marsh bulrush, and seashore paspalum (*Paspalum vaginatum*) are also prominent species in brackish marshes (Craig et al. 1987). Other dominants of the intermediate (slightly brackish) marshes include soft-stemmed bulrush, southern blue flag (*Iris virginica*), and switchgrass (Shafer et al. 2007). Species present in southern brackish marshes but not in northern marshes include seashore paspalum, torpedo grass (*Panicum repens*), California bulrush (*Schoenoplectus californicus*), water hemp (*Amaranthus australis*), alligatorweed (*Alternanthera philoxeroides*), southern blue flag, spider lilies (*Hymenocallis crassifolia, H. occidentalis*), southern swamp lily (*Crinum americanum*), white-top sedge (*Dichromena colorata*), deer pea (*Vigna luteola*), Gulf cordgrass, giant cutgrass (*Zizaniopsis miliacea*), and cabbage palm (Tiner 1993a).

On the West Coast, California bulrush, cattails, Olney's three-square, pickleweed, alkali bulrush, Pacific silverweed (*Argentina egedii*), seaside arrow-grass, Lyngbye's sedge (*Carex lyngbyei*), Baltic rush, and tufted hair-grass (*Deschampsia cespitosa*) are important brackish species (Josselyn 1983; Seliskar and Gallagher 1983; Culberson 2001). In British Columbia, the last five species are also dominant brackish species along with meadow barley (*Hordeum brachyanatherum*) (MacKenzie and Moran 2004). In the Pacific Northwest, low mesohaline to oligohaline marshes in the middle to upper estuary may also include Douglas water hemlock (*Cicuta douglasii*), Douglas aster (*Symphyotrichum subspicatum*), seaside angelica (*Angelica lucida*), Pacific lady fern (*Athyrium filix-femina*), common spike-rush (*Eleocharis palustris*), marsh pea (*Lathyrus palustris*), Pacific water-parsley (*Oenanthe sarmentosa*), silverweed, seaside arrow-grass, arctic plantain (*Plantago macrocarpa*), and water parsnip (*Sium suave*) (Christy and Brophy 2007). In Alaska, several sedges (*C. rariflora, C. ramenskii, C. glareosa, C. lyngbyaei*) may be found in brackish coastal meadows along with a few grasses (*Dupontia fisheri, Calamagrostis canadensis, C. deschampsioides, Arctophila fulva, Poa eminens*), Pacific silverweed, black crowberry (*Empetrum nigrum*), and two willows (*Salix frucesens, S. ovalifolia*) (Jorgenson and Dissing 2010; Jorgenson and Roth 2010). Brackish species along rivers in the James Bay region include spike-rushes, chaffy sedge, mare's-tail (*Hippurus vulgaris*), alkali bulrush, soft-stemmed bulrush, common three-square, various sedges, and rushes (Glooschenko and Martini 1978).

a

b

Figure 5.22. Examples of southeastern brackish marshes: (a) North Carolina black needlerush (*Juncus roemerianus*) marsh and (b) Louisiana wiregrass (*Spartina patens*) marsh.

The constant flow of freshwater into estuaries from spring through fall from melting sea ice, snow and glaciers, and surface water runoff create brackish conditions along the Alaskan coast where salt marshes would form in other regions (Jon Hall, pers. comm. 2011). Along Alaska's Pacific Coast where major uplifting is occurring, salt and brackish marshes grade imperceptibly into virtually nontidal freshwater marshes that may be flooded by extreme tides. These meadows were once estuarine marshes but their elevations have been raised above that of most tides by tectonic activity. Some of the salt marsh species still predominate, such as Lyngbei's sedge and many-flower sedge (*C. pluriflora*) while shrubs like Alaska bog willow (*Salix fuscescens*) and sweet gale may colonize the upper portions of the meadows near the edge of coastal forests (Tande 1996; Jorgenson et al. 2010).

Tidal Fresh Marshes

Beyond the limit of average saltwater penetration in coastal rivers, tidal fresh marshes replace the brackish marshes. Here, the tide still influences water levels, but the water is typically fresh (less than 0.5 ppt salinity), although subject to occasional pulses of saltwater during dry periods. Tidal freshwater marshes and swamps are common along river systems with extensive drainages, lacking dams. They are best developed on low flat landscapes—drowned river valleys—where they form along the upper tidal reaches of coastal rivers. Consequently, they reach their fullest extent on the coastal plain from New Jersey to northern Florida on the Atlantic Coast and from northwest Florida to Louisiana on the Gulf Coast, and along the Alaskan coast (Hall 2009; Leck et al. 2009). In the North Atlantic, rivers with substantial amounts of tidal fresh marshes include the Saint Lawrence (QC), Saint John (NB), Merrimack and North Rivers (MA), Connecticut (CT), Hudson (NY), Great Egg Harbor, Mullica, and Maurice Rivers (NJ), Delaware (NJ and PA), Nanticoke (MD and DE), and Choptank, Patuxent, Pocomoke, and Potomac Rivers (MD). Flooding of adjacent wetlands is influenced by river overflows (e.g., spring freshets) as well as by tides. A recent book—*Tidal Freshwater Wetlands* (Barendregt et al. 2009)—offers an in-depth examination of these marshes in North America and Europe.

With little or no saltwater stress, hundreds of hydrophytic plant species can become established. Herbaceous marsh plants usually predominate immediately upstream of brackish marshes. Tidal fresh marshes include some of the most diverse plant communities in North America (Figure 5.23). It is important to emphasize that the change from slightly brackish (oligohaline) marshes to strictly freshwater marshes is relatively inconspicuous as numerous freshwater plants tolerate mild salinity levels (<3 ppt) in late summer. The best indicators for separating the two marshes might be the absence of smooth cordgrass and the presence of scattered tree saplings in the marsh near the upland border.

Hydrophytic trees and shrubs form tidal swamps along the upland border of the tidal fresh marshes and on natural levees adjacent to the river, while in other places marshes lay behind tidal swamps. As one moves upriver, these wet forests sometimes appear to gradually advance toward the river from the backmarsh, eventually forming a solid band of trees from riverbank to upland. What is responsible for these varied patterns? Tidal fresh marshes are subjected to occasional saltwater intrusion during droughts and perhaps more frequently in late summer during dry years, so the pattern of marshes and swamps may be related to the frequency and duration of these salt pulses. Or, is it attributed to rising sea level, hurricanes, or simply sedimentation rates and topographic differences? Discovery of the remains of a sedge-alder swamp below today's marshes led scientists to conclude that a combination of factors including rising sea levels, increased sediment, eutrophication, and salinity changes led to demise of

Figure 5.23. Examples of tidal fresh marshes: (a) diverse community along the St. Lawrence River (QC); (b) classic zonation evident along Crosswicks Creek (NJ); (c) mixed perennials with seedlings of wild rice (*Zizania aquatica*) in the spring (NJ); and (d) wild rice marsh in bloom (NJ).

the swamp (Leck and Simpson 1987; Orson et al. 1992; Schuyler et al. 1993; Leck and Crain 2009). Swamps tend to be at higher elevations than the marshes, so sedimentation patterns likely play a role in swamp formation. Direct evidence of this is found on riverbanks of the marshes in which wetland forests occupy the higher linear natural levees while marshes dominate the low-lying areas behind. While swamp trees appear to be able to handle acute salinity pulses during drought or hurricanes, chronic exposure of trees to salinities greater than 2 ppt appears to promote their conversion to marshes (Hackney et al. 2007).

Tidal fresh marshes occur in zones known for their high turbidity (Baldwin et al. 2009). High sedimentation coupled with natural disturbances (e.g., ice scouring, erosion, wrack deposition, and herbivory) create an environment ripe for annual plants (Leck et al. 2009; Figure 5.24). A mixture of annual and perennial herbs characterizes many of these marshes, with annuals in particular providing an interesting dynamic to marsh composition over time (Table 5.9). Annual seed production and seedbanks play important roles in determining species composition and abundance in these plant communities (Leck and Simpson 1994, 1995). While it is possible to produce a list of likely plants for an area based on past observations, predicting their abundance is virtually impossible due to the variable response of annual plants. Tidal fresh marshes are unique among tidal wetlands in that

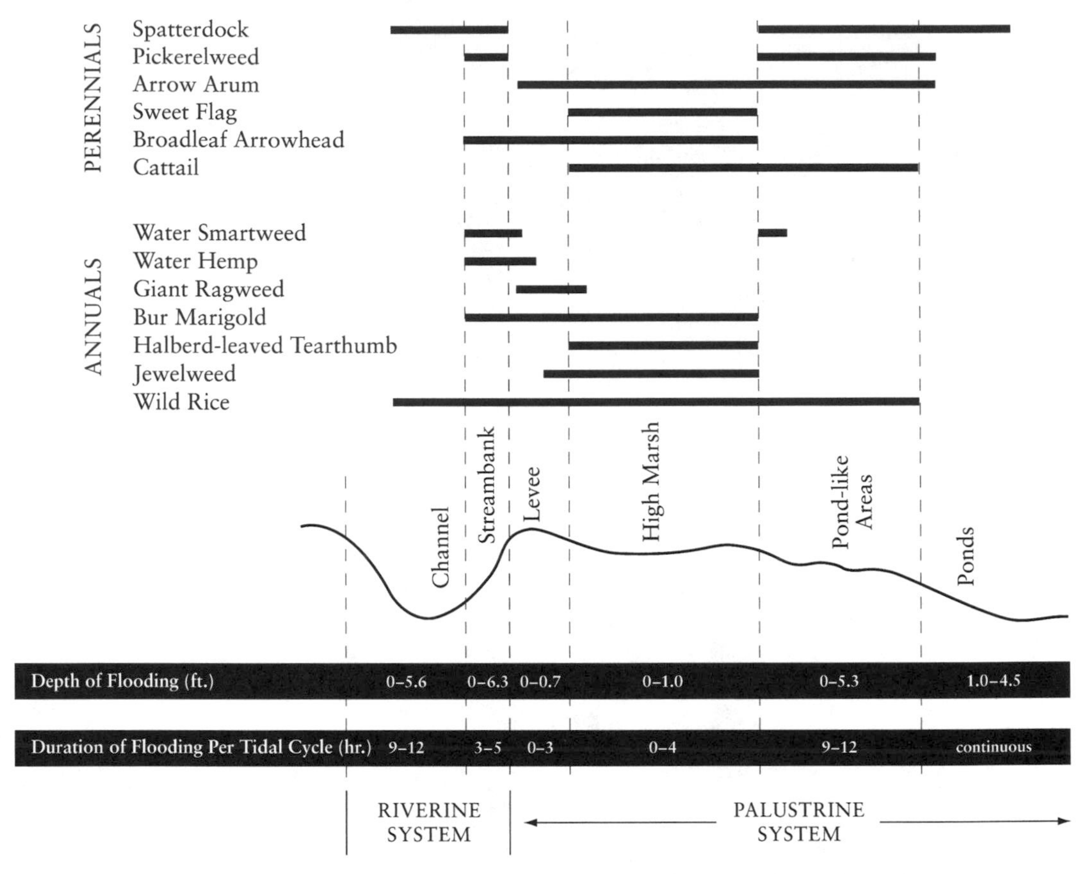

Figure 5.24. Generalized plant zonation in a Delaware River tidal fresh marsh. (Redrawn from Simpson et al. 1983a)

they exhibit a marked seasonal change in vegetation patterns (dominance and aspect). In winter, much of the marsh resembles a mudflat at low tide. In early spring, seedlings of annual plants cover the high marsh, while spatterdock (a perennial herb) dominates the low marsh. By late spring and early summer, other perennials including broad-leaved types (arrow arum, pickerelweed, and big-leaved arrowhead) and sweet flag may predominate (Figure 5.23c). Later, taller herbs that are mostly annuals visually dominate tidal fresh marshes. In the Mid-Atlantic region, these species include smartweeds, tearthumbs (*Polygonum arifolium, P. sagittatum*), jewelweed (*Impatiens capensis*), water hemp, wild rice, and bur-marigold (*Bidens laevis*) (McCormick and Ashbaugh 1972; Tiner 1985a; Leck and Simpson 1995; Tiner and Burke 1995; Leck et al. 2009; Perry et al. 2009).

Like their salt marsh counterparts, tidal fresh marshes often exhibit a marked zonation of species in relation to tidal flooding—low marsh and high marsh (Figure 5.24). A regularly flooded, low marsh is evident along the water's edge, with an irregularly flooded, high marsh above. Along the Merrimack River, the low marsh was flooded for 16 hours per day (Caldwell and Crow 1992). Low marsh species include wild rice, arrowheads (*Sagittaria latifolia, S. cuneata, S. graminea, S. rigida*), spatterdock (*Nuphar luteum*), arrow arum, pickerelweed, water hemp, common three-square, spike-rushes, soft-stemmed bulrush, and sometimes swamp loosestrife (*Decodon verticillatus*). Just above this zone, wild rice, arrow

Table 5.9. Annual herbs of tidal fresh marshes in the Northeast

Common Name (*Scientific name*)
Sensitive joint vetch (*Aeschynomene virginica*)*
Water hemp (*Amaranthus cannabinus*)
Giant ragweed (*Ambrosia trifida*)
Pink ammania (*Ammannia latifolia*)
Nodding bur-marigold (*Bidens cernua*)
Devil's beggar-ticks (*Bidens frondosa*)
Bur-marigold (*Bidens laevis*)
Pygmyweed (*Crassula aquatica*)
Fragrant flatsedge (*Cyperus odoratus*)
Dodder (*Cuscuta gronovii*)
Overlooked hedge hyssop (*Gratiola neglecta*)
Jewelweed (*Impatiens capensis*)
Bearded sprangletop (*Leptochloa fusca*)
False pimpernel (*Lindernia dubia*)
Marsh dayflower (*Murdannia keisak*)
Clearweed (*Pilea pumila*)
Fall panic grass (*Panicum dichotomiflorum*)
Halberd-leaved tearthumb (*Polygonum arifolium*)
Pinkweed (*Polygonum pensylvanicum*)
Dotted smartweed (*Polygonum punctatum*)
Arrow-leaved tearthumb (*Polygonum sagittatum*)
Mock bishopweed (*Ptilimnium capillaceum*)
Swamp dock (*Rumex verticillatus*)
Annual salt marsh aster (*Symphyotrichum subulatum*)
Wild rice (*Zizania aquatica*)

Note: *denotes federally threatened species.

arum, and big cordgrass are still present but plant diversity increases and includes many of the plants listed in Table 5.8. Other tidal fresh marsh species common in the northeastern United States include sweet flag (*Acorus calamus*), bur-marigold, sneezeweed (*Helenium autumnale*), spotted Joe-Pye-weed (*Eupatoriadelphus maculatus*), rice cutgrass (*Leersia oryzoides*), Walter millet (*Echinochloa walteri*), tearthumbs, dotted smartweed (*Polygonum punctatum*), broad-leaved cattail, tussock sedge (*Carex stricta*), marsh fern, and royal fern (*Osmunda regalis*). Sandy or gravelly areas may have Parker's pipewort (*Eriocaulon parkeri*), riverbank quillwort (*Isoetes riparia*), dwarf St. John's-wort (*Hypericum mutilum*), bog rush (*Juncus pelocarpus*), golden-pert (*Gratiola aurea*), and others (Ferren et al. 1981). Some halophytic species extend their distribution into these marshes including salt marsh sedges, marsh straw sedge (*Carex hormathodes*), Baltic rush, prairie cordgrass, silverweed, narrow-leaved cattail, water hemp, rose mallow, annual salt marsh aster (*Symphyotrichum subulatum*), salt marsh bulrush, and marsh orach.

Shrubs, such as alders (*Alnus* spp.), swamp rose, buttonbush (*Cephalanthus occidentalis*), willows (*Salix* spp.), silky dogwood (*Cornus amomum*), false indigo (*Amorpha fruticosa*), and southern arrow-wood (*Viburnum dentatum*), plus saplings of red maple (*Acer rubrum*), green ash (*Fraxinus pennsylvanica*), and black willow (*Salix nigra*) may be scattered in clusters at higher elevations within these marshes, especially near the marsh-upland border. Poison ivy may grow here as an erect shrub over 6 feet (2 m) tall. These and other shrubs may form tidal shrub swamps, but these swamps are not common when compared with the tidal fresh marshes and tidal forests. Buttonbush often characterizes the wettest shrub swamps, while speckled alder (*Alnus rugosa*), smooth alder (*A. serrulata*), and seaside alder (*A. maritima*) become dominant species along some rivers. The latter species is rare, with restricted distribution in the eastern United States where it forms tidal shrub thickets on Maryland's Eastern Shore (e.g., Marshyhope Creek). Sweet flag usually characterizes the herb stratum along with jewelweed, arrow arum, rice cutgrass, and other herbs (Harrison and Stango 2003). Climbing hempweed may cover much of the shrub canopy in some tidal shrub swamps. In Delaware and Maryland, wax myrtle forms a dense shrub thicket along the landward edge of many tidal fresh marshes and may also be found along tidal creeks. Groundsel-bush and creeping spike-rush (*Eleocharis fallax*) are common associates in the former landscape position, whereas swamp rose and marsh fern are characteristic of the latter (Harrison and Stango 2003).

Tidal fresh marshes along the St. Lawrence River in Quebec may possess many of the species found along the Merrimack River (Table 5.8) along with endemic plants that include varieties of several species: tufted hairgrass (*Deschampsia cespitosa* var. *intercotidalis*), wild rice (*Zizania aquatica* var. *brevis*), hairy willow-herb (*Epilobium ciliatum* var. *ecomosum*), poison hemlock (*Cicuta maculata* var. *victorinii*), fringed gentian (*Gentianopsis crinita* var. *victorinii*), water horehound (*Lycopus americanus* var. *laurentianus*), overlooked hedge hyssop (*Gratiola neglecta* var. *glaberrima*), square-stemmed monkeyflower (*Mimulus ringens* var. *colpophilus*), three-lobe beggar-ticks (*Bidens tripartita* var. *orthodoxa*), and sneezeweed (*Helenium autumnale* var. *fylesii*) (Baillargeon 1981 reported in Glooschenko and Grondin 1988). More than 60 other species may occur in these very diverse marshes. Common three-square is a particularly abundant species.

Regional Differences. Tidal fresh marshes are common along many southern rivers and many of the species of northeastern marshes dominate here as well (Figure 5.25). Two types of tidal marshes occur: 1) the typical marsh dominated by rooted emergent; and 2) the floating marsh that forms in shallow water, especially along slow-moving rivers and tidal embayments (e.g., tidal lakes). Giant cutgrass (*Zizaniopsis miliacea*), maidencane, cattails (*Typha latifolia, T. domingensis*), and other freshwater species predominate, plus others that occurred in brackish marshes are still common or locally dominant, including lance-leaved arrowhead (*Sagittaria lancifolia*), common reed, salt hay grass, narrow-leaved cattail, big cordgrass, deer pea, sand cordgrass, alligatorweed, saw grass, arrow arum, pickerelweed, common three-square, smartweeds, soft-stemmed bulrush, Walter millet, and wild rice. Other characteristic tidal fresh marsh species are jointed spike-rush (*Eleocharis equisetoides*), wood reed (*Cinna arundinacea*), umbrella sedges (*Cyperus* spp.), sedges, swamp dock (*Rumex verticillatus*), bur marigold, beggar-ticks (*Bidens* spp.), rice cutgrass (*Leersia oryzoides*), water primroses (*Ludwigia* spp.), marsh dayflower (*Murdannia keisak*),

Figure 5.25. Louisiana freshwater tidal marsh dominated by southern wild rice (*Zizaniopsis miliacea*).

water parsnip, water hemlock (*Cicuta maculata*), tearthumbs, American cup-scale (*Sacciolepis striata*), jewelweed, false nettle (*Boehmeria cylindrica*), marsh fern, royal fern, climbing hempweed, soft rush, sweet flag, marsh pennywort (*Hydrocotyle umbellata*), mock bishop-weed, asters, butterweed (*Senecio glabellus*), cardinal flower (*Lobelia cardinalis*), and marsh eryngo (*Eryngium aquaticum*). Shrubs and scattered trees may occur within these marshes including many species found in more northern marshes. Southern additions to the community include water locust (*Gleditsia aquatica*), swamp or Carolina willow (*Salix caroliniana*), bald cypress, southern swamp lily, and marsh spider lily (*Hemerocallus crassifolia*). In California's Sacramento-San Joaquin Delta, tule or bulrushes (*Schoenoplectus acutus, S. californicus*), Olney three-square, cattails, common reed, and arroyo willow (*Salix lasiolepis*) are common species (Atwater et al. 1979). Associates include water smartweed (*Polygonum amphibium* var. *emersum*), giant bur-reed, Pacific silverweed, and hedge bindweed (*Calystegia sepium*). In the Pacific Northwest, some tidal fresh marshes are included among "surge plain communities," which are flooded or saturated by high tides (Kunze 1994). Dominant marsh species include European bur-reed (*Sparganium emersum*), Lyngbye's sedge, slough sedge (*Carex obnupta*), pale spike-rush (*Eleocharis macrostachya*), marsh horsetail (*Equisetum fluviatile*), hard-stemmed bulrush, common three-square, narrow-leaved cattail, broad-leaved cattail, and marsh marigold (*Caltha palustris* ssp. *asarifolia*).

Floating marshes are common in the Mississippi River Delta where they represent 60 to 70 percent of Louisiana's freshwater wetlands (Sasser et al. 2009). Dominant flotant species include maidencane, alligatorweed, bull-tongue, spike-rushes (e.g., *E. baldwinii, E. parvula*), American frog-bit (*Limnobium spongia*), water pennywort (*Hydrocotyle ranunculoides*), water hyacinth (*Eichhornia crassipes*), and water lettuce (*Pistia stratiotes*). Floating marshes also occur along the shores of many other southern rivers. Actively forming marshes in the Mississippi Delta may be colonized by several species depending on the successional stage including delta bulrush (*Schoenoplectus deltarum*), big arrowhead, bull-tongue, soft-stemmed bulrush, cattails, rice cutgrass, giant cutgrass, American frog-bit, dotted smartweed, looseflower water-willow (*Justicia ovata* var. *lanceolata*), elephant-ear (*Colocasia esculenta*), river seedbox (*Ludwigia leptocarpa*), marsh fern, peat mosses, wax myrtle, and black willow (Shaffer et al. 1992; White 1993; Evers et al. 1998; Sasser et al. 1996, 2009). Mats comprise more than 75 percent organic matter (by dry weight) and vary in thickness. Some mats float continuously over water, while others float in summer and some are occasionally inundated. Vertical fluctuations in the mat range from about an inch (3 cm) to 3.6 feet (110 cm) during the year (Sasser et al. 2009).

Tidal Freshwater Swamps

Forested wetlands often border tidal fresh marshes and eventually replace them as the dominant alluvial wetland (Figure 5.26). Most tidal swamps in the northeastern United States are flooded for shorter periods than are the low-lying marshes, although the wetter swamps may possess microtopography that allows for longer inundation in the hollows (depressions) than on the hummocks, or may have elevations low enough for daily tidal flooding. Of all the coastal wetlands, the tidal swamps are the least studied. The uppermost tidal swamps often go relatively unnoticed, since they are virtually indistinguishable from neighboring nontidal floodplain swamps. Upon closer inspection, however, one might find that their composition differs somewhat, but such investigations have not been conducted.

In most of the North and Middle Atlantic region, dominant trees consist of red maple

Figure 5.26. Changes from marsh to forested wetland can be abrupt or more gradual: (a) abrupt change along Barren Creek; (b), (c), and (d) show changes along Maryland's Choptank River within a distance of four miles. Note the encroachment of tidal swamp near the upland edge (b); expanding its coverage (c); and replacing the marsh (d).

and green ash (*Fraxinus pennsylvanica* var. *subintegerrima*), with black gum (*Nyssa sylvatica*) becoming increasingly important as one moves south (Table 5.10; Figure 5.27). Along the St. Lawrence River, crack willow (*Salix fragilis*), silver maple (*Acer saccharinum*), and American elm (*Ulmus americana*) are significant species (Couillard and Grondin 1986; Martin Jean, pers. comm. 2004). Along the St. John River in New Brunswick, silver maple swamps are characteristic of the tidal floodplain. Green ash, black ash (*F. nigra*), and red maple are the most abundant trees in tidal swamps on the Hudson River while 223 other plant species have also been recorded (Westad and Kiviat 1986). Other arborescent species include black willow and ironwood (*Carpinus caroliniana*); and in more southerly areas, bald cypress, sweet bay (*Magnolia virginiana*), river birch (*Betula nigra*), loblolly pine, and sweet gum (*Liquidambar styraciflua*). Less common trees are sycamore (*Platanus occidentalis*), persimmon (*Diospyros virginiana*), and Atlantic white cedar (*Chamaecyparis thyoides*).

Tidal forests have a rich shrub understory that may include buttonbush, willows, swamp rose, silky dogwood, highbush blueberry (*Vaccinium corymbosum*), wax myrtle, groundsel-bush, alders, southern

Table 5.10. Examples of tidal swamps in the Mid-Atlantic region

Dominance Type (Location)	Associates
Red maple (*Acer rubrum*) (New Jersey)	Trees: Sweet bay (*Magnolia virginiana*), American holly (*Ilex opaca*), Black gum (*Nyssa sylvatica*), Sweet gum (*Liquidambar styraciflua*) Shrubs: Swamp azalea (*Rhododendron viscosum*), Sweet pepperbush (*Clethra alnifolia*), Common elderberry (*Sambucus nigra* ssp. *canadensis*), Serviceberry (*Amelanchier canadensis*), Southern arrowwood (*Viburnum dentatum*) Herbs: Cinnamon fern (*Osmunda cinnamomea*), Jewelweed (*Impatiens capensis*), Sensitive fern (*Onoclea sensibilis*), Soft rush (*Juncus effusus*), Smartweed (*Polygonum* sp.), Arrow arum (*Peltandra virginica*), Tall meadow-rue (*Thalictrum pubescens*), Three-way sedge (*Dulichium arundinaceum*) Others: Poison ivy (*Toxicodendron radicans*), peat mosses (*Sphagnum* sp.), other mosses
Red maple–Sweet Gum (New Jersey)	Trees: Sweet bay Shrubs: Sweet pepperbush, Southern arrowwood, Fetterbush (*Eubotrys racemosa*), Common elderberry, Swamp azalea, Swamp cottonwood (*Populus heterophylla*), Willow (*Salix* sp.) Herbs: Arrow arum, Sensitive fern, Sweet flag (*Acorus calamus*), White grass (*Leersia virginica*), Narrow-leaved cattail (*Typha angustifolia*), Sedges (*Carex* spp.)
Red maple (Delaware)	Trees: Green ash, Sweet gum Shrubs: Spicebush (*Lindera benzoin*) Herbs: Sedges Vines: Grape (*Vitis* sp.), Trumpet creeper (*Campsis radicans*), Japanese honeysuckle (*Lonicera japonica*)
Green ash–Red maple (Delaware)	Shrubs: Highbush blueberry (*Vaccinium corymbosum*), Smooth alder (*Alnus serrulata*) Herbs: Sedges, Wood reed (*Cinna arundinacea*) Others: Common greenbrier (*Smilax rotundifolia*), Poison ivy, mosses
Red maple–Green ash (Maryland)	Trees: Black gum, Sweet bay Shrubs: Winterberry (*Ilex verticillatus*), Highbush blueberry, Southern arrowwood Herbs: Wood reed (*Cinna arundinacea*), Net-veined chain fern (*Woodwardia areolata*), Jewelweed (*Impatiens capensis*), Sedges, Tall meadow-rue (*Thalictrum pubescens*), White grass, Goldenrod (*Solidago* sp.) Others: Poison ivy, Grape, Common greenbrier
Green ash–Black gum (Maryland)	Trees: Sweet gum, River birch (*Betula nigra*), Red maple Shrubs: Winterberry Herbs: Lizard's tail (*Saururus cernuus*) Vines: Cross vine (*Bignonia capreolata*)
Sweet gum–Black gum (Maryland)	Trees: Black willow (*Salix nigra*) Shrubs: Spicebush, Smooth alder, Common elderberry Herbs: False nettle (*Boehmeria cylindrica*), Jewelweed, Virginia chain fern (*Woodwardia virginica*), Sensitive fern, Lizard's tail, Goldenrod Others: Poison ivy, Common greenbrier, Blackberry (*Rubus* sp.), Dodder (*Cuscuta gronovii*)
Bald cypress (*Taxodium distichum*) (Maryland)	Trees: Red maple, Green ash Shrubs: Southern arrowwood, Smooth alder Herbs: Lizard's tail, False nettle, Wood reed, Turtlehead (*Chelone glabra*), Sedge, Clearweed (*Pilea pumila*), Bugleweed (*Lycopus* sp.) Vines: Virginia creeper (*Parthenocissus quinquefolia*)

Table 5.10. (*continued*)

Dominance Type (Location)	Associates
Bald cypress–Red maple–Green ash (Maryland)	Trees: American elm (*Ulmus americana*) Shrubs: Smooth alder, Winterberry, Southern arrowwood, Pawpaw (*Asimina triloba*), Silky dogwood (*Cornus amomum*), Swamp rose (*Rosa palustris*) Herbs: Sensitive fern, Jewelweed, Lizard's tail, Fringed sedge (*Carex crinita*), Cardinal flower (*Lobelia cardinalis*), Turtlehead, Arrow-leaved tearthumb (*Polygonum sagittatum*), Jack-in-the-pulpit (*Arisaema triphyllum*) Others: Poison ivy, Common greenbrier, Japanese honeysuckle, Grape
Loblolly pine–Wax Myrtle (*Pinus taeda–Morella cerifera*) (Maryland)	Trees: Sweet gum Herbs: Cinnamon fern, Royal fern (*Osmunda regalis*), Sensitive fern Vines: Poison ivy, Virginia creeper, Common greenbrier, Trumpet creeper, Grape

Sources: Personal observations; Tiner and Burke 1995.

arrowwood, sweet pepperbush (*Clethra alnifolia*), spicebush (*Lindera benzoin*), common winterberry (*Ilex verticillata*), swamp azalea (*Rhododendron viscosum*), and fetterbush (*Eubotrys racemosa*). Less common shrubs are black haw (*Viburnum prunifolium*), red chokeberry (*Photinia pyrifolia,* formerly *Aronia arbutifolia*), common elderberry (*Sambucus nigra* ssp. *canadensis,* formerly *S. canadensis*), and maleberry (*Lyonia ligustrina*). Vines are often seen climbing up shrubs and trees, including common greenbrier (*Smilax rotundifolia*), laurel-leaved greenbrier (*S. laurifolia*), climbing hempweed, poison ivy, Virginia creeper (*Parthenocissus quinquefolia*), Japanese honeysuckle (*Lonicera japonica*), and trumpet creeper (*Campsis radicans*). In Maryland and Delaware, cross vine (*Bignonia capreolata*) and American or oak mistletoe (*Phoradendron leucarpum,* formerly *P. flavescens*), a parasitic epiphyte on deciduous trees, may be present (Tiner and Burke 1995).

Many herbs can be found in tidal swamps, especially in the hollows and in more open canopy swamps. Some of the more frequently occurring species include skunk cabbage (*Symplocarpus foetidus*), jewelweed, cardinal flower, lizard's tail (*Saururus cernuus*), royal fern, cinnamon fern (*Osmunda cinnamomea*), sensitive fern (*Onoclea sensibilis*), halberd-leaved tearthumb (*Polygonum arifolium*), and tussock sedge. Less abundant species include wood reed, marsh horsetail, arrow-leaved tearthumb (*P. sagittatum*), manna grasses (*Glyceria* spp.), blue flag (*Iris versicolor*), arrow arum, rose mallow, and other smartweeds.

Regional Differences. Tidal influence extends long distances up numerous southern rivers, so tidal swamps are common features on the coastal plain (Figure 5.27). The wettest of southeastern tidal freshwater forested wetlands are cypress-gum swamps (bald cypress, water gum, and swamp black gum or swamp tupelo [*Nyssa biflora*]). Other southern tidal swamps are dominated by sweet bay, water locust, red maple, elms, ashes, cabbage palm, river birch, black willow, water hickory (*Carya aquatica*), laurel oak (*Quercus laurifolia*), ironwood, loblolly pine, and sweet gum (Table 5.11). Less common trees include sycamore, persimmon, Atlantic white cedar, tulip poplar (*Liriodendron tulipifera*), and the invasive Chinese tallow (*Triadica sebifera*). Locally, Atlantic white cedar may co-dominate with sweet bay.

a

b

c

Figure 5.27. Tidal swamps are common on the Atlantic Coastal Plain: (a) red maple–green ash (*Acer rubrum–Fraxinus pennsylvanica*) swamp (NJ); (b) tidal swamp behind regularly flooded marsh (DE); and (c) water tupelo (*Nyssa aquatica*) swamp (VA).

Table 5.11. Some trees and shrubs reported in tidal forested wetlands in southern swamps

Species	Location				
	MD	VA	GA	FL	MS
Trees					
Green ash (*Fraxinus pennsylvanica*)	X	X	X*	—	X*
Red maple (*Acer rubrum*)	X	X	X	X	X
Swamp tupelo (*Nyssa sylvatica biflora*)	X	X	X	X	X
Sweet bay (*Magnolia virginiana*)	X	X	—	X	X
Ironwood (*Carpinus caroliniana*)	X	X	X	X	—
Atlantic white cedar (*Chamaecyparis thyoides*)	X	—	—	—	X
Wax myrtle (*Morella cerifera*)	X	—	X	X	X
Black gum (*Nyssa sylvatica*)	—	X	—	—	—
American elm (*Ulmus americana*)	—	X	—	X	—
Slippery elm (*Ulmus rubra*)	—	X	—	—	—
American beech (*Fagus grandifolia*)	—	X	—	—	—
Sweet gum (*Liquidambar styraciflua*)	—	X	X	X	X
Bitternut hickory (*Carya cordiformis*)	—	X	—	—	—
Swamp chestnut oak (*Quercus michauxii*)	—	X	—	—	—
American holly (*Ilex opaca*)	—	X	—	—	X
Bald cypress (*Taxodium distichum*)	—	X	X	X	X
Water tupelo (*Nyssa aquatica*)	—	—	X	X	X
Black willow (*Salix nigra*)	—	—	X	—	—
Pumpkin ash (*Fraxinus profunda*)	—	X	—	X	—
Carolina ash (*Fraxinus caroliniana*)	—	—	—	X	X*
Water-elm (*Planera aquatica*)	—	—	X	X	—
Cabbage palm (*Sabal palmetto*)	—	—	—	X	—
Laurel oak (*Quercus laurifolia*)	—	—	X	X	X
Loblolly pine (*Pinus taeda*)	—	—	—	X	X
Swamp bay (*Persea palustris*)	—	—	X	X	X
Swamp titi (*Cyrilla racemiflora*)	—	—	—	—	X
Buckwheat tree (*Cliftonia monophylla*)	—	—	—	—	X
Water oak (*Quercus nigra*)	—	—	X	—	X
Darlington's oak (*Quercus hemisphaerica*)	—	—	—	—	X
Southern magnolia (*Magnolia grandiflora*)	—	—	—		X
Persimmon (*Diospyros virginiana*)	—	—	—	X	X
Live oak (*Quercus virginiana*)	—	—	—	X	X
Eastern redbud (*Cercis canadensis*)	—	—	—	—	X
Fringe-tree (*Chionanthus virginicus*)	—	—	—	—	X
River birch (*Betula nigra*)	—	—	—	X	—
Water locust (*Gleditsia aquatica*)	—	—	—	X	—
Shrubs					
Swamp azalea (*Rhododendron viscosum*)	X	—	—	—	—
Common winterberry (*Ilex verticillata*)	X	X	X	—	—
Arrowwood (*Viburnum dentatum*)	X	—	X	—	—
Sweet pepperbush (*Clethra alnifolia*)	X	X	—	—	X
Spicebush (*Lindera benzoin*)	X	X	—	—	—
Highbush blueberry (*Vaccinium corymbosum*)	X	X	X	—	X*
Virginia sweetspire (*Itea virginica*)	X	—	X	—	—
Fetterbush (*Eubotrys racemosa*)	X	—	X	—	—
Silky dogwood (*Cornus amomum*)	X	—	—	—	—
Serviceberry (*Amelanchier canadensis*)	—	X	—	—	—

Table 5.11. (*continued*)

Species	Location				
	MD	VA	GA	FL	MS
Shrubs (*continued*)					
Common elderberry (*Sambucus nigra* ssp. *canadensis*)	—	—	X	—	—
Buttonbush (*Cephalanthus occidentalis*)	—	—	X	X	—
Stiff dogwood (*Cornus foemina*)	—	—	X	X	—
Southern wild raisin (*Viburnum nudum*)	—	—	X	—	—
Possumhaw (*Ilex decidua*)	—	—	X	—	—
Southern alder (*Alnus serrulata*)	—	—	X	—	—
Groundsel-bush (*Baccharis halimifolia*)	—	—	X	—	—
Deerberry (*Vaccinium stamineum*)	—	—	—	—	X
Elliott's blueberry (*Vaccinium elliottii*)	—	—	—	—	X
Hawthorn (*Crataegus* sp.)	—	—	—	X	X
St. Andrew's cross (*Hypericum hypericoides*)	—	—	—	—	X
Dahoon (*Ilex cassine*)	—	—	X	—	X
Large gallberry (*Ilex coriacea*)	—	—	—	—	X
Inkberry (*Ilex glabra*)	—	—	X	—	X
Georgia holly (*Ilex longipes*)	—	—	—	—	X
Yaupon (*Ilex vomitoria*)	—	—	—	—	X
Maleberry (*Lyonia ligustrina*)	—	—	—	—	X
Flameleaf sumac (*Rhus copallinum*)	—	—	—	—	X
Saw palmetto (*Serenoa repens*)	—	—	—	—	X
Farkleberry (*Vaccinium arboreum*)	—	—	—	—	X

Sources: MD = Maryland, Nanticoke River (Baldwin 2007); VA = Virginia, Pamunkey River (Rheinhardt 2007); GA = Georgia, Savannah River (Duberstein and Kitchens 2007); FL = Florida, Lower Swannee River (Light et al. 2007); MS = Mississippi, Atlantic white cedar swamp (Keeland and McCoy 2007).
Notes: * Species not specified for this site. *Cyrilla, Cliftonia, Cercis,* and *Chionanthus* are short trees that occur also in shrub forms, while *Crataegus* typically occurs as trees.

Associated shrubs include buttonbush, willows, bluestem palmetto (*Sabal minor*), wax myrtle, groundsel-bush, smooth alder, red bay (*Persea borbonia*), southern red cedar, water elm (*Planera aquatica*), and, rarely, titi (*Cyrilla racemiflora*). Common vines may be present, such as climbing hempweed, pepper-vine (*Ampelopsis arborea*), and poison ivy. The herbaceous layer may consist of sand cordgrass, royal fern, marsh fern, saw grass, sedges, umbrella sedges, lizard's tail, marsh dayflower, arrow arum, arrow-leaved tearthumb, rice cutgrass, water hemlock, large-fruit beggar-ticks (*Bidens coronata*), bur marigold, southern blue flag, butterweed, and asters. *Ecology of Tidal Freshwater Forested Wetlands of the Southeastern United States* (Conner et al. 2007a) provides detailed descriptions of coastal plain tidal swamps and their ecology.

Only remnants of tidal freshwater swamps exist in California. Species that occupy such sites include Oregon ash (*Fraxinus latifolia*), willows (*S. exigua, S. goodingii, S. lasiolepis*), box-elder (*Acer negundo*), Fremont cottonwood (*Populus fremontii*), Southern California walnut (*Juglans californica*), valley oak (*Quercus lobata*), common elderberry, California wild rose (*Rosa californica*), California blackberry (*Rubus ursinus*), California wild grape (*Vitis californica*), western white clematis (*Clematis ligusticifolia*), and California dutchman's pipe (*Aristolochia californica*) (J. Collins, pers. comm. 2011).

In the Pacific Northwest, tidal floodplains may be represented by Oregon ash with

Table 5.12. Common vascular aquatic bed plants found in tidal waters of the Northeast

Habitat	Common Name (*Scientific name*)
Marine waters	Eel-grass (*Zostera marina*)
Brackish waters	Widgeon-grass (*Ruppia maritima*), Sago pondweed (*Stuckenia pectinata*), Clasping-leaved pondweed (*Potamogeton perfoliatus*), Curly pondweed (*P. crispus*), Horned pondweed (*Zannichellia palustris*), Slender naiad (*Najas flexilis*), Eurasian water milfoil (*Myriophyllum spicatum*)
Slightly brackish and fresh waters	Wild celery (*Vallisneria americana*)
Fresh waters	Southern Naiad (*Najas guadalupensis*), Waterweeds (*Elodea* spp.), Coontail (*Ceratophyllum demersum*), Duckweeds (*Lemna* spp., *Spirodela polyrhiza*), Pondweeds (*Potamogeton epihydrus, P. pusillus, P. foliosus*), Bullhead lily (*Nuphar lutea* ssp. *variegata*), White water lily (*Nymphaea odorata*), South American elodea (*Egeria densa*), Marsh mermaid-weed (*Proserpinaca palustris*), Water shield (*Brasenia schreberi*), Fanwort (*Cabomba caroliniana*), Floating-heart (*Nymphoides aquatica*), Water pennywort (*Hydrocotyle ranunculoides*), Bladderworts (*Utricularia* spp.), Mosquito-fern (*Azolla caroliniana*)

stinging nettle (*Urtica dioica*), balsam poplar (*Populus balsamifera* ssp. *trichocarpa*) with red osier dogwood (*Cornus sericea* ssp. *sericea*) and jewelweed, and Sitka spruce with red alder (*Alnus rubra*) and slough sedge, whereas tidal shrub swamps are characterized by species including Sitka willow (*Salix sitchensis*), Pacific willow (*S. lucida* ssp. *lasiandra*), dune willow (*S. hookeriana*), and red osier dogwood (Kunze 1994; Christy and Brophy 2007). Associated shrubs may include Pacific crabapple, Douglas or rose spirea (*Spiraea douglasii*), Nootka rose (*Rosa nutkana*), and salmonberry (*Rubus spectabilis*).

Coastal Aquatic Beds

Aquatic beds typically grow in open water ranging from marine embayments and estuarine coastal ponds to small pools within tidal marshes and impoundments. They are represented by colonies of three types of plants: floating-leaved rooted, free-floating, and submerged (underwater). They include both algae and vascular plants. In northern areas, marine or estuarine aquatic beds may be exposed to air during low tides, especially spring low tides. Strong offshore winds may expose these beds in microtidal embayments (e.g., Barnegat Bay, NJ). The algal beds of rocky shores are another type of coastal aquatic bed, although an intertidal one growing on exposed shores (see discussion under Rocky Shores). Like other coastal wetlands, the composition of the aquatic beds dramatically changes from marine waters to tidal fresh waters, with a similar, corresponding increase in diversity (Table 5.12).

A few introduced (nonnative) aquatics have become established in some estuaries. Water chestnut (*Trapa natans*), a Eurasian species, covers much of the nearshore waters of the tidal fresh reach of Hudson River and can be found in fresh to slightly brackish waters of upper Chesapeake Bay. Hydrilla (*Hydrilla verticillata*) from Southeast Asia has colonized the Potomac River, whereas Eurasian water-milfoil (*Myriophyllum spicatum*) and South American elodea (*Egeria densa*) are widespread throughout the Northeast and elsewhere.

Regional Differences. In southern marine waters and high-salinity estuarine bays, submerged aquatic beds are represented by eelgrass, widgeon-grass, turtle-grass (*Thalassia testudinum*), manatee-grass (*Cymodocea filiformis*), shoal-grass (*Halodule wrightii*), and Engelmann's sea-grass (*Halophila engelmannii*). Brackish and tidal freshwater aquatics are many of the same species that dominate more northern waters. Water hyacinth and water lettuce are two other species of note due to their invasive nature, while floating beds of alligatorweed, water pennywort, mosquito-fern (*Azolla caroliniana*), American frog-bit, maidencane, and bladderworts are also common in southern tidal freshwaters. Sea grasses on the Pacific Coast include two species of eel-grass (*Z. marina, Nanozostera japonica*), two species of widgeon-grass (*R. maritima, R. cirrhosa*), horned pondweed, three surfgrasses (*Phyllospadix torreyi, P. scouleri, P. serrulatus*), sago pondweed (*Stuckenia pectinata*) and other pondweeds (*Potamogeton* spp.) in fresher waters.

Further Readings

Nonvegetated Tidal Wetlands

The Ecology of Intertidal Flats of North Carolina: A Community Profile (Peterson and Peterson 1979)

The Ecology of New England Tidal Flats: A Community Profile (Whitlatch 1982)

The Biology of Soft Shores and Estuaries (Little 2000)

The Biology of Rocky Shores (Little et al. 2009)

Vegetated Tidal Wetlands

BROAD RANGE

Wet Coastal Ecosystems (Chapman 1977)

The Ecology of a Salt Marsh (Pomeroy and Wiegert 1981)

"Comparative ecology of tidal freshwater and salt marshes" (Odum 1988)

Tidal Freshwater Wetlands (Barendregt et al. 2009)

CANADA

"Salt marshes of Canada" (Glooschenko et al. 1988)

EAST COAST

The Ecology of New England High Salt Marshes: A Community Profile (Nixon 1982)

The Ecology of the Mangroves of South Florida: A Community Profile (Odum et al. 1982)

The Ecology of Tidal Freshwater Marshes of the United States East Coast: A Community Profile (Odum et al. 1984)

The Ecology of Irregularly Flooded Salt Marshes of the Northeastern Gulf of Mexico: A Community Profile (Stout 1984)

The Ecology of Regularly Flooded Salt Marshes of New England: A Community Profile (Teal 1986)

Tidal Salt Marshes of the South Atlantic Coast: A Community Profile (Wiegert and Freeman 1990)

Ecology and Management of Tidal Marshes—A Model for the Gulf of Mexico (Coultas and Hsieh 1997)

Ecology of Tidal Freshwater Forested Wetlands of the Southeastern United States (Conner et al. 2007a)

PACIFIC COAST

"Beach and salt marsh vegetation of the North American Pacific Coast" (Macdonald and Barbour 1974)

"Plant and animal communities of Pacific North American salt marshes" (Macdonald 1977)

The Ecology of Southern California Coastal Salt Marshes: A Community Profile (Zedler 1982)

The Ecology of San Francisco Bay Tidal Marshes: A Community Profile (Josselyn 1983)

The Ecology of Tidal Marshes of the Pacific Northwest Coast: A Community Profile (Seliskar and Gallagher 1983)

"Tidal marsh plants of the San Francisco Estuary" (Baye et al. 2000)

6 Tidal Wetlands as Wildlife Habitat

Like plants, animals have adapted to the rigors of the tidal wetland environment. Resident sessile animals have to cope with whatever the environment brings their way. They must develop adaptations to flooding, varying salinities, alternating wetting and drying, temperature changes, and oxygen deficits (e.g., Levin and Talley 2000). Most wetland animals are mobile and can usually avoid undesirable conditions, returning when conditions improve. Some animals may visit these wetlands seasonally, at favorable tide conditions for food, for protection from predators, or to avoid adverse weather conditions (coastal storms). Most birds common to northern wetlands migrate south to avoid harsh winters. Terrestrial animals may frequent wetlands when not inundated. The salt marsh snail (*Melampus bidentatus*) and the marsh periwinkle (*Littoraria irrorata,* formerly *Littorina irrorata*) both climb the stems of smooth cordgrass (*Spartina alterniflora*) during flood tides to avoid inundation and predation from fishes and crabs (Figure 6.1). Adults of the salt marsh snail may migrate within salt marshes to higher levels during spring tides and lower levels during neap tides (Price 1984). The ribbed mussel (*Geukensia demissa*) closes its shell at low tide to avoid exposure, and then opens it at high tide for feeding. The salt marsh killifish or mummichog (*Fundulus heteroclitus*) can survive some out-of-water exposure by breathing air, allowing it to survive in tidal pools with oxygen deficiencies (Halpin and Martin 1999). Mites may inhabit the hollow stems of grasses for protection. Prior to marsh flooding, fiddler crabs return to their burrows where they are virtually inactive under low oxygen conditions before emerging at low tide to feed (Vernberg and Vernberg 1972). Ants (e.g., *Crematogaster*) may use their large heads to close the opening of their nests when tides flood the marshes (Mendelssohn and Batzer 2006). Some benthic animals of mud and sandflats (e.g., lugworms) can live in anoxic substrates by excavating U-shaped burrows and pumping water through the burrows at high tide, which stores enough oxygen to survive stressful conditions at low tide (Little 2000). Other animals have developed internal mechanisms that detoxify hydrogen sulfide, and some have symbiotic bacteria in their gills that can extract energy from hydrogen sulfide.

Tidal wetlands serve many habitat functions. They may be feeding areas, resting or loafing areas, breeding grounds, refuges, or permanent homes. The vegetation and aquatic beds provide cover—refugia—for prey species and transient juveniles of predators (Zimmerman et al. 2000). The ebb and flow of the tides create quite varied and challenging environmental conditions. When flooded at high or spring tides, the wetland becomes an aquatic ecosystem providing fishes and marine invertebrates access to the riches of the marsh or swamp. At low tide, the wetland is exposed to air creating a more terrestrial habitat where spiders

Figure 6.1. Marsh periwinkles (*Littoraria irrorata*) climbing smooth cordgrass stems to avoid flooding and predation at high tide in a Georgia salt marsh. (Jan Mackinnon)

and insects may abound. Consequently, most tidal wetlands may be viewed as semi-aquatic, amphibious habitats, neither water nor land, but with properties of both depending on the moment. Food web interactions can be quite complex between aquatic and terrestrial organisms (Figure 6.2). Tidal flats are homes for benthic invertebrates and feeding grounds for fishes and waterfowl at high tide and for shorebirds and wading birds at low tide. Marshes are residences for some mammals and reptiles. While northern marshes are particularly important breeding and feeding grounds for many birds in summer, southern marshes offer year-round habitat. Tidal swamps are used by many species, much like their nontidal counterparts (forested wetlands), providing nesting habitat for colonial nesting birds, breeding habitat for a long list of migratory songbirds (neotropical migrants), and food and shelter for many terrestrial animals. In this chapter I introduce the use of tidal wetlands by various animals; more detailed treatments are found in other sources listed as citations or as additional readings. The absence of salt in the water makes tidal freshwater wetlands suitable habitat for many wildlife. Consequently, most of the frogs, toads, salamanders, reptiles, and mammals living in nontidal wetlands in the coastal zone can be found in tidal freshwater wetlands. The focus of this chapter will be on animals of salt and brackish marshes since tidal freshwater wetlands are habitats for thousands of animals typical of nontidal wetlands.

Invertebrate Habitat

Since tidal wetlands include nonvegetated flats, beaches, rocky shores, frequently flooded grasslands, and wet forests, they undoubtedly provide habitats for thousands of invertebrates. The environment, which is alternately flooded and exposed, creates conditions that at times provides habitat for aquatic species and at other times for more terrestrial species. Swimming invertebrates such as grass shrimp (*Palaemonetes*), juvenile penaeid shrimp (*Farfantepenaeus* and *Litopenaeus,* formerly *Penaeus*), and blue crab (*Callinectes sapidus*) utilize these wetlands at high tide, while other estuarine aquatic species may reside in the lower areas of the intertidal zone or in tidal pools or ponds where conditions are regularly or permanently flooded (worms, mussels, clams, and crabs). Terrestrial invertebrates including insects, mites, and spiders thrive in the upper intertidal zone where drier conditions exist for much of the time, but they can access lower elevations at low tide. Although variations exist, several macroinvertebrates are common in many salt marshes along the Atlantic and Gulf Coasts (Table 6.1; Figure 6.3). Mudflats may harbor more invertebrates

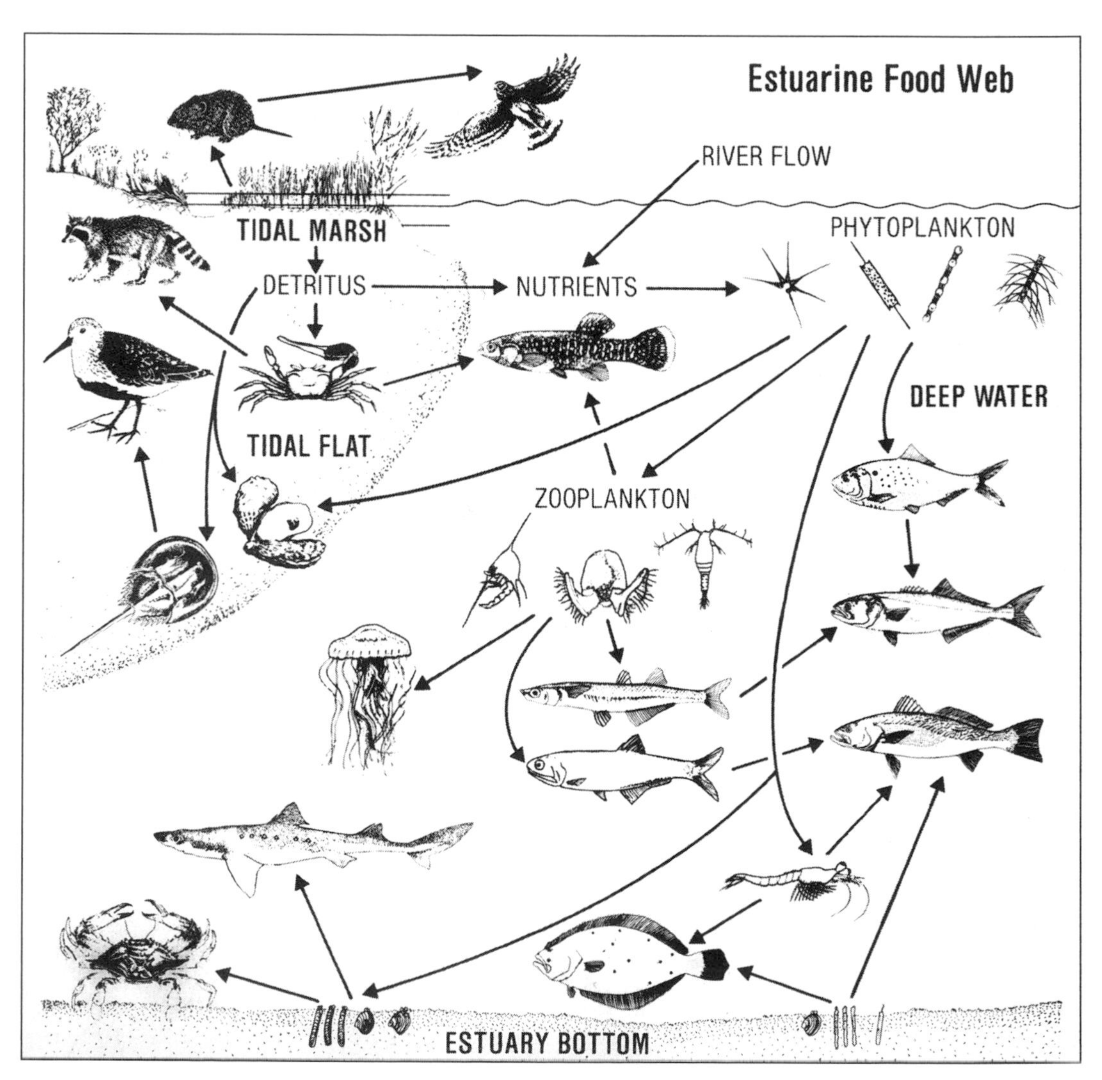

Figure 6.2. This generalized food web shows the complexity of interactions between plants and animals in the estuarine environment. Primary producers capture energy from the sun and convert it into biomass that serves as food for some animals, which are the food for others, and so forth. (Bryant and Pennock 1988; courtesy of University of Delaware, College of Earth, Ocean, and Environment and the Delaware Sea Grant College Program)

Table 6.1. Some characteristic invertebrates in Atlantic and Gulf Coast salt marshes, excluding insects and arachnids that occupy the marsh at low tide

State (Source)	Characteristic Species
Massachusetts (Dexter 1947)	Snails: Salt marsh snail (*Melampus bidentatus*), Periwinkles (*Littorina littorea, L. obtusata, L. saxatilis*), Mud snail (*Ilyanassa obsoleta*) Bivalves: Ribbed mussed (*Geukensia demissa*), Blue mussel (*Mytilis edulis*) Crabs: Green crab (*Carcinus maenas*) Others: Isopod (*Philoscia vittata*), Amphipods (*Orchestia, Gammarus*), Barnacles (*Balanus balanoides*)
Connecticut (Tiner 1974; Fell et al. 1982)	Snails: Salt marsh snail, Rough periwinkle (*L. saxatilis*), Snails (*Odostomia, Hydrobia*), Mud snail Bivalves: Ribbed mussel, Eastern oyster (*Crassostrea virginica*) Crabs: Mud fiddler crab (*U. pugnax*) Others: Isopods (*Philoscia, Jaera, Porcellio*), Amphipods (*Orchestia, Gammarus, Ampithoe*), Nudibranch (*Alderia modesta*), Nematodes, Polychaete worms (*Arabella iricolor, Nereis succinea*), Anemone (*Haliplanella luciae*), Barnacle (*Balanus*)
Virginia (Wass 1963; Wass and Wright 1969)	Snails: Salt marsh snail, Marsh periwinkle (*Littoraria irrorata*) Bivalves: Ribbed mussel, Carolina marsh clam (*Polymesoda caroliniana*) Crabs: Fiddler crabs (*Uca pugilator, U. pugnax, U. minax*), Wharf or marsh crabs (*Armases [Sesarma] reticulum, A. cinereum*) Others: Barnacles (*Chthalamus*), Amphipod (*Orchestia*)
Georgia (Teal 1962; Kneib 1984; Covi and Kneib 1995)	Snails: Salt marsh snail, Marsh periwinkle, Coastal marsh snail (*Littoridina tenuipes*) Bivalves: Ribbed mussel Crabs: Fiddler crabs (3 spp.), Wharf crabs (2 spp.), White-clawed mud crab (*Eurythium limosum*) Others: Polychaete worms (*Capitella capitata, Laeonereis culveri, Manayunkia aestuarina, Neanthes [Nereis] succinea, Streblospio benedicti, Fabricia* sp.), Oligochaete worms (3 spp.), Isopod (*Cyathura carinata*), Amphipod (*Uhlorchestia spartinophila*)
Northeastern Gulf of Mexico (Subrahmanyam et al. 1976; Heard 1982)	Snails: Olive nerite (*Neritina usnea*), Hydrobiid snail (*Littoridinops palustris, Heleobops* sp.), Marsh periwinkle, Salt marsh snail, Horn shells (*Cerithidea* spp.) Bivalves: Florida marsh clam (*Polymesoda maritima*), Carolina marsh clam, Dall's marsh clam (*Cyrenoida floridana*), Ribbed mussel, Eastern oyster (*Crassostrea virginica*) Crabs: Striped hermit crab (*Clibanarius vittatus*), Blue crab (*Callinectes sapidus*), White-clawed mud crab, Marsh crabs (*Armases* spp.), Fiddler crabs (*U. pugilator, U. panacea, U. longisignalis*) Others: Worms (*Oligochaeta, Polychaeta, Nematoda*), Salt marsh barnacle (*Chthamalus fragilis*), Amphipods (*Corophium, Gammarus, Grandidierella, Melita, Orchestia*), Isopod (*Cyathura polita*), Tanaids (*Halmyrapseudes bahamensis* and *Hargeria rapax*)

Sources: Heard (1982) contains descriptions of macroinvertebrates found in brackish marshes of the northeastern Gulf of Mexico, while a list of southern California salt marsh invertebrates can be found in Talley and Levin (1999).

Notes: List does not include migrants coming in at high tide such as grass shrimp (*Palaemonetes pugio*), penaeid shrimp, and blue crab (*Callinectes sapidus*).

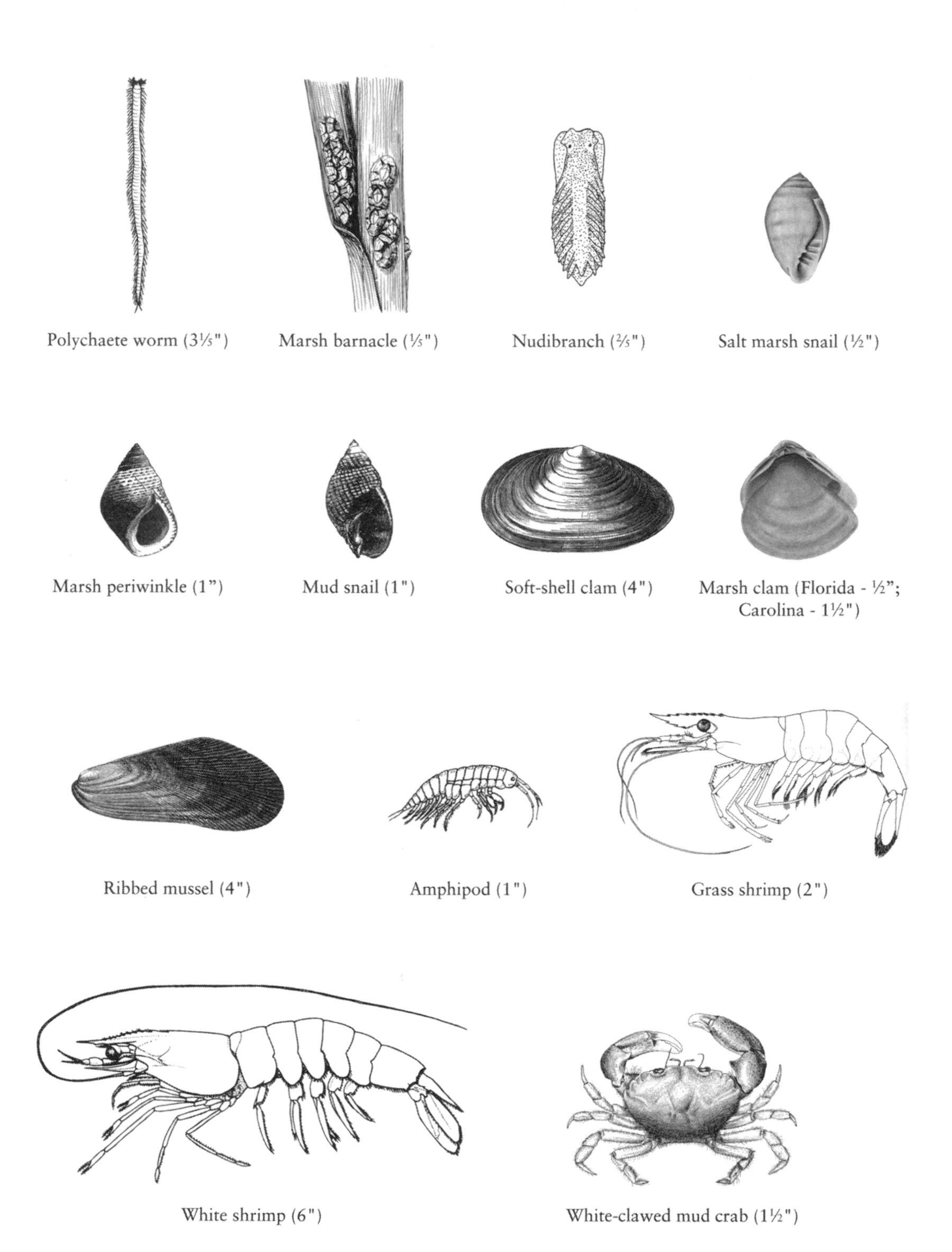

Figure 6.3. Some invertebrates living in or feeding in salt marshes. (Arnold 1901: ribbed mussel, soft-shelled clam, mud snail, marsh periwinkle, fiddler crab, green crab, and blue crab; Heard 1982: polychaete worm, salt marsh snail, marsh clam, salt marsh barnacle, grass shrimp, white shrimp, wharf crab, white-clawed crab, and black-clawed crab; Wiegert and Freeman 1990: salt marsh grasshopper; other drawings by the author)

Black-clawed mud crab (1½")

Marsh (wharf) crab (1")

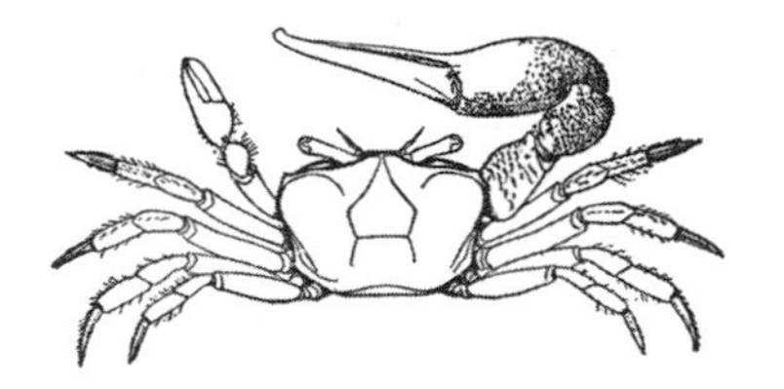

Fiddler crab (1½")

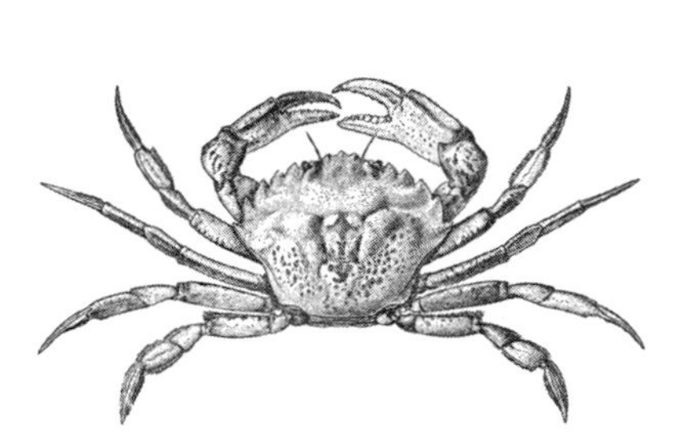

Green crab (to 4")

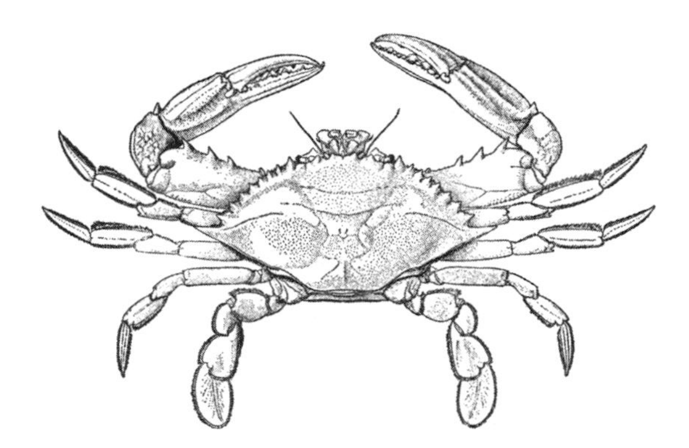

Blue crab (10+")

Greenhead (1⅛")

Salt marsh mosquito
(½"; striping on legs not shown)

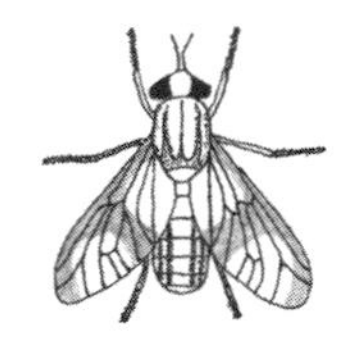

Deer fly (½")

Salt marsh grasshopper (⅔–¾")

Wolf spider (Lycosidae; ½")

than sandflats in New England do—62 different species including 31 polychaete worms, 8 bivalves/snails, and 6 amphipods were recorded in the former, whereas only 30 different species were found in the latter including 7 bivalves and 6 amphipods (Heck et al. 1995).

Most bivalves (clams) are found in the subtidal zone, but 25 species occupy habitats including the littoral zone from the Bay of Fundy to Cape Hatteras (Gosner 1971). Most of these species are found in the lowest tidal wetlands—the intertidal flats—in addition to deeper waters. The more familiar tidal flat bivalves include the hard clam or quahog (*Mercenaria mercenaria*), soft-shell clam (*Mya arenaria*), surf clam (*Spisula solidissima*), and razor clams (*Ensis directus, Tagelus gibbus*). The blue mussel (*Mytilis edulis*) forms enormous shellbeds in the intertidal zone of northern waters, while the eastern oyster (*Crassostrea virginica*) builds intertidal reefs along southern shores (see Figure 5.4). The most abundant bivalve in salt marshes is the ribbed mussel, which also inhabits low-salinity Gulf Coast marshes along with Rangia clams (*Rangia cuneata, R. flexuosa*), Carolina marsh clam (*Polymesoda caroliniana*), and Florida marsh clam (*Cyrenoida floridana*) (Bishop and Hackney 1987; Moore 1992). In a Connecticut salt marsh, the ribbed mussel was most numerous along the water's edge in the tall-form smooth cordgrass zone and inhabited in lesser numbers the wetter portion of the high marsh dominated by salt grass (Tiner 1974). Similarly, in a Georgia salt marsh, 46 percent of the ribbed mussels occurred in the tall *Spartina*-creek head zone, which represented only 6 percent of the marsh, while the medium-height *Spartina*-levee and short-form *Spartina*-low elevation marshes that collectively represented 37 percent of the marsh accounted for 32 percent of the mussels (Kuenzler 1961). The relationship between ribbed mussels and smooth cordgrass may be an example of facultative mutualism in which both organisms benefit from co-habitation. The presence of mussel beds helps stabilize shorelines and protect smooth cordgrass from erosion, while adding nutrients to the soils through deposition of pseudofeces that stimulate cordgrass root production. Meanwhile, the cordgrass stems provide the mussels with some means of protection from predators and a place for them to climb to avoid burial by sediments (Bertness 1984b; Bertness and Grosholz 1985). Detrital material derived from smooth cordgrass also comprises part of the diet of this filter-feeding bivalve (Peterson et al. 1985; Kreeger et al. 1988).

Sixty species of gastropods (snails and nudibranchs) have been reported in the intertidal zone from Nova Scotia to North Carolina (Gosner 1971). Of these, the salt marsh snail is most abundant in the northern salt marshes, whereas the marsh periwinkle occurring from Long Island south dominates smooth cordgrass marshes from Virginia south. The former is an air-breathing snail that is especially well adapted to life in the intertidal zone, while the marsh periwinkle possesses gills (McMahon and Russell-Hunter 1981). Other salt marsh snails of northern marshes include rough periwinkle (*L. saxatilis*), common periwinkle (*Littorina littorea*), smooth periwinkle (*L. obtusata*), mud snail (*Ilyanassa obsoleta*), salt marsh hydrobe (*Spurwinkia salsa*), other hydrobid snails (*Hydrobia totteni, H. truncata*), a dusky snail (*Amnicola winkleyi*), and the snail (*Odostomia* sp.) (Tiner 1974; Davis et al. 1982). Other gastropods of southern marshes include Spartina hydrobe (*Onobops jacksoni*) and salt marsh hydrobe (Pung et al. 2008). In Mississippi low-salinity marshes, several species of snails occupy the drier high marsh, for example, Florida marsh snail (*Detracia floridana*), a hydrobid snail (*Littoridinops palustris*), Spartina hydrobe, salt marsh snail, amber snail (*Succinea ovalis*), another air-breathing snail (*Vertigo ovata*), and a terrestrial slug (*Deroceras laeve*) (Bishop and Hackney 1987; Moore 1992).

Densities of the salt marsh snail can be extremely high when they congregate for breeding. In a Connecticut salt marsh, densities exceeding 5,000 snails per m^2 were observed during May and June when they congregated for breeding (Tiner 1974), while 1 to 4 million per acre (0.4–1.6 million/ha) have been reported for New Jersey marshes (Alpaugh and Ferrigno 1973). Egg-laying and hatching are synchronized with spring tides (Russell-Hunter et al. 1972). Three or four cycles of egg-laying may take place during early summer in Massachusetts. Peak hatching of each set of eggs occurs two weeks later, thereby allowing the next spring tides to transport their planktonic larvae (veligers) to estuarine waters. The salt marsh snail is a major winter food for black ducks in the northeastern United States where it comprised 64 percent of the total dry weight of foods eaten in New Jersey salt marshes (Costanzo and Malecki 1989). On southern marshes, densities of the marsh periwinkle above 150 individuals per m^2 have been reported in North Carolina marshes, while in Georgia, healthy marshes may support up to 558 individuals per m^2 and some marsh sites experiencing die-off have had densities of more than 2,600 individuals per m^2 (Stiven and Hunter 1976; Silliman et al. 2005).

Crustaceans are among the more readily observed invertebrates in coastal wetlands. While most marine crustaceans are associated with subtidal waters, many inhabit the intertidal zone. Eighteen species of isopods, 92 species of amphipods (beach fleas and scuds), and 20 crab species have been listed as occurring in the littoral zone from Nova Scotia to Cape Hatteras (Gosner 1971). Some crabs including the familiar blue crab enter the marshes at high tides to feed. Fiddler crabs (*Uca*) and wharf crabs (*Armases*) are salt marsh residents, whereas the mud crabs (*Rhitropanopeus, Panopeus,* and *Eurypanopeus*) occupy oyster reefs and muddy bottoms but are also common in southeastern salt marshes (Kneib 1984). Three species of fiddler crabs occur in salt marshes—sand fiddler (*Uca pugilator*), mud fiddler (*U. pugnax*), and red-jointed fiddler (*U. minax*). The last species is more prevalent in brackish and tidal fresh marshes from Massachusetts to Texas and can even be found in nontidal wetlands (Moore 1992), yet it was the dominant fiddler in wet pockets of sandy-peaty soils located within the short-form smooth cordgrass zone of Georgia's salt marshes (Teal 1958). The density of crab burrows can be extremely high as 300 burrows per m^2 have been counted in South Carolina (McCraith et al. 2003). Fiddlers play an important role in the ecology of salt marshes. Their burrows improve soil drainage, increase oxidation of soils, enhance decomposition of belowground biomass, and increase available nitrogen through their feces, thereby stimulating productivity of smooth cordgrass (Bertness 1985). Interestingly, the advancement of smooth cordgrass into softer sediments appears to eventually benefit fiddler crabs by providing a more stable substrate for their burrows (Bertness and Miller 1984). The relationship between smooth cordgrass and fiddler crabs therefore benefits both parties—a form of mutualism. Another fiddler crab (*U. spinicarpa*) has been reported in low-salinity tidal marshes along the Gulf Coast where it occurs with other crabs including wharf crab (*Armases cinereum*) and mud crab (*Rhithropanopeus harrisii*) (Moore 1992). Amphipods (*Gammarus, Orchestia, Uhlorchestia*), isopods (*Philoscia, Jaera, Cyathura*), ostracods, and copepods are common small crustaceans of salt marshes. The distribution and abundance of these and other invertebrates on the marshes are affected by multiple factors including predation, competition, selective larval settlement and mortality, physical conditions of the environment, and unpredictable or cyclical disturbances (Kneib 1984).

Insects, mites, and spiders are common and extremely abundant in coastal wetlands. Tidal freshwater wetlands should support more of these species than salt marshes do given the lack of salt stress and the greater diversity of plants including an abundance of flowering herbs. While biting insects including salt marsh mosquitoes (*Aedes sollicitans, A. cantator, A. taeniorhynchus;* also referred to by genus *Ochlerotatus*), deer flies (*Chrysops* spp.), greenhead flies (*Tabanus nigrovittatus*), and no-see-ums (*Culicoides* spp.) get most of our attention for obvious reasons, grasshoppers (*Conocephalus, Orchelimum*), planthoppers (e.g., *Prokelisia marginata, Fieberiella florii*), ants (e.g., *Myrmica emeryana*), crickets (e.g., *Allonemobius sparsalus, A. fasciatus*), aphids (e.g., *Dactynotus* sp.), and fly larvae of many species (e.g., *Apallates, Bezzia, Chaetopsis, Dasyhelea, Diaphorus, Dimecoenia, Laphystia, Limonia, Megaselia, Orthocladius, Paracleius, Parydra, Pelastoneurus, Polytrichophora, Stomoxys*) can also be found in salt marshes (Tiner 1974; Davis 1978; Rey and McCoy 1986). Hundreds of insect species occur in the variety of tidal wetlands across the United States. More than 200 species were collected in North Carolina salt and brackish marshes while more than 500 species were found in Gulf Coast marshes (Davis and Gray 1966; Rey and McCoy 1997). Numbers of insect species are likely to be less in northern salt marshes; about 50 species were reported in an urban salt marsh in Connecticut (Pupedis 1997). Some butterflies of salt marshes include Aaron's skipper (*Poanes aaroni*), broad-winged skipper (*P. viator*), and salt marsh skipper (*Panoquina panoquin*), whereas monarch butterflies (*Danaus plexippus*) migrate along the coast and can be found feeding on seaside goldenrod (*Solidago sempervirens*) and other coastal plants in late summer (Opler and Malikul 1992; Tiner 2005; Table 6.2). The maritime ringlet (*Coenonympha nipisiquit*), a Canadian endangered species, and the salt marsh copper (*Lycaena dospassosi*) are the only two butterflies that are exclusive to Canada's maritime salt marshes (New Brunswick Maritime Ringlet Recovery Team 2005). The maritime ringlet's larva is a green caterpillar that blends in with its prime host plant—salt hay grass (*Spartina patens*), while the orange-colored adult feeds on nectar mostly from the flowers of sea lavender (*Limonium carolinianum*). Salt grass (*Distichlis spicata*) is an important host plant for other butterflies including salt marsh skipper, obscure skipper (*Panoquina panoquinoides*), and wandering skipper (*Panoquina errans*), a southern California species. Adults of the salt marsh copper also rely on nectar from sea lavender, but the caterpillars favor silverweed (*Argentina* spp.), which is a host plant of other salt marsh butterflies (e.g., dorcas copper [*Lycaena dorcas*] and the purplish copper [*L. helloides*], a Pacific Coast species). The caterpillar of the short-tailed or maritime swallowtail (*Papilio brevicauda*) feeds on scotch lovage (*Ligusticum scoticum*) and other plants in the Canadian Maritimes (Pyle 1995). In southern marshes, caterpillars of great southern white (*Ascia monuste*) and eastern pigmy blue (*Brephidium pseudofea*) feed on sea rocket (*Cakile* spp.) and glassworts (*Salicornia* spp.), respectively, while adults of both species obtain nectar from flowers of saltwort (*Batis maritima*) (Butterflies and Moths of North America at www.butterfliesandmoths.org). A number of moths occur in northeastern salt marshes, and they appear to depend on plants of the upper marsh border. Prairie cordgrass (*Spartina pectinata*) is the host plant for the many-lined cordgrass moth (*Chortodes enervata*) and the Spartina borer (*Spartiniphaga inops;* a species of special concern in MA and CT). Groundsel-bush (*Baccharis halimifolia*) provides food for caterpillars of the Baccharis borer (*Oidaematophorus balanotes*), dark-edged Eusarca moth (*Eusarca fundaria*), and the straight-lined Pero moth (*Pero zalissaria*), while seaside goldenrod (*Solidago sempervirens*) is the host plant for the seaside goldenrod stem borer (*Papaipema duovata;* also a species of special concern in CT) (Dale Schweitzer, pers. comm. 2012). Dragonflies, damselflies, and other freshwater aquatic species such as water

Table 6.2. Some butterflies reported from North American tidal wetlands

Common Name (*Scientific name*)	Tidal Habitat	General Range
Short-tailed swallowtail (*Papilio brevicauda*)	Grassy sea cliffs	Canadian Maritimes
Anise swallowtail (*Papilio zelicaon*)	Sea-level tidelines	Pacific Coast
Palamedes swallowtail (*Pterourus palamedes*)	Coastal swamps	Southeast
Great southern white (*Ascia monuste*)	Beaches, salt marshes	Southeast
Cloudless giant sulphur (*Phoebis sennae*)	Seashore	Southern California and Gulf Coast
Bronze copper (*Hyllolycaena hyllus*)	Edges of salt marshes	Northeast and Maritimes
Purplish copper (*Epidemia helloides*)	Tidal marshes	Pacific Coast
Dorcas copper (*Epidemia dorcas*)	Salt marshes	Maine north
Acis hairstreak (*Strymon acis*)	Beaches	South Florida
Martial hairstreak (*Strymon martialis*)	Mangroves	South Florida
Eastern pygmy blue (*Brephidium isophthalma*)	Salt marshes, tidal flats	Southeast
Little metalmark (*Calephelis virginiensis*)	Salt marshes	Southeast
Little wood satyr (*Megisto cymela*)	Salt bays, brackish streamsides	Atlantic and Gulf Coasts
Mangrove skipper (*Phocides pigmalion*)	Mangroves	South Florida
Manuel's skipper (*Polygonus manueli*)	Tidal flats	South Florida
Rare skipper (*Problema bulenta*)	Brackish and tidal fresh marshes	South Atlantic
Woodland skipper (*Ochlodes sylvanoides*)	Tidewater marshes	Pacific Coast
Saffron skipper (*Poanes aaroni*)	Salt, brackish, and tidal fresh marshes	Middle and South Atlantic
Broad-winged skipper (*Poanes viator*)	Salt marshes	Atlantic and Gulf Coasts
Saw-grass skipper (*Euphyas pilatka*)	Saw-grass tidal marshes	Southeast
Salt marsh skipper (*Panoquina panoquin*)	Salt marshes	Connecticut to Florida and Mississippi
Obscure skipper (*Panoquina panoquinoides*)	Salt marshes	Gulf Coast
Wandering skipper (*Panoquina errans*)	Salt marshes, beaches	Southern California
Long-winged skipper (*Panoquina ocola*)	Salt marshes	Gulf Coast; also New Jersey

Source: Pyle 1995.

boatmen, backswimmers, water beetles, and water-striders can be found in brackish and tidal fresh marshes and some can be found in salt marsh pools (e.g., Davis et al. 1982). Mite species of salt marshes include members of several genera (*Ceratozetes, Leptus, Balaustium, Microtrombidium*), and several genera of spiders have been reported as well (*Grammonota, Ceraticelus, Pardosa, Pirata, Lycosa, Schizocosa, Clubiona*) (Tiner 1974).

Given that a variety of plant communities occur within salt marshes, one would expect at least some variation in animal distributions within the marshes as well. One example of this difference was observed in a Connecticut salt marsh where invertebrates of five major plant communities were studied (Table 6.3). Insect communities associated with black needlerush (*Juncus roemerianus*) marshes on the Gulf Coast appear to have more in common with Mid-Atlantic freshwater marshes than with Atlantic Coast salt marshes (Moore 1992).

Tidal creeks provide habitats for aquatic invertebrates (jellyfish, comb jellies, shrimp, green crabs, and blue crabs) and fishes as well as access to the marshes at high tide. Grass shrimp (*Palaemonetes pugio*) are probably the most abundant macrocrustacean using flooded salt marshes along the Atlantic and Gulf Coasts, while penaeid shrimp and juvenile blue crabs are also important visitors to marshes (Peterson and Turner 1994).

Commercially significant shellfish (crabs, shrimp, mussels, and clams) depend on coastal wetlands. Penaeid shrimp use estuary–wetland complexes along the South

Table 6.3. More abundant invertebrates associated with five plant communities in a Connecticut salt marsh

Plant Community	Common Invertebrates
Smooth cordgrass, tall form (*Spartina alternifora*) (38–41 species)	Salt marsh snail (*Melampus bidentatus*), Planthoppers (Delphacid nymphs, *Prokelisia marginata*), Ribbed mussel (*Geukensia demissa*), Rough periwinkle (*Littorina saxatilis*), Amphipod (*Orchestia grillus*), Spider (*Grammonota* spp.), Nudibranch (*Alderia modesta*), Nematodes, Anemone (*Haliplanella luciae*), Polychaete worm (*Arabella iricolor*), and Carabid beetle (*Bembidion laterale*) (*Note:* Fiddler crabs were observed in this zone, but avoided detection in plots.)
Salt hay grass (*Spartina patens*) (26 species)	Salt marsh snail, Isopod (*Philoscia vittata*), Spiders (*Ceraticelus emertoni, Cornicularia* sp., and *Clubiona* spp.), Amphipod (*O. grillus*), Mites (*Erythraeidae* and *Ceratozetes* sp.), and Planthoppers (Delphacid nymphs and *Fieberiella florii*)
Salt grass (*Distichlis spicata*) (39 species)	Salt marsh snail, Nudibranch (*Alderia*), Isopod (*Philoscia*), Ribbed mussel, Spiders (*Exigonidae, Ceraticelus emertoni, Cornicularia* sp., and *Lycosidae*), Amphipod (*O. grillus*), Nematodes, Mite (*Camisia* sp.), Planthoppers (Delphacid nymphs), and Unidentified beetle (*Coleptera*)
Black grass (*Juncus gerardii*) (46–55 species)	Isopod (*Philoscia*), Salt marsh snail, Amphipod (*O. grillus*), Nudibranch (*Alderia*), Spiders (*Cornicularia* sp., *Ceraticelus emertoni,* and *Lycosidae*), Plantbug (*Hemiptera*), and Mite (*Ceratozetes* sp.)
High-tide bush–Black grass (*Iva frutescens–Juncus gerardii*) (33 species)	Aphid (*Dactynotus* sp.), Isopod (*Philoscia*), Salt marsh snail, Mites (*Ceratozetes* sp. and *C. emertoni*), and Amphipod (*O. grillus*)

Source: Tiner 1974.

Atlantic and Gulf Coasts as nursery grounds where juveniles spend their time feeding in tidal creeks and marshes (Wiegert and Pomeroy 1981). Here they grow to 2 inches (5 cm) long before returning to deeper estuarine waters (Anderson 1970). In fact, there appears to be a direct relationship between the area of tidal marshes and the yield of the shrimp harvest offshore (Turner 1992). Postlarval brown shrimp enter the estuaries in late winter and early spring to access the marshes and grow to subadult size by May when the postlarval white shrimp begin entering the estuaries to begin their use of the marshes (Conner and Day 1987). This is a great example of resource partitioning by similar species that minimizes competition. Oysters and blue crabs spend virtually their entire lives (except as plankton) in estuaries, although female crabs do migrate offshore for spawning (Williams 1965). Low-salinity (oligohaline) marshes and associated waters are vital nursery grounds for juvenile blue crabs (Posey et al. 2005). Louisiana marsh ponds larger than 820 feet (250 m) wide with direct connections to waterways had high densities of blue crabs in the fall (Rozas and Minello 2010). Ponds lacking such connections had few animals, presumably due to lack of access, greater environmental stress, and predation. While most people would not associate lobsters with tidal wetlands, a couple of connections can be made: 1) young lobsters have been found in eel-grass beds on Cape Cod; and 2) inshore populations use salt marsh creekbanks as nursery grounds (Able et al. 1988; Heck et al. 1989).

The world's largest population of horseshoe crabs (*Limulus polyphemus*) congregates in spring on the beaches of Delaware Bay to mate and lay eggs in the sand (Walls et al. 2002; Figure 6.4). These eggs represent a major food source for thousands of migrating shorebirds, providing needed energy for birds such as the red knot that are making a 20,000-mile (32,190 km) return flight to the Arctic from their wintering grounds in Tierra del Fuego at the tip of South America. The survival of the red knot depends on this egg supply; recent declines in horseshoe crabs due to overharvesting for bait has greatly reduced the availability of this food and may be responsible for the nearly catastrophic drop in red knot numbers observed in the Delaware Bay region. (*Note:* Concern about the impact of overharvesting horseshoe crabs on shorebirds led several states including New Jersey, Delaware, and Maryland to regulate the fishery.)

Salt marshes and seagrass beds serve as corridors for the movement of blue crabs (Micheli and Peterson 1999). Benthic macroinvertebrates were more abundant and had higher diversity in oyster reefs separated from these habitats. Artificial oyster reefs placed next to salt marshes and seagrass beds received more blue crab predation than reefs separated from them. Molting blue crabs hide in seagrass beds until their shells harden (Ryer et al. 1990). Larger populations and higher diversity of fishes and invertebrates were found in eel-grass beds than in adjacent nonvegetated bottoms (Mattila et al. 1999). Eel-grass beds are settlement areas for spat of bay scallops (*Pecten irradians*) and blue mussels, and nursery grounds for rock crabs (*Cancer*), starfishes (*Asterias*), and hermit crabs (*Pagarus*) (Roman et al. 2000; Basquill 2003). Higher numbers of fishes and crustaceans occur along marsh shorelines than along bulkheaded or rip-rap shores (Peterson et al. 2000).

Differences in exposure and submergence that create distinctive plant zonation along rocky shores have also influenced invertebrate communities. Terrestrial animals (insects, spiders, isopods, and mites) occupy the barren zone where flooding is infrequent. Common and rough periwinkles (*Littorina littorea, L. saxatilis*) frequent the black zone represented by cyanobacteria (blue-green algae). A zone of barnacles (*Balanus balanoides,*

Figure 6.4. Horseshoe crabs mating on the beaches of Delaware Bay. (Greg Breese, U.S. Fish and Wildlife Service).

Chthamalus fragilis) occurs between the black zone and the fucoid zone. Blue mussels can also be found here along with dog whelks (*Nucella lapilla*), Atlantic limpets (*Acmaea testudinalis*), chitons, common whelks (*Buccinium undatum*), smooth periwinkle, and other periwinkles. The fucoid zone dominated by rockweeds (*Fucus* spp., *Ascophyllum nodosum*) is inhabited by numerous marine invertebrates including copepods, mites, amphipods, a variety of worms, encrusting bryophytes, hydroids, and small anemones plus a few starfishes and green sea urchins (*Strongylocentrotus drobachiensis*). The lowest intertidal zone exposed for only short periods of time contains the most diverse assemblage of invertebrates including rock, jonah, and green crabs (*Cancer irroratus, C. borealis, Carcinus maenas*), brittle stars, starfishes, green sea urchins, tunicates, sponges, common periwinkles, limpets, sea slugs, amphipods, isopods, copepods, encrusting bryophytes, hydroids, anemones, and an assortment of marine worms (Gosner 1971; Davis and Browne 1996). These and other invertebrates are food staples for many fishes. For example, young pollock spend two years feeding in this zone and use the rockweed zone as nursery grounds (Rangeley and Kramer 1995).

Sandy beaches are difficult environments for animals for the same reasons they pose problems for plants. Animals tend to be more abundant in the wave-washed lower beach, although the upper beach may support amphipods, collembolans, flies, and other invertebrates beneath rotting seaweeds washed up on the beaches. Predatory tiger beetles (*Cicindela* spp.) including the federally threatened northeastern beach tiger beetle (*C. dorsalis dorsalis*) feed on the former invertebrates. Ghost crabs (*Ocypode quadrata*) can be found on southern beaches from New York south. Two invertebrates of particular note on Atlantic sandy beaches are the mole crab (*Emerita talpoidea*) and coquina clams (*Donax* spp. from Virginia south). Both species live in the wet sands of the "surf zone" in rather large numbers. As the tide changes, they migrate up and down the slope by burrowing in the sand. Mole crabs can bury themselves in 1.5 seconds, while the coquina clam can do so in 5.6 seconds (Little 2000). (*Note:* After 50+ years, I still remember the thrill my brother Gordon and I had trying to catch mole crabs on the beach at Barnegat Lighthouse in New Jersey. Their expert burrowing skills definitely outmatched our ability to dig them out. I think we caught only one or two that afternoon.)

Tidal sandflats are home to oligochaete and polychaete worms, amphipods, moon snails (*Polinices*), razor clams, and Gemma clams, whereas mudflats support soft-shell clams, Baltic clams (*Macoma balthica*), mud snails, amphipods, and polychaete worms (Whitlatch 1982; Figure 6.5). Mud snails are the most obvious mudflat species as hundreds can often be observed at low tide. At high tide, crustaceans such as sand shrimp (*Palaemonetes vulgaris*), grass shrimp, green crabs, blue crabs, lady crabs (*Ovalipes ocellatus*), and spider crabs (*Libinia*) migrate across the flats in search of food. Mudflats support high densities of the marine worms, amphipods, mud snails, and bivalves (e.g., soft-shell clam and Baltic clam) that are important food for migrating shorebirds, resident birds, and coastal fishes. Juvenile blue crab growth was greater in nonvegetated mud and sandflats in the upper part of the York River in Virginia than in similar habitats or in seagrass beds lower in the river, which underscores the importance of brackish intertidal flats to this species of major commercial importance (Seitz et al. 2005).

Fish Habitat

The sheltered location, warmer water, and rich aquatic productivity of estuaries have undeniably fortified their use as primary nursery grounds for many species of marine fishes worldwide. The abundance of food plus the lack of large predators make tidal

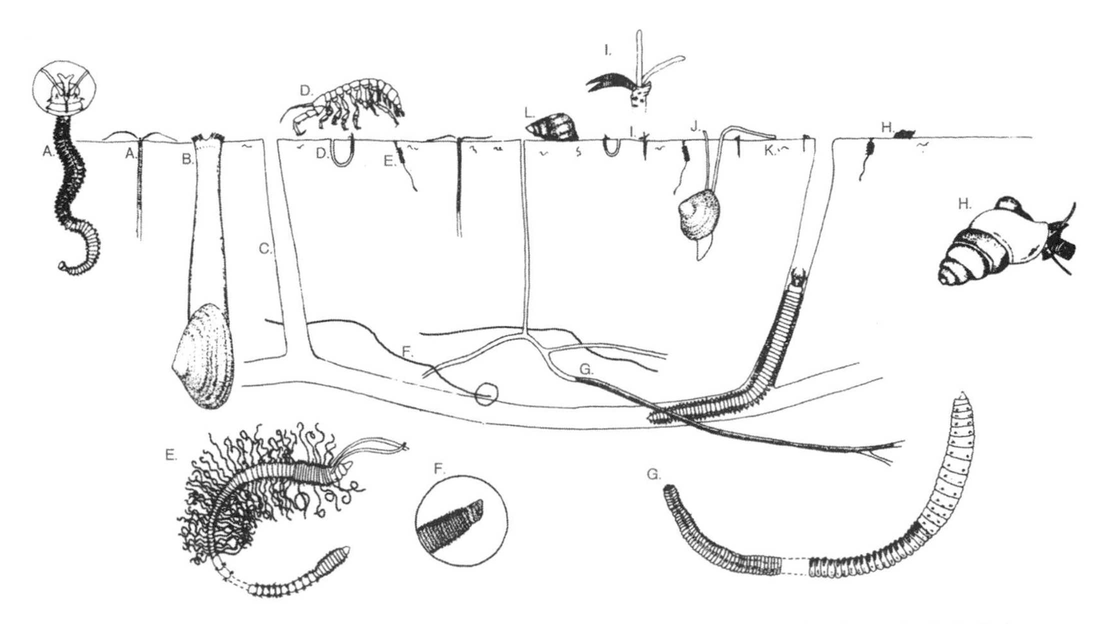

Figure 6.5. Some invertebrates common to New England mudflats. Suspension feeder: soft-shelled clam (*Mya arenaria,* A); burrowing omnivore: nereid polychaete worm (*Nereis virens,* C); Burrowing deposit feeders: polychaete worms (*Tharyx* sp., E; *Lumbrinereis tenuis,* F; *Heteromastus filiformis,* G, and oligochaete worm, K; Surface deposit feeders: polychaete worms (*Polydora ligni,* A; *Streblospio benedicti,* I); mud snail (*Ilyanassa obsoleta,* L); hydrobid snail (*Hydrobia totteni,* H); amphipods (*Corophium* spp., D); and Baltic clam (*Macoma balthica,* J). (Whitlatch 1982)

creeks in salt and brackish marshes prime feeding grounds and refuges for juvenile fishes (Boesch and Turner 1984; Moore 1992). Four factors may be responsible for the latter functions: 1) vegetation; 2) shallow depth; 3) physio-chemical environment; and 4) turbidity (Deegan et al. 2000). Submerged aquatic beds also offer protection from predators for the young of estuarine-spawning fishes (e.g., shad, herring, and striped bass) as well as for small resident fishes (e.g., sheepshead minnow, killifishes, mosquitofish, and silversides). Many fish species use estuaries for only the young-of-year and juvenile portion of their lives, spending their adult time offshore (Figure 6.6). They require a variety of habitats to complete their life cycle.

Fish abundance may vary directly with density of aquatic vegetation (Homer 1988), and the high marsh appears to be important spring and summer spawning and nursery areas for resident salt marsh fishes (Talbot and Able 1984). Interestingly, seasonal salinity changes in the estuary allow use of the same location by freshwater fishes and marine species at different times, with freshwater species using more of the estuary during the spring freshets. They will be replaced by marine species in summer during low flows and maximum saltwater intrusion (Rogers et al. 1984; Moore 1992). Fish diversity appears to be higher in the freshwater tidal portion of coastal rivers and in the lower reaches of estuaries where salinity is more stable than in the middle reaches where salinity is more variable (Moore 1992). (*Note:* See Table 6.4 for scientific names of fishes.)

About two-thirds of the U.S. commercial fish landings are dependent on estuaries, with higher dependencies in certain areas such as the Chesapeake Bay region and the Gulf of Mexico where 97 percent of the fish harvest is estuarine-dependent (McHugh 1966, 1975; Gunter 1967; Tiner and Burke 1995). The diversity of habitats (marshes, tidal flats, seagrass beds, creeks, shallow water, and deep water) and the resources (food and cover) available probably make estuaries attractive to this wide variety of species. Along the Atlantic Coast, menhaden, bay anchovy, weakfish, bluefish, spotted sea trout, croaker, spot, black drum, summer flounder, winter flounder, mullet, northern puffer, and striped bass are among the most significant estuary–marsh dependent species. Major tributaries of Chesapeake Bay provide spawning grounds for about 90 percent of the East Coast population of striped bass (Berggren and Lieberman 1977; Tiner and Burke 1995).

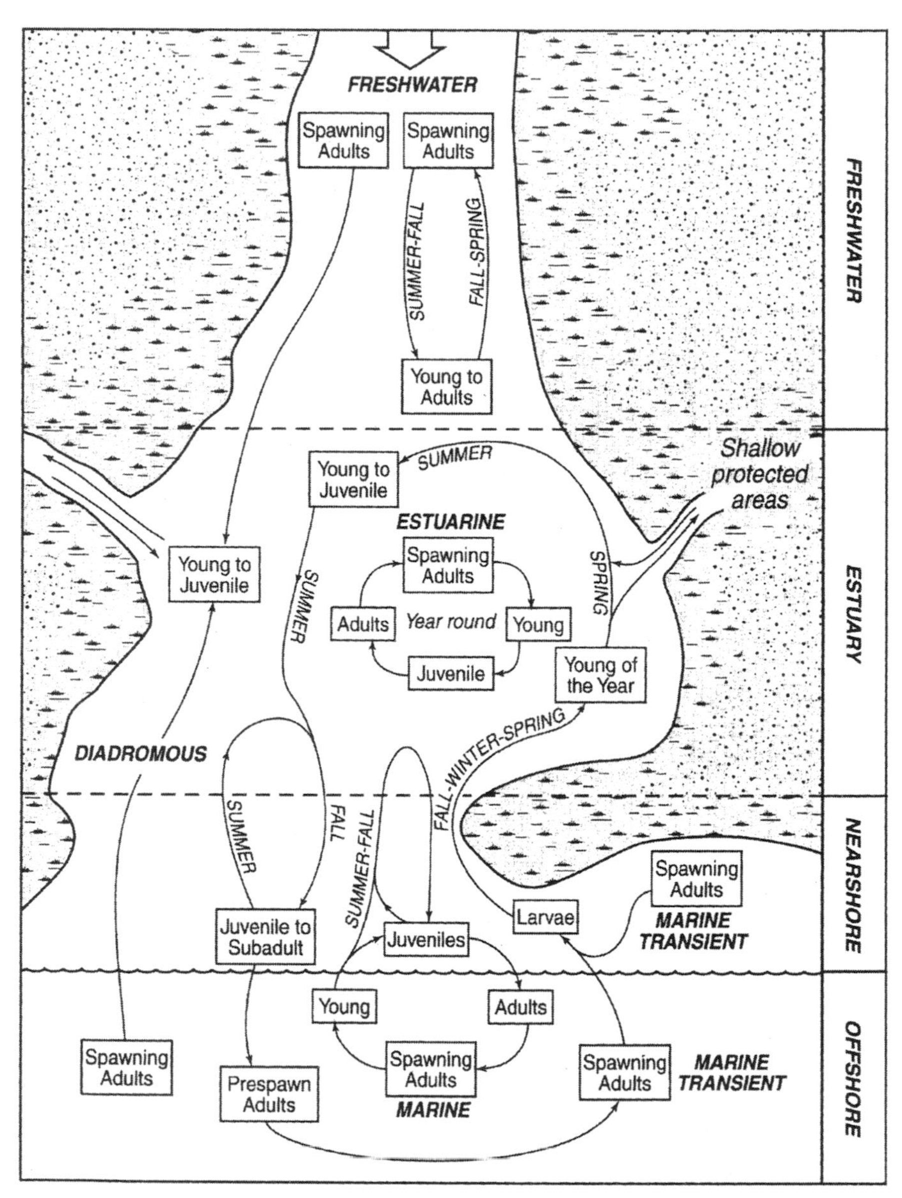

Figure 6.6. Temporal and spatial use of estuaries by marine, estuarine, freshwater, and estuarine-dependent marine nekton. (Figure 1 from Deegan et al. 2000; *Concepts and Controversies in Tidal Marsh Ecology*; copyright Kluwer Academic Publishers; reprinted by kind permission from Springer Science+Business Media B.V.)

An abundance of forage fishes and macroinvertebrates provide a wealth of food to support the growth of young predaceous fishes such as bluefish and weakfish. Abandoned former rice fields in South Carolina that are now brackish marshes and waters are preferred over the main river channel as spawning and nursery grounds by blueback herring (Osteen et al. 1989). On the Pacific Coast, the juvenile stage of chum and chinook salmon (*Oncorhynchus keta* and *O. tshawytscha*) spend two to four months in estuaries feeding in tidal creeks at high tide where they add 3 to 6 percent of their body weight per day (Simenstad et al. 1982). Tidal marshes and creek areas may also offer juvenile salmon protection from predators.

More than 200 species are associated with salt marshes in eastern North America from Hudson Bay to Texas (Nordlie 2003). Of these, 9 percent are permanent residents, 18 percent are marine nursery species, 6 percent are diadromous fishes (using both fresh and salt water during their life cycle), 52 percent are marine transients, and 15 percent are freshwater transients. Permanent residents (various killifishes and sticklebacks) have the widest range of salt tolerance, followed by marine nursery species. Striped killifish, mummichog, fourspine stickleback, threespine stickleback, sheepshead minnow, Atlantic silverside, and common eel are among the more common species in Atlantic Coast salt marsh creeks, and they may also frequent neighboring eel-grass beds (Figure 6.7; Table 6.4). Species utilizing these marshes and tidal creeks as nursery grounds include winter flounder, tautog, sea bass, alewife, menhaden, bluefish, mullet, sand lance, and striped bass (Teal 1986). Diadromous species include sea-run populations of freshwater fishes (i.e., brook, brown, and rainbow trout) as well as anadromous fishes that migrate from the ocean to freshwater to spawn (e.g., salmon, herring, rainbow smelt, and striped bass). The marsh creek edges are the places where most salt marsh collections are made and where juveniles of commercially and recreationally important fish and shellfish are captured (Montague and Wiegert 1991). In southern waters, predatory fishes including snook, tarpon, red drum, croaker, sea trout, and kingfish feed in tidal creeks as do sharks, rays, and needlefish. More than 80 species of fishes have been captured in creeks and open waters of black needlerush marshes along the northeastern Gulf Coast (Subrahmanyam and Drake 1975; Stout 1984).

Fishes using the salt marsh interior can be divided into three groups based on their habitat utilization: 1) interior marsh residents; 2) interior marsh users; and 3) edge marsh users (Peterson and Turner 1994). The first group stays in the marsh at low tide, retreating to marsh pools or burrowing in the mud. They may be considered true marsh residents, and in Louisiana they include bayou and diamond killifishes, sheepshead minnow, and sailfin molly. The second group contains resident fishes that use the marsh at high tide and return to the tidal creek at low tide and includes the gulf killifish and possibly the rainwater killifish. Marsh edge users don't move too far into the marsh interior and are represented by some resident species, such as naked and darter gobies, saltmarsh topminnow, and tidewater silverside; and transient species, such as spotted sea trout and possibly mullet. The marsh creek edge is a shallow water habitat offering some vegetative cover and plenty of food through zooplankton, phytoplankton, and detrital particles. Transient fishes occupy deeper water at low tide and move to the marsh edge at high tide where they find small fishes (forage species) to feed upon. In Delaware Bay, striped bass were more abundant around marsh creek mouths than in upper portions of creeks (Tupper and Able 2000). Here around low tide they found concentrations of prey species (e.g., blue crab, grass shrimp, sand shrimp, mummichog, bay anchovy, and Atlantic silverside).

Mummichogs are the epitome of a salt marsh–dependent fish. They spend their entire life in the marshes. In winter, they

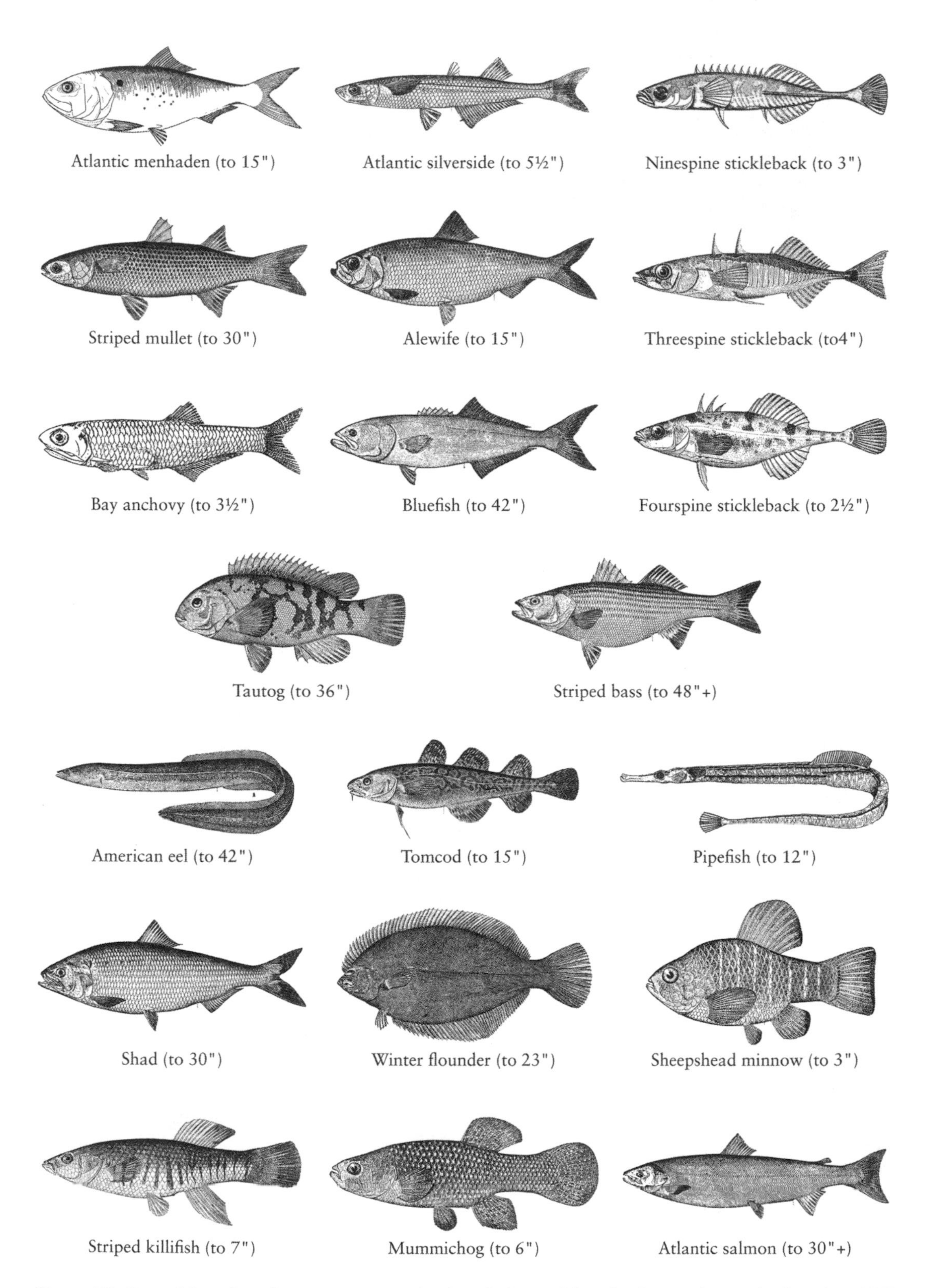

Figure 6.7. Some fishes of northeastern estuaries that benefit from tidal wetlands. (Bigelow and Schroeder 1953)

Table 6.4. Examples of fishes found in tidal creeks in Atlantic and Gulf regions

Region	Species
Gulf of Maine	Blueback herring (*Alosa aestivalis*), Hickory shad (*A. mediocris*), Alewife (*A. pseudoharengus*), Atlantic menhaden (*Brevoortia tyrannus*), Atlantic herring (*Clupea harengus*), American sand lance (*Ammodytes americanus*), American eel (*Anguilla rostrata*), Fourspine stickleback (*Apeltes quadracus*), Threespine stickleback (*Gasterosteus aculeatus*), Blackspotted stickleback (*G. wheatlandi*), Ninespine stickleback (*Pungitius pungitius*), Lumpfish (*Cyclopterus lumpus*), Sea snail (*Liparis atlanticus*), Mackerel scad (*Decapterus macarellus*), Mummichog (*Fundulus heteroclitus*), Striped killifish (*Fundulus majalis*), Atlantic cod (*Gadus morhua*), Atlantic tomcod (*Microgadus tomcod*), Pollock (*Pollachius virens*), Red hake (*Urophycis chuss*), White hake (*U. tenuis*), Rock gunnel (*Pholis gunnellus*), Inland silverside (*Menidia beryllina*), Atlantic silverside (*M. menidia*), Tidewater silverside (*M. peninsulae*), White perch (*Morone americana*), Striped bass *(M. saxatilis*), Striped mullet (*Mugil cephalus*), Sea raven (*Hemitriperus americanus*), Grubby (*Myoxocephalus aeneus*), Longhorn sculpin (*M. oxtodecimspinosus*), Shorthorn sculpin (*M. scorpius*), Wrymouth (*Cryptacanthodes maculatus*), Rainbow smelt (*Osmerus mordax*), Butterfish (*Peprilus tricanthus*), Sea lamprey (*Petromyzon marinus*), Yellowtail flounder (*Pleuronectes ferrugineus*), Winter flounder (*Pseudopleuronectes americanus*), Windowpane flounder (*Scopthalmus aquosus*), Bluefish (*Pomatomus saltatrix*), Atlantic salmon (*Salmo salar*), Brown trout (*S. trutta*), Brook trout (*Salvelinus fontinalis*), Atlantic mackerel (*Scomber scombrus*), Northern sennet (*Sphyraena borealis*), Northern pipefish (*Syngnathus fuscus*), Cunner (*Tautogolabrus adspersus*)
Southern New England	Mummichog, Striped killifish, Threespine stickleback, American eel, Fourspine stickleback, Grubby, Northern pipefish, White mullet (*Mugil curema*), Atlantic silverside, Winter flounder, Sheepshead minnow (*Cyprinodon variegatus*), Atlantic menhaden, Striped mullet, Blueback herring, Ninespine stickleback, Shiner (*Notropis* sp.), Black sea bass (*Centropristis striata*), Rainwater killifish (*Lucania parva*), Seaboard goby (*Gobiosoma ginsburgi*), Bay anchovy (*Anchoa mitchilli*), Butterfish, Tautog (*Tautoga onitis*), Alewife, Cunner, Sand lance, Striped bass, Bluefish
Delaware Bay	Blueback herring, Hickory shad, Alewife, American shad (*Alosa sapidissima*), White catfish (*Ameiurus catus*), Bay anchovy, Silver perch (*Bairdiella chrysura*), Atlantic menhaden, Atlantic herring, Weakfish (*Cynoscion regalis*), Sheepshead minnow, Common carp (*Cyprinus carpio*), American gizzard shad (*Dorosoma cepedianum*), Mummichog, Striped killifish, Threespine stickleback, Naked goby (*Gobiosoma bosci*), Eastern silvery minnow (*Hybognathus regius*), Channel catfish (*Ictalurus punctatus*), Spot (*Leiostomus xanthurus*), Pumpkinseed (*Lepomis gibbosus*), Atlantic silverside, Atlantic croaker (*Micropogon undulatus*), White perch, Striped bass, Golden shiner (*Notomigonus crysoleucas*), Striped cusk-eel (*Ophidion marginatum*), Toadfish (*Opsanus tau*), Summer flounder (*Paralichtys dentatus*), Yellow perch (*Perca flavescens*), Black drum (*Pogonias cromis*), Bluefish, Striped searobin (*Prionotus evolans*), Winter flounder, Windowpane flounder (*Scophthalmus aquosus*), Northern pipefish, Hogchoker (*Trinectes maculatus*), Striped hake (*Urophycis regia*)
South Carolina	Mummichog, Bay anchovy, Spot, White mullet, Atlantic silverside, Striped anchovy (*Anchoa hepsetus*), Striped mullet, Pinfish (*Lagodon rhomboides*), Striped killifish, Spotfin mojarra (*Eucinostomus argenteus*), Pigfish (*Orthopristis chrysoptera*), Naked goby, Toadfish, Tonguefish (*Symphurus* sp.), Sheepshead minnow, Threadfin shad (*Dorosoma petenese*), Southern flounder (*Paralichthys lethostigma*), Planehead filefish (*Stephanolepis hispidus*), Striped blenny (*Chasmodes bosquianus*), Northern puffer (*Sphaeroides maculatus*), Atlantic menhaden, Atlantic needlefish (*Strongylura marina*), Atlantic bumper (*Chloroscombrus chrysurus*), Darter goby (*Ctenogobius boleosoma*), Striped burrfish (*Chilomycterus schoepfi*), Feather blenny (*Hysoblennius hentz*), Northern pipefish, bluefish, Atlantic Spanish mackerel (*Scomberomorus maculatus*)
Florida—Gulf Coast	Bay anchovy, Spot, Longnose killifish (*Fundulus similis*), Gulf killifish (*Fundulus grandis*), Sheepshead minnow, Diamond killifish (*Adinia xenia*), Spotfin mojarra, Inland silverside, Pinfish, Marsh killifish (*Fundulus confluentus*), Sailfin molly (*Poecilia latipinna*), Striped mullet, Rainwater killifish, Clown goby (*Microgobius gulosus*)
Louisiana	Atlantic croaker, Gulf menhaden (*Brevoortia patronus*), Spot, Sea catfish (*Galeichthys felis*), Sand seatrout (*Cynoscion arenarius*), Inland silverside, Striped mullet, Gafftopsail catfish (*Bagre marinus*), Bay anchovy, Atlantic threadfin (*Polydactylus octonemus*), Southern puffer (*Sphaeroides nephelus*), Hogchoker, Silver perch, Rough silverside (*Membras martinica*), Naked goby, Gulf killifish, Diamond killifish, Bayou killifish (*Fundulus pulvereus*), Darter goby, Sailfin molly, Rainwater killifish, Sheepshead minnow, Saltmarsh topminnow (*Fundulus jenkinsi*), Longnose Killifish Pinfish, Mosquitofish (*Gambusia affinis*), Star drum (*Stellifer lanceolatus*), Speckled worm eel (*Myrophis punctatus*), Clown goby, Blackcheek tonguefish (*Symphurus plagiusa*), Red dum (*Sciaenops ocellatus*), Spotted seatrout (*Cynoscion nebulosus*), Southern flounder

Sources: Gulf of Maine: Neckles and Dionne 1999; Southern New England: Werme 1981; Howes and Goehringer 1996; Raposa and Roman 2001a; Roman et al. 2002; Delaware Bay: Able et al. 2001; South Carolina: Bretsch and Allen 2006; Florida Gulf Coast: Perret 1971; Rogers and Herke 1985; Subrahmanyam and Coultas 1980; Louisiana: Gosselink 1984; Peterson and Turner 1994.

seek out salt marsh pools with organic and fine-textured soils for burrowing where winter temperatures will remain above 33.8°F (1°C), while other pools may freeze (Raposa 2003). They also spawn at high spring tides so eggs are deposited on the marsh where they later hatch when submerged. Fry are carried to pools where they grow for six to eight weeks before moving to tidal creeks and ditches (Warren and Fell 1995). Mummichogs are a vital link in the estuarine food web as they are part of the diet of many predatory fishes, birds, and other wildlife. Through this process, energy is transferred from the marsh to the adjacent waters and other habitats.

Eel-grass beds are important nursery grounds for both fishes and crabs. The highest densities of juvenile cod in Newfoundland were found in eel-grass beds (Gregory 2004), and they supported higher numbers of fish species than nearby nonvegetated bottoms in Chesapeake Bay and in New York's Great South Bay (Briggs and O'Connor 1971; Orth and Heck 1980). Aquatic beds appear to be safe havens for juveniles, helping them avoid predation. Nighttime use was generally greater for most species in the Kouchibouguac estuary, New Brunswick, with tomcod, winter flounder, and sand shrimp more abundant at night and threespine stickleback and cunner predominant during the day (Joseph et al. 2004). The juxtaposition of seagrass beds and tidal marshes may increase fish productivity. Pinfish were more than twice as abundant in North Carolina salt marshes adjacent to shoal-grass and eel-grass beds as in salt marshes adjacent to nonvegetated bottoms (Irlandi and Crawford 1997). Growth of pinfish in enclosures placed in marsh–seagrass bed areas was far greater than for pinfish placed in other locations—they were more than 90 percent heavier. Pinfish are a major prey species in southern estuaries for commercially important species including summer flounder, bluefish, and Spanish mackerel. Abundance of fourspine stickleback, seaboard goby, and oyster toadfish were higher in eel-grass beds adjacent to salt marshes, while eel-grass beds near beaches supported more Atlantic silversides, winter flounder, and grass shrimp (Raposa and Oviatt 2000).

The upper estuary with its lower salinity (0–10 ppt) is considered the critical zone for larval stages of most anadromous fish species based on observations in Chesapeake Bay and the Hudson River (Dovel 1971, 1981). Striped bass migrate from the sea to oligohaline and tidal freshwaters to spawn. Other anadromous species using low-salinity waters as nursery grounds include alewife, American shad, blueback herring, gizzard shad, Atlantic sturgeon, shortnose sturgeon, Atlantic tomcod, white perch, and yellow perch (semi-anadromous: migrating from brackish water to spawn in freshwater) (Odum et al. 1984; Able and Fahey 1998). The first three species and white perch also use these areas for spawning from spring to early summer. These lower salinity regions are major nursery grounds for bay anchovies and hogchokers; they spend their first summer feeding in these areas (Lippson 1973). Resident fishes of freshwater tidal wetlands and waters include chain pickerel, carp, goldfish, golden shiner, spottail shiner, silvery minnow, white catfish, channel catfish, brown bullhead, banded killifish, mummichog, mosquitofish, tessellated darter, largemouth bass, pumpkinseed, and bluegill (Good et al. 1975; Hastings and Good 1977; Odum et al. 1984). While fishes and swimming invertebrates (collectively called "nekton") were more abundant in the salt marsh systems of Virginia, nekton diversity was greater in tidal fresh marshes (Yozzo and Smith 1998). Interestingly, estuarine species dominated both habitats. Low-salinity marshes in Louisiana may be important nursery areas for commercially important gulf menhaden (Rozas and Minello 2010).

Bird Habitat

Hundreds of avian species, including both migratory and resident birds, depend on tidal wetlands, and many wildlife refuges have been established to protect and manage these habitats (Table 6.5). Many species are frequently observed (Figure 6.8), while others are rare or occasional visitors. Coastal wetlands and estuaries in most of the northeastern United States and eastern Canada provide feeding grounds virtually year-round for some species, while South Atlantic and Gulf Coast marshes provide year-round habitat for many species and serve as vital overwintering grounds for many North American waterfowl. Salt and brackish marshes are common coastal habitats from New Hampshire south, forming an almost continuous green band behind barrier islands from Long Island south. Rich tidal flats intermixed with these and more northern marshes provide a muddy shore rich in invertebrates that supports high populations of North American migratory birds along the Atlantic Flyway. Tidal wetlands are valuable feeding grounds and resting areas, which are critical for successful bird migration between their southern wintering habitat and northern breeding grounds. They also serve as breeding habitat for many species in summer, with marsh ponds and tidal creeks providing important summer foraging habitat for waterfowl, wading birds, and shorebirds. Bird use of the vegetated interior portion of the marsh is more limited (Erwin et al. 2006). Sparrows, grackles, crows, and red-winged blackbirds may be more prominent there. Numerous studies have demonstrated relationships between tidal marsh plant communities and bird distribution (e.g., Burger et al. 1982; Craig and Beal 1992; Reinert and Mello 1995; DeLuca et al. 2004).

Fall and spring are peak seasons for migratory birds to concentrate in and around coastal wetlands. Some particularly important stopover places (staging areas) for migrants to feed and rest during their travels along the Atlantic Flyway are Lake St. Pierre and Cap Tourmente (QC), Bay of Fundy (NB and NS), Sandy Hook and Cape May (NJ), Delaware Bay (NJ and DE), and Chesapeake Bay (MD and VA). Near the start of their southward migration, snow geese feed heavily on common three-square marshes along the St. Lawrence River in Quebec. Up to one million birds, or 30 percent of the world's population of snow geese, stage in Cap Tourmente marshes during migration. Here they consume about 35 percent of the common three-square production (Glooschenko and Grondin 1988). Later, they move south where they graze heavily on smooth cordgrass, denuding hundreds of acres of salt marsh in the Delaware Bay–southern New Jersey area each winter. From July to November, 35 species of shorebirds migrate to the Bay of Fundy to fatten up for their long journey to Central and South America wintering grounds, and 24 of these species frequent mudflats or salt marshes (Table 6.6; Figure 6.9; Hicklin 1987). From 50 to 95 percent of the world population of semipalmated sandpipers feed here on mudflat invertebrates, doubling their weight in preparation for their 3,700-mile (6,000 km) flight to South America (Mawhinney et al. 1993). An estimated 1 to 2 million shorebirds feed on these mudflats before continuing their southward journey. Delaware Bay is the largest spring migration staging area on the East Coast for shorebirds returning to northern breeding grounds with 800,000 to 1.5 million birds recorded each spring (U.S. Fish and Wildlife Service 1997). Six species make up 95 percent of these shorebirds—semipalmated sandpiper, red knot, ruddy turnstone, sanderling, dunlin, and short-billed dowitcher. Horseshoe crab eggs are a prime food source and in the spring, Delaware Bay shores contain the largest concentration of these eggs in the eastern United States. Twenty-two species of shorebirds were observed feeding in tidal inlets in the southeastern United States during spring and fall migrations (Harrington 2008). Seven shorebirds showed a preference for feeding in the intertidal zone of coastal inlets from North Carolina to Florida: black-bellied plover, piping plover, red knot,

Figure 6.8. Some common birds seen in and around salt marshes. (Mike Baird: black-crowned night heron, brant, great blue heron, greater yellowlegs, and willet; Andrew Cruz: great egret; Steve Hillebrand: osprey; Chelsi Hornbaker: tree swallows; Lee Karney: canvasback; Dave Menke: snow goose; Laura Meyers: glossy ibis, clapper rail, marsh wren, seaside sparrow, and saltmarsh sharp-tailed sparrow; Glen Smart: black duck; U.S. Fish and Wildlife Service: double-crested cormorant and snowy egret)

Table 6.5. Examples of bird use of tidal wetlands in the northeastern United States and eastern Canada

Season and Use	General Bird Type	Species	General Wetland Type
Fall or Spring feeding	Shorebird	Semipalmated plover	Beach, flat
		Greater yellowlegs	Beach, marsh
		Lesser yellowlegs	Beach, marsh
		Semipalmated sandpiper	Flat, marsh
		Least sandpiper	Flat, marsh
		Short-billed dowitcher	Flat, marsh
	Raptor	Peregrine falcon	Beach, marsh
Summer breeding	Shorebird	Piping plover	Beach
		Willet	Marsh, beach
		Spotted sandpiper	Beach, flat
	Waterfowl	Wood duck	Tidal swamp
	Wading bird	Glossy ibis	Marsh
		Least bittern	Marsh
		American bittern	Brackish-fresh marsh
		Green heron	Marsh
	Other waterbird	King rail	Brackish-fresh marsh
		Black rail	Marsh
		Moorhen	Brackish-fresh marsh
		Common tern	Beach
		Least tern	Beach
		Black skimmer	Beach
	Raptor	Osprey	Marsh
	Songbird	Marsh wren	Brackish-fresh marsh
		Yellow warbler	Fresh marsh, shrub swamp
		Prothonotary warbler	Forested swamp
		Common yellowthroat	Fresh marsh, shrub swamp
		Nelson's sharp-tailed sparrow	Salt marsh, fresh marsh
Summer feeding	Wading bird	White ibis	Marsh, flat
		Yellow-crowned night heron	Marsh, flat
		Black-crowned night heron	Marsh, flat
		Tri-colored heron	Marsh, flat
		Little blue heron	Marsh, flat
		Snowy egret	Marsh, flat
		Great egret	Marsh, flat
	Other waterbird	Laughing gull	Beach, flat, marsh
		Roseate tern	Beach
	Songbird	Bobolink	Fresh marsh
Winter feeding	Shorebird	Black-bellied plover	Beach, flat
		Piping plover	Beach
		Ruddy turnstone	Beach
		Red knot	Beach, flat
		Sanderling	Beach
		Dunlin	Beach, flat
	Waterfowl	American black duck	Salt marsh
		Snow goose	Salt-brackish marsh
	Other waterbird	Virginia rail	Salt marsh
	Raptor	Rough-legged hawk	Salt marsh
		Short-eared owl	Salt marsh
	Songbird	Sedge wren	Marsh

Table 6.5. *(continued)*

Season and Use	General Bird Type	Species	General Wetland Type
Winter feeding		Savannah sparrow	Beach, marsh
		Snow bunting	Beach
Year-round resident	Shorebird	American oystercatcher	Beach, flat
	Waterfowl	Mallard	Marsh
		American black duck	Marsh
	Wading bird	Great blue heron	Marsh, flat
		American bittern	Marsh
	Other waterbird	Clapper rail	Salt marsh
		Virginia rail	Brackish-fresh marsh
		Herring gull	Beach, flat, marsh
		Great black-backed gull	Beach, flat, marsh
	Raptor	Northern harrier	Marsh
	Songbird	Saltmarsh sharp-tailed sparrow	Salt marsh
		Seaside sparrow	Marsh
		Song sparrow	Marsh, shrub swamp
		Swamp sparrow	Fresh marsh
		Red-winged blackbird	Marsh
		Boat-tailed grackle	Salt marsh
		American goldfinch	Fresh marsh

Notes: Fall and spring use is by migratory species. Year-round residency mainly applies to wetlands in the southern parts of this region, generally from Long Island, New York, south. Most winter-use species also use coastal wetlands during migration. "Marsh" includes salt marsh, shallow ponds, and tidal creeks.

ruddy turnstone, snowy plover, western sandpiper, and Wilson's plover (Harrington 2008). Other species observed feeding there during annual migrations were American avocet, American oystercatcher, dunlin, greater yellowlegs, lesser yellowlegs, long-billed dowitcher, least sandpiper, marbled godwit, sanderling, short-billed dowitcher, semipalmated plover, spotted sandpiper, whimbrel, and willet. Four species were observed nesting on coastal beaches—piping plover, snowy plover, Wilson's plover, and American oystercatcher.

Wintering areas for northern-breeding waterfowl, waterbirds, and shorebirds are critical not only for survival but also for their future reproductive success. While many birds travel to South Atlantic and Gulf Coast marshes and tropical regions for winter, many waterfowl overwinter in northeastern U.S. estuaries and ice-free estuaries in the Maritimes where they find open water or salt marshes (Smith et al. 1989c). Among the more common winter visitors to northeastern estuaries are common eider, greater and lesser scaup, brant, common merganser, red-breasted merganser, hooded merganser, old squaw, common goldeneye, bufflehead, American wigeon, canvasback, mallard, American black duck, green-winged teal, northern pintail, northern shoveler, scoters, great cormorant, double-crested cormorant, horned grebe, common loon, red-throated loon, and tundra swan (e.g., Boyle 1991). About 75 percent of the western Atlantic population of brant overwinters in Long Island bays and New Jersey estuaries where they feed on eelgrass and sea lettuce (U.S. Fish and Wildlife Service 1997). American black ducks overwinter in coastal marshes from Maine to South Carolina (mostly in Mid-Atlantic states). Some may overwinter in more northern marshes, like in Nova Scotia, during mild winters (McAloney 1981). In northern marshes, freezing of mudflats during exposure at low tide causes black duck to shift feeding to ice-free zones near the low-tide line (Hartman 1963). Marsh and tidal flat invertebrates (e.g., snails, mussels, clams, and amphipods) make up the majority of their diet. They feed mainly at dawn and dusk (crepuscular) and on bright moonlit evenings (Lewis and Garrison 1984). Chesapeake Bay is the winter home of about one-third of the waterfowl using the Atlantic Flyway (Tiner and Burke 1995), especially for mallard, canvasback, and ruddy duck. Delaware Bay is particularly important for overwintering snow geese. North Carolina and South Carolina tidal marshes support the highest numbers of many wintering waterfowl along the Atlantic Flyway including gadwall, American wigeon, green-winged teal, northern pintail, and redhead. Coastal marshes along the Gulf and Pacific Coasts are vital wintering grounds for waterfowl and shorebirds along other flyways. Half of the winter duck population of Mississippi Flyway is found in coastal marshes and estuaries from Alabama through Louisiana, while Texas marshes are vital wintering grounds for more than half of Central Flyway's waterfowl (Wilson et al. 2002). Seventy-seven percent of the North American population of redheads overwinter in the Laguna Madre in Texas where they feed on shoal grass that covers much of the shallow water zone (Tunnell et al. 2002). More than 50 percent of the ducks that winter in Canada use the Fraser River estuary (British Columbia) and other coastal marshes—they are recognized as internationally important habitats for migratory birds (Glooschenko

Table 6.6. Migratory shorebirds using the Bay of Fundy mudflats and salt marshes

Species	Habitat
Semipalmated plover	Mudflats, sandflats and salt marshes
Killdeer**	Open fields and salt marshes
Black-bellied plover	Mudflats, sandflats, beaches, and salt marshes
Common snipe**	Freshwater and salt marshes, pastures, and mowed hayfields
Spotted sandpiper**	Beaches and salt and freshwater marshes
Greater yellowlegs	Salt marshes, mudflats, and beaches
Lesser yellowlegs	Salt marshes, mudflats, and beaches
Willet**	Salt marshes and mudflats
Red knot	Salt marshes and mudflats
Pectoral sandpiper	Salt marshes
White-rumped sandpiper	Beaches, mudflats, and salt marshes
Baird's sandpiper	Beaches, mudflats, and salt marshes
Least sandpiper	Beaches, mudflats, and salt marshes
Curlew sandpiper*	Beaches and mudflats
Dunlin	Beaches and mudflats
Semipalmated sandpiper	Beaches, mudflats, and salt marshes
Western sandpiper*	Beaches, mudflats, and salt marshes
Sanderling*	Beaches, mudflats, and salt marshes
Short-billed dowitcher*	Beaches, mudflats, and salt marshes
Long-billed curlew*	Salt marshes and mudflats
Stilt sandpiper*	Beaches and salt marshes
Marbled godwit*	Beaches and salt marshes
Hudsonian godwit	Beaches, mudflats, and salt marshes
American avocet*	Salt marshes

Source: Hicklin 1987.
Notes: *indicates rare species in this area; **indicates species breeding in the region. Major fall migrants are black-bellied plovers, dunlins, and semipalmated sandpipers.

Figure 6.9. Swarming shorebirds put on a most interesting display of synchronized group movement off St. Mary's Point (NB) in late summer. Two of many displays observed during the course of a half hour.

et al. 1988). Coastal wetlands from California into Mexico support high concentrations of shorebirds and brant in winter (e.g., Page et al. 1992, 1999; Pacific Flyway Council 2002).

Wetland animals and plants provide food for birds at other times as well. In northern areas, rockweed beds provide amphipods and periwinkles for black ducks and eider ducks in the spring. Eider ducklings even feed in the smooth cordgrass marshes at high tide (Reed and Moisan 1971). Adult diving ducks feed on aquatic invertebrates, fish, shellfish, and submerged aquatic vegetation (e.g., eel-grass). Dabbling or puddle ducks are surface feeders dipping into the water to feed on plant and microscopic animal life. Some coastal wetland plants that provide important food for waterfowl include smooth cordgrass, widgeon-grass, eel-grass, bulrushes, spike-rushes, arrow arum, Walter millet, rice cutgrass, wild rice, spatterdock, pondweeds, and tape-grass (see Table 7.2). Fish, aquatic invertebrates, and amphibians are also part of the diet of some species. Northern harrier (marsh hawk) and osprey can be frequently observed flying over salt marshes in search of food: rodents, rabbits, and birds for the former and fish for the latter. For some birds, adult use of coastal marshes changes when they are feeding young. For example, adult white ibis typically feed in salt marshes, but when feeding young, they travel to freshwater wetlands (DeSanto et al. 1997). When the young fledge, they fly to freshwater marshes, swamps, and impoundments to feed while adults feed in salt marshes (Petit and Bildstein 1986).

Coastal wetlands provide breeding habitat for many birds in summer. Upper beaches are vital nesting grounds for least, common and roseate terns, and piping plover (federally endangered). Herring gull, great black-backed gull, laughing gull, black skimmer, and American oystercatcher nest on sandy islands and shores. The first four species may also nest in salt marshes. American black ducks are the main breeding waterfowl in salt marshes like those of the St. Lawrence Estuary where widgeon-grass ponds are abundant (Reed and Moisan 1971). Other breeding waterfowl that use tidal marshes include mallard, gadwall and Canada goose. The low salt marsh dominated by the tall form of smooth cordgrass is the prime habitat for the clapper rail, a secretive bird that feeds heavily on fiddler crabs. (Listen closely and you can hear the clucking of this rail—*chit-chit* or *kek-kek*—from Massachusetts south in summer.) Its West Coast relative, the light-footed clapper rail, a federally endangered species, lives in similar places in southern California marshes. Pacific cordgrass at least 2 feet (60 cm) tall is required for nesting as the nests are constructed in a way that allows the nest to float upward with rising tides (Zedler 1993). In the northeastern United States, marsh wrens also nest in the low marsh zone. Nelson's and saltmarsh sharp-tailed sparrows, seaside sparrows, willets, gulls, red-winged blackbirds, and boat-tailed grackle are other birds nesting in Atlantic salt marshes (DeRagon 1988; Boyle 1991; Hanson 2004). The short-form smooth cordgrass zone appears to be an important plant community for breeding red-winged blackbirds, seaside sparrows, and sharp-tailed sparrows with the first two species nesting only in this part of the salt marsh, while the last feed extensively there (Reinert et al. 1981). The common tern and American oystercatcher now nest in salt marshes where their former nesting grounds (dunes) have been developed or disturbed by human traffic (Boyle 1991). Many ospreys now nest on artificial structures with platforms raised well above the salt and brackish marshes. Most birds breed at higher elevations adjacent to the marsh and feed in salt marshes. Heronries for colonial-nesting species are established on treed shores and islands in estuaries (e.g., New York Harbor Islands and Pea Patch Island, DE). Great blue herons, night-herons (black-crowned night and yellow-crowned night), egrets (snowy, great, and cattle), and glossy ibis seek out food in tidal creeks and flats along coastal marshes, while barn swallows and tree swallows feed on airborne insects flying above the marshes. Mute swans nest in coastal ponds in such numbers as to become nuisance species (e.g., in Rhode Island). Gulls nest on marsh islands, dredged material disposal islands, dunes, and beaches. More than 60 species of birds use black needlerush marshes along the Gulf Coast, but clapper rail and seaside sparrows may be the only species nesting in this cover (Stout 1984). Other species nest in shrubs and trees in and around these marshes. In Barataria Bay in Louisiana, shrub thickets of black mangrove provide nesting habitats for some colonial nesting birds (egrets, herons, and ibises) and even brown pelican have nested in these areas (Conner and Day 1987).

Common reed marshes, especially the wetter ones, are used by a number of birds like other marshes. Thirty-three species of birds have been observed breeding in New Jersey reed marshes including many that typically breed in other types of marshes (Kane 2001): pied-billed grebe, night-herons, herons, glossy ibis, egrets, bitterns, rails, American coot, ducks, geese, swamp sparrow, common yellowthroat, marsh wren, red-winged blackbird, common grackle, and northern harrier. These marshes are particularly important for state-endangered and threatened birds such as the pied-billed grebe, night-herons, and American bittern. Overwintering pheasants seek cover in reed marshes.

A host of birds breed in tidal fresh marshes and swamps including typical forest-nesting neotropical migrant species like warblers, thrushes, vireos, and other passerines as well as marsh-nesting species such as green heron, rails (sora, Virginia, king, and black), and bitterns (American and least). Wood ducks nest in tree cavities as well as nesting boxes placed on wooden posts in open water by wildlife managers. Since most wood ducks winter south of Maryland, they are typically summer residents in the Northeast. Red-winged blackbirds and marsh wrens may be the most common birds in tidal fresh marshes. In late summer, bobolinks congregate in tidal fresh marshes to feed on the grain of wild rice. Other freshwater marsh birds include swamp sparrows and song sparrows, yellowthroats, yellow warblers, goldfinches, and many other songbirds plus several species of ducks, Canada geese, great blue heron, and green heron. Belted kingfishers are a common sight along tidal swamp creeks.

In southern regions, many of the species that are seasonal migrants in northern marshes are year-round residents, and various wading birds nest in huge rookeries including ibises (white and glossy), egrets (cattle, snowy, and great), herons (tri-colored, little blue, and black-crowned) (Moore 1992). While the tricolor heron may be the most common heron along the South Carolina–Georgia coast, great egrets and yellow-crowned night herons may be most abundant in brackish and tidal fresh marshes (Sandifer et al. 1980).

Reptile and Amphibian Habitat

Birds are the most frequently observed animals of tidal marshes; however, other vertebrates also live in or frequent these wetlands. These often less conspicuous animals include amphibians, reptiles, and mammals. A few reptiles are found in salt and brackish marshes, but they are more common in freshwater tidal habitats (Table 6.7; Figure 6.10). Snakes are uncommon in northern salt marshes, although an occasional garter or ribbon snake (*Thamnophis*) may be found hunting in the upper edges of northern salt marshes. In contrast, southern salt marshes have resident salt marsh snakes (*Nerodia clarkii*): Atlantic salt marsh snake (subspecies *taeniata*) on the South Atlantic Coast and the Gulf salt marsh snake (subspecies *clarkii*) on the Gulf Coast. Another subspecies, the mangrove snake (*N. clarkii compressicauda*) occurs in Florida's mangrove swamps. This snake lives among the roots of mangroves and eats fish exclusively. While most snakes shake their tails to attract prey, the mangrove snake is unique in using flicks of its curled-up tongue as the lure (Hansknecht 2008). Fresher conditions upstream in estuaries provide more suitable habitats for snakes and other reptiles. Mud and rainbow snakes (*Farancia abacura, F. erytrogramma*) may be found in fresh to low-salinity wetlands from Virginia south. Water snakes are common from eastern Canada south: the eastern water snake (*Nerodia sipedon*) that ranges northward into eastern Canada (along the Atlantic Coast from North Carolina) and the banded water snake (*Nerodia fasciata*) in southern marshes and swamps. They are non-venomous snakes that resemble rattlesnakes and are frequently killed for this reason. The only poisonous snake likely to be found in tidal wetlands from Virginia south is the cottonmouth or water moccasin (*Agkistrodon piscivorous*), although other rattlesnakes (*Crotalus* spp.) and copperhead (*A. contortrix*) are common in maritime forests and might travel into local tidal wetlands on occasion.

The American alligator (*Alligator mississippensis*) and the American crocodile (*Crocodylus acutus*) are the largest reptiles found in North American tidal wetlands. Alligators can be found mostly in tidal fresh and brackish marshes and swamps (less common in salt marshes) from North Carolina south, while crocodiles live in saltwater and mangrove swamps around Florida Bay. Alligators lay their eggs in a mound made of marsh herbs, while American crocodiles lay theirs in holes or mounds of vegetation. An interesting fact about these crocodilians is that the sex of their embryo is determined by temperature during incubation. One study found female alligators developed at 86°F (30°C) and 94.1°F (34.5°C) during days 21 to 45, whereas a temperature of 91.4°F (33°C) produced males (Lang and Andrews 1994). Young hatch in about two months.

Diamondback terrapins (*Malaclemys terrapin*) frequent tidal marsh creeks, nest in sandy areas above the salt marsh, and reportedly burrow in mud at night (Montague and Wiegert 1991). In northern coastal waters, they overwinter in the muddy bottoms of the tidal creeks and shallow embayments. From mid-November through December, terrapins begin migrating from open water to tidal creeks where they hibernate in natural depressions that are 5 to 8 feet (1.5–2.4 m) or more deep at low tide or in burrows dug into sides of eroding creekbanks (Yearicks et al.

1981). They seem to prefer slumping banks that lack roots and therefore make for easier digging. While many overwinter in single burrows, some terrapins hibernate in groups beneath eroding banks until April or May when they become active. Other turtles can be found in and around tidal marshes. Snapping turtles (*Chelydra serpentina*), more typical of brackish and freshwater areas, may also be occasionally observed in salt marshes along the Atlantic and Gulf Coasts. The Alabama red-bellied turtle (*Pseudemys alabamensis*, a federally endangered species) and Florida cooter

Table 6.7. Amphibians and reptiles reported from North American tidal marshes

Region	Species	Habitat
East Coast	Southern chorus frog (*Pseudacris nigrita*)	Salt marsh, varied
	Spotted chorus frog (*Pseudacris clarkia*)	Salt marsh
	Little grass frog (*Pseudacris ocularis*)	Salt marsh, fresh marsh
	Green tree frog (*Hyla cinerea*)	Fresh marsh
	Gray tree frog (*Hyla versicolor*)	Fresh marsh
	Pine woods tree frog (*Hyla femoralis*)	Varied
	Eastern narrow-mouthed toad (*Gastrophryne carolinensis*)	Varied
	Southern leopard frog (*Rana sphenocephala*)	Fresh marsh
	Pickerel frog (*Rana palustris*)	Fresh marsh
	Pig frog (*Rana grylio*)	Fresh marsh
	Common snapping turtle (*Chelydra serpentine*)	Salt marsh, fresh marsh
	Spotted turtle (*Clemmys guttata*)	Fresh marsh
	Painted turtle (*Chrysemys picta*)	Fresh marsh
	Florida cooter (*Pseudemys concinna*)	Fresh marsh
	Florida redbelly turtle (*Pseudemys nelsoni*)	Fresh marsh
	Striped mud turtle (*Kinosternon baurii*)	Fresh marsh
	Eastern mud turtle (*Kinosternon subrubrum*)	Salt marsh, fresh marsh
	Green anole (*Anolis carolinensis*)	Varied
	Slender glass lizard (*Ophisaurus attenuatus*)	Salt marsh
	Salt marsh snake (*Nerodia clarkii*)	Salt marsh
	Northern water snake (*Nerodia sipedon*)	Salt marsh, fresh marsh
	Mississippi green water snake (*Nerodia cyclopion*)	Fresh marsh
	Graham's crayfish snake (*Regina grahami*)	Fresh marsh
	Black swamp snake (*Seminatrix pygeae*)	Fresh marsh
	Mud snake (*Farancia abacura*)	Salt marsh, fresh marsh
	Eastern indigo snake (*Drymarchon corais*)	Varied
	Eastern racer (*Coluber constrictor*)	Varied
	Rough green snake (*Opheodrys aestivus*)	Fresh marsh
	Eastern rat snake (*Elaphe obsoleta*)	Varied
	Common king snake (*Lampropeltis getula*)	Varied
	Eastern diamond-backed rattlesnake (*Crotalus adamanteus*)	Varied
	Timber rattlesnake (*Crotalus horridus*)	Salt marsh, fresh marsh
	American crocodile (*Crocodylus acutus*)	Salt marsh, fresh marsh
	American alligator (*Alligator mississippiensis*)	Salt marsh, fresh marsh
West Coast	Pacific pond turtle (*Emys marmorata*)	Fresh marsh
	West Mexican water snake (*Nerodia valida*)	Fresh marsh
	Common garter snake (*Thamnophis sirtalis*)	Fresh marsh
	Ring-necked snake (*Diadophis punctatus*)	Fresh marsh

Source: Greenberg and Maldonado 2006.
Note: East Coast includes the Atlantic and Gulf Coasts.

Figure 6.10. Reptiles are not well represented in tidal marshes: (a) American alligator (*Alligator mississippensis*) frequents tidal fresh marshes and coastal impoundment; (b) American crocodile (*Crocodylus acutus*) occurs in saltwater and mangrove swamps from the tip of Florida into Colombia and Ecuador; (c) diamond-backed terrapin (*Malaclemys terrapin*) is the most common turtle found in estuaries; and (d) Gulf salt marsh snake (*Nerodia clarkii clarkii*). (a: Adam Mackinnon; b: Tomás Castelazo; c: Ryan Hagerty, U.S. Fish and Wildlife Service; d: Kelly Jones)

(*Pseudemys floridana*) frequent Gulf Coast black needlerush marshes and fresh tidal marshes. Sea turtles nest on ocean beaches from Virginia south, with the loggerhead turtle (*Caretta caretta*), a federally threatened species, being the most abundant of those nesting in the United States. Archie Carr National Wildlife Refuge (Atlantic Coast of Florida) contains the most important nesting area in the western hemisphere with as many as 1,000 nests per mile reported (www.nmfs.noaa.gov/pr/species/turtles/loggerhead.htm).

Amphibians are virtually absent from salt and brackish marshes as salamanders and frogs tend to be restricted to freshwater environments. One exception is the southern leopard frog (*Rana utricularia*), which has been reported in brackish marshes with salinities as high as 21 ppt but more typically less than 1 ppt (Pearse 1936). On occasion, one might expect to find a transient frog or toad from the adjacent upland or freshwater wetland along the upper edges of a salt or brackish marsh. Frogs and amphibians characteristic of a region are likely inhabitants of tidal freshwater wetlands.

Mammal Habitat

A variety of mammals have been observed in North American tidal marshes (Table 6.8).

Small mammals such as red-backed vole (*Clethrionomys gapperi*), meadow vole (*Microtus pennsylvanicus*), white-footed mouse (*Peromyscus leucopus*), meadow jumping mouse (*Zapus hudsonius*), and rice rat (*Oryzomys palustris*) may be found in salt and brackish marshes, but they are more abundant in tidal freshwater wetlands. In the oligohaline marshes of Rhode Island, mammal runs may cover 20 percent of the marsh floor (Leck and Crain 2009). Small mammals feed on plants, invertebrates, and fish. Rice rats are major egg predators on long-billed marsh wrens and seaside sparrows in southern marshes where they build their nests in smooth cordgrass and black needlerush stands near the nests of these birds (Montague and Wiegert 1991). Marsh rabbits (*Sylvilagus palustris*), cotton mice (*Peromyscus gossypinus*), and cotton rats also occur in southern coastal marshes. Muskrats (*Ondatra zibethicus*) inhabit brackish and tidal fresh marshes as evidenced by the abundance of their houses in cattail, three-square, and bulrush marshes throughout the Northeast. Meadow voles and rice rats often inhabit the muskrat houses (Harris 1953). Nutria (*Myocastor coypus*), a South American introduction, is common from Maryland into North Carolina and along the Gulf Coast where it has successfully displaced muskrats from freshwater marshes into less favored brackish and saline habitats (Wilson 1968). The round-tailed muskrat (*Neofiber alleni*) occupies the high salt marsh at Merritt Island in Florida. Raccoon (*Procyon lotor*), red fox (*Vulpes fulva*), gray fox (*Urocyon cineroargenteus*), and white-tailed deer (*Odocoileus virginianus*) frequent coastal wetlands as do mink (*Mustela vison*) and river otter (*Lontra canadensis*). Raccoons feast on the eggs of sea turtles, diamond-backed terrapins, and marsh and land birds.

Larger mammals also inhabit tidal wetlands. Perhaps the most well-known are the wild ponies of Chincoteague made famous by Marguerite Henry's children's book *Misty of Chincoteague*. The ponies are actually wild horses (*Equus caballus*) that live on Assateague Island (in Virginia and Maryland; Figure 6.11). The ponies live in social groups called "bands" (Keiper 1990). On Assateague Island, harem bands consist of one stallion, one to several mares, and their colts and fillies, while groups of a few young males or mixed bands of young females and males also form. Stallions defend their harems rather than a territory. Wild horses and hogs (*Sus scrofa*) can be found in salt marshes surrounding sea islands on the North Carolina and Georgia coasts, while wild horses also live on Texas barrier islands. These horses are descendants of horses brought to the Americas by the Spanish and may have come to the islands with explorers and early settlers or from shipwrecks of Spanish galleons. The horses feed on a variety of dune and marsh grasses with smooth cordgrass being an important part of their diet (Wood et al. 1987). Their grazing of cordgrass can negatively affect marsh use by nesting birds such as laughing gull and forster's tern, while increasing use by a greater variety of species including willet and least sandpiper (Levin et al. 2002). Also by removing vegetation such grazing may make resident marsh crabs more susceptible to predation by fish at high tide.

Deer trails can be found through many tidal marshes. In Nova Scotia, white-tail deer feed on rockweeds in the intertidal zone in winter (Davis and Browne 1996). On Assateague Island, sika deer (*Cervus nippon*), an introduction from eastern Asia, frequent the back-barrier salt marshes where they graze on cordgrasses and seaside goldenrod (Keiper 1990). Moose (*Alces alces*) have been observed feeding on salt hay grass in the Maritime provinces of Canada (Alan Hanson, pers. comm. 2004). Beaver (*Castor canadensis*) may on rare occasions take up residence in tidal freshwater wetlands. Black bear (*Ursus americanus*) inhabit southern coastal plain swamps including tidal swamps, and the Louisiana

Table 6.8. Mammals reported from North American tidal marshes

Region	Species	Habitat
East Coast	Virginia opossum (*Didelphis virginiana*)	Salt marsh, varied
	Cinereus shrew (*Sorex cinereus*)	Salt marsh
	Southeastern shrew (*Sorex longiostris*)	Fresh marsh
	Northern short-tailed shrew (*Blarina brevicauda*)	Salt marsh, fresh marsh
	Least shrew (*Cryptotis parva*)	Salt marsh, fresh marsh
	Eastern mole (*Scalopus aquaticus*)	Varied
	Swamp rabbit (*Sylvilagus aquaticus*)	Salt marsh,
	Marsh rabbit (*Sylvilagus palustris*)	Salt marsh, fresh marsh
	Eastern harvest mouse (*Reithrodontomys humulis*)	Salt marsh, varied
	Meadow vole (*Microtus pennsylvanicus*)	Salt marsh
	Long-tailed vole (*Microtus longicaudus*)	Fresh marsh
	Townsend's vole (*Microtus townsendii*)	Salt marsh, fresh marsh
	Red-backed vole (*Clethrionomys gapperi*)	Salt marsh, fresh marsh
	Muskrat (*Ondatra zibethicus*)	Salt marsh, fresh marsh
	Marsh rice rat (*Oryzomys palustris*)	Salt marsh, fresh marsh
	Round-tailed muskrat (*Neofiber alleni*)	Salt marsh
	Eastern woodrat (*Neotoma floridana*)	Salt marsh, varied
	White-footed mouse (*Peromyscus leucopus*)	Varied
	Meadow jumping mouse (*Zapus hudsonius*)	Salt marsh, fresh marsh
	House mouse (*Mus musculus*)*	Salt marsh, varied
	Black rat (*Rattus rattus*)*	Salt marsh, varied
	Brown rat (*Rattus norvegicus*)*	Salt marsh, varied
	Nutria (*Myocastor coypus*)*	Salt marsh, fresh marsh
	Gray fox (*Urocyon cinereoargenteus*)	Varied
	Red fox (*Vulpes vulpes*)	Varied
	Coyote (*Canis latrans*)	Varied
	Raccoon (*Procyon lotor*)	Salt marsh, fresh marsh
	American mink (*Mustela vison*)	Salt marsh, varied
	Northern river otter (*Lontra canadensis*)	Varied
	Striped skunk (*Mephitis mephitis*)	Salt marsh, varied
	Domestic cat (*Felis catus*)*	Salt marsh, varied
	Hog (*Sus scrofa*)*	Salt marsh, varied
	Horse (*Equus caballus*)*	Salt marsh, varied
	White-tailed deer (*Odocoileus virginianus*)	Salt marsh
	Sika deer (*Cervus nippon*)*	Salt marsh, varied
West Coast	Virginia opossum*	Salt marsh, varied
	Ornate shrew (*Sorex ornatus*)	Salt marsh
	Vagrant shrew (*Sorex vagrans*)	Salt marsh
	Marsh shrew (*Sorex bendirii*)	Varied
	Desert shrew (*Notiosorex crawfordi*)	Salt marsh, shrub swamp
	Broad-footed mole (*Scapanus latimanus*)	Varied
	Brush rabbit (*Sylvilagus bachmani*)	Varied
	Salt marsh harvest mouse (*Reithrodontomys raviventris*)	Salt marsh
	Western harvest mouse (*Reithrodontomys megalotis*)	Salt marsh
	California vole (*Microtus californicus*)	Salt marsh
	House mouse*	Salt marsh, varied
	Black rat*	Salt marsh, varied
	Brown rat*	Salt marsh, varied
	Red fox*	Varied
	Coyote	Varied
	Raccoon	Salt marsh, fresh marsh
	Striped skunk	Salt marsh, varied
	Domestic cat*	Salt marsh, varied
	Elk (*Cervus elaphus*)	Salt marsh, varied

Source: Expanded slightly from Greenberg and Maldonado 2006.
Note: Species marked with * are introduced, nonnative to the region.

Figure 6.11. Wild horses (*Equus caballus*) attract much visitor attention at several barrier island refuges and parks along the Atlantic and Gulf Coasts.

subspecies (*U. americanus luteolus*) is a federally threatened species. Coastal brown bear (*U. arctos,* a federally threatened species in the lower 48 states), the largest land carnivore in North America, feeds on spawning salmon in tidal streams along the Pacific Coast from Washington into Alaska. This diet has made them much larger than their inland relatives—the grizzly bear (*U. arctos horribilis*). Upon awakening from hibernation, coastal brown bears travel to intertidal mudflats to feed on clams and brackish plants before heading inland for the June salmon run (Smith and Partridge 2004). One bear may harvest as many as 100 clams during low tide (Troyer 2005). The intertidal environment is particularly important to female bears with cubs and single small bears.

Marine mammals also frequent tidal wetlands. Harp, hooded, and gray seals (*Pagophilus groenlandicus, Cystophora cristata, Halichoerus grypus*) plus whales and dolphins are common in the coastal waters of the Gulf of St. Lawrence. Seals use a variety of intertidal wetlands as "haul-outs" for resting, birthing, and nursing: rocky shores, mudflats, sand bars, and beaches. Harbor seals (*Phoca vitulina*) are year-round residents of coastal waters from Maine north, and seasonally they use intertidal beaches and rocky shores in southern New England and New York as haul-outs from September through late May prior to the pupping season when they return again to northern waters to give birth (Katona et al. 1993). Gray seals, year-round residents in southern New England, congregate in large groups on coastal beaches where females birth pups from late December into mid-February. Muskeget Island off of Nantucket hosts the largest population of gray seals on the East Coast during the pupping season (Dawicki 2009). On the Pacific Coast, sea lions (Family Otariidae) and harbor seals use similar wetlands as haul-outs. Bottlenose dolphins (*Tursiops truncatus*) feed and make high use of estuaries and salt marsh creeks in southern waters. An estimated 3 to 7 percent of annual production of North Inlet, South Carolina, is needed to support its dolphin population (Young and Phillips 2002). On occasion, dolphins can be seen herding schools of fishes up onto the banks of salt marshes where they are easy prey. The West Indian manatee (*Trichechus manatus*), a familiar marine mammal of Florida's coastal waters, is known to feed on smooth cordgrass in addition to its typical diet of submerged aquatic plants that include widgeon-grass (*Ruppia maritima*) and sago pondweed (*Stuckenia pectinatus*) (Rathbun, et al. 1990; Walkup 2007).

Rare Species Habitat

Given their history of human use and the fact that many coastal wetlands have been destroyed, significantly altered, or grossly polluted, some of the remaining wetlands provide the last refuge for rare and endangered wildlife. Federally endangered and threatened species that frequent tidal wetlands are listed in Table 6.9 (Figure 6.12), while Table 6.10 lists some species of concern in northeastern states or eastern Canadian provinces (see Greenberg et al. 2006 for species and subspecies of concern in several states elsewhere). Many of these provincial and state rare species are so designated because these locations are at the limits of their geographic range.

A Special Note on Common Reed as Wildlife Habitat

Common reed or *Phragmites* (its genus name by which it is often referred to; *Phragmites australis*) is widely recognized as the number one invasive plant threatening estuarine wetlands in the northeastern United States. While a native subspecies of common reed exists (*Phragmites australis* ssp. *americanus*), it is not particularly abundant today as the nonnative species has displaced it along with other native halophytic species. In particular, common reed has outcompeted typical salt marsh plants in areas where tidal flow has been significantly restricted, where fill has been deposited in wetlands, or where excessive nutrients have been introduced from upland sources. *Phragmites* is a good disturbance indicator as it readily colonizes exposed soils in the coastal zone and even inland areas along highways (Marks et al. 1994). Plant diversity usually declines with the invasion of common reed as this species typically forms monotypic stands, especially in brackish waters (Meyerson et al. 2000). Changes in plant composition alter the habitat use by some to many species. There is general agreement that pure *Phragmites* stands generally yield poorer quality wildlife habitat than the marshes they replace, although they may be important habitat for some species (Roman et al. 1984; Kiviat 1987). The tall, dense reeds restrict wildlife movement and also adversely affect hydrology with negative impacts on aquatic species.

More than 50 species of birds have been found in common reed marshes; however, no birds depend solely on these wetlands (Meyerson et al. 2000). Common birds using *Phragmites* marshes include marsh wren, red-winged blackbird, and swamp sparrow. Northern harriers have been found nesting in pure common reed marshes or those mixed with poison ivy in New Jersey and New York, respectively (Dunne 1984; England 1989). Ringed-necked pheasant and American bittern have also been observed in common reed stands (Tiner, pers. obs.). Muskrats, raccoons, and other wildlife frequent these marshes.

From the aquatic organism perspective, marsh flooding provides access for fishes and swimming (nektonic) invertebrates, and anything reducing this connection will have a negative impact on their use of the marshes. Common reed is known to accelerate the buildup of the marsh surface and to reduce drainage density by filling in small ditches and creeks, thereby restricting access to the marshes by fishes and transient shellfish (Weinstein and Balletto 1999). Reducing the frequency of tidal flooding has obvious negative impacts on aquatic species. Also small pools on the marsh surface at low tide are important for resident fishes. Fish and shellfish density in *Phragmites* stands vary with hydrology and wetland geomorphology. High stem density and litter accumulation may reduce tidal flow rates, leading to

Table 6.9. U.S. federally endangered and threatened plants and animals found in tidal wetlands, excluding fishes and sea turtles that nest on southern beaches

Common Name (*Scientific name*)	Status	Habitat (Primary state or U.S. range)
Plants		
Seabeach amaranth (*Amaranthus pumilus*)	T	Beach (MA–SC)
Sensitive joint-vetch (*Aeschynomene virginica*)	T	Brackish and tidal fresh marsh (NJ–NC)
California sea blite (*Suaeda californica*)	E	Salt marsh and beach (CA)
Suisun thistle (*Cirsium hydrophilum hydrophilum*)	E	Brackish marsh (CA)
Soft bird's-beak (*Cordylanthus mollis mollis*)	E	Salt and brackish marsh (CA)
Salt marsh bird's-beak (*Cordylanthus maritimus maritimus*)	E	Salt and brackish marsh (CA)
Invertebrates		
Northern beach tiger beetle (*Cicindela dorsalis dorsalis*)	T	Beach (MA–VA)
Puritan tiger beetle (*Cicindela puritana*)	T	Tidal fresh beach (MA)
Reptiles		
American crocodile (*Crocodylus acutus*)	T	Mangrove (South Florida Bay)
Atlantic salt marsh snake (*Nerodia clarkii taeniata*)	T	Salt marsh (FL)
Birds		
Western snowy plover (*Charadrius alexandrinus nivosus*)	T	Beach (WA–CA)
Piping plover (*Charadrius melodus*)	T	Beach (ME–SC)
Cape Sable seaside sparrow (*Ammodramus maritimus mirabilis*)	T	Brackish to fresh marsh (FL)
Wood stork (*Mycteria americana*)	E	Salt to fresh marsh; mangrove (FL)
Roseate tern (*Sterna dougallii dougallii*)	E	Beach (NC north)
	T	Beach (NC south)
California least tern (*Sterna antillarum browni*)	E	Beach (CA)
California clapper rail (*Rallus longirostris obsoletus*)	E	Salt marsh (CA)
Light-footed clapper rail (*Rallus longirostris levipes*)	E	Salt marsh (CA)
Steller's eider (*Polysticta stelleri*)	E	Tidal flat, salt marsh, eel-grass bed (AK)
Whooping crane (*Grus americana*)	E	Tidal marsh (TX)
Mammals		
West Indian manatee (*Trichechus manatus*)	E	Estuarine aquatic beds (FL)
Lower Keys marsh rabbit (*Sylvilagus palustris hefneri*)	E	Salt to fresh marsh (FL)
Rice rat (*Oryzomys palustris natator*)	E	Salt to fresh marsh; mangrove (FL)
Florida salt marsh vole (*Microtus pennsylvanicus dukecampbelli*)	E	Salt marsh (FL)
Salt marsh harvest mouse (*Reithrodontomys raviventris*)	E	Salt marsh (CA)
Louisiana black bear (*Ursus americanus luteolus*)	T	Tidal swamps (MS–TX)

Source: U.S. Fish and Wildlife Service.
Notes: E = endangered; T = threatened. This list should contain most of the presently designated species.

a reduction in the depth of tidal flooding. In a Hudson River brackish marsh, common mummichogs, herrings, grass shrimp, and blue crabs were captured primarily near the creekbanks whereas only a few individuals were collected from the marsh interior (Hanson et al. 2002). While fishes use accessible parts of common reed marshes, a greater abundance of mummichogs were found in neighboring smooth cordgrass marshes. A similar study in Alloway Creek (NJ) found the lack of water on the reed marsh at high tide to be a major impediment for mummichogs (Hagen et al. 2007).

Light-footed clapper rail

California least tern

Piping plover

Wood stork

Whooping crane

Salt marsh harvest mouse

West Indian manatee

Figure 6.12. Some federally endangered wildlife species found in coastal wetlands. See also American crocodile (Figure 6.10b). (Steve Hillebrand: whooping crane; Gene Nieminen: piping plover; Jim P. Reid: manatee; Scott Streit: California least tern and light-footed clapper rail; California Department of Water Resources: salt marsh harvest mouse)

Here, former reed marshes that had been treated with a combination of herbicides (Rodeo) and controlled burning provided productive mummichog habitat similar to smooth cordgrass marshes. Consequently, while reed marshes do provide habitat for fish and wildlife, most do not appear to be used by as many species or support the abundance of individuals that the native tidal marshes do.

Further Readings

"The ecological distribution of spiders in non-forest maritime communities at Beaufort, North Carolina" (Barnes 1953)

"Zonal and seasonal distribution of insects in North Carolina salt marshes" (Davis and Gray 1966)

Life and Death of the Salt Marsh (Teal and Teal 1969)

The Biology of Estuarine Animals (Green 1971)

Table 6.10. Examples of rare, threatened, and endangered plants and animals frequenting coastal wetlands in the northeastern United States and eastern Canada

	State or Provincial Concern/Rare	Habitat
Plants		
Seaside gerardia (*Agalinis maritima*)	NS, ME, NH	Salt marsh
Gaspé arrow-grass (*Triglochin gaspensis*)	PEI	Salt marsh
Parker's pipewort (*Eriocaulon parkeri*)	QC, NB, ME, MA, CT	Tidal fresh marsh
River bulrush (*Schoenoplectus fluviatilis*)	MA, PA	Tidal fresh marsh
Eaton's beggar-ticks (*Bidens eatonii*)	ME, MA, CT, NJ	Tidal fresh marsh
Estuarine beggar-ticks (*Bidens hyperborea*)	NS, ME, MA, NY	Brackish marsh
Long's bitter-cress (*Cardamine longii*)	ME, NH, MA, RI, CT, NJ, DE, MD	Brackish to fresh marsh
Seabeach sandwort (*Honckenya peploides*)	NH, MD, CT	Beach
Bathurst aster (*Symphyotrichum subulatum* var. *obtusifolius*)	NB	Beach, Salt marsh
Gulf of St. Lawrence aster (*Symphyotrichum laurentianum*)	CN-threatened	Beach, Salt marsh
Victorin's gentian (*Gentianopsis procera* ssp. *macounii* var. *victorinii*)	QC	Tidal fresh marsh
Victorin's water hemlock (*Cicuta maculata* var. *victorinii*)	QC	Tidal fresh marsh
Hooded arrowhead (*Sagittaria calycina* ssp. *spongiosa*)	QC, ME, MA, NY, PA	Tidal fresh marsh
High-tide bush (*Iva frutescens*)	ME, NH	Salt marsh
Salt marsh sedge (*Carex recta*)	PEI, NB, ME, MA	Salt marsh
Bigelow's glasswort (*Salicornia bigelovii*)	ME, NH, NY	Salt marsh
Annual salt marsh aster (*Symphyotrichum subulatum*)	ME, NY	Salt marsh
Annual salt marsh pink (*Sabatia stellaris*)	MA, CT, NY	Salt marsh
Lizard's tail (*Saururus cernuus*)	CT, RI	Tidal fresh wetland
Golden club (*Orontium aquaticum*)	CT, MA, RI, NY, PA	Tidal fresh marsh
Scotch lovage (*Ligustrum scoticum*)	CT, NY	Salt marsh
Slender sea purslane (*Sesuvium maritimum*)	NY, MD	Salt marsh, beach
Sea milkwort (*Glaux maritima*)	NJ, MD	Salt marsh
Sweet grass (*Hierochloe odorata*)	DE	Salt marsh
Elongated lobelia (*Lobelia elongata*)	DE	Brackish to fresh marsh
Prairie cordgrass (*Spartina pectinata*)	DE	Tidal fresh marsh
Slender blue flag (*Iris prismatica*)	NS, ME, NH, MY, PA, MD	Brackish to fresh marsh
Seabeach knotweed (*Polygonum glaucum*)	MA, CT, NY, NJ, DE, MD	Beach, Salt marsh
Sea ox-eye (*Borrichia frutescens*)	MD	Salt marsh
Animals		
Maritime ringlet (*Coenonympha nipisiquit*)	CN-endangered (QC, NB)	Salt marsh
Short-tailed swallowtail (*Papilio brevicauda*)	QC, NB, NS, LB	Salt marsh
Maritime copper (*Lycaena dospassosi*)	QC, NB	Salt marsh
Spartina borer (*Spartiniphaga inops*)	MA	Salt to brackish marsh
Common tern (*Sterna hirundo*)	NH	Beach
Least tern (*Sterna antillarum*)	ME, NY	Beach
Least bittern (*Ixobrychus exilis*)	QC, NY	Tidal fresh marsh
American bittern (*Botaurus lentiginosus*)	CT	Brackish to fresh marsh
Diamondback terrapin (*Malaclemys terrapin*)	MA, RI, DE	Salt marsh

Notes: CN = Canada for federal species of concern; LB = Labrador. U.S. federally endangered and threatened species are not listed; see previous table.

"The relationship of marine macroinvertebrates to salt marsh plants" (Kraeuter and Wolf 1974)

Animals of the Tidal Marsh (Daiber 1982)

"Seasonal abundance and diversity of spiders in two intertidal marsh plant communities" (LaSalle and de la Cruz 1985)

Habitat Management for Migrating and Wintering Waterfowl in North America (Smith et al. 1989c)

"The role of tidal marshes in the ecology of estuarine nekton" (Kneib 1997)

Concepts and Controversies in Tidal Marsh Ecology (Weinstein and Kreeger 2000)

The Edge of the Sea (Carson 1998)

Baylands Ecosystem Species and Community Profiles: Life Histories and Environmental Requirements of Key Plants, Fish and Wildlife (Goals Project 2000; for San Francisco Bay)

Terrestrial Vertebrates of Tidal Marshes: Evolution, Ecology, and Conservation (Greenberg et al. 2006)

"Animal communities in North American tidal freshwater wetlands" (Swarth and Kiviat 2009)

See also readings listed at end of Chapter 5.

7 Functions and Values of Tidal Wetlands

Today tidal wetlands are universally regarded as valuable natural resources by scientists, and most North Americans probably share this view. Wetlands are the vital link between land and water, a location that endows them with many properties that support fish and wildlife and produce many services that benefit people. Tidal wetlands can be among the world's most productive natural ecosystems. Wetlands temporarily store potentially harmful floodwaters, buffer and stabilize shorelines, help cleanse natural waters, reduce siltation in navigable waters, and yield natural products for human use and consumption, all while providing vital fish and wildlife habitat. This is a win-win situation for both people and wildlife. For these reasons, laws have been passed and policies issued in the United States and Canada to protect wetlands, or at least control their exploitation, whereas most upland habitats have not been accorded similar protection.

Wetland functions and wetland values are quite different in meaning, although they are frequently used interchangeably to underscore the importance of wetlands to people (Table 7.1). Functions are activities that wetlands perform whether or not people consider them to be important. Hence, functions are value-neutral. Nonetheless, values are the direct result of people's perspectives—their views on how good or useful something is. Values change over time, while functions do not (unless changed by natural forces or human actions).

Changing Attitudes about Tidal Wetlands

For natural resources, their value or the perception of their value by the masses is largely based on the nature of the society or culture and varies according to the needs of the society. Generally speaking, the closer one's lifestyle is tied to nature, the greater the appreciation of natural resources. As the majority of a society moves away from this dependency, the less valuable the resources appear in the view of the populace—perhaps until such time as the resource becomes scarce, rare, or threatened with extinction. Despite public attitudes, natural resources in one way or another are the foundation supporting all societies.

Before European colonization, Native Americans relied on natural resources for all of their needs—food, fiber, and shelter. For tribes along the coast, tidal marshes provided a wealth of food in the form of fish, shellfish, terrapins, waterfowl, and mammals; materials for clothing and adornment; and items for trade (e.g., wampum made from the shells of quahogs and whelks) (Kraft 1986). "Middens"—piles of shells, bones, and other refuse—located in and adjacent to today's marshes provide ample evidence of this life-supporting relationship.

Table 7.1. Major functions of tidal wetlands and some of their values

Function	Values
Water storage	Flood- and storm-damage protection, water source during dry season (freshwater wetlands), peat deposits, fish and shellfish habitat, waterfowl and waterbird habitat, recreational boating, fishing, shellfishing, waterfowl hunting, nature photography, aesthetic appreciation
Nutrient retention and cycling	Water-quality renovation, increases in plant productivity and aquatic productivity, decreases in eutrophication, pollutant abatement, global cycling of nitrogen, sulfur, methane, and carbon dioxide, peat deposits, carbon sequestration, greenhouse gas reduction
Sediment retention	Water-quality renovation, reduced sedimentation of waterways, pollution abatement (contaminant retention)
Provision of substrate for plant colonization	Shoreline stabilization, reduction of flood crests and water's erosive potential, plant-biomass productivity, organic export, aquatic productivity, fish nursery grounds, bird breeding habitat, other fish and wildlife habitat, trapping, hunting, fishing, nature observation, production of timber (mangroves and tidal swamps), production of salt hay, scientific study, environmental education, nature photography, aesthetic appreciation

Hunter-gatherer societies are solely dependent on natural resources and likely hold these lands in the highest regard. Native Americans had a healthy respect for what we call the environment and, perhaps better than any culture that followed, revered human kinship with nature and its wildlife.

When America was first settled by Europeans, Colonists initially were as dependent as the natives were on the available natural resources. They were farmers, hunters, trappers, fishermen, and gatherers living from whatever nature could provide and they could grow. They probably held the Old World utilitarian attitude toward natural resources—they were to be used for the sole benefit of mankind. Tidal marshes provided free pastures for livestock grazing and winter fodder that helped farmers keep some livestock through the harsh New England winters. Their fledgling agrarian society depended on livestock to do labor that was not possible for people to do by hand (e.g., haul timber and remove stumps to clear forests for farmland). Salt hay was a critical food for their livestock and therefore vital to the farmer's livelihood and their society as a whole. In 1879, Reverend Allen Brown, a New Jersey historian, referred to salt marshes as "natural privileges" since they produced annual hay crops without any cultivation (Moonsammy et al. 1987).

Later, as more forest was cleared for agriculture, upland grasses and clover provided pastures for livestock, and the growing society became less dependent on salt hay and salt marshes. With further increases in population and the beginning of the industrial revolution, American society became more dependent on technology to support people's livelihood. It became easier to dike and drain marshes, converting them to cropland as was done in the Netherlands and England since the 11th century. Vegetables, wheat, corn, and other crops could be produced on diked lands to help feed America's rising population. As population increased and cities boomed, demand also increased for more land for agriculture, more real estate for homes, industry, commerce, and ports to handle maritime commerce. Values changed to the point that tidal wetlands were looked upon as unused or unproductive lands that should be reclaimed and put to "better use." Passage of the Swamp and Overflow Land Acts by the U.S. Congress in 1849, 1850, and 1860 is stark evidence of the negative view and the reclamation value of wetlands that prevailed in that era. Through these acts, the federal government gave fifteen states title to millions of acres of wetlands provided they were drained and put into agricultural production. During this time frame, coastal waterways were polluted by disposal of sewage and industrial wastes directly into rivers in urban areas, producing foul-smelling waters and contaminating fishes and shellfishes (e.g., Crawford et al. 1994). This water quality degradation further reduced the value of tidal wetlands to the burgeoning society.

From Colonial times, there was also widespread concern about tidal marshes breeding disease-carrying mosquitoes and as such they were viewed as public health hazards. In 1766, a local public health ordinance for Baltimore—"Act to Remove a Nuisance in Baltimore Town"—demanded that property owners fill "a large miry marsh giving off noxious vapor and putrid effluvia" (Vileisis 1997). Filling was one solution to the problem, but perhaps the most widely used technique to rid society of these menaces was drainage. How-to textbooks were written on the subject including titles such as *Draining for Profit and Draining for Health* (Waring 1867) and *Engineering for Land Drainage: A Manual for the Reclamation of Lands Injured by Water* (Elliott 1912). Waring devoted an entire chapter (Chapter IX) to salt marsh reclamation and another (Chapter X) to malarial diseases in which he remarked, "If there is any fact well established by satisfactory experience, it is that thorough and judicious draining will entirely remove the local source of the miasma which produces these diseases" (216). In referring to New Jersey's Hackensack marshlands, Waring writes: "Its area is divided among many owners, and, while ninety-nine acres in every hundred are given up to muskrats, mosquitoes, coarse rushes and malaria, the other one acre may belong to the owner of an adjacent farm who values the salt hay which it yields him. . . . The inherent wealth of the land is locked up, and all of its bad effects are produced, by the water with which it is constantly soaked or overflowed. Let the waters of the sea be excluded, and a proper outlet for the rain-fall and the upland wash be provided, both of which objects may, in a great majority of cases, be economically accomplished, and this land may become the garden of the continent. Its fertility will attract a population, (especially in the vicinity of large towns,) which could no where else live so well nor so easily" (192). His views reflect the prevailing attitude toward wetlands from the late 1700s to the mid-1900s and helped pave the way for wetland destruction. Increasing the value of these lands was considered a noble contribution in helping the nation grow and prosper. The U.S. Department of Agriculture actively supported tidal marsh and other wetland reclamation projects to create farmland across the country (e.g., Nesbit 1885; Smith 1907). After numerous failures of tidal marsh reclamation projects in agricultural areas, however, some economists began to question the merits of these types of large-scale government-financed projects. For example, Harrison and Kollmorgen (1947) suggested that it might be more prudent for the government to focus on building up the land in the Mississippi Delta through conserving the soil that is being transported from the river directly into the Gulf of Mexico.

Despite the overwhelmingly negative view of wetlands during these early times, naturalists such as Henry David Thoreau and William Bartram revered wetlands as special places where one could get closer to God (Vileisis 1997). The great American landscape architect Frederick Law Olmsted saw the aesthetic quality of salt marshes and re-created one in his urban park in Boston—the Back Bay Fens—which was perhaps the nation's first wetland restoration project. John Charles Van Dyke, an art history professor at Rutgers College, wrote eloquently about nature and was the first to write a book about the beauty of American deserts in *The Desert* (1901; http://southwest.library.arizona.edu/vand/body.1_div.2.html). His *Nature for Its Own Sake* attempted to draw public attention to the raw beauty of nature "regardless of human meaning or use" (Van Dyke 1898). Van Dyke claimed that tidal marshes have been denigrated—"Because they cannot be utilized to advantage, they have been regarded with some contempt by mankind and the preacher, the orator, and the poet have always paralleled them with human stagnation or vileness. But they do not deserve such odious comparisons. "Humble and peaceful under the falling sunlight, they have their share of universal glory, and were constructed by nature for a useful purpose. They are the fortifications of the coast, keeping back the sea, and growing strong vegetation to prevent the wear of water on the land. . . . [T]hey are far from being the pestilent congregation of vapors and malaria which fancy usually pictures them" (249, 251). He then described their natural beauty, comparing their "picturesque qualities" to those offered by the dunes and meadows of Holland. Charles Townshend, author of *Sand Dunes and Salt Marshes,* promoted the value of tidal marshes in their natural state and urged that the marshes be protected from development (Townshend 1913). Another champion of coastal marshes was Percy Viosca Jr. of the Louisiana Department of Conservation. After witnessing the impact of construction of levees along the Mississippi, navigation canals, and drainage on Louisiana's wetlands, Viosca wrote about the value of these wetlands to wildlife and fisheries, the need to restore tidal flow to brackish marshes, and the need to cease the seemingly endless development of Gulf Coast wetlands (Viosca 1928). Later natural history writers including Rachel Carson (1947) were alarmed by the wanton destruction of tidal wetlands and through their writings continued the effort to stimulate public appreciation for the natural values of coastal wetlands and support for wetland conservation.

The post–World War II building boom brought much harm to Mid-Atlantic salt marshes as many dredge-and-fill developments converted marshland to residential development (see Chapter 8). These tremendous losses did not go unnoticed as wildlife management agencies began reporting on wetland losses including the effect of ditching on marsh ecology and the value of tidal marshes (e.g., Bourn and Cottam 1950; Steenis et al. 1954). Particularly noteworthy are a series of reports prepared by the U.S. Fish and Wildlife Service (1959, 1965) that addressed accelerating destruction of the nation's coastal wetlands. These were state assessment reports covering wetland status and trends from 1955 to 1959 and 1959 to 1964. (*Note:* These studies were the precursor to the National Wetlands Inventory, which commenced in the mid-1970s.)

The 1950s also marked a time when tidal marshes began to receive widespread attention by scientists as subjects for scientific research. In 1953, the University of Georgia established the Marine Institute at Sapelo Island to study tidal marsh ecology from a multidisciplinary perspective. John Teal, senior author of the classic environmental awareness book *Life and Death of the Salt Marsh,* was one of its early researchers. In March 1958, the institute held the first scientific conference on salt marshes to advance their recognition as a subject for general scientific study and for bringing together the science of other disciplines to understand the salt marsh ecosystem. Among the more notable attendees were E. S. Barghoorn (one of the preeminent paleontologists of his time), V. J. Chapman (author of a book on salt marshes of the world), J. L. McHugh (author of a classic paper on estuarine dependency of coastal fisheries), E. T. Moul (aquatic plant ecologist), H. T. Odum (a pioneer in ecosystem ecology), L. R. Pomeroy (marsh productivity expert), H. M. Raup (botanist and plant geographer), A. C. Redfield (author of a highly cited article on salt marsh formation), J. A. Steers (British salt marsh expert), and J. M. Teal (Figure 7.1). Scientists from the Marine Institute and other researchers working in marshes from Rhode Island to Louisiana made significant contributions

Figure 7.1. Scientists attending the first scientific conference devoted to salt marshes at Sapelo Island, Georgia, in March 1959. (Courtesy of the Marine Institute, University of Georgia; see ScholarWorks@UMass for list of participants)

to our understanding of tidal wetlands and their vital link to the estuary and coastal fisheries from the 1950s into the 1970s. They also began to transfer this scientific information to popular books and nature periodicals. One that caught my eye in 1970 was an article by Eugene Odum (1961)—"The Role of Tidal Marshes in Estuarine Production"—published in a wildlife magazine by New York State Department of Conservation. In 1961, Connecticut College published a public information bulletin—Connecticut's Coastal Marshes—A Vanishing Resource—that contained articles on the values of tidal wetlands, recent threats, and a call to action (Barske 1961; Darling 1961; Goodwin 1961; Niering 1961; Pough 1961; Rankin 1961). This may represent one of the first forays of research scientists into environmental activism. A short time later, one of the contributors, William Niering wrote *The Life of the Marsh*—an introduction to North American marshes in which he described these wetlands, their wildlife, their functions, and addressed the question—wetlands or wastelands? (Niering 1966). The 1960s also produced *Life and Death of the Salt Marsh* by John Teal and Mildred Teal (1969), a best seller that like Rachel Carson's *Silent Spring* (1962) raised public awareness about an environmental disaster—in this case, drawing attention to the plight of tidal marshes and their ecological significance.

These efforts coupled with growing environmental activism by concerned citizens, plus the firsthand observations of accelerated destruction of these natural vistas, helped sway public opinion. In the 1960s and 1970s, American public opinion started to shift toward wetland conservation. Most people began to better understand and appreciate tidal wetlands and their functions. The destruction of salt marshes probably played a significant role in advancing the environmental movement on a broader scale. With newly acquired knowledge, people started to hold coastal wetlands in high regard and were concerned about the future well-being of these natural resources. Environmental activists ("conservationists") brought the plight of tidal marshes to the attention of politicians on both the Atlantic and Pacific Coasts (e.g., Connecticut Conservation Association and Save the San Francisco Bay Association). Massachusetts enacted the first law to protect salt marshes in 1963; and in 1965, California passed a law establishing a public agency to regulate filling of the San Francisco Bay. By the mid-1970s, all northeastern states had passed laws to protect coastal wetlands from exploitation. Later, other coastal states followed the lead of New England states and passed similar laws. Tidal wetlands had finally regained the respect they had earlier received—public opinion had come full circle. People once again had a strong connection to these marshlands and demanded that they be protected from uncontrolled development.

During the past 400 years, the basic functions of tidal wetlands did not change, only people's view of their "worth" to society changed. Today most people better understand and appreciate the values of natural lands to our society and are motivated to promote protection, conservation, and restoration of these precious resources.

In the rest of this chapter, I provide an overview of tidal wetland functions and values, excluding the provision of habitat for plants and animals, which have been discussed in the two preceding chapters. By intent, it is not an exhaustive review of a topic that is worthy of a book unto itself. Attention is given to salt and brackish marshes because function and values of tidal freshwater wetland have been thoroughly described elsewhere (Barendregt et al. 2009). For more information on the functions of particular wetland types, consult references cited in the accompanying text and publications listed in Further Readings at the end of this chapter.

Ecosystem Productivity

The estuary and its associated tidal wetlands are especially valued because they link the sea with the river and stand between land and water. Their landscape position has endowed tidal wetlands with many unique properties that other lands do not possess. The ebb and flow of tides acting upon the constant seaward flow of freshwater create an environment where some of the highest levels of natural productivity can be achieved. Upon these fertile soils, vegetation takes root and flourishes. Primary producers are the foundation of the estuarine food web—vascular plants, macro- and micro-algae, and plankton. These plants are food for grazing marsh invertebrates (e.g., amphipods, isopods, worms, and mollusks), detritivores (e.g., killifishes), and herbivorous birds that are all or part of the diet of other animals. All or most estuarine animals ultimately depend on wetland flora in one form or another and on organic matter contributed by coastal watersheds (Figure 7.2).

Wetland plants provide food for wildlife either directly or indirectly. Living plant parts (e.g., roots and leaves) may be grazed by herbivores including snow geese, muskrats, nutria, and insects such as grasshoppers. Planthoppers and aphids suck sap from plant conductive tissue, while the seeds of some plants are food for other animals (Table 7.2). Direct grazing of salt marsh plants is more limited than for their freshwater counterparts. In southern marshes, 7 percent of the annual production of smooth cordgrass (*Spartina alterniflora*) is consumed by two insects (a grasshopper, *Orchelimum fidicinum* and a planthopper, *Prokelisia marginata*) (Smalley 1960; Marples 1966; Kraeuter and Wolf 1974). Eutrophication (nutrient enrichment of marshes and estuaries) appears to increase direct grazing. Grazing reduces smooth cordgrass production in the eutrophic marshes by 50 to 75

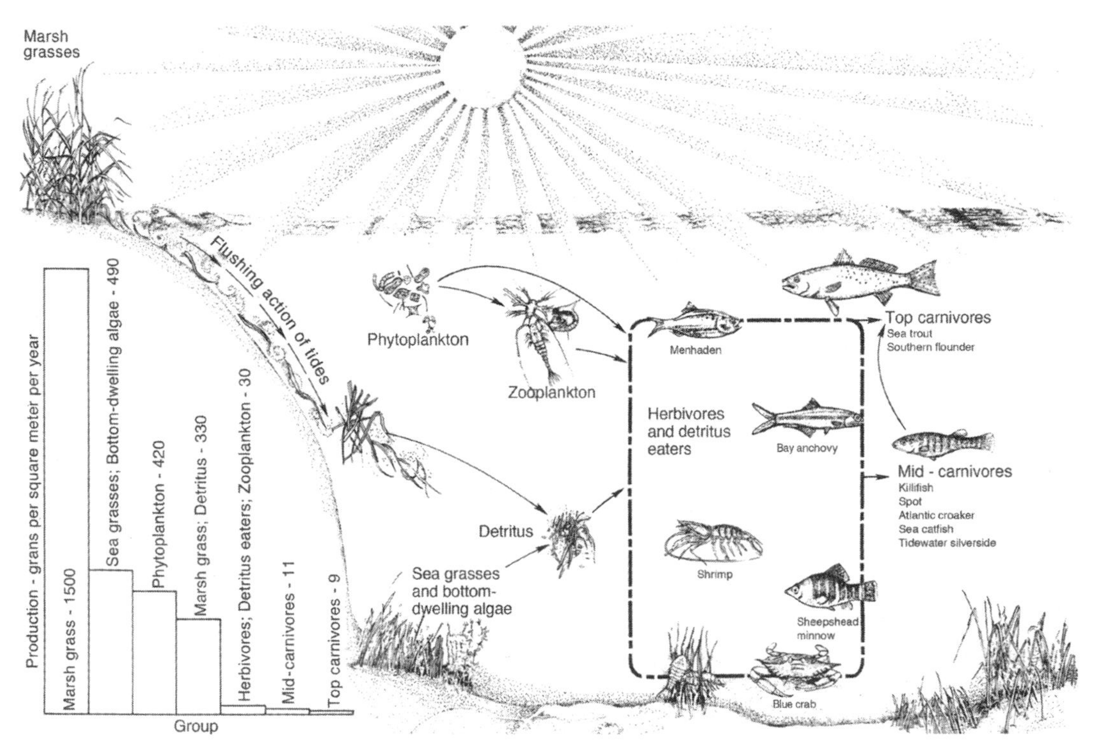

Figure 7.2. Plants (plankton, sea grasses, benthic algae, and marsh plants) are the foundation of the estuarine food web. Relative productivity of these plants is shown in the bar graphs along with the estimated production of consumers. Production in pounds per acre can be computed by multiplying each number by 10. (Gosselink 1980)

percent of what it would be in the absence of herbivory, while little impact from insect herbivory was noted in salt marshes surrounded by natural habitat (Bertness et al. 2008; see Chapter 4 for more discussion of grazing impacts). Fungi are important decomposers of dead shoots of tidal marsh plants. In Georgia salt marshes, fungi biomass may account for 3 percent of the living cordgrass biomass in summer and 28 percent in winter (Newell and Porter 2000). This biomass serves as food for marsh periwinkles, amphipods, and undoubtedly other invertebrates.

The main food value of tidal wetland plants to estuarine organisms, however, comes from the breakdown of the leaves and stems that are shed during the growing season or that fall to the ground or into the water with senescence in the late summer and fall. These plant parts are broken down into small fragments by invertebrates ("shredders") and natural processes to form "detritus" (particulate organic matter; Figure 7.3). The dead leaves provide a substrate for growth of bacteria and fungi that enrich this matter (Odum and de la Cruz 1967; Squiers and Good 1974; Kreeger and Newell 2000). While decomposing on-site, the detritus supports marsh-dwelling invertebrates (primary consumers). These animals provide food for larger organisms such as shrimp, killifishes, silversides, clapper rails, black ducks, egrets, raccoons, and various rodents. When exported to the adjacent estuary, detritus provides food for both aquatic animals and burrowing organisms (the "infauna") of intertidal and subtidal substrates forming the foundation of an intricate estuarine food web (see Figure 7.2). Animals such as zooplankton, crabs, shrimp, worms, and forage fishes

Table 7.2. Some wildlife food plants that occur in tidal wetlands

Life Form, Common Name	Plant Parts	Animals
Aquatic plants		
Widgeon-grass	Seeds, stems, leaves	Waterfowl, coot, swans, geese, brant, shorebirds, manatees, muskrat, and nutria
Eel-grass	Leaves, seeds, and rootstocks	Waterfowl and brant
Spatterdock	Seeds	Waterfowl
Pondweeds	Seeds, leaves, and tubers	Waterfowl, shorebirds, muskrat, and manatee
Tape-grass	Seeds, leaves, and tubers	Waterfowl, coot, swans, fish, and turtles
Water lily	Seeds and rootstocks	Ducks
	Plants	Muskrat
Graminoids		
Smooth cordgrass	Roots	Snow geese and muskrat
	Seeds	Birds, small mammals, and manatee
Salt hay grass	Roots and rhizomes	Nutria
	New foliage	Geese
Alkali-grasses	Foliage	Geese
Bulrushes	Seeds	Waterfowl, geese, swans, rails, sandpipers, songbirds, small mammals, and muskrat
	Rootstocks and stems	Waterfowl, geese, muskrat, and rabbits
Spike-rushes	Stems, roots, and seeds	Waterfowl
Saw grass	Seeds	Ducks and shorebirds
	Tubers	Muskrat and nutria
Salt grass	Seeds	Ducks, sora rail, and small mammals
	Rhizomes	Ducks, geese, muskrat, and nutria
	Stems	Muskrat
Black needlerush	Leaves and rhizomes	Nutria and muskrat
Walter's millet	Seeds	Waterfowl, sora rail, and songbirds
Rice cutgrass	Seeds	Waterfowl, shorebirds, sora rail, songbirds, and small mammals
	Rhizomes	Waterfowl
	Plants	Muskrat
Wild rice	Seeds	Waterfowl, coot, rails, and songbirds
Cattail	Tubers	Muskrat and waterfowl
	Stems and leaves	Muskrat
Switchgrass	Rhizomes	Deer (winter)
	Seeds	Birds and small mammals
Forbs		
Arrow arum	Seeds	Waterfowl, rails, and muskrat
Arrowheads	Seeds and tubers	Ducks
	Tubers and plants	Muskrat
Bur-reeds	Seeds	Waterfowl, marsh birds, and shorebirds
	Stems and foliage	Muskrat
Pickerelweed	Seeds	Waterfowl and muskrat
Glassworts	Seeds, stems, and leaves	Geese
Smartweeds	Seeds	Waterfowl and songbirds
Swamp milkweed	Nectar	Butterflies
	Roots	Muskrat
Seaside goldenrod	Nectar	Butterflies (especially for fall migrating monarchs)
Water hemp	Seeds	Waterfowl

Table 7.2. (*continued*)

Life Form, Common Name	Plant Parts	Animals
Shrubs		
Alders	Bark	Deer (winter)
	Twigs and foliage	Moose, muskrat, beaver, and rabbits
	Seeds	Songbirds
	Buds and catkins	Woodcock and grouse
Carolina wolfberry	Berries	Whooping crane (winter)
Sweet pepperbush	Foliage	Deer
Dangleberry	Berries	Birds and mammals
Highbush blueberry	Berries	Birds and mammals
Buttonbush	Nectar	Bees
	Seeds	Waterfowl and shorebirds
	Foliage	Deer
Wax myrtle and Bayberry	Fruits	Songbirds (fall and winter)
Red cedar	Fruits	Birds, rabbits, raccoons, and other mammals (winter)
	Foliage	Deer (winter)
Trees		
Sweet bay	Foliage and twigs	Deer
	Seeds	Songbirds, mice, and gray squirrels
White pine	Bark	Mammals
	Foliage	Deer and rabbits
	Seeds	Songbirds and rodents
Loblolly pine	Seedlings	Deer and rabbit
	Seeds	Birds and small mammals
Oaks	Acorns	Deer, bear, squirrels, other mammals, turkey and waterfowl
Sweet gum	Twigs and buds	Deer (winter)
	Seeds	Birds, rodents, and squirrels
Green ash	Twigs	Deer and beaver (winter)
	Seeds	Rabbits, rodents, and wood duck
	Foliage	Rabbits and rodents
Atlantic white cedar	Foliage	Deer (winter)
	Bark	Rodents
Red maple	Twigs and bark	Deer and mice
	Stump sprouts	Deer
Black willow	Buds and flowering catkins	Birds and rodents
	Twigs and foliage	Deer
	Bark	Rodents
Vines		
Greenbriers	Berries	Birds and mammals
	Foliage	Deer and rabbits
Poison ivy	Foliage	Deer
	Berries	Songbirds
Virginia creeper	Berries	Deer, squirrels, and songbirds
	Foliage	Deer

Sources: USDA Forest Service fire effects information at http://www. fs.fed.us/database/feis/plants; Martin et al. 1961; Butzler and Davis 2006.
Note: Small mammals include rodents and rabbits.

eat detritus or graze upon the bacteria, fungi, algae, diatoms, and protozoa growing on its surfaces. The ribbed mussel attains 50 to 80 percent of its diet from *Spartina*-derived organic matter (Peterson et al. 1985; Langdon and Newell 1990). Forage fishes such as anchovies, sticklebacks, menhaden, killifishes, and silversides, plus grass shrimp, are primary food for commercial and sport fishes, especially bluefish, flounder, weakfish, and perch. Invertebrates and/or fishes are important in the diet of most birds frequenting the coastal marshes. Blue crabs are the primary winter food for the endangered whooping crane that uses Texas salt marshes as wintering grounds (Butzler and Davis 2006). And, of course, penaeid shrimp, blue crabs, and many fishes are consumed by people. So in simple terms, tidal wetlands may be regarded as aquatic farmlands where great volumes of food are produced to support estuarine and marine organisms that ultimately provide a variety of seafood for people.

Depending on local conditions, much of this material may remain in the marsh or be flushed out to the estuary by a process called "outwelling" (Odum 1980). The degree to which tidal marshes export organic matter is difficult to quantify and has been a topic of much debate (e.g., Haines 1976; Nixon 1980). Forty-five percent of the organic matter (net primary production) of a mesotidal Georgia salt marsh was reportedly flushed

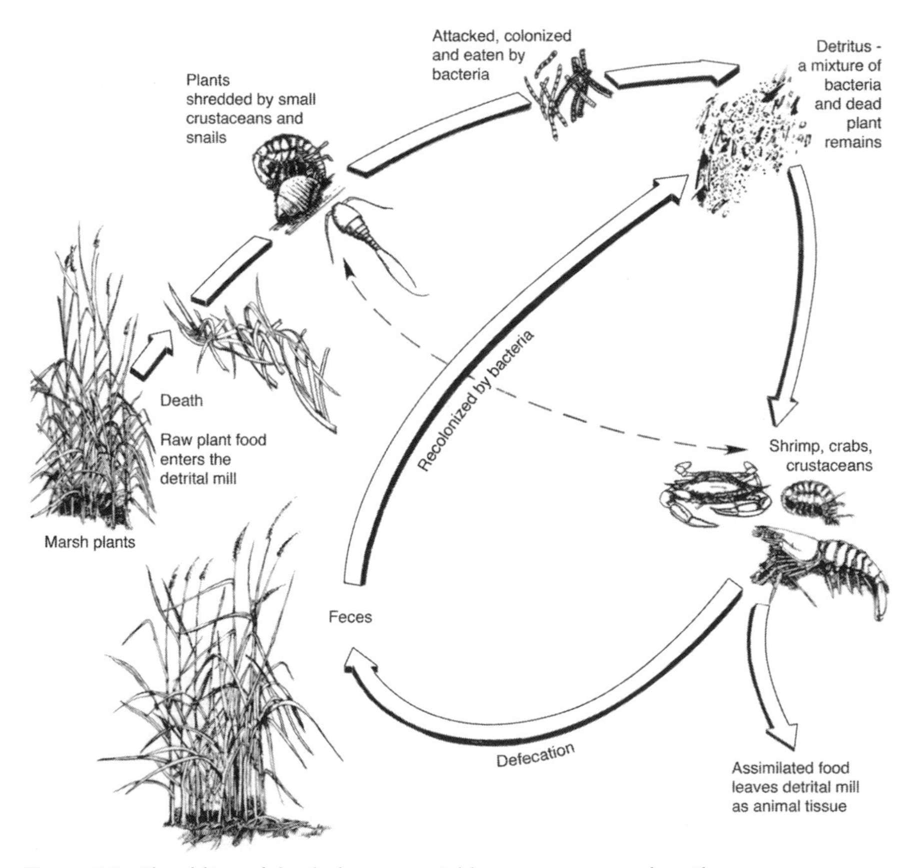

Figure 7.3. Shredding of dead plant material by crustaceans and snails creates more surface for bacteria and fungi, which produce enriched plant material called "detritus" for a number of small detritus-feeding animals including sheepshead minnow. The waste products of these animals are recolonized by bacteria and fungi repeating the cycle. (Gosselink 1980)

into the adjacent estuary, and a more recent study estimated that nearly 19 percent of marsh plants are exported (Teal 1962; Duarte and Cebrián 1996). Outwelling should be greater in marshes with high tide range or open access to the ocean (Odum et al. 1979; Odum 2000). In macrotidal areas with turbid waters, export of detritus from estuaries may make an even bigger contribution to estuarine–nearshore environments than they do elsewhere because the total primary production from the marshes represents a higher proportion of the organic matter in these waters where planktonic algae may be reduced by turbidity. For example, in the Bay of Fundy, the detritus from salt marshes reportedly comprises 53 percent of the primary production of the Cumberland Basin, with intertidal benthic diatoms and phytoplankton contributing 31 percent and 16 percent, respectively (Gordon and Cranford 1994). The detritus from salt marshes supports the animals of tidal flats (the main food for shorebirds) as well as zooplankton (copepods and mysid shrimp), which are important food species for estuarine forage fishes. The interdependence of the tidal wetlands and aquatic ecosystems should be evident. After 10 years of study, French scientists concluded that the salt marshes of Mont Saint-Michel Bay contributed organic matter to the diet of all tidal flat invertebrates examined and that salt marsh production enhanced productivity of the bay ecosystem (Lefeuvre et al. 2000). It should be clear that organic matter is exported from tidal marshes to tidal creeks and nearby estuarine waters and that the quantity exported to the ocean by outwelling depends on the interchange between the estuary and the ocean (Childers et al. 2000). The availability of particulate organic carbon may reach its highest level in northern marshes in late summer and fall as plants die back, but more southern marshes may have a fairly constant crop of dead biomass available for export (Woodwell et al. 1977; Roman and Daiber 1989). Regardless of marsh location, major storms appear to export substantial quantities of organic matter from marshes to the estuary (Pickral and Odum 1977; Hackney and Bishop 1981; Chalmers et al. 1985; Roman and Daiber 1989).

Keep in mind that animals utilizing marsh resources also transport carbon to the estuary and other places through the food web. Small young-of-the-year mummichogs are important prey species for many estuarine predators (e.g., blue crabs, grass shrimp, summer flounder, and fish-eating birds and mammals). These and other marsh fishes and invertebrates are an integral part of an energy transfer network ("trophic relay") transporting marsh productivity to the estuary (Rountree 1992; Kneib 1997, 2000; Deegan et al. 2000; Smith et al. 2000). Another point worth emphasizing is that this transfer of energy between the marsh and estuarine nekton is focused along the marsh–water edge. The edge is the most accessible portion of the marsh for these species, while birds and mammals can access other areas of the marsh more easily. Consequently the marsh edge and channels with submerged aquatic vegetation have been referred to as "hot spots" of production transfer between the marsh and the estuary (Kneib 2000; Zimmerman et al. 2000).

Although estuaries receive nutrients from the land via upstream watersheds, an interesting twist occurs in the Pacific Northwest where a significant transfer of energy takes place from the estuary to the upland. Juvenile salmon feed in tidal marshes where they gain weight before migrating to the ocean, while adult salmon return to spawn in natal freshwater streams. Salmon are important to these freshwater ecosystems as well as to adjacent forests. Since bears eat the salmon and return to woodlands, bear droppings fertilize these often nutrient-poor habitats, thereby linking these habitats to marine and estuarine systems as well (e.g., Helfield and Naiman 2006). Adult Pacific salmon die after spawning, so their carcasses enrich natal waters with marine-produced nutrients. This contribution appears to facilitate more rapid growth of both freshwater fishes and young salmon and the linkage may be vital to sustaining productive freshwater ecosystems in this region (Wipfli et al. 2003).

Tidal wetlands may also enhance productivity of aquatic beds. Salt marshes and mangroves intercept land-derived nitrogen (nitrate) and thereby lower nitrate levels of coastal waters, providing an "ecological subsidy" to neighboring seagrasses (Valiela et al. 2004). This is especially important to seagrasses because they are highly sensitive to nitrate: the more salt marsh or mangrove area in the vicinity, the higher the seagrass production.

Finally, one must appreciate the export of marsh organic matter in the form of larval and juvenile fishes and crustaceans that are using the marsh as nursery and feeding grounds. While much of the documentation of these relationships comes from studies in the southeastern United States (e.g., Pomeroy and Wiegert 1981; Peterson and Howarth 1987; Deegan et al. 1990), similar studies in the Pacific Northwest (e.g., Kistritz 1978; Seliskar and Gallagher 1983), Chesapeake Bay (Stribling and Cornwell 1997), Massachusetts (Deegan and Garritt 1997), and the Canadian Maritimes (e.g., Hatcher and Patriquin 1981) have demonstrated the significance of marshes to estuarine food webs elsewhere.

Plant Productivity

Given differences in species composition and the effects of climate and other factors, some tidal marshes annually yield from 2 to more than 10 tons of organic matter per acre (Table 7.3). The latter value is more organic matter than is produced by the richest cornfields in North America. In southern coastal watersheds, salt marsh grasses may generate 80 percent of the estuary's net primary productivity, while phytoplankton and microalgae each contribute 10 percent and support a complex detritus-based food web (Wiegert et al. 1981). Tidal freshwater wetlands are almost always more productive than similar nontidal communities (Whigham 2009).

Marsh plants do not produce equal amounts of such matter, and even the same species produces more in some locations than others due to differences in climate, tides, local conditions (e.g., hydrogeomorphic setting, tidal flushing, anaerobiosis, salinity, and nutrient enrichment), and other factors (Nixon and Oviatt 1973; Steever et al. 1976; Turner 1976; Kirwan et al. 2012; Figure 7.4). Individuals of a given species tend to produce more aboveground biomass in southern marshes due to the longer growing season and temperature (e.g., Kirwan et al. 2009). Differences in production within the same marsh for a given species are reflected in plant height, with plants in the low marsh yielding more biomass than the same species in the high marsh. Brackish and tidal fresh marshes may produce more aboveground biomass than salt marshes. A study of Georgia marshes found highest standing biomass in brackish marshes (around 1,700 g/m^2), while tidal fresh marshes and salt marshes produced 1,400 g/m^2 and 1,000 g/m^2, respectively (Wieski et al. 2010). The higher biomass of the

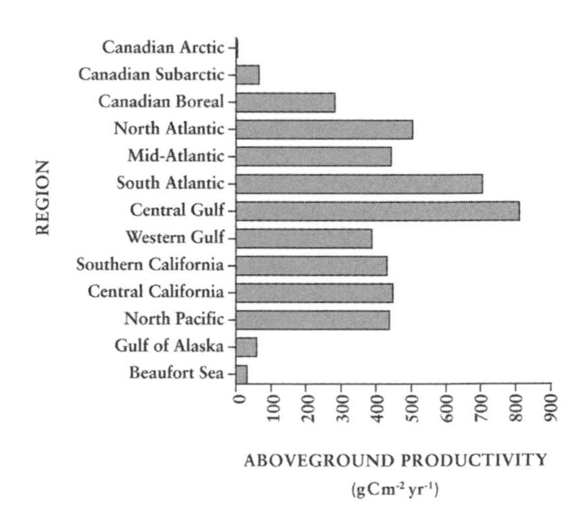

Figure 7.4. Marsh productivity varies with species and climate, with the most productive marshes occurring in the central Gulf of Mexico and the South Atlantic. (Redrawn from Mendelssohn and Morris 2000)

Table 7.3. Examples of annual net production of vascular plants from tidal marshes in various locations

Species	Location	Net Production (g dry wt/m^2)	Sources
Smooth cordgrass, tall form (*Spartina alterniflora*)	Nova Scotia	514–900	Livington and Patriquin 1981; Patriquin 1981
	Maine	431	Linthurst and Reimold 1978
	Massachusetts	1,320	Valiela et al. 1976
	Rhode Island	433–1,383	Nixon and Oviatt 1973
	New York	664–1,119	Udell et al. 1969
	New Jersey	735–1,700	Good 1972; Sugihara et al. 1979
	Delaware	1,487	Roman and Daiber 1984
	North Carolina	1,300	Stroud 1976
	Georgia	1,520–3,700	Gallagher et al. 1980; Dai and Wiegert 1996
	Florida	700	Kruczynski et al. 1978
	Louisiana	702–2,645	Kirby and Gosselink 1976
Smooth cordgrass, short form	Maine	246	Linthurst and Reimold 1978
	Massachusetts	420–664	Valiela et al. 1976; Roman et al. 1990
	Rhode Island	430	Nixon and Oviatt 1973
	Connecticut	250	Steever 1972
	New York	341–660	Udell et al. 1969
	New Jersey	444–590	Good 1972; Sugihara et al. 1979
	Delaware	654–916	Roman and Daiber 1984
	North Carolina	330	Stroud 1976
	South Carolina	209–694	Morris and Haskin 1990
	Georgia	1,105–1,300	Gallagher et al. 1980; Dai and Wiegert 1996
	Florida	130	Kruczynski et al. 1978
Salt hay grass (*Spartina patens*)	Nova Scotia	371–577	Gordon et al. 1985; Patriquin 1981
	New Brunswick	379	Connor and Chmura 2000
	Maine	912	Linthurst and Reimold 1978
	Rhode Island	430	Nixon and Oviatt 1973
	Connecticut	300	Steever 1972
	New York	424–547	Udell et al. 1969
	New Jersey	535–618	Good 1972; Sugihara et al. 1979
	Delaware	962–1,147	Linthurst and Reimold 1978; Roman and Daiber 1984
	Virginia	805	Wass and Wright 1969
	Georgia	176–980	Linthurst and Reimold 1978
	Louisiana	2,000–4,159*	Hopkinson et al. 1980
Salt grass (*Distichlis spicata*)	Connecticut	359	Steever 1972
	New York	523–774	Udell et al. 1969
	New Jersey	613–670	Good 1972, Sugihara et al. 1979
	Delaware	785–1,142	Linthurst and Reimold 1978; Roman and Daiber 1984
	Georgia	128–458	Linthurst and Reimold 1978
	Louisiana	700–2,881*	Hopkinson et al. 1980
Black grass (*Juncus gerardii*)	Maine	244–644	Linthurst and Reimold 1978
	Connecticut	570	Steever 1972
	Delaware	560	Linthurst and Reimold 1978
Seaside plantain (*Plantago maritima*)	New Brunswick	222	Connor and Chmura 2000

Table 7.3. (*continued*)

Species	Location	Net Production (g dry wt/m^2)	Sources
Chaffy sedge (*Carex palaecea*)	Nova Scotia	1,307	Patriquin 1981
Creeping bent grass (*Agrostis stolonifera* var. *palustris*)	Nova Scotia	430	Patriquin 1981
Bluejoint (*Calamagrostis canadensis*)	Nova Scotia	1,319	Patriquin 1981
Common reed (*Phragmites australis*)	Connecticut-Delaware	980–2,642 (tidal fresh) 727–3,663 (brackish)	Meyerson et al. 2000
	Rhode Island	900	Nixon and Oviatt 1973
	New York	2690	Harper 1918
	Delaware	1,380–2,940	Linthurst and Reimold 1978; Roman and Daiber 1984
	Louisiana	700–2,364*	Hopkinson et al. 1980
Black needlerush (*Juncus roemerianus*)	North Carolina	688–1,900	Kuenzler and Marshall 1973; Woerner and Hackney 1997
	Georgia	2,200	Pomeroy et al. 1981
	Florida	130–700**	Kruczynski et al. 1978
	Louisiana	1,200–3,295*	Hopkinson et al. 1980
Big cordgrass (*Spartina cynosuroides*)	Georgia	2,176	Linthurst and Reimold 1978
	Louisiana	398–1,767*	Hopkinson et al. 1980
Wild rice (*Zizania aquatic*)	New Jersey	1,437	Good et al. 1975
Spatterdock (*Nuphar luteum*)	New Jersey	605	Good et al. 1975
Narrow-leaved cattail (*Typha angustifolia*)	New Jersey	830	Good et al. 1975
Arrow arum (*Peltandra virginica*)	New Jersey	1,262	Good et al. 1975
	New Jersey	591	Whigham and Simpson 1976
	Virginia	423	Doumlele 1981

Notes: *Brackish marsh community.
**Lower estimates are from plants in the high marsh (short form of black needlerush).
Within-species differences reflect differences in climate, hydrologic, and soil conditions, and locations within marshes. Different sampling procedures also affect recorded productivity values. Plant density is also a contributing factor to differences. Values are for dry weight of aboveground biomass; some have been rounded off from actual values. Most of these measurements are probably underestimates since they do not account for leaf loss during the growing season; belowground biomass is even greater than these values (Long and Mason 1983). Note that 450 g dry wt/m^2 is equivalent to approximately 2 tons/acre.
Comment: These examples underscore the need for a standardized method so that better comparisons can be made; see Table 7.4, Shew et al. 1981 and Morris 2007 for comparison of various methods.

brackish marsh may have been due, in large part, to the retention of older leaves (from the past several years) by black needlerush (*Juncus roemerianus*), the dominant brackish species (Eleuterius and Lanning 1987).

Many researchers have evaluated plant productivity of tidal marshes across the country (Table 7.3), but unfortunately no standard method is agreed upon, and so a variety of methods were employed. An early study evaluating several approaches to determine aboveground productivity found estimates ranging from 750 to 2,600 g/m²/year for the same marsh depending on the method used (Kirby and Gosselink 1976). Site-to-site comparisons would be more meaningful if researchers employed a standard evaluation technique. Some common methods include end-of-season harvest as well as those presented by Smalley (1958), Wiegert and Evans (1964), and Milner and Hughes (1968). Use of the Wiegert and Evans method may yield production values twice as high as those from the Smalley method, and three to six times that derived from end-of-season harvest (Reimold and Linthurst 1977; Marinucci 1982; Table 7.4). All these methods tend to underestimate aboveground productivity because of the difficulty in accounting for turnover in live material between sampling periods (Turner 1976). Since the 1970s, other techniques including non-destructive procedures (no harvest) have been developed to predict net primary productivity (Hsieh 1996; Daoust and Childers 1998; see Morris 2007 for further discussion of methods).

Eutrophication in estuaries may stimulate higher productivity of certain species. For example, sewage-enriched portions of Rhode Island estuaries produced the highest standing crop values for the tall form of smooth cordgrass (Nixon and Oviatt 1973). Marshes along the Providence River and upper Narragansett Bay produced 39 percent more biomass than lower bay marshes and 107 percent more than Block Island Sound marshes. Increased production of all forms of smooth cordgrass was observed in a sewage-enriched North Carolina marsh where net production of tall, medium, and short forms increased by 33 percent, 26 percent, and 80 percent, respectively (Marshall 1970).

Belowground biomass may be many times higher than aboveground biomass. Reported values of underground biomass

Table 7.4. Comparison of techniques for estimating productivity for some common marsh plants along the Atlantic Coast

State	Species	Productivity Estimates by Method (g/m²)				
		Peak standing crop	Milner and Hughes (1968)	Smalley (1958)	Valiela et al. (1975)	Wiegert and Evans (1964)
Maine	Black grass (*Juncus gerardii*)	644	634	1,940	1,940	4,027
	Smooth cordgrass (*Spartina alterniflora*)	431	431	758	758	1,602
	Salt hay grass (*Spartina patens*)	912	912	3,523	3,523	5,833
Delaware	Salt grass (*Distichlis spicata*)	856	864	1,274	1,191	2,017
	Black grass	524	524	884	775	1,540
	Salt hay grass	807	522	980	1,241	2,753
Georgia	Salt grass	395	283	1,258	988	4,378
	Salt hay grass	946	705	1,674	1,028	3,925

Source: Reimold and Linthurst 1977.

range from near equal amounts to over eight times higher than aboveground production (Good et al. 1982; Roman and Daiber 1984; Teal 1986). Fringe marshes in northern New England were found to have lower belowground to aboveground biomass ratios than more expansive meadow marshes (4.8 to 6.9), with belowground biomass averaging roughly 1,400 g/m^2 and 1,900 g/m^2 in fringe and salt meadow marshes, respectively (Morgan et al. 2009). Seasonal changes in belowground biomass occur, with a decrease from spring to mid-summer in response to aboveground growth and an increase from mid-summer to fall as perennial plants build up a reserve for the next year's growth (Connor and Chmura 2000).

While emphasis has been placed on the vascular plants, the role of algae in contributing to estuarine productivity should not be ignored (Table 7.5). For example, microalgae in short-form smooth cordgrass marshes and mudflats made up 45 percent and 22 percent, respectively, of the total primary production at North Inlet estuary, South Carolina (Pinckney and Zingmark 1993). In Georgia salt marshes, benthic algae annually produced 200 grams of carbon per square meter, while annual primary production of salt marsh algae in southern California marshes ranges from 185 to 341 grams of carbon per square meter depending on vascular plant cover (Pomeroy 1959; Zedler 1980). The latter rates represent 0.8 to 1.4 times that of associated vascular plants. Microalgae are also important for stabilizing mudflats, while fungi provide similar benefits for sandflats (Underwood and Patterson 1993; Meadows et al. 1994). A study of Mississippi salt marsh ecosystems found that benthic and planktonic algae appeared to be more important than vascular plants to most estuarine consumers (Sullivan and Moncreiff 1990). Regional differences in the significance of algae in food-web support may be attributed to differences in predominant plant species (black needlerush vs. smooth cordgrass), hydrology, and climatic conditions. Algae are an important food source for many estuarine invertebrates and omnivorous fishes. In most cases, from a productivity standpoint, salt marsh grasses produced more primary productivity per unit area than other biota in a South Carolina marsh–estuarine ecosystem (Table 7.6), whereas in southern California, algal production was nearly equal to or up to 1.4 times that of vascular plants (Zedler 1980; see Table 7.5).

Water Quality Renovation

Coastal wetlands like their inland counterparts play a significant role in improving

Table 7.5. Primary productivity of microalgae growing beneath vascular halophytes in U.S. tidal marshes

State	Productivity (g carbon/m^2/yr)	BMP/VPP x 100%	Source
Massachusetts	105	25	Van Raalte 1976
Delaware	61–99*	33	Gallagher and Daiber 1974
South Carolina	98–234*	12–58	Pinckney and Zingmark 1993
Georgia	150–200	25	Pomeroy 1959; Pomeroy et al. 1981
Mississippi	28–151**	10–61	Sullivan and Moncreiff 1988
Texas	71	8–13	Hall and Fisher 1985
California	185–341	76–140	Zedler 1980

Sources: Pomeroy et al. 1981; Sullivan and Currin 2000.

Notes: *Higher values were beneath short-form smooth cordgrass with lower value under the tall form.

**Lowest value (28 g C/m^2) was under dense, tall-form black needlerush (3.9 ft [1.2 m] tall), which lowered light penetration; algae production was twice as high (57 g C/m^2) beneath smooth cordgrass (2 ft [0.6 m] tall), and three times higher under salt grass (88 g C/m^2).

BMP/VPP is the ratio of benthic microalgae production to vascular plant aboveground production. Where ranges are given, data are from algae under different vascular plants.

Table 7.6. Primary production (g carbon/m^2/yr) from different biota and habitats in a South Carolina marsh–creek ecosystem

Habitat	Phytoplankton	Microbenthic	Macrobenthic	Marsh Grass	Total
Tidal Creek	265 (24%)	400 (36%)	450 (40%)	0	1,115
Oyster Reef	0	400 (34%)	790 (66%)	0	1,190
Tall Marsh	0	400 (14%)	290 (11%)	2,078 (75%)	2,768
Mid-marsh	0	400 (37%)	20 (2%)	666 (61%)	1,086
Short marsh	0	400 (12%)	10 (<1%)	2,888 (88%)	3,298

Source: Dames et al. 1991.
Note: Percentage of total (%) also given.

water quality. Since they form in areas of active sedimentation, their vegetation enhances this process, thereby reducing turbidity. This is especially important for aquatic life and for reducing siltation of ports, harbors, and rivers.

Salt and brackish marshes are important for improving water quality. Nutrient cycling in the Great Sippewissett Salt Marsh on Cape Cod, Massachusetts, has been described (Teal 1986), and a detailed treatment of wetland biochemistry has been published (Reddy and DeLaune 2008). Algae and nitrogen-fixing bacteria on the roots of marsh grasses fix nitrogen from the atmosphere, making nitrogen available to marsh plants and increasing productivity (Figure 7.5). Other bacteria convert other forms of nitrogen under aerobic and anaerobic conditions. Nitrification (biological oxidation of ammonium to nitrate) occurs when marsh soils are partly aerated such as at low or ebb tides, whereas denitrification (microbial respiration of nitrate and nitrite to produce nitrous oxide or nitrogen gas) takes place under saturated, anaerobic conditions. Evapotranspiration increases water uptake creating more aerated soil conditions that may enhance nitrification rates. The denitrification rate in a Georgia salt marsh was 10 times higher in tall smooth cordgrass than in the short-form stands, and fiddler crab burrows provided more oxidized sediment to facilitate this (Dollopf et al. 2005). Nitrification and denitrification were strongly correlated to each other. An earlier study in a Massachusetts salt marsh found the highest denitrification rates in muddy creek bottoms and pannes, the next highest in the low marsh, and the lowest rates in the short-form smooth cordgrass marsh (Kaplan et al. 1979). The higher rates in the creek bottom might be related to increased nitrate levels in groundwater, but the high rate in pannes was unexplainable. In northern areas, marshes retain inorganic nitrogen and export nitrogen as particulate organic matter (detritus and living cells) mainly in the fall. These processes are vital to supporting estuarine organisms. Tidal marshes may serve as important processors of nitrate from runoff as there appears to be a positive correlation between eel-grass beds (sensitive to nitrate pollution) and extensive salt marshes in New England estuaries (Valiela et al. 2000; Wigand et al. 2004). Organic-rich salt marshes more than a few football fields wide will denitrify most of the nitrate from groundwater sources, providing a buffer for eutrophication of coastal waters (Teal and Howes 2000). Georgia tidal marshes remove or store between 13 to 32 percent of the nitrogen exported from their contributing watersheds (Loomis and Craft 2010). Marshes act as sinks for phosphorus and many trace elements by storing them in their soils (e.g., phosphorus sorption); they may also be a sink for nitrogen, which involves more complicated processes to unravel. Nutrient cycling takes place throughout the year as northern marshes of the St. Lawrence estuary have been shown to be year-round nutrient processors (Poulin et al. 2009).

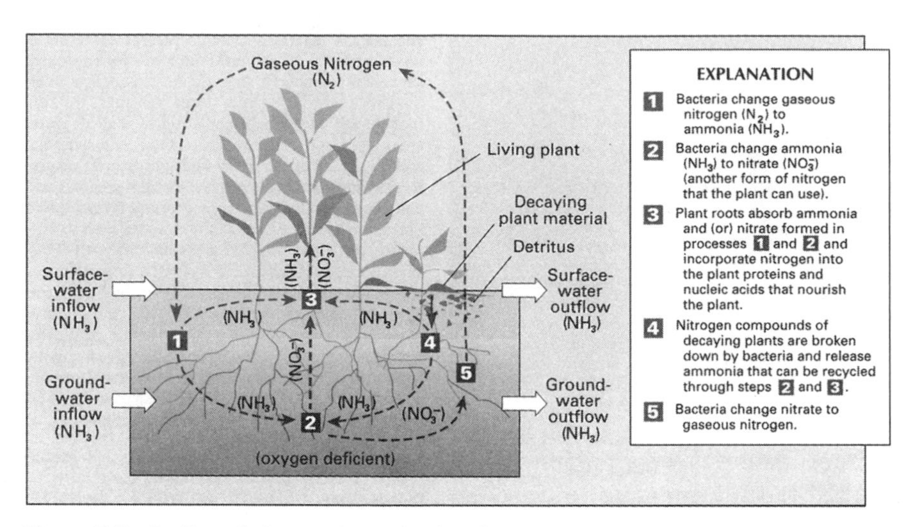

Figure 7.5. Cycling of nitrogen in wetlands. (Carter 1996)

Tinicum Marsh, a 512-acre (207 ha) tidal fresh marsh across from the Philadelphia International Airport, is one of the world's best-known natural wetland examples of water quality renovation. On a daily basis, the marsh removes 4.9 tons of phosphate (ortho-phosphate), 4.3 tons of ammonia (ammonium nitrogen), and 138 pounds of nitrate from Darby Creek (a tributary of the Delaware River), while adding 20 tons of oxygen to the creek (Grant and Patrick 1970). Aquatic life clearly benefits from this process, especially since the creek receives discharges from three sewage treatment plants that add nutrients and further increase oxygen demands. Studies of other tidal freshwater wetlands on the Delaware River found that vegetation utilized nutrients (nitrogen and phosphorus) and concentrated heavy metals (copper, lead, and nickel) in their tissues during the growing season when the estuarine waters are most stressed, but later released the latter when plants died back and plant litter decomposed (e.g., Whigham and Simpson 1976; Simpson et al. 1983b). The effectiveness of marsh vegetation on nutrient and heavy metal removal may be more seasonal in nature in these marshes where litter decomposes rapidly, although some storage occurs in underground parts of perennial plants. Litter decomposition may be slower in salt marshes. For example, in a Louisiana marsh, organic matter was accumulating in the soil at more than 400 g/m^2 thereby retaining much of the inorganic nitrogen taken up by smooth cordgrass (DeLaune et al. 1981). Marsh soils retain more insoluble metals such as lead, which may be entering tidal waters from non-point source runoff (Simpson et al. 1983b). All tidal wetlands play an important role in mitigating the impacts of heavy metal inputs from upland and upstream ecosystems (Simpson and Good 1985).

Coastal Flood and Storm Protection

While most tidal wetlands are frequently flooded for relatively short periods, heavy onshore winds and storm tides bring excessive amounts of water (i.e., storm surges) into estuaries that may stay there for longer periods. Heavy rains can cause coastal rivers to swell and overflow their banks, flooding the uppermost zones of tidal freshwater wetlands as well as contiguous nontidal wetlands. Coastal wetlands also absorb storm energy by 1) reducing the area of open water that can promote wave formation; 2) increasing drag on water motion thereby lowering storm surge height; and 3) absorbing wave energy. Storm surge may be reduced by up to 10.5 inches per mile (16.6 cm/km) depending on the vegetation type: 2.6 to 3.1 inches per mile (4.0–4.9 cm/km) by marshes and 10 inches per mile (15.8 cm/km) by mangroves (Lovelace 1994; Day et al. 2007; Krauss et al. 2009). The presence of tidal wetlands reduces the risk and level of flood damages to private property (Gedan et al. 2011).

Lost tidal wetlands translate into lost storm protection for coastal states. It has been reported that storm surge along the Louisiana coast is reduced by 3 inches for every mile of marsh (4.7 cm/km; Louisiana Coastal Wetlands Conservation and Restoration Task Force and the Wetlands Conservation and Restoration Authority 1998). Resource economists have estimated the value of storm protection provided by wetlands along the Atlantic and Gulf Coasts at $23.2 billion annually (Costanza et al. 2008). The wide expanse of coastal wetlands along many coasts provides a type of horizontal levee that reduces wave and wind energy and thereby lowers storm damage. These natural protection areas are more effective than constructed levees that are prone to failure, require maintenance, and promote development in low-lying areas behind them. With rising sea levels, additional stress will be placed on man-made levees worldwide. In the tropics, mangrove forests function as bioshields protecting villages from tsunamis, cyclones, and other major storms; reducing mangroves has increased potential for the loss of life and damage from these storms. Along India's Bay of Bengal coast, villages with wide belts of mangroves experienced fewer deaths from a 1999 super cyclone than did villages with few or no mangroves (Das and Vincent 2009). The 2004 tsunami that devastated much of the southeastern Asian coast caused less damage in communities where mangrove forests were preserved (Danielsen et al. 2005). Estimates for the value of Thailand's mangroves for this function alone amount to $2,340 per acre ($5,850/ha) (Barbier 2007).

Shoreline Stabilization

Many coastal shores, except rocky ones, are susceptible to erosion. The position of wetlands between water and erodable uplands helps protect these lands from erosion. It must be recognized, however, that vegetated wetlands tend to form in sheltered areas, so vegetation once established has the potential to stabilize shorelines in the low to moderate-energy conditions. Wave height appears to be an important factor influencing the occurrence of vegetation along shorelines (Roland and Douglass 2005). The value of salt marsh for shoreline protection is recognized (King and Lester 1995; Gedan et al. 2011). Marsh vegetation helps reduce wave heights by slowing the water's velocity, diminishing shear stress of the surface by increasing substrate elevations, and stabilizing the substrate with its root mass (Leonard and Luther 1995; Daehler and Strong 1996; Nepf 1999; Langlois et al. 2003). Wave reduction is even significant when marsh vegetation is completely submerged (Neumeier and Ciavola 2004). Enhanced sedimentation from the presence of marsh plants can be rapid and substantial averaging 0.2 inches per year (4.3 mm/yr) with peak values of 3.2 inches per year (82 mm/yr) (Boorman 1996; Wolanksi 2007).

Broad expanses of marshes as well as narrow fringing marshes help protect shorelines by dampening wave action (Knutson et al. 1982; Shafer et al. 2003; Barbier et al. 2008; Morgan et al. 2009). Ship-generated waves can be reduced by 25 to 50 percent after traveling 492 feet (150 m) (Sorenson 1973). Salt marshes in Great Britain have been found to lower wave heights by 60 percent (average) and dissipate total wave energy by 82 percent (average) compared with only a 15 percent reduction and 29 percent energy dissipation by sandflats (Möller et al. 1999). Similar results were reported for northern

New England where boat-generated wave height was reduced by 63 percent after traveling across 23 feet (7 m) of salt marsh compared with only a 33 percent reduction across a nonvegetated area (Morgan et al. 2009). Vegetation, microtopography, and marsh width are key factors operating to further reduce wave energy. Smooth cordgrass stems may reduce wave heights by 71 to 91 percent and wave energy by 92 to 100 percent, while its roots bind the soil (Wayne 1976; Knutson et al. 1990). Black needlerush offers more frictional resistance than the short form of smooth cordgrass and is therefore more effective at reducing wave heights and tidal surges (Miller 1988).

Shoreline erosion is a significant problem in Chesapeake Bay where nearly 25,000 acres (~10,120 ha) of land were eroded between the mid-1800s and 1947 (Slaughter 1967); more recently, some islands have disappeared (Kearney and Stevenson 1991). Since marsh vegetation dampens wave action and stabilizes substrates, marsh creation can be an effective nonstructural solution to shoreline erosion problems in some places (e.g., low-energy shorelines). Examination of wetlands constructed for this purpose—"living shorelines"—has found that marsh widths of 33 feet (10 m) can stabilize eroding shorelines (Knutson et al. 1990). Such shorelines are being constructed throughout the coastal zone to protect private property (e.g., Piazza et al. 2005; Swann 2008). The U.S. Army Corps of Engineers is using dredged material to restore shoreline marshes along Poplar Island (MD). The state of Maryland, which has promoted tidal marsh creation as a nonstructural means of reducing shoreline erosion for decades, passed the Living Shoreline Protection Act in 2008. This act requires living shorelines to be the first defense against shoreline erosion, with some exceptions. It is important to emphasize that there are limits as to how much plants can help protect shorelines, largely related to wave action and energy. For example, shoreline marshes created where the fetch was greater than 5.6 miles (9 km) were quickly eroded (Knutson et al. 1981).

Marshes, of course, are not the only vegetated wetland type that stabilize shorelines. Woody plants of tidal swamps, including mangroves, are effective for this function. The roots of tidal swamp trees growing along coastal rivers have helped stabilize banks and minimize erosion for years. While trees cannot prevent erosion, they do help bind the soil (lowering its erosion potential) and dampen wave energy through friction that reduces current velocity and floodwater's erosive potential. The overall effect of vegetation whether herbaceous or woody is a decrease in soil erosion potential—the more persistent the vegetative cover, the better.

Provision of Natural Products

Since the arrival of humans in the coastal zone, tidal wetlands have provided a bounty of food, especially waterfowl, fishes, shellfishes, furbearers, and edible plants. As the population grew and demand for these resources increased, some of these resources were overharvested, while others were degraded by development and disposal of society's wastes. For example, in the late 1800s and early 1900s, market-hunting of wading birds (e.g., herons and egrets) for food and for feathers for the millinery trade decimated their numbers to the point that some species were on the verge of extinction. Outlawing such hunting in the 1900s allowed these species to recover. Harvesting shellfish and fish beyond sustainable yields put their populations in jeopardy, and pollution largely from industrial sources and municipal sewage contaminated shellfish beds and made some fish unsuitable for consumption.

Despite the abuses of the past, we still harvest natural resources to variable degrees. Sport and commercial fishing for coastal species generate billions of dollars in revenues (Table 7.7) and jobs for millions of people. Likewise, shellfish (e.g., shrimp, crabs, oysters, soft-shell clams, quahogs, and mussels) are harvested commercially and recreationally. Penaeid shrimp harvests along the Gulf of Mexico have been found to be proportional to the area of tidal marsh landward of the shrimping area (Turner 1992), the more marsh, the more shrimp caught. Surprisingly, shrimp landings have increased in the northern Gulf of Mexico despite heavy losses of marshes to subsidence. Scientists identified four possible reasons for this: conversion of marsh to open water creates more edge habitat, longer flooding periods allow more time for shrimp to access available marsh resources, expansion of the saltwater zone creates more shrimp habitat, and shortened connections to the Gulf (Zimmerman et al. 2000).

Tidal marshes are vital to sustaining local and regional fisheries (e.g., Kritzer and Hughes 2009; Figure 7.6). Of the commercially important estuarine-dependent fishes, menhaden generates the most revenue, with harvests valued at $93 million and $91 million in 2007 and 2008, respectively (Pritchard 2008). Shrimp and blue crab landings during this time yielded even higher returns. The South Atlantic and Gulf Coast shrimp fisheries generated over $1 billion in harvest, with about 90 percent caught in the Gulf region. Blue crab harvest yielded about $150 million annually. These and other estuarine fisheries support a major seafood industry involving all sectors of some local economies ranging from fishing boats, fuel, commercial processing plants, and fish markets to restaurants. Important recreational estuarine fishes include striped bass (about 2 million caught annually in 2007 and 2008), bluefish (about 8 million), spotted sea trout (about 15 million), spot (about 14 million), and mullet (8 million) (Pritchard 2008). Marine worms and macroalgae, mainly rockweeds, Irish moss, and dulse, are harvested from tidal flats and rocky shores for commercial purposes in Canada and Maine. The marine worm harvest in the United States was valued at $8 million in 2007 and $11 million in 2008 (Pritchard 2008). Besides recreational fishing, many coastal residents and visitors go crabbing and clamming.

Hunting for waterfowl and other animals living in wetlands (e.g., deer and rabbits) remains a popular sport for an estimated 1.1 million licensed duck hunters in the

Table 7.7. Domestic commercial landings of some fishes and invertebrates utilizing coastal wetland ecosystems

	2007 Statistics		2008 Statistics	
Species	1000s of pounds	1000s of dollars	1000s of pounds	1000s of dollars
Bluefish	7,663	2,737	6,148	2,579
Atlantic croaker	20,303	8,818	18,768	8,695
Winter flounder (Atlantic/Gulf Regions)	5,900	12,320	5,192	9,934
Menhaden	1,483,701	92,718	1,341,413	90,725
Mullet	11,846	6,980	13,174	7,181
Striped bass	7,383	15,883	7,072	15,256
Spot	5,721	4,322	2,889	1,861
Blue crab, hard	146,027	138,413	155,340	160,863
Blue crab, soft	2,135	6,845	2,011	5,367
Shrimp, South Atlantic	21,141	43,585	22,963	47,624
Shrimp, Gulf	225,154	367,028	188,295	363,136
Soft shell clam	3,947	24,348	3,818	21,649
Horseshoe crab	2,131	1,224	1,736	910
Worms	936	8,160	808	11,108

Source: Pritchard 2008.

United States (U.S. Fish and Wildlife Service and U.S. Census Bureau 2006). Trapping provides some people with substantial income in various coastal communities, with muskrat, nutria, and alligator being the main target species. Demand for fur has declined over the past couple of decades for a number of reasons including the economic downturn and changes in fashions. Most of the U.S. fur harvest comes from Louisiana, which accounts for 40 percent of the nation's wild fur harvest. Nutria, an introduced rodent from South America, had destroyed more than 100,000 acres of Louisiana coastal marsh by 1998. In an effort to reduce their population, the state created a marketing campaign to increase demand for nutria meat and fur and actually paid trappers a bounty for nutria tails. While this increased nutria harvest to more than 350,000 animals, this is still far short of the target of trapping one million nutria to reduce marsh and crop damage (Association of Fish and Wildlife Agencies; www.fishwildlife.org/furbearer_cases.html). Muskrat is served as a delicacy under the euphemistic name "marsh rabbit" in local restaurants on the Delmarva Peninsula. Other furbearers sought by trappers include mink, raccoon, and otter. Louisiana with its abundance of coastal wetlands is the leading state in harvest of living resources through hunting, fishing, and trapping. A statewide assessment of the economic contributions of these activities in 2006 provides some perspective on harvest values (Table 7.8).

Horseshoe crabs are extremely valuable to both fishermen and the medical industry. Fishermen use the female crabs as bait for catching eels. In Delaware Bay, the site of the largest populations of horseshoe crabs, recent harvesting of millions of crabs for bait reduced the population of horseshoe crabs by 70 percent. This is believed to have caused the huge drop in the red knot population by limiting the number of crab eggs available for feeding during their migration. When the red knot return from South America, they fly across the Atlantic directly to Delaware Bay to feed on horseshoe crab eggs laid in the intertidal beaches. Here the birds gain energy and build up fat reserves for continuing their flight to Arctic breeding grounds. Since the State of New Jersey issued a moratorium on horseshoe crab harvest in 2008, and the State of Delaware

Figure 7.6. Estuaries produce fish and shellfish for human consumption: seafood market in Quebec touting some products of the St. Lawrence Estuary.

Table 7.8. Contributions of major fish and wildlife resources to Louisiana's economy in 2006

Activity	Retail Sales ($)	Total Economic Effect ($)	Jobs Supported	State and Local Tax Revenues ($)
Migratory bird hunting	93 million	154 million	2,043	10.6 million
Saltwater sport fishing	472 million	757 million	7,733	45.6 million
Commercial marine, finfish fishing	342 million	455 million	5,107	32.3 million
Commercial marine, shellfishing	1.4 billion	1.9 billion	21,238	134.6 million
Alligator harvest, wild	11.9 million	21.9 million	149	1.1 million
Fur harvest, pelts	124,637	229,144	3	11,808
Nutria control*	1.9 million	2.6 million	21	128,067

Source: Southwick Associates, Inc. 2008.
Notes: Although the study did not separate coastal wetlands from other resources, coastal wetlands play a major role in contributing to these figures. * State paid $5 per nutria tail through its Coastal Nutria Control Program (375,683 tails total).

reduced its harvest levels, horseshoe crab numbers have risen dramatically. There has not yet been a corresponding increase in red knots as there may be a lag time in response (Associated Press 2010).

In 1956, Dr. Fred Bang of the Woods Hole Marine Biological Laboratory discovered that the blood of horseshoe crab clotted when exposed to certain pathogenic bacteria (Marine Biological Laboratory, Woods Hole, MA; www.mbl.edu/marine_org/images/animals/Limulus/blood/bang.html). The blue blood of horseshoe crabs contains copper and limulus amebocyte lysate, an important clotting agent that is vital for testing intravenous drugs for bacterial contamination and identifying infections caused by spinal meningitis, E. coli, certain cancers, and blood clots. Crabs harvested for medicinal purposes are returned to the estuaries within 24 hours after laboratory technicians extract up to one third of their blood. Mortality from this practice may be as high as 15 percent (Public Broadcasting Service 2008).

The number of wetland plants that are edible is significant (Table 7.9). As with all foods, the palatability depends on one's taste, and the plants listed may not produce marketable products for the everyday American consumer. They are edible and nutritious nonetheless, and some may actually be quite tasty and already part of many people's diets (e.g., blackberries and blueberries). Some plants have parts that may be boiled and mashed and mixed with flour and/or cold water to form a puree to which salt and pepper or sometimes sugar may be added for taste. Others have foliage or seeds that can be added to boiled water to create starchy soups, while berries can be used to make fruit or sweet soups. The roots and/or seeds of many species can be dried and converted into flour for bread-making. In some cases, preparation may be time-consuming. Roots may have to be dried for a long time (e.g., Jack-in-the-pulpit), or dried kernels of acorns have to be boiled to remove unpalatable substances. Cooking involves roasting roots or boiling foliage, young shoots, or other parts. Some plants require two or more boilings to remove bitter taste, and in some cases, harmful substances. While many people may have eaten wild rice since it is commercially available, cattail is another plant that everyone should try: collect some young flower spikes (before you see pollen), boil them, and serve them with butter. Eat them like you would corn on the cob for the center is a hard woody core. Surprisingly, even the inner bark of some trees can be eaten. Native Americans stripped the bark of trees and ate the inner bark (e.g., American elm and white pine; White 1913).

Table 7.9. Examples of edible tidal wetland plants

Common Name (*Scientific name*)	Use
Seaweeds	
Irish moss (*Chondrus crispus*)	Soup
Dulse (*Palmaria palmata*)	Raw vegetable, sun-dried vegetable, pan-fried chips, soup, salad
Nori or Laver (*Porphyra* spp.)	Wrap for sushi, soup
Flowering aquatics	
Spatterdock (*Nuphar lutea*)	Seeds: soup, nuts, breakfast cereal
Bur-reeds (*Sparganium* spp.)	Roots: cooked vegetable
Pondweeds (*Potamogeton* spp.)	Roots: cooked vegetable
Water lotus (*Nelumbo lutea*)	Seeds: nuts, breakfast cereal
White water lily (*Nymphaea odorata*)	Roots and seeds: cooked vegetable
Water-shield (*Brasenia schreberi*)	Roots: cooked vegetable
Salt and brackish herbs	
Arrow-grasses (*Triglochin* spp.)	Seeds: breakfast cereal, breadstuff
Silverweeds (*Argentina* spp.)	Roots: raw or cooked vegetable
*Sea-blite (*Suaeda* spp.)	Branches and foliage: cooked vegetable
Glassworts (*Salicornia* spp.)	Young stems: raw vegetable
Goosefoots (*Chenopodium* spp.)	Seeds: breadstuff Foliage: cooked vegetable
Marsh orach (*Atriplex patula*)	Young leaf tips: raw vegetable
Amaranth (*Amaranthus* spp.)	Foliage: cooked vegetable Seeds: breadstuff
Walter millet (*Echinochloa walteri*)	Seeds: breadstuff, meal
Salt marsh bulrush (*Schoenoplectus robustus, S. maritimus*)	Tubers: vegetable, flour
Common reed (*Phragmites australis*)	Roots: cooked vegetable Young shoots: breadstuff and cooked vegetable Seeds: breakfast cereal, breadstuff
Scotch lovage (*Ligusticum scoticum*)	Vegetable (substitute for celery)
Freshwater herbs	
Cattail (*Typha* spp.)	Young flower spikes (before bloom): puree, raw or cooked vegetable Pollen: breadstuff Roots/new shoots from rootstock: raw or cooked vegetable
Docks (*Rumex* spp.)	Leaves: puree Seeds: breadstuff
Fireweed (*Chamerion angustifolium*)	Young shoots: puree
Manna-grass (*Glyceria* spp.)	Seeds: soup, breadstuff
Wild rice (*Zizania aquatica*)	Seeds: soup, breakfast cereal, breadstuff
Violets (*Viola* spp.)	Roots: soup
Arrowheads (*Sagittaria* spp.)	Roots: cooked vegetable
Water plantain (*Alisma* spp.)	Roots: cooked vegetable
Flatsedges (*Cyperus* spp.)	Roots: cooked vegetable
Arrow arum (*Peltandra virginica*)	Roots and seeds: cooked vegetable
Golden club (*Orontium aquaticum*)	Roots and seeds: cooked vegetable
Beach pea (*Lathyrus japonicus*)	Young peas: cooked vegetable
Bugleweeds (*Lycopus* spp.)	Roots: cooked vegetable
Pickerelweed (*Pontederia cordata*)	Seeds: nuts
Soft-stemmed bulrush (*Schoenoplectus tabernaemontani*)	New shoot tip of rootstock: raw vegetable, thirst-quencher Roots and pollen: breadstuff
Jack-in-the-pulpit (*Arisaema triphylum*)	Roots: breadstuff
Sweet flag (*Acorus calamus*)	Roots: boiled/sugared candy Young shoots: raw vegetable

Table 7.9. (*continued*)

Common Name (*Scientific name*)	Use
Shrubs	
Elderberry (*Sambucus nigra* ssp. *canadensis*)	Pith: puree Berries: soup
Highbush blueberry (*Vaccinium corymbosum*)	Berries: fruit, soup, pie filling, add to muffins, breads, and pancakes
Cranberry (*V. oxycoccos*)	Berries: fruit, relish, juice, add to muffins and breads
Vines and trailing plants	
Greenbrier (*Smilax* spp.)	Young leaves and sprouts: puree
Swamp dewberry (*Rubus hispidus*)	Berries: soup
Groundnut (*Apios americanus*)	Roots: cooked vegetable
Hog peanut (*Amphicarpa bracteata*)	Underground seeds: cooked vegetable

Notes: Reference to roots includes tubers; foliage is often best when young. An asterisk (*) marks plants whose foliage needs to be boiled more than two times to increase palatability.

In fact, the name "Adirondack" means "tree-eaters" (Fernald and Kinsey 1943). While some natural berries are gathered locally, blueberries and cranberries are now grown commercially for mass production.

Although not as widespread an activity as it once was, salt hay farming continues in places (e.g., shores of the Gulf of St. Lawrence, Massachusetts Bay, Long Island Sound, and Delaware Bay). Some salt marshes have been harvested for more than 200 years (Figure 7.7). This activity can be practiced with or without diking the marshes. The principal salt hay species are black grass (*Juncus gerardii*), salt hay grass (*Spartina patens*), and salt grass (*Distichlis spicata*). In Colonial times, salt hay was winter food for horses, cattle, and sheep, and the number of livestock was typically dependent on the acres of salt hay available for farmers. The location of many New England towns was determined by their proximity to these natural haylands (Nixon 1982). Livestock grazing in tidal wetlands was commonplace in Colonial times but is rare today (Gedan et al. 2009). Sheep grazing of marshes still occurs at L'Isle Verte (an island in the St. Lawrence River) and the meat of the lambs is highly prized (Gail Chmura, pers. comm. 2005). Salt hay has also been used as packing material for glassware and pottery, roping for cast iron pipes, and butcher's wrapping paper (Moonsammy et al. 1987). Today, salt hay is preferred as weed-free mulch and for animal bedding. Rotational mowing of salt hay (e.g., once every two years) for commercial purposes in New England may have a few long-term effects including reduced total nutrients in soil, although this did not appear to affect plant productivity (Buchsbaum et al. 2009). The process of haying creates conditions that favor one species over another. Haying appears to favor salt hay grass over smooth cordgrass and observations of higher densities of salt hay grass in hayed marshes versus reference marshes suggest that haying stimulates tilling in salt hay grass. Shorebirds were attracted to hayed marshes for feeding, causing short-term reductions of some marsh invertebrates.

Some tidal marsh plants are harvested for basket-making and wreath-making. From northern Maine to Nova Scotia, sweet grass (*Hierochloe odorata*) is still collected by Native Americans and woven into baskets or burned as incense. It has also been used for spiritual or medicinal purposes. In South Carolina, purple muhly grass (*Muhlenbergia filipes*), a grass occurring in sand dunes and along the edges of salt marshes, is the main plant used to make low-country coiled baskets, while black needlerush originally used for the core bundle of the basket may still be used in combination with other native plants including cabbage palm (*Sabal palmetto*) and white oak (*Quercus alba*) for decorative purposes (Dufault et al. 1993). These baskets are prized by many and sell for hundreds of dollars. Wreaths of sea lavender (*Limonium*) called "sea heather wreaths" are made by local crafters. This plant is also used in dried flower arrangements. In many places in the northeastern United States, overharvesting has been a problem and picking sea lavender has been prohibited in some locales.

Other plants harvested from tidal wetlands include intertidal algae. Knotted wrack (brown algae) is used for fertilizer, algin, and for packing lobsters. Dulse (red algae) is widely regarded as an excellent source of vitamins and nutrients. Irish moss (red algae) offers the widest variety of uses from medicinal purposes (anti-coagulant, anti-inflammatory, expectorant, emollient, and tonic) to food products (tea) to cosmetics (skin softener). It is made into an extract (carrageenan powder) that is used as a thickener in food preparation (chocolate milk, ice cream, cheese, jellies, soups, pudding, and pie filling), for clarifying beer, and as a gel for toothpaste and hand creams.

Salt was another natural product collected for use and regional trade by Native

a

b

Figure 7.7. Harvesting salt hay has a long tradition in New England: (a) haying a Massachusetts salt marsh in the early 1900s and (b) haycock at high tide portrayed on early postcard. (Courtesy of Hampton Historical Society and Lane Memorial Library, Hampton, NH)

Americans—the Ohlone tribe—in the San Francisco Bay area well before the Spanish colonized California (Collins and Grossinger 2004). The region's climate with long dry summers produced hypersaline ponds and pannes in salt marshes that yielded concentrations of natural salt through solar evaporation. When Europeans settled the area, the Ohlone continued to work the salt ponds to provide salt for Spanish missions. From the mid-1800s, salt ponds were enlarged for commercial purposes by leveeing salt marsh-pond complexes and using windmill pumps and later electric pumps to move water through the system and regulate pond salinities.

Carbon Storage and Greenhouse Gas Reduction

Increases in carbon dioxide, methane, chlorofluorocarbons, and other "greenhouse gases" released by human activities have contributed to a rise in the Earth's temperatures. Like all plants, tidal marsh plants fix carbon dioxide during photosynthesis and convert carbon to plant matter. A waterlogged environment (wetland) favors accumulation of this matter in the soil when the plant dies, thereby creating organic soils or enriching wet mineral soils rather than oxidizing the material, which is typical of terrestrial environments (Chapter 3). Consequently, many wetlands, especially coastal wetlands, store more carbon per unit area of soil than upland (dryland) soils can store (Brevik and Homburg 2004). Through this carbon sequestration, wetlands help reduce the amount of carbon released to the atmosphere and contribute to global warming. High plant productivity and anaerobic soil conditions in coastal marsh soils produce higher rates of carbon storage than upland soils produce, and storage is continuous (Hussein et al. 2004). In fact, carbon storage of salt marshes and mangroves is higher than in peatlands and they release only negligible quantities of greenhouse gases (Chmura et al. 2003). The presence of sulfides in seawater suppresses the activity of methane-producing bacteria (methanogenesis), making salt marshes less likely to release methane during anaerobic decomposition and yielding further benefits to mitigate the greenhouse effect (Chmura 2009). Salt marshes also appear to be the best natural carbon sinks as they store more carbon per unit area in their soils than any other habitats—about 84 to 91 times more than tropical forests and 18 to 150 times more than temperate forests (Table 7.10; Choi and Wang 2004; Pidgeon 2009). Globally, tidal salt marshes store more than 430 teragrams (1 Tg = 10^{12} grams) of carbon in the upper 20 inches (50 cm) of the soil, while North American estuarine wetlands have been estimated to sequester 10.2 Tg of carbon per year (Chmura et al. 2003; Bridgham et al. 2006). These wetlands store carbon at rates 10 times that of other wetland ecosystems on a per area basis due to high sedimentation rates, high soil organic carbon content, and constant burial due to sea-level rise. For the United States, tidal wetlands represent 1 to 2 percent of the nation's carbon sink (Chmura 2009). New England salt marshes likely store much more carbon than southern marshes because they possess peat deposits that can be more than 20 feet (6 m) thick (Redfield 1972). In contrast, southern salt marshes with some exceptions are typically organic-rich mineral soils (e.g., loams and clays). Georgia's freshwater tidal marshes had higher levels of organic carbon in their soils than did brackish and salt marshes; salt marshes had the lowest levels (Loomis and Craft 2010). Restoration of tidal wetlands can add significantly to the world's carbon reserves. For example, if all the dykelands (former tidal marshes) in the Bay of Fundy were restored to salt marsh, they would sequester enough carbon to amount to 4 to 6 percent of Canada's targeted reduction of 1990-level emissions of carbon dioxide (Connor et al. 2001). While vegetated wetlands store more carbon than nonvegetated portions of estuaries, mudflat

Table 7.10. Comparison of estimated carbon reserves and long-term carbon accumulation in major ecosystems around the world

Ecosystem	Carbon per Unit ($g\ C/m^2$)		Total Global Area ($x\ 10^{12}\ m^2$)	Global Carbon Stock ($x\ 10^{15}\ g\ C$)		Long-term Rate of Carbon in Sediment ($g\ C/m^2/v$)
	Plants	Soil		Plants	Soil	
Tidal salt marshes	—	—	Unknown, but 0.22 reported	—	—	210
Mangroves	7,990	—	0.152	1.2	—	139
Seagrass meadows	184	7,000	0.3	0.06	2.1	83
Kelp forests	120–170	—	0.02–0.4	0.009–0.02	—	Not reported
Wetlands	4,286	72,857	3.5	15	225	20
Tundra	632	12,737	9.5	6	121	0.2–5.7
Deserts/semi-deserts	176	4,198	45.5	8	191	0.8
Croplands	188	8,000	16	3	128	Not reported
Forests, tropical	12,045	12,273	17.6	212	216	2.3–2.5
Forests, temperate	5,673	9,615	10.4	59	100	1.4–12.0
Forests, boreal	6,423	34,380	13.7	88	264	0.8–2.2
Savannas and grasslands, tropical	2,933	11,733	22.5	66	264	Not reported
Grasslands and shrublands, temperate	720	23,600	12.5	9	295	2.2

Source: Pidgeon 2009 from several references.

Table 7.11. Annual rates of carbon storage in substrates of marine and estuarine ecosystems

Tidal Habitat	Carbon Storage
Salt marsh	151 $g\ C/m^2/yr$
Mangrove	139 $g\ C/m^2/yr$
Seagrass meadows	83 $g\ C/m^2/yr$
Nonvegetated estuaries*	45 $g\ C/m^2/yr$
Open continental shelf	17 $g\ C/m^2/yr$

Source: Duarte et al. 2005.
Note: *Includes mudflats.

and estuary bottoms also store considerable quantities—about half that of seagrasses (Table 7.11).

Scientific and Archeological Research

Wetlands provide unique opportunities for scientific study and archeological research. Because wetlands are subject to a variety of stresses and contain a variety of habitats, they are excellent natural laboratories for teaching and studying ecology including plant and animal adaptations to flooding, soil saturation, differences in salinity, and other variables. The University of Georgia's Marine Institute at Sapelo Island was established to investigate salt marsh ecology and to provide students and others with an opportunity to learn about coastal resources. The university with support from the National Science Foundation hosted the first conference devoted to salt marshes. The March 25–28, 1958, Salt Marsh Conference brought together some of the world's most prominent scientists who had been studying salt marshes in their own specialized area to share their knowledge with researchers from other disciplines (Ragotzkie et al. 1959; see Figure 7.1). Organizers also intended to advance recognition of salt marshes as a subject for scientific study. At the time, Professor Valentine Jackson Chapman (University of Auckland, New Zealand) was in the process of completing the first monograph about salt marshes (Chapman 1960). Since then countless studies, many of which have been cited throughout this primer (Figure 7.8), have been undertaken in a multitude of disciplines that have widened our understanding of these interesting ecosystems. Scientific journals devoted to publishing information on coastal ecosystems include *Estuaries, Estuarine, Coastal and Shelf Science, Estuaries and Coasts,* and *Journal of Coastal Research* while others like *Limnology and Oceanography, Wetlands, Wetlands Ecology and Management, Aquatic Botany, Conservation Biology, Ecology, Journal of Ecology, Science,* and *Restoration Ecology* also contain findings from tidal wetland studies.

Besides ecological, physiological, hydrologic, and physiographic studies, tidal marshes provide researchers with opportunities to conduct paleobotanical studies that have helped scientists discover regional vegetation patterns and reconstruct past climates. For example, analyses of sediment cores from Piermont Marsh, a brackish marsh along the Hudson River in New York, have allowed researchers to reconstruct the history of past climate change and vegetation patterns, assess human impact on the environment, and estimate changes in marsh sedimentation rates (Pederson et al. 2005). Archeologists have also found value in studying tidal marshes. In early times, the abundance of fish and wildlife in tidal wetlands supported tribes of Native Americans in coastal regions across North America. Permanent and seasonal camps were located near tidal marshes. Rising sea level has buried their artifacts—preserving them and making them available for archeological investigations. Through these studies, we have a better understanding of Native American culture and how they used tidal wetland and other coastal resources. In Coquille River salt marshes in Oregon, archeologists have found hearths, stone and bone tools, wooden weirs, and shell deposits buried beneath marsh peat (Byram and Witter 2000). Most of the

Figure 7.8. Georgia salt marsh study site and boardwalk used by various researchers including John Teal during his classic study of energy flow in a salt marsh. (Wade Shelton, University of Georgia)

weir artifacts have been determined to be approximately 600 to 1,200 years old. Note that the age of the oldest weir (3,370 years old) seems to correlate with estimates of the time when postglacial sea-level rates slowed. The oral history of the Coquille Indians mentions surviving the great flood from the ocean. In the Southeast, shell middens have been located in salt marshes (many are islands in the marsh) and shell rings on marsh islands. These middens contain the remains of fish, shellfish, and turtles and give archeologists information on the diet of coastal tribes, while the shell rings offer similar information plus evidence of their cultural events, possibly seasonal feasts by their structure and added remains of fancy serving ware (Saunders 2007). In southern regions, tribes may have occupied these sites year-round as opposed to the more seasonal use of coastal marshes in northern areas (Russo 1991; Marquardt 1996).

Aesthetics, the Arts, and Recreation

A drive to most beaches will reveal coastal waterways and marshlands in route. Driving from the mainland to the barrier island beaches, most people are awestruck by the openness of the verdant plains lying behind the sand dunes with the ocean in plain view. The natural beauty of the marshes varies with seasons, green in the peak of summer and yielding a golden glow in the fall, sometimes with hints of red coming from salt marsh pannes dominated by glassworts. Salt marshes provide unique natural vistas where the rivers meet the sea or freshwater mixes with saltwater, and they have inspired many artists past and present (Figure 7.9).

Testimony of this can be found at any coastal art show where paintings and photographs of the variety of coastal wetlands including rocky shore, tidal marshes, and beaches will often dominate the selections. Since many people can't afford to have a marsh view from their backyard, they will bring home artwork showing these vistas to decorate their living room, family room, or study. One of the more famous painters of salt marshes is 19th-century American luminist and landscape painter Martin Johnson Heade. He is noted for panoramic paintings showing salt hay stacks in Northeast marshes, mainly along Massachusetts' North Shore (Figure 7.10, cover). A collection of his work can be viewed online at Hay in Art (www.hayinart.com/000163.html).

Before the mass production of plastic and cork-bodied decoys, decoy carving was a tradition among waterfowl hunters throughout North America. Wooden decoys were carved by hand from pine or cedar. Decoys, of course, were used to lure ducks and shorebirds to areas where hunters were waiting. In many cases, huge numbers of decoys were set in water. Today wooden decoys are widely appreciated as art with numerous decoy festivals and local, national, and international competitions held annually. Some museums are dedicated to collecting, displaying, and telling the history of decoy carving. The Ward Museum of Wildfowl Art (Salisbury, MD) maintains perhaps the largest collection of antique decoys with examples from across the country. The museum has sponsored the Ward World Championship Wildfowl Carving Competition for more

Figure 7.9. An art class painting a salt marsh landscape on Cape Cod.

Figure 7.10. Martin Johnson Heade painting *Sunlight and Shadow: The Newbury Marshes.* (John Wilmerding collection; image courtesy of National Museum of Art, Washington, DC)

than 40 years. Art books introducing the history of carving and showing examples of their works have been published (e.g., Fleckenstein 1983). Decoys that sold for $20 or less for a dozen back in the 1920s are now prized by collectors and are worth hundreds and sometimes thousands of dollars each. Even reproductions may go for $1,000 or more.

The raw beauty and vastness of southern salt marshes have stirred emotions in writers (Figure 7.11). "Marshes of Glynn" by Sidney Lanier, a 19th-century Georgian poet, is the most famous poem about marshes. It was one of several poems featured in his *Hymns of the Marshes* immortalizing the salt marshes of Glynn County, Georgia (http://theotherpages.org/poems/lanier01.html). In his view, the marshes were expressions of the greatness of God and places that captivate the human spirit. James Dickey, poet in residence at the University of South Carolina and author of the acclaimed novel *Deliverance,* wrote a lesser known poem—"The Salt Marsh"—about the swaying grasses of his beloved low-country marshes (www.eclectica.org/v1n5/salt_marsh.html).

Although many coastal wetlands are surrounded by private property, many government-owned lands are open for public use and many others can be accessed by boat. Federal, state, and provincial wildlife refuges; wildlife management areas; and parks, as well as local parks provide recreational opportunities such as fishing, hunting, boating (including canoeing and kayaking), nature photography, nature observation, and hiking. These and other activities generate substantial income for coastal communities. More than 40 percent of adult Americans visit the coast or estuaries at least once a year for recreation (Leeworthy and Wiley 2001). For many, the sheer natural beauty of expansive marshes with their winding creeks and open bay waters in the background offer an undeniably breathtaking view and feeling of solitude. In forested regions, coastal wetlands provide one of the few places where such views can be had. Nationwide, nearly 88 million people participate in wildlife-related recreation, spending $122 billion (U.S. Department of the Interior 2006). Saltwater fishing is enjoyed by almost 8 million people, generating roughly $9 billion in expenditures. More than 12 million hunters spend more than $22 billion. These numbers pale in comparison with the 71 million "wildlife watchers" in the country. While most of these people observe wildlife in their

Figure 7.11. An inspiring view of one of Georgia's many salt marshes. (Jan Mackinnon)

backyards, many enjoy watching birds in coastal marshes and beaches. Fifteen million people watch waterfowl and 12 million like to observe waterbirds. Ten percent of the nation's population (229 million who are 16 years or older) take trips to watch wildlife. Going to the beach alone may contribute between $6 billion and $30 billion to the U.S. economy each year, yet recreational fishing may yield about $15 billion annually and coastal wildlife viewing adds between $4.9 billion and $40 billion dollars each year (Pendleton 2008). Although heavy visitation may adversely affect the quality of natural areas and their use by wildlife, experiences with nature (connecting people with nature) should have a positive effect on public appreciation for wetlands and other natural resources and build support for continued protection, preservation, and restoration efforts.

Increased Real Estate Value

Anyone who has bought or sold a home has heard that the most important factor in real estate is location, location, location. Since tidal wetlands offer a natural view that is enjoyed by virtually everyone, properties with a view of the salt marsh should be higher valued than neighboring properties lacking this view. Properties actually bordering the marsh and those with boat access to navigable water should rate even higher. In fact, their value is so great that the costs of such properties are beyond the budgets of most people. One study found that beachfront properties were valued 207 percent higher than similar properties two blocks away and that bayfront properties were 73 percent more valuable than nearby properties without this view (Major and Lusht 2004).

Economic Value

During the early Colonial period of North America, salt marsh hay was a main fodder for livestock and early settlers depended on this hay to feed their livestock through the cold winters (an acre of salt marsh could produce enough hay to get one cow through the winter; Hatvany 2001). Salt marshes were arguably the most valuable natural resource in the Colonies. Land prices in coastal agrarian communities were affected by the availability of salt marshes. For example, from the 1720s to mid-1800s, salt marshes were rated higher or equal in value to cleared land on Prince Edward Island

(PEI) (Hatvany 2001). Salt marshes were the only open land in the Canadian Maritimes, and they were so valuable that when the island was first divided into 67 lots, each lot was designed to contain some salt marsh or water access (Figure 7.12). In the early 1800s, PEI land values per acre were $0.50 for backlands, $1.00 for frontlands, and $5.00 for marsh and cleared lands. Salt marsh was five to ten times more valuable than non-cleared lands. Marsh owners had both economic power and social status. By the mid-1800s, however, more land had been cleared for upland hay production and horse-drawn reapers and rakes were used to harvest this hay in marked contrast to the manual labor required to harvest salt hay. Changes in technology had changed the value of PEI salt marshes, at least in economic terms.

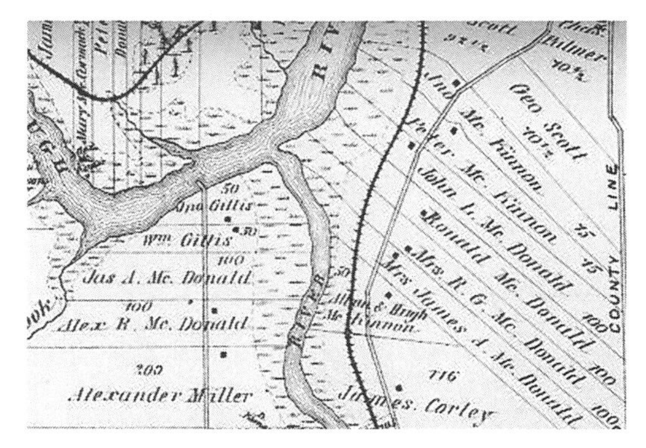

Figure 7.12. Portion of an 1880 map showing lot configuration along the Hillsborough and Pisquid Rivers on Prince Edward Island. These rectangular lots all contain a portion of tidal marsh providing each settler access to the river (transportation), the salt marshes (marsh hay), good bottom land, and uplands away from the river. (Map provided by Matthew Hatvany)

As mentioned at the beginning of this chapter, functions continue to operate whether or not they are valued by society. But since values are set by society, values may change as society changes. In early times, salt marshes were highly valued by the general public in those agrarian societies since their livelihood largely depended on wetlands. Then, with a change in culture to an industrial society that needed to support an ever-increasing population, values changed. Salt marshes became viewed by the societal leaders and the public-at-large as worthless lands, more of a public nuisance as they produced hordes of mosquitoes that carricd malaria and other life-threatening diseases. Consequently, tidal wetlands were ditched, drained, filled, or manipulated for a wide range of purposes (Chapter 8).

In the past, the value of a good or service was based on what someone might pay for the product or, in terms of land, what one could sell it for or earn from it by growing crops, raising livestock, harvesting timber, or extracting minerals. For wetlands in their natural state, one might consider the monetary value of what might be termed by-product production, that is, what you could reap by catching fish, crabbing, hunting, trapping, and gathering berries and other natural products. This type of valuation undervalues natural resources that are not harvested and sold for profit. The conventional pricing system also seems to make tidal wetlands more valuable as the surrounding area becomes more developed. This reflects the typical landowner's perspective because it represents the direct monetary value to him or her. The value of tidal wetlands and other natural resources, however, extends well beyond the individual property owner to society-at-large because the functions provided by these wetlands benefit all (e.g., shoreline and flood protection, water quality renovation, recreation, and aesthetics). This is the failure of a strictly market-based assessment. The conventional pricing system (real estate value) clearly ignores the natural value of tidal wetlands as habitats for fish and wildlife and for helping to maintain a healthy estuarine ecosystem.

In the 1970s, scientists began to address this shortcoming and devised an economic valuation system that included an analysis of the life support role tidal wetlands play. Instead of an annual by-product value of $100 per acre, they concluded that the annual value of an acre of tidal marsh was $50,000 to $80,000 (Table 7.12) and that in estuaries subject to human-induced nutrient loading, the value for waste assimilation would generate even higher values (Gosselink et al. 1973, 1974). This was perhaps the beginning of what eventually emerged as the discipline of ecological economics that attempts to value the non-market values of natural resources—"natural capital" and "environmental services." The Gosselink report also referenced another method of analysis suggested by economics professor Herman Daly. He recommended listing all the functions that tidal marshes perform and then determining how much it would cost to provide those functions with an alternate means. This "least-cost alternative" approach would likely generate values that would dwarf those published in the Gosselink report.

Using more sophisticated methods to estimate the economic value of wetlands, resource economists have determined that the world's tidal wetlands may produce environmental services worth about $3 billion annually (Table 7.13). A northeastern United States example of the use of these modern assessment techniques valued New Jersey's estuarine wetlands at over $6,000 per acre annually, for a total resource value between $1.1 and $1.2 billion (Table 7.14; Costanzo et al. 2006). The state's beaches have a higher per acre value ($42,147)

Table 7.12. Estimated economic value of an acre of tidal marsh–estuary

Function	Annual Return ($)	Income-capitalized Value (at 5% interest) ($)
Commercial and sport fisheries	100	2,000
Potential for aquaculture		
Moderate oyster culture	630	12,600
Intensive culture of oyster beds	1,575	32,000
Waste treatment (assimilation)		
Secondary	280	5,600
Phosphorus removal	950	19,000
Adjusted tertiary	2,500	50,000
Total life-support value	4,100	82,000

Source: Gosselink et al. 1973, 1974.
Note: Estimate based on the earliest known report on economic values of tidal marshlands.

Table 7.13. Estimated total economic value of tidal wetlands by continent

Continent	Salt/Brackish Marsh ($)	Nonvegetated Sediment ($)	Mangrove ($)
North America	29,810,000	550,980,000	30,014,000
Latin America	3,129,000	104,782,000	8,445,000
Europe	12,051,000	268,333,000	–0–
Asia	23,806,000	1,617,518,000	27,519,000
Africa	2,466,000	159,118,000	84,994,000
Australasia	2,120,000	147,779,000	34,696,000
Total	73,382,000	2,848,575,000	185,667,000

Source: Schuyt and Brander 2004.
Notes: Unvegetated sediment wetlands are mostly represented by coastal wetlands. Values of tidal freshwater wetlands (non-mangrove) were not included in these figures.

Table 7.14. Economic value of New Jersey's saltwater (estuarine) wetlands and beaches

Function	Saltwater Wetland Value/acre ($)	Beach Value/acre ($)
Disturbance prevention	1 or 310	27,276
Waste assimilation	6,090 or 5,413	—
Habitat refugium	230 or 201	—
Recreation/aesthetics	26	14,847
Cultural/spiritual	180	24
Total annual value/acre	6,527 or 6,131	42,147

Total for All Saltwater Wetlands (based on 190,520 acres) = $1,243,524,040 or $1,168,078,120

Total for All Beaches (based on 7,837 acres) = $330,306,039

Source: Costanza et al. 2006.

due to their protection of shorelines and recreational usage, but have a lower total resource value (just over $330 million annually) due to their lesser extent.

Given recent hurricane damage on the Gulf Coast and predictions for more frequent and severe hurricanes related to global climate change, economists have calculated the economic value of storm protection services provided by coastal wetlands in each U.S. state (Costanza et al. 2008). Based on this work, U.S. tidal wetlands produce storm protection benefits worth $23.2 billion per year. The economic value of storm protection services lost as a result of historic losses of wetlands for Louisiana was estimated at $816 million per year prior to Hurricane Katrina, while Hurricane Katrina's destruction of wetlands caused an additional loss of $34 million in storm protection annually. The present value of Louisiana coastal wetlands for storm protection alone is estimated to be $28.3 billion. If current projections related to climate change suggesting an increase in hurricane frequency and intensity are correct, then the value of coastal wetlands becomes even greater. Wetland conservation and restoration, therefore, appear to be quite cost-effective operations.

While these monetary values are impressive, they are not the sole reason that wetlands should be protected. Maintaining a healthy, productive ecosystem that supports self-sustaining coastal fisheries, provides the vital links for migratory birds moving between breeding grounds and winter habitat, and produces many other values to society (e.g., flood protection, shoreline stabilization, water quality renovation, and recreational opportunities) yields the greatest benefits for the most people and supports wetland conservation and restoration efforts.

Further Readings

Discovering the Unknown Landscape: A History of America's Wetlands (Vileisis 1997)

Restoration of an Urban Salt Marsh: An Interdisciplinary Approach (Casagrande 1997)

Tidal Freshwater Wetlands (Barendregt et al. 2009)

The Management of Natural Coastal Carbon Sinks (Laffoley and Grimsditch 2009)

See other references at the end of Chapter 5.

8 Extent, Threats, and Human Uses of North American Tidal Wetlands

Tidal wetlands are located along the shores of oceans, estuaries, and coastal rivers worldwide. Their current extent has been shaped by natural processes and greatly modified by human development. This chapter introduces the distribution of estuarine wetlands globally and in much of North America (excluding northern Canada), describes the current status of tidal wetlands in this region, and examines common threats and historic trends in various subregions. The probable future of the region's tidal wetlands is addressed in the last chapter (Chapter 12).

Distribution

Worldwide, salt marshes typically occur above 25° latitude in the northern and southern hemispheres, while mangrove swamps ("mangals") dominate the tropics in-between (Figure 8.1). By some estimates, roughly 160 million acres (65 million ha) of tidal wetlands exist throughout the world (Table 8.1; Schuyt and Brander 2004). North America may account for more than a third (38%) of the world's salt and brackish marshes, 37 percent of the world's nonvegetated wetlands, and only 4 percent of the mangrove swamps.

United States

Wetland extent for the United States comes from the National Wetlands Inventory (NWI), U.S. Fish and Wildlife Service, which has produced maps and digital data for the entire coastline of the conterminous United States and Hawaii. Only estimates are available for Alaska (Hall et al. 1994). According to the best figures available, more than 10 million acres (4.05M ha) of tidal wetlands are currently present in the United States (Table 8.2). The distribution of wetlands in the conterminous United States can be seen online at ScholarWorks@UMass (http://scholarworks.umass.edu/umpress/).

An irregular coastline with numerous embayments plus a high tidal range (macrotidal) has produced ideal conditions for the formation of extensive tidal flats and intertidal rocky shores in Maine. Surprisingly, it has the most acreage of marine wetlands in the United States with almost 70,000 acres (28,330 ha) mapped (Table 8.2). Alaska is second-ranked with an estimated 48,600 acres (19,670 ha) (Hall et al. 1994) followed by Florida, Washington, California, Massachusetts, and Oregon. Estuarine wetlands in the nation's largest state—Alaska—have been estimated at 2,131,900 acres (862,700 ha): 1,771,700 acres (717,000 ha) nonvegetated and 360,200 acres (145,800 ha) vegetated (Hall et al. 1994). Approximately half (50.7%) of these wetlands occur in the western region from the Aleutian Islands and Bristol Bay to Norton Sound and the Bering Strait, while nearly one-third (31.7%) are found in the south-central region along the Pacific Ocean from Yakutat west to the end of the

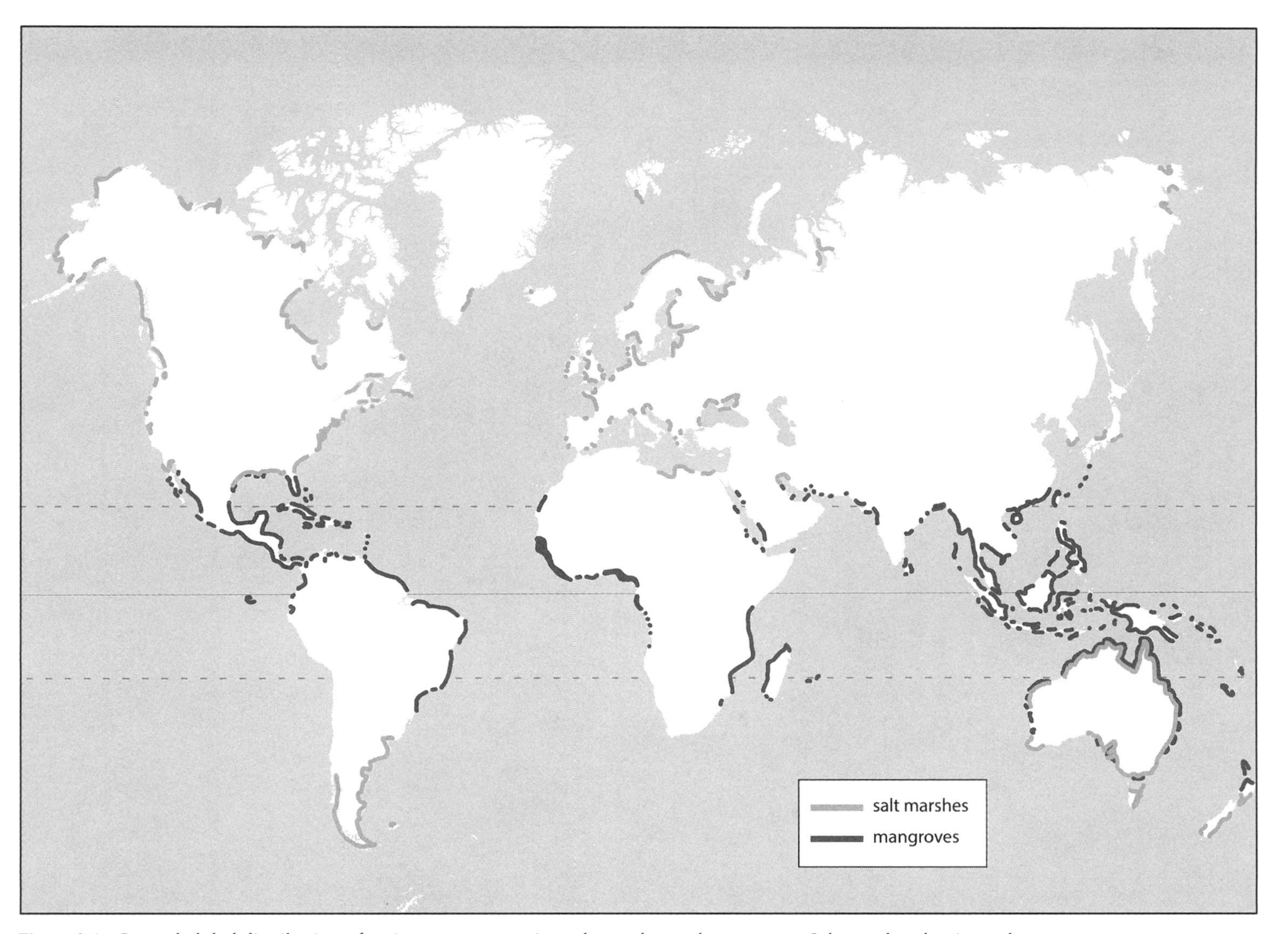

Figure 8.1. General global distribution of major areas supporting salt marshes and mangroves. Salt marshes dominate the coasts at mid to high latitudes, while mangroves occupy tropical and subtropical shores. (Adapted from: Boorman 2003; Chapman 1977; data from Geoscience Australia, NASA Earth Observatory, and U.S. Fish and Wildlife Service's National Wetlands Inventory)

Table 8.1. Estimated tidal wetland area by continent

Continent	Salt/Brackish Marsh acres (ha)	Nonvegetated Sediment acres (ha)	Mangrove acres (ha)
North America	6,363,000 (2,575,000)	41,780,000 (16,906,000)	1,260,000 (510,000)
Latin America	4,218,000 (1,707,000)	22,790,000 (9,223,000)	10,440,000 (4,224,000)
Europe	1,236,000 (500,000)	5,866,000 (2,374,000)	— (—)
Asia	2,538,000 (1,027,000)	19,800,000 (8,011,000)	3,556,000 (1,439,000)
Africa	1,203,000 (487,000)	11,450,000 (4,632,000)	9,108,000 (3,686,000)
Australasia	1,139,000 (461,000)	11,470,000 (4,641,000)	5,567,000 (2,253,000)
Total	16,750,000 (6,758,000)	113,100,000 (45,788,000)	29,930,000 (12,112,000)

Source: Schuyt and Brander 2004.
Note: Nonvegetated sediment wetlands are mostly from the coastal zone.

Table 8.2. Acreage of tidal wetlands in the United States

State	Marine Wetlands	Estuarine Nonvegetated	Estuarine Vegetated	Freshwater Tidal Wetlands	Total
Alabama	670	3,383	28,116	1,804	33,973
Alaska	48,600	1,771,700	360,200	NA	2,180,500
California	27,923	79,671	95,919	14,799	218,312
Connecticut	—	6,509	12,279	1,920	20,708
Delaware	622	4,880	78,202	11,448	95,152
Florida	31,536	141,066	1,241,144	95,039	1,508,785
Georgia	8,848	14,044	369,184	96,326	488,402
Hawaii	3,413	140	1,872	767	6,192
Louisiana	5,354	24,456	1,704,840	569,961	2,304,611
Maine	69,816	53,678	29,497	14,680	167,671
Maryland	722	23,672	224,542	45,851	294,787
Massachusetts	22,305	16,188	49,222	5,129	92,844
Mississippi	1,609	2,771	54,795	17,209	76,384
New Hampshire	886	3,280	6,017	854	11,037
New Jersey	4,224	5,156	203,557	43,632	256,569
New York	4,983	7,143	29,018	5,297	46,441
North Carolina	9,452	32,699	258,802	16,979	317,932
Oregon	16,399	36,236	16,105	16,471	85,211
Pennsylvania	—	55	—	1,396	1,451
Rhode Island	930	3,488	3,800	177	8,395
South Carolina	5,295	31,602	346,374	182,691	565,962
Texas	8,094	223,105	389,591	22,400	643,190
Virginia	4,377	144,499	205,690	89,518	444,084
Washington	26,073	47,440	163,028	19,792	256,333
Total U.S.*	302,131	2,676,861	5,871,794	1,274,140	10,124,926

Sources: Hall et al. 1994 for Alaska based on statistical sampling; other estimates from the U.S. Fish and Wildlife Service's National Wetlands Inventory data, September 2009 and Massachusetts DEP.
Notes: To convert acreage to hectares multiply by 0.4047. * District of Columbia had 230 acres of tidal freshwater wetland.

Aleutian Islands. The remaining wetlands reside in southeast Alaska (10.8%, Juneau region and the Alexander Archipelago) and along the Arctic Ocean (6.8%, the northern region including Chukchi Sea and Beaufort Sea). Louisiana, second-ranked in estuarine wetland area, has the most estuarine marsh area—nearly 1.7 million acres (688,000 ha). Florida with its abundance of mangroves along Florida Bay and extensive salt marshes along the Gulf of Mexico's northeast coast is next ranked with almost 1.4 million acres (566,600 ha). Florida possesses the greatest extent of estuarine scrub-shrub and forested wetlands (mangrove swamps) in the nation due to its subtropical climate. Other states with more than 300,000 acres (121,400 ha) of estuarine wetlands include Texas, Georgia, South Carolina, and Virginia. The arid to semi-arid climate along the south Texas coast and the length of its coast has given the state the most area of estuarine tidal flats in the coterminous United States. Estuarine aquatic beds are most abundant in Florida (seagrasses) and Washington (algae-covered rocky shores and eel-grass). Freshwater tidal wetlands are most extensive in Louisiana with nearly 570,000 acres (230,700 ha). South Carolina is second-ranked with more than 180,000 acres (72,840 ha) due to the abundance of tidal swamps, while Florida and Georgia possess nearly 100,000 acres (40,470 ha), closely followed by Virginia (90,000 acres [36,420 ha]). No data on tidal freshwater marshes are available for Alaska.

Regionally, the Gulf Coast possesses more than half of the nation's estuarine and tidal freshwater wetlands (58% and 54%, respectively) with Louisiana alone containing 2 million acres (809,400 ha) of tidal marshland (Figure 8.2). The North and Middle Atlantic Regions have 44 percent of the nation's marine intertidal habitats due largely to Maine's huge tidal range and irregular shoreline. Some states have more tidal wetland than others, mainly because of the extent of their coastlines (Figure 8.3). In the North Atlantic Region, Maine has the greatest abundance of estuarine wetlands, which can be attributed to its irregular shoreline and high tidal range, while Massachusetts with its wealth of salt and brackish marshes is second-ranked. For the Mid-Atlantic Region, Virginia and Maryland are the leaders in estuarine wetlands reflecting the dominance of Chesapeake Bay with its vast shoreline and numerous tributaries. New Jersey with its long coastline of barrier islands and the north shore of Delaware Bay is, however, not far behind. Nearly equal amounts of estuarine wetlands occur in Georgia and South Carolina, with North Carolina having only slightly less. The Mississippi Delta gives Louisiana the edge over Florida in the Gulf Region, while Puget Sound and its algae-covered shores helps Washington account for more estuarine area than occurs in California. Many of California's estuarine wetlands have been diked or filled.

Canada

Data for Canada's tidal wetlands are more limited than for the United States (Table 8.3). The extent of tidal wetlands in Canada is not known, but estimates are available for the Maritime Provinces, where salt marshes may be dominated by either high or low marsh in various regions. On Prince Edward Island, high marsh is the predominant type making up 60 percent of the marshes along the Northumberland shore and 52 percent of the Gulf of St. Lawrence shore (Wells and Hirvonen 1988). Nova Scotia tends to have mixed dominance with low marsh prevailing on most shores (61% low for Northumberland shore to northeast Cape Breton and 57% low along the Atlantic shore from Cape Breton to Shelburne), while high marsh is more prevalent along the Bay of Fundy. No comparable data were available for New Brunswick.

Threats to Tidal Wetlands

Coastal wetlands are dynamic ecosystems subject to change brought about by natural forces as well as by human actions

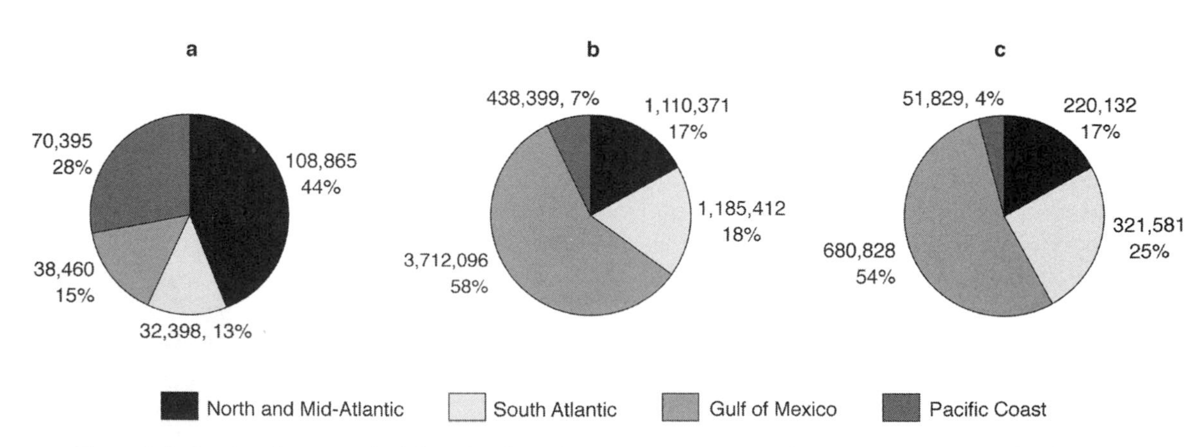

Figure 8.2. Regional extent of tidal wetlands in the conterminous United States: (a) marine wetlands; (b) estuarine wetlands; and (c) tidal freshwater wetlands. (Compiled from U.S. Fish and Wildlife Service 2009 data and Massachusetts Department of Environmental Protection data—an interactive version of this map that provides more detail can be seen online at ScholarWorks@UMass (http://scholarworks.umass.edu/); look under University of Massachusetts Press)

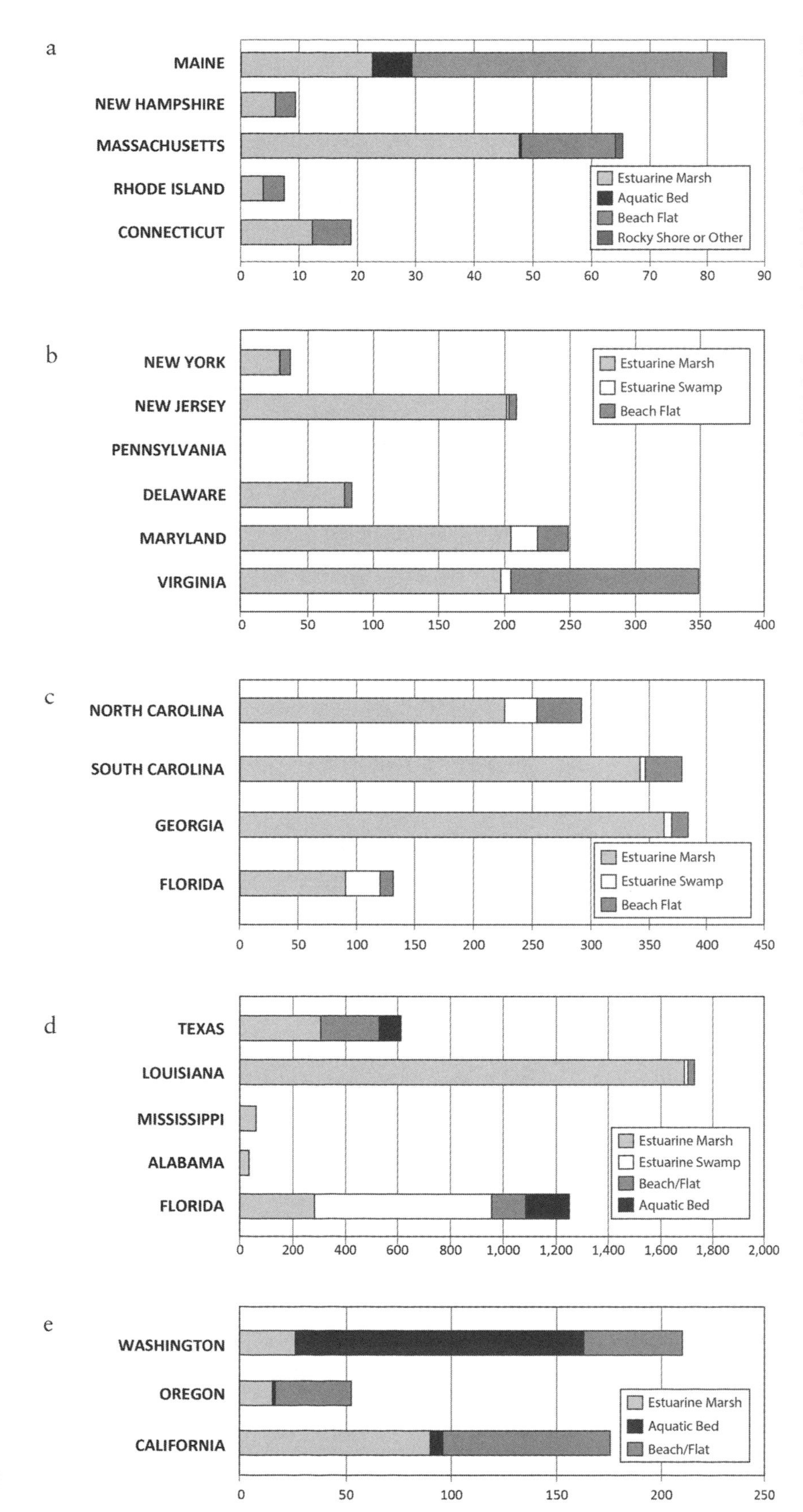

Figure 8.3. Extent of major estuarine wetland types in the conterminous United States: (a) North Atlantic, (b) Mid-Atlantic, (c) South Atlantic, (d) Gulf of Mexico, and (e) Pacific. Area is given in thousands of acres. Note: Pennsylvania has only 55 acres of estuarine wetlands. Miami-Dade County, Florida was included in Gulf totals. (Compiled from U.S. Fish and Wildlife Service 2009 data and Massachusetts Department of Environmental Protection data)

Table 8.3. Area of tidal wetlands in eastern Canadian provinces

Province	Marine Beach Acreage (ha)	Estuarine Marshland Acreage (ha)	Estuarine Tidal Flat Acreage (ha)
Quebec	Unknown	22,338 (9,040)	1,798,186 (727,700)
Prince Edward Island	3,994 (1,617)	15,420-16,911 (6,243–6,847)	107,564 (43,548)
Nova Scotia	6,501 (2,632)	30,732 (12,442)	507,555 (205,488)
New Brunswick	4,826 (1,954)	20,931 (8,474)	213,813 (86,564)

Sources: Quebec: Gratton and Dubreuil 1990; St. Lawrence Centre 1996; PEI: Hanson and Calkins 1996; Hanson, pers. comm. 2000; PEI Department of Environment, Energy, and Forestry 2003; Nova Scotia and New Brunswick: Hanson and Calkins 1996.
Notes: Quebec also has 9,227 acres (3.734 ha) of tidal freshwater wetlands.

(Table 8.4). Natural processes shaping these wetlands include rising sea level, coastal subsidence, isostatic (postglacial) rebound, sedimentation and erosion, hurricanes and coastal storms, and herbivory (see Chapters 2 and 3 for discussion). While natural processes have been operating forever to create and shape coastal wetlands, human development has had a more devastating effect on these wetlands in just the past two centuries. The coastal zone with its accessible water-based transportation and the wealth of food resources is among the world's most desirable places for human habitation. Most major cities are located on the coast. The location of tidal wetlands in this zone has made them vulnerable to coastal development worldwide. Some major human actions that have altered and continue to affect wetlands in some areas around the globe are summarized below. While the book has focused on non-mangrove tidal wetlands, it is important to recognize that mangrove swamps are probably among the most threatened natural ecosystems worldwide as they are being cut down for timber, converted to aquaculture (shrimp and fish farms) or agriculture, and filled for development. At least 35 percent of the world's mangrove forests were destroyed from 1980 to 2000, with annual loss rates estimated in the 2 to 3 percent range (Valiela et al. 2001).

Filling

Deposition of fill has been the principal cause of tidal wetland loss in many, if not most, areas of the United States, especially in ports, urban centers, and resort areas. Filling could be performed for a number of purposes including waste management (i.e., landfills and dredged spoil), public health concerns, or creation of real estate. Deposition of fill for development or disposal grounds of material dredged from navigable waterways elevates wetland surfaces above the tides and may make the land suitable for development (e.g., buildings, playgrounds, golf courses, and roadways). Dredge-and-fill development with canals providing waterfront lots was a common practice in the post–World War II building boom along the U.S. coast, which lasted until the Corps began regulating such projects in the 1970s (Figure 8.4). Road and railroad construction across tidal wetlands has both direct and indirect effects on wetlands. While the roadbeds destroy a certain portion of wetlands through filling, the more significant impact on tidal wetlands is their effect on hydrology of upstream wetlands. In the 1930s, filling tidal pools with sand was one approach to control mosquito populations on Long Island (Conard 1935). Worldwide, mangrove swamps continue to be filled for coastal development and to be harvested for timber (Polidoro et al. 2010). Wetland filling and upland development result in a loss of water storage capacity and increased runoff, respectively. In coastal watersheds, this can cause more frequent pulses of freshwater into estuaries and groundwater intrusion that can alter salinity gradients, change

Table 8.4. Some impacts to coastal wetlands from natural processes or activities and human-induced disturbances

Natural Processes or Activities	Possible Impacts
Sea-level rise and coastal plain subsidence	Submergence (loss) of wetland, change from vegetated wetland to tidal flat/open water, change from tidal flat to open water, change in shorelines, landward migration of marshes if suitable space is available, increase in salinity with changes in plant communities
Coastal processes (wave action, currents, and tides)	Erosion, sedimentation, wetland loss or gain, smothering of vegetation (e.g., from deposition of tidal wrack or overwash sediments), and changing shorelines
Deltaic soil compaction	Same as for sea-level rise and coastal plain subsidence, except migration
Hurricanes and other storms	Sedimentation (including overwash deposits), erosion, vegetation impacts, wetland loss or gain, saltwater intrusion, and changing shorelines
Ice scour	Erosion, vegetation removal, creation of open water in marsh
Grazing by animals (e.g., waterfowl, fur-bearers, and invertebrates)	Loss of vegetation (denuded areas) and vegetation changes
Insect infestations	Loss of vegetation and vegetation changes
Disease outbreaks	Loss of vegetation and vegetation changes
Beaver dams	Reduce tidal flowage to tidal freshwater wetlands, and vegetation changes
Fire from lightning	Vegetation impacts
Droughts	Brown marsh syndrome, vegetation changes, and saltwater intrusion
Human-induced Disturbances	**Possible Impacts**
Filling for development (port, industrial, commercial, residential, resort, etc.)	Loss of wetland and altered hydrology
Disposal of dredged material or garbage	Loss of wetland or change in plant community depending on amount of spoil
Construction of jetties and groins	Loss of wetland, changes in wetland type, altered hydrology, shoreline changes, and changes in wildlife use
Armoring shorelines (e.g., bulkheads and rip-rap)	Loss of wetland and stops landward migration of tidal wetland
Construction of docks	Shading of vegetation and disturbance to wildlife
Dredging for marinas and residential development (canal development)	Loss of wetland and degraded water quality
Drainage for mosquito control	Altered hydrology, increased salinity, vegetation changes, and local subsidence
Diking (for many purposes)	Altered hydrology, vegetation changes, localized subsidence, loss or diminished estuarine exchange, and loss of wetland
Road and railroad crossings	Loss of wetland, altered hydrology, and vegetation changes
Installation of tide gates	Altered hydrology, salinity changes, and changes in vegetation and aquatic life
Runoff from farms, lawns, or other developed areas (nonpoint source pollution)	Eutrophication, vegetation changes, and changes in aquatic life
Discharge of industrial or municipal wastewater	Water pollution, fish kills, degraded water quality, increased algal blooms, hypoxia, and changes in vegetation and aquatic life

Table 8.4. (*continued*)

Human-induced Disturbances	Possible Impacts
Oil spills	Vegetation die-off, changes in aquatic life, substrate contamination, death for oiled wildlife, fish kills, and water pollution
Deepening channels and dredging canals for navigation	Increase salt water intrusion, altered hydrology, shoreline erosion, vegetation changes, and change in aquatic life
Damming of tidal rivers for power generation, water supply, or other purposes	Reduction in sediment load, loss of wetland, altered hydrology, altered salinity regimes, vegetation changes, and changes in aquatic life (e.g., fish migration)
Water withdrawals	Altered hydrology, possible subsidence, altered salinity regimes, and corresponding changes in vegetation and aquatic life
Diversion of river flows	Altered hydrology, increased salinity (at least seasonally), and corresponding changes in aquatic life
Groundwater withdrawals	Subsidence and changes in vegetation
Oil and gas withdrawals	Possible subsidence, corresponding changes in vegetation and aquatic life, and pollution from spills and leaks (see oil spills)
Prescribed burning for wildlife management	Loss of soil organic matter and vegetation changes
Timber harvest	Vegetation changes, may facilitate conversion to other uses
Log storage	Topographic changes, vegetation changes, and habitat degradation
Marine aquaculture	Change in current patterns, disturbance to waterbirds, and possible loss of wetland (via conversion to open water)
Harvest of baitworms	Disturbance to waterbirds and changes in invertebrate density
Plant introductions	Invasive species replacement of native flora and altered hydrology
Animal introductions	Replacement of native fauna
Grazing by livestock	Soil compaction and changes in vegetation
All-terrain vehicle traffic	Vegetation impacts, disturbance of nesting birds, destruction of nests, and incidental kill of young birds
Beach renourishment	Change in sand composition and changes in invertebrate usage
Noise and light pollution (nighttime)	Unknown effect on wildlife, probable displacement of sensitive species; latter disturbance disorients nesting sea turtles
Plant collection	Loss or reduction of local populations (overharvest, e.g., sea lavender)
Nitrogen inputs from runoff	Change in vegetation, plant productivity, and aquatic life
Spraying of pesticides	Aquatic organism kills

Notes: With human impacts, changes are mostly one-directional (i.e., losses). Where a significant change in vegetation occurs, a change in wildlife use will likely occur.

sediment pH, and stress benthic organisms in small estuaries (Welsh et al. 1978).

Diking, Impoundment Construction, and Leveeing

Diking of tidal wetlands for agricultural purposes has been a common practice in much of North America. In the Colonial times, many northeastern U.S. salt marshes were diked and managed for salt hay production, whereas in the southeastern United States, tidal swamps were cleared and diked for rice culture. On the West Coast, Pacific Northwest marshes were diked to create mostly pasture and hayland and some cropland, while tidelands in San Francisco Bay were diked for producing salt. Likewise, it was standard agricultural practice in eastern Canada to dike salt marshes to improve the land for farming or upland hay fields (e.g., New Brunswick, Nova Scotia, and Quebec), and such diking eliminated many, if not most, of the original tidal marshes. Tidal wetlands have been diked for other reasons. In Maine, tidal embayments have been diked to create lobster pounds for holding harvested lobsters for later sale. Because of concerns over breeding salt marsh mosquitoes, many tidal marshes in Florida urban areas have been impounded for mosquito

Figure 8.4. Dredge and fill projects were common land development practices from the 1950s to the 1970s: (a) active filling of Long Island tidal wetlands in 1965 and (b) dredge and fill development along the New Jersey shore. (a: Al Dole, U.S. Fish and Wildlife Service)

control. Wildlife managers have impounded tidal marshes along all coasts to provide habitat for migratory waterfowl and for hunting preserves (see Figure 9.4). In the Southeast, many of these areas are former rice impoundments (Figure 8.5). These and other impoundments negatively impact estuarine fishes and invertebrates by restricting access to the marshes (see Montague et al. 1987 for a review of the ecological effects of coastal marsh impoundments). A South Carolina assessment of impounded versus natural tidal marshes revealed significant differences beyond the obvious vegetation difference (submergent versus emergent species) and restricted access by macroinvertebrates and fishes, such as the timing of maximum nutrient export (spring for former due to open tide gates, and summer in natural marsh), less diverse and abundant benthic fauna in the former, poor water quality in summer in the impoundments (low dissolved oxygen), and higher waterfowl use in managed impoundments (DeVoe and Baughman 1987).

The practice of impounding tidal wetlands for waterfowl management and other purposes has come under increasing scrutiny in the United States during the past four decades as the value of tidal marshes to estuarine and marine life has been more fully understood and appreciated. Diking of tidal marshes was a common wildlife management practice from the late 1930s until the 1980s when recognition of the value of tidal wetlands in their natural state and the emergence of federal and state wetland protection programs forced wildlife managers to re-evaluate this practice. Since then, impounding U.S. tidal marshes for waterfowl habitat has been greatly curtailed. Concern about the impact of diked marshes on rare and endangered species habitat has been another issue, especially on the West Coast. Furthermore, with today's concern over rising sea levels and the impact on tidal wetlands, wildlife managers are considering abandonment of certain diked marshes and restoration of tidal marshes as vital parts of their long-term waterfowl management

Figure 8.5. Aerial view showing diked wetlands (former rice fields) and natural marsh along the Combahee River (SC).

strategies. While much of the diking has been stopped in the United States, these practices continue around the globe. For example, Mexican salt marshes are being diked and converted to shrimp aquaculture (Alonzo-Pérez et al. 2003), while mangrove swamps are diked and converted to agriculture or aquaculture operations in many places (e.g., Barbier and Cox 2004; Barbier and Sathirathai 2004; Kathiresan and Qasim 2005).

Levees are simply huge dikes intended to protect lands behind them from flooding so that they can be converted to other uses. Converting large sections of the Hackensack Meadowlands in New Jersey to dryland in the 1800s was made possible by constructing levees across the marshes. The effect of levees has been most pronounced along the Mississippi River in the East and the Sacramento–San Joaquin Rivers in the West (Figure 8.6). Today, much of the Mississippi's sediment travels directly into deep waters of the Gulf of Mexico and is no longer available for marsh accretion. This lack of sediments has significantly reduced the ability of marshes to sustain themselves and directly contributes to the enormous loss of tidal wetlands that the Mississippi Delta is experiencing. Recently, some of the Mississippi River water has been diverted to the Atchafalaya River, which is helping to build coastal wetlands in a new delta. California's Sacramento–San Joaquin Delta once contained the greatest extent of tidal wetlands on the West Coast. Levee construction for livestock grazing and crop production prevents deposition of riverborne sediments in the former marshes, so the sediments now pass through the delta and directly into San Francisco Bay. Increased sedimentation in the bay has required more dredging to maintain navigation (Goals Project 1999).

Figure 8.6. Aerial view of leveed marshes in the Sacramento–San Joaquin Delta.

Excavation

Dredging of coastal marshes has converted them to deepwater habitats for ports and marinas. Construction of the Atlantic and Gulf intercoastal waterways from New Jersey to Florida to Texas have destroyed tidal wetlands mainly by dredging the channels and depositing the fill directly on the marshes or in confined disposal areas constructed in the marshlands (Figure 8.7). In places these waterways disrupted local biota where lagoons having different salinities were connected to provide a direct corridor for boat travel. Harbor and channel dredging can also increase saltwater intrusion upstream affecting both plant and animal life and affecting water levels upstream. Dredging of the Lewes and Rehoboth Canal in Delaware in 1917 raised the tidal range in the adjacent salt marshes by 32 inches (80 cm) (Carey 1996). Dredging canals across Louisiana's coastal wetlands for oil and gas development has greatly disrupted natural hydrologic conditions and facilitated saltwater penetration upstream. Secondary impacts include erosion of marshes from boat traffic along these and other waterways and invasion of spoil sites by common reed (*Phragmites australis*) and other invasive species. While channel dredging has moved the salt wedge farther upstream in most estuaries, dredging of channels and passes to open blind estuaries such as the Laguna Madre in Texas have reduced salinities in the lagoon from excessively hyperhaline (over 100 ppt) to moderately hypersaline (40–80 ppt) (Tunnell et al. 2002).

a

b

Figure 8.7. Dredged spoil areas along waterways include open marsh disposals and diked containment areas: (a) constructing a diked disposal area (NJ) and (b) open marsh and diked disposal areas (SC). (a: U.S. Fish and Wildlife Service)

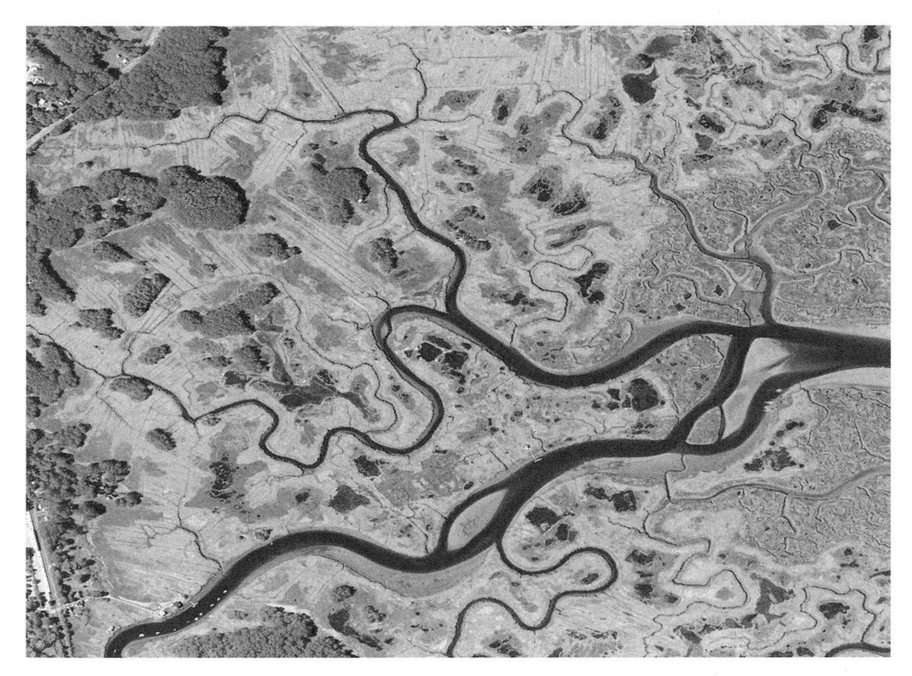

Figure 8.8. Ditched and nonditched sections of marsh are evident in this Massachusetts salt marsh. Ditched areas (light gray) are mowed for salt hay annually.

Excavation for resource extraction (mining) has had minimal direct effects on tidal wetlands; however, indirect effects can be substantial. With the discovery of phosphate in South Carolina's tidal rivers, marshes, and adjacent uplands in the late 1800s, mining took place in rivers and adjacent uplands and may have affected some marshes. Today Florida's inland wetlands are major sources of phosphate for use in fertilizers. Also in the late 1800s, hydraulic mining for placer gold on the western slopes of the Sierra Nevada in California caused heavy sedimentation in the Sacramento and San Joaquin Rivers, which both created extensive marshlands and impeded navigation (Kelley 1959). Required dredging operations used the excavated sediments to construct levees around tidal wetlands to convert them to agricultural land.

That said, some types of marsh excavation may have beneficial effects. Marsh restoration techniques that involve excavating marshes to re-establish shallow ponds for colonization by widgeon-grass help compensate for pools eliminated by mosquito-control grid-ditching.

Ditching

Tidal marshes were ditched for two main reasons: initially for salt hay production and later for mosquito control. The former use probably did not produce as many ditches as the latter and besides promoting salt hay grass (*Spartina patens*) over smooth cordgrass (*Spartina alterniflora*), the salt hay operation appears to have had minimal effect on salt marsh ecology (Buchsbaum et al. 2009). In the northeastern United States, many, if not most, of the remaining salt marshes are ditched (Figure 8.8). Most were grid-ditched in the 1930s for mosquito control and to keep people employed during the Depression. Extensive ditching dries out the marsh and drains depressions (e.g., pools and ponds) intercepted by ditches. While altered hydrology is the most obvious impact, ditching also influences sedimentation patterns, soil composition, interior marsh elevations, vegetation patterns, and wildlife use. A recent study of ditched and non-ditched portions of a Massachusetts salt marsh found that 1) the interior section of ditched marshes flooded before the tide overflowed the creekbanks of non-ditched areas; 2) sediments were coarser and had less organic matter in the ditched interiors than in the non-ditched interiors; and 3) interior sections of ditched marshes had lower elevations than the inner portion of non-ditched marshes (LeMay 2007).

Lowering of water tables has allowed plants that typically occupy the marsh–upland border to colonize more interior portions of the high marsh. High-tide bush (*Iva frutescens*) colonized these sites and the berm created by material excavated from the ditches (Bourn and Cottam 1950). Vegetation changes may be rapid as well as dramatic. For example, within five years of completing extensive ditching in a New Jersey salt marsh, shrub cover increased from 21 to 84 percent, converting the salt hay grass meadow to an estuarine shrub thicket (Burger and Shisler 1978). Meanwhile an adjacent marsh with only interior ditches showed no change in vegetation. Connecting ditches to estuarine waters increased drainage thereby favoring the growth of salt bushes.

The effect of ditching on wildlife use is variable depending on the animals of interest and the marsh location (see Clarke et al. 1984 for numerous references). Mosquito ditching focused largely on draining marsh pools, thereby eliminating widgeon-grass (*Ruppia maritima*) and valuable feeding areas for waterfowl and shorebirds. Ditched marshes are used by breeding birds, but densities of most breeding birds (e.g., red-winged blackbirds, seaside sparrows, sharp-tailed sparrows, and marsh wrens) were found to be higher on unditched Rhode Island marshes (Reinert et al. 1981). Bird densities on unditched marshes with pools in Massachusetts were high for all guilds (Clarke et al. 1984). Ditching also eliminated short-form smooth cordgrass habitat (a preferred habitat for sharp-tailed sparrows), forcing the sharp-tailed sparrow to shift its activity to tall-form smooth cordgrass (mostly for feeding) and the high marsh (for breeding) in ditched marshes. Unditched marshes with a greater number of ponds provide more habitat for waterfowl, wading birds, shorebirds, gulls, and terns during spring and summer (Reinert et al. 1981; Clarke et al. 1984; Adamowicz and Roman 2005). Similar findings have been observed for West Coast marshes (Resh et al. 1980). For example, ditching for mosquito control altered the spatial patterns of primary productivity in salt marshes as well as the distribution and abundance of resident passerine birds and macroinvertebrates (Balling and Resh 1983; Collins and Resh 1985).

Altered Tidal Regimes and Freshwater Flows

Besides ditching, deepening harbors, and dredging channels, other human activities adversely affect wetland hydrology and ecology. These activities include building structures across wetlands and coastal waters including jetties, groins, levees, roads and railroads, and of course, dams. Rerouting freshwater flows also significantly alters estuarine processes and ecology.

Construction of groins alters nearshore current patterns, affecting natural sediment transport and delivery to beaches and tidal wetlands, which in turn causes increased erosion or sedimentation depending on the location. Salinity patterns in back-barrier bays may also be altered, with significant impacts to estuarine biota. Jetties built to stabilize shifting tidal inlets between barrier beaches have the same effect. The shift of Assateague Island south of the stabilized inlet is a vivid example of this. Stabilization of inlets and dredging operations can significantly impact tidal wetlands by altering water levels. Inlet stabilization for coastal ponds disrupted hydrologic regimes (tidal flushing) and changed the ecology of the ponds (Lee 1980). Stabilization of the Indian River inlet (DE) increased the tidal range in both the Indian River and Rehoboth Bay by 31 inches (77 cm) and 15 inches (38 cm), respectively (Carey 1996).

Levee construction along the Mississippi River and Sacramento–San Joaquin Rivers has altered river flows and floodplain development with major adverse impacts to deltaic wetlands. Tidal inlets of most if not all of the estuaries of southern California have been altered to either increase or decrease tidal exchange (Grossinger et al. 2011). Tidal wetland filling of estuaries decreases the tidal prism of the estuary and also affects circulation patterns (Davis and Zarillo 2003).

Construction of roads and railroad beds across tidal marshes destroys parts of these marshes by filling, but their greatest impact is often the reduction of tidal flows to upstream marshes (see Chapter 11). Causeways serve as barriers restricting sheet flow across the marshes from spring and storm tides, and many roadways do not have adequate-sized culverts to pass sufficient tidal water upstream. Even where a combination of causeways and short-span bridges cross the marshes, upstream tidal flow is altered to some degree. Where tidal flow is significantly restricted, upstream salinity gradients and hydrologic regimes change and the natural salt marsh vegetation is replaced by common reed, brackish marsh, or essentially freshwater marsh. In some cases, the roads have served as a dike creating freshwater impoundments on the upstream side. In other cases, tidal flow has been intentionally controlled by the installation of tide gates to prevent flooding of low-lying upstream areas. This act reduces access of estuarine fishes and invertebrates to the impounded marsh. In the northeastern United States, common reed has replaced former salt marsh vegetation in many of these places (Roman et al. 1984). Where tidal flow was completely severed, common reed, cattails (*Typha* spp.), purple loosestrife (*Lythrum salicaria*), and other freshwater plants have become established. In more extreme cases, tidal flushing is so poor that all the water coming in does not flow out on the ebb tide, leading to waterlogging of the marsh with accompanying anaerobic conditions. The Nonquit Salt Marsh (Dartmouth, MA) apparently suffered this fate when its culvert became partly clogged. Between 60 to 80 percent of the tidal waters entering the marsh stayed there at low tide, while in most marshes only 10 percent of the volume remains (Howes and Goehringer 1996). More than 60 percent of the marsh vegetation died back, resulting in extensive denuded areas. In the arid West, tidally restricted areas become hypersaline, brackish, or freshwater nontidal wetlands depending on the amount of upland runoff they receive. If they are subject to desiccation, oxidation of the sediments can lead to acidification and the mobilization of metals, especially iron, giving the areas a rust or orange coloration. Depressions in diked saline wetlands may develop very low pH, producing often barren or sparse, low-diversity vegetation (Madrone Associates et al. 1983; Goals Project 2000).

Significant groundwater withdrawals may cause local subsidence as extraction of oil and gas has done along the Gulf Coast (see below). These activities may lower marsh surfaces and make less water available for discharge to wetlands. The greatest loss of brackish marsh at Blackwater National Wildlife Refuge (MD) is situated near the center of a groundwater withdrawal area (Stevenson et al. 2000). Withdrawals correspond with a two-to-three-times rise in relative sea-level rise for this area.

Construction of dams along rivers emptying into the sea has altered flow regimes, adversely affecting the ecology of estuaries. Decreased freshwater flow allows ocean-derived saltwater to penetrate farther upstream (similar to what happens during droughts), while increased flow has the opposite effect—moving the salt line farther downstream. In the Bay of Fundy region, at least 25 of 44 major rivers have tidal barriers of some kind (Wells 1999). These barriers include dikes (also spelled dykes), aboiteaux, causeways, dams, and wharves. These structures have induced drastic changes in the coastal environment by altering the hydrology and water chemistry in former tidal areas, shortening the length of tidal rivers, altering freshwater discharges to the coastal waters, eliminating salt marsh and other tidal wetlands, and restricting the movement of fish and aquatic invertebrates.

While many rivers have been dammed, only a few have had their flows re-routed to other rivers (e.g., Santee River in South Carolina). These hydrologic alterations will cause corresponding shifts in salinity and wetland vegetation patterns and reduce sediment inputs into coastal waters as dams collect much of the sediments arriving from upstream watercourses. Damming of rivers negatively impacts accretion rates in estuarine marshes by retaining sediments that would otherwise flow downstream and settle out in estuaries. Lowered sedimentation in the Otter Point Creek (MD) ecosystem due to dam construction has stopped deltaic wetland formation and changed plant communities (Hilgartner and Brush 2006). Dams also alter migration patterns for anadromous aquatic species unless measures are undertaken to accommodate them (i.e., fishways). The alteration of freshwater inflows by river damming and diversion is also a major concern for the San Francisco Bay Estuary (Schubel 1992). Damming has caused upstream migration of saline and brackish conditions affecting many aspects of the estuarine ecosystem including the plant communities of tidal marshes (Collins and Foin 1992; Jassby et al. 1995; Contra Costa County 2010). Upper limits of coastal streams have been dammed to create impoundments for various uses.

Nutrient Enrichment

Pollution has degraded water quality in, and downstream of, many urban and agricultural watersheds. Given the amount of development in coastal watersheds, nutrient loading has become a serious problem in coastal waters (Nixon et al. 1986; Valiela et al. 1990). Excess nutrients, especially nitrogen, in surface runoff promote eutrophication and algal blooms in affected waters. The location of tidal wetlands between upland development and coastal waters place them in a position to receive excess nutrients from both land and water (Caffrey et al. 2007). Nutrient loading may also facilitate the spread of common reed in salt marshes and induce changes in plant community composition in low-salinity tidal wetlands (Bertness et al. 2002; Silliman and Bertness 2004; Crain 2007; see Chapter 4, the Nutrient Availability section, for additional discussion). Current levels of nutrient enrichment in coastal waters may have a negative impact on salt marsh elevations by reducing underground biomass (roots and rhizomes) and carbon buildup (Turner et al. 2009). Eutrophication can increase the likelihood for disease outbreaks that can devastate eel-grass beds and adversely affect other biota (Valiela et al. 1990; Tyrrell 2004). Nutrient enrichment of estuaries has also caused hypoxia (oxygen-depleted waters) and changes in plankton communities that impact fish and invertebrate abundance (e.g., Breitburg et al. 2009; Glibert 2010).

Oil and Gas Exploration

More than 160,000 oil and gas wells have been constructed in Louisiana's coastal wetlands (Lyles and Namwamba 2005). Eighteen percent of the nation's oil production, valued at $6.3 billion annually, and 24 percent of the nation's gas production, generating $10.3 billion per year, originates in, is transported through, or processed in these wetlands. Oil and gas extraction along the Gulf Coast is causing subsidence of the marshes further exacerbating the effects of sea-level rise (Morton and Bernier 2010). Dredging canals through Louisiana's marshland to access production sites has greatly altered the area's hydrology. Over 8,000 miles of canals have been excavated through the state's marshes for navigation and installation of oil and gas pipelines (Turner 1987; Figure 8.9). As much as 43 percent

Figure 8.9. Aerial view of Louisiana marshes showing extensive canalization and oil facility (lower center).

of Louisiana's marsh loss appears to be occurring in areas where petroleum is being extracted (Morton et al. 2002, 2003). The effects of the petroleum industry on Gulf Coast wetlands have been described in detail (Ko et al. 2004; Ko and Day 2004). Filling of wetlands along the Arctic Coastal Plain for well sites, access roads, and pipeline corridors has impacted an unknown amount of tidal wetland. Drilling for oil may also have unintended consequences such as oil contamination from leaks or discharges, with the impact of the blowout of the offshore drilling rig Deepwater Horizon on April 20, 2010, being an excellent example.

Contamination

Contaminants from industrial, commercial, and urban sources entering the water may be adsorbed to waterborne soil particles that eventually settle out in tidal wetlands and bottoms. Heavy metals including zinc, copper, lead, cadmium, nickel, and mercury may be found in tidal wetland sediments or in leachate oozing out of landfills constructed in tidal wetlands. Depending on concentration levels, such contamination may adversely affect aquatic animal life in wetlands and pose health risks to people consuming contaminated fishes (Rai 2008). Degraded water quality from industrial sources has been a major environmental problem for estuaries in urban areas since the late 1800s (e.g., Newark Bay, NJ, for example; Crawford et al. 1994). More than 200 years of industrial activity in this area produced some of the country's most contaminated sites (7 Superfund sites) in the Hackensack Meadowlands and vicinity. For example, direct discharges from a mercury-processing plant into Berry Creek produced mercury levels in tidal marshes that led the U.S. Environmental Protection Agency to designate the Ventron/Velsicol site as a Superfund National Priority List site (U.S. Fish and Wildlife Service 2007). It is one of the worst mercury-contaminated sites in the country. Mining of mercury and its use in gold extraction during the Gold Rush era in California produced contaminated runoff that eventually led to an accumulation of mercury in tidal sediments of the Sacramento–San Joaquin Delta. The occurrence of methyl mercury (MeHg, a form of mercury that bioaccumulates) in wetland soils has necessitated pre-restoration analysis of mercury at potential tidal wetland restoration sites in the San Francisco Bay region and identified the need for post-restoration monitoring of MeHg as well (Grenier et al. 2010).

Many insecticides have been used in attempts to control mosquitoes (e.g., Paris green, DDT, dieldrin, chlordane, lindane, malathion, and Abate; Daiber 1986). Spraying marsh with an organochloride pesticide (DDT) from the 1950s to the 1970s for controlling mosquito populations had negative effects on the reproductive success of birds of prey such as the ospreys and eagles, and fish-eating brown pelicans (Carson 1962). DDT accumulation through the food chain led to concentrations high enough to interfere with calcium production, causing a decrease in egg shell thickness that led to egg breakage and an eventual decline in bird populations. Eliminating the use of this insecticide (banned in the United States in 1972) eventually allowed populations of the affected birds to recover (Daiber 1986).

Oil pollution from boat or industrial spills, ruptured pipelines, or offshore oil rigs can have devastating impacts on wetland plants and wildlife (e.g., Mendelssohn et al. 1990). Depending on the size of the spill and its timing, hundreds of birds and large stands of marsh vegetation can be killed. Marsh recovery takes many years, well after vegetation has re-colonized the site, as oil remains in the marsh substrate and poses problems for burrowing animals. In 1969, a barge ran aground in Buzzards Bay, Massachusetts, spilling 200,000 gallons of oil. Initially, high oil content in sediment reduced crab density and caused heavy overwinter mortality, and after seven years marsh recovery was still incomplete (Krebs and Burns 1977). The degree and duration of damage from oil spills is dependent on several factors: 1) type of oil spilled; 2) quantity and duration of the spill; 3) seasonal, oceanographic, and meteorological conditions; 4) nature of the exposed biota; 5) habitat and substratum; 6) geographic location; and 7) type of spill cleanup used (Clark and Finley 1977). NOAA has published reviews of oil spill impacts for subarctic fucoid beds, mussel reefs, and intertidal mudflats (Fukuyama et al. 1998). The Exxon Valdez spill in Prince William Sound, Alaska, destroyed an estimated 5,800 metric tons of *Fucus* biomass by oiling and the cleanup with high-pressure hot water, with the latter possibly causing more harm than the oil. Fucoid recovery from oil spills may be further hampered by the effect on grazers or on its competitors (e.g., green algae and mussels). The discharge from the blowout of the Deepwater Horizon oil drilling rig off the coast of Louisiana spewed up to 4.9 million barrels (206,000,000 US gallons; 779,000 cubic meters) of oil into the gulf with disastrous impacts to fish, shellfish, birds, sea turtles, and marine mammals; the coastal wetlands; and people and businesses dependent on these resources. To call it an "oil spill" is a euphemism as it discharged 12,000 to 19,000 barrels (over 500,000 gallons) of oil per day (McNutt 2010). Nearly 550 miles of Gulf Coast shorelines (including marshes) were oiled from Louisiana to Florida (www.fws.gov/home/dhoilspill/). The assessment of environmental damages is under way, and the full impact of the event will not be realized for years to come (Mendelssohn et al. 2012).

Other Wetland Degradation

The quality of the remaining tidal wetlands is degraded in many ways from obvious sources such as minor filling, ditching, and pollution and in more subtle ways from adjacent land use, increased human traffic, noise pollution, and local runoff that can negatively impact vegetation and wildlife (e.g., DeLuca et al. 2004). Degraded water quality in some of our estuaries stresses plant life, giving the edge to disturbance-tolerant species like common reed. Invasive species have altered the composition and structure of some tidal wetlands along the all coasts (Table 8.5). For example, along the Atlantic Coast, common reed alters the salt marsh landscape by converting the natural low- to medium-height grassland to a tall grass monoculture and disrupts drainage patterns by fragmenting microhabitats and building up a thick mat of dead reeds. All these changes tend to produce a negative impact on plant diversity, bird use, and fish habitat in most cases (Benoit and Askins 1999; Weinstein and Balletto 1999; Meyerson et al. 2000; Able and Hagan 2003). Despite these changes, reed marshes may still provide breeding, feeding, and shelter for many birds and the wetter ones may serve as habitat for mummichog, grass shrimp, and blue crabs (Marks et al. 1994;

Table 8.5. Examples of recognized invasive species growing in estuarine wetlands of the United States

Common Name (*Scientific name*)	State or Region of Occurrence
Plants	
Common reed (*Phragmites australis*)	Northeast
Purple loosestrife (*Lythrum salicaria*)	Northeast
Perennial pepperweed (*Lepidium latifolium*)	New England, California
Water hyacinth (*Eichhornia crassipes*)	South Atlantic, Gulf Coast
Redtop (*Agrostis stolonifera*)	Oregon
Reed canary grass (*Phalaris arundinacea*)	Oregon
Curly dock (*Rumex crispus*)	Oregon
Brass-buttons (*Cotula coronopifolia*)	Oregon
Salt marsh sand spurrey (*Spergularia salina*)	Oregon
Smooth cordgrass (*Spartina alterniflora*)	Puget Sound, California
Dense flower cordgrass (*Spartina densiflora*)	Puget Sound, California
Common or English cordgrass (*Spartina anglica*)	Puget Sound, California
Salt meadow cordgrass (*Spartina patens*)	Puget Sound, California
Japanese eel-grass (*Zostera japonica*)	Puget Sound
Chinese tallow (*Triadica sebifera*)	Gulf Coast
Tamarisks (*Tamarix* spp.)	Southern California
Rabbitfoot grass (*Polypogon monspeliensis*)	California
Ripgut brome (*Bromus diandrus*)	California
Oppositeleaf Russian thistle (*Salsola soda*)	California
Ice plant or Hottentot fig (*Carpobrotus edulis*)	California
Australian saltbush (*Atriplex semibaccata*)	California
Animals	
Green crab (*Carcinus maenas*)	Atlantic Coast
Asian shore crab (*Hemigrapsus sanguineus*)	New England, Puget Sound
Mitten crab (*Eriocheir sinensis*)	South Atlantic, Mississippi Sound, Puget Sound
Green porcelin crab (*Petrolisthes armatus*)	South Atlantic, Gulf
Atlantic ribbed mussel (*Geukensia demissa*)	Puget Sound
Atlantic gem clam (*Gemma gemma*)	Puget Sound
Northern quahog (*Mercenaria mercenaria*)	Puget Sound
Purple varnish clam (*Nuttalia obscurata*)	Puget Sound
Asian date mussel (*Musculista senhousia*)	Puget Sound, South Atlantic
Mediterranean mussel (*Mytilus galloprovincialis*)	Puget Sound, South Atlantic
Atlantic oyster drill (*Urosalpinx cinerea*)	Puget Sound
Asian mud snail (*Batillaria attramentaria*)	Puget Sound
Nutria (*Myocastor coypus*)	Maryland to Texas

Sources: Massachusetts Office of Coastal Zone Management 2002; Adamus et al. 2005; Ray 2005; California Invasive Plant Council 2006; Whitcraft et al. 2007; Nahkeeta Northwest Services 2008; Vanderhoof et al. 2009; Fetscher et al. 2010.

Kane 2001; Hanson et al. 2002; Neuman et al. 2004). Where invasive species are involved in marsh formation on intertidal flats, the marsh itself may be interpreted as a form of wetland degradation. On the Pacific Coast, the colonization of mudflats by introduced cordgrasses is reducing already declining shorebird feeding habitat (mudflats) and changing their algae-based food web to a detritus-based one (Page et al. 1999; Levin et al. 2006).

Sea-Level Rise

Rising ocean levels accompanying climate change threaten tidal wetlands due mainly to increased hydroperiod (frequency and duration of flooding) and salinity (see Chapter 12 for details). Higher water levels will submerge lower marshes, converting them to tidal flats or shallow bottoms and changing high marsh plant communities to low marsh. Saltwater will move farther upstream replacing freshwater species with more salt-tolerant species typical of brackish marshes. Nontidal areas may also become tidally influenced with possible increased flooding and corresponding changes in plant community composition. The presence of hardened shoreline (bulkheads, seawalls, and rip-rap) or dikes along many marshes will prevent their natural landward migration, further reducing the extent of tidal wetlands. Rates of mineral and organic accretion within tidal wetlands will be critical to their capacity to adapt to sea-level rise (Cahoon et al. 2006).

Other Threats

Many other human activities negatively impact the world's tidal wetlands. European tidal marshes have been used for livestock grazing for centuries, but such use is not widespread in North America today. Timber harvest in tidal swamps may alter vegetation patterns. Mangroves are harvested around the world, causing vegetation changes including conversion to nonvegetated wetlands. Building seawalls and bulkheads along coastal marshes will prevent natural marsh migration landward. Development and human activities in coastal wetland buffers have an adverse effect on wildlife use by some species. A noticeable decline in beach-nesting birds such as piping plovers and least terns has occurred in the New York Bight due to coastal development and predation by raccoons, skunks, foxes, cats, and dogs (U.S. Fish and Wildlife Service 1997). Beach-nesting birds are vulnerable to nest destruction by off-road vehicle use, beach-walking, and unleashed pets. Boat wake may be accelerating erosion of coastal marshes in some areas.

Human actions can also affect populations of animals with devastating consequences for tidal marshes. For example, snow goose feeding behavior may have been disrupted by human exploitation of coastal resources that led to increased grazing pressure on remaining wetlands. Destruction of mid-latitude tidelands (the winter feeding grounds for snow geese) and expansion of agricultural fields in the coastal zone may have caused snow geese to shift some of their feeding activity to nutrient-rich farmland. This shift allowed the geese to triple their populations, leading to subsequent denuding of Arctic marshes by their grubbing activity and secondary effects (Gedan et al. 2009). Likewise, overharvesting of blue crabs and diamond-backed terrapins has reduced snail predator populations to the point at which snail grazing can have a devastating impact on smooth cordgrass during times of drought (Silliman et al. 2005). Nitrogen-loading in the coastal zone, and enrichment of marsh grasses in urban areas, may be triggering increased insect herbivory as evidenced by a nearly 40 percent reduction in productivity of Narragansett Bay salt marshes attributed to insect grazing (Bertness et al. 2008).

Tidal Wetland Trends in the Conterminous United States

Similar data for Canada and Alaska are lacking, so only trends in the conterminous United States (lower 48 states) are summarized. Data for tidal wetland trends in the United States come mainly from a series of studies undertaken by the U.S. Fish and Wildlife Service's National Wetlands Inventory Program since the 1980s (Frayer et al. 1983; Dahl and Johnson 1991; Dahl 2000, 2011). The value of these studies is that they apply the same definitions and data collection and analysis techniques for predicting changes, so the trends are more representative of the changes than comparing these findings with the results from studies employing other methods. Since these studies commenced in the 1980s, other sources had to be consulted to report on earlier trends.

In the 1920s nearly 10 million acres (4.05 M ha) of tidal wetlands may have existed in the lower 48 states (Gosselink and Baumann 1980; this estimate probably focused on vegetated wetlands). From 1922 to 1954, the annual loss rate was estimated at 0.2 percent per year and from 1954 to 1974, the rate more than doubled to 0.5 percent. Postwar development, population growth, and virtually uncontrolled construction in the coastal zone placed the "worthless" tidal wetlands at risk. The first U.S. Fish and Wildlife Service national wetland trends study estimated only 5.6 million acres (2.27 M ha) for 1954 and reported a total loss of 372,000 acres of estuarine vegetated wetlands from the mid-1950s to the mid-1970s (Frayer et al. 1983). This total amounted to 7.6 percent of the estuarine vegetated acreage that was present in the mid-1950s. The study period represented a time when wetland regulations either did not exist or were in their infancy, and many coastal wetlands were filled or dredged at will for residential and other development or were used as dumping grounds for landfills and dredged material (see Figures 8.4 and 8.7). Since then, such activities have been strictly regulated in most areas, chiefly through a combination of state wetland laws and federal regulations related to water quality (see Chapter 9). By the mid-1980s, wetland loss rates had been cut by more than half. From 1974 to 1983, estuarine wetland losses amounted to 59,265 acres (23,994 ha) for a 0.12 percent annual loss rate (Tiner 1991). During this time, estuarine vegetated wetlands declined by 70,799 acres (28,650 ha) or nearly 7,900 acres per year (3,197 ha/yr)—a 1.5 percent loss during that decade. Most of the estuarine vegetated wetlands were transformed to deepwater habitat by a combination of factors including sea-level rise, coastal subsidence, erosion, and dredging (Table 8.6). Nearly equal amounts of estuarine vegetated wetland were converted to agriculture and urban development. While most of the wetland acreage was unchanged, representing generally stable conditions, the dynamic movement of estuarine marshes and swamps between deepwater and tidal flats is evident in these findings. Further declines in estuarine wetland losses were recorded in the succeeding decades. From 1986 to 1997, the marine and estuarine wetland losses totaled 10,400 acres (4,209 ha) for an annual loss of only 945 acres (Table 8.7; Dahl 2000). Forty-three percent of the losses were attributed to urban and rural development (24% and 19%, respectively), while other upland accounted for 30 percent, development 13 percent, agriculture 1 percent, and conversion to freshwater wetlands 14 percent. Findings from 1998 to 2004 showed a 58,700-acre (23,715 ha) drop in marine and estuarine wetlands, which is quite shocking given the period is only a 6-year interval (Table 8.7; Stedman and Dahl 2008b). It represents an annual loss of 9,950 acres, which is an order of magnitude greater than the loss rate during the prior decade. Ninety-six percent of the loss was due to conversion to open water, 1.7 percent to urban and rural development, 2 percent to other upland, and 0.3 percent to freshwater wetlands. What is responsible for this loss to open water? Is it a combination of sea-level rise, coastal subsidence, and marsh erosion, which are well-documented causes for Louisiana marsh loss? The problem is particularly focused in the Gulf region where more than twice as much loss occurred in estuarine vegetated wetlands as along the Atlantic Coast. Since the findings represent conditions prior to Hurricane Katrina, the hurricane is not the cause. Given that the report did not make any comparisons with previous data, the question remains while federal agencies search for answers. The difference may in part be attributed to study design as there was a change in the allocation of plots used in this study (Thomas Dahl, pers. comm. 2010).

The latest national report on wetland trends identified a net loss of estuarine vegetated wetlands of 110,900 acres (44,800 ha) from 2004 to 2009 (Dahl 2011). While modest gains in nonvegetated types and estuarine forested/scrub-shrub wetlands were reported (18,300 acres [7,406 ha] and 6,000 acres [2,428 ha], respectively), emergent losses were enormous (111,500 acres [45,120 ha]). Eighty-three percent of the loss of emergent types was attributed to deepwater habitat and 16 percent to non-vegetated tidal wetlands, presumably due to sea-level rise, erosion from coastal storms, subsidence, and other coastal processes. Only 1 percent of the estuarine marshes was converted to upland.

Human Use of Tidal Wetlands

Humans have had a tremendously adverse impact on wetlands of all kinds for thousands of years. From early times, perhaps beginning in Mesopotamia 6,000 years ago, civilizations around the globe have drained wetlands for agricultural purposes to support a growing population (Butzer 2002; Hatvany 2003). Egyptians, Greeks, and Romans drained wetlands along the Mediterranean Sea, while wetlands along

Table 8.6. Causes of estuarine vegetated losses and gains in the conterminous United States between 1974 and 1983

Nature of Change	Estimated Acreage (%SE)
Losses to	
Marine intertidal wetland	1,614 (39.3)
Estuarine nonvegetated wetland	6,077 (21.2)
Palustrine wetland	11,577 (48.0)
Deepwater habitat	57,004 (12.7)
Agricultural land	3,980 (79.9)
Urban development	4,502 (34.5)
Other land*	7,126 (18.9)
Gains from	
Marine intertidal wetland	1,428 (88.2)
Estuarine nonvegetated wetland	5,271 (23.8)
Palustrine wetland	148 (58.1)
Deepwater habitat	13,095 (20.1)
Agricultural land	276 (98.2)
Other land (non-urban)*	863 (35.2)
No change	4,762,035 (4.2)

Source: Tiner 1991.

Notes: Since the numbers were generated by statistical sampling, the standard error is given: %SE represents the standard error as a percentage of the estimated total; in general when the SE is 25% or less, the estimate is considered reliable. These results were derived from a trends analysis of the nation's wetlands; no sample plots were located in the coastal region of California, Oregon, Pennsylvania, and Washington. * Other land includes rangeland, forests, strip mines, quarries, and land being converted to an unknown use, for example. To convert acreage values to hectares multiply by 0.4047.

Table 8.7. Estimated changes in estuarine and marine wetlands for two decades from 1986 to 2004

Wetland Type	Acres 1986 (D 2000)	Acres 1998 (D 2000)	Change Acres	Acres 1998 (S&D 2008)	Acres 2004 (S&D 2008)	Change Acres
Marine intertidal	133,100	130,900	−2,200	135,970	134,180	−1,860
Atlantic Coast	NA	NA	NA	105,130	105,160	+30
Gulf Coast	NA	NA	NA	30,830	28,950	−1,890
Estuarine nonvegetated	580,400	580,100	−300	699,390	704,660	+5,270
Atlantic Coast	NA	NA	NA	287,920	287,680	−290
Gulf Coast	NA	NA	NA	411,470	416,980	+5,510
Estuarine vegetated	4,623,100	4,615,200	−7,900	4,950,430	4,885,460	−64,970
Atlantic Coast	NA	NA	NA	1,842,320	1,822,780	−19,540
Gulf Coast	NA	NA	NA	3,108,110	3,062,680	−45,430
Total change	—	—	−10,400	—	—	−61,560

Source: Dahl 2000 = D 2000, Stedman and Dahl 2008b = S&D 2008.
Notes: Data also shown for the Atlantic and Gulf coasts for 1998–2004. Standard errors for these numbers are not reported since one of the original documents was incomplete in this regard. The difference in estimates for 1998 for the two studies related to a change in sampling design. NA, not analyzed. To convert acreage values to hectares multiply by 0.4047.

the Ganges (India) and Yellow (China) Rivers were diked and drained for agriculture. Mexico and Mesoamerica freshwater wetlands were similarly utilized. By AD 800, the French, Germans, Dutch, and English all had developed techniques for draining coastal wetlands.

General Overview

When Europeans settled America, they established communities in places along the coast with good water access, and up coastal rivers as far as the ships could travel, which was often in the tidal freshwater zone (Baldwin et al. 2009). These new communities served as major places of commerce and later led to port and urban expansion often at the expense of tidal marshlands. They were focal points for the growth of a nation. In choosing areas for settlement, colonists also sought out places with tidal marshes because they did not have the ability or time and effort required to convert forests to farmland. The marshes offered them free food for livestock and a wealth of fish and game for their tables. European settlers brought with them the knowledge of marsh exploitation and later the technology to continue these practices in the New World when they needed more land to farm. Once diked, marshes provided lands suitable for crop production as well as for livestock grazing. In many areas along the North Atlantic Coast, tidal marshes were the only naturally level lands with fertile soils—in stark contrast to upland pine and spruce forests on acidic, nutrient-poor soils.

During Colonial times, marsh conversion to agriculture may have been the first significant human impact to tidal wetlands, with diking of marshes to create salt hay farms being most active in the northeastern United States and to create upland pastures and agricultural land in the Canadian Maritimes. In the southeastern United States, tidal swamps were converted to plantation rice fields, whereas tidal marshes along the Pacific Coast from California to British Columbia were diked and drained for agriculture starting around the mid-1800s. Later, rivers were dammed, thereby reducing tidal flow, eliminating tidal wetlands and their plants and animal life, and cutting off access to upstream spawning grounds for anadromous fishes. Dams were built across the upper reaches of many tidal rivers to power local mills in New England and the Mid-Atlantic states.

As the population in U.S. coastal regions grew, the society moved from an agrarian base to an industrial one that included mechanized farming. Mechanization of activities that once required huge investments of human and/or draught animal labor (e.g., land clearing, agriculture, and production of manufactured goods) greatly increased the capacity of humans to modify the natural environment in a rapid fashion and allowed the population to rise at rates never before witnessed in human history. Attitudes toward tidal wetlands changed as a society as the percentage of the population actually working on farms dropped and coastal rivers became polluted (see discussion in Chapter 7, Changing Attitudes about Tidal Wetlands). Instead of seeing wetlands as valuable resources to support livestock and livelihood, wetlands were viewed as "wastelands"—unproductive lands that could be made productive through "reclamation" to better serve society's needs. The population and businesses in coastal cities were growing and in the view of civic leaders and real estate developers, tidal wetlands represented unused acreage ripe for development. In the 1800s, inventions of the clamshell dredge, draglines, and similar mechanized equipment made it relatively effortless to modify huge tracts of wetlands. The expansion of all major seaports involved filling and dredging of coastal marshlands and bay bottoms, with dredged spoil typically deposited on wetlands or fill material brought in from nearby uplands to make substrates suitable for construction. Some examples in the Northeast are Boston, New York City–Newark, Philadelphia, Baltimore, and Washington, DC. Port expansion also required deepening navigation channels and harbors to support an ever-growing shipping industry with larger vessels. Tidal wetlands were convenient places to dispose of dredged materials and the sites could later be converted to marketable real estate. Wetland destruction was actively supported by the U.S. Congress that passed the Swamp Lands Act of 1849–1860, giving 15 states including six coastal states (Alabama, California, Florida, Louisiana, Mississippi, and Oregon) possession of 16 million acres (6.5M ha) of tidal and nontidal marshes and swamps provided they drain them for agriculture or other purposes (Hibbard 1965; U.S. Geological Survey 2006). Nearly all of these once public lands eventually ended up in private ownership. Throughout the U.S. coastal zone, marshes were dredged to create marinas to meet the increasing demands for boat access by commercial and recreational fishermen and boaters. The growth of metropolitan areas in the coastal zone, surrounding suburban communities, and seaside resort towns led to further destruction of coastal wetlands for commercial and residential real estate, industrial facilities, and airports (e.g., Logan, JFK, Newark, and Philadelphia). Many of the remaining wetlands were ditched or impounded and sprayed with pesticides to reduce mosquito populations that posed public health risks. In the 1930s, the federal government also was responsible for grid-ditching millions of acres of tidal and other wetlands across the country for mosquito control through the Civilian Conservation Corps, established in the 1930s to put thousands of people to work during America's Great Depression, and to help address a public health concern. Today, finding a salt marsh from Massachusetts to Delaware without mosquito ditches would be difficult. Other tidal wetlands were impounded for waterfowl management or aquaculture. Water quality of estuaries and associated wetlands was degraded by various forms of pollution ranging from raw sewage to heavy metals and other contaminants from manufacturing operations to nutrients from runoff. The hydrology of many tidal marshes was altered by road or railroad crossings and by diverting river water for commercial and industrial uses. In some areas, extracting groundwater for public water supplies and farmland irrigation or oil and gas development caused coastal subsidence, leading to drowning of coastal marshes. Invasive species that were intentionally or accidentally introduced by people have replaced or threaten native biota. The cumulative impact of civilization on coastal wetlands has been devastating. In the United States alone, more than half of the original (pre-settlement) tidal wetlands have been lost (Kennish 2001).

Today, 53 percent of the U.S. population lives in the nation's coastal counties, representing an increase of nearly 28 percent from 1980 levels (Crossett et al. 2004). While the dredging and filling of tidal wetlands has been significantly reduced through wetland and water quality protection laws in the United States and some Canadian provinces, threats still remain. Urbanization continues to negatively impact coastal wetlands in more subtle ways. Coastal development is introducing more nitrogen into coastal waters through runoff from lawns, golf courses, and farms, and this is degrading the adjacent tidal wetlands. In southern New England, nitrogen loading is causing smooth cordgrass to replace salt hay cordgrass, and common reed to invade lower elevations displacing the diverse assemblage of upper marsh plants including black grass (Bertness et al. 2002). This has reduced plant diversity fivefold and has lowered bird diversity in tidal marshes (Benoit and Askins 1999).

Regional Review

An abbreviated history of wetland uses and trends in each region follows. The discussion is not intended to be a comprehensive treatment but should give the reader a historic perspective and some insight into the current state of affairs regarding tidal wetlands. More attention is given to the northeastern United States and eastern Canada since those regions are in large part the focus of this book. Although the discussion focuses on wetland losses, it should be noted that since the latter part of the 20th century, governments at the state, federal, and in some cases, provincial level have taken legislative action to protect or regulate uses of wetlands, while both public agencies and private organizations have worked cooperatively in the last few decades to restore damaged or degraded tidal marshes (see Chapters 9 and 11).

CANADA'S MARITIME PROVINCES

This region includes the provinces of New Brunswick, Nova Scotia, and Prince Edward Island. It was settled by French Acadians in the 17th century. "Reclamation" of tidal marshes (conversion to cropland) was a common practice in their homeland. Diking in the Bay of Fundy region began at Annapolis Royal and Port Royal in the 1630s (Milligan 1987; Nova Scotia Department of Agriculture and Fisheries 2003). Marshes were leveed and ditched; one-way water-control gates ("aboiteaux") were then built to allow freshwater to flow out at low tide, while preventing saltwater from coming into the "dyked lands" (as it is spelled in the Maritimes) on rising tides. After two to four years of letting snow and rain wash out the salt, these dyked lands were ready to plant. They were converted to farmland for growing crops such as wheat, oats, barley, rye, peas, corn, flax, and hemp and to gardens planted with beets, carrots, parsnips, onions, chives, shallots, herbs, salad greens, cabbage, and turnips. With plenty of farmland made this way, there was little need to clear forests for agricultural purposes. Interestingly, the Acadians were sometimes referred to as "defricheurs d'eau" (clearers of water) because they converted marshes ("water") to land rather than clearing forests. Like in Colonial America, undiked marshes, including the seaward portions of the diked marshes, were cut for salt hay to sustain livestock through winter. Salt hay was harvested and stacked to dry on wooden platforms ("staddles") that were built above the high tide mark.

Marsh diking continued until the early 1900s. At that time, hay was worth $28 per ton, making hayfields high-valued real estate. Interestingly, the elite social club in Amherst, Nova Scotia, was called the Marshland Club. In the late 1920s, when oil and machines replaced hay and livestock for farming, hay prices dropped greatly to about $7 per ton making dyked lands less valuable property. With dikes falling into disrepair and the 18-year cycle of high tides nearing, the federal government passed the Maritime Marshland Rehabilitation Act in 1948 and initiated a program to restore the deteriorating dikes (Milligan 1987; Nova Scotia Department of Agriculture and Fisheries 2003). Today, 232 miles (373 km) of dikes protect 44,480 acres (18,000 ha) of converted lands (including hayland, some cropland, and residential and commercial buildings) in Nova Scotia and 37,070 acres (15,000 ha) in New Brunswick (Figure 8.10).

Since the 1700s, more than 65 percent of the original Maritime tidal marshes have been converted (Lynch-Stewart 1983; MacKinnon and Scott 1984). Only 20 percent of the famous Tantramar Marshes on the upper Bay of Fundy remain in an unaltered state. More recently, "tidal barrages" were constructed to control tides in rivers to protect upstream "fertile dykelands" from flooding. In the Bay of Fundy region, at least 25 of 44 major rivers have tidal barriers of some kind (Wells 1999), including dikes (dykes), aboiteaux, causeways, dams, and wharves. They affect 72 percent of the major rivers in New Brunswick and 46 percent in Nova Scotia. These structures have induced drastic changes in the coastal environment by altering the hydrology and water chemistry in former tidal areas, shortening the length of tidal rivers, altering freshwater discharges to the bay, eliminating salt and other tidal wetlands, and restricting the movement of fish and aquatic invertebrates. Today, more than 82,220 acres (33,275 ha) of land are "protected by dykes" in the Canada's three maritime provinces, representing a loss of 75 to 90 percent of the region's historic tidal marshes; these former tidal wetlands are protected as farmlands through the Agricultural Marshland Conservation Act. Ducks Unlimited has built freshwater impoundments on these dykelands to benefit waterfowl. From 15 to 35 percent of the original tidal wetlands remain in the Fundy region and many are significantly degraded by tidal restrictions (Lynch-Stewart 1983; MacKinnon and Scott 1984; Percy 1996b). In 1984, a large tidal

Figure 8.10. Much of New Brunswick's tidal marshes have been diked for agriculture: natural and diked marshes are evident in this image. (Service New Brunswick)

barrier was constructed across Nova Scotia's Annapolis River to run a tidal-power generating station (Wells 1999). Proposals have been made to dam other portions of the Bay of Fundy to create tide-generated energy. These proposals are currently inactive; if these projects are constructed the extent of wetland alteration and adverse impact to wildlife (especially migratory shorebirds and waterfowl) will be enormous.

Urbanization, industrialization, and coastal development have posed more recent threats to the remaining coastal marshes. Salt marshes around Halifax, Nova Scotia, and St. John, New Brunswick, have been converted to house lots and industrial sites (Glooschenko et al. 1988). More recently in New Brunswick, housing development has focused on the Northumberland Strait between Bouctouche and Shemogue (Lee Swanson, pers. comm. 2004). Road and railroad crossings have altered tidal regimes and have adversely affected upstream marshes in many places. Sea-level rise is also causing marsh loss and landward migration (i.e., marine transgression of the land). In Nova Scotia, dead spruce trees can be found along the edges of salt marshes (Davis and Browne 1996).

ST. LAWRENCE REGION

The St. Lawrence Region was first colonized by the French in the late 17th century (Rousseau 1967). Like neighboring Colonial settlements in New England, marshes were a vital resource sustaining livestock through the long, cold winters and their late springs. From the Colonial period to the mid-20th century, marsh grasses were used for forage, while tidal wetlands served as pastures for grazing domesticated animals. Marshes were mowed at low tide, then as the tide rose and lifted the cut grass, a team of men, women, and children using rope and a horse-drawn cart would work together to encircle and pull the floating grass up to the high tide line where it was loaded in a cart for transport to upland fields where it was dried (Farnham 1888 cited in Hatvany 2003). Where the salt hay could not be easily brought to shore, hay was stacked on staddles for removal in the winter when the marsh was frozen and could support the weight of draft animals. Smooth cordgrass was used as thatch to protect the salt hay from rain and snow. Salt hay was widely harvested in the St. Lawrence until the mid-1940s (Matthew Hatvany, pers. comm. 2011).

By 1850, the best agricultural land in the St. Lawrence Valley had been settled, limiting the next generation of non-inheriting farm children to marginal land in the Canadian Shield, finding industrial employment in the urban centers of Quebec and New England, or going West, like many New Englanders were doing, to find land in the prairies (Hatvany 2003). Canada's first permanent school of agricultural was inaugurated in 1859 at Sainte-Anne-de-la-Pocatière in reaction to this agricultural dilemma. In response to the dearth of good land, the school launched a large-scale marsh drainage project to expand available agricultural land. Based on salt marsh diking techniques then in use in France, the project had the desired outcome of encouraging local and regional farmers in the St. Lawrence Valley to dike the tidal marshes as a means of increasing cropland and intensifying production. The first dike, about 6 feet wide at the base, 5 feet high, and 1 mile long (1.8 m x 1.5 m x 1.6 km), was built on the high marsh with tidally operated sluice gates (aboiteaux) to permit drainage of the marshes at low tide. With industrialization and increased market production of agricultural goods, diking and drainage accelerated in the early 1900s and was substantially aided by an influx of provincial funding in the 1920 and 1930s. That capital provided the means to initiate large-scale communal (multifarm) dikes. The prevailing view of marshlands held at this time by engineers and government developers (technocrats) was utilitarian—the marshes were underutilized in their natural state and were considerably more valuable when converted to arable land. To buttress this argument, scientists claimed that through sedimentation and biological succession new marsh would naturally form on the seaside of diked marshes approximately every 20 years. This paradigm, however, based on French examples where marshes like Mont-St.-Michel have been diked multiple times since the Middle Ages, did not hold true for Quebec. Insufficient rates of sedimentation in the St. Lawrence estuary inhibited the formation of new marsh outside diked areas (Hatvany 2003).

At the height of an industrial period in the 1970s, 32 percent (about 2,500 acres [1,012 ha]) of the original tidal marshes on the south shore of the St. Lawrence from Monte-des-Ours to Pointe-a-la-Loupe had been transformed (Reed and Moisan 1971). Agricultural "reclamation" claimed 27 percent (2,081 acres [842 ha]), while industrial development converted 5 percent (415 acres [168 ha]). A study of more recent changes in St. Lawrence wetlands from Cornwall to Pointe-des-Montes (including nontidal wetlands) identified losses of 9,111 acres (3,687 ha) between 1950 and 1978. While agricultural conversion represented a significant cause of loss (33.5%), the majority of recent losses were attributed to urban development (landfill = 26.5%, residential development = 11.8%, industry = 9.5%, clearing = 9.5%, and services = 9.2%) (St. Lawrence Centre and Université Laval 1991).

With the rise of the environmental movement in the 1970s and a growing ecological consciousness in the public during the postindustrial period, pressure mounted on agricultural interests and the government to protect St. Lawrence wetlands. As a result, the last officially sanctioned diking ended in 1980. While small losses of wetland continue in some urban areas, there has been nearly a 10 percent increase in tidal wetlands (Table 8.8). This gain is largely attributed to restoration projects and government economic incentives (tax exemptions) to encourage farmers to convert diked land to marsh (Hatvany 2009). A 2011 government report offers a very detailed account of wetland trends along much of the St. Lawrence River from 1970 to 2002 (Jean and Létourneau 2011). Analyzing aerial photographs, these researchers found that from 1990–91 to 2000–2, 430 acres (174 ha) of low marsh colonized former open water areas in the lower estuary, while 178 acres (63 ha) of marshes established on tidal flats and 67 acres (27 ha) of marshes became tidal flats in the upper estuary. During this time significant changes in wetland type were also detected: for the lower estuary, 178 acres (72 ha) of low marsh became high marsh and 897 acres (363 ha) of high marsh

Table 8.8. Changes in St. Lawrence tidal wetlands from 1990–2002

Section of Estuary	Wetland Acreage (ha)			
	1990–1991	2000–2002	Change in area	Percentage change
River estuary (freshwater)	6,306 (2,552)	7,411 (2,999)	+1,105 (+447)	+17.5
Québec-Lèvis	2,350 (951)	2,350 (951)	0 (0)	0
Middle estuary (brackish water)	7,717 (3,123)	8,103 (3,279)	+386 (+156)	+5.0
Maritime estuary*	3,603 (1,458)	4,033 (1,632)	+ 430 (+174)	+11.9
Total	19,980 (8,084)	21,900 (8,861)	+1,920 (+777)	+9.6

Source: Hatvany 2009.
*= Partial coverage.

became low marsh; and for the upper estuary, 15 acres (6 ha) went from low marsh to high marsh and 736 acres (298 ha) changed from high to low marsh. Currently, erosion from natural and anthropogenic sources as well as expansion of invasive species, including common reed and purple loosestrife, constitute important threats to the future biodiversity of St. Lawrence wetlands (Jean et al. 2002; Hatvany 2009).

NEW ENGLAND AND MID-ATLANTIC REGION

Some people may think that wetlands were always regarded as wastelands and that only recently have we begun to appreciate their natural bounty, but this is not the case. As mentioned elsewhere, in Colonial times salt marshes were highly valued for their salt hay. In the 1600s, they were vital resources for an agrarian society building a new nation. These natural grasslands in an otherwise forested region provided winter feed and bedding for livestock with no need to plow, seed, or control undesirable weeds. The amount of salt hay available determined how many cattle a farmer could keep through winter. Salt hay grass, salt grass, and black grass were the main species, with the latter being preferred (Nixon 1982). Settlers moved out from established population centers like Plimouth Plantation to create farms in areas with extensive marshes because of the availability of this natural fodder (Russell 1976). Coastal towns grew up in these locations across New England and the Mid-Atlantic region. Following English common law (e.g., Magna Carta of 1215), tidelands were typically held in the public trust by the government for use by the people. Long Island salt marshes were owned by towns and each year they sold the rights to hay the marshes to the highest bidders (Kavenagh 1980). Salt marshes were so valuable that some farmers actually converted freshwater wetlands to salt hay meadows (Eliot 1748; Clark and Patterson 1985).

Initially the marshes were simply used as grazing meadows for horses and cattle and for the harvest of salt hay. In some cases, marshes were ditched to increase production of salt hay (Rozsa 1995). Later, some marshes were diked to control flooding, increase productivity of salt hay, and to make conditions easier for harvest. Diking of salt marshes was a common practice for salt hay farming from southern New England to Delaware Bay.

As America's population grew, forests were cleared for pasture and cultivation and eventually demand for salt hay declined. It is interesting to note that forests in coastal regions were first cleared with the help of livestock that were supported by salt hay (Casagrande 1997b). Since there was plenty of livestock feed available from upland grasses and clover, some coastal farmers began to employ "reclamation" techniques from the Old World to convert tidal marshes to farmland to grow crops. Dutch and English settlers learned these techniques from their former countrymen as the Netherlands and England had undertaken huge tidal wetland reclamation projects for centuries. By the 11th century, the Dutch had diked salt marshes in the Netherlands to protect uplands from tidal flooding, and later, with the development of the windmill and power pumps, they converted the diked marshes to fertile agricultural lands (Sebold 1992). As early as 1675, the Dutch began diking marshes in New Castle, Delaware. In 1788, the New Jersey legislature passed a law to promote the formation of "meadow companies" that would organize the diking and draining of privately owned marshes (Weinstein et al. 2000). In the late 1800s, the U.S. Department of Agriculture began actively promoting conversion of salt marshes to cropland. Crops including wheat, corn, hay, oats, strawberries, potatoes, tomatoes, and other garden vegetables grew well in these former marshlands.
The Delaware Bay area was a focal point for marsh diking. By 1885, two-thirds of the salt marshes in New Castle County, Delaware, and more than 20,000 acres in New Jersey (mostly in Cumberland and

Salem counties), had been reclaimed (Sebold 1992). In the early 1900s, the lack of cooperation among landowners, high maintenance costs, the Great Depression, and other factors led to a failure to maintain the Delaware Bay meadows. Some were converted to waterfowl impoundments, while others were used for trapping, hunting, flood protection, and salt hay production. In the 1940s, there was a movement to build impoundments for mosquito control and this lasted until the 1960s when the public became concerned about tidal wetland preservation (Weinstein et al. 2000).

Salt marshes were home to biting insects such as mosquitoes, no-see-ums (midges), and greenhead flies, which were no friends of man or beast. By the late 1800s, mosquitoes were identified as vectors for malaria and other diseases, thereby posing a threat to human health and livestock. As early as 1675, marshes were diked in New Castle County, Delaware, to get rid of malaria-carrying mosquitoes while producing agricultural crops. From 1895 to 1917, the Connecticut legislature passed three laws focused on eliminating mosquito-breeding habitat (e.g., "Nuisance Arising from Swampy Places" law and "An Act Providing for the Elimination of Mosquito Breeding Places and Areas"). Towns were mandated to maintain mosquito ditches and tide gates (Casagrande 1997b). During the Great Depression of the 1930s, many salt marshes were ditched through a federal program to keep people working during this economic crisis. Based on the belief that salt marsh mosquitoes bred in pools, the federal government used the Civilian Conservation Corps to grid-ditch marshes (Figure 8.11) and to drain "pools" in the marsh interior. These pools usually contained heavy growths of widgeon-grass, a valuable waterfowl food plant. As it turned out, the salt marsh mosquitoes did not actually breed in these pools, but rather laid their eggs in moist soil around shallower depressions that are flooded by exceptionally high tides (O'Meara 1992). (*Note:* Under optimal conditions, adults develop in 4 to 5 days after hatching.) The ditches made the salt marshes drier and eliminated much valuable waterfowl habitat. By 1940, nearly 90 percent of the salt marshes in the Northeast had been ditched (Bourn and Cottam 1950). This promoted the growth of high-tide bush in many areas and converted short-form smooth cordgrass depressions to salt hay grass turf (Bourn and Cottam 1950; Miller and Egler 1950; Rozsa 1995).

As industrial centers grew in the Northeast, farming became less important in the region (i.e., the nation's farming center eventually relocated to the Midwest) and the salt marshes became less important to the majority of people. At the beginning of the U.S. industrial revolution in the 1700s, some coves and marshes (e.g., Connecticut's central and western shores) were dammed to create millponds where tidal energy was harnessed through a combination of tide gates and sluiceways (Rozsa 1995). People became more removed from the land and became more dependent on technology (e.g., for jobs as well as for bringing food to the masses from agricultural regions). They moved from rural areas to urban areas for better paying jobs and also to gain more leisure time than farming allowed. Salt marshes were no longer the vital economic resource they were in Colonial times. Instead, they became viewed as places for development where ports could be expanded and factories built, or as valuable real estate that could be filled and sold as house-building lots. Significant filling probably commenced in the early 1800s and continued into the early 1900s. For example, over 4,942 acres (2,000 ha) of greater Boston's tidal marshes and flats were filled for urban and industrial development mostly between 1830 and 1930 (Seasholes 2003; Figure 8.12). Much of Newark, New Jersey, was created from filled salt marsh (Crawford et al. 1994). Not all people held a negative view of wetlands. Coastal wetlands were still viewed as important by salt hay farmers, fishermen, and sportsmen, but fewer

Figure 8.11. In the 1930s, extensive ditching of tidal wetlands in the Northeast was done by a civilian workforce in a government-sponsored back-to-work program: (a) workers digging tidal ditches and (b) grid-ditched marsh lined with peat blocks. (Connecticut Department of Environmental Protection)

people worked in these occupations or took part in waterfowl hunting, so the average American's view of coastal wetlands was not a positive one, especially when mosquitoes were identified as disease vectors. Nonetheless, salt hay remained an important commodity into the early 1900s.

After World War II, the U.S. population and economy soared, creating a boom in residential, industrial, and port development from the 1950s to the 1970s. One result of this growth was accelerated coastal wetland destruction. Wetlands were filled for airports and commercial development in major cities (e.g., Boston, New York City, and Washington, DC) and filled with material dredged from neighboring bays and marshes to create real estate for housing developments (e.g., Ferrigno et al. 1973; Sugihara et al. 1979; Tiner 1985b, Daiber 1986; Stalter and Lamont 2002; Hartig et al. 2003). Other wetlands were simply used as dumping grounds for household and commercial wastes. Most of the landfills in New York City were built on tidal wetlands (Walsh and LaFleur 1995). During the 1950s and 1960s, the Hackensack Meadowlands in New Jersey was being filled with about 10,000 tons of garbage per day—enough to fill an average-sized football stadium (Figure 8.13; Sullivan 1998; U.S. Fish and Wildlife Service 2007). By the 1980s, the cumulative national losses of these wetlands resulted in an estimated loss of fisheries worth $208 million annually (National Marine Fisheries Service 1983). Construction of roads across wetlands filled wetlands but also had significant indirect impacts on upstream marshes. Undersized culverts and tide gates beneath roads and railroad beds restricted tidal flow and created conditions favoring the growth of common reed (Roman et al. 1984). An estimated 20 percent of salt marshes in New Hampshire are tidally restricted (Burdick et al. 1997). The upper reaches of many coastal rivers have been impounded.

Historic coastal geodetic survey maps from the late 1700s and early 1800s have been compared with current wetland data to determine historic losses of salt marsh for New England excluding Connecticut (Bromberg and Bertness 2005). Thirty-seven percent of the region's original salt marsh has been lost, with Rhode Island losing the highest percentage—53 percent—since 1832. Massachusetts lost 41 percent (with most of the loss in the greater Boston area—losing 81 percent of its wetlands since 1777), while New Hampshire lost 18 percent and Maine less than 1 percent since 1851.

Another study used aerial photointerpretation and similar map evaluation techniques to assess changes in estuarine marshes in selected areas of Massachusetts

a

b

Figure 8.12. Boston and vicinity: (a) circa 1775 and (b) today. (a: Frothingham 1849)

(Boston Harbor, Cape Cod and the Islands—Nantucket, Martha's Vineyard, and the Elizabeth Islands) from 1893 to 1995. Over 15,000 acres (6,070 ha) of marsh were lost during that period, with nearly 7,000 acres (2,833 ha) in gains resulting in a net loss of 8,285 acres (3,353 ha) (Carlisle et al. 2005). From 1893 to 1952, losses averaged 90 acres per year (36.4 ha/yr), while annual losses increased to 134

Figure 8.13. Changes in the Hackensack Meadowlands (NJ) from 1889 to 1995. While subjected to drainage, diking, and road construction since Colonial times, filling of the Meadowlands to create real estate accelerated after World War II. Wetland conversion slowed since the 1980s and today wetland restoration efforts are under way. (Tiner et al. 2002)

acres (54.2 ha) from 1952 to 1971 and then, with state salt marsh regulations in place, declined to 19 acres per year (7.7 ha/yr) from 1971 to 1995. Boston Harbor lost nearly half (47%) of its late 19th-century wetlands between 1893 and 1952 and 62 percent of its wetlands during the 102-year period. The other nearby areas experienced less estuarine marsh loss during this time: Cape Cod losing 23 percent and the islands 18 percent. Most of the marsh losses from 1952 to 1971 in Boston Harbor were due to filling for real estate, while conversion to open water or other wetlands (e.g., tidal flats) was the major cause of loss during the most recent period. For Cape Cod, the latter causes were most significant during both periods, whereas filling and drainage were the major causes of estuarine marshes loss on the islands since the 1950s.

Heaviest losses due to filling and dredging probably occurred in New Jersey and New York where such activities created thousands of acres of real estate for residential housing after World War II (Figure 8.14; see also Figure 8.4). Since state laws and strengthened federal regulations have been enacted, coastal wetland losses have diminished dramatically. Significant changes in the wetland loss rate are documented for New Jersey, Delaware, Maryland, and the Chesapeake Bay ecosystem (Table 8.9). From the mid-1970s to the mid-1980s, coastal wetland losses in New England have been minimal, ranging from no detectable losses in the Neponset River estuary (MA) and the Cobscook Bay/St. Croix River estuaries (ME) to an 18-acre (7.3 ha) loss in the Casco Bay estuary in Maine (Foulis and Tiner 1994 d,e,f; Foulis et al. 1994; Tiner et al. 1998). Wetland trends for several Maryland counties show similar results for the 1980s. St. Mary's County had no detectable losses of estuarine vegetated wetlands, while Charles County lost only 1.8 acres (0.7 ha), Calvert County 8.4 acres (3.4 ha), and Dorchester County 145.6 acres (58.9 ha) (Foulis and Tiner 1994 a,b,c; Foulis et al. 1995).

Human-caused losses are not the only threat facing northeastern coastal wetlands. Given the level of protection for today's tidal wetlands through various regulations (see Chapter 9), natural processes are the main threat to these wetlands. A recent study documented trends from the 1930s to 1995 for five northeastern marsh systems significant to migratory waterfowl: Nauset Marsh (MA), Forsythe National Wildlife Refuge (NJ), and three marshes of the Virginia Coastal Reserve (Erwin et al. 2004). All areas, except one, experienced net losses during the study interval, with accelerated losses noted from the 1970s to the 1990s. Most of the loss was at the marsh–open water boundary, while Forsythe NWR lost 10 percent of its marsh area due to creek widening and pond expansion. Marsh ponds at Nauset Marsh increased by 100 percent, but 7 percent of the marsh was lost to overwash. Winter ice action may have been responsible for the rise in ponds, but winter herbivory by snow geese at Forsythe NWR may have been another contributing factor. One of the Virginia marshes (Mockhorn Island) increased in size as the marsh expanded into neighboring tidal flats, with most of the growth noted during the last 27 years, whereas the other two Virginia sites experienced small net losses.

The recent effects of sea-level rise and coastal subsidence have been dramatic, especially on Maryland's Eastern Shore (see Figure 12.12). For example, a U.S. Fish and Wildlife Service wetland trends study for Dorchester County found that from 1981–82 to 1988–89, nearly 1,065 acres (431 ha) of estuarine wetlands and 214 acres (87 ha) of nontidal wetlands changed due to sea-level rise: 1,000 acres (405 ha) of estuarine forests with a full canopy of pine changed to 300 acres (121 ha) of stressed pines (canopy mixed with living and dead trees), 667 acres (270 ha) of dead estuarine forests, and 34 acres (14 ha) of salt marsh (Foulis et al. 1995). In addition, 12 acres of dead estuarine forest changed to salt marsh and 17 acres (7 ha) of salt marsh became subtidal habitat. Besides these changes, more than 200 acres (>81 ha) of nontidal forest came under tidal influence and was classified as estuarine forested wetland in 1989. Saltwater intrusion into freshwater streams is also affecting aquatic life with an obvious decline in freshwater fishes (Love et al. 2008). Significant deterioration of tidal marshes due to sea-level rise has been reported for both Chesapeake Bay

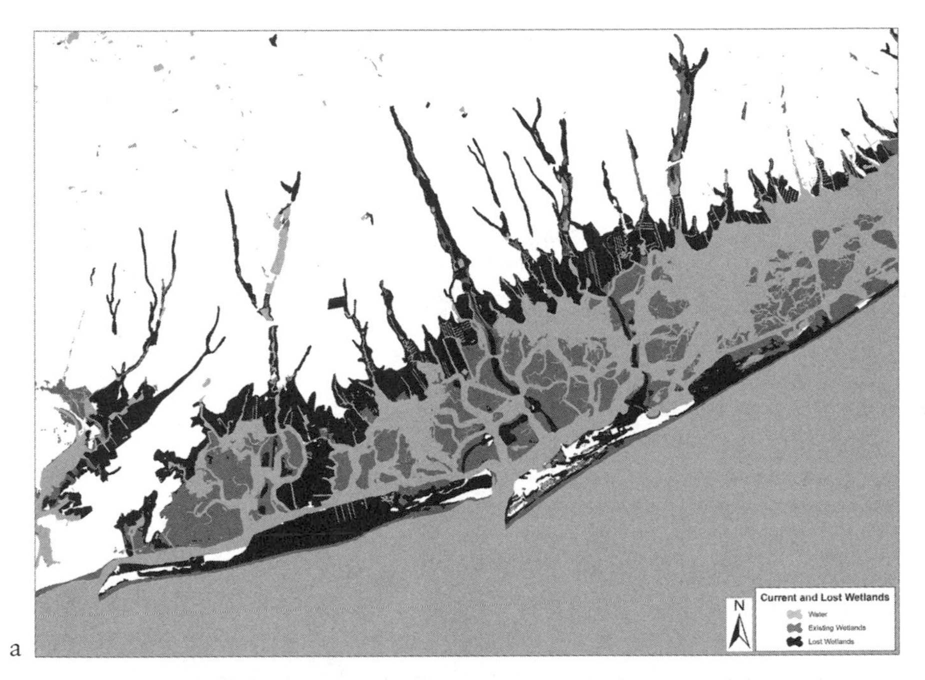

a

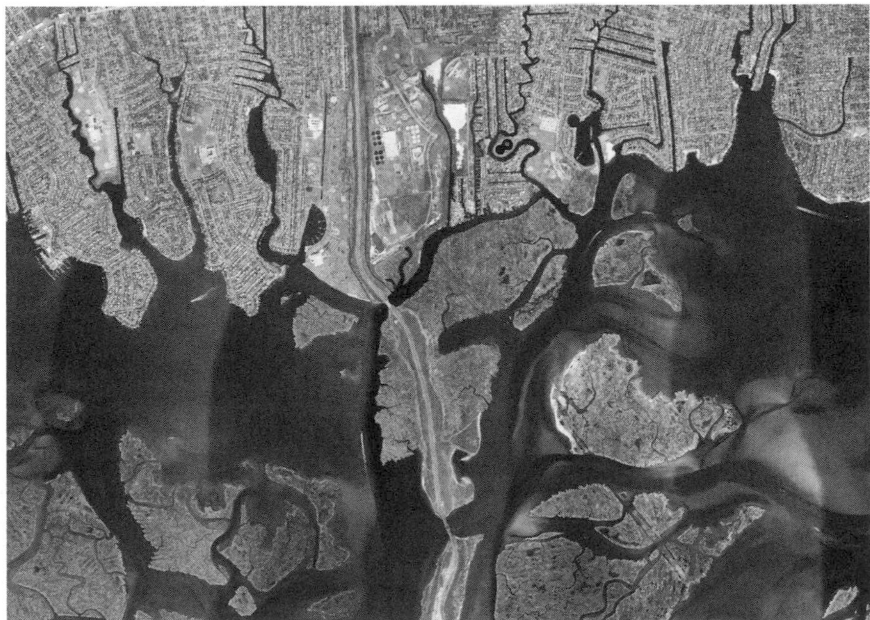

b

Figure 8.14. Historic losses of tidal wetlands in Nassau County, New York: (a) map highlighting filled areas and (b) filled former salt marshes now residential areas and current marshes. (Tiner et al. 2012)

Table 8.9. Estimated trends for coastal wetlands in northeastern United States

Location	Time Interval (years)	Acreage Lost (hectares)	Percentage Lost (%)	Source
Connecticut tidal marshes	1880–1970 (90)	6,242 (2,526)	30	Rozsa 1995
Jamaica Bay, NY marsh islands	1924–1974 (50)	507 (205)	20	Hartig et al. 2002
	1974–1994 (20)	531 (215)	27	Hartig et al. 2002
	1994–1999 (5)	220 (89)	15	Hartig et al. 2002
Long Island, NY tidal marshes	1900–2004	23,650	50	Tiner et al. 2012
Northern New Jersey tidal marshes	1925–1976 (51)	23,300 (9,429)	75	Tiner 1985b
New Jersey tidal marshes	1953–1973 (20)	61,677 (24,960)	23	Ferrigno et al. 1973
Cape May County, NJ estuarine emergent wetlands	1977–1984 (7)	221 (89)	0.3	Smith and Tiner 1993
	1984–1991 (7)	108 (44)	0.2	Smith and Tiner 1993
Delaware estuarine marshes	1938–1973 (35)	8,252 (3,339)	9	Daiber et al. 1976
	1954–1971 (17)	7,548* (3,055)	na	Lesser 1971
	1956–1981 (25)	3,878 (1,569)	6	Tiner & Finn 1986
	1973–1979 (6)	140 (57)	0.1	Hardisky & Klemas 1983; Tiner 1985a
Maryland estuarine vegetated	1955–1978 (23)	10,025 (4,057)	8	Tiner & Finn 1986
	1982–1989 (7)	671 (272)	6	Tiner et al. 1994
Chesapeake Bay estuarine vegetated wetlands	1956–1979 (23)	12,585 (5,093)	9	Tiner and Finn 1986
	1983–1989 (6)	1,145 (463)	0.5	Tiner et al. 1994

Notes: In the 1970s, New York, New Jersey, Maryland, and Virginia passed state laws to protect tidal wetlands and the U.S. federal government strengthened wetland protection under the Clean Water Act. Natural processes are now the leading cause of tidal wetland change in these states. * Based on 440 acres/year loss rate for the designated period.

and Delaware Bay (Kearney et al. 2002). For the latter, in 1984 only 25 percent of the marshes showed signs of significant stress, yet by 1994 the percentage rose to 54 percent. Severe to complete degradation was observed in about 20 percent of the Chesapeake Bay marshes and about 14 percent of the Delaware Bay marshes, with roughly 50 percent and 40 percent of their wetlands showing signs of slight to moderate degradation, respectively. Ongoing changes are also occurring in tidal fresh marshes as sea-level rises (e.g., low marsh replaces high marsh). An assessment of the status of tidal fresh marsh in the upper Delaware River estuary found that although the marsh acreage did not change appreciably from the late 1970s to the late 1990s, there was a noticeable increase in low marsh at the expense of high marsh—from 9 to 34 percent of the marshes (Field and Philipp 2000).

SOUTH ATLANTIC REGION

Instead of diking salt marshes for salt hay and crop production, southern plantation owners focused on tidal swamps, mainly cypress swamps. Using slave labor, they could first harvest the valuable timber products, then through diking convert these tidelands to rice production (Mattoon 1915; Hilliard 1978; Carney 2001; Edelson 2007).

Rice fields were also built in oligohaline (slightly brackish) waters where freshwater could be maintained (Hilliard 1975). From 1800 to 1860, southeastern states produced 90 percent of the U.S. rice crop (Reimold 1976). Along the Atlantic Coast, rice cultivation took place from Cape Fear River (NC) to the Ogeechee River (GA), with South Carolina being the most productive area (DeVoe and Baughman 1987; Figure 8.15). After rice became unprofitable (lack of a labor source with the abolition of slavery), the diked impoundments were either abandoned or maintained and managed for winter waterfowl habitat. Most of South Carolina's coastal impoundments are in private ownership (Tufford 2005) and many are used as hunting preserves. Federal and state wildlife agencies acquired coastal wetlands for waterfowl management. Dikes along former rice fields were maintained or restored, and new impoundment acreage was created by diking salt and brackish marshes.

Significant tidal wetland losses occurred in the region from the 1950s to the 1970s—a time of virtually unregulated wetland exploitation (Koneff and Royle

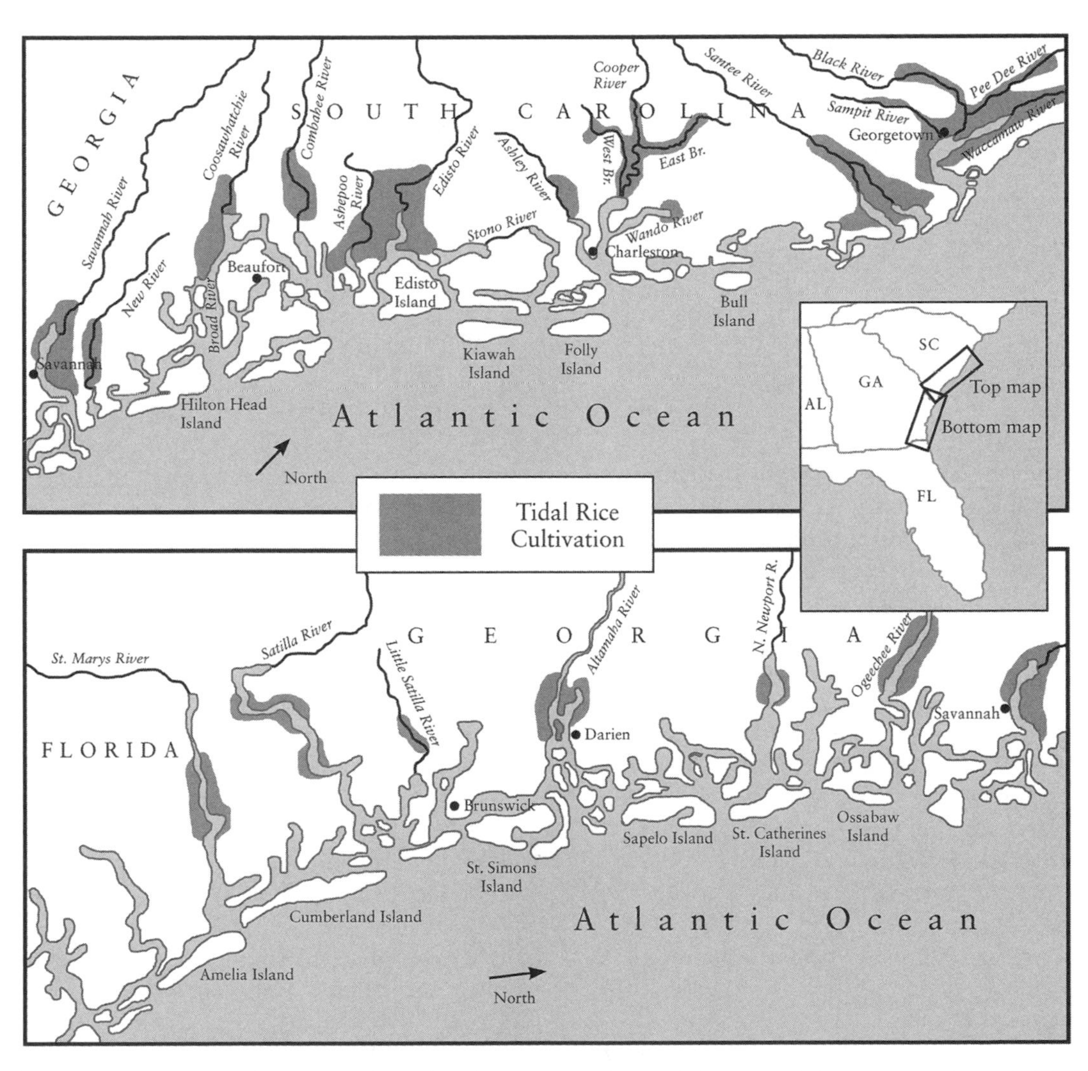

Figure 8.15. Location of tidal rice plantations in South Carolina and Georgia. (Redrawn from Carney 2001)

2004). More than 25,950 acres (10,500 ha) of estuarine marsh was lost from Virginia to Florida during this period, with the region from Charleston, South Carolina, to Savannah, Georgia, suffering the heaviest losses. Much of this loss can be attributed to wetland filling for port development (Charleston and Savannah), residential and resort development (e.g., Hilton Head Island), and disposal of dredged material to maintain the intercoastal waterway and harbors, or to the construction of waterfowl impoundments.

Later, strengthened wetland protection through federal and state regulations slowed but did not necessarily eliminate losses of estuarine wetlands. For example, permitted alterations in the early 1970s accounted for 70 percent of the salt marsh losses in North Carolina between 1970 and 1984 (Stockton and Richardson 1987). Total losses during this period represented 2 percent of the state's salt marshes. During the 1980s, however, the future of the state's salt marshes improved as the rate of marsh loss significantly declined. This was also the case for the South Atlantic region as a whole, as the region experienced an estimated net gain of 494 acres (200 ha) of estuarine marsh from the 1980s to the 1990s (Koneff and Royle 2004). Despite regulations, estuarine wetlands in North Carolina appeared to be altered more frequently than did their freshwater counterparts in the 1990s (Kelly 2001).

Installation of a tide gate across a main distributary of the Savannah River in 1977 caused saltwater to move farther upstream in the river (Pearlstine et al. 1993; Tufford 2005). Marsh vegetation in the Savannah National Wildlife Refuge was adversely affected by this and in 1991 the tide gate was removed. A major diversion of freshwater took place in South Carolina when in 1942 water from the Santee River was diverted to the Cooper River to generate hydropower (Bradley et al. 1990). This action led to dramatic shoaling issues of Charleston Harbor requiring millions of dollars for annual maintenance dredging. To relieve this problem, water was returned to the Santee in 1985. Salinity changes in the Cooper River went from 30.1 parts per thousand (ppt) to 16.8 ppt to 22.0 ppt over this time span (Kjerfve and Magill 1990). Significant changes in tidal vegetation will occur along each river in the future.

FLORIDA

Large-scale drainage of South Florida wetlands began in 1881 with a peak alteration from 1960 to 1970 when the population increased by 35 percent (Patterson 1986). Dredge-and-fill operations were a common practice to create developable land in Florida. Everglades wetlands were reduced by 65 percent, 2,508 acres (1,015 ha) of mangroves were destroyed in Collier County, and 5,300 acres (2,145 ha) of mangroves and uplands on Marco Island were converted by dredge-and-fill operations to finger-canal housing developments (Kushlan 1986; Patterson 1986). Interestingly, the U.S. Army Corps of Engineers' 1976 denial of a permit for mangrove filling on Marco Island was its first major decision to exert its authority over dredge-and-fill activities in wetlands under the Rivers and Harbors Act. This action was instrumental in helping pave the way to better protection for the nation's wetlands (see Figure 9.3). Tremendous population growth in Florida's coastal counties has taken a great toll on estuarine wetlands because of the desire for waterfront property. For example, 81 percent of Tampa Bay's seagrass beds and 44 percent of its mangrove swamps have been destroyed, and 86 percent of the mangroves and 30 percent of the seagrass beds in the Indian River estuary on Florida's east coast are gone (Durako et al. 1988; see Raabe et al. 2012 for a historical perspective on the change from marshes to mangroves in Tampa Bay). Marsh-breeding mosquitoes are a nuisance and are viewed as a public health hazard; yet, more people live on Florida's coast, so mosquito control has been and continues to be an important force altering salt marshes. These activities have included grid-ditching, application of pesticides, and diking and flooding of salt marshes. From 1954 to 1972, many of the Indian River marshes were impounded for mosquito control, which began in the 1920s (Platts et al. 1943). By 1972, 34,820 acres (14,090 ha) of marshes were impounded including nearly all (95%) of Brevard County's salt marshes (Montague and Wiegert 1991). Today, these wetlands are freshwater to slightly brackish impoundments and many are also managed for waterfowl. From the 1950s to the 1970s, peninsular Florida lost 19,270 acres (7,800 ha), with southwest Florida experiencing heavy loss of estuarine marshes (Koneff and Royle 2004). During the next decade (1970s–1980s), 2,700 acres (1,093 ha) of Florida's estuarine vegetated wetlands were filled for urban development (Frayer and Hefner 1991). Another 900 acres (364 ha) became deepwater habitat and 1,000 acres (405 ha) were lost to unspecified development. Despite these changes, estuarine vegetated wetlands experienced a net loss of 700 acres (283 ha) during this decade mainly due to gains from nonvegetated wetlands (1,700 acres [688 ha]) and deepwater habitats (1,900 acres [769 ha]). Since the 1970s and 1980s, federal and state regulations and policies have virtually protected estuarine emergent wetlands from further human alterations. From the 1980s to 1990s, one study reported a net gain of 2,224 acres (900 ha) of estuarine marsh in peninsular Florida (Koneff and Royle 2004).

GULF COAST REGION

The first settlers grazed livestock in Gulf Coast wetlands during the dry season (Reimold 1976). Later, earthen levees were constructed to make more marsh available for cattle and to control flooding, especially along the Mississippi River. Since the early 1900s, waterfowl management practices involved creating impoundments, installing weirs (low dams of steel or wood), blasting potholes, burning vegetation, and planting preferred food plants in coastal marshes (Chabreck 1976).

By the late 1970s, nearly 12 percent of the tidal marshes (about 10,000 acres [4,047 ha]) in the Mississippi Sound were dredged or filled, whereas 22 percent of the Mobile Bay marshes (3,944 acres [1,596 ha]) were destroyed by dredging operations alone (Stout 1984). From 1955 to 1979, Alabama lost 29 percent of its estuarine marsh and 69 percent of its coastal freshwater marsh due to development (industrial, commercial, and residential), erosion, and subsidence (Roach et al. 1987). Construction of the causeway across Mobile Bay as well as bridge-causeway crossings over three rivers (Blakely, Apalachee, and Tensaw) have altered the hydrology by restricting freshwater flows into the bay and reducing movement of saltwater upstream, resulting in a change in vegetation and sedimentation rates (Fearn et al. 2005). More than 8,500 acres (3,440 ha) of coastal wetlands were lost in Mississippi since 1930, mostly due to industrial and urban development. Since 1973, Mississippi tidal wetlands have been regulated under the Coastal Wetlands Protection Act and losses due to human activities have been greatly reduced (Demas and Demcheck 1996b). From 1979 to 1988, there was no loss of estuarine marsh (McPherson 1996). Recent casino development along the Gulf Coast has filled some tidal marshes (Shafer et al. 2007).

Louisiana has suffered the greatest coastal wetland loss of any state in the nation (Table 8.10). Its location at the mouth of the Mississippi River—the Mississippi Delta—produced a vast expanse of tidal marshes. As much as 4.5 million acres (1.8 million ha) may have been present before European settlement (Smith 1993). Since deltaic sediments compact over time, contributing to natural subsidence, sediment deposition on the floodplain generally maintained and created marshes for thousands of years (Törnqvist et al. 2008; Blum and Roberts 2009). However, when the Mississippi River was leveed after the disastrous 1927 flood,

Table 8.10. Persistent changes in wetlands of coastal Louisiana from 1932 to 2010

Basin	Area of Land Loss	Area of Land Gain	Total Change
Atchafalaya Delta	16.49	29.95	+13.46
Barataria	421.71	5.91	–415.80
Breton Sound	160.87	6.03	–154.84
Calcasieu-Sabine	198.94	9.02	–189.92
Mermentau	134.37	6.87	–127.50
Mississippi River Delta	152.02	23.74	–128.28
Pontchartrain	179.25	8.31	–170.94
Teche-Vermillion	64.61	3.85	–60.76
Terrebone	459.99	10.42	–449.57
Total	1,788.24	104.11	–1,684.13*

Source: Couvillion et al. 2011.
Notes: Values are square miles; to compute km^2 multiply values by 2.59. All regions experienced net losses except the Atchafalaya Delta. The report indicated a net loss of 1,883 square miles based on somewhat different calculations. The 2010 wetland area is 75% of the 1932 total. * Totals may differ slightly from sums due to computer round-off procedures.

sediment that would normally replenish the delta marshes was transported directly into the deep waters of the Gulf of Mexico. This, plus the construction of upstream dams, robbed the marshes of sediment they needed to raise their elevations to keep pace with rising sea level and seasonal flooding. The protection provided by the levees also encouraged more development of coastal wetlands. In the Lake Pontchartrain area alone, construction of continuous levees along the Mississippi River may be responsible for a loss of 90,000 acres (36,420 ha) of wetlands (Lopez 2009). Waterborne transportation was promoted by channelizing the Mississippi River and its distributaries making New Orleans a major U.S. port. Degradation of tidal marshes was further exacerbated by oil and gas production, which accelerated coastal subsidence (Morton et al. 2002, 2003; Morton and Bernier 2010), and by dredging canals through the marshes to access drilling sites in the marshes and to install pipelines (Turner 1987; see Figure 8.9). Canal digging did more than simply destroy the marsh through excavation. The dredged material was placed directly on marshes, creating spoil banks that restricted drainage, interrupted the natural flow of the tides (e.g., preventing sheet flow across marshes), and prevented sedimentation from floods. Also the canals were subject to erosion, and over time they have greatly expanded in width. The canals also caused a salinity change in the marshes by moving saltwater farther upstream into freshwater marshes. Salt-sensitive species died back, leaving huge open water areas in their place. Some canals routed freshwater runoff away from the marshes into estuarine lakes and bays (Gosselink 1984). The Mississippi River Gulf Outlet, a canal dredged through the Breton Sound Basin in 1963, increased salt intrusion that killed thousands of acres of freshwater forested wetlands (Day et al. 2007). All the above factors (natural and human-induced) are operating in complex ways to cause extensive losses in coastal Louisiana. Many waterfowl and wildlife impoundments have also been constructed in the tidal marshes, while others were built to raise crawfish. Since European settlement, Louisiana may have lost one-quarter to one-half of its salt, brackish, and freshwater coastal marshes, while only 25 to 50 percent of its intermediate marshes remain (Smith 1993). Louisiana's recent loss of coastal wetland comprised 67 percent of the nation's total loss (Johnston et al. 1995). From 1978 to 1990, the loss was 290,432 acres (117,500 ha), representing an annual loss rate of 24,203 acres (9,795 ha) or 38 square miles (98 km^2). Between 1956 and 1978, net wetland losses were even greater—661,700 acres (267,800 ha), for a yearly loss of 30,000 acres (12,140 ha) or 47 square miles (121 km^2). Hurricanes Katrina and Rita converted an estimated 217 square miles (562 km^2) of wetland to open water (Barras 2006). The state is expected to lose 500 square miles (1,295 km^2) over the next 50 years (Barras et al. 2003) due to ongoing shifts in the ecosystem. Despite these enormous losses, coastal marshes are expanding in Atchafalaya Bay due to sedimentation brought in by re-routing Mississippi floodwaters to the Atchafalaya River Basin (Figure 8.16; Demas and Demcheck 1996a). While heavy losses continue, it appears that the phenomenal rate of wetland loss observed during the 1970s has slowed markedly. The most recent assessment indicates that Louisiana lost wetlands at a rate of 16.57 square miles per year (42.9 km^2) from 1985 to 2010 (Couvillion et al. 2011). This rate is equivalent to losing a wetland the size of a football field every hour.

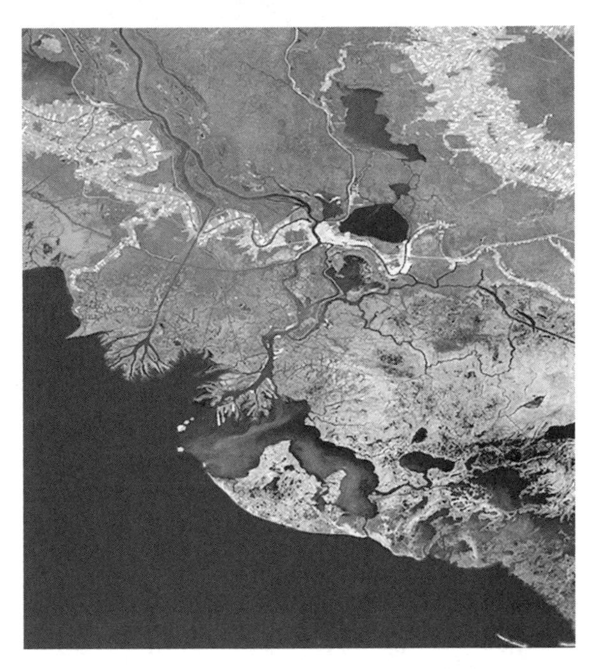

Figure 8.16. New deltas are forming in Atchafalaya Bay, one at the mouth of the Atchafalaya River (on right) and the other at the Wax Lake Outlet.

From the mid-1950s to 1992, Texas lost nearly 10 percent of its estuarine wetlands with marshes falling by 8 percent and tidal flats by 13 percent (Moulton et al. 1997). During this time, a slight increase (8%) in estuarine scrub-shrub wetlands was detected. Most of the marsh loss was attributed to conversion to open water (19,931 acres [8,066 ha]). The bulk of this loss was observed from Freeport to Port Arthur where groundwater withdrawal and oil and gas extraction are causing faulting and land subsidence, which probably induced drowning and erosion of salt marshes. The rest of the lost marshes were converted to freshwater marshes (9,238 acres [3.738 ha]), reservoirs (7,023 acres [2,842 ha]), or upland (6,291 acres [2,546 ha]). Four causes were responsible for the loss of estuarine tidal flats: other upland development (15,805 acres [6,396 ha]), conversion to estuarine marsh (14,376 acres [5,818 ha]), rural development (4,079 acres [1,651 ha]), and conversion to freshwater marsh (3,686 acres [1,492 ha]). The other upland development category includes deposition of dredged material along the Gulf Intracoastal Waterway and ship channels as well as road, levee, and other filling. Nearly all of the gain in estuarine shrub swamps (2,403 acres [973 ha]) came from estuarine marsh. A wetland trend study for Galveston Bay found that by 1989, more than 15,000 acres (6,070 ha) of estuarine marsh present in the 1950s had been converted to open water or intertidal flats (White et al. 1993). This area is experiencing significant subsidence due to sea-level rise. Some of this subsidence may result from human activities, mainly withdrawal of groundwater, but also through oil and gas extraction and sulfur mining around salt domes on the Texas coast. Active surface faults may also be contributing to subsidence in this area.

CALIFORNIA

Tidal wetlands in California have been subject to filling for urban development and to diking for a variety of purposes including creating pastures, arable lands, evaporation ponds for salt production, waterfowl impoundments, and to prevent oil spills from reaching estuaries. In addition, the hydrology of many tidal wetlands has been severely compromised by major freshwater diversions, levee construction, channel deepening, road crossings, and other development (see Josselyn 1983 and Atwater et al. 1979 for reviews of the history of San Francisco Bay estuary tidal wetlands). During the California Gold Rush (beginning in 1849), the greater San Francisco Bay region possessed an estimated 543,600 acres (220,000 ha) of tidal marshes: 345,900 acres (140,000 ha) in the Sacramento–San Joaquin Delta and 197,700 acres (80,000 ha) in San Francisco, San Pablo, and Suisan Bays (Gilbert 1917; Atwater et al. 1979; Goals Project 1999; Drexler et al. 2009). Agriculture was the first main use of these tidal marshes with marshes along San Pablo and Suisun Bays and the Sacramento–San Joaquin Delta diked for cropland or pasture. Most of this diking took place between 1860 and 1910, stimulated by passage of the Swamp Lands Act in which the federal government gave marshes and swamps to a number of states on the provision that the states reclaim these lands for agriculture. In the late 1800s, wetlands along the south portion of San Francisco Bay were diked for salt production, with about 36,000 acres (14,570 ha) converted to evaporation ponds (Goals Project 1999; see Collins and Grossinger 2004 for review of the history of South Bay wetlands). New marshes also formed at this time as increased sediment input from hydraulic mining in the Sierra Nevada Mountains caused shoaling of embayments, allowing marshes to colonize former mudflats (Gilbert 1917). As much as one-third of the San Francisco estuary marshes were created by this process (Atwater et al. 1979). The current acreage of tidal wetlands in the San Francisco Bay estuary amounts to about 40,000 acres (16,000 ha), of which only 16,000 acres (6,475 ha) are historic marshland (Goals Project 1999; Figure 8.17). Consequently, only 8 percent of the bay's historic marshes remain. In the Sacramento–San Joaquin Delta only a few marshes remain as islands in a virtual sea of farmland that characterizes the delta (Drexler et al. 2009). Construction of roads and railroads across wetlands has restricted tidal flow across many California tidal wetlands, thereby reducing tidal flushing. Development around these wetlands has further impaired their quality. Urban wetlands suffer from disposal of wastewater, increased freshwater inflow from street drains, oil spills, and human-induced sediment deposition. Southern California's coastal wetlands have been greatly modified with large portions of many tidal marshes filled for development. In the mid- to late 1800s, there were about 49,400 acres (20,000 ha) of estuarine habitats. An atlas of historic coastal wetlands of Southern California graphically shows how these wetlands have been greatly modified since the 1800s (Grossinger et al. 2011). Filling of San Diego Bay commenced in 1888 and accelerated just before World War II (Smith 1977). By the 1970s, approximately 27 percent of the bay had been filled with material dredged from bay bottom and most of the filled land was occupied by salt marshes and tidal flats.

PACIFIC NORTHWEST REGION

Only a small percentage of the tidal wetlands in the Pacific Northwest remain unchanged (Seliskar and Gallagher 1983). Upon settlement of the region, tidal wetlands were used in their natural state as pastures, as was the case elsewhere (Nesbit 1885). Since the 1800s, approximately two-thirds (68%) of Oregon's tidal wetlands may have been destroyed with estimated losses of 50,436 acres (20,410 ha) (Good 2000; Table 8.11). In non-urban areas, most of the estuarine wetlands were diked ("reclaimed") and partly drained to create pastures and hayland. Diking occurred at various times. For example, in Washington from 1880 to 1940,

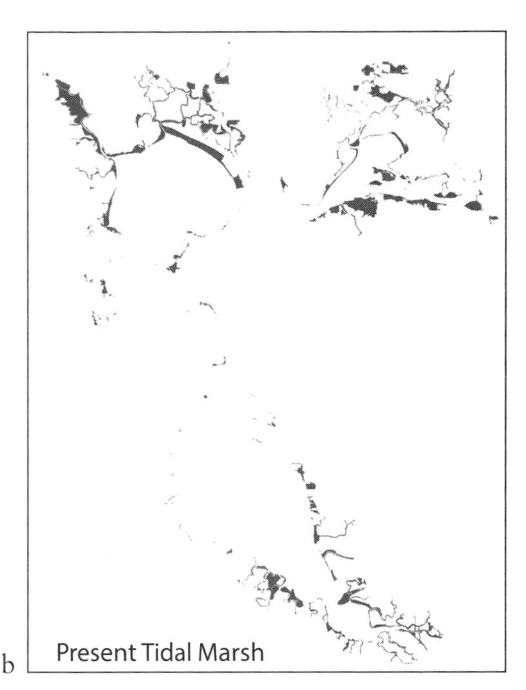

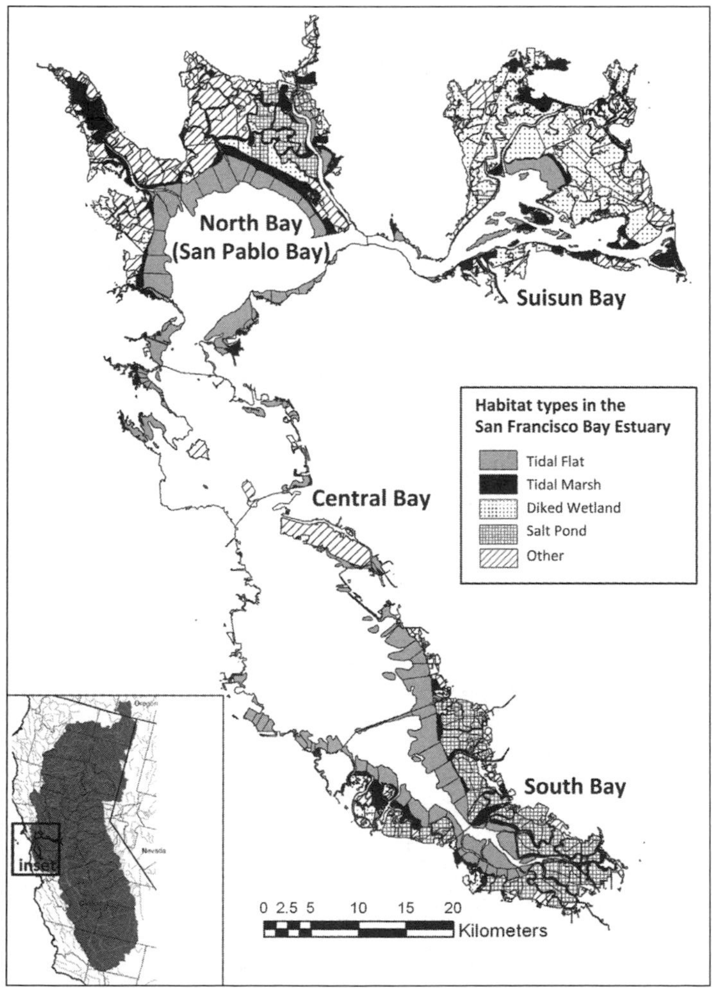

Figure 8.17. Tidal wetlands of the San Francisco Bay Estuary: (a) in the late 1800s; (b) today; (c) composite. (Goals Project 1999; courtesy of the San Francisco Estuary Institute—see their EcoAtlas for color maps at: www.sfei.org/ecoatlas/)

Snohomish River wetlands were diked, whereas significant losses of wetlands in Willapa Bay and Grays Harbor took place between 1933 to 1974 and 1960 to 1973, respectively (Boulé et al. 1983). Dredge-and-fill operations in urban areas have caused the loss of many wetlands in river deltas (Everett—Snohomish River; Seattle—Duwamish River; Tacoma—Puyallup River; Table 8.12). Logging of forests and subsequent erosion led to increased sedimentation in Coos Bay, which necessitated dredging operations to maintain navigable waterways. Dredged material was deposited in the marshes. Only 20 percent of the Skagit River delta marshes of Puget Sound remain, while 60 percent of the Willapa Bay tidal marshes still exist. For the Duwamish River (Seattle area), over half of the tidal swamps were destroyed between 1854 and 1908 (Blomberg et al. 1988). From 1908 to 1940, dredging to create navigation channels for deep-draft ships took place. Dredged material was deposited on the remaining tidal swamps and on 60 to 70 percent of the tidal marshes, flats, and shallow water zones. By 1985, only 2 percent of the original estuarine habitat remained. Until the 1970s, many of Oregon's tidal wetlands were used to store logs floated downriver by log drives. The history of Puget Sound tidal wetlands has been painstakingly reconstructed and detailed through analysis of historic maps (Collins and Sheikh 2005). Today's wetland acreage in the sound amounts to 17 to 19 percent of their extent at European settlement, with Whidbey Basin losing the most acreage (nearly 37,000 acres [15,000 ha])—only 20 percent of the historic acreage remains—and with most sub-basins losing more than half of their tidal wetland acreage. While 70 percent of Oregon's tidal marshes may have been lost, destruction of tidal swamps has been even greater with 90 to 95 percent lost since the 1850s (Brophy 2005). Since 1870, more than half of the tidal marshes and swamps in the lower Columbia River have been lost due to diking, draining, filling, dredging, and flow regulation (Jennings et al. 2003).

Table 8.11. Changes in extent of tidal marshes and swamps in Oregon's estuaries since 1870

Estuary	1870 Acreage	1970 Acreage	Acreage Lost, Filled, or Diked	Percentage Loss, 1870–1970 (%)
Alsea	1125	460	665	59
Chetco	9	4	5	56
Columbia	46,200	16,150	30,050	65
Coos Bay	5,087	1,727	3,360	66
Coquille	4,876	276	4,600	94
Necanicum	147	132	15	10
Nehalem	2,095	524	1,571	75
Nestucca	2,365	205	2,160	91
Netarts	244	228	16	7
Rogue	74	44	30	41
Salmon	551	238	313	57
Sand Lake	471	462	9	2
Siletz	675	274	401	59
Siuslaw	2,002	746	1,256	63
Tillamook	4,158	884	3,274	79
Umpqua	2,419	1,201	1,218	50
Yaquina	2,114	621	1,493	71
Total	74,612	24,176	50,436	68

Source: Good 2000.
Note: To convert acres to hectares multiply by 0.4047.

Table 8.12. Wetland losses in Washington estuaries

Estuary	Tidal Wetland Acreage		Acres Lost	Percentage Lost (%)
	Historic	Circa 1980s		
Grays Harbor	48,190	33,610	14,580	30.3
Skagit River	7,166	2,965	4,201	58.6
Snohomish River	9,637	2,471	7,166	74.4
Duwamish River	4,448	445	4,003	89.9
Puyallup River	2,471	10	2,461	99.6
Nisqually River	1,409	1,013	396	28.1
Skokomish River	519	346	173	33.3

Source: Simenstad and Thom 1992.
Note: To convert acres to hectares multiply by 0.4047.

BRITISH COLUMBIA

Like similar wetlands in the U.S. Pacific Northwest, many tidal marshes were diked for agriculture in the Fraser River Delta. More recently, estuarine wetlands have been negatively impacted by residential, shipping, and industrial development (Glooschenko et al. 1988). Eighty percent of the wetlands in the Fraser River Delta have been lost or severely degraded (Rubec 1994). From 1967 to 1982, about 7 percent of the tidal and riparian marshes in the southwestern Fraser Lowland were converted to urban and agricultural land, but 51 percent of the wetlands were protected for conservation and recreational purposes (Pilon and Kerr 1984). Port development at Prince Rupert and the Kitimat Estuary destroyed coastal wetlands. About 60 percent of the estuarine marshes in the Strait of Georgia have been lost (Levings and Thom 1994).

ALASKA

Coastal wetlands in Alaska remain largely intact due to their remoteness, and they probably experience more change due

Figure 8.18. Alaskan tidal marsh dominated by Ramensk's sedge (*Carex ramenskii*) near Anchorage. (Jerry Tande)

to coastal processes and tectonic activity (Figure 8.18). Natural changes in the central part of the Yukon–Kuskokwim Delta accounted for small losses of some tidal wetland types (0.5% of tidal flats and 0.3% of brackish sedge meadows) and minor gains in slightly brackish sedge meadow (0.7%) and brackish drained pond barrens (0.5%) from 1948 to 2008 (Jorgenson and Dissing 2010). Like other coastal cities, port development and transportation projects have negatively impacted tidal wetlands in Anchorage and Juneau, while marinas and docking facilities have altered tidal wetlands in smaller coastal towns. Anchorage's Westchester Lagoon (an estuarine embayment) was dammed in the 1970s to create a freshwater lake. Between 1950 and 1990, 137 acres (55 ha) of estuarine wetlands were filled in Anchorage, yet despite these losses, the area experienced a 98 percent increase (a 781-acre [311 ha] gain) in estuarine marsh area (U.S. Fish and Wildlife Service 1993). This enormous gain was attributed to land subsidence resulting from the 1964 earthquake—the surfaces of palustrine emergent wetlands dropped into the intertidal zone initiating conversion to salt marshes.

A rather unique impact to tidal wetlands takes place in the Eagle River Flats where the military uses the area for artillery practice (G. Tande, pers. comm. 2010).

Further Readings

From Marsh to Farm: The Landscape Transformation of Coastal New Jersey (Sebold 1992)

Discovering the Unknown Landscape (Vileisis 1997)

"The full circle: a historical context for urban salt marsh restoration" (Casagrande 1997b)

Saving Louisiana? The Battle for Coastal Wetlands (Streever 2001)

Marshlands: Four Centuries of Environmental Change on the Shores of the St. Lawrence (Hatvany 2003)

Gaining Ground: A History of Landmaking in Boston (Seasholes 2003)

History, Ecology, and Restoration of a Degraded Urban Wetland (Clemants et al. 2004)

The Hackensack Meadowlands Initiative (U.S. Fish and Wildlife Service 2007)

"Centuries of human-driven change in salt marsh ecosystems" (Gedan et al. 2009)

Human Impacts in Salt Marshes: A Global Perspective (Silliman et al. 2009)

Salt Marshes: A Natural and Unnatural History (Weis and Butler 2009)

9 Tidal Wetland Conservation and Management

After more than two centuries of being considered wastelands and sites in need of "reclamation," wetlands are now recognized as valuable natural resources that must be conserved. Wetland conservation involves a combination of efforts that result in protecting, enhancing, restoring, and creating wetlands. Tidal wetlands were the first wetlands to receive protection through state regulations. Since the 1950s and 1960s, the status of coastal wetlands in the United States has greatly improved due to environmental laws, regulations, policies, expanded wetland acquisition, increased public awareness and participation in conservation, and more recently to wetland restoration initiatives (see Table 9.1 for timeline of important events in tidal wetland conservation). In the past two decades, Canada's tidal wetlands have been receiving more attention from the regulatory perspective. Federal and state/provincial governments of the United States and Canada strive to have native peoples, nongovernmental organizations, industries, local communities, and the general public participate in their wetland conservation, environmental education, and public outreach efforts. In this chapter I briefly describe tidal wetland protection through acquisition, regulation, and government policies as well as describing wetland management for wildlife and mosquito control, with emphasis on activities in the United States (for more information visit agency websites). In later chapters I address other aspects of wetland conservation including wetland identification, delineation, functional assessment, restoration, creation, and monitoring.

Public Trust Doctrine

The public trust doctrine has a long history dating back to Roman times. Under Roman law, the air, the rivers, the seas, and the seashores were not capable of being owned by private parties. They were public resources to which everyone had right of access.

In Colonial times, America's coastal waters and their adjacent tidelands were recognized as important public resources in accordance with the public trust doctrine of English law. When the Magna Carta of 1215 was signed, British tidal marshes became common property as the King of England could no longer grant private use of these wetlands. In 1667, tidal marshes were reinstated as the King's property but with the condition that they be held in trust for the good of the people (Kavenagh 1980). The public trust doctrine meant that while the Crown held transferable title to such lands, the public had the right to navigate, fish, and hunt fowl in these areas. Even when such lands were granted to private individuals, there was a presumption of priority given to public usage. For example, when the Massachusetts Bay Colony in the mid-1600s issued title to tidal lands to individuals for private wharf construction

Table 9.1. Timeline of numerous significant events impacting tidal wetland conservation

Date	Event
1849–1860	Swamp Lands Act gave land to 15 states (excluding the original 13 states) to be drained and reclaimed (LA, AL, AR, CA, FL, IL, IN, IA, MI, MS, MO, OH, WI, MN, and OR), and thereby promoted destruction of tidal wetlands.
1899	Rivers and Harbors Act enacted giving U.S. Army Corps of Engineers responsibility for regulating dredging, filling and construction of structures in navigable waters. (*Note:* Although not originally used to conserve wetlands, it later became the first federal legal instrument for preventing dredging and filling of tidal wetlands.)
1903	President Theodore Roosevelt issued an executive order to establish Pelican Island (FL) as the "first national bird reservation;" over 50 more reservations were authorized similarly by 1909.
1929	Migratory Bird Conservation Act authorized acquisition of wetlands for waterfowl habitat (no funding provided).
1933	President Franklin D. Roosevelt established the Civilian Conservation Corps to help relieve unemployment caused by the Great Depression, putting young men to work on restoring the nation's natural resources through public works (e.g., forestry, soil erosion control, flood control, and other projects); unfortunately for tidal wetlands, one of the major public work projects involved ditching of salt marshes for mosquito control.
1934	Migratory Bird Hunting Stamp Act required hunters to purchase a duck stamp for waterfowl hunting annually, and revenues were dedicated to purchase or lease waterfowl habitat (including wetlands).
	Fish and Wildlife Coordination Act required federal agencies (Agriculture and Commerce) to provide technical assistance to states and other federal agencies to protect and increase game and fur-bearing animals and to study the effects of pollution on wildlife resources. (*Note:* This act was later amended to require that federal regulatory agencies consult with federal and state fish and wildlife agencies on the impact of proposed work in wetlands. See 1946.)
1946	Fish and Wildlife Coordination Act amended to require federal agencies to consult with the U.S. Fish and Wildlife Service and state Fish and Wildlife agencies for projects requiring federal permit or license; objective is to prevent loss of and damage to wildlife.
1954	U.S. Fish and Wildlife Service published *Wetlands of the United States* (Circular 39) to report on the current status of wetlands in coterminous U.S. with emphasis on wetlands important to waterfowl.
1958	Fish and Wildlife Coordination Act amended to require that fish and wildlife resources be given equal consideration when planning federal water resource development projects; development agencies must consult with the U.S. Fish and Wildlife Service, National Marine Fisheries Service, and the state wildlife agency on the impacts to fish and wildlife resources of proposed federal projects or projects requiring federal permits.
1959	New York passed the Long Island Wetlands Act, authorizing the state to assist towns and counties in preserving tidal wetlands through cooperative agreements, technical assistance, and cost-share reimbursable funding (law later repealed and included in the Tidal Wetlands Act of 1973).
1960s	U.S. Fish and Wildlife Service published reports on threats to tidal and nontidal wetlands; these reports alerted the public to the rapid destruction of tidal wetlands and may have stimulated public concern about accelerated wetland losses.
	Scientific papers produced by the University of Georgia's Sapelo Island Marine Institute were instrumental in documenting the value of salt marshes to estuarine fisheries and increasing the public's knowledge of tidal marsh ecology.
1963	Massachusetts passed first state wetland protection law in the nation to protect salt marshes—the Jones Act.
	Recreation Coordination and Development Act established Land and Water Conservation Fund that provided funds to states for outdoor recreation projects and to federal agencies for land acquisition (funds used to purchase wetlands).
1965	California passed law that established the San Francisco Bay Conservation and Development Commission as a temporary agency to address public concern for the bay status; the commission became permanent in 1969 and now regulates all filling and dredging activities in the bay.

Table 9.1. *(continued)*

Date	Event
1966	National Wildlife Refuge System Administration Act—organic act for the National Wildlife Refuge System (amended in 1997).
1967	New Hampshire passed law to protect coastal wetlands.
1968	Land and Water Conservation Fund Act established a program to provide funding to states and local governments to acquire lands (including wetlands) for outdoor recreation and open space.
	Rhode Island passed law to protect coastal wetlands.
1969	National Environmental Policy Act required federal agencies to prepare environmental impact statements for proposed projects.
	Connecticut passed tidal wetlands protection law including a moratorium on coastal wetland alteration until mapping of jurisdictional wetlands was completed.
	California made the San Francisco Bay Conservation and Development Commission permanent and adopted the San Francisco Bay Plan as state law.
1970	Maryland and New Jersey passed legislation to protect tidal wetlands and initiate wetland mapping.
	Georgia passed state law to protect coastal marshlands (including mudflats and bottoms of tidal waters).
	Court decision—Zabel v. Tabb—supported U.S. Army Corps of Engineers position to deny a permit to fill mangroves on environmental grounds; Corps begins to seriously consider environmental impact in Section 10 permit program.
1971	Oregon enacted the Fill Law (now the Removal-Fill Law) to regulate filling in estuaries.
	Washington passed Shoreline Management Act, which included a permit program and development of local master plans for shoreline conservation.
	North Carolina passed dredge-and-fill law requiring permit for such activities in estuarine waters, tidelands, and marshlands, and state-owned lakes.
1972	Federal Water Pollution Control Act (later renamed Clean Water Act) amendments gave Corps of Engineers and Environmental Protection Agency authority to regulate deposition of fill material and other pollution in waters of the U.S. (Section 404); Section 401 gave water quality certification authority to states, and it requires any applicant for a federal permit or license involving discharges into navigable waters to obtain a water quality certificate from the state to ensure that such discharge complies with state water quality standards.
	Coastal Zone Management Act provided federal funding to states to establish coastal zone management programs to protect and conserve coastal wetlands and other resources.
	Endangered Species Act enacted to protect ecosystems that support endangered species; many such species require wetlands for survival.
	Virginia passed tidal wetland law.
	California Coastal Commission established (by voter initiative on Proposition 20) to protect, conserve, restore, and enhance environmental and human-based resources in the coastal zone.
1973	Delaware, New York, and Mississippi enacted laws to protect coastal wetlands (the Delaware law also protected contiguous nontidal wetlands, 400 acres or more in size; the Mississippi law referred only to public-owned tidal wetlands).
	Oregon law established the Oregon Statewide Planning Program, which among other things classified estuaries as natural, conservation, or development types and required preparation of estuary plans to address estuarine resources (including tidal wetlands) that are now part of city and county comprehensive plans.
1974	U.S. Fish and Wildlife Service initiated plans to conduct a national wetlands inventory.
	Court decision—United States v. Holland—affirmed the U.S. EPA's position that the Corps of Engineers had the authority to regulate deposition of fill in wetlands above the mean high water mark and that intertidal wetlands were intended to be regulated under the federal Clean Water Act.

Table 9.1. (*continued*)

Date	Event
	Nation's first estuarine sanctuary established in South Slough Estuary, Coos Bay, Oregon.
	North Carolina passed Coastal Area Management Act that provided regulation over tidal wetlands, allowing impacts only from projects deemed to have overriding public benefit and appropriate mitigation.
	California passed Suisun Marsh Preservation Act that required formulation of a marsh protection plan.
1975	Court decision—Natural Resources Defense Council vs. Callaway—determined that wetlands are part of the waters of the U.S. and that the Corps of Engineers must regulate activities impacting wetlands under Section 404 of the Clean Water Act; the Corps began regulating tidal wetlands above the mean high water mark as well as most nontidal wetlands.
	Corps of Engineers and EPA published proposed wetland definition for regulation in the Federal Register (July 25, 1975).
	U.S. Fish and Wildlife Service proposed initial wetland classification system (undergoes several iterations prior to final publication in 1979).
	Maine passed coastal wetlands act.
1976	Corps of Engineers denied permit to fill mangroves at Marco Island—the agency's first major decision to exercise its regulatory authority over dredge-and-fill developments.
	U.S. Fish and Wildlife Service used interim wetland classification to begin national wetlands inventory; the National Wetlands Inventory (NWI) would serve as primary source of wetland information for many states, especially those that did not conduct their own wetland inventory.
	Washington became first state to receive federal approval of its coastal zone management program.
	Oregon adopted statewide planning requirements for estuary planning at the local level.
	California enacted the California Coastal Act to provide long-term protection for the state's coastline.
1977	Corps of Engineers and EPA finalized wetland definition based on review (July 19, 1977).
	President Carter issued Executive Orders 11990 (Protection of Wetlands) and 11988 (Floodplain Management) requiring all federal agencies to minimize wetland destruction or degradation and avoid development on floodplains when carrying out their duties including providing technical assistance.
	Clean Water Act amendments added provisions to delegate regulatory authority to states and directed the U.S. Fish and Wildlife Service to provide technical assistance to states in developing regulatory programs for regulating the discharge of dredge-and-fill materials into U.S. waters including wetlands and also authorized the U.S. Fish and Wildlife Service to conduct a national inventory of wetlands.
	South Carolina and Alaska passed state coastal management acts.
1978	Louisiana and Florida passed state coastal resources management act.
	South Carolina published rules and regulations for permitting in critical areas of the coastal zone.
1979	U.S. Fish and Wildlife Service published final version of its wetland classification system, *Classification of Wetlands and Deepwater Habitats of the United States,* for conducting national wetlands inventory (later adopted as the national standard for wetland inventories and for reporting on the status of the nation's wetlands by the Federal Geographic Data Committee).
	Oregon amended Removal-Fill Law to require mitigation for wetland alteration.
1981	Court decision—Avoyelles Sportsman's League, Inc. vs. Alexander Marsh (in Louisiana)—established three criteria (hydrophytic vegetation, hydric soils, and wetland hydrology) for identification of federally regulated wetlands.
1982	Coastal Barrier Resources Act designated undeveloped barrier beaches and related habitats as part of the Coastal Barrier Resources System; such areas are ineligible for federal financial assistance.

Table 9.1. *(continued)*

Date	Event
1984	U.S. Fish and Wildlife Service published national report, *Wetlands of the United States: Current Status and Recent Trends,* documenting wetland status, trends, and values and reporting that over half of the wetlands in the lower 48 states have been lost.
	Florida enacted wetlands protection act.
1985	U.S. Supreme Court decision—United States vs. Riverside Bayview Homes (Ohio)—affirmed Corps of Engineers jurisdiction in wetlands adjacent to navigable waters.
1986	Emergency Wetlands Resources Act authorized purchase of wetlands through the Land and Water Conservation Fund, required the Secretary of Interior to report to Congress at 10-year intervals on the status and trends of the nation's wetlands, and established completion dates for the National Wetlands Inventory (September 30, 1998, for lower 48 states; September 30, 1990, for Alaska and non-contiguous areas; funding not provided).
	North American Waterfowl Management Plan signed by U.S. and Canada; initiated joint ventures to restore and protect waterfowl habitat (Mexico became a signatory in 1994).
1987	Corps of Engineers published its wetland delineation manual as "guidance" for determining areas subject to jurisdiction under Section 404 of the Clean Water Act; use of the manual was discretionary.
1988	National Wetlands Policy Forum (government–business–private citizens–academics–bipartisan group) recommended that the federal government adopt a policy of "no-net-loss of wetlands."
	EPA published its wetland delineation manual for determining jurisdictional wetlands.
	Maine's Natural Resources Protection Act includes provisions for regulating uses of wetlands including coastal wetlands.
1989	Federal interagency wetland delineation manual published; developed by four federal agencies (Corps of Engineers, EPA, Fish & Wildlife Service, and Soil Conservation Service) to serve as the technical standard for wetland identification (January 10, 1989).
	Corps of Engineers and EPA issued a memorandum of agreement adopting the interagency manual as the national standard for identifying and delineating wetlands subject to Clean Water Act jurisdiction (January 20, 1989); first mandated standard for delineation of federally regulated wetlands—before this, each Corps district had its own local procedures for identifying jurisdictional wetlands leading to inconsistencies across the country.
	President George H. W. Bush's administration announced federal policy of no-net-loss of wetlands.
	North American Wetlands Conservation Act provided matching funds for organizations and individuals to acquire, restore, and enhance wetlands.
1990	Coastal Wetlands Planning, Protection, and Restoration Act established National Coastal Wetlands Conservation Grant Program providing funds for acquisition, restoration, and enhancement of these wetlands.
1991	EPA proposed revisions to interagency manual; thousands of comments recommended rejecting this proposal since the requirements would significantly limit the number of wetlands subject to federal regulation.
1992	Energy and Water Development Appropriations Act prohibited the Corps of Engineers from using the interagency manual; Corps adopted its 1987 wetland delineation manual as standard for making jurisdictional determinations.
1993	National Research Council (NRC) created Committee on Wetland Characterization to conduct scientific review of wetland delineation practices used by the federal government and to summarize current understanding of wetland functions.
	President Clinton's White House Office of Environmental Policy issued the report *Protecting America's Wetlands: A Fair, Flexible, and Effective Approach* that reaffirmed President Bush's policy of no-net-loss wetlands and laid out plans for a national wetland policy.

Table 9.1. *(continued)*

Date	Event
1995	National Academy Press published *Wetland Characteristics and Boundaries* detailing the National Research Council's findings on the science behind wetland delineation; recommended, among other things, that the federal government develop a new wetland delineation manual based on experiences learned since 1987.
2000	Corps of Engineers published revisions of nationwide permits.
	Estuaries and Clean Waters Act authorized development of a national estuary habitat restoration strategy, provision of grants for such restoration (Title I of the Act is referred to as the Estuary Protection Act), and expanded cooperative efforts to restore Chesapeake Bay.
	National Research Council published report on wetland mitigation concluding that wetlands lost are not being adequately replaced by mitigation.
2002	National Oceanic and Atmospheric Administration established the Coastal and Estuarine Land Conservation Program that provides funds to states with approved coastal zone management programs to acquire wetlands and other coastal lands of significance.
2004	President George W. Bush announced program to increase wetland restoration and established goal of three million acres over the next 5 years.
2005	Corps of Engineers begins to develop regional supplements to its wetlands delineation manual to present more science-based techniques for delineating wetlands, including more hydrology indicators and explicit guidance on problem situations (regional supplements are produced over the next decade).
2008	Maryland passed Living Shorelines Protection Act requiring landowners to employ nonstructural alternatives (living shorelines) for shoreline erosion control unless can demonstrate that hardened measures are required (formalized existing regulations into a law; nonstructural measures were previously recommended).
2013	Corps of Engineers plans to propose a new wetland delineation manual incorporating experiences learned since the 1980s.

Notes: Laws mentioned are federal laws unless stated otherwise. Note that there are other events that benefited tidal wetland wildlife, such as the Audubon Society's efforts in the late 1800s to fight against the slaughter of birds for feathers for women's fashions, and the first of many wildlife conservation acts (e.g., Lacey Act, Weeks-McLean Act, and the 1918 Migratory Bird Treaty Act that prohibited market hunting of migratory birds and importation of such birds), but they are not included in this timetable because they address species and not habitat.

to benefit maritime commerce, the public's right to use such lands for traditional purposes and as access to other lands was retained (Archer et al. 1994). Over time, however, this priority public right was forgotten or ignored as coastal wetlands were used for non-water-dependent purposes and public access and uses were denied. With increased public interest in environmental protection during the 1980s and 1990s, the public trust has received more attention. In a 1988 court case, the U.S. Supreme Court reaffirmed the application and importance of the public trust doctrine to coastal resources management (Phillips Petroleum v. Mississippi; see Slade et al. 1997 for a review of this doctrine as it relates to managing resources in coastal states).

Conservation

Wetland conservation can be accomplished in a variety of ways including acquisition, regulations, or landowner-incentive programs. Since tidal wetlands are now recognized as one of the world's most highly regarded natural resources, they receive better protection than most of their nontidal counterparts. Table 9.2 lists some laws that benefit tidal wetlands; some of these laws

Table 9.2. Some U.S. federal laws that provide benefits to tidal wetlands

Name of Law (Year enacted)	Key Points
Rivers and Harbors Act (1899)	The oldest federal environmental law prohibits discharge of refuse into navigable waters and tributaries without a permit from the U.S. Army Corps of Engineers; Section 10 prohibits excavation of or deposition of fill in these waters and lands below the normal high water mark including wetlands.
Migratory Bird Conservation Act (1929)	Authorized the federal government through the U.S. Department of the Interior to acquire by purchase or rent migratory bird habitat subject to approval by the Migratory Bird Conservation Commission (federal–state commission) that was created under this act.
Migratory Bird Hunting and Conservation Stamp Act (1934)	Commonly called the "Duck Stamp Act"; generates funding for wetland acquisition through the sale of stamps to hunters, stamp collectors, and others interested in contributing to habitat acquisition; funds deposited in the Migratory Bird Conservation Fund to be used for acquisition of migratory bird refuges.
Clean Water Act (1948)	Formerly Federal Water Pollution Control Act; established to reduce pollution of nation's waterways including the improvement of water quality for fish and aquatic life; administered by U.S. Environmental Protection Agency; Section 404 of the 1972 amendments regulates filling of wetlands while Section 401 gave water quality certification authority to states; the 1977 amendments included a provision authorizing $6 million for the Secretary of the Interior to complete the National Wetlands Inventory by December 31, 1981; the 1987 amendments established the National Estuary Program to develop and implement conservation and management plans to protect estuaries and restore their biological, chemical, and physical integrity.
Fish and Wildlife Coordination Act (1958)	Allows U.S. Fish and Wildlife Service and state wildlife agencies to comment on federal permits and federally funded actions; requires other federal agencies to fully consider fish and wildlife impacts of proposed development.
National Wildlife Refuge System Administration Act (1966)	An organic act that recognized and supported expansion of the national wildlife refuge system, which includes many coastal wetlands at strategic locations for migratory birds; administered by U.S. Fish and Wildlife Service.
Estuary Protection Act (1968)	Authorized the Secretary of Interior to study and inventory estuaries of the U.S. and determine whether such areas should be acquired for their protection; promoted increased cooperation among federal agencies to conserve estuaries and acquire wetlands through cost-sharing agreements.
National Environmental Policy Act (1969)	Requires all federal agencies to conduct environmental impact assessments of major actions significantly affecting the quality of the human environment including programs and projects (e.g., unavoidable impacts and alternatives actions) and to seek review comments from other agencies and the public.
Coastal Zone Management Act (1972)	Strives to improve conservation and management of nation's coastal resources; established the National Estuarine Research Reserve System to protect areas for long-term research, water-quality monitoring, education, and coastal stewardship; administered by National Oceanic and Atmospheric Administration.
Coastal Barrier Resources Act (1982)	Designated various undeveloped coastal barrier habitats for inclusion in the Coastal Barrier Resources System; such areas are ineligible for direct or indirect federal financial assistance that might support development (e.g., flood insurance); administered by U.S. Fish and Wildlife Service.
Emergency Wetlands Resources Act (1986)	To promote conservation of the nation's wetlands; required the Secretary of Interior to establish a national wetlands priority conservation plan to assist decision-makers in identifying wetlands warranting priority consideration for federal and state acquisition using federal land and water conservation funds; requires the U.S. Fish and Wildlife Service to continue its National Wetlands Inventory Project (mapping) and update its national wetlands status and trends report at 10-year intervals; also encouraged states to develop state wetlands priority plans as part of their statewide comprehensive outdoor recreation plan.

Table 9.2. (*continued*)

Name of Law (Year enacted)	Key Points
North American Wetlands Conservation Act (1989)	Established North American Waterfowl Management Plan and tripartite agreement on wetlands between U.S., Canada, and Mexico to promote wetland conservation and to fund conservation projects in the signatory countries; the U.S. Fish and Wildlife Service administers the conservation grant program that supports long-term protection, restoration, and/or enhancement of wetlands and associated upland habitats.
Coastal Wetlands Planning, Protection and Restoration Act (1990)	Provides funds for planning and implementing projects to create, restore, and enhance wetlands in Louisiana, and through the National Coastal Wetlands Conservation Grant Program provides matching grants to other coastal states for similar activities plus acquisition.
The Estuary Restoration Act (2000)	Promotes restoration of estuarine habitats, developed a national strategy for estuary restoration titled "A National Strategy to Restore Coastal and Estuarine Habitat" (Restore America's Estuaries 2002), and authorizes funding for restoration projects.

Notes: Recognize that the focus of most of these laws is much broader than wetland protection. The dates given refer to the original law, as amendments have been passed for many laws.

are discussed below, while information on others can be obtained from the administrating agency. In addition, various presidents have proclaimed executive orders that have supported tidal wetland conservation. For example, on March 14, 1903, President Theodore Roosevelt issued an order establishing Pelican Island as the first federal bird reservation. This was the first of numerous executive orders setting aside lands for preservation, including many tidal wetlands. The federal bird reservations formed the foundation of what is today the National Wildlife Refuge System. On May 24, 1977, President Jimmy Carter signed an order to protect wetlands that required each federal agency to minimize destruction, loss, and degradation of wetlands from its activities and to preserve and enhance wetlands. President Barack Obama's executive order to restore Chesapeake Bay issued on May 12, 2009, attempts to reduce pollution of the bay and restore its habitats (including tidal wetlands) and living resources. Federal policies have also helped maintain and improve the status of tidal wetlands. For example, President George H. W. Bush's proclamation of a federal policy of "no-net-loss" of wetland functions has been a keystone for wetland conservation since 1989 and is a policy tool embraced by states for wetlands and other resources of interest as well as by Canada and some of its provinces (Rubec and Hanson 2009).

Protection through Acquisition

Government agencies and private environmental organizations are actively acquiring wetlands to protect their functions and to preserve their values for our benefit and for the benefit of future generations. Acquisition of land may be by direct purchase (fee simple) or through buying conservation easements. Fee-simple purchases create government-owned lands that may be protected or managed to promote wildlife and other natural resources, whereas purchasing easements are used to protect wetlands on private property by promoting conservation and restricting development on such lands. The easement may be valid for a certain number of years or may be a perpetual easement. Fee-simple acquisition provides the best protection as easements require monitoring to ensure that the wetlands remain unaltered. The federal government actively purchases wetlands and adjacent lands to expand National Parks, National

Wildlife Refuges (U.S.), National Wildlife Areas (Canada), and National Seashores or to establish new ones (Figure 9.1). State and provincial governments have done likewise, establishing wildlife management areas or public seashores that contain tidal wetlands. In the United States, federal grants are also awarded to state and nonprofit organizations for coastal wetland acquisition through various programs. Private nongovernmental organizations such as The Nature Conservancy and Ducks Unlimited purchase wetlands through funds secured by private donations from individuals and corporations. This section will focus on acquisition efforts in the United States, recognizing

a

b

Figure 9.1. Both the United States and Canada are protecting tidal wetlands through acquisition: (a) Rachel Carson National Wildlife Refuge (ME) and (b) Bic National Wildlife Park (QC).

Figure 9.2. The duck stamp program provides funding to support wetland conservation in addition to promoting the beauty of waterfowl through the arts.

that many similar efforts are under way in Canada.

The first National Wildlife Refuge in the United States was established at Pelican Island, Florida, home to one of Florida's largest rookeries for colonial nesting egrets. At the time of its creation in 1903, the island's birds were being hunted indiscriminately for their plumage and other parts used in the fashion industry to adorn women's hats, shoes, and fans. President Theodore Roosevelt recognized the need to stop this slaughter and established by Executive Order Pelican Island as the first National Bird Reservation. This marked the unofficial beginning of what later became the National Wildlife Refuge System (NWRS) (www.fws.gov/refuges/centennial/pdf2/pelicanIsland_reffalt.pdf). By the time he left office, President Roosevelt had established 53 such areas. Today, the NWRS provides a network of sites along major flyways connecting breeding grounds with overwintering areas where possible. Many tidal wetlands are included in the NWRS because they provide vital feeding and resting habitats for migratory birds, overwintering grounds, and breeding habitat for migratory species.

In 1929, Congress passed the Migratory Bird Conservation Act, which was designed to provide yearly funding of $1 million to acquire wetlands that were important for migratory birds. Later that year, the stock market crashed, plunging the country into the Great Depression. At that point, funding for wetland acquisition dwindled (Bolen 2000). Continued concern about the destruction of waterfowl habitat in the United States led to passage of the Migratory Bird Hunting Stamp Act ("Duck Stamp Act") in 1934. Since then waterfowl hunters have been required to purchase a duck stamp (Figure 9.2). Ninety-eight cents out of every dollar from duck stamp sales goes to the Migratory Bird Conservation Fund to purchase or lease wetland habitat, which by 2002 had purchased more than 5 million acres of habitat (U.S. Fish and Wildlife Service 2002). Duck stamps are also purchased by stamp collectors and by non-hunters to support wetland conservation as well. Canada has a similar program administered by Wildlife Habitat Canada, a nonprofit foundation, but that program has a more general wildlife habitat orientation than the U.S. program. State, provincial, and local governments also acquire wetlands to serve as public parks, natural areas, forests, or wildlife management areas. The first two levels of government mentioned also may have duck stamp or wildlife habitat stamp programs to generate funds to preserve valuable waterfowl habitat, including coastal wetlands.

Several more recent federal laws have facilitated wetland acquisition for wildlife through federal–state, federal–nonfederal, and multinational partnerships to support wetland conservation. Brief descriptions are given below. (*Note:* More complete summaries of these and other wildlife acts are published online at www.fws.gov/laws/lawsdigest/resourcelaws.htm.)

The Endangered Species Act of 1973 offers some protection for species designated as federally threatened or endangered. Designation of critical habitat for such species dependent on tidal wetlands may protect such areas from federally financed projects or other projects receiving some federal funding. All federal agencies must ensure that any action authorized, funded, or carried out by the agency is not likely to jeopardize the continued existence of an endangered or threatened species, or to result in destruction or adverse modification of a critical habitat of a species. The act also allows for purchase of critical habitat through use of Land and Water Conservation Funds.

The Emergency Wetlands Protection Act was passed in 1986 to promote wetlands conservation for the public benefit and to help fulfill international obligations in various migratory bird treaties and conventions. To accomplish this, the act 1) authorizes the purchase of wetlands from Land and Water Conservation Fund monies; 2) requires the Secretary of the Interior to establish a national wetlands priority conservation plan; 3) requires states to include wetlands in their comprehensive outdoor recreation plans; and 4) transfers funds from import duties on arms and ammunition to the Migratory Bird Conservation Fund to acquire wetlands.

Another law promoting wetland conservation is the North American Wetlands Conservation Act passed in 1989. This act has several objectives including that it 1) encourages partnerships among public agencies and other interests to protect, enhance, restore, and manage an appropriate distribution and diversity of wetland ecosystems and other habitats for migratory birds and other fish and wildlife in North America; 2) maintains current or improves distributions of migratory bird populations; and 3) sustains an abundance of waterfowl and other migratory birds consistent with the goals of the North American Waterfowl Management Plan and the international obligations contained in the migratory bird treaties and conventions and other agreements with Canada, Mexico and other countries.

The Coastal Wetland Planning, Protection, and Restoration Act of 1990 established a plan to restore wetlands, mainly in Louisiana, funded through excise taxes on fishing equipment and motorboat and small engine fuels. While the act emphasizes planning and implementing wetland restoration on a large-scale for Louisiana (the state that has lost the largest amount of tidal wetlands), it also establishes the National Coastal Wetlands Conservation Grants Program (a federal/nonfederal cost-sharing grant program) to fund wetland acquisition and restoration in coastal states, except Louisiana. This grant program has helped acquire thousands of acres of wetlands since its inception. The act also allocates 15 percent of the available appropriations in each fiscal year for wetlands conservation projects authorized by the North American Wetlands Conservation Act in coastal states. As of 2010, about $183 million in grant monies have been awarded to 25 coastal states and to one U.S. territory to acquire, protect, or restore more than 250,000 acres (100,000 ha) of coastal wetland ecosystems. Typically, between $13 million and $17 million in grants are awarded annually through a nationwide competitive process.

Besides its initiative to improve management of coastal resources, the Coastal Zone Management Act of 1972 required the National Oceanic and Atmospheric Administration (NOAA) to establish a national system of protected estuarine sanctuaries for the purposes of long-term research, public awareness, and education. During the 1970s, NOAA designated five estuarine sanctuaries: South Slough in Coos Bay (OR), Sapelo Island (GA), Rookery Bay (FL), Apalachicola Bay (FL), and Elkhorn

Slough in Monterey Bay (CA). Today this system, now called the National Estuarine Research Reserve System, contains 28 reserves.

The National Park Service (NPS) has established 10 national seashores that contain tidal wetlands: Assateague Island (VA), Canaveral (FL), Cape Cod (MA), Cape Hatteras (NC), Cape Lookout (NC), Cumberland Island (GA), Fire Island (NY), Gulf Islands (two districts: FL and MS), and Padre Island (TX). Other NPS lands possessing tidal wetlands include the following national parks: Acadia (ME), American Samoa, Biscayne (FL), Everglades (FL), and Virgin Islands, plus Cape Krusenstern National Monument (AK), Virgin Islands Coral Reef National Monument, and Boston Harbor Islands National Recreation Area.

Joint ventures have been established to conserve, restore, and enhance waterfowl, shorebird, waterbird, and landbird habitat across North America. In the East, two joint ventures are operating along the Atlantic Coast: Eastern Habitat Joint Venture addresses habitat in six eastern Canadian provinces, while the Atlantic Coast Joint Venture does the same for U.S. coastal states from Maine to Florida (including Puerto Rico). These joint ventures have developed implementation plans that address continental and regional waterfowl population and habitat goals for conservation of breeding, migration, and wintering habitat. They have also created federal–state/provincial–private partnerships to achieve these objectives. Ducks Unlimited and The Nature Conservancy are among the private partners. Other joint ventures that include coastal wetlands are the Gulf Coast Joint Venture, San Francisco Bay Joint Venture, and Pacific Coast Joint Venture. Although not formally part of the joint ventures, similar planning efforts are under way to conserve waterbird and shorebird habitat (e.g., Southeast Waterbirds Conservation Plan—Hunter et al. 2006; North Atlantic Shorebird Management Plan—Clark and Niles 2000; Pacific Northwest—Thomas et al. 2004). The Western Hemisphere Shorebird Reserve Network was established to develop partnerships that recognize and protect vital wetlands in the Americas for shorebird migration. Delaware Bay was the first designated reserve.

Nonprofit environmental organizations have established nature preserves to meet some of their resource objectives. These groups include the National Audubon Society (and its state chapters), The Nature Conservancy, the Conservation Fund, Trustees for Reservations, Massachusetts Audubon Society, nature trusts of the Maritime provinces, and local land trusts. Ducks Unlimited purchases wetlands to preserve or enhance their value as waterfowl habitat. Some of these organizations purchase private lands for conservation and later sell the land to government to replenish the funds they have available for buying land.

While acquired lands may be largely protected from development, they remain subjected to damage from outside forces such as water, air, and noise pollution; invasive species; and development encroachments on their borders, which can adversely affect their habitat quality. Some coastal wetlands managed for waterfowl may be diked and impounded, which alters their natural characters and functions. Tidal wetlands are subjected to the effects of rising sea level, which may ultimately lead to their elimination in areas where low-lying lands are not available for their landward migration (see Chapter 12).

Conservation through Regulation

Since the cost of acquiring wetlands far exceeds the dollars available for such purposes, and since some landowners would rather keep their wetlands in private ownership, other means have been taken to conserve the remaining wetlands and to protect the public values they provide (see Chapter 7). This gap is largely filled by wetland regulations that affect use of wetlands on both public and private property. These land-use regulations have the broadest effect on protecting wetlands from development. A combination of federal and state laws and regulations and, in some cases, local ordinances, are used to minimize coastal wetland conversion in the United States. The U.S. Congress and many state legislatures have passed laws to prevent uncontrolled loss and degradation of wetlands. Regulations promulgated to implement these laws spell out what types of activities are controlled or exempted and the specific procedures that must be followed if public agencies, private organizations, or private landowners want to alter wetlands.

From the Colonial period to the 1960–70s, U.S. coastal wetlands had virtually no protection, except where they had been acquired by government agencies or interested organizations for conservation purposes or were regulated under Section 10 of the federal Rivers and Harbors Act, which at the time, was administered with a focus on preventing obstructions to navigation and not for wetland protection. Consequently, losses were tremendous (Chapter 8). With increased knowledge of coastal marsh functions, especially regarding marine fisheries, the public's growing awareness of wetland values, and concern over accelerating wetland losses in the 1950s and 1960s, federal and state governments in the United States enacted laws or promulgated regulations to curtail coastal wetland destruction. By the mid-1970s, all northeastern coastal states had passed laws ("tidal wetland acts") to protect tidal wetlands. These laws may represent the first U.S. laws passed to restrict, on a large scale, uses of natural resources on private property due to their public benefits (e.g., fish spawning and nursery grounds, waterfowl habitat, flood protection, and shoreline stabilization). Similar laws to regulate uses of nontidal freshwater wetlands on public and private lands were later enacted by most of these same northeastern states and by several others across the country.

Federal laws that provide protection to wetlands are not wetland-specific, but are laws designed to protect navigability or water quality of the nation's waters. Since most wetlands are considered part of the "waters of the United States," they are granted some, but not exclusive, protection from certain types of development. In contrast to the federal laws, specific wetland protection laws have been passed by state legislatures. These laws recognize the multitude of functions performed by wetlands and their value to society, and are designed to either protect wetlands or at least to conserve them. To gain local control over activities in wetlands that might not be covered by federal and state regulations, some counties, municipalities, and towns have established ordinances to regulate development in and around wetlands. Examples include Falmouth's (MA) town wetland regulations, Sarasota County's (FL) coastal setback code, and Kauai County's (HI) shoreline setback and coastal protection ordinance.

FEDERAL WETLAND REGULATION

Two federal laws give the U.S. government authority to regulate activities in wetlands: 1) the Rivers and Harbors Act of 1899; and 2) the Clean Water Act of 1972 and amendments. The purpose of the Rivers and Harbors Act was to regulate construction of piers, wharfs, bridges, dams, dikes, causeways, and other actions that could obstruct navigation. Section 10 of this act (also known as the "Refuse Act") prohibits filling or excavation within navigable waters. Wetlands at or below mean high water level in navigable waters were included within the U.S. Army Corps of Engineers jurisdiction. Filling of these wetlands now required permission from the Corps. In the process of fulfilling one of its prime missions—to maintain navigability of the nation's waters—the Corps filled many acres of tidal wetlands to dispose of dredged material from harbor and waterway maintenance projects. Wetlands were used as dumping grounds, and the Corps was exempt from permit requirements since it was the regulatory authority. In 1946 Congress passed amendments to the Fish and Wildlife Coordination Act of 1934 (FWCA) that forced the Corps to at least consider environmental impacts of their projects. These amendments required any federal agency proposing to impound, divert, dredge, or otherwise alter streams or other waters to consult the U.S. Fish and Wildlife Service (FWS) and the corresponding state wildlife agency for an assessment of fish and wildlife impacts of their project. Following 1958 amendments to FWCA, comments from the wildlife agencies regarding wildlife impacts, and recommendations to mitigate the adverse impacts must be given full consideration but such mitigation was not a requirement for project authorization by Congress. Proposed federal projects had to include a cost-benefits analysis and an assessment of wildlife costs and benefits. Prior to litigation, the Corps was more single-minded in their mission (i.e., maintaining navigability) when it came to watercourses, and wetlands appeared to receive only minimal attention. This attitude would change as public interest in and support for environmental protection increased.

In 1967, the Corps used environmental impact concerns from the FWS, several state agencies, private citizens, and others to deny a permit to dredge and fill tidal wetlands in Boca Ciega Bay, Florida. This decision was extremely controversial since the Corps had issued these types of permits on a routine basis beforehand. At that time, the ecological values of mangroves had been discovered, and that knowledge was used by natural resource agencies and concerned citizens to support the stoppage of the era of wanton filling of mangrove swamps. The developers sued the Corps and won the case in district court; however, the decision was appealed to the U.S. Court of Appeals for the Fifth Circuit which reversed the lower court's decision—Zabel v. Tabb, 430 F.2d 199 (5th Cir. 1970, cert. denied 39 U.S.L.W. 3356). The court cited the FWCA and the National Environmental Policy Act of 1969 to support the Corps' position and to affirm Corps authority to deny the permit for environmental reasons (Goldstein 1971). This was the first time the Corps used its regulatory power under Section 10 solely for conservation of fish and wildlife resources.

In the 1970s, when the Environmental Protection Agency (EPA) was established, most of the activities covered by the Federal Pollution and Control Act of 1972 (now called the Clean Water Act) were assigned to them. A few provisions, however, were given to the Corps of Engineers. The one most relevant to wetlands was Section 404, which dealt with permitting the disposal of fill material in the nation's waters. Congress gave the Corps this responsibility in part because they already had a nationwide permit program in place for navigable waters in accordance with the Rivers and Harbors Act. This assignment was controversial as the Corps did not have a good track record for safeguarding aquatic resources. To watch over the Corps administration of this provision, Congress delegated oversight authority to the EPA giving them veto power over Corps permits on environmental grounds. At this time, the Corps did not embrace its new environmental responsibilities. Initially the Corps limited its jurisdiction to lands and waters below the ordinary high water mark, but EPA contended that the Clean Water Act Amendments included wetlands above mean high water. A court case in the U.S. District Court for the Middle District of Florida (United States v. Holland 373 F. Supp. 665) centered on the Corps' authority to regulate a dredge-and-fill development in shallow water and mangroves. In March 1974, the court ruled in favor of the EPA interpretation that the Corps had the authority to regulate deposition of dredged or fill material in wetlands above mean high water and that intertidal wetlands were intended to be regulated under the Clean Water Act. The Corps still refused to regulate filling of high marsh and only required a permit for dredge-and-fill developments in the high marsh when the canals were to be connected to navigable waters. By this time, the high marsh was damaged beyond repair, so high marsh destruction continued without permit. In April 1974, the Corps published new regulations that limited the extent of Section 404 jurisdiction to traditional navigable waters, outraging environmental agencies and organizations. The National Resources Defense Council (NRDC), the National Wildlife Federation, and others sued the Corps for narrowing the scope of the Clean Water Act to navigable waters. In 1975, the U.S. Court of Appeals (2nd District) found that Congress did not intend the law to be restricted to navigable waters, and the Corps regulations did not meet the intent of the law and must be revised accordingly (NRDC v. Callaway 392 F. Supp. 685). The Corps reluctantly expanded the scope of the regulations to all waters of the United States including wetlands (Figure 9.3). For details of the conflict between the Corps and environmental groups, see Vileisis (1997).

Today, individuals seeking to dredge, fill, build structures, or otherwise alter coastal wetlands and many inland wetlands must apply for a federal permit from the U.S. Army Corps of Engineers. If an activity involving a discharge of dredged or fill material represents a new use of the wetland, and the activity would result in a reduction in reach or impairment of flow or circulation of regulated waters, including wetlands, the activity is not exempt from the permit process. Both conditions must be met in order for the activity to be considered nonexempt. In general, any discharge of dredged or fill material associated with an activity that converts a wetland to upland is not exempt, and requires a Section 404 permit (see Table 9.3 for exemptions).

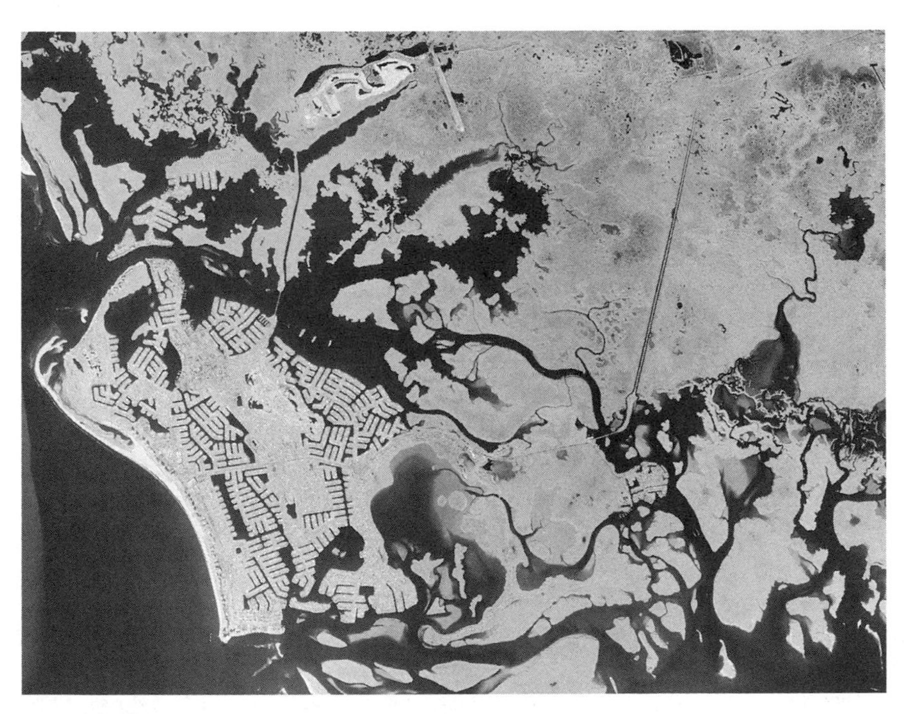

Figure 9.3. In 1976, the Corps, for the first time, denied filling of thousands of acres of mangrove at Marco Island (FL)—this was a major action by the Corps in exercising its new authority to regulate dredge and fill operations in accordance with recent court interpretation on the Corps regulatory responsibilities. This type of development had proceeded unabatedly before this time. The Corps' decision was a major victory for environmentalists and for tidal and other wetlands.

Table 9.3. Exempt activities under Section 404(f) of the Clean Water Act

Exempt Activities
Plowing
Seeding
Cultivating
Harvesting food, fiber, and forest products
Minor drainage (including removal of blockages)
Upland soil and water conservation practices
Maintenance (not construction) of drainage ditches
Construction and maintenance of irrigation ditches
Construction and maintenance of farm or stock ponds
Construction and maintenance of farm and forest roads following approved best management practices
Maintenance of dams, dikes, levees, and similar structures

Note: Exempt if they are part of an ongoing farming, ranching, or forestry operation.

"Wetlands Law and Policy: Understanding Section 404" (Connolly et al. 2005) provides an in-depth review of the Section 404 Program.

STATE REGULATIONS

Many states had enacted laws to regulate dredging operations or drainage practices prior to the 1960s, however, they were not focused on protecting wetlands. In 1962, the Massachusetts Department of Natural Resources was directed by the legislature to investigate and report on the state's coastal wetlands in terms of location, ownership, and value to fish and wildlife. The report published the following January (Commonwealth of Massachusetts 1963) summarized wildlife, shellfish, historic, scenic, and fisheries values and identified threats from development. These findings were enough to convince the state legislature to enact the first law to specifically protect wetlands—in this case salt marshes—from development (the Jones Act, effective May 22, 1963). By 1975, all states from Maine through North Carolina had enacted legislation to protect tidal wetlands, mainly salt and brackish marshes, in some form. These laws were necessary because the states clearly recognized the contributions of tidal wetlands in providing nursery grounds for commercially and recreationally important fishes and habitat for shellfish, waterbirds, and other wildlife; in protecting private property from shoreline erosion and floods; and protecting other values in the public interest. Tidal wetlands were defined by hydrology, by connectivity (past and present) to tidal waters, by elevation, and by vegetation (Table 9.4). These laws have received widespread public support and have withstood numerous court challenges addressing claims of a "taking" of private property without payment of just compensation.

In some states, local governments (e.g., town conservation commissions) administer these and other wetland regulations with state agency oversight, whereas in other states, state committees or agencies run the wetland program and issue permits. Today, virtually all coastal states regulate activities in tidal wetlands to some degree. Fifteen coastal states have state regulatory programs that include both tidal and nontidal wetlands, while eight states have state programs only for coastal wetlands (Environmental Law Institute 2008). States lacking a state regulatory program (e.g., Alabama and Texas) use authority given to them by Section 401 of the federal Clean Water Act to approve or deny any activity that requires a federal permit or license to ensure that such activity complies with the state's water quality standards. Most western states use this authority to regulate inland wetlands. In addition to its 401 program, California regulates wetlands within its coastal zone through the Coastal Zone Management Act of 1972 and is currently developing new policy (the California Wetland and Riparian Area Protection Policy) to regulate wetlands and riparian areas statewide under its Water Quality Control Act of 1969 (J. Collins, pers. comm. 2011). Some states also have regulatory authority along the borders of their tidal wetlands. For example, the State of Maryland regulates all lands within 1,000 feet (305 m) of the mean high water mark

Table 9.4. Examples of wetland definitions in some state laws

State	Definition (Source and year)
Maine	"all tidal and subtidal lands including all areas below any identifiable debris line left by tidal action, all areas with vegetation present that is tolerant of salt water and occurs primarily in a salt water habitat, and any swamp, marsh, bog, beach, flat or other contiguous lowland subject to tidal action or normal storm flowage at any time excepting periods of maximum storm activities." —Alteration of Coastal Wetlands, Maine Public Law 595, Title 38, Article 5, 1975
Connecticut	"those areas which border on or lie beneath tidal waters, such as, but not limited to banks, bogs, salt marsh, swamps, meadows, flats, or other low lands subject to tidal action, including those areas now or formerly connected to tidal waters, and whose surface elevation is at or below an elevation of one foot above local extreme high water; and upon which may grow or be capable of growing some, but not necessarily all, of the following: . . . [list of many salt marsh indicator species]." —Connecticut Tidal Wetlands Act, Connecticut Public Act of 1969, Number 695
Delaware	"those lands above the mean low water elevation including any bank, marsh, swamp, meadow, flat or other low land subject to tidal action in the state of Delaware . . . along an inlet, estuary or tributary waterway or any portion thereof, including those areas which are now or in this century have been connected to tidal waters, whose surface is at or below an elevation of two feet above local mean high water, and upon which may grow or is capable of growing any but not necessarily all of the following plants: . . . [list of indicator plants] and those lands not currently used for agricultural purposes containing four hundred (400) acres or more of contiguous non-tidal swamp, bog, muck, or marsh exclusive of narrow stream valleys where fresh water stands most, if not all, of the time due to high water table, which contributes significantly to ground water recharge, and which would require intensive artificial drainage using equipment such as pumping stations, drain fields or ditches for the production of agricultural crops." —Wetlands Act, Title 7, Delaware Code, Chapter 66
Virginia	"all that land lying between and contiguous to mean low water and an elevation above mean low water equal to the factor 1.5 times the mean tide range at the site of the proposed project in the county, city or town in question; and upon which is growing on July 1, 1972, or grows thereon subsequent thereto, any one or more of the following:" . . . [listing of representative salt marsh and tidal freshwater wetland species]." —Virginia Wetlands Act of 1972, Virginia Code, Sections 62.1-13.1 to 62.1-1-13.20
Georgia	"any marshland or salt marsh in the State of Georgia, within the estuarine area of the State, whether or not the tide waters reach the littoral areas through natural or artificial water courses. Marshlands shall include those areas upon which grow one, but not necessarily all, of the following: . . . [list of indicator species presented]. The occurrence and extent of salt marsh peat at the undisturbed surface shall be deemed to be conclusive evidence of the extent of a salt marsh or a part thereof." —Coastal Marshlands Protection Act of 1970, Official Code of Georgia Annotated, Section 12-5-281 (*Note:* Chapter 391-4-12 rules for "Coastal Marshland Protection" published by the Georgia Department of Natural Resources specified that marsh elevations lie within "a tide-elevation range from five and six-tenths (5.6) feet above mean tide level and below.")
Mississippi	"all publicly owned lands subject to the ebb and flow of the tide, which are below the watermark of ordinary high tide; all publicly owned accretions above the watermark of ordinary high tide and all publicly owned submerged water-bottoms below the watermark of ordinary high tide." —Coastal Wetlands Protection Law," Mississippi Laws 1973, Chapter 385

Notes: Some definitions were later amended. The Mississippi definition pertains only to publicly owned land; the Delaware definition includes some nontidal wetlands.

or from the landward edge of tidal wetlands under the Critical Areas Act (1984).

CANADIAN PROVINCIAL REGULATIONS

The federal government of Canada has policies to encourage wetland conservation but has relied on its provincial governments to regulate wetland uses. Most of the Maritime provinces (New Brunswick, Newfoundland, Nova Scotia, and Prince Edward Island) have enacted general environmental laws that include wetland protection and provisions requiring individuals, organizations, or companies seeking to alter coastal wetlands to obtain a permit from the provincial government. The watercourse and wetland alteration regulations in New Brunswick apply to work within 98.4 feet (30 m) of a wetland that is 2.5 acres (1 ha) or larger (New Brunswick Department of Natural Resources and Energy and Department of Environment and Local Government 2002; Table 9.5). All coastal wetlands are considered provincially significant and the government does not support any proposed development within 98.4 feet (30 m) of such wetlands with two exceptions: development that rehabilitates or enhances a degraded area, and projects deemed necessary for the greater public good. The Environmental Act and corresponding Wetland Designation Policy in Nova Scotia (Nova Scotia Environment and Labour 2006) recognizes freshwater wetlands and salt marshes as critical ecosystems providing a suite of environmental and societal services: 1) maintain and improve water quality and quantity; 2) reduce impacts and damage from flooding and storm surges; 3) provide habitat for wildlife and other wetland-dependent species; 4) provide opportunities for recreation and education; and 5) generate products that create opportunities for economic development. The policy established a wetland permit program for filling, draining, flooding, or excavating wetlands. The policy excludes tidal mudflats, tidal ponds, man-made wetlands created for wastewater or stormwater treatment, and former salt marshes that are now mainly used for agriculture covered under the Agricultural Marshland Conservation Act. The Prince Edward Island (PEI) regulations administered by the Department of Environment, Energy, and Forestry and promulgated under PEI's Environmental Protection Act (1989) require permits for watercourse alterations and include a 60-foot (18.3 m) setback for subdivision development along

Table 9.5. Activities considered "alterations" subject to regulation under New Brunswick's Clean Water Act

Regulated Activities
Any changes made to existing structures in the watercourse or wetland including repairs, modifications or removal, whether the water flow in a watercourse or wetland is altered or not
The operation of machinery on the bed of a watercourse other than at a recognized fording place
The operation of machinery in or on a wetland
Any deposit or removal of sand, gravel, rock, topsoil or other material into or from a watercourse or wetland or within thirty meters of a wetland or the bank of a watercourse
Any disturbance of the ground within thirty meters of the bank of a watercourse, except grazing by animals, the tilling, plowing, seeding and harrowing of land, the harvesting of vegetables, flowers, grains and ornamental shrubs and any other agricultural activity prescribed by regulation for the purposes of this paragraph, that occur more than five meters from a wetland or the bank of a watercourse
Any removal of vegetation from the bed or bank of a watercourse
Any removal of trees within thirty meters of the bank of a watercourse
Any removal of vegetation from a wetland or from within thirty meters of a wetland except the harvesting of vegetables, flowers, grains and ornamental shrubs and any other agricultural activity prescribed by regulation for the purposes of this paragraph, that occur more than five meters from a wetland

Source: New Brunswick Department of Natural Resources and Energy and Department of Environment and Local Government 2002.
Note: An "alteration" to a watercourse or wetland is a temporary or permanent change made at, near, or to a watercourse or a wetland, or to water flow in a watercourse or wetland.

wetlands, watercourses, beaches, and dunes; also, no building can be built within 75 feet (22.9 m) of a wetland or watercourse. PEI's Department of Community and Cultural Affairs administers the province's Planning Act that requires wetland buffers for all buildings, subdivisions, and sewage disposal systems (Prince Edward Island 2003).

Mitigation

The U.S. federal regulatory (Section 404) program now uses a triage sequence of evaluation steps to determine whether a proposed project should be permitted:

1) avoidance (Does the project have to be located in a wetland?);
2) minimization (Can the impact of the project on the wetland be reduced to an insignificant level?); and
3) mitigation or compensation (If impacts can't be avoided or minimized, then the wetland alteration must be compensated for).

In 1989, the U.S. government established a policy of no-net-loss of wetlands, while the Canadian government adopted a similar policy but applies it only to federal projects. Several states and provinces have established similar policies for evaluating wetland or watercourse alteration permits. Compensation has been interpreted to mean that for every acre of wetland loss, an acre or more of new wetland would have to be established; or in terms of function, for every unit of functional capacity lost, an equal or greater value of functional capacity has to be created or restored. Wetland type, functions, and timing of the project are factors influencing compensation ratios.

Mitigation may involve several approaches, and the approach taken is determined by the regulatory agency: 1) in-kind replacement of wetland lost (at some ratio, 1:1, 2:1, etc.); 2) out-of-kind replacement (restoring or constructing a different type of wetland, at some ratio); 3) purchasing credits from an established wetland mitigation bank (a created or restored wetland established to serve as a means of facilitating compensation for an unavoidable alteration; may be a government-run bank or a privately run operation); 4) payment of an in-lieu fee (to be used for wetland conservation or to support agency work in restoration; typically restricted to government and nonprofit organizations); and 5) wetland protection (acquisition of a wetland that is currently unprotected). Of these, the last approach is usually considered the least desirable since most wetlands are "protected" by regulations, but many fall outside current jurisdiction that are also worthy of protection through acquisition. This approach also does not meet the goals of "no-net-loss." Compensation ratios may vary based on the type of wetland altered and/or the type of mitigation offered (e.g., King and Price 2004). For example, the Province of New Brunswick has recommended the following ratios for mitigation: 2:1 for restoration, 3:1 for creation, 4:1 for enhancement, and 10:1 for preservation which must be coupled with some restoration, creation, or enhancement. The State of New Jersey requires mitigation ratios of 2:1 for restoration or creation and a minimum of 27:1 for preservation in accordance with the Freshwater Wetlands Protection Act. The State of New Hampshire requires 2:1 for tidal wetland restoration, 3:1 for creation, and 15:1 for preservation of upland buffer area in addition to requiring mitigation for disturbances of the tidal buffer zone (New Hampshire Department of Environmental Services 2009).

When mitigation is to be achieved by restoration, establishment (creation), or enhancement, such projects require extensive planning and monitoring (see Chapter 11) and may involve preparation of adaptive management and long-term management plans. These projects may be performed on-site (property where the wetland alteration is taking place) or off-site, but should be within the watershed and in locations where lost functions can be replaced. In-kind replacement (same wetland type as altered) is preferred, but out-of-kind replacement may be permitted, especially if such types have suffered significant losses regionally in the past. On-site mitigation is typically accomplished by the permittee, while off-site mitigation may be through an authorized mitigation bank or by the permittee.

Mitigation banking is a fairly recent effort to provide opportunities to compensate for authorized wetland destruction and to address common failure problems associated with piecemeal, on-site efforts to create small parcel wetlands (e.g., Brown and Veneman 1998; Committee on Mitigating Wetland Losses 2001; Turner et al. 2001;

Table 9.6. Examples of mitigation banks in tidal wetlands

Mitigation Bank (Location)	Description and Source
Richard P. Kane Mitigation Bank (NJ)	A 240-acre wetland bank in the Hackensack Meadowlands involving a public–private partnership to restore a 200-acre common reed marsh to a more diverse intertidal plant community and 20 acres of forested wetland; treated common reed with herbicide, lowered elevations by grading, constructed numerous tidal channels, planted native marsh grasses, installed netting to goose-proof the site, removed berms, and built dike to protect neighboring properties from flooding; contaminant issues; use of credits are restricted for transportation projects by New Jersey's four state transportation agencies; construction initiated in 2010 and area planted in 2011. —Mogensen and McBrien 2010
Julie J. Metz Wetlands Bank (VA)	First entrepreneurial wetland mitigation bank in Virginia; included preservation of nearly 200 acres of fresh tidal wetlands, adjacent uplands, and creation of 19 acres of nontidal wetlands. To date, it has provided mitigation for 16 projects. —Wetland Studies and Solutions, Inc. 2010
Sandy Island (SC)	Established by the South Carolina Department of Transportation (SCDOT) to expedite permitting of transportation projects; involved preservation of Sandy Island and parts of three other plantations; purchased with contribution from The Nature Conservancy who will be responsible for long-term management; contains uplands and tidal and nontidal wetlands. —Hall 2001b
Huspa Creek (SC)	Also established by SCDOT; 250-acre partially impounded tidal marsh where tidal flow was restored by breaching the dike in 13 places; extent of open water has been reduced; salt marsh vegetation has recolonized intertidal areas. —Hall 2001b
Little Pine Island Mitigation Bank (FL)	Entrepreneurial bank on state lands (Matlacha Pass Aquatic Preserve) to enhance 1,507 acres of wetlands by removing exotic plants (mostly *Melaluca*) and restoring 48.3 acres of mosquito ditches and associated berms to tidal wetlands. —www.dep.state.fl.us/water/wetlands/docs/mitigation/permits/little_pine_island/pdf
Blue Heron Slough Conservation and Mitigation Bank (WA)	Plan to establish 344-acre site by the Port of Everett and Wildlands, Inc., as a joint fisheries conservation and wetland mitigation bank; restoring tidal flow to former tidal wetland (diked and drained in the early 1900s); re-establish tidal mudflat and marsh and rehabilitate degraded wetlands; constructing an internal slough network, reshaping elevations, re-establishing native vegetation, removing a tide gate, breaching the dike in four locations, and lowering dike in five areas to allow for high tide flooding. —U.S. Army Corps of Engineers and Washington State Department of Ecology 2010
Wilbur Island (OR)	Oregon's first coastal mitigation bank, an entrepreneurial venture; 162-acre site involving removal of dikes, disabling ditches, and re-establishing tidal channels to restore tidal flow and native vegetation. —http://statelands.dsl.state.or.us/DSL/news/pr0905_wilbur_mb.shtml

U.S. Government Accounting Office 2005). Banks may include restored, created, enhanced, and, in some cases, protected wetlands and adjacent habitats (Table 9.6). As a condition for permit approval, developers may be required to purchase credits from an approved mitigation bank to compensate for wetland losses from their proposed projects. Credits are often based on ratios in accordance with the nature of the compensation (i.e., restoration, creation, enhancement, or preservation). Regulatory agencies have established guidance for bank creation, use, and operation (e.g., national guidance from U.S. EPA at www.epa.gov/wetlands/guidance/mitbankn.html and State of Virginia guidance for tidal wetland mitigation banks at www.mrc.virginia.gov/regulations/bankguide.shtm). To use a bank, the developer may have to demonstrate that 1) all mitigative actions (including avoidance and minimization) have been incorporated into the proposed project; 2) the project must be water-dependent; 3) the project needs to be constructed in a wetland; and 4) its public and private benefits are deemed overwhelming (Virginia Marine Resources Commission and Virginia Institute of Marine Science 1997). Mitigation banks have been created either by government agencies, nonprofit organizations, private entrepreneurs, or public-private partnerships. By 2006, more than 400 mitigation banks were in place, with Louisiana having the most (Environmental Law Institute 2006). While mitigation banking has been gaining support as agencies strive to achieve no-net wetland loss, the approach has not been embraced by all (e.g., Brown and Lant 1999; Turner et al. 2001; Bauer et al. 2004; Environmental Law Institute 2006; Salzman and Ruhl 2007; Austen and Hanson 2008; Bourriaque 2008). A main concern over mitigation banks is the use of preservation as compensation for the loss of natural wetlands. While preservation is an admirable objective, preservation alone will not accomplish the goal of no-net wetland loss; however, combining it with restoration and creation makes for a more reasonable approach.

Policy and Planning

Governments can create policies to strengthen wetland protection. A general wetland policy adopted recently by the United States and Canada has been called "no-net-loss" (The Conservation Foundation 1988; Lynch-Stewart 1992). This refers to the attempt to achieve no-net-loss of wetlands and their functions over time from development activities. In the United States, the policy, first established by President George H. W. Bush's Administration in 1989, is applied in the regulatory process and includes activities on both public and private lands, while various federal agencies such as the National Park Service have embraced no-net wetland loss for their land management activities. The United States, however, still lacks a comprehensive national policy for wetland conservation, despite a proposal for such policy by the Clinton Administration in 1993 (White House Office of Environmental Policy 1993). Canada has embraced a federal wetland conservation policy since 1991.

Other government policies include providing federal funding to support state and local programs to regulate wetlands, while others focus on wetland acquisition (mentioned earlier), restoration, or public education. Both the United States and Canada have federal policies that promote the establishment of estuarine and marine sanctuaries: the Coastal Zone Management Act of 1972 and the 1996 Oceans Act, respectively. These areas are protected and may be important sites for research. States may have designated zones in and around coastal wetlands (e.g., buffer zones) as critical resource areas (e.g., Massachusetts' areas of critical environmental concern or Maryland's critical areas for Chesapeake and Atlantic Coastal Bays) and such designation may have resulted in the preparation of comprehensive management plans.

There are literally hundreds of programs across the United States and Canada that address coastal resources benefiting tidal wetlands. The few programs summarized below give the reader an idea of the kinds of activities going on to improve the status of tidal wetlands. For specific information on activities in a state or province, consult coastal resource agency websites.

Nongovernmental organizations, including conservation organizations and local watershed associations, are also involved in planning efforts that benefit coastal wetlands. For example, The Nature Conservancy, an international organization that has been instrumental in protecting wetlands through acquisition, has hosted workshops that bring together conservation-minded groups and individuals to identify resources at risk and to advance preservation of biodiversity (e.g., Hudson River estuary—Shirer et al. 2005; northern Gulf of Mexico—Beck et al. 2000).

U.S. POLICIES AND PROGRAMS

Coastal Zone Management. In 1972, Congress passed the Coastal Zone Management (CZM) Act. This act established a federal program that creates state–federal partnerships to improve management of coastal resources, especially for protecting estuaries and coastal wetlands. This law includes the Great Lakes as estuaries and coastal wetlands for purposes of administering this program. Grants to the states support establishment of new coastal zone management programs that seek to protect and conserve coastal resources through various means. If states do not have state laws to regulate tidal wetlands, they may use CZM funds to develop similar regulatory programs or other approaches to achieve the goal of tidal wetland protection. Federal grants to states through the CZM program also support tidal wetland acquisition and restoration projects and fund public education programs to increase public awareness of the values of coastal resources. Another provision of the act—federal consistency—gives states a significant role in federal decision-making. Federal activities in the coastal zone must comply with state coastal zone management policies. Consequently, federal actions including dredging projects, beach nourishment, and offshore wind power projects are subject to state approval. One study of the CZM program's effectiveness evaluated CZM programs for 23 states and a few U.S. possessions (Good et al. 1998; Table 9.7).

Table 9.7. General findings and recommendations from a 1998 review of the Coastal Zone Management Act effectiveness

Study Findings
1. The importance of estuary and coastal wetland protection is relatively high for most states and the nation as a whole.
2. The potential effectiveness of state coastal management programs in protecting estuaries and coastal wetlands looks good "on paper."
3. Outcome effectiveness of state coastal management programs in protecting estuaries and coastal wetlands gets moderate to high ratings for states with sufficient data.
4. The overall performance of state coastal management programs in protecting estuaries and coastal wetlands is relatively good for states with sufficient data.
5. Management of nontidal freshwater wetlands needs CZM attention.
6. Nonregulatory wetland restoration is an underutilized tool in CZM.
Recommendations
1. Establish a national outcome monitoring and performance evaluation system. (Including wetland loss permitted v. violation, gains through mitigation, and data on acquisition and nonregulatory restoration)
2. Improve nontidal freshwater wetland management by expanding coastal zone boundaries as necessary to encompass all "coastal" wetlands.
3. Establish national CZM policy goals for wetland restoration policy including: a) no-net-loss of wetland area and function in the short term implemented through regulatory programs; and b) a gain in wetland area and function over the long term implemented through nonregulatory restoration programs.

Source: Good et al. 1998, 1999.

A number of new approaches to improve wetland protection came out of state coastal zone management programs (Good et al. 1999). One of the more innovative means of improving coastal zone protection—special area management planning—focuses on resources within a specific geographic area and brings in regional planning and dispute resolution practices to address conflicts between stakeholders with different interests. Another reported CZM creation was requiring compensatory mitigation for wetland loss, which was incorporated in Oregon's CZM policy in 1976. Later, Oregon established the mitigation banking concept to address the issue of compensation for small projects. Other novel approaches to resource management include

- integrated land and water use planning (e.g., OR and NH),
- public–private partnerships for environment-development dispute resolution (CA),
- use of water-dependency tests for shoreline development, exclusion zones for major facilities (DE),
- shoreland buffers (NH, MD, and NJ),
- coastal habitat restoration (LA, CT, and DE),
- GIS-based methods for wetland evaluation and restoration planning (NC and WA), and
- use of federal consistency standards in lieu of a separate state wetland permit program (SC).

The CZM Act also provides funds to coastal states to acquire lands and to operate and manage reserves dedicated to supporting research, education, stewardship, system-wide monitoring, and graduate studies.

National Estuary Program. The National Estuary Program was established through the Clean Water Act Amendments in 1987 as a community-based program to restore and maintain the water quality and ecological integrity of estuaries designated as nationally significant. The U.S. EPA, the lead federal agency, has developed partnerships with other government agencies at all levels, nonprofit organizations, academia, business and environmental groups, and private individuals. To date, 28 national estuary programs (NEP) are in existence across the country. Each NEP has drafted a comprehensive conservation and management plan (CCMP) addressing specific actions required to improve and protect water quality, habitat, and living resources. Actions include measures to combat leaking septic systems, stormwater runoff, and invasive species; monitoring indicators of ecological health; and disseminating information for public education. The program has produced coastal condition reports that assess the condition of NEP estuaries. A companion program—Climate Ready Estuary Program—now works with NEP and other coastal managers to 1) assess coastal vulnerability to climate change; 2) develop and implement adaptation strategies; 3) engage and educate stakeholders; and 4) share lessons learned from coastal managers. EPA offers grants and technical assistance for conducting vulnerability assessments, developing adaptation plans and monitoring plans, and related projects. For additional information on both these programs, visit the EPA website (www.epa.gov/nep and www.epa.gov/cre, respectively).

Coastal Barrier Resources Act. The Coastal Barrier Resources Act (CBRA) is a unique federal policy that encourages conservation of coastal barriers and associated aquatic habitats by restricting federal expenditures that encourage development, such as federal flood insurance through the National Flood Insurance Program. The U.S. Fish and Wildlife Service has produced maps identifying approximately 3.1 million acres (1.2 M ha) of land and associated aquatic habitat that are federally designated coastal barriers—what are now part of the John H. Chafee Coastal Barrier Resources

System (www.fws.gov/habitatconservation/coastal_barrier.html). CBRA is a free-market approach to conservation. These areas can be developed, but the federal government does not underwrite the investments, so if a house built in a designated coastal barrier is destroyed by floods or other events, reconstruction costs are borne solely by the property owner. CBRA has saved more than $1 billion to date while discouraging development of coastal barriers.

Federal–State–Private Partnerships. Federal–state–private partnerships have been developed to protect important coastal resources. Coastal America is a nationwide partnership designed to address coastal environmental issues from the public and private sectors (www.coastalamerica.gov/). This cooperative seeks to protect, preserve, and restore the U.S. coast through leveraging capabilities, expertise, and funding among the partners. A major focus of the effort is to restore tidal flow to altered salt marshes. To date, nearly 700 non-federal organizations have initiated more than 700 restoration and protection projects in 26 states, 2 territories, and the District of Columbia. In addition, a network of coastal ecosystem learning centers has been created to raise public awareness and to increase public stewardship for coastal and marine resources. On a smaller scale, federal–state–private partnerships have been developed to bring attention to unique coastal resources in certain estuaries. One example of this is the ACE Basin Project (Ashepoo–Combahee–Edisto river basin) in South Carolina (www.acebasin.net/). In 1988, Ducks Unlimited, The Nature Conservancy, the South Carolina Department of Natural Resources, the U.S. Fish and Wildlife Service, and private landowners formed the ACE Basin Task Force. Later Westvaco Corporation, the Lowcountry Open Land Trust, and Nemours Wildlife Foundation joined the group, which is dedicated to maintaining the natural character of the basin by promoting wise resource management on private lands and by protecting strategic tracts by conservation agencies and organizations. This basin has been designated by The Nature Conservancy as one of its Last Great Places and has also been added to the national estuarine research reserve program.

Federal–State Agreements. Various government agencies may cooperate to protect coastal resources of common interest. For example, the Chesapeake Bay Agreement in 1983 committed the states of Maryland, Virginia, and Pennsylvania, the U.S. Federal Government, District of Columbia, and the Chesapeake Bay Commission to developing a bay-wide policy to protect wetlands. This resulted in the adoption of a multi-agency wetland policy that seeks to achieve a net gain in wetland acreage and function by 1) protecting existing wetlands; 2) rehabilitating degraded wetlands; 3) restoring former wetlands; and 4) creating artificial wetlands. Included within the policy, among other actions, are recommendations for strengthening regulations, improving mapping and monitoring, eliminating government programs counterproductive to wetland protection, and establishing private sector protection incentives (Chesapeake Bay Executive Council 1988).

State Coastal Policies. Some states have enacted legislation that establishes policies for conserving coastal resources including tidal wetlands. One example is Rhode Island's Coastal Resources Management Act passed in 1971 "to preserve, protect, develop and where possible restore coastal wetlands." The Coastal Resources Management Council was established to develop a coast-wide management plan and to promulgate regulations to accomplish the objectives of the law (Olsen and Lee 1993). With the aim of promoting wise use of coastal resources, these types of laws establish mechanisms to identify and articulate resource issues (habitat loss, degradation, fragmentation, pollution, overharvest, etc.), to develop specific management plans, to increase public awareness of coastal issues, to provide opportunities for public involvement, and to promulgate regulations. Similar laws and programs have been established in all U.S. coastal states and territories, typically with support from NOAA (http://coastalmanagement.noaa.gov/mystate/welcome.html).

CANADIAN POLICIES AND PROGRAMS

National Wetland Conservation Policy. The Canadian Federal Government has adopted an explicit national policy for wetland conservation (Canadian Wildlife Service 1991). This tool is a multidimensional approach including seven strategies to promote wetland conservation: 1) develop public awareness; 2) manage wetlands on federal lands and waters and in all federal programs (goal of no-net-loss for all federal agencies and crown corporations); 3) promote wetland conservation in federally protected areas; 4) enhance cooperation with provinces and nonprofit conservation organizations; 5) conserve wetlands of significance to Canadians (Table 9.8); 6) ensure a sound scientific basis for policy; and 7) promote international action (through various treaties and international agreements). Although the Canadian policy does not include a national wetland regulatory program, it does require mitigation for federal projects on federal lands and waters (e.g., National Wildlife Areas and Migratory Bird Sanctuaries), in areas where the continuing loss or degradation has reached critical or severe levels regardless of federal ownership, and where wetlands are designed as ecologically or socioeconomically important to the region. A separate document—the Federal Policy on Wetland Conservation Implementation Guide for Federal Land Managers (Lynch-Stewart et al. 1996)—explains how federal agencies are to implement the policy under the Canadian Environmental Assessment Act and specifies mitigation sequencing (avoidance, minimization, and compensation). Detailed information on Canadian wetland policies and no-net-loss

Table 9.8. Wetlands of significance to Canadians

Wetlands of significance to Canadians are:

(a) "exemplary" or "characteristic" of the wetlands dominant or rare within each of Canada's twenty wetland regions and the full range of wetland forms and types;

(b) "strategic" or "essential" to meeting a goal or objective specific to a wetland function (e.g., a marsh essential to the maintenance of a migratory bird population).

Wetlands are considered strategically significant for a number of factors including:

Wetland Quality: The wetland enhances water quality directly, in a related groundwater system, in a watershed in general, or in a domestic or other water source.

Toxics: The wetland acts naturally to retain toxic substances, thereby improving local or regional soil or water quality.

Water Quantity: The wetland enhances watershed water storage capacity, thereby affecting flood peaks and seasonal water releases.

Habitat: The wetland provides a range of valuable wildlife habitats in terms of quality, quantity, and/or diversity.

Wildlife: The population of some species of wildlife is dependent on the wetland.

Endangered Species: The wetland is habitat for any endangered species as defined by the Committee on the Status of Endangered Wildlife in Canada (COSEWIC).

Human Use: A significant portion of the economic benefits derived from consumptive uses of the wildlife, forest products, peat, rice, or other wetland natural resources is dependent on the wetland site or complex.

Recreation: Various non-consumptive and other recreational values are derived from the wetland.

Economic: The wetland has a range of significant economic uses or potentials based on the resources and values present.

Education and Research: The wetland is used or has potential for education, scientific monitoring, or research.

Uniqueness: The wetland is a unique or highly representative example of an unusual ecosystem.

Quantity: Based on regional or national thresholds established for specific forms of wetlands, the wetland is a significant component of the remaining baseline area and critical to the maintenance of local, regional, or national biodiversity.

Source: Canadian Wildlife Service 1991.

guidelines is published and available online (Lynch-Stewart 1992; Lynch-Stewart et al. 1999).

Like the United States, Canada has designated national marine conservation areas. Such areas are to be managed for sustainable use and contain smaller protected areas. These conservation areas are protected from ocean dumping, undersea mining, and oil and gas exploration and development. Traditional fishing is permitted but is managed for the conservation of the ecosystem. In eastern Canada, shallow water zones of Magdalen Islands, Gaspé Bay, and Kouchibouguac are included in the system.

Provincial Policies. In 2002, the Province of New Brunswick established a wetland conservation policy that takes a multipronged approach to wetland protection. It establishes a provincial goal of no-net-loss of wetlands, seeks to prevent destruction of provincially significant wetlands, promotes and develops wetland education and awareness programs to build support for wetland conservation, and aims to improve private stewardship of wetlands and increase acquisition of wetlands through enhanced cooperation among various levels of government and the private sector (New Brunswick Department of Natural Resources and Energy and Department of Environment and Local Government 2002). This type of broad-based policy should serve as a model for other provinces. All coastal marshes are recognized as provincially significant wetlands and are therefore accorded the highest degree of protection. All provincially significant wetlands are listed and mapped, and the information is publicly available. The Department of Environment developed a wetland and watercourse alteration permit program to administer the province's Clean Water Act (see Table 9.5). The province has also adopted a Coastal Areas Protection Policy that restricts development in salt marshes and within a 98.4-foot (30 m) buffer (New Brunswick Department of the Environment and Local Government 2002).

In 2011, the Province of Nova Scotia released a new conservation policy to provide direction and a framework for conservation and management of wetlands (Province of Nova Scotia 2011). The overall goal is "to prevent net loss of wetland in Nova Scotia through wetland conservation practices that integrate the need for wetland protection with the need for sustainable economic development, now and in the future." This policy applies to all freshwater and certain tidal wetlands (i.e., salt marshes and coastal saline ponds), with certain exceptions including wetlands on federal lands, small wetlands (<100 m^2), wetlands created for wastewater and stormwater treatment, wetlands created on uplands (ponds; not compensatory wetlands), and wetlands designated as "Marshlands" (agricultural land) under the Agricultural Marshland Conservation Act.

In 2003, Prince Edward Island adopted a wetland conservation policy that also attempts to achieve no-net-loss of wetlands (Prince Edward Island 2003). The policy now requires that the Department of Fisheries, Aquaculture, and Environment must apply the standard three-step mitigation sequencing approach when evaluating projects (avoidance, minimization, and compensation) to ensure no-net-loss of wetlands and their functions.

International Agreements

In 1918, the United States and Great Britain signed an international treaty to protect birds migrating between the United States and Canada. Later agreements with Mexico, Japan, and Russia extended similar protection to migratory species across their borders. In 1986, the United States and Canada developed a strategy—the North American Waterfowl Management Plan—to protect, restore, and manage important waterfowl wetlands along continental flyways and to bring public and private interests together to address wetland conservation needs. In

1994, Mexico signed on to the agreement. The plan is revised periodically. One of the initiatives of the plan was the creation of joint ventures in which private organizations and public agencies would work in partnership to conserve waterfowl, shorebird, and other migratory populations and habitat. Joint ventures have been established for all major waterfowl flyways. For the coastal regions of North America, the main joint ventures are 1) the Atlantic Coast Joint Venture, which focuses on black ducks and other species from Florida northward (including Puerto Rico); 2) the Gulf Coast Joint Venture, covering the region from Alabama through Texas; and 3) the Pacific Coast Joint Venture from northern California through Alaska.

Canada and the United States are signatories of an international wetland policy called the Convention on Wetlands of International Importance Especially Waterfowl Habitat (commonly called "Ramsar Convention" after the city in Iran where the first convention was held in 1971). Canada and the United States did not sign on to the treaty until 1981 and 1987, respectively. Participating countries agree to include wetland conservation in their natural resource planning efforts, promote wise use of wetlands (i.e., maintain their ecological character), and designate wetlands of international importance in their country following specified criteria (Table 9.9). A list of internationally important wetlands is maintained. As of May 2010, the list contained 1,888 wetlands including 37 sites in Canada and 26 in the United States (www.ramsar.org). The sites may contain both wetlands and uplands. Appearance on the list does not require any specific management or conservation action, but it does serve to give special attention to wetlands deemed internationally important, thereby increasing public awareness of wetland values and status. As of 2010, 15 coastal wetland complexes in Canada and 10 in the United States have been listed as wetlands of international importance (Appendix A).

Table 9.9. Criteria for designation of wetlands of international importance following Ramsar Convention

GROUP A. Sites containing representative, rare or unique wetland types.
Criterion 1. Wetland contains a representative, rare, or unique example of a natural or near-natural wetland type found within the appropriate biogeographic region.
GROUP B. Sites for conserving biological diversity.
Criterion 2. Wetland supports vulnerable, endangered, or critically endangered species or threatened ecological communities.
Criterion 3. Wetland supports populations of plant and/or animal species important for maintaining the biological diversity of a particular biogeographic region.
Criterion 4. Wetland supports plant and/or animal species at a critical stage in their life cycles, or provides refuge during adverse conditions.
Criterion 5. Wetland regularly supports 20,000 or more waterbirds.
Criterion 6. Wetland regularly supports one percent of the individuals in a population of one species or subspecies of waterbird.
Criterion 7. Wetland supports a significant proportion of indigenous fish subspecies, species or families, life-history stages, species interactions and/or populations that are representative of wetland benefits and/or values and thereby contributes to global biological diversity.
Criterion 8. Wetland is an important source of food for fishes, spawning ground, nursery and/or migration path on which fish stocks, either within the wetland or elsewhere, depend.

Source: www.ramsar.org

Management

While tidal wetlands have been used for many purposes (Chapters 7 and 8), resource management of their vegetation and hydrology has been performed for a few purposes, mainly for waterfowl habitat and mosquito control, and more recently for endangered species habitat and invasive species control. By the 1950s, wildlife management agencies and public health departments were busily impounding tidal wetlands mainly to provide more valuable waterfowl or muskrat

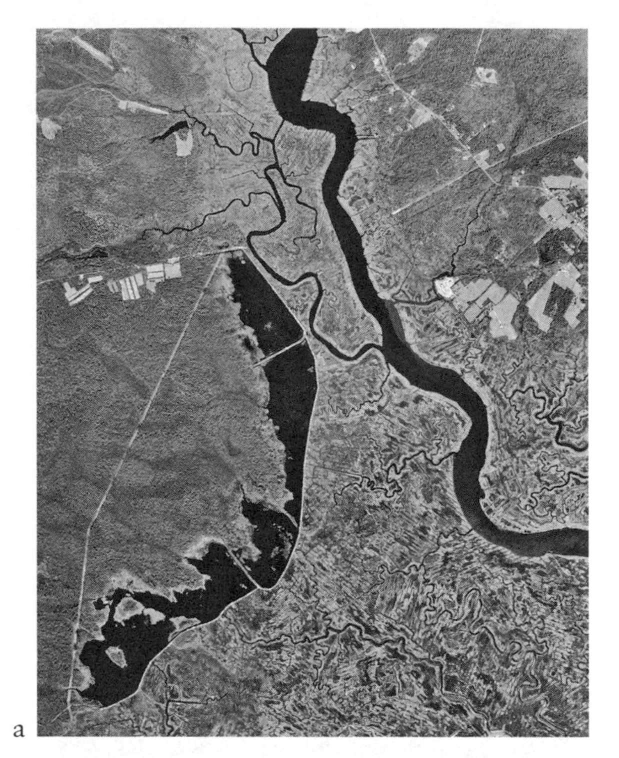

a

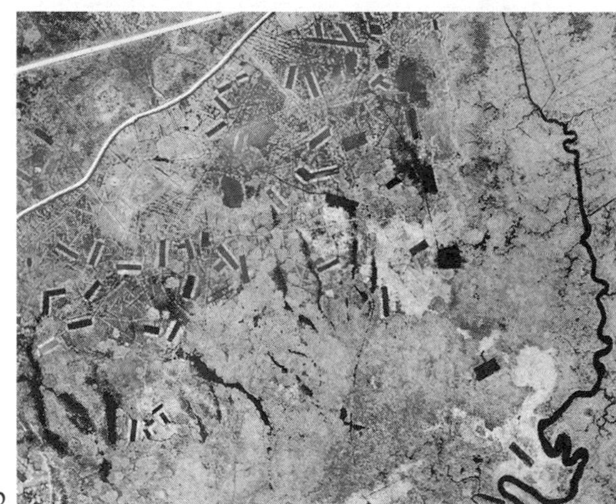

b

Figure 9.4. Waterfowl impoundments and creating dug-out ponds in tidal marshes were common wildlife management practices: (a) New Jersey and (b) Maryland.

habitat, or to control salt marsh mosquito populations (Newson 1968). Water management practices may directly or indirectly affect tidal wetlands. Wetland restoration is a type of management practice; it is addressed in Chapter 11.

Wildlife Management

In the 1930s, wildlife management as practiced today was in its infancy. Prior to that time, natural resources were generally viewed as virtually limitless or inexhaustible (pioneer days) or as materials that could be freely used without restriction (an inalienable right) as long as its use did not harm other people. Early efforts of wildlife researchers focused on management techniques, wildlife control, food habitats, diseases, and life history data, while little attention, at least through research studies, addressed the plight of marshland destruction and the effect on waterfowl (Alexander 1962). After wildlife scientists studying waterfowl food habits identified certain wetland plants important for ducks, waterfowl managers used this information to justify building brackish to freshwater impoundments in coastal marshes with water-control structures to regulate water levels for growing these duck food plants or for excavating shallow open water bodies in tidal wetlands (Valentine 1984; Figure 9.4). A government publication titled *Flyways: Pioneering Waterfowl Management in North America* (Hawkins et al. 1984) offers an interesting historical account of the evolution of waterfowl management across the continent. Seeing the need to protect wetlands from draining and filling, public wildlife agencies and private environmental organizations started acquiring wetlands as part of their wildlife management activities. Tidal wetlands were recognized as vital habitats for waterfowl and shorebird species migrating along the coast between northern breeding grounds and southern wintering grounds, while many southern marshes were prime overwintering areas for ducks, geese, and wading birds. Wildlife management practices have had a significant impact on tidal wetlands as wildlife agencies have both protected tidal wetlands and altered them for the benefit of waterfowl.

Waterfowl management has been applied to many coastal wetlands to enhance their value for ducks. Wildlife managers have used weirs, tide gates, and other structures—often in combination with dikes—to create hydrologic conditions that benefit waterfowl and furbearers. The main technique applied was impounding marshes and installing water-control structures to regulate water levels for promoting growth of desirable food plants (e.g., submerged aquatic vegetation, especially widgeongrass). This usually meant opening and closing the control structure to limit or prevent daily tidal flows to the marshes. This action restricted access to the marshes by fishes, shrimp, crabs, and other aquatic species, while changes in vegetation affected bird use. Although benefiting waterfowl and shorebirds, impoundments altered high marsh habitat and thereby reduced this habitat for marsh specialists such as rails and sparrows (Mitchell et al. 2006; Table 9.10). Marsh impoundments have also been constructed for furbearer management. While the benefits of natural tidal marshes were recognized, widespread opinion held that a high marsh produced few benefits compared with the more productive low marsh, and that its wildlife value could be improved by impoundment. The only commercial fishery that might have been threatened by high marsh impoundment was believed to be the white shrimp fishery and cautions were raised for that industry, although decreased access to impounded

Table 9.10. Effects of dikes and impoundments on vegetation structure and possible impacts on marsh avifauna

Effect of Structure	Marsh Appearance	Impact	Guild(s) Potentially Affected (typical species)
Marsh ponds not subject to tides	Ponds retain water year-round; ratio of open water to emergent vegetation increases, submerged vegetation increases	Positive	Species that feed or loaf on shallow to moderately deep pools (waterfowl, coots, herons, egrets, terns)
	Marsh surface flooded year-round; access to estuarine invertebrates limited	Negative	Ground-foraging or ground-nesting species (seaside sparrow, sedge wren)
Timing of drawdowns controllable, not dependent on droughts; wetland manager actively conducts seasonal drawdowns	Open mudflats may be available during spring or fall migration; summer drawdowns may enhance production of seed-producing annual plants; open water may be available throughout winter	Positive	Species feeding on flats (shorebirds, waterfowl, gulls)
	Marsh surface may be flooded year-round; access to estuarine invertebrates limited	Negative	Ground-foraging or ground-nesting species (seaside sparrow, sedge wren)
	Availability of open water habitat may be limited during drawdowns	Negative	Species that feed or loaf on shallow to moderated deep ponds (waterfowl, coots herons, egrets, terns)
Flood-tolerant vegetation dominates; wetland manager implements a deepwater management regime	Tall reeds (cattails or common reed) increase	Positive	Species that nest in tall dense vegetation (red-winged blackbird, boat-tailed grackle, marsh wren, common yellowthroat)
	Short marsh vegetation distribution decreases	Negative	Species associated with grassland habitats (seaside sparrow, Nelson's sharp-tailed sparrow)

Source: Adapted from Mitchell et al. 2006.

marshes by estuarine-dependent species is a common concern (Herke and Rogers 1989; Herke et al. 1992). Given the widespread use of impoundments for waterfowl management, some wildlife managers were interested in seeking ways to maintain some linkage between the impoundments and the estuary to minimize the negative impact on estuarine life. In Louisiana marshes, studies showed that 117 species of fishes and 12 crustaceans used the marshes as nursery and migrated between the Gulf of Mexico and the marshes (Herke and Rogers 1984). Moreover, there was not a time when migration by some species did not occur (e.g., cold weather or low water periods in hot weather fueled emigration from the marsh to avoid these stresses). Studies demonstrated that the number of species using the weired marsh (closed or slotted weirs) was less and the catch of most species was substantially lower than from the adjacent unaltered marsh (Herke and Rogers 1989; Rogers et al. 1994). From a weir design standpoint, the height of the weir crest did, however, made a big difference in allowing for better or no access.

Because of the trade-offs involved in marsh impoundments—favoring waterfowl or muskrat at the expense of effectively eliminating access by estuarine fishes and invertebrates and reducing export of organic matter—this traditional practice has met increased opposition from environmentalists and fisheries biologists. Marsh impoundments have altered thousands of acres of tidal marshes from Maine to Texas and along the West Coast. This was a common practice prior to the passage of tidal wetland protection laws by states and strengthened federal regulation under the Clean Water Act. Today faced with the problem of rising sea levels, wildlife managers are developing plans that will determine when to abandon such impoundments and initiate actions to restore a more natural marsh.

To counter accelerating submergence of Louisiana coastal marshes, wetland managers have developed a technique called "pond terracing." It involves using

Figure 9.5. Terraces in shallow water of a Louisiana coastal marsh. (U.S. Fish and Wildlife Service)

Figure 9.6. Seasonal burning of the marshes is done to promote the growth of desirable waterfowl food plants and in some cases to manage invasive species such as common reed (*Phragmites australis*). (U.S. Fish and Wildlife Service)

dredged material to build narrow strips of marsh within shallow open water areas. Although the pattern of created marsh is entirely artificial looking (Figure 9.5), it does create more edge marsh that provides valuable habitat for fish and waterbirds, and the activity seems to promote growth of submerged aquatic vegetation (food for waterfowl). Bird density was almost four times higher in terraced ponds as compared with unterraced ponds, and species richness was also greater (O'Connell and Nyman 2010). Wading birds made heavy use of the marsh edge habitat of terraced ponds during their nesting season. The abundance of submerged aquatic vegetation in shallow water (<1 m deep) next to the terraces appears to benefit dabbling ducks. While fish and swimming invertebrate (nekton) densities were similar in terraced and unterraced ponds, nekton density was more than two times greater at marsh edges than in open water. To increase nekton biomass over that of unterraced ponds, terraced ponds may need to have more than 25 percent of the pond occupied by terraces (Rozas and Minello 2001).

Native Americans saw that after lightning set southern marshes ablaze, vegetation re-growth attracted significant numbers of waterfowl for hunting, and therefore they set seasonal fires themselves to accomplish the same results (Yancey 1964). Lightning-induced fires are common in coastal marshes along the Gulf and South Atlantic Coasts. Today wildlife managers use fire management (prescribed burning) as a means of controlling undesirable plants and promoting the growth of wildlife food plants that attract waterfowl (Nyman and Chabreck 1995; Figure 9.6). For example, burning of brackish marshes where salt marsh bulrush and salt hay grass co-exist may favor the growth of the former whose seeds and tubers are preferred food by waterfowl. Marsh burns are also initiated to support waterfowl hunting, or by trappers seeking easier access to marshes, or by ranchers trying to improve marshes for livestock grazing. In the 1960s, up to one million acres (404,700 ha) of marsh were burned annually (Hoffpauer 1968). In 2002, the U.S. Fish and Wildlife Service alone burned 61,280 acres (24,800 ha) of coastal marsh from Virginia to Texas (Mitchell et al. 2006). Two types of burns are used in brackish marshes: 1) wet or cover burns (when the soil is saturated to the surface or flooded) to remove the aboveground vegetation; and 2) dry burns (during very dry periods) to convert a marsh with a low density of three-square or bulrush to one with a higher density of these plants by damaging the roots of salt hay grass. In Louisiana, fall or winter burns facilitate growth of Olney's three-square, while spring burns promote salt hay grass (Chabreck 1981). Where the marsh peat is dry for some depth, the burn may create shallow water bodies (Yancey 1964). Most studies on prescribed burning have evaluated the effect of fire on plants; the impact of burns on birds and marsh invertebrates has received little attention (Mitchell et al. 2006; Table 9.11).

Intentional or accidental introduction of nonnative species has presented problems to natural ecosystems around the globe. While most nonnative species are innocuous, the

Table 9.11. Impacts of prescribed burns on marsh structure and birds

Effect of Prescribed Burn	Marsh Appearance/Condition	Bird Impact	Guild(s) Potentially Affected (typical species)
Reduces or removes herbaceous vegetation	Vegetation height and density - temporary reduction or loss of standing vegetation	Positive	Waterfowl, gregarious, visual surface feeders (snow goose and blackbirds)
Inhibit woody species	Lack of shrub cover, perches (live or dead) absent, standing canopy removed	Negative	Species nesting in or foraging from shrubs (colonial waterbirds, belted kingfisher, peregrine falcon, swamp sparrow, American bittern)
Reduces or removes surface of last year's grasses	Marsh surface exposed, no cover for small mammals; rodent surface-tunnels removed	Negative	Species that nest in/on this material or that nest on mats of previous years' litter grasses; species that prey on rodents (black rail, black tern, Forster's tern, least bittern, northern harrier, short-eared owl)
		Positive	Species that forage on ground are no longer constrained by dense vegetation (possibly seaside sparrow)
Peat or root burns expose underlying mineral soil	Open water more abundant; emergent vegetation replaced by shallow to moderately deep ponds; long-term inundation inhibits re-vegetation	Negative	Species that use emergent vegetation in unbroken marsh or that nest or forage on the ground (sedge wren, marsh wren, red-winged blackbird, common yellowthroat, seaside sparrow, other sparrows)
	Ponds remain as open water, submerged vegetation increases; or pond bottoms exposed due to tides or drought	Positive	Species using broken, hemi-marsh or mudflats for feeding or loafing (American coot, pied-billed grebe, waterfowl, waders, shorebirds)
Fire suppression	Buildup of dense emergent vegetation and litter, colonization by woody species; small ponds obscured	Variable	Change in bird use depending on vegetation type

Source: Mitchell et al. 2006.

ones causing harm to the environment, economy, or public health are designated as invasive. From the environmental perspective, invasive species threaten natural biodiversity by displacing native plants and animals, creating monotypic plant communities, and disrupting natural food webs. Natural resource agencies at federal and state levels have established programs to control or eradicate species that are recognized as invasive. Federal agencies are collaborating to implement a national invasive species management plan (National Invasive Species Council 2008). The plan involves five strategic goals: 1) prevention; 2) early detection and rapid response; 3) control and management; 4) restoration of native species and habitat conditions; and 5) collaboration among international, federal, state, local, and tribal governments and private organizations and individuals. At the state level, natural resource agencies have often established multi-organizational teams and developed statewide management plans or policies that include listing of noxious species, prohibiting sale of invasive aquatic species, measures to address ballast water treatment, programs for boat checks to keep aquatic invasive species out of state waters, and public outreach campaigns. The invasive species that pose the greatest threats to tidal wetlands have been noted in Chapter 8 (see Table 8.5). Common reed is the major plant species of concern along the Atlantic Coast, while Chinese tallow is invading numerous brackish marshes along the Gulf Coast, and nonnative cordgrasses are causing problems in West Coast estuarine wetlands. Controlling these species is a management practice that may also be viewed as a restoration project. Although common reed stands expand mainly through rhizomes, a recent study has shown that seed production appears to be the primary means of spreading the species throughout an estuary. Seed dispersal allowed common reed to invade brackish marshes in both developed and undeveloped (forest) surroundings (McCormick et al. 2010). Controlling seed production may be more important for managing this invasive species than originally believed.

Nutria, a South American rodent, was introduced in Louisiana by the fur industry in the 1930s (Genesis Laboratories Inc. 2002). Accidental and intentional releases of captive nutria led to the establishment of feral populations along the Gulf Coast and in other locations including coastal Maryland, Virginia, North Carolina, and Washington State. In just twenty years, the Louisiana population was estimated at 20 million animals and by 1962, nutria had replaced muskrat as the leading furbearer in Louisiana. At Blackwater National Wildlife Refuge (MD), the nutria population grew from about 250 animals in 1968 to between 35,000 and 50,000 by the late 1990s (www.fws.gov/blackwater/nutriafact.html). A series of cold winters and a decline in their food source took a toll on most of the nutria in subsequent years, but eradicating the remaining estimated 8,300 nutria cost the government about $2 million. By 2004 the refuge was reportedly nutria-free (Fahrenthold 2004). Nutria adversely affect tidal marshes by their intensive grazing, which creates enormous barren areas (eat-outs) that eventually erode and become shallow open water.

Some barrier islands from Maryland to Texas have populations of feral livestock (horses, sheep, goats, cattle, and/or hogs). The presence of "wild ponies" in salt marshes and on coastal islands is a major tourist attraction in this region. Since they have both direct and indirect effects on marshes (e.g., grazing and trampling, Turner 1987; Levin et al. 2002), natural resource management of islands owned by federal and/or state governments (e.g., Assateague—MD, Chincoteague—VA, Currituck Banks—NC, Cumberland Island—GA, and Padre Island—TX) usually involves regulating the horse population to some degree. Herd management includes annual roundups of wild ponies for auction to the public (e.g., Chincoteague);

immunization and contraception measures; euthanizing starving, ill, or disease-carrying horses; and restricting range through fencing (Taggart 2008). Such practices are necessary to reduce the impacts of the horses and maintain healthy populations.

Since passage of the federal Endangered Species Act, special attention is given to these species when assessing wetlands. Since California wetlands have been decimated over time, the remaining wetlands contain several federally endangered species that are benefiting from wetland restoration. For example, tidal wetland alterations for transportation and flood-control projects in San Diego required mitigation for impacts to three endangered species—two birds and one plant: California least tern (*Sterna antillarum browni*), light-footed clapper rail (*Rallus longirostris levipes*), and salt marsh bird's-beak (*Cordylanthus maritimus maritimus*). In this case, 28 acres of tidal wetlands were created by restoring elevations and creating open water areas at a former dredged disposal site (Zedler 1997). The largest tidal marsh restoration project on the West Coast is located in South San Francisco Bay. Project goals are targeting the recovery of many threatened or endangered species, including the California clapper rail (*Rallus longirostris obsoletus*), California least tern, salt marsh harvest mouse (*Reithrodontomys raviventris*), and multiple plant species (Marcus 2000; Philip Williams & Associates et al. 2006; Joshua Collins, pers. comm. 2011).

Mosquito Control

Prior to the 1880s, malaria was believed to be caused by foul-smelling air ("miasma") emanating from the decomposing organic matter of swamps and rivers. A Cuban physician, Carlos Juan Finlay, observed a relationship between malarial outbreaks and peak production of the female mosquito, and in 1881 he presented a paper on this connection. This linkage was initially rejected as absurd (how could a small insect cause this much suffering?); however, Dr. Finlay spent the following years conducting experiments to prove his theory. Later, Dr. Walter Reed of the U.S. Army performed highly controlled experiments that confirmed without question that mosquitoes were vectors for malaria (Tan and Sung 2008). Since then, governments have sought to control mosquito populations in heavily populated areas by a variety of means including application of pesticides, ditching tidal marshes, and in some cases, impounding tidal marshes to eliminate mosquito-breeding habitat (see SWS Wetland Concerns Committee 2009 for review).

Tidal marsh ditching was perhaps the most widespread technique used that altered the physical structure and hydrology of tidal marshes, while spraying of a variety of pesticides had significant negative effects on wildlife (see Rachel Carson's book *Silent Spring*). Tidal marshes were partly drained to get rid of shallow marsh ponds that were believed to be the breeding grounds for the salt marsh mosquito. In the 1930s, many wetlands in the Northeast were grid-ditched for mosquito control (see Figures 8.8 and 8.11). This work was done by the Civilian Conservation Corps (CCC), part of President Franklin D. Roosevelt's plan to put young men to work during the Great Depression. The CCC drained marsh pools as well as the marshes themselves. The loss of the pools represented significant loss of valuable waterfowl habitat since the pools often contained widgeon-grass, a preferred waterfowl food. Although these areas were believed to be prime mosquito-breeding areas, later scientific research discovered that the pools were not so.

In the late 1960s, waterfowl biologists involved with mosquito control developed a marsh management technique that attempted to bring back the pools and improve the hydrology of the grid-ditched tidal marshes—Open Marsh Water Management (OMWM) (e.g., Ferrigno and Jobbins 1968; Candeletti 2007; State Mosquito Control Commission 2008). OMWM consists of creating deepwater pools in known mosquito-breeding areas in the marsh along with a series of connecting ditches to provide access for killifish that consume mosquito larvae in the pools and in the marshes at high tide (Lent et al. 1990; Figure 9.7). The system can be closed (not connected directly to estuarine waters via ditches) or open. Construction of the ponds is done by excavation no deeper than 3 feet (0.9 m) including shallow areas for specific benefits: 12 to 18 inches (30–46 cm) for migratory birds, 2 to 3 inches (5–8 cm) for shorebirds, and 7 to 25 inches (18–64 cm) for widgeon-grass production. Existing ditches in the area are plugged with marsh soil or marine plywood (Taylor 1998). The biggest problem in constructing this network is disposal of excavated material, yet most of this material can be used for ditch plugging. Rotary ditching equipment broadcasts the spoil over a large area thereby minimizing the impact. OMWM has both positive and negative effects on the marsh and its wildlife (Table 9.12). While OMWM appears to be effective in reducing mosquito populations by some accounts (Meredith and Lesser 2007; James-Pirri et al. 2009), others have questioned the effectiveness and net benefits of these practices (Rochlin et al. 2009, Potente 2010a). Concerns have been raised about impacts to marsh animals (especially marsh-interior birds such as seaside and sharp-tailed sparrows), structural changes in the marsh surface, and altered hydrology—rerouting of tidal waters through a new network of ditches (Mitchell et al. 2006; Pepper 2008; Greller 2010; Pepper and Shriver 2010; Potente 2010b; Riepe 2010). The creation of ponds may actually hamper the ability of the high marsh to sustain itself in the face of rising sea level (Gedan and Bertness 2010).

Tidal marsh management for mosquito control started in the early 20th century in San Francisco Bay and elsewhere on the West Coast. The main approach was not grid-ditching the marshes or OMWM but instead involved constructing ditches

a

b

Figure 9.7. Open marsh water management for mosquito control: (a) close-up of pool and (b) aerial view of a salt marsh pockmarked with OMWM projects adjacent to ditched marsh.

to connect natural drainage channels into suspected mosquito breeding areas, such as marsh ponds. While most of these ditches are less than 3 feet (1 m) wide and less than 3 feet (1 m) deep, the cumulative effect of ditching can greatly increase the overall extent or density of drainage networks. This plus the spoil piles left along the ditches can severely disrupt the hydrology of the marshes (Collins et al. 1986). Since most ditches tend to trap tidal-borne sediment,

Table 9.12. Potential ecological effects of open marsh water management on Atlantic Coast tidal marshes

Negative Effects	Positive Effects
Loss of salt hay grass habitat for sparrows; loss of short-form smooth cordgrass	Reduction of mosquito breeding sites
Fragmentation of inner marsh with pools and radial ditches; increased erosion and marsh loss due to sea-level rise	Increases forage fish populations and enhanced waterbird (wading birds, shorebirds) feeding habitats
Compaction of emergent marsh from use of heavy equipment on the marsh surface	Restored hydrology from ditch plugging
Change in plant community structure; invasion by salt shrubs and common reed due to elevation changes	More perches and nesting habitat for passerines (marsh sparrows and wrens) and wading birds

Source: Adapted from Mitchell et al. 2006.

they have to be periodically cleaned out, producing additional spoil on the marshes.

Besides ditching and application of pesticides, tidal wetlands have been impounded to eliminate mosquito-breeding habitat since the salt marsh mosquito does not lay eggs in water. While successful in mosquito control, the impoundments also eliminated the use of the marshes by estuarine aquatic species. Studies in the 1960s found that seasonal flooding of Florida mosquito impoundments had the same effect on mosquitoes as year-round flooding (Clements and Rogers 1964). These findings led to development of a technique called "rotational impoundment management" (RIM). Reconnecting the impoundments to tidal waters by installing culverts with water-control structures through the dikes allows for the management of tidal exchange. In Florida's Indian River Lagoon, impoundments are flooded from May to October (mosquito-breeding season) and opened to tidal flow thereafter (Poulakis et al. 2002). This management practice has increased nekton diversity and allowed recreationally important marine and estuarine fishes to utilize the impounded marshlands, while controlling mosquito breeding (Gilmore et al. 1982; Rey et al. 1990).

Water Management

Freshwater flows to estuaries may be managed in ways that purposely or indirectly affect tidal wetlands and salt balance in estuaries. In the Mississippi Delta, wetland restoration may involve diverting freshwater to certain basins (e.g., Barataria Bay and Breton Sound) to increase sedimentation for wetland establishment and, in some cases, oyster production (Chatry et al. 1983; Lane et al. 1999). The receiving basins will experience reductions in salinity with impacts to plant and animal communities (e.g., Boshart 1998; de Mutsert 2010). Such water diversions have helped rebuild brackish tidal marshes by lowering salinity and adding sediment that raises marsh elevations in the Atchafalaya River estuary (DeLaune et al. 2003). Damming rivers to provide freshwater for agricultural, municipal, and industrial uses diverts freshwater away from estuaries, causing saltwater intrusion upstream. The impact of freshwater diversion is particularly harmful to salt marshes, and "inverse or reverse estuaries" in arid or semi-arid coastal regions where high evapotranspiration and temporary closure of inlets already exerts enormous salt stress on vegetation. In these areas, extreme soil salinities exceeding 100 ppt create salt barrens devoid of vascular plants (>117 ppt for California marshes; Purer 1942). Salt marsh vegetation in these areas benefits from periodic inputs of freshwater. To counter the loss of these inputs by river diversion for municipal uses, the City of Corpus Christi, Texas, has been experimenting with discharging secondary-treatment wastewater onto salt marshes as a means of reducing hypersaline and dry conditions while adding nutrient-rich freshwater to the lower Nueces River (Alexander and Dunton 2006). This operation would also eliminate direct discharge of treated wastewater into the river and reduce wastewater treatment costs by eliminating tertiary treatment. While this discharge did increase plant cover, biomass, and diversity at the discharge point, long-term monitoring is necessary to ensure that nitrogen loading does not have negative impacts on habitat quality. Along the Central California coast, most notably in San Francisco Bay and Tomales Bay, recognition of the value of tidal marshes for upstream flood control (National Park Service and California State Lands Commission 2007) and downstream navigation (Dedrick 1979) has been increasing. In essence, estuarine marshes at low tide function as floodplains for rivers and streams, and the tidal prism of marshes helps to maintain the depths and widths of associated intertidal and subtidal channels.

Livestock Management

Former tidal wetlands (now diked) are used for pasture in some parts of North America, but management of natural salt marshes for livestock grazing is not a major issue in the United States. Salt marsh grazing is, however, a common practice in northern Europe that has received some attention from ecologists. Studies suggest that light to moderate grazing may be beneficial for these marshes by increasing plant diversity and species richness, while heavy grazing reduces these attributes and decreases populations of marsh invertebrates (Andresen et al. 1990; Bouchard et al. 2003). Moderate grazing by sheep has been suggested as a possible tool for maintaining high diversity in French salt marshes, and light grazing by cattle has been recommended as a best management practice in northern Germany. The applicability of these findings to North American salt marshes has not been evaluated.

Further Readings

Conservation of Tidal Marshes (Daiber 1986)
Statewide Wetlands Strategies (World Wildlife Fund 1992)
Our National Wetland Heritage (Kusler and Opheim 1996)
Discovering the Unknown Landscape (Vileisis 1997)
Impacts of marsh management on coastal-marsh bird habitats (Mitchell et al. 2006)
Wetlands Deskbook (Strand and Rothschild 2010)
A Sustainable Chesapeake (Burke and Dunn 2010)
Tidal Marshes of Long Island, New York (Potente 2010b)

10 Wetland Identification, Mapping, Delineation, and Functional Assessment

Wetland identification and delineation are essential parts of wetland regulations because they establish the limits of governmental jurisdiction. Once a wetland is identified on a property and targeted for alteration, an assessment of wetland functions is often required to determine lost functions and appropriate mitigation—to insure no-net-loss of functions. Assessment techniques vary depending on the scale of interest—landscape-level to project-area evaluation. The former is usually done to aid in watershed, countywide or townwide planning, or to gain perspective on the significance of a project's impacts on a larger area. Site-specific procedures address project impacts to a particular wetland and can be used to determine where environmental impacts can be minimized and to consider mitigation options.

Identification and Delineation

Wetland identification is performed for a number of reasons including: 1) to map wetlands for natural resource conservation and planning including field verification of map accuracy; 2) to assess the presence of wetland and general boundaries on property planned for purchase; and 3) to identify wetlands on a property where construction may impact a wetland that may require a federal, state, or provincial permit. The level of effort and the need for documentation varies with these purposes. The first effort—wetland mapping—involves interpretation of aerial imagery (e.g., aerial photography, digital imagery acquired by aircraft, or satellite imagery). Some field checking is performed to verify accuracy of the interpretation of general wetland boundaries and the wetland type. The second purpose (wetland determination) generally does not require extensive analysis and documentation of site characteristics as the intention is to identify the existence of a wetland and note its approximate limits (e.g., how much of the property contains wetland and is there sufficient non-wetland to build the proposed project). The third objective for wetland identification requires evaluation of regulatory criteria for verifying the presence of wetland and may involve marking a definitive boundary for the regulated wetland (wetland delineation). In some cases, states have produced tidal wetland maps that define legal boundaries, but such boundaries are subject to field review when development is proposed and usually are refined at that time.

The degree of wetness and seasonal variability in hydrology makes wetland identification relatively easy or quite challenging. Drier-end wetlands, including seasonally saturated and temporarily flooded meadows and forests and partly drained wetlands, are difficult to identify especially through remote sensing, because they are dry for most of the year and their vegetation is not much different from that of neighboring moist fields and woodlands. The longer a wetland is wet, usually the easier it is to identify. This is true even when the soils are not visibly wet at the time of inspection because wetland hydrology produces soil properties and vegetation that are generally quite different from that of adjacent uplands. Since tidal wetlands are frequently flooded they possess properties that rank them among the easiest wetlands to identify through remote sensing and by field investigations. Salt marshes should be the easiest to recognize purely from a vegetation standpoint as they support many halophytic plants that do not grow in freshwater or in adjacent uplands. Wetland identification of some freshwater tidal wetlands gets a bit trickier as they possess plants that can grow in non-wetlands. For the most part, however, they should be readily identifiable because 1) the frequency of flooding in these wetlands usually promotes the growth of plant species that occur only in wetlands (obligate hydrophytes) or that grow in wetlands more than not (facultative wetland hydrophytes); 2) tidal flooding typically leaves evidence of "wetland hydrology" (e.g., water-carried debris, water marks, or water-stained leaves); and 3) these wetlands are always associated with a water body (contiguous with tidal embayments, rivers, and streams); they are not completely surrounded by upland. Consequently, the vegetation of most tidal wetlands is distinctly different from that of neighboring lands. That is undoubtedly why most of the early state tidal wetland laws relied on indicator plants to identify the presence and limits of wetlands (Table 10.1). The upper limits of some tidal wetlands, however, are more challenging where they grade gently into low-lying fields, forests, or nontidal wetlands.

Identification and Delineation through Aerial Interpretation

Since most tidal wetlands are easily recognized by their vegetation and adjacency to a water body, they can be readily identified on remote-sensed imagery (e.g., aerial photographs or satellite imagery). Tidal wetland maps can be produced in an efficient and cost-effective manner. When some states passed laws to protect tidal wetlands, they initiated tidal wetland inventories that, for the most part, involved a combination of photointerpretation of aerial images and field checking (ground-truthing). Because of the national significance of tidal and other wetlands, the U.S. Federal Government has been mapping the nation's wetlands through the U.S. Fish and Wildlife Service, National Wetlands Inventory program (NWI) since the mid-1970s. Wetlands along the St. Lawrence River in Quebec have been inventoried with tidal marshes classified to dominant species and the results published in an environmental atlas (St. Lawrence Centre and Université Laval 1991; Létourneau and Jean 1996). While most tidal wetlands are easily identified on aerial imagery, the timing of the imagery and certain wetland types can complicate the process (Table 10.2; Figure 10.1). Wetland boundaries established through remote sensing should be considered approximate as boundary delineation is dependent on scale and image quality as well as on site characteristics (e.g., topographic relief and distinctiveness of the adjacent plant community). Wetland boundaries for wetland plant communities with unique photo-signatures such as salt marshes should be accurate; however, identifying their upper boundaries when grading into low-lying meadows, or separating forested wetlands from neighboring upland forests in flat terrain like that of the coastal plain, is more difficult. The best and most accurate boundaries are delineated through field inspections, and that is why government wetland permit programs typically require field delineations.

Identification to Verify or Produce Tidal Wetland Maps

When designed for regulatory purposes, wetland maps are essentially zoning maps showing regulated lands (i.e., some state wetland maps). When developed for natural

Table 10.1. Plants specifically referenced in state tidal wetland laws in some northeastern states for wetland identification

Plant Species	State Law Citing Species			
	NH	CT	NY	DE
Symphyotrichum tenuifolium (*Aster tenuifolius*)	—	—	—	X
Plantago maritima (*P. oliganthos*)	—	—	—	X
Pluchea odorata (*P. purpurascens*)	—	—	—	X
Ptilimnium capillaceum	—	—	—	X
Spartina patens	X	X	X	X
Distichlis spicata	X	X	X	X
Juncus gerardii	X	X	X	X
Spartina alterniflora	X	X	X	X
Salicornia maritima (*S. europaea*)	X	X	X	X
Salicornia bigelovii	X	X	X	X
Salicornia depressa (*S. virginica*)	—	—	—	X
Limonium carolinianum	X	X	X	X
Schoenoplectus maritimus (*Scirpus maritimus, S. paludosus*)	X	X	—	—
Schoenoplectus robustus (*Scirpus robustus*)	—	X	—	—
Schoenoplectus torreyi (*Scirpus torreyi*)	—	—	—	X
Spergularia salina (*S. marina*)	X	X	—	—
Spergularia canadensis	X	—	—	—
Iva frutescens	X	X	X	X
Eleocharis parvula	X	—	—	—
Eleocharis halophila	X	—	—	—
Schoenoplectus pungens (*Scirpus americanus*)	X	X	—	X
Agrostis stolonifera var. *palustris* (*A. palustris*)	X	X	—	—
Suaeda linearis	X	—	—	X
Suaeda maritima	X	—	—	X
Atriplex patula	X	—	—	X
Triglochin maritima	X	—	—	—
Solidago sempervirens	X	—	—	X
Spartina pectinata	—	X	X	—
Panicum virgatum	—	X	—	X
Typha angustifolia	—	X	X	X
Typha latifolia	—	X	X	X
Eleocharis rostellata	—	X	—	—
Hierochloe odorata	—	X	—	—
Spartina cynosuroides	—	—	X	—
Baccharis halimifolia	—	—	X	X
Hibiscus moscheutos (*H. palustris*)	—	—	X	X
Zostera marina	—	—	—	X
Ruppia maritima	—	—	—	X
Stuckenia pectinata (*Potamogeton pectinatus*)	—	—	—	X
Fucus vesiculosus	—	—	—	X

Notes: Plants are listed by current scientific name with the one listed in the law given in parentheses (to species level, with typos corrected). Where genus only was referenced, the appropriate species have been marked. Common names vary from law to law. See Chapter 5 for common name.

Table 10.2. Major limitations of mapping both tidal and nontidal wetlands through photointerpretation

1. *Target mapping unit* (tmu). A *tmu* is an estimate of the minimum-sized wetland that the producer is attempting to consistently map. It is not the smallest wetland shown on the maps. The tmu for wetlands generally varies with the scale of the aerial photography used, wetland type, project design, and funding.
2. *Use of spring imagery*. Where spring photography is used, aquatic beds and nonpersistent emergent wetlands may be under-represented; these areas are typically classified as open water, unless vegetation is observed during field investigations. Low-lying scrub-shrub wetlands such as buttonbush swamps may be submerged during springtime, avoiding photo-detection; they too would be included within mapped open water bodies. In some instances, flooded emergents may be misclassified as scrub-shrub wetlands due to misinterpretation of photo-signature. The low marsh in freshwater tidal areas is often dominated by nonpersistent vegetation; these areas may be mapped as tidal flats.
3. *Use of leaf-on imagery*. The canopy of deciduous trees makes it extremely difficult to separate all but the wettest forested wetlands from upland forests. In some areas, such as the Pacific Northwest, spring photography is difficult to acquire due to cloud cover, so leaf-on photography was used for wetland mapping. In Alaska, most of the aerial photography is acquired in midsummer, which results in conservative mapping of forested wetlands. While posing problems for forested wetland interpretation, growing season photography usually improves interpretation of aquatic beds and nonpersistent emergent wetlands.
4. *Detection of forested wetlands*. These are among the more difficult types to photointerpret; they are conservatively mapped. Forested wetlands on glacial till are difficult to photointerpret, so many of these wetlands may be under-represented. Since temporarily flooded and seasonally saturated forested wetlands are among the most difficult to identify on the ground as well as through photointerpretation, many of these wetlands are missed. This limitation is a common problem along the coastal plain and in glaciolacustrine plains. In southeast Alaska, subtle differences in the photo-signatures of evergreen forested wetlands and evergreen forested uplands result in both errors of omission and commission.
5. *Upland inclusions*. Small upland areas may occur within delineated wetlands due to minimum mapping size and on the coastal plain due to the complexity of wetland-upland interspersion. Field inspections and/or use of larger-scale photography are necessary to refine wetland boundaries for site-specific assessments.
6. *Delineating the limits of estuarine and tidal waters*. Delineation of the break between the estuarine and riverine (tidal) systems and the oligohaline (slightly brackish) segment of estuaries are approximate as they are based on available reports and/or limited field checking. In Maine, the irregular rocky coastline complicates delineation of the boundary between the marine and estuarine systems, with more open, exposed embayments classified as marine and more sheltered ones and those receiving significant inputs of freshwater typed as estuarine.
7. *Mapping of intertidal flats*. Since aerial photos are not always captured at low tide, all intertidal flats are not visible. Boundaries of these nonvegetated wetlands may be approximated from coastal and geodetic survey maps and topographic maps when necessary. Flats in sandy areas tend to shift over time, so boundaries may be quite different than depicted on the various maps.
8. *Mapping zones within coastal wetlands*. Identification of high marsh (irregularly flooded zone) vs. low marsh (regularly flooded zone) in estuarine wetlands is conservative; the photo-signatures of these zones may not be distinctive; separating the short form of smooth cordgrass from the tall form may be challenging depending on the scale of the imagery and seasonality of the image.
9. *Classification of water regimes*. Water regime designations are based on photo-signatures coupled with limited field verification; they are therefore approximate. Long-term hydrologic studies are required to accurately describe the hydrology of any particular wetland.
10. *Mapping of linear wetlands* (long, narrow). Because these follow drainage-ways and stream corridors, they may or may not be mapped depending on project objectives. In any event, there are limits to which one can identify and delineate linear features on aerial imagery.
11. *Mapping of partly drained wetlands*. Even though these may qualify as wetland if they are wet for long enough times and often enough, many of these wetlands may not appear on various wetland maps because they are difficult to identify on aerial imagery.
12. *Currentness of the data*. All wetland maps are dependent on the date of the imagery used to produce the data. Activities of humans (e.g., filling and drainage) or of beavers may have caused changes in wetlands since the aerial photos were taken, so maps do not show losses or gains in wetlands since that date.
13. *Aerial imagery reflects temporal wetness conditions* during the specific year and season when they were taken; if taken during a dry season or a dry year, many wetlands will be missed and not mapped.
14. *Mapping of drier-end wetlands* (e.g., seasonally saturated and temporarily flooded) are conservatively mapped since they are the most difficult to identify on-the-ground as well as through remote sensing.
15. *Accuracy of wetland boundaries*. The mapped boundaries may be somewhat different than if based on detailed field observations, especially in areas with subtle changes in topography (e.g., pitch pine lowlands, coastal plain flatwoods, or glaciolacustrine plains).

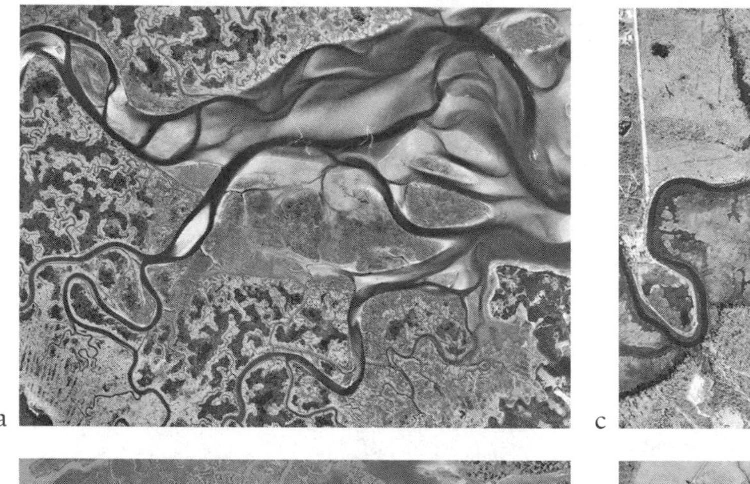

a

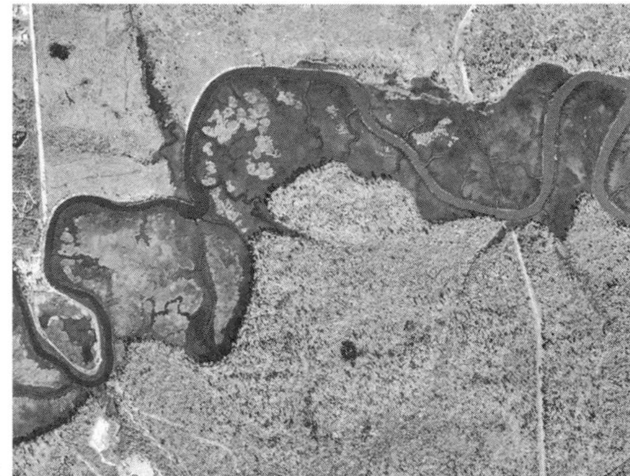

c

b

d

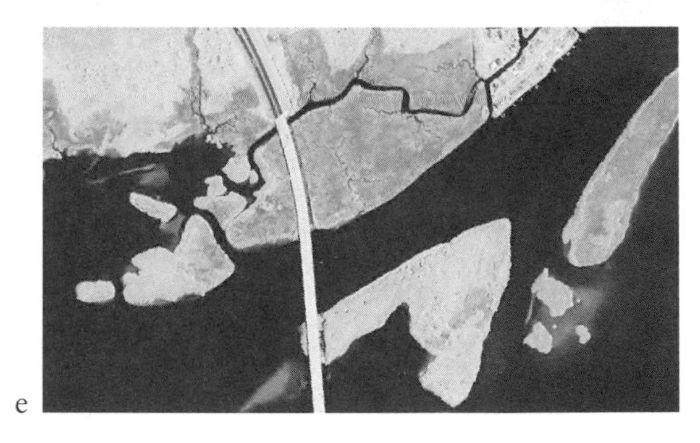

e

Figure 10.1. Examples of aerial imagery showing different tidal wetlands: (a) Massachusetts salt marsh and tidal flat (low marsh: medium gray areas along whitish to light gray sand flats; high marsh: lighter gray areas; salt pannes and pools: dark gray and black areas within the high marsh); (b) salt marsh and salina on the mainland side of a South Carolina sea island (low marsh: medium gray areas on upper left; high marsh: darker gray; salina: whitish to light gray areas); (c) change from salt marsh to brackish marsh along a South Carolina tidal creek (smooth cordgrass, *Spartina alterniflora:* smooth dark areas on right; black needlerush, *Juncus roemerianus:* lighter areas on left); (d) tidal freshwater marsh and swamp along a Maryland tidal river (marsh: light to medium gray, smooth-textured areas along river bend on left with forested swamp behind and upstream; forested wetland: dark gray, coarse-textured areas); and (e) black needlerush marshes and mangroves (mostly *Rhizophora mangle*) along the estuarine section of a Florida tidal river (marsh: darker gray areas; mangroves: lighter gray smooth or coarse-textured on this black-and-white image but red on color infrared image).

Note: For color images see ScholarWorks@UMass (http://scholarworks.umass.edu/); look under University of Massachusetts Press.

resource planning, wetland maps are advisory—a tool that can be used to identify important natural resources for conservation and where permits may be required for development (e.g., National Wetlands Inventory maps). Even states without tidal wetland laws have mapped tidal wetlands largely because of their importance to coastal fisheries. All mapping projects require some level of fieldwork to aid in wetland interpretation and to verify map accuracy.

In most cases, states producing tidal wetland maps have conducted limited field inspections as a routine part of the photo-interpretation and mapping process. Many state laws require that tidal wetlands be identified by the presence of certain species typical of salt and brackish marshes, so field identification of tidal wetlands requires plant identification skills. Field verification involves checking enough wetlands on the ground to feel confident about the interpretation of a particular photo-signature. To accomplish this, a small percentage of the tidal wetlands in a state are usually field checked, and the level of information collected will vary according to agency needs.

The tidal mapping program in Connecticut was unique among other state tidal wetland inventories in that it was done on the ground. The limits of all the state's estimated 15,000 acres (6,070 ha) of tidal wetlands were walked by field personnel who identified the boundaries on the ground using vegetation as the determining factor. The following plants were listed in the act as the major indicators of tidal wetlands: "salt meadow grass (*Spartina patens*), spike grass (*Distichlis spicata*), black grass (*Juncus gerardii*), saltmarsh grass (*Spartina alterniflora*), saltworts (*Salicornia europaea* and *Salicornia bigelovii*), sea lavender (*Limonium carolinianum*), saltmarsh bulrushes (*Scirpus robustus* and *Scirpus paludosus* var. *atlanticus*), sand spurrey (*Spergularia marina*), switch grass (*Panicum virgatum*), tall cordgrass (*Spartina pectinata*), high-tide bush (*Iva frutescens* var. *oraria*), cattails (*Typha angustifolia* and *Typha latifolia*), spike rush (*Eleocharis rostellata*), chairmaker's rush (*Scirpus americana*), bent grass (*Agrostis palustris*), and sweet grass (*Hierochloe odorata*)." Later this list was expanded to include plants of freshwater tidal wetlands, so that the law would be applied to all tidal marshes and swamps. Where upland forest bordered the salt and brackish marshes, wetland identification was easy and the boundaries were clear-cut. Where the borders became less clear, a combination of subtle differences in plant community, changes in topography, and the presence of tidal debris were used to locate the upper limit of these wetlands. The boundaries of Connecticut's tidal wetlands were marked with wooden stakes or blazes on trees with these locations recorded on large-scale aerial photographs in the field for later use in establishing official wetland boundaries through a public hearing process.

When checking the draft versions of federal wetland maps on the ground, NWI personnel use a combination of properties to identify wetlands. Many wetlands, both tidal and nontidal, are readily identified by their characteristic vegetation or obvious signs of wetness or recent flooding (e.g., water-carried debris—tidal wrack) and their landscape position (e.g., floodplain or shoreline). Where tidal vegetation and site wetness are not obviously wetland, landscape position (the proximity of the site to a tidal water body) and relative elevation are good clues for the possible occurrence of a tidal wetland. Along the Atlantic Coastal Plain, the overall low topographic relief often makes it difficult to determine the upper limits of tidal wetlands because salt marshes gently grade into nontidal wetlands with virtually no difference in dominant vegetation. Switchgrass (*Panicum virgatum*) may occupy the upper high marsh as well as the adjacent fields. In these cases, examination of soil properties is typically used to separate wetland from upland. This field verification technique has been described as the "primary indicators method" to identify wetland (Tiner 1993b). The rapid wetland identification method relies on unique properties—primary indicators—to identify wetlands and delineate their boundaries (Tables 10.3 and 10.4). Since these properties are unique to wetlands, they are powerful indicators in their own right for identifying wetlands not subject to hydrologic alteration. If obligate wetland vegetation is found, there is no need to examine soils. If such vegetation is absent (e.g., when facultative-type species predominate), then the focus shifts to the soils, looking for the presence or absence of indicators of hydric soils. Once the area is identified as wetland,

Table 10.3. List of primary indicators of wetlands

Vegetation Indicators of Wetland

- VI. OBL species comprise more than 50 percent of the abundant species of the plant community. (*An abundant species is a plant species with 20 percent or more areal cover in the plant community.*)
- V2. OBL *and* FACW species comprise more than 50 percent of the abundant species of the plant community.
- V3. OBL perennial species collectively represent at least 10 percent areal cover in the plant community and are evenly distributed throughout the community and not restricted to depressional microsites.
- V4. One abundant plant species in the community has one or more of the following morphological adaptations: pneumatophores (knees), prop roots, hypertrophied lenticels, buttressed stems or trunks, and floating leaves. (*Note:* Some of these features may be of limited value in tropical U.S., e.g., Hawaii.)
- V5. Surface encrustations of algae, usually blue-green algae, are materially present. (*Note:* This is a particularly useful indicator of drier wetlands in arid and semi-arid regions including high marsh salinas.)

Soil Indicators of Wetland

- S1. Organic soils (peats or mucks) present.
- S2. Histic epipedon (e.g., organic surface layer 8–16 inches thick) present.
- S3. Sulfidic material (hydrogen sulfide, odor of "rotten eggs") present within 12 inches of the soil surface.
- S4. Gleyed* (low chroma) horizon or dominant ped faces (chroma 2 or less with mottles *or* chroma 1 or less with or without mottles) present immediately (within 1 inch) below the surface layer (A- or E-horizon) *and* within 18 inches of the soil surface.
- S5. Non-sandy soils with a low chroma matrix (chroma of 2 or less) within 18 inches of the soil surface *and* one of the following present within 12 inches of the surface:
  - a. iron and manganese concretions or nodules; or
  - b. distinct or prominent oxidized rhizospheres along several living roots; or
  - c. low chroma mottles.
- S6. Sandy soils with one of the following present:
  - a. thin surface layer (1 inch or more) of peat or muck where a leaf litter surface mat is present; or
  - b. surface layer of peat or muck of any thickness where a leaf litter surface mat is absent; or
  - c. a mineral surface layer (A-horizon) having a low chroma matrix (chroma 1 or less and value of 3 or less) greater than 4 inches thick; or
  - d. vertical organic streaking or blotchiness within 12 inches of the surface; or
  - e. easily recognized (distinct or prominent) high chroma mottles occupy at least 2 percent of the low chroma subsoil matrix within 12 inches of the surface; or
  - f. organic concretions within 12 inches of the surface; or
  - g. easily recognized (distinct or prominent) oxidized rhizospheres along living roots within 2 inches of the surface.

Source: Adapted from Tiner 1993b.

Notes: The presence of any of these characteristics in an area that has not been significantly drained typically indicates wetland. The upper limit of wetland is determined by the point at which none of the indicators are observed. OBL = obligate hydrophytes (>99% of time in wetland) and FACW = facultative hydrophytes (67–99% of time in wetlands).

*Gleyed colors are low chroma colors (chroma of 2 or less in aggregated soils and chroma 1 or less in soils not aggregated; plus hues bluer than 10Y) formed by excessive soil wetness; other non-gleyed low chroma soils may occur due to 1) dark-colored materials (e.g., granite and phyllites); 2) human introduction of organic materials (e.g., manure) to improve soil fertility; and 3) podzolization (natural soil leaching process in acid woodlands where a light-colored, often grayish, E-horizon or eluvial-horizon develops below the A-horizon; these uniform light gray colors are not due to wetness). (These definitions differ from those later developed for identifying hydric soils; see Vasilas et al. 2010.)

Table 10.4. Steps for using the primary indicators method to verify the presence of wetlands and their boundaries for wetland mapping projects

Step 1.	Walk project site and identify different plant communities that are not significantly drained for evaluation (*see footnote for significantly drained sites). When identifying a plant community, consider both overstory and understory species and landscape position. Go to Step 2.
Step 2.	In each homogeneous plant community, determine visually whether any primary vegetation indicators of wetland are present. If necessary, representative sampling plots may be established. The following plot sizes are recommended: 30-foot radius circular plot for woody plants, and 5-foot radius circular plot for herbaceous plants. Expand plot size appropriately in high-diversity communities. If any primary vegetation indicator is present, the area is wetland; go to Step 4. If no such indicators are present, go to Step 3.
Step 3.	Examine soil properties by digging a 1-foot diameter hole that is about 2 feet deep, as necessary, and look for primary soil indicators of wetland. If any are present, the area is wetland. If soil indicators are not present, then the area is not a wetland. Go to Step 4.
Step 4.	Repeat Steps 2 and 3 for each remaining plant community. When all communities have been identified as wetland or non-wetland, go to Step 5.
Step 5.	Delineate boundaries between wetland and non-wetland plant communities. The limits of wetland are established by the point where primary indicators are lacking. By identifying several points between these plant communities, a relationship will be established that often correlates the wetland boundary with a specific elevation or contour. Use these points to identify a contour that delimits the wetland boundary, follow that contour between the two plant communities, and check periodically to ensure that relationship is still holding true.

Source: Tiner 1993b.
*Significantly drained sites should be evaluated based on criteria established by the appropriate regulatory authority. The criteria may require installing and monitoring groundwater observation wells over a multiyear period and comparing with a hydrology standard for that particular wetland type in a specific region of the country. Alternatively, drainage models may be developed for specific wetland types in certain soils to determine whether the apparent drainage is sufficient to effectively eliminate wetland functions of concern to the regulatory agency. (Ditching in salt and brackish marshes typically is not sufficient to convert them to non-wetlands as the ditches allow tides to reach farther inland.)

the next question is: Is the wetland tidal? In many cases, signs of tidal debris or evidence of bidirectional flow (e.g., observing recent water marks or exposed wet soil left by ebbing tide) are used. While this procedure is clearly not rigorous, it is a practical approach and adequate for natural resource inventories. Regulatory matters require more effort.

Identification and Delineation of Regulatory Boundaries

In the 1980s, federal agencies (U.S. Army Corps of Engineers and the U.S. Environmental Protection Agency, and later these two agencies with the U.S. Fish and Wildlife Service and the USDA Natural Resources Conservation Service) developed standardized techniques for identifying and delineating wetlands potentially subject to federal jurisdiction through the Rivers and Harbors Act and Clean Water Act (Environmental Laboratory 1987; Sipple 1988; Federal Interagency Committee for Wetland Delineation 1989; see Tiner 1999 for review of these methods). Since wetland determinations for regulatory compliance require more documentation, these methods use a combination of indicators (hydrophytic vegetation, hydric soils, and wetland hydrology) for wetland identification—the "three-parameter test" (Figure 10.2).

Regulation of wetlands has been controversial because it provides for government review and requires agency approval of private land development. The controversy reached a head in the early 1990s when the federal government began using a consistent science-based approach for identifying wetlands potentially subject to federal jurisdiction (Tiner 1999). The government was criticized for expanding its jurisdiction without adequate public review. Subsequently, the federal government provided funds to the National Research Council to do an independent review of federal wetland delineation methods and their regulatory application. While the NRC report found that the methods being employed were generally acceptable practices, the NRC team made numerous suggestions to improve the manuals and the delineation process (National Research Council 1995). One of their recommendations was to better incorporate regional differences into the manual by developing regional supplements that address locally valid wetland indicators. The Corps examined this issue and recommended development of regional supplements to the Corps wetlands delineation manual (Wakeley 2002).

Since 2005, the Corps of Engineers has been preparing regional supplements to its 1987 manual for improving wetland identification. These supplements build on more than 20 years of experience delineating wetlands on the ground, and they attempt to identify additional reliable indicators specific to particular geographic areas that may not be applicable nationwide. While the federal method still requires evidence of three parameters in most cases, it recognizes and provides guidance on situations where less than three parameters can be used to verify wetlands (e.g., disturbed areas or problematic wetlands). Tidal wetlands are readily identified following these procedures as there are plenty of indicators listed that are found in these wetlands (Table 10.5). The only shortcoming of this approach is that it still requires evidence of two or more parameters, while most tidal wetlands as well as the wetter nontidal wetlands can be simply identified by vegetation alone from a practical standpoint. Nonetheless, this is probably worthwhile to ensure that investigators do a more thorough examination of sites to be regulated under federal law, provide more documentation to regulators, and besides, even the boundaries of many tidal wetlands still require evaluation of soil properties where clear vegetation and topographic breaks are not evident. The Corps' wetland delineation manual is currently being updated and may be published for peer review in 2013.

While field investigations are typically required to identify the presence and limits of jurisdictional wetlands, wetland and soil maps may be used to help locate possible

Figure 10.2. Biologist evaluating vegetation, soils, and hydrology to make a wetland determination.

Table 10.5. Three factors for identifying wetlands subject to federal jurisdiction and some indicators that might be used for identifying tidal marshes

Factor	Typical Indicators
Hydrophytic vegetation	Rapid test: all dominants are OBL and/or FACW and no FACU or UPL dominants
	Dominance test: more than 50% of dominants are OBL, FACW, or FAC
	Prevalence index (PI) test: PI ≥3.0
Hydric soils	Histosol (A1 = organic soil), Histic epipedon (A2), Hydrogen sulfide (A4 = sulfidic odor within 12 inches), Muck presence (A8), 1 cm (0.5 in) muck (A9), 2.5 cm (1 in) muck (A10), Thick dark surface (A12), Alaska gleyed (A13), Sandy mucky mineral (S1), 2.5 cm (1 in) mucky peat or peat (S2), 5 cm (2 in) mucky peat or peat (S3), Sandy gleyed (S4), Sandy redox (S5), Loamy gleyed matrix (F2), Depleted matrix (F3), Anomalous bright loamy soils (F20)
Wetland hydrology	Surface water (A1), High water table (A2), Saturation (A3 = within one foot of the surface), Water marks (B1), Sediment deposits (B2), Drift deposits (B3), Algal mat or crust (B4), Iron deposits (B5), Water-stained leaves (B9), Aquatic fauna (B13), Hydrogen sulfide odor (C1), Oxidized rhizospheres on living roots (C3), Presence of reduced iron (C4), Other (not specified in the manual = fiddler crab burrows)

OBL = obligate wetland plants (nearly always in wetlands)
FACW = facultative wetland plants (usually in wetlands; occasionally in non-wetlands)
FAC = facultative species (found equally in wetlands and non-wetlands)
FACU = facultative upland plants (occasionally in wetlands; usually in non-wetlands)
UPL = obligate upland plants (nearly always in non-wetlands)

Source: Rabenhorst 2001; U.S. Army Corps of Engineers 2010; Vasilas et al. 2010.
Notes: The rapid hydrophytic vegetation test should suffice for identifying most tidal marshes, but evaluation of other hydrology indicators may be necessary for locating upper boundaries. Only primary hydrology indicators are listed as there should be no need to look for secondary hydrology indicators in most tidal wetlands, with the possible exceptions of those in more arid climates. A larger suite of soil indicators and hydrology indicators will be necessary to identify tidal freshwater swamps and upper edges of marshes. The alpha-numeric codes listed in parentheses are the indicator references used in the Corps supplements.

wetlands prior to such work. Two useful sources available nationwide are National Wetlands Inventory (NWI) "maps" from the U.S. Fish and Wildlife Service and U.S. Department of Agriculture soil "maps." The term "maps" is referenced in quotations because these sources now post their geospatial data online for the public to view and from which custom maps can be made rather than producing a set of hardcopy maps for distribution as was done previously. The NWI data can be accessed via a "wetlands mapper" posted at www.fws.gov/wetlands/Data/Mapper.html, while the soil data are available through a "web soil survey" mapping tool at http://websoilsurvey.nrcs.usda.gov/app/HomePage.htm. Examples of these maps covering the same area are shown in Figures 10.3 and 10.4, respectively (for summaries of these data sources, see Tiner 1999). The NWI data can be superimposed on the latest aerial image as well as recent historic images through a link to Google Earth. Some states have produced wetland maps explicitly for regulatory purposes. The wetlands and boundaries shown on these maps represent the "official" regulatory boundaries, but in many cases are subject to field review and modification as necessary. A new wetland information website—Wetlands One-Stop—established by Virginia Tech, the Association of State Wetland Managers, and the U.S. Fish and Wildlife Service posts multiple sources of wetland geospatial data that can be linked to a host of base maps and aerial images (aswm.org/wetland-science/wetlands-one-stop-mapping). The site also

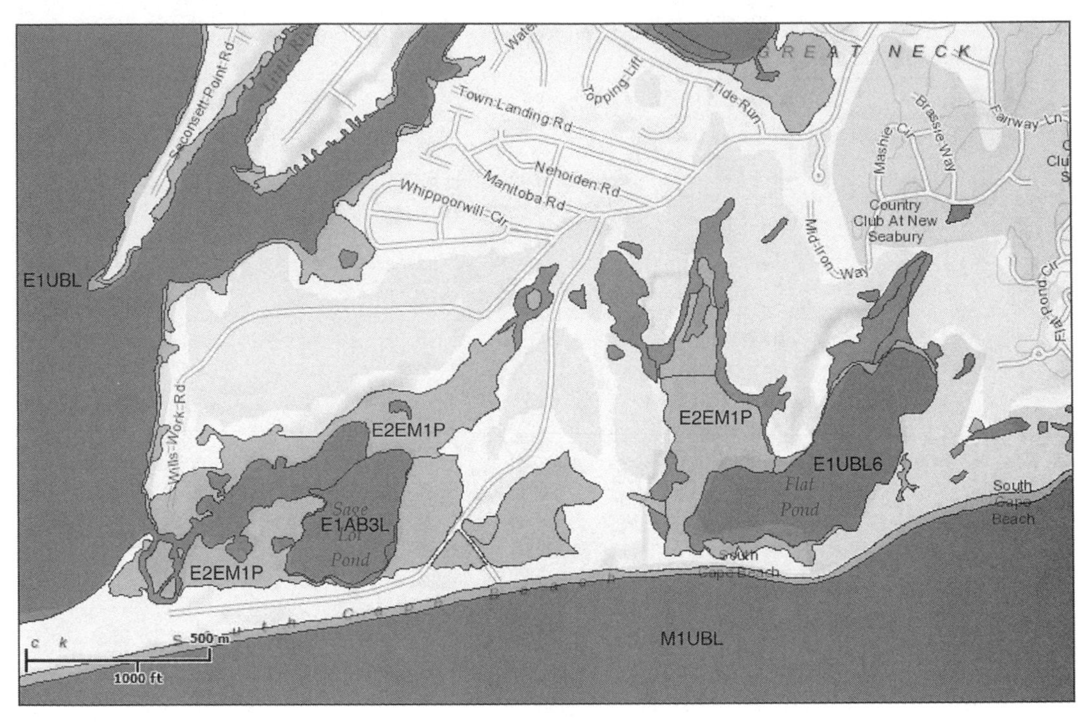

Figure 10.3. Portion of a National Wetlands Inventory map from the "Wetlands Mapper." The data may be displayed on maps as shown or on aerial imagery. Codes: E = estuarine; M = marine; 1 = subtidal; 2 = intertidal; AB3 = aquatic bed, rooted vascular; EM1 = emergent, persistent; UB = unconsolidated bottom; L = subtidal; P = irregularly flooded; 6 = oligohaline. (U.S. Fish and Wildlife Service)

Figure 10.4. Soil survey data for the site shown in the previous figure (10.3). Tidal marshes are represented by map units 66A (Ipswich, Pawcatuck, and Matunuck peats) and 54A (Freetown and Swansea mucks), while units 610 are beaches. (USDA Natural Resources Conservation Service)

has links to state websites where additional wetland data can usually be accessed.

Assessment

Once wetlands have been identified through inventories or on-the-ground techniques, an evaluation of their values, functions, or condition (health) may be performed. A large number of assessment techniques have been developed for these purposes and others are under development. The National Park Service has created an ecological assessment methods database that reviews more than 80 approaches including many dealing with wetlands (www.websitefororg.com/OldWebsites/NPS/GeneralSurveyFrameset.htm). Wetland assessment methods have been summarized by various scientists over the past two decades (Bartoldus 1999; Danielson and Hoskins 2003; Carletti et al. 2004; Fennessy et al. 2004, 2007). Assessments of wetland values emphasize the role of wetlands in protecting property, human health and welfare, and harvestable commodities, while wetland functional assessments attempt to address the natural functions of wetlands whether they are valued by society or not. Wetland condition assessments may be considered a report card on wetland health or quality, and when they are repeated over time they represent a type of monitoring (see Chapter 11). Such assessments may address the condition of features such as vegetation, soil, channel morphology, and buffer zone, but often include analysis of a wetland's departure from "reference" (least disturbed condition) based on vegetation, hydrology, soils, and wildlife. The degree of impact (severe, moderate, relatively little, or no) can be interpreted as indicators of wetland health or quality—the more degraded the less healthy or lower the quality. Degraded wetlands can be evaluated for restoration potential (see next chapter).

Most wetland assessments, whether for function, values, or condition can be performed at three scales—regional (national, state, or ecoregion), watershed (catchment or basin), and site. The larger the area of interest, the more assessment methods rely on remotely sensed data (e.g., interpreted from aerial imagery) and/or statistical (probability-based) sampling. Some regional and watershed-based wetland assessments depend solely on remotely sensed (or mapped) geospatial data, for example, the preliminary watershed-based wetland functional assessments conducted for watersheds and larger geographic areas by the U.S. Fish and Wildlife Service (see further discussion under Inventory-Based Methods below). Others use probabilistic sampling to identify wetlands for in-field assessment as exemplified by the national wetland condition assessment conducted by the U.S. Environmental Protection Agency, which used a probabilistic sampling design to identify 900 wetlands for on-the-ground evaluation (http://water.epa.gov/type/wetlands/assessment/survey/index.cfm). In 2013, this survey will provide a statistically valid assessment of the condition or quality of the nation's wetlands and rank the relative predominance of key stressors at national and regional scales. Two types of tidal wetlands will be evaluated—emergent and scrub-shrub. Site-scale assessment focuses on evaluation of individual wetlands and comparison with reference wetlands (either least disturbed wetlands or a range of types along a disturbance gradient) through making detailed measurements of environmental variables utilizing field-sampling techniques (e.g., Brinson and Rheinhardt 1996; Findlay et al. 2002; Wigand et al. 2010). Even these assessments often make use of remotely sensed information to evaluate landscape-level indicators (e.g., land-use activities within a certain distance of the wetland and the extent of vegetated buffers).

Regardless of their focus, assessment methods can be divided into three types: 1) inventory-based methods; 2) rapid assessment methods; and 3) model-based methods. Many so-called rapid assessment methods are actually developed after "model" development and although the application of the method is rapid, the creation of such methods often involves a huge investment of time and resources in model development. Examples of their application to tidal wetlands are given below. Some of these methods focus on wetland functions, while others address values, and a few combine the two to evaluate "functional values." Still others have been developed to describe wetland condition or health. Best professional judgment or field studies collecting data may form the basis of the evaluations.

Inventory-Based Methods

These types of assessment methods focus on wetland characteristics obtained from classification and inventory. They are desktop approaches usually applied to large geographic areas such as watersheds or estuarine drainage basins. The underlying principle is that one or more wetland properties can be used to predict wetland functions. These methods usually require application of remote sensing techniques, perhaps coupled with additional map interpretation and sometimes review of other data to develop a set of variables correlated to specific wetland functions. These methods are often applied at the landscape-level to present a broadscale view of wetland functions (level one assessment according to U.S. EPA guidelines; www.epa.gov/owow/wetlands/pdf/techfram_pr.pdf).

One of the earliest inventory-based methods for tidal wetlands was prepared by the Audubon Society of New Hampshire (Cook et al. 1993). The "coastal method" was intended to be used for townwide inventories of tidal wetlands. It was designed to educate local conservation commissioners, planners, and others on the functions of tidal wetlands and to provide guidance for inventorying and collecting site-specific information on particular wetlands and for using that information in natural resource planning, management, and restoration. The "coastal method" begins with an examination of existing National Wetlands Inventory (NWI) maps. Once wetlands are identified on maps and classified as either coastal/back-barrier marsh, estuarine meadow marsh, or estuarine fringe marsh, evaluation units are identified for examination on the maps and aerial photographs and through field inspection. Nine "functions" are identified for assessment:

- ecological integrity (related to level of human development in and around the marsh),
- shoreline anchoring,
- storm surge protection,
- wildlife/finfish/shellfish habitat,
- water quality maintenance,
- recreation potential,
- aesthetic quality,
- education potential, and
- noteworthiness (other features of local or regional significance).

Some of these "functions" are actually better characterized as values. For each function and each evaluation unit (section of the marsh), a series of questions must be answered and a functional index assigned (0.1, 0.5, or 1.0) for each question to determine the average functional index for each function. For each marsh, the findings for each function are graphically displayed to show the percentage of the marsh providing a certain level of function (from 0.0 to 1.0). The user then selects the best management option for the unit based on a combination of high or low average functional index ratings.

In the mid-1990s, the U.S. Fish and Wildlife Service's Northeast Region (FWS-NE) began exploring use of its NWI data for landscape-level functional assessment. The effort came about through a joint effort with the Massachusetts Executive Office of Environmental Affairs, which was starting a watershed-based wetland restoration program. That program aimed to identify potential wetland restoration sites for selected watersheds and to determine the likely benefits that could be gained from each site. For using NWI data for this purpose, the FWS-NE recognized the need to expand the information in its NWI database to provide more parameters that could be used to predict wetland functions. Abiotic descriptors for landscape position, landform, water flow path, and water body type (LLWW descriptors) were created to provide a set of hydrogeomorphic properties. These descriptors could be used in combination with existing NWI characteristics (ecological system, vegetation, water regime, and special modifiers for human impacts) to predict wetland function. Correlations between the characteristics in the expanded NWI database (NWI+) can be used to predict 11 functions for watersheds (Table 10.6 summarizes these correlations for tidal wetlands; Tiner 2003b, 2010, 2011a). The approach, called "watershed-based preliminary assessment of wetland functions" (W-PAWF), uses relationships between characteristics in the NWI+ database and functions based on a review of the literature along with the best professional judgment of wetland scientists to predict high or moderate levels of performance for wetlands at the landscape level (Tiner 2003b). Given the overall value of tidal wetlands, it is no surprise that they rank high in nearly every function, except streamflow maintenance. The FWS-NE has used this approach to produce functional assessments for several watersheds that include tidal wetlands: Casco Bay (ME), Peconic River (NY), Hackensack River (NJ), Nanticoke River (DE/MD; Tiner 2005b), and Coastal Bays (MD). It has also been applied to other geographic areas containing coastal marshes: the entire states of Connecticut, New Jersey, Rhode Island, Delaware, and Massachusetts; Long Island, New York; two coastal counties of South Carolina (Horry and Jasper); nine coastal counties of Georgia; and coastal areas in Mississippi and Texas (Tiner 2010). Reports including summaries of wetlands by type and wetland functions, and maps of significant wetlands, have been produced for these areas (Figure 10.5). These reports are posted online at: http://library.fws.gov/wetland_pubs.html or www.fws.gov/northeast/wetlands/publications.html. Maps are posted at Wetlands One-Stop.

Rapid Assessment Methods

These methods involve on-site assessment for evaluating wetlands on a given property or within a local area. They may also be used in a probabilistic sampling study designed to predict wetland condition or quality for a large geographic area. Field investigations are conducted to identify characteristics associated with a variety of functions. Before going to the field, the rapid assessment method may include reviewing maps or doing an inventory-based assessment. Most methods involve scoring a number of variables, and some methods recommend evaluating reference wetlands beforehand, to develop a standard for comparison. Some methods were designed for general planning purposes, while most were created for site-specific wetland assessment. The ones created for on-site use typically involved considerable fieldwork in the planning and design stage; given this level of effort they can be categorized as model-based types.

At least two rapid assessment methods have been created for tidal wetlands in the northeastern United States: one for New England and the other for the Mid-Atlantic region. The New England rapid assessment method was developed by coastal biologists with the Massachusetts Coastal Zone Management Program (Carullo et al. 2007); suggested modifications are based on evaluation of 81 southern New England tidal marshes (Wigand et al. 2011). A total of 25 metrics cover vegetation, soils, plant communities and marsh landscape, natural lands, and on-site marsh landscape disturbances (Table 10.7). The southern New England assessment found 14 percent of the study wetlands to be in good condition, 11 percent in poor condition, and 75 percent in moderate condition. The researchers felt that a combination of vegetation communities and marsh landscape features, on-site disturbances, and watershed natural lands appeared to be the best metrics

Table 10.6. Correlations between wetland functions and characteristics for tidal wetlands in the eastern United States used for preliminary assessments

Function	Potential	Characteristics
Coastal storm surge detention	High	Tidal wetlands (excluding diked types)
	Moderate	Diked tidal wetlands, Nontidal wetlands contiguous to estuarine wetlands (may be subject to flooding by storm tides)
Nutrient transformation	High	Estuarine vegetated wetlands, Estuarine reefs, Marine aquatic beds (rooted vascular), Freshwater tidal vegetated wetlands that are seasonally flooded or wetter
	Moderate	Freshwater tidal vegetated wetlands that are temporarily flooded-tidal
Carbon sequestration	High	Estuarine vegetated wetlands (excluding aquatic beds), Freshwater vegetated tidal wetlands that are seasonally flooded and wetter
	Moderate	Estuarine aquatic beds, Estuarine and freshwater tidal flats (organic or mud, not sand), Shallow fresh tidal ponds
Sediment and other particulate retention	High	Tidal vegetated tidal wetlands, Marine aquatic beds (rooted vascular)
	Moderate	Estuarine nonvegetated wetlands (excluding rocky shore), Marine unconsolidated shores, Marine reefs, Freshwater tidal nonvegetated (fringe) wetlands, Shallow fresh tidal ponds
Bank and shoreline stabilization	High	Tidal vegetated wetlands (except island wetlands), Estuarine nonvegetated wetlands (irregularly flooded), Marine and Estuarine rocky shores
	Moderate	Tidal nonvegetated wetlands (regularly flooded; excluding island types), Marine unconsolidated shores, Marine and Estuarine reefs (when along a shoreline)
Fish and shellfish habitat	High	Estuarine marshes (excluding irregularly flooded *Phragmites*), Tidal unconsolidated shores (regularly flooded), Marine and Estuarine reefs, Marine and Estuarine aquatic beds, Tidal freshwater marshes (excluding temporarily flooded-tidal *Phragmites*), Freshwater swamps semipermanently flooded-tidal
	Moderate	*Phragmites*-dominated tidal marshes (regularly flooded and irregularly flooded but excluding marsh interior), Estuarine tidal swamps mixed with emergent vegetation (excluding *Phragmites*), Freshwater tidal swamps mixed with emergent vegetation seasonally flooded and wetter (excluding *Phragmites*), Shallow tidal ponds (≧1 acre)
Waterfowl and waterbird habitat	High	Estuarine marshes (including *Phragmites* if regularly flooded), Tidal unconsolidated shores (excluding temporarily flooded-tidal), Marine and Estuarine reefs, Marine and Estuarine aquatic beds, Marine and Estuarine rocky shores (excluding jetties and groins), Tidal freshwater semipermanently flooded-tidal wetlands, Freshwater tidal marshes and mixed types (seasonally flooded-tidal or wetter; excluding *Phragmites*), Riverine tidal unconsolidated shores (excluding temporarily flooded-tidal), Shallow freshwater tidal ponds associated with semipermanently flooded-tidal wetland
	Moderate	Estuarine and freshwater tidal *Phragmites* wetlands contiguous with water body (excluding temporarily flooded-tidal), Estuarine mixed shrub-emergent wetlands, Other shallow tidal ponds (≧1 acre)

Table 10.6. *(continued)*

Function	Potential	Characteristics
	Wood Duck	Freshwater tidal swamps along rivers and streams
Other wildlife habitat	High	Tidal vegetated wetland complex >20 acres, Tidal wetlands 10–20 acres with two or more vegetated classes (excluding *Phragmites*)
	Moderate	Other tidal vegetated wetlands
Unique, uncommon, or highly diverse plant communities	Regionally Significant	Estuarine wetlands regularly flooded (Northeast), Slightly brackish and tidal fresh marshes and shrub swamps, Freshwater tidal cypress swamps (Mid-Atlantic), Estuarine aquatic beds (not algal)
	Locally Significant*	Urban wetlands, Oyster and mussel reefs, Freshwater tidal swamps

Source: Tiner 2003b, 2011a.
Notes: These correlations are probably applicable throughout the U.S. coast zone. * Depends on local abundance of type. The conservation of biodiversity function from an earlier version was changed to provision of habitat for unique, rare, and highly diverse wetland plant communities to highlight this aspect of biodiversity.

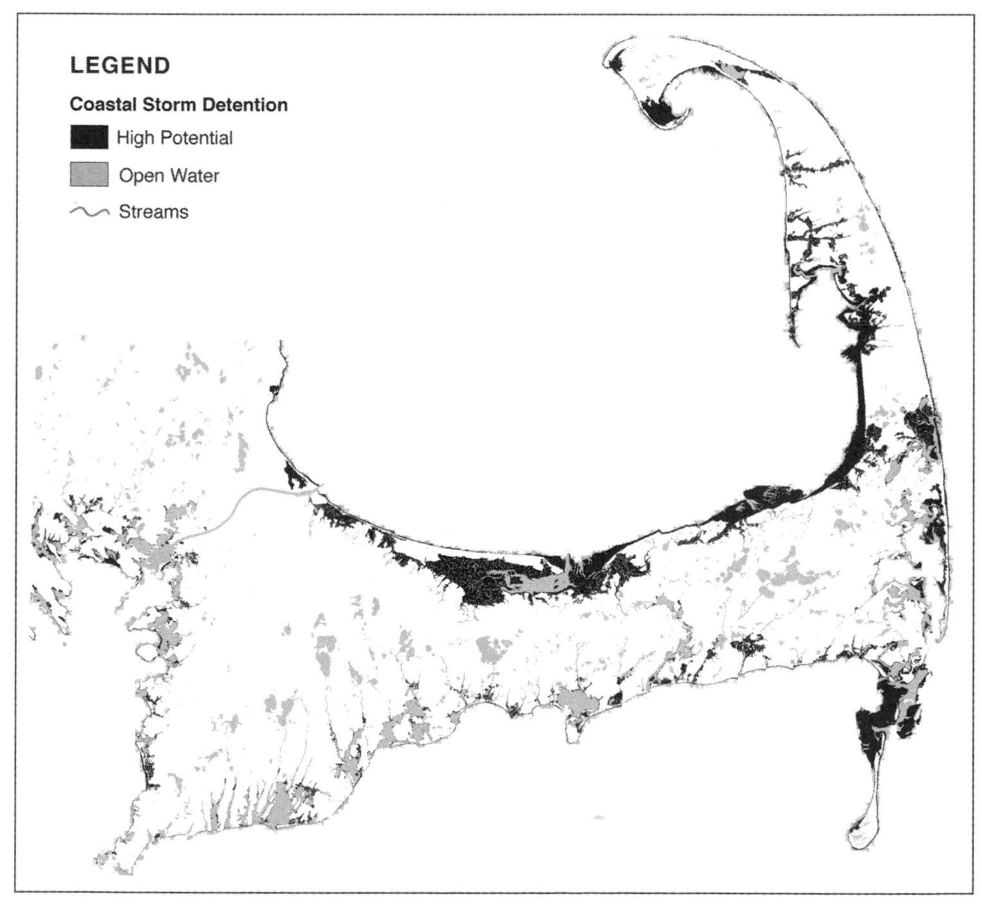

Figure 10.5. Map showing significant wetlands for coastal storm surge detention for Cape Cod, Massachusetts. (U.S. Fish and Wildlife Service)

for a rapid assessment as they were quick and inexpensive to acquire and produced condition findings similar to those from intensive assessments (Wigand et al. 2011).

The Mid-Atlantic tidal wetlands rapid assessment method (MidTRAM) uses a combination of qualitative and quantitative data to rate 20 metrics addressing hydrology, habitat, and the wetland buffer (Table 10.8; Jacobs et al. 2010). Data are collected along transects and within 820.2 feet (250 m) of the wetland edge. Scores for individual wetlands are compared with condition scores from a set of regional reference wetlands. The method also describes other characteristics including plant composition, stability of the assessment area (i.e., healthy and stable, beginning to deteriorate and/or some fragmentation, and severe deterioration and/or severe fragmentation), depth of the organic layer, and salinity. A two-person team should be able to complete one site assessment in two hours. Upon completion of the assessment, the team gives a qualitative rating of disturbance to the wetland (minimal, moderate, or high) based on

Table 10.7. Metrics adapted from the New England rapid assessment method for estuarine marshes

Attribute	Metrics
Vegetation	Average sums of the following: brackish border species, low marsh species, high marsh species, invasive species, short form *Spartina alterniflora,* short form *S. alterniflora* of high marsh, and salt marsh obligates plus a count of unique species
Soil	Average depth of blows and average of fiber volume of survey points
Vegetation communities and marsh landscape	Average area of the following habitats: brackish border, high marsh, invasive habitat, and low marsh, number of marsh habitats, sum of average area of low marsh, high marsh, and marsh border, and percent cover by invasives in the assessment unit
Natural lands	Number of natural habitats within 0.6 miles (1 km) of the assessment unit and sum of the area of natural habitats divided by the number of natural habitats
Onsite marsh landscape disturbance	Scores for the following: percentage of upland edge occupied by barrier to migration, extent of diking, extent of ditching, extent of filling and fragmented marsh, extent of flushing, and type/extent of stressors

Source: Carullo et al. 2007; Wigand et al. 2011.
Note: Vegetation data are derived from point-intercept sampling.

Table 10.8. Metrics used in the Mid-Atlantic tidal wetland rapid assessment method (MidTRAM)

Attribute	Condition Metrics
Buffer/Landscape	Percentage of the AA's perimeter with 16.4 feet (5 m) of buffer, average buffer width, percentage of developed land within 820.2 feet (250 m) from edge of the AA, landscape condition within 820.2 feet (250 m) surrounding the AA (i.e., nativeness of vegetation, disturbance to substrate, and extent of human visitation), and percentage of landward perimeter of wetland within 820.2 feet (250 m) that has physical barriers preventing wetland migration inland (barriers to landward migration)
Hydrology	Presence of ditches, fill or wetland fragmentation from anthropogenic sources, dikes or other tidal flow restriction, and localized sources of pollution in the AA
Habitat	Soil-bearing capacity, vegetative obstruction, number of plant layers (floating or aquatic, short, medium, tall, or very tall), percentage of co-dominant invasive species, and percent cover by invasive species in the AA

Source: Jacobs et al. 2010.
Notes: The portion of the wetland being evaluated is called the assessment area (AA). The term "buffer" is used to represent natural or semi-natural (undeveloped) land around the wetland.

Table 10.9. Metrics planned for use in U.S. EPA's rapid assessment for reporting on the condition and stress on the nation's wetlands

Attribute	Metric (Examples of field indicators)
Buffer	Condition: Percentage of assessment area having buffer (not developed in any way), buffer width (using 100 m as most effective) Stress: Stress to the buffer zone (numerous field indicators identified for each of the following categories of stressors—hydrological, habitat/vegetation, residential/urban/commercial, and agricultural)
Hydrology	Condition: None identified since it is reflected by other condition attributes Stress: Alterations to hydroperiod (ditches, dams, dikes/levees, culverts, tidal restriction, evidence of plant die-off, tiles, pumps/siphons, stormwater inputs, and encroachment by upland plants in assessment area), stress to water quality (point sources, sedimentation/pollutants, eutrophication, mining impacts, and salinity)
Physical structure	Condition: Topography complexity (micro- and macro-relief), patch mosaic complexity (mosaics—diversity and interspersion) Stress: Habitat/substrate alterations (subsidence, sediment splays, dredging, grazing, plowing/disking, soil compaction, fill, garbage, ruts, off-road vehicle/mountain bike trails, and fire lines)
Biological structure	Condition: Vertical complexity (number of and cover of plant strata), plant community complexity (diversity of plants with ≧10% cover) Stress: Percent cover of invasive plants (strata with ≧10% cover), vegetation disturbance (stressors indicated within three categories—human use/management, excessive grazing, and fire)

Source: U.S. Environmental Protection Agency 2011.
Notes: Examples of indicators are identified. For stressors, severity is rated as severe, moderately severe, or not severe.

general guidance and best professional judgment. MidTRAM has been applied to 53 randomly selected tidal wetlands as part of a watershed-wide assessment for the St. Jones River watershed in Delaware. Fifty percent of the tidal wetlands in the St. Jones River watershed were rated as minimally or not stressed, 32 percent as moderately stressed, and 18 percent as severely stressed (Rogerson et al. 2010). (*Note:* MidTRAM utilized some metrics from the California Rapid Assessment Method—CRAM—see Collins et al. 2008 for details.)

The U.S. Environmental Protection Agency is initiating a national wetland condition assessment with the results scheduled for publication in 2013. To evaluate overall condition and stress on the nation's wetlands, the EPA has created a rapid assessment method that examines key characteristics for randomly selected study areas (each 1.2 acres [0.5 ha] in size with a 328-foot [100 m] buffer). Four attributes of condition and stress are recognized: buffer, hydrology, physical structure, and biological structure. Rapid assessment metrics have been developed for each attribute (Table 10.9).

Model-Based Methods

Model-based assessment methods are resource-intensive (time and expense), requiring much up-front work in collecting sufficient field data to develop and test the models. Field data from a large number of wetlands are required to serve as the foundation for creating indices, which are then used to predict functions or levels of functions. Two common model-based methods are index of biotic integrity (IBI) and hydrogeomorphic (HGM) approach.

INDEX OF BIOTIC INTEGRITY (IBI)

Created as a biological alternative to traditional chemical-based assessment of water quality, the IBI approach was originally developed to assess river and stream quality based on fish communities and was later expanded to include macroinvertebrates and algae (www.epa.gov/bioindicators/html/ibi-metrics.html). Watershed health is evaluated by comparing scores of a few to many biological indices recorded within a particular watershed with a set of reference values derived from detailed field investigations along a disturbance gradient. Although the IBI method has been applied to benthic invertebrates, phytoplankton, and tidal fish in Chesapeake Bay, the approach has not been used to assess tidal wetlands, with perhaps two exceptions—Massachusetts and Delaware. A tidal marsh evaluation procedure developed by the Massachusetts Office of Coastal Zone Management employs the concept of IBI (Carlisle et al. 2004a). The method does not have a formal name, but it involved creating a set of indices to determine the ecological integrity of tidal wetlands based on the plant and macroinvertebrate responses to anthropogenic development along a disturbance gradient. In this assessment, two indices described ecological condition: plant community index (PCI) and invertebrate community index (ICI) (Table 10.10), while three other indices address stressors: impervious surface, anthropogenic nitrogen, and land use index (a multiparameter index). The ecological indices are like the IBI approach in that they can be used as biological indicators of wetland quality (or biotic integrity). For the PCI index, fixed interval sampling along six transects is recommended with the last sampling point located at the upper edge of the wetland. Sampling for the ICI is performed in the tidal creek and along the creekbank in the smooth cordgrass low marsh. A wetland condition index is calculated by summing the PCI and ICI values. In a subsequent investigation, researchers expanded the ecological indices to include assessments of avifauna and nekton, and revised the indicators for plant and invertebrate communities (Carlisle et al. 2004b). For this application, undisturbed wetlands were compared with tidally restricted wetlands (pre- and post-restoration conditions). Given the level of

Table 10.10. Ecological indicators used for evaluating Massachusetts salt marshes

Biotic Feature	Indicators
Plant community	Taxa richness, invasive species, nutrient affinity, salinity tolerance, persistent standing litter, opportunistic species, habitat affinity
Plant community (revised)	Abundance (% cover), taxa richness, abundance of invasives (% cover), abundance of low marsh grass (% cover), abundance of high marsh grasses (% cover), habitat affinity value, and *Phragmites* height
Invertebrate community	Total taxa richness, percentage of predators, percentage representation of dominant taxa group, percentage representation of dominant trophic group, percentage abundant, percentage rare, percentage *Palaemonidae* shrimp, percentage of introduced species, community taxa similarity index, community trophic similarity index
Invertebrate Community (revised)	Abundance, taxa richness, marine taxa abundance, number of crabs, shrimp abundance, number of amphipods and isopods, community composition
Avifauna	Abundance, taxa richness, number of wetland-dependent species, number of neotropic migrants, abundance of aerial foraging species, number of resident species, avian community composition
Nekton	Abundance, taxa richness, mummichog abundance, non-mummichog abundance, green crab relative biomass, mummichog length, number of transient species, nekton community composition

Source: Carlisle et al. 2004a, b.
Notes: A Bray-Curtis similarity index was calculated from the variables. Plant community and invertebrate community variables were modified for the second application of these indices, hence the revised listing.

effort required for this type of assessment, it has not been applied to more than a few study sites.

The Delaware Department of Natural Resources and Environmental Control, Water Resources Division used the IBI approach in evaluating bird habitat in a wetland condition assessment for the St. Jones River watershed (Rogerson et al. 2010). An index of marsh bird community integrity (IMBCI) was developed from visual and audio detections within a predetermined distance from a sampling point and from bird response to callback tapes for secretive birds (rails and bitterns). Each species detected was then scored by foraging habitat (generalist, marsh facultative, marsh specialist), nesting substrate (non-marsh nester, marsh vegetation nester, marsh groundnester), breeding range, and conservation status. Number of obligate marsh bird species was also factored into the index.

HYDROGEOMORPHIC APPROACH (HGM)

The HGM approach for wetland assessment was developed mainly for regulatory purposes—initially to help determine appropriate mitigation and to evaluate the success of wetlands restored or created for mitigation purposes (Brinson 1993). Other possible applications include evaluating and designing wetland restoration projects. Guidebooks are prepared for various wetland types based on field investigations of reference wetlands. Such work provides reference values or measurements that are applied to a set of indices for determining the relative functions of wetlands. Guidebooks for tidal wetlands have been prepared for the Hackensack Meadowlands, Mississippi–Alabama coasts, Northwest Gulf of Mexico and Oregon estuaries (Shafer et al. 2002; The Louis Berger Group, Inc. 2004; Adamus 2006; Shafer et al. 2007). Once developed, the HGM method can be employed as a rapid assessment tool or for more long-term assessments (Kleindl et al. 2010).

Thirteen variables were selected for evaluating northwest Gulf Coast tidal wetlands (Table 10.11). These variables are used in developing functional capacity index (FCI) for five functions (wave energy attenuation, biogeochemical cycling, nekton utilization potential, provision of habitat for tidal marsh-dependent wildlife, and maintenance of characteristic plant community structure and composition). An example of the FCI for biogeochemical cycling follows:

$$FCI = [\, V_{HYDRO} \times V_{COVER} \times V_{LANDUSE}]^{1/3}$$

where V_{HYDRO} is the value of the variable for hydrologic regime, V_{COVER} is the value of the variable for percentage of native emergents, and $V_{LANDUSE}$ is the value for adjacent land use.

When used for evaluating a mitigation site, functions are evaluated before and after project completion. The change in the FCIs gives an estimate of the functional loss or gain. Several steps are involved: 1) characterize the project area; 2) identify red flags; 3) define wetland assessment area (WAA—the area to be evaluated); 4) collect data needed to measure or estimate variables; 5) analyze data (multiply the FCI for each function by the total size of the WAA to calculate the total number of functional capacity units [FCU] for each function); and 6) apply results to project assessment and compare FCUs pre-project with FCUs post-project. Red flags include known locations of state/federal rare or endangered species, rare or unusual plant communities, unique geological features, flood-prone areas, wildlife refuges or management areas, public parks, marine or estuarine sanctuaries, archeological sites, hazardous waste sites, designated groundwater aquifers, and areas protected by various acts, international treaties, or coastal zone management plans.

The State of Oregon has produced a hydrogeomorphic guidebook for evaluating tidal wetlands (Adamus 2005). It contains what may be considered by some to be a rapid assessment method that incorporates information gathered from prior evaluation of a large subset of individual wetlands. The rapid assessment technique involves

Table 10.11. Variables and their reference standards used to evaluate functions performed by tidal marshes in Mississippi and Alabama

Variable	Reference Standard	Functions
Total % cover of native emergents	Cover >70%	Wave energy attenuation, Biogeochemical cycling, Provision of wildlife habitat, Maintenance of characteristic plant community
Mean height of tallest herb strata	Mean height at least 100 cm	Provision of wildlife habitat
% cover by woody plants	Cover ≦5%	Maintenance of characteristic plant community
% cover by invasive or exotic species	Cover ≦5%	Maintenance of characteristic plant community
Wetland plant indicator status	Sum of cover class midpoints of OBL and FACW/sum of cover midpoints of FAC and FACU species = 1 or less	Maintenance of characteristic plant community
Aquatic edge	Well-developed tidal drainage network, very narrow fringe	Nekton utilization potential, Provision of wildlife habitat
Hydrologic regime	Free exchange of tidal waters	Biogeochemical cycling, Nekton utilization potential
Nekton habitat diversity	Five or more habitats (low marsh, high marsh, intertidal creek, subtidal creek, pond, mudflat, submerged aquatic vegetation and oyster reef)	Nekton utilization potential
Wildlife habitat diversity	Lack tall herbs or tall vegetation restricted to a narrow fringe (<10 m)	Provision of wildlife habitat
Mean marsh width	Mean marsh width ≧100 m	Wave energy attenuation
Wave energy exposure	One or more shorelines located directly along water's edge of estuary or a bay	Wave energy attenuation
Adjacent land use	>95% of perimeter is bounded by undeveloped naturally vegetated areas or open water	Biogeochemical cycling
Wetland patch size	<0.04 hectare	Provision of wildlife habitat

Source: Shafer et al. 2007.
Notes: Names of variables and functions have been simplified for this table. Index values range from 0.0 to 1.0; requirements for the standard are listed.

analyzing site characteristics based on direct field observations, direct measurement, and review of aerial photos or maps. The guidebook includes a thorough literature review, geospatial data, Excel spreadsheets, and standard forms (Adamus 2005, 2006; Table 10.12). Using this approach, a single wetland may be evaluated in one day. The method also includes an option for users to place a value on specific functions (e.g., primary production/export, nutrient cycling/pollutant processing/sediment stabilization, invertebrate habitat, anadromous fish use, resident and marine fish use, waterfowl and shorebird use, and habitat for other wildlife) using professional judgment.

Since the model-based approaches are labor intensive during the model formulation and calibration stages, only a few have been developed for tidal wetlands. Moreover, the need for such rigorous assessments may lessen given the fact that tidal wetlands are widely recognized as being among the most valuable wetlands and are generally considered off-limits to development (as opposed to nontidal wetlands). Model-based approaches may, however, be more useful for planning and evaluating success of wetland restoration sites.

Hybrid Methods

A couple of methods have combined landscape-level assessment with on-site evaluations. Oregon's rapid wetland assessment protocol (ORWAP) is one such hybrid assessment approach that produces scores of wetland functions, values, ecological condition, wetland stress, and sensitivity (Adamus et al. 2009). The development of this method required extensive fieldwork that produced data to formulate a standardized scoring system. It begins with an in-office review of existing geospatial information (maps and aerial photographs) to answer landscape-level questions. This is followed by a field assessment of the site in which a number of indicators are evaluated including the specific types of wetlands present, wetland type of conservation concern, tidal/nontidal hydroconnectivity, groundwater, outflow duration, outflow confinement, inlet and outlet, island, upland edge complexity, wetland size, wetland size

Table 10.12. Wetland site and landscape context properties considered in Oregon's rapid assessment method for tidal wetlands

Indicators of Risks to Tidal Wetland Integrity	Rapid Indicators of Estimating Functions (estimated) (*continued*)
100 foot (30.5 m) buffer	Types of freshwater sources entering wetland
Chemical inputs (toxicity, contaminants)	Wetland width
Nutrient inputs (source, loads)	Maximum width of adjoining tidal flat
Sediment inputs	Number of shorebird roosts within 1.5 miles (2.4 km)
Soil disturbance	Part of uninhabited island
Effect of diking	Fetch
Foot traffic	Vegetation forms in wetland
Boat traffic (proximity)	Percentage of border bounded by alder
Home distance	Presence of eel-grass
Road proximity	Soil texture
Invasives (plants and animals)	Tidal marshes percent of estuary's acreage
Instability (vegetation, historic changes in wetland, etc.)	Estuary's connection to the ocean
	Dominance of undiked marshes in the estuary
Direct Indicators of Wetland Integrity	
Channel proportions (width and depth)	Rapid Indicators Requiring Aerial Photos or Measuring Equipment
Vegetation (numerous parameters derived from plot sampling)	Percentage of land within 1.5 miles (2.4 km) that are nontidal wetlands, water bodies or cropland/pasture on flat terrain
Rapid Indicators of Estimating Functions (estimated)	Percentage of surrounding area that is developed or persistently bare ground
Flooding	Internal channel complexity
Shading	Number of internal channel exits
Bare substrate (including pools and pannes)	Number of internal channel junctions
Pannes	Salinity difference in wetland pool/panne versus adjoining estuary (refractometer)
Transition angle along wetland edge	
Upland edge	
Large woody debris in channel and wetland	
Length of driftwood along marsh–upland boundary	
Length of fish-accessible nontidal tributaries entering wetland	

Note: Indicator names have been simplified; see Adamus 2006 for details.

uniqueness in watershed, historic hydrologic connectivity, contributing area present, nonvegetated surface in the contributing area, upslope storage, transport from upslope, native versus nonnative species, known water quality issues in the input water, and known water quality issues below the wetland. After completing the forms and entering the results into a spreadsheet, wetlands are assigned standardized scores for 16 functions (e.g., water storage, sediment retention and stabilization, phosphorus retention, nitrate removal and removal, thermoregulation, carbon sequestration, and several habitat functions) and 14 related values that are then used to compute relative scores for grouped services, such as water quality support, fish habitat support, aquatic support, terrestrial support, public use and recognition, and provisioning services (e.g., timber harvest and other commercial products). This method was modified in 2010 (Adamus et al. 2010).

Another hybrid method has been adopted for use in Nova Scotia for regulatory purposes to assess the condition and functions of the wetlands at a proposed project site and the affected watershed (Nova Scotia Environment 2011). The method called NovaWET was initially developed for teaching practicing environmental scientists and provincial agency personnel how to evaluate the functions of wetlands at a given project site and in a watershed context. The method utilizes a desktop assessment to gain a watershed perspective on wetland functions and a rapid assessment for a site-specific evaluation. It is intended to provide Nova Scotia Environment (NSE) with basic information for wetland alteration or environmental assessment applications on project site wetlands, the surrounding landscape, and the contributing watershed to help evaluate the significance of wetlands in a project area that may be affected by proposed alterations. The information gives NSE pertinent information for making their determination to approve or deny the project and for determining appropriate compensation for adverse impacts. A NovaWET assessment is required for projects where proposed impacts are 0.5 ha or more. NovaWET is essentially a two-step process involving off-site assessment using maps and aerial imagery (modeled after the U.S. Fish and Wildlife Service's watershed-based method; Tiner 2003b) and on-site evaluation to confirm or modify the off-site interpretation and to record site-specific information on wetland characteristics and indicators of wetland functions in the project area. The on-site portion of this method was adapted chiefly from Minnesota's rapid assessment method (MnRAM 3.2; Minnesota Board of Water and Soil Resources 2008) but was significantly expanded with contributions from other assessment methods (Collins et al. 2008; Adamus et al. 2009; North Carolina Department of Environment and Natural Resources 2009), from method reviews (Fennessy et al. 2004; Hanson et al. 2008), and earlier versions of NovaWET created by Ralph Tiner. The end result of the analysis is a characterization of the wetland condition and likely functions, the condition of the wetland buffer, the relationship between wetlands in a project area and neighboring wetlands and water bodies, and a general assessment of the contributing watershed. The method does not generate a numeric score or compare the subject wetland with similar wetlands of the type, because thus far no detailed studies of wetlands in Nova Scotia have been performed to develop data for a representative suite of reference wetlands. It does, however, identify red flags (critical wetland functions or watershed conditions) that may likely preclude project approval by NSE.

Further Readings

Wetlands: Characterization and Boundaries (Committee on Characterization of Wetlands 1995)

Wetland Indicators: A Guide to Wetland Identification, Delineation, Classification, and Mapping (Tiner 1999)

11 Tidal Wetland Restoration, Creation, and Monitoring

Over the past 25 years, much has been learned about wetland restoration and creation (e.g., Kusler and Kentula 1990; National Research Council 1992, 1994, 2001; Zedler 2000, 2006; Teal and Peterson 2005; Falk et al. 2006; Mitsch 2006). Wetland regulations and government regulatory programs are largely responsible for promoting wetland restoration and creation as compensation for permitted wetland losses (see Mitigation in Chapter 9). Monitoring of mitigation projects has been performed to determine compliance with permit requirements. Many government agencies have also initiated proactive wetland restoration programs independent of the regulatory process to simply restore wetlands because they provide valued environmental services to society. Consequently, government agencies, cooperating institutions and organizations, and private consulting firms have gained hands-on experience in wetland restoration, creation, and monitoring. In this chapter I describe in general terms several types of projects, project planning, data needs, and criteria that may be used for tracking progress. I also provide examples of specific projects and their accomplishments to illustrate real-world applications. The discussion focuses on restoration and creation and monitoring for those types of projects. Other forms of monitoring may be conducted to track ecological changes in coastal wetlands in response to climate change (e.g., sea-level rise and changes in regional precipitation) or tectonic activity, but these are not emphasized here.

Defining Restoration, Creation, and Enhancement

Classifying projects as restoration, creation, and enhancement projects may seem at first glance to be rather obvious and unnecessary, but in many cases the differences are subtle. The terms need definition and some examples would be useful before discussing the subjects in more detail and addressing project planning, goal setting, and design monitoring.

Wetland restoration can be separated into two types: re-establishment and rehabilitation (Figure 11.1). Re-establishment involves bringing a wetland back at a site where it formerly existed; in other words, restoring a former wetland. This type of restoration will produce an increase in wetland extent (area) as well as functions. An example would be restoring tidal flooding to a former wetland that had been diked, drained, and previously converted to cropland. Another example would be removing fill from former tidal wetland that had been filled for a parking lot.

In contrast, rehabilitation deals with improving conditions for an existing wetland that is degraded. Although this action does not increase wetland area, it seeks to restore lost or diminished functions, leading to a net gain in wetland functions.

a

b

c

Figure 11.1. Three examples of potential restoration projects: (a) diked former tidal wetland now pasture (Bandon Marsh National Wildlife Refuge, OR); (b) filled wetland lacking structures; and (c) common reed marsh (*Phragmites australis*) has established on this former salt marsh whose tidal connection is altered by a causeway and an undersized culvert. (a: Roy Lowe, U.S. Fish and Wildlife Service)

A common example of rehabilitation is restoring tidal flow to a tidal wetland with limited tidal influence due to tide gates or undersized culverts.

Wetland creation involves building a wetland where one did not exist since Colonial times, for example. It is not the site of a former wetland, but either upland (dryland) or water. Constructing a wetland in those locations seeks to increase wetland extent and functions. Wetlands can be created by excavating an upland (dryland) area to an elevation subject to frequent tidal flooding or by depositing sufficient fill in water to raise the bottom to the level of the intertidal zone. Creating a marsh by adding fill to a mudflat is typically considered wetland creation, while it could be viewed as a type of enhancement since the mudflat is technically a wetland. Sometimes the line is blurry between what constitutes wetland restoration or creation. For example, if the shallow bottom to be filled was a former marsh that was lost due to coastal subsidence and erosion, creating marsh there would be considered wetland restoration, although the restoration practices would be more typical of those used in creation projects.

Wetland enhancement is the act of modifying a wetland to attain a level of a desired function above or different from that which the wetland type is normally capable of reaching. The targeted function may even be a function that the wetland did not perform. The wetland is altered in some way to satisfy the management objective. It may change the wetland to another type (e.g., tidal marsh to aquatic bed) or affect the wetland's hydrology, vegetation, wildlife use, or other properties. Enhancement could be applied to a pristine wetland or a degraded one. In either case it modifies the wetland's characteristics for the benefit of one or more functions at the expense of other functions. Examples include diking of natural tidal wetlands to create waterfowl impoundments or burning marsh vegetation to encourage the growth of "desirable" plants for waterfowl or muskrat. Whether enhancement is good or bad is in the eyes of the beholder or resource manager.

The difference between rehabilitation and enhancement may be subtle or quite apparent. As mentioned above, rehabilitation refers to improving the level of performance of a degraded wetland by removing or minimizing the effect of some type of significant alteration or disturbance (rehabilitating impaired functions or reducing the extent of degradation). In contrast, enhancement is promoting one or more wetland functions over other functions that a wetland is now performing. The objective of enhancement is not to restore a lost or degraded function, but to amplify certain desired functions (e.g., waterfowl habitat or stormwater detention), which typically results in a different wetland type in terms of vegetation and hydrology, for example. Oftentimes, wetland enhancement requires installing structures to increase water levels over those that normally occur at the site. The presence of a dike and water-control structure that converts a salt marsh to a brackish marsh and pond is an example of enhancement, while the installation of a weir in a ditch or a ditch plug is a comparative example of restoration for the salt marsh. The latter action aims to restore the original hydrology by negating the impact of the drainage ditch (the disturbance factor) and restoring a tidal hydrology more like that of a natural marsh. In the first scenario, the dike will amplify water levels over those that would normally occur in a salt marsh and thereby increase its capacity to provide one or more functions, such as providing open water for waterfowl at the expense of other functions (e.g., habitat for sharp-tailed sparrows and clapper rails).

Agency Collaborations

While most created wetlands and some restoration may be performed by private consultants for mitigation purposes, tidal wetland restoration projects are often government-financed projects requiring investments of funds and personnel from multiple agencies and organizations. Since the 1990s, the U.S. Federal Government and most, if not all, coastal states have been actively engaged in restoring tidal marshes. The magnitude of these projects and the need for permits necessitates interagency cooperation and participation by other organizations to assist with public outreach and monitoring. Wetland restoration therefore tends to be a collaborative effort with agencies, organizations, private contractors, and individuals working together for the common good. The National Marine Fisheries Service, Fish and Wildlife Service, Natural Resources Conservation Service, Corps of Engineers, and Environmental Protection Agency are common participants in proactive tidal wetland restoration at the federal level in the United States, while the last two agencies require restoration or creation through the federal permitting process. Following oil spills and natural resources damage assessments, multiple agencies (federal, state, and tribal) are involved in restoration as required under the Oil Pollution Act of 1990. State natural resource agencies typically collaborate with federal agencies on restoration projects related to both mitigation and proactive initiatives.

In 1991, the U.S. government formed the Coastal America Partnership consisting of representatives from 9 federal agencies for the purpose of promoting and coordinating protection, preservation, and restoration of coastal resources (www.coastalamerica.gov/index.php). Today, 16 federal agencies are working with state and local governments and private organizations to accomplish these objectives. In 1999, Massachusetts Executive Office of Environmental Affairs worked with the U.S. EPA and the Gillette Company to establish the Massachusetts Corporate Wetlands Restoration Partnership in which environmentally responsible companies could work openly with government agencies to support wetland restoration initiatives. Within six months, 17 corporations joined the partnership and contributed more than $1 million for projects. Since 2000, Coastal America has expanded this Massachusetts initiative to the national and international levels. New England states and Canada's Maritime Provinces are coordinating their restoration activities through the Gulf of Maine Council's Habitat Restoration Working Group (http://restoration.gulfofmaine.org/). Similar multi-agency initiatives are under way elsewhere (e.g., Louisiana and San Francisco Bay).

Goals and Objectives

From the outset, project goals and objectives must be established and clearly defined. For tidal wetlands, the basic goal may be self-evident—restoring or creating a tidal wetland like other relatively undisturbed wetlands in the locale, or returning a wetland to its pre-disturbance condition. Under ideal circumstances, the end result should be a self-sustaining wetland that will function like unaltered or minimally impacted wetlands in the locale and be subject to the same natural processes that create, maintain, and reshape wetlands over time. "Like" is the operative word for we cannot expect to produce an exact replica of a natural marsh in either form or function in the short-term. There are too many uncontrollable variables involved in time and space. The term "functional equivalency" has received criticism in part because evaluations focus on structure (condition at a point in time) and assume an unproven linear relationship with function (Zedler and Lindig-Cisneros 2000). While restored or created wetlands may appear to be structurally similar to natural marshes from a vegetation standpoint in a few years, equivalency in performance of multiple functions (whether the wetland is a fully working estuarine ecosystem) requires more effort to evaluate and time (Edwards and Proffitt 2003). Most evaluations are not carried out over a long enough time to evaluate equivalency in any regard. For example, it took 20 years for salt marsh snail (*Melampus bidentatus*) populations in a restored wetland to reach densities comparable with those in a local reference wetland (Warren et al. 2002). (See Project Success, Compliance, or Progress later in this chapter.)

For designing the project and evaluating its success, specific objectives are needed. Some objectives include restoring tidal flow (fully or partially depending on concerns for local flooding of property), establishing a specific wetland plant community, managing invasive species, creating habitat for certain wildlife (e.g., waterfowl or endangered species), or producing an area that provides one or more other wetland functions or valued services (e.g., flood storage, shoreline stabilization, or water-quality improvement). Besides recovering lost functions and values, wetland restoration may help reduce a public health and safety concern. For example, in Connecticut, tidal marsh restoration is recognized as a mosquito control technique for diked and drained marshes (Rozsa 2005).

Basic Considerations in Project Planning

In all cases involving existing wetlands, the level of functional impairment or degradation should be evaluated and documented at the outset of project planning to clearly separate significantly altered wetlands (i.e., where a prior disturbance or alteration has caused significant impairment of one or more of the functions of that wetland type) from altered wetlands with relatively insignificant functional changes. Restoration efforts should ideally be targeting the significantly altered wetlands where the most benefits can be achieved, given cost considerations and feasibility. An example of the difference might be illustrated by considering hydrologic restoration for one of two tidally restricted wetlands. Each site has diminished tidal flow due to undersized road culverts with seemingly healthy marsh on the seaward side of the road. One of these wetlands is dominated by the invasive, non-native common reed (*Phragmites australis*) while the other is represented by a mixture of typical salt and brackish marsh plants with common reed limited to a narrow band along the upland. The severity of the tidal restriction seems to be greater in the former marsh and restoring its hydrology should therefore yield more significant habitat improvement. Of course, all cases are not that straightforward. Some basic questions to consider in planning wetland restoration or creation are given in Table 11.1. Besides being vital for project design, answering these questions will facilitate establishment of measurable parameters for monitoring purposes and evaluating project success. Also it must be emphasized that wetland restoration and creation projects are likely to need permits from regulatory agencies and, in some cases, environmental impact statements (EIS) may be required (see EIS for Giacomini wetland restoration project; National Park Service and California State Lands Commission 2007).

Table 11.1. Fundamental questions to consider in planning tidal wetland restoration or creation

1. What type of tidal wetland is desired? (Specifying this may be required by regulatory agencies as a condition for approval of a construction project impacting wetlands.)
2. How do existing government wetland regulations apply to the proposed restoration or creation project? (Consult federal, state, and local agencies for specifics; permits are often required for proactive restoration projects.)
3. Where are suitable restoration or creation sites located and are they available for use? (This step involves locating available sites most suitable for restoration/creation and securing landowner approval.)
4. Is the intended project a restoration, creation, or enhancement project? (Relates to the reason for initiating the project—mitigation or proactive restoration, creation, or enhancement—and is vital for accomplishment reporting; used to help assess how the nation is faring re: "no-net-loss of wetlands.")
5. What should the project size be? (This is dictated by the regulators for mitigation projects and by site conditions for restoration projects, although there are other considerations, e.g., cost for proactive restoration.)
6. If the project is a mitigation project, should the project be located on-site, off-site in the same watershed, or off-site outside the watershed? (This will be answered by the regulatory process.)
7. What hydrologic conditions need to be established? (The proportion of low marsh to high marsh depends on the project goals.)
8. What plant communities are desired? (Considers lost wetlands either at proposed development site or historic types; for latter, review literature and reference wetlands.)
9. Is planting or seeding required to promote such communities? (Check local wetlands for composition of existing wetland plant communities and use local plant materials for planting.)
10. What faunal species and kinds of animal use are desired?
11. How will the proposed project improve wetland functions over existing habitat? (Requires an assessment of functions of existing site and comparison with expected functions from project.)
12. What is an acceptable risk of structural failure (e.g., dike, weir, ditch plug, or water control device) that would require repair at some frequency (e.g., 5, 10, 25, 50, or 100 years)? (The best projects are designed to be self-sustaining with no structures requiring maintenance.)
13. What other risk factors exist that may compromise project success? (Consider factors such as invasive plant species; release of possible contaminants such as methyl mercury; nuisance animals including geese, nutria, and other heavy grazers; erosive potential of site; sea-level rise; excessive flooding during wet years; sedimentation rates; and adjacent land use; may have to install devices to reduce herbivory, e.g., goose netting).
14. What features could be monitored, at what frequency, and what criteria will be used to determine project success?
15. How will site be managed in the long-term?
16. Does the restoration/creation plan have a provision for adaptive management? (This may affect success criteria.)

Wetland Restoration

Regulatory agencies usually prefer wetland restoration over wetland creation for mitigation projects because former wetlands are located in landscape positions that at least were once suitable for wetland establishment. The "raw materials" for success may be there (e.g., former marsh substrates and natural seedbanks). Restoring wetlands (both tidal and nontidal) has also become a major conservation initiative engaged by natural resource agencies and organizations. Since many tidal wetlands have been diked or crossed by roads and railroads causing a restriction of tidal flow and changes in vegetation and wildlife use, there are literally thousands of opportunities to restore tidal wetlands across the country. Note that restoration can sometimes be accomplished naturally without human investment. For example, storms can breach dikes around former tidal or tidally restricted wetlands, thereby re-establishing or improving tidal flowage. In the Bay of Fundy, such passive restoration has led to the re-establishment of salt marsh on former agricultural land ("reclaimed" marshland). The recovering marsh has attained plant cover and productivity similar to natural marshes (MacDonald et al. 2010). Of particular interest for this marsh was an increase in pool density over unaltered wetlands—13 percent of marsh versus 5 percent, respectively. This should yield significant benefits for certain wildlife, especially waterfowl.

Common Types of Restoration and Challenges

Pre-existing conditions dictate whether the wetland restoration effort will be a re-establishment project or a rehabilitation project (Table 11.2). Former tidal wetlands include filled areas (now dry land, such as a dredged spoil island), dredged areas (now deepwater habitat), and nontidal freshwater wetlands that were once tidal. Coastal wetlands may be restored in the former two areas by removing fill material or rebuilding elevations in dredged areas, respectively. The latter is not practical in most situations unless part of a dredging material disposal project, but may be practiced on a smaller scale through the infilling of ditches. The objective of a re-establishment project is to put a coastal wetland back on the landscape where one previously existed. This restored wetland will not, of course, be exactly the same as the original, yet it should, if successful, possess many of the same properties and should function similarly in the long-term to neighboring tidal wetlands. Rehabilitation projects seek to recover lost or damaged functions. Common examples of this type are a tidally restricted wetland that may now be a freshwater wetland lacking any tidal flow and a brackish marsh with significantly reduced tidal flowage (e.g., Burdick et al. 1997). By providing a tidal connection for the former or expanding the opening of an existing culvert for the latter, significant tidal flow can be returned to these wetlands. This may sound simple enough, and it really is in undeveloped areas, but there are issues to consider. For example, in developed areas, homes or other buildings have often been built along the edges of coastal wetlands complicating the restoration process. People living there will be concerned that restoring tidal flow will lead to flooding of their homes, septic system failure, or contamination of their drinking water wells. This is a major constraint for wetland restoration in developed areas, but it should not be a limiting factor for homes built on high ground. Yet, where the homes were constructed on filled land (former marshes), the situation may represent a roadblock for restoration. In these cases, hydrologic evaluations must be conducted to determine allowable limits of tidal flow restoration. Devices such as self-regulating tide gates or electric tide gates may be necessary to provide protection from storm tide flooding, while improving tidal flow to the marsh.

Other factors may challenge coastal wetland restoration. Drained or impounded marshes may have undergone considerable subsidence since they were cut off from the tides. Sedimentation is eliminated or drastically reduced by diking. Organic or organic-rich soils of diked marshes rapidly decompose when aerated (increased microbial activity) or are eroded by permanent inundation. Return of full tidal flow to such areas may create shallow water habitats rather than intertidal marshes. These water bodies should fill in with time and may eventually support wetland vegetation if sufficient sediments are available for accretion. Great Harbor Marsh in Connecticut had subsided 2 feet (60 cm) below its original elevation before the tide gate was destroyed by a hurricane in the 1950s (Rozsa 1995). Since then, smooth cordgrass (*Spartina alterniflora*) has established a low marsh

Table 11.2. Coastal wetland restoration types

Restoration Type	Conditions Necessitating Restoration	Examples of Restorative Action
Re-establishment	Filled	Remove fill to restore elevations subject to tidal flooding; plant desirable species
	Effectively drained	Restore hydrology through blocking ditches; plant desirable species if necessary
	Excavated	Restore elevations suitable for recolonization by tidal wetland vegetation by adding clean dredged or fill material; plant desirable species
	Impoundment	Remove or breach dikes; if necessary install water control devices to allow tidal flow sufficient to restore tidal wetland vegetation and create new tidal creeks if original ones are filled; plant desirable species
	Tidal flow eliminated	Remove tide gates or install water-control devices to allow tidal flow sufficient to restore tidal wetland vegetation; or open or remove dikes; or install culvert
Rehabilitation	Partly drained	Restore hydrology through blocking ditches or installing weirs
	Tidal restriction	Remove tide gates or install water-control devices to allow tidal flow sufficient to restore hydrology; or remove all or portions of dikes to permit tidal flow; or enlarge culvert
	Partial filling	Remove fill and restore suitable elevations to support tidal wetland vegetation growth; plant desirable species if necessary
	Exotic and invasive species	Control or manage by various means (herbicides, biological control, mowing, pulling by hand, etc.) and plant desirable species if necessary; may involve restoring tidal hydrology to former salt marshes
	Chemical contamination	Eliminate the pollution source and restore natural chemistry; remove contaminated soils if necessary; plant desirable species
	Vegetation removal (e.g., timber harvest of tidal trees)	Plant desirable species
	Interior marsh collapse (ponding)	Restore elevations suitable for recolonization by tidal wetland vegetation by adding clean dredged or fill material; plant desirable species

Table 11.3. Situations faced by tidal wetlands or former tidal wetlands requiring restorative action: possible solutions and concerns

Problem	Source	Solution and Concerns
Restricted tidal flow	Undersized or damages culvert; tide gates; completely cut off from tides; causeways	Solution: Increase culvert size or replace with a bridge; repair damaged structure; install adjustable or self-regulating tide gates; reconnect wetland to tides by installing culvert; replace part of causeway with bridge Concern: Possible flooding of low-lying development surrounding wetland and salt intrusion into shallow wells (*Note:* In some cases, this is due to intentional diking, in which case, dikes need to be breached or completely removed; subsidence. See permanent flooding below.)
Elevation above tides	Dredged or other fill material	Solution: Remove material to elevation subject to tidal action; consider possible rebound of compressed peat; will need to transplant vegetation Concern: Hazardous materials on-site, under-excavation and over-excavation (final elevation incorrect); need to transplant or seed to stabilize substrates (need to get plants at suitable elevations); need to guard against colonization by invasive species; need to protect transplants/seedlings from goose grazing
Permanent flooding	Diked impoundments with or without water-control structures	Solution: Breach dike in places or completely remove dike Concern: Subsidence of original organic substrates and new sedimentation rates
Shift in plant species	Increase freshwater flow from various sources	Solution: Identify sources and divert flow elsewhere, restore salinity
Invasive species	Urban environment, disturbance along marsh edges	Solution: Various controls based on the target species; biological control would be best; in some cases, this problem is solved by restoring a more natural hydrologic regime and salinities Concern: Effect of control on nontarget organisms
Habitat fragmentation	From causeways and development	Solution: Reconnect through bridging or create greenway corridors to reconnect fragmented parcels with larger wetlands; daylighting streams that now go beneath developed areas (e.g., pavement and buildings)
Erosion	Loss of protective barrier; lack of sediments and rising sea levels	Solution: Create artificial berm at seaward end, deposit fill material behind barrier, and transplant species at appropriate elevations Concern: Sustainability given rising sea level; will marsh be able to accrete at rate needed to keep pace with rising seas; protect transplants from geese and other heavy grazers
Chemical contamination	Oil spills, industrial waste disposal, and discharge of wastewater	Solution: Remove or clean up pollution source; if natural bioremediation is not possible, remove or clean up contaminated substrate, replace with clean fill (if necessary) to proper grade, and transplant or seed species Concern: Release of contaminants into the estuarine ecosystem and their effect on biota; protect transplants from goose grazing and erosion

a 

b 

Figure 11.2. Aerial views of two potential tidal marsh restoration sites. Site a: road crossing a salt marsh with a single culvert for tidal creek and scouring basins on both sides of the road—evidence of insufficient opening; common reed is present on both sides but interesting, more abundant on right side. Site b: road crossing tidal marsh with a drastic change in vegetation from one side compared with the other (right: typical salt marsh; left: note presence of more woody plants especially encroaching from the margins).

community in the area originally dominated by high marsh species. Soil chemistry properties of marshes also change with drainage, producing an increase in acidity, iron oxides, and jarosite. Restoring tidal flow needs to take these soil conditions into consideration (Portnoy 1999). Catastrophic events such as tropical storms and hurricanes or long-term droughts may pose unanticipated problems after basic restoration work has been completed. Natural temporary closure of inlets, increased flooding during wet years, and increased sedimentation from watershed land-use activities further complicate the restoration process (e.g., Zedler and West 2008). To address these and other post-restoration problems, all plans for restoration and creation projects should include an adaptive management component for making adjustments to the project based on problems noted during monitoring. Monitoring should be a mandatory requirement for any restoration, creation, or enhancement project.

Tidal wetland restoration may address a number of issues (Table 11.3), with altered hydrology perhaps being the most widespread problem and the one that causes other problems such as vegetation change (e.g., spread of invasives) and limited exchange between restricted marshes and the estuarine ecosystem. Signs of altered hydrology may include

- a distinct vegetation change on the upstream side of a road,
- scour pools on one or both sides of culverts,
- marsh bank slumping (undersized culverts),
- similar scouring at bridge crossings,
- collapsed or clogged culverts,
- poorly installed culverts (opening too high or too low to pass normal tides, or angled upward or downward),
- causeways with narrow bridge openings (upstream channel much narrower than seaward channel), or
- the presence of tide gates and similar structures (Figure 11.2).

Even for a project requiring removal of fill, the main objective is to restore hydrology to the site by creating elevations suitable for tidal flooding and the establishment of tidal wetland vegetation. The ebb and flow of tides are the lifeblood of coastal marshes, permitting nutrient exchange and allowing access to marshes by aquatic organisms. Without such flow, the marshes become freshwater marshes, disconnected from the sea, and many such areas have been or are being invaded by nonnative common reed or other unwanted species. These conditions may also increase local mosquito populations (e.g., impounded salt hay farms are one of the largest sources of salt marsh mosquitoes in New Jersey; Slavin and Shisler 1983) and increase fire hazards in densely populated areas (i.e., burning of common reed). In the northeastern United States, extensive grid-ditching for mosquito control drained many natural widgeon-grass pools that were important for waterfowl. Plugging of ditches with an earthen plug or installing weirs at appropriate locations will create shallow pannes and pools. They may also be re-established by excavating shallow depressions in the high marsh. (*Note:* Since diked wetlands are among the most common sites for tidal wetland restoration, the journal *Restoration Ecology* has devoted an entire issue to dike/levee breach restoration: volume 10, number 3, September 2002; it contains several case studies, discusses performance criteria, and offers recommendations for project planning, implementation, and monitoring.) Unique challenges face restoration ecologists when confronted with contaminated marshes and restoration of tidal swamps. The former require careful removal of pollutants either through bioremediation or other means. Tidal swamp restoration projects will often involve recreating microtopography and drainage patterns, planting woody species, and long-term monitoring to chart the trajectory of forest development, identify problems along the way, and take necessary actions (adaptive management) to ensure that a forest becomes established.

Over the past several decades, nonnative common reed has become more widespread in coastal wetlands and elsewhere (Chambers et al. 1999; Saltonstall 2002; see Table 4.7 for differences between invasive and native forms). Altered hydrology with its profound decrease in salinity has promoted the invasion of this species in the northeastern United States. Decreased water quality (nutrient enrichment) also has contributed to its success (Bertness et al. 2002). While this species has displaced native plant species and their fauna, ecosystem functions of food chain support (detrital material), water quality renovation, sediment detention, and shoreline stabilization may still be maintained (Chambers et al. 1999), especially where common reed marshes are not tidally restricted. From the wildlife management perspective, however, there remains great concern over the spread of this aggressive species and its adverse impact on native wildlife. Universal agreement does not exist on this or on the extent to which marshes dominated by common reed marshes should be restored (Kane 2001). Some of these marshes may provide good wildlife habitat and any efforts to control this species should evaluate current habitat use by fish and wildlife beforehand. Since the early 1990s, efforts have increased to reduce or eliminate common reed (Tiner 1998). This is accomplished by restoring salt marsh hydrology (e.g., expanding culvert openings, removing fixed tide gates, or installing controllable tide gates) or by diverting the flow of incoming freshwater from the marsh edge to the sea. Many examples can be found where salt marsh vegetation has returned upon restoration. The pace of recovery can be quite rapid with noticeable changes occurring within one year. For example, just one year after reintroducing tidal flow to a restricted New England marsh dominated by common reed, the abundance of salt marsh grasses increased, the height and density of common reed declined, and the diversity of fishes and motile crustaceans (nekton) was comparable to that of the adjacent unaltered marsh (Roman et al. 2002). Despite the negative impacts mentioned above, common reed has been suggested as a possible tool for combating brackish marsh loss (Stevenson et al. 2000). Its high productivity and ability to form thick mats on the marsh surface could help stabilize sediment-starved marshes subject to extreme subsidence.

The use of potential restoration sites by rare, threatened, and endangered species offers unique challenges for restoration projects. For example, in San Francisco Bay, tidal flats and marshes invaded by nonnative cordgrasses have become nesting habitat for the federally endangered California clapper rail. The highest rail concentrations in the bay area are located in South Bay where the heaviest infestations of the invasive cordgrasses occur (Casazza 2010). This has created an interesting dilemma for organizations interested in protecting the endangered species and those focused on eradicating invasive plants.

While tidal restoration projects may simply re-establish tidal flow, some projects involve reconfiguration of the drainage network, removal of fill, and adjustment of elevations, and perhaps even an element of wetland creation depending on the degree of past alteration (Figure 11.3). Restoration projects have been initiated in estuaries across the country (Table 11.4). Given the extent of restoration under way in Louisiana and in San Francisco Bay, these efforts are briefly summarized in the following paragraphs. Information on other wetland restoration projects can be obtained via the web at www.coastalamerica.gov/tex/cwrpprojlist.html and from other sources (e.g., Yozzo and Titre 1997; agency websites).

Since Louisiana is losing more coastal wetland than any other state, much recent attention has been given to restoring its wetlands. The Coastal Wetlands Planning, Protection, and Restoration Act (1990) required development of a comprehensive plan to restore Louisiana's wetlands and authorized expenditure of millions of dollars annually. State officials have requested $8.5 billion over a 20-year period to continue restoration efforts (U.S. Government Accounting Office 2007). Wetland restoration is particularly complicated due to the relationship between human use, management of the Mississippi River (e.g., levees, channel dredging, and dams), and petroleum extraction in coastal wetlands (including dredging of thousands of miles of canals for pipelines and navigation). Detailed reviews of these conditions and implications for restoration have been prepared, and readers interested in these efforts should consult "Scientific Assessment of Coastal Wetland Loss, Restoration, and Management in Louisiana" (Boesch et al. 1994); "Coast 2050: Toward a Sustainable Coastal Louisiana" (Louisiana Coastal Wetlands Conservation and Restoration Task Force and the Wetlands Conservation and Restoration Authority 1998); *Saving Louisiana? The Battle for Coastal Wetlands* (Streever 2001); Louisiana—Ecosystem Restoration Study (U.S. Army Corps of Engineers 2004); and Drawing Louisiana's New Map: Addressing Land Loss in Coastal Louisiana (Committee on the Restoration and Protection of Coastal Louisiana 2006).

Given the complexity of these conditions and the sediment available from the Mississippi River, attempts to restore these coastal marshes are particularly challenging (e.g., Turner 1997). A number of conventional restoration techniques (e.g., reflooding agricultural lands, removing spoil banks, deposition of dredged material, and backfilling canals) as well as new approaches are being tried (e.g., breaching levees to initiate crevasse-splay formation and terracing) to address Louisiana's unique conditions (Turner and Streever 2002). Most techniques use different approaches to direct sediments to sinking marshes, while backfilling canals seeks to prevent further erosion of marshes, limit saltwater intrusion, and restore more natural hydrology. One technique recreates the natural fluvial process of the Mississippi River where the river breaks through its levee and deposits sediments through the opening. Here an opening (crevasse) is cut through a levee allowing alluvial sediments to flow into shallow water or subsiding marshes, thereby creating a mudflat (splay) suitable for colonization by marsh plants. Another technique called "terracing" involves depositing dredged material in a series of linear

Figure 11.3. Series of photographs showing growth of smooth cordgrass (*Spartina alterniflora*) plantings in just a single season (2009) at the Poplar Island restoration site (MD): (a) May 4; (b) June 2; (c) July 6; and (d) September 21. This project involves rebuilding a lost island and associated wetlands with dredged material. (Griff Evans)

Table 11.4. Examples of tidal wetland restoration projects on the Atlantic and Pacific Coasts of the United States

Project Type	Location	Initial Condition	Action Taken
Rehabilitation and Re-establishment	Hatches Harbor, MA	200-acre (80.9 ha) diked former salt marsh	Installed box culverts with adjustable tide gates, progressively increased openings until maximum allowable tidal exchange was achieved
Rehabilitation	Potter Pond, RI	5-acre (2.04 ha) tidally restricted estuarine pond	Replaced culverts
Rehabilitation and Re-establishment	Galilee Salt Marsh, RI	~100 acres (40.5 ha) degraded *Phragmites* marsh	Reconstructed channel, installed two box culverts with self-regulating tide gates, and removed dredged material
Rehabilitation	Barn Island, CT	52-acre (21 ha) diked former salt marsh (waterfowl impoundment–oligohaline cattail *Typha* marsh)	Installed culverts
Rehabilitation	Arthur Kill, NY	Salt marsh contaminated by oil spill	Excavated trenches and vacuumed oil from soil; fertilized shoreline to accelerate microbial breakdown of oil; planted vegetation
Rehabilitation	Delaware Bay, NJ/DE	Diked salt hay farms and *Phragmites*-dominated wetlands	Breached dikes and constructed primary and secondary channels and allowed smaller creeks to develop naturally; controlled *Phragmites* through herbicide treatment (aerial spraying) and prescribed burns (to remove *Phragmites* mat)
Re-establishment	Bogue Banks, NC	Salt marsh filled with dredged material	Created living shoreline; removed fill and planted vegetation
Rehabilitation and Re-establishment	Tomales Bay, CA	560-acre (227 ha) diked former tidal marsh (wet pastures with some estuarine wetlands)	Removed levees, tide gates, culverts, agricultural infrastructure and manure deposits; filled in ditches, recreated tidal sloughs and creeks; reoriented creeks to historic alignment; removed invasive species
Re-establishment	Siuslaw River Estuary, OR	12-acre (4.8 ha) diked area (former tidal swamp)	Breached dike, filled in ditches, removed culverts, created meandering tidal channel, and planted tidal shrubs and trees (saplings)

Note: See Appendix B for detailed profiles of these and others, including assessment of results.

ridges (approximately 3.5 feet [about 1 m] high) in shallow water to capture sediments sufficient to create low marsh. Applying moderate amounts of sediments to sinking marshes through hydraulic dredging has produced positive results (e.g., higher plant cover, improved aeration and nutrient availability, and decreased sulfide) over marshes receiving heavy, little, or no sediment-slurry (Mendelssohn and Kuhn 2003; Slocum et al. 2005; Stagg 2009; Stagg and Mendelssohn 2010). The best results were achieved by depositing just enough material to raise elevations sufficiently to reduce stress from submergence. Efforts to restore Louisiana's coastal wetlands have been funded through the Coastal Wetland Planning, Protection, and Restoration Act. As of June 2007, $1.7 billion dollars have been spent on 147 restoration projects, with costs per acre ranging from $9,000 for marsh grass planting to $54,000 for barrier island restoration (U.S. Government Accounting Office 2007). Costs for backfilling canals appear to be considerably less, estimated at $486 to $1,377 per acre (Turner et al. 1993, 1994).

Another large-scale restoration project involves San Francisco Bay's salt ponds that were diked for commercial salt production from 1850 to the 1920s (Collins and Grossinger 2004). In 2003, state and federal agencies and private foundations purchased 15,100 acres (6,111 ha) of these ponds for habitat protection and restoration. Before transferring the ponds to government agencies, Cargill Inc. was required to reduce salinities of the ponds. Introduction of bay water to some of the ponds began in 2004. In 2006, breaching of levees connected 800 acres (324 ha) of the ponds to the bay, and in 2008, design and construction began for habitat restoration, recreation, and flood protection. Restoring salt marsh will produce more habitat for endangered species such as the California clapper rail and salt marsh harvest mouse. Since the salt ponds provide habitat for shorebirds, the agencies are facing the challenge of striking a balance between restoring salt marsh and preserving existing uses of salt pond habitats. The South Bay Salt Pond Restoration Project, like restoration of Louisiana's coastal wetlands, is an ongoing effort (for the latest information, visit www.southbayrestoration.org/track-our-progress/).

Wetland Creation

Tidal marshes and swamps can be created by excavating dryland or filling shallow water or nonvegetated tidal flats to elevations suitable for establishment of wetland vegetation (halophytes in estuary and hydrophytes in tidal freshwaters). Although wetland creation is usually not the preferred alternative for mitigation, creating tidal wetlands in places where they did not exist can be successful and beneficial. For example, tidal wetlands have been created to protect shorelines from erosion as an alternative to hardened structures (e.g., bulkheads and rip-rap), to restore eroded wetlands along coastal islands (could also be viewed as re-establishment), or as a beneficial use of dredged material that solves the problem of disposing of such materials by navigation projects. Since these wetlands are usually not constructed to mitigate for losses of tidal wetlands elsewhere, the concern about functional equivalency may not be paramount. When built for mitigation, however, functional assessments should be done to determine the appropriate ratio for compensation, and such ratios should be significantly greater than 1:1 due to a number of factors (e.g., lost services, time needed to "mature," and difficulty of achieving full performance of the multitude of functions performed by natural wetlands).

Considerations for Wetland Creation Projects

Tidal wetland creation projects must be sited in or adjacent to tidal waters for rather obvious reasons. This location, however, will not guarantee success as other factors need to be considered. Several issues that must be addressed in any tidal wetland creation project include:

- attaining proper elevations to ensure desired tidal flooding and drainage (e.g., too low would result in permanent flooding, while too high would not receive tidewater);
- stabilizing substrate for sites constructed in shallow water;
- shifting tidal access during construction (increased erosion could uproot plants, while excessive sedimentation could smother plants);
- amending soil with fertilizers and organic matter to drive microbial activity;
- planting species at appropriate elevations;
- finding local sources of seeds and transplants (must be local ecotype to avoid introducing types not adapted to local conditions; may need to store plants from tidal wetlands that were destroyed by construction projects; must not contain invasives);
- controlling spread of invasive species (must have a plan to eliminate invasives);

- protecting transplants and seedlings from grazing by geese and other animals and if possible from disease outbreaks; and
- connecting site with other tidal wetlands (to maximize seed dispersal into constructed wetland from natural wetlands, especially for annual species) (Seneca and Broome 1992; Zedler 1992; Tiner pers. obs.).

Some critical elements to successful marsh creation are listed in Table 11.5.

Table 11.5. Some critical elements for successful creation of tidal marshes

1. Implement careful planning and site selection (consider physical, chemical, and biotic conditions; boat traffic and wave and current energy).
2. Allow time to secure necessary permits before optimum planting dates.
3. When seeding, be sure to get correct elevations prior to restoring tidal flow.
4. Create gentle marsh slope (1–3%) and make wetland as wide as possible (dissipates wave energy).
5. Include creeks in marsh design to facilitate tidal exchange of nutrients and access to the marsh by aquatic organisms.
6. Sand substrates are easier to work with than are finer sediments or organic deposits.
7. Fertilization may be beneficial, especially since nitrogen is limiting at most sites (phosphorus may also be limiting); ammonium sulfate and triple superphosphate are best sources of these nutrients.
8. Select plants from surrounding area (local ecotypes are best adapted to these conditions) and use best horticultural practices to achieve successful plantings.
9. Include exclusion devices (fencing and netting) to protect planting/seedlings from foot and vehicle traffic and predation by geese and other herbivores.
10. Make sure that plants receive adequate sunlight; shading from trees along creeks may be a problem at some sites.
11. Monitor sites frequently during first few years; need to make sure that vegetation is well-established; litter (tidal wrack) may smother plants; need for fertilizer and other soil amendments can be evaluated; removal of invasive species. After five years you should know if project is on the right trajectory or needs reworking.

Source: Seneca and Broome 1992.

Planting desirable vegetation is a basic component of most wetland creation projects since bare ground is ripe for erosion and colonization by undesirable species, notably invasive plants. Planting should be done where specific plant species are desired to perform certain functions (e.g., food for wildlife or nutrient uptake) to mitigate for lost wetlands, where no natural seedbank (or imported seedbank) exists, and for many shrub and forested wetland restoration projects initiated to mitigate for destroyed wetlands of these types. While plantings are routinely done for creation projects, they may also be done for some restoration projects. For example, planting of acorns is a common technique for re-establishing southern bottomland hardwoods forests and may be useful for restoring tidal swamps. Seeding of created sites may also be warranted for soil stabilization. The importance of planting or seeding local ecotypes cannot be overstated as studies have shown considerable ecotypic variation within salt marsh species (Seliskar et al. 2002; Proffitt et al. 2003; Travis et al. 2004; Howard 2010). Planting different ecotypes in the same area should produce plant communities with greater adaptability to changing environmental conditions.

Creating tidal wetlands often involves converting shallow bottoms and tidal flats to marshes, so a trade-off occurs in aquatic habitats. Beginning in the 1970s, techniques for salt marsh creation were developed in an effort to make productive use of dredged material as opposed to direct disposal or construction of diked containment areas in salt marshes, which was common practice (Woodhouse et al. 1972, 1974; Garbisch et al. 1975). During the 1970s, the U.S. Army Corps of Engineers conducted the Dredged Material Research Program that sought to identify productive uses of dredged material and ways of minimizing negative impacts of conventional disposal methods. Using dredged material to create marshes and islands in marshes for nesting waterbirds were some of the program's accomplishments (e.g., Buckley and McCaffrey 1978; Landin et al. 1978; Lunz et al. 1978; Parnell et al. 1978; Soots and Landin 1978). At least nine marshes were constructed by the program in various locations around the country including James River (VA), Buttermilk Sound (GA), Apalachicola Bay (FL), Galveston Bay (TX), San Francisco Bay (CA), and Columbia River (OR). The James River site was a created freshwater tidal marsh that was planted at considerable cost, yet within six months the site was nearly completely covered by volunteer native species (Lunz et al. 1978). Plantings of smooth cordgrass, common three-square (*Schoenoplectus pungens*), and others were virtually eliminated by Canada goose grazing, erosion, and competition from more than 75 species of volunteer native plants. Unfortunately, some of the early wetlands created for this purpose bore little resemblance to natural wetlands: they were diked, rectangular blocks—not the best design to withstand tides and currents. Some created wetlands were simply washed away over the years. Techniques have improved, and wetland creation through use of dredged material has produced at least some productive tidal wetlands that are comparable in aboveground biomass and appear to be on trajectories to develop soil properties and benthic infauna similar to natural marshes (e.g., Craft et al. 1999). While the dredge-related creation projects are usually expensive, they also address the problem of dredge material disposal by increasing marsh habitat. Material used in these projects should be relatively clean and free of contaminants.

For three decades or more, the State of Maryland has promoted tidal marsh creation as an alternative soft-engineering approach for protecting tidal shores from erosion (Tiner and Burke 1995), while other states have more recently embraced this approach for shoreline erosion control. This approach involved building fringing wetlands along eroding shores—"living shorelines"—instead of creating hardened shorelines of bulkheads or rip-rap. These living shorelines work best in areas of low to moderate wave energy as they are doomed for failure in high-energy locations. Projects often require constructing a protective berm at the seaward end of the proposed wetland, then making a gentle grade with fill upslope to the natural bank, and finally transplanting seedlings of several species at appropriate elevations for growth, reproduction, and survival (Figure 11.4; see Hardaway et al. 2010 for guidelines). The protective berm may be created from natural biodegradable materials (e.g., fiber logs) or from rock materials (e.g., rip-rap or gabion), depending on the location's susceptibility to erosion. Creating oyster reefs or other intertidal/subtidal rocky habitats at the water's edge of these projects is beneficial. Living shorelines provide a nonstructural alteration to shoreline protection that provides better habitat than rip-rap and bulkheads, while helping improve water quality. In 2008 the Maryland legislature passed the Living Shoreline Protection Act, which requires landowners seeking to protect their property against erosion to use living shorelines, with some exceptions (e.g., areas where the state has indicated that structural alternatives are necessary).

Questions remain about the long-term viability of created wetlands, such as: How long will it take to achieve parity with natural wetlands, or will they ever reach this state? Are they sustainable without human intervention? How will they respond to rising sea levels? A 25-year study of constructed smooth cordgrass marshes in North Carolina found that 1) the plant community established quickly and had productivity equivalent to nearby natural wetlands in 5 to 10 years; 2) within 15 to 25 years, the density and diversity of benthic invertebrates were greater in the constructed wetland; and 3) soil bulk density decreased over time in the constructed marsh but, after 25 years, still had lower carbon and nitrogen storage values than the 2,000-year-old natural marsh (Craft et al. 1999; Craft and Sacco

Figure 11.4. Living shorelines are a more natural solution to shoreline erosion that becomes part of the estuarine ecosystem as opposed to structural alternatives (e.g., bulkheads, seawalls, and rip-rap), which harden shorelines: (a) before; (b) during construction; (c) after construction; and (d) project plan. (Courtesy of Virginia Institute of Marine Science; Hardaway et al. 2009)

2003). Another long-term study found that the vegetation of a created North Carolina brackish marsh attained aboveground biomass similar to natural wetlands in 3 years for smooth cordgrass and in 9 years for big cordgrass (*Spartina cynosuroides*) and black needlerush (*Juncus roemerianus*), while soil development proceeded more slowly (Craft et al. 2002). The researchers estimated that it would take an additional 30 years for the constructed marsh to achieve nitrogen levels and 90 years to achieve organic carbon levels and soil porosity equivalent to natural marshes. The uppermost zone characterized by salt hay grass (*Spartina patens*) would need more than 200 years to replicate natural conditions. Comparison of a 12-year-old tidal marsh constructed by excavating upland with natural marshes showed many similarities and a few differences, namely, in sediment organic carbon at depth, mature saltbush (*Iva* and *Baccharis*) density (lower in the constructed marsh), and use by birds (related to saltbush density differences) (Havens et al. 2002). Studies of Texas marshes documented use of both natural and constructed marshes by the same aquatic species, but found lower densities of nekton (fishes and macroinvertebrates) and benthic infauna (mainly polychaete worms) in the constructed marshes (Minello and Webb 1997), while the infauna of a created 3-year-old North Carolina salt marsh were different from that of the neighboring natural marshes due to substrate differences (Moy and Levin 1991). Although constructed wetlands are becoming more productive habitats, they are not yet equivalent to natural marshes. This is not surprising since most natural marshes are at least a couple of thousand years old, and the more recent ones hundreds of years old.

Data Needs for Project Planning and Developing a Monitoring Plan

Three types of information may be needed to aid in designing a restoration, creation, or enhancement project and a corresponding monitoring plan: 1) characteristics and conditions of suitable reference wetlands; 2) existing or baseline conditions (prior to project construction); and 3) as-built conditions (after construction). At the outset of any planning exercise for restoration, it is important to know what conditions you are seeking to restore, which requires some sense of what an unimpaired wetland is or what the pre-disturbance condition was. State wetland reports published by the U.S. Fish and Wildlife Service or state natural resource agencies, and native plant community descriptions published by state natural heritage programs, offer useful information on the types of plants expected in these habitats, but they do not provide specific data on densities of individual species and other variables (e.g., soil and hydrology) necessary for project planning. Such data must be collected at reference sites. Vegetation data and soil properties for reference wetlands can often be compiled by a single site visit during the peak of the growing season.

The term "reference wetlands" has been used in various contexts to describe pristine wetlands of a particular type, or a series of wetlands of a common type representing a range of environmental conditions along a disturbance gradient from pristine or minimally impaired to highly modified. The pristine type may represent the wetland that existed in pre-settlement times or a minimally disturbed wetland. This type of reference is often used as the standard for assessing biological condition of wetlands when applying indices of biotic integrity (e.g., U.S. EPA 1998; see Chapter 10). While both types of reference have been used for evaluating restoration projects, most projects probably select minimally impacted wetlands in the local area as their reference. They represent the most realistic targets given current land uses and environmental conditions in the general vicinity of the proposed project. Knowing and documenting baseline conditions at the proposed restoration site are the critical steps in designing any project. For enhancing or restoring existing wetlands, it is crucial to know the level of impairment or degradation, so that one can properly design the project and later measure success. As mentioned earlier, this is also needed to determine whether a proposed project is either a restoration or an enhancement activity. Only projects that attempt to restore pre-disturbance conditions or something similar for significantly altered wetlands are considered restoration projects, whereas projects seeking to change the hydrology to something different from the pre-disturbance water regime are enhancement projects. A baseline assessment for a potential salt marsh restoration might include comparing the vegetation, substrate, and salinity of the proposed restoration site to that of reference salt marshes. This type of comparative analysis should yield information on the effect of the known disturbance on the subject wetland that can be used to determine significance of the impairment and for planning restoration. A regional analysis of restoration sites in the Gulf of Maine ecosystem found that degraded wetlands (sites for potential restoration) had more brackish species, lower tidal ranges, and reduced salinity compared with typical salt marshes used as reference sites (Konisky et al. 2006).

Establishing and Monitoring Reference Sites

Data from reference wetlands are particularly important in geographic areas where wetlands have not been well studied and reported in the literature, or for wetland types that are little studied in general. The purpose of reference wetlands is to gain a better understanding of the variability among wetlands of a common type in terms of plant composition, hydrology (especially water table fluctuations and flooding), and soil conditions, and to then use that information for designing wetland restoration and creation projects and for establishing measurable and comparable objectives. Reference wetlands permit comparisons between the functions of a restored or created wetland and a naturally functioning one of similar type. Even if a good foundation of scientific information on plant communities, soil types, and hydroperiods (wetland hydrology dynamics) for a given wetland type exists, evaluating reference types will provide specific information on local characteristics and temporal variations. Monitoring of the hydrology of reference wetlands of even well-studied types is useful for evaluating how well a particular wetland restoration or creation project is responding to local conditions and whether the site's hydrology is truly similar to that of local wetlands of the subject type.

Basic site analysis should be performed at a number of reference sites for major wetland types likely to be restored or created to gain a better local understanding of the characteristics of these areas. Once established and characterized, reference sites can then be monitored to track the performance of a particular restored or created wetland. No magic number can be given as to the number of reference sites per local wetland type that should be studied, but probably two or three nearby reference sites should be sufficient for comparison and to track how well the restored or created wetland is mimicking the hydrology. Eventually, comparison of vegetation and other properties will also be possible. Reference sites should be similar to the targeted type, but do not have to be exactly the same in all respects. A suite of reference sites or a reference database (e.g., Brophy 2009; Brophy et al. 2011) may facilitate evaluation of project results. In selecting appropriate reference sites for potential salt marsh creation projects in New Hampshire, researchers used their experience in the region; reviewed the literature; conducted estuary-wide sampling to measure north–south orientation, surface slope, average fetch, and wetland length and width; and then applied statistical analyses (principal components analysis) to find suitable matches between proposed creation sites and available reference wetlands (Short et al. 2000). It is important to emphasize that reference wetlands are not restricted to "unaltered or pristine natural wetlands," since it is well recognized that many wetlands have some history of human disturbance and that "pristine" wetlands are not necessarily the target condition for restoring an emergent or shrub wetland due to ambient environmental conditions and adjacent land uses (e.g., in an area of moderate or poor water quality). Reference wetlands for a given project ideally should be based on the wetland type desired, be contained in the local area, and be subjected to the same types of external influences (e.g., water quality and adjacent land use) as the wetland to be restored.

When tidal forested wetland is the targeted type, reference wetlands should include a suite of wetlands representing the successional stages that the restored or created wetland will pass through on its way to becoming a forested wetland. Forested wetlands cannot be created/restored in a short time, for obvious reasons, hence the precursor of the forested wetland, such as a wet meadow planted with tree saplings, may be the initial type produced. The reference wetland for this stage of development could be a previously harvested forested wetland in succession. Such sites would be useful for evaluating intermediate stages of the restoration trajectory. The process of establishing a forested wetland may take 20 years or more for the trees to reach sufficient height and canopy closure to begin looking like a forest and functioning as a forested wetland. Some researchers predicted that it would take 35 to 50 years for forested wetlands in the Mid-Atlantic region created for mitigation to achieve vegetation and wildlife similar to mature forests (Perry et al. 1998, 2000; Perry 2010). Others have said restoration of forested wetlands may take a lifetime (Mitsch and Gosselink 2007). The hydrology of such wetlands will change over time as more tree coverage occurs and rates of evapotranspiration increase, thereby lowering water tables more rapidly and more deeply in summer. To evaluate success of forested wetland restoration projects, it may be satisfactory to determine whether the wetland is on a trajectory that suggests it is making progress toward becoming a forested wetland. In most cases, these results should be apparent in 10 years, but monitoring in some form should be continued at least until canopy closure.

In any restoration or creation project, it should not be expected that the vegetation will ever look exactly like the reference sites except perhaps for very simple monotypic or low diversity wetland communities such as cattail marshes. The bottom-line should be that the hydrology of the restored or created wetland is similar to that of the reference wetlands, and that typical species of the targeted type are present. Initially, the vegetation may look quite different, but over time, it is expected that the vegetation will more closely resemble that of the reference sites, provided invasive species are controlled as necessary. Performance curves or trajectory models can be created by plotting the changes in functions or measured indicators over time. These curves can be used to determine 1) levels of functions that can be achieved by both natural and created/restored wetlands in a particular locale; 2) whether or when a project has attained a level of function comparable to reference wetlands; and 3) how long it takes for restoration/creation projects to reach a desired level of function (Kentula et al. 1992). For example, trajectory models were built from data collected on salt marsh creation projects in the Great Bay Estuary (NH/ME) and compared with values from natural unaltered wetlands (Morgan and Short 2002; Figure 11.5).

Documenting Pre-Existing Conditions

A prerequisite to monitoring is the establishment of baseline conditions in terms of pre-existing conditions (before project construction). It is vital to know what these conditions are so that gains or

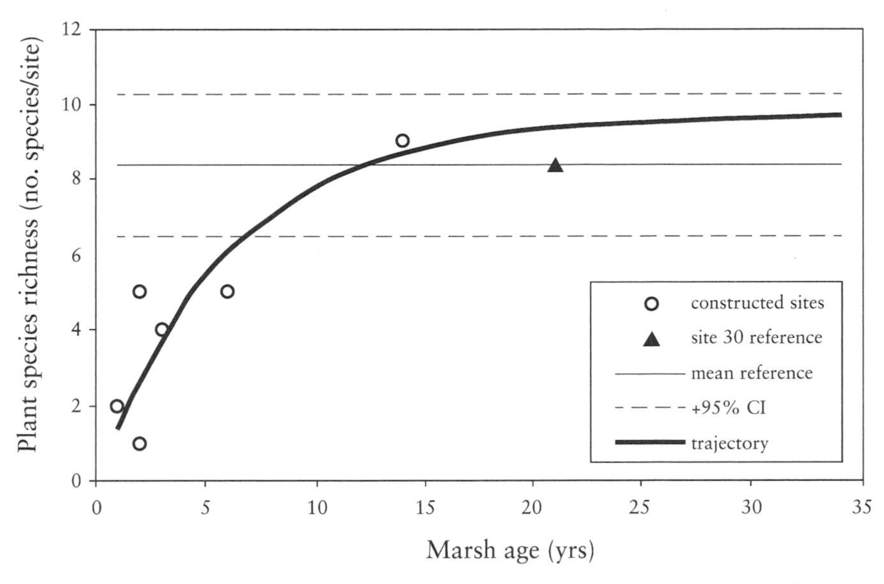

Figure 11.5. Graph showing trajectory of plant species richness in created marshes compared with a natural marsh. (Redrawn from Morgan and Short 2002)

improvements can be documented. These conditions should include answers to questions such as:

1) Is the area presently a wetland?
2) If not an existing wetland, is the site a former wetland (potential restoration site) or an upland (creation site)?
3) If a former wetland, what are the soils, current hydrology, and existing vegetation?
4) If an existing wetland, why does it warrant restoration?
5) For restoration projects, what was the pre-disturbance condition that is the target for restoration (project goal)?
6) What is the condition of the surrounding land (landscape context), and what threats does it pose or opportunities does it provide for project success?

Documenting As-Built Conditions

Once the project is constructed, the "as-built conditions" should be recorded. It is not uncommon for construction to deviate from project plans for any number of reasons. Documenting as-built conditions provides information on the configuration of the restored or created wetland (e.g., elevations) and other pertinent site conditions (e.g., soil characteristics and the locations of plantings or seeded areas). Projects should have an "as-built" plan showing elevations, plantings, seeded areas, and other factors relevant for monitoring and evaluating project success.

Wetland Monitoring

Site-specific monitoring can be initiated for three main purposes: 1) to establish pre-project site conditions for use in designing restoration, creation, or enhancement projects (baseline monitoring); 2) to track the progress and success or failure of projects (project monitoring); or 3) to evaluate changes in natural wetlands over time (research-based monitoring or simply research). Monitoring can also be used to determine how wetlands are changing due to both natural processes and human-induced action (wetland trend analysis). That type of monitoring (or surveillance) is applied to large geographic areas rather than individual wetlands and is beyond the scope of this chapter.

Baseline monitoring provides vital information on existing site conditions for determining pre-existing conditions (including the level of impairment) and for planning specific actions needed for restoration. Every wetland restoration, creation, or enhancement project should have a monitoring plan to establish explicit performance criteria for meeting project goals and objectives, to assess progress, to identify problems needing immediate correction, and to evaluate project success or failure after some period of time. Knowledge gained from project monitoring also aids in designing future projects. Of course, monitoring requires a considerable expenditure of time in the planning and design process and in the implementation, analysis, and reporting phases. While project monitoring documents what is happening at the site and patterns of change over time, it does not address why. To answer the why question(s), scientific research with testable hypotheses may be required. Research-based monitoring may be applied to a specific project to answer project-specific questions or may be independent of any project and focused on addressing ecological processes aimed to help advance our knowledge of wetland development, especially in the face of climate change and other factors. Salt marsh research monitoring programs are under way at National Estuarine Research Reserves in Maine, Rhode Island, Virginia, North Carolina, Georgia, and Oregon, and at numerous National Wildlife Refuges to assess their ecological condition in the context of climate change and/or various management activities. Examples of monitoring protocols for assessment of natural wetland processes and patterns are available (e.g., Roman et al. 2001; Moore 2009; Jorgenson et al. 2010). These protocols typically have focused on the marsh itself and tend to ignore the fact that many marshes will be advancing landward with rising seas. Consequently, sampling transects should be extended into adjacent lowlands, and vegetation-soil plots established therein to monitor the rate of and nature of marine transgression.

Wetland restoration projects initiated as mitigation for permitted alterations should require some monitoring based on performance criteria set forth by the regulatory agency. Proactive (non-mitigation) projects may not have this requirement, yet monitoring is just as important, since they are publicly financed projects intended to increase environmental services provided by wetlands. If designed properly, a monitoring plan will yield useful information to help improve project design both for the site monitored and for future projects. For example, the results of monitoring sites with different creation practices may provide answers to concerns such as how much organic matter should be added to sites for best results, what the appropriate levels of fertilization are, what the maximum spacing of transplants should be for successful re-colonization by various species, and under what circumstances planting is critical rather than optional. The following sections provide general guidance for any wetland restoration, creation, or enhancement project, with reference to special considerations for tidal wetlands where necessary. Examples of metrics for project monitoring are provided since they are critical for evaluating the success or failure of projects and for identifying where adaptive management (mid-course correction) may be required to get the project back on track. Although research studies may be required to determine why certain things are happening at a project site and to help improve the success of future projects, this discussion does not address research needs. For monitoring wetland mitigation sites, follow guidance from applicable regulatory agencies.

Project Success, Compliance, or Progress

Project success can be evaluated by monitoring certain properties or processes to determine whether the project is on the right trajectory, when corrective action is needed, and when the project can be deemed a success or in compliance with regulatory requirements. Since mitigation projects are expected to compensate for damage to existing wetlands of known type and function, establishing reasonable and realistic measures to evaluate project success is of the utmost importance. Although I use the word "success" in this chapter for convenience, the use of that word for evaluating a project has been criticized for a number of reasons including: 1) it implies a finality—a yes or no end-result, or that all of the objectives have been met; 2) many restoration and creation projects are too young to evaluate their ultimate outcome; and 3) most projects do not meet all the intended objectives and yet they may not be failures (Zedler and Callaway 2000). These scientists suggest that for mitigation projects, "compliance" is more appropriate than "success" because the project either complies with the permit requirements or it does not. The word "progress" is recommended for proactive restoration to indicate the project is on the right trajectory. In assessing the success of any restoration/creation project, it must be realized that habitat development takes time. Restoring a wetland is not like restoring a damaged car. Perhaps it is more like the recovery of a person from a serious accident or illness where months or years of physical therapy or treatment are required and the outcome may not be 100 percent complete. While the results of restoration are often quite impressive, with significant re-establishment of vegetation in a relatively short time, it will take much more time to perform all or most functions at the same level of natural unaltered ones. Projects deemed in compliance with regulatory requirements are not necessarily fully functioning wetlands as the criteria for evaluation are more short-term and oriented for the regulatory program versus the ultimate goal of establishing a self-sustaining fully functioning wetland. The latter requires long-term monitoring to determine. A few studies have shown that 10 to 20 years after restoring tidal flow to former diked marshes, the vegetation of these marshes was similar to natural marshes (Morgan and Short 2002; Warren et al. 2002; Williams and Orr 2002), while others have suggested that it may take 100 years or more to attain similar vegetation (Crooks et al. 2002; Thom et al. 2002). The latter situations involved diked marshes where subsidence was significant and post-restoration accretion slow. Restored and created wetlands in urban areas may never reach functional parity with natural wetlands given the modified landscape, altered hydrology, and presence of invasive species (Zedler and Callaway 1999). New Hampshire scientists developed a procedure for determining the degree of success of restoring or creating wetlands by comparing the values for each indicator versus the mean values for all reference sites (Figure 11.6; Short et al. 2000).

Metrics

If the primary goal of projects is to create or restore a tidal wetland, assessing hydrology, vegetation, accumulation of organic matter in the soil surface layer, salinity, and other metrics should provide adequate metrics for most projects to evaluate whether the project has successfully produced a restored or created vegetated wetland. These metrics are typically monitored for a limited time to ensure compliance with regulatory requirements. To evaluate functional equivalency will take more time and effort, such as carefully designed, long-term research studies (e.g., Williams and Zedler 1999). For all projects (e.g., mitigation, proactive restoration, or wetland management), explicit and measurable objectives should be established so that monitoring can be conducted to assess whether the project has successfully accomplished all or most of its stated objectives.

This section provides a review of techniques and procedures for documenting measurable properties of both reference wetlands and restored, created, and enhanced

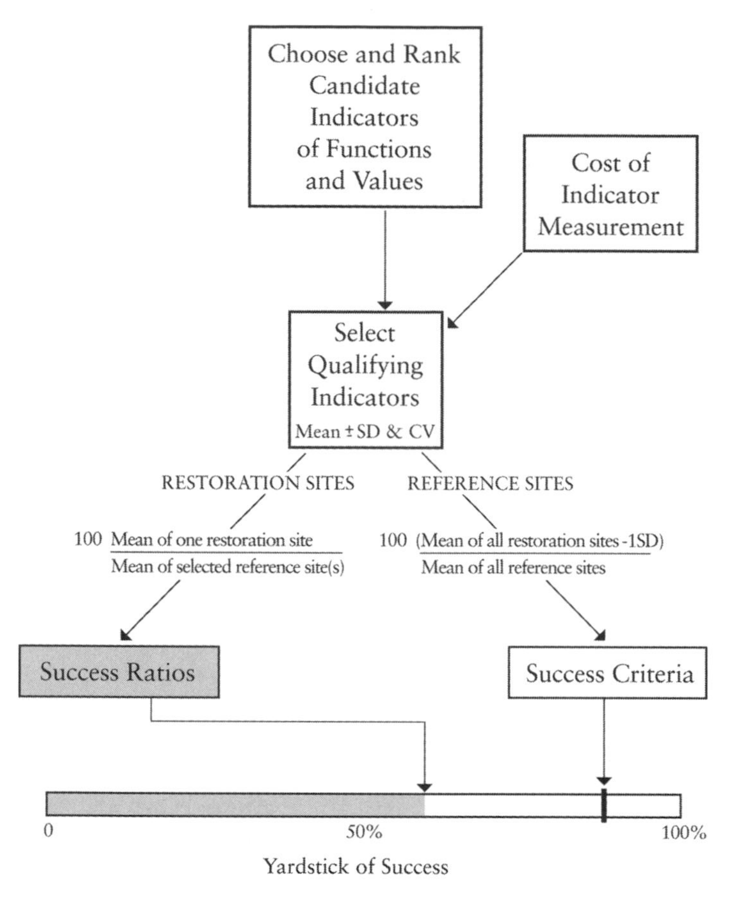

Figure 11.6. Flow diagram showing process for developing and conducting a monitoring program to evaluate project success. (Redrawn from Short et al. 2000)

wetlands in lieu of more comprehensive research-based studies designed to evaluate wetland functions and better understand ecological processes (see Zedler and Callaway 2000, 2003). As mentioned previously, a comparison of the project wetland with natural reference wetlands is recommended and pre- and post-project construction is often required. The discussion addresses several attributes and possible metrics: wetland hydrology, hydrophytic vegetation (including the success of plantings and seeding), accumulation of organic matter, salinity, development of hydric soil properties for created sites, wildlife use, and wetland extent for mitigation projects (Table 11.6). Regulatory and natural resource agencies have developed more specific guidance for restoration planning and monitoring (see later section for examples).

HYDROLOGY

Restoration of tidal wetlands will usually involve plugging ditches to restore pannes (natural depressions in the marsh), repairing broken culverts, replacing undersized culverts with larger ones, removing tide gates, or replacing them with automated or self-regulating tide gates to allow for increased water exchange to the extent possible given site conditions. In other cases, causeways or dikes may need to be removed or breached to allow more tidal flowage. Even restoration of sites where fill is removed (e.g., parking lots and dredged spoil sites) aims to restore hydrology by creating appropriate elevations within the tidal zone. Monitoring the hydrology of these types of projects is quite different from that of nontidal wetlands, since the hydrology of tidal wetlands is driven by surface water, namely the tides—their frequency and duration. While one could measure the changes in water tables, more emphasis should be placed on ensuring that the project area is sufficiently flooded by spring tides and dewatered thereafter. An existing culvert may be sufficient to pass the daily tides, but may greatly limit the penetration of spring tides which are vital to maintain the salt–fresh water balance upstream and to sustain salt or brackish marsh plant communities.

Monitoring the hydrology of restored tidal wetlands will at a minimum require assessing the tidal exchange during a spring tide, and for tidal floodplains during river high-water events, while more sophisticated methods (i.e., data loggers that record hourly water levels) may be employed for large-scale projects. For projects with restricted culverts, for example, the best time for observing the spring tide can be determined by consulting the "tide tables" prepared by the U.S. Department of Commerce (http://tidesandcurrents.noaa.gov/) for the upcoming year to identify the day of the highest predicted spring tides when you can observe virtually the entire period of a rising and falling spring high tide. Ideally, the slack water period should be around 6:00 or 7:00 AM with the tide beginning to rise around 9:00 AM, peaking around noon, and reaching low tide around 6:00 PM. This is the most convenient time for

Table 11.6. Some metrics to consider for monitoring and evaluating success of tidal marsh restoration and creation projects

Attribute	Method of Sampling	Metrics
Vegetation	*Photographs from fixed locations	Plant cover General species composition
	*Plots (within different zone)	*Species composition *Percent cover of species Height of species (*for *Phragmites* restoration sites) Density of species (*for *Phragmites* restoration sites) Signs of reproduction (flowers/fruits) *Signs of dead plants Signs of disease Evidence of herbivory *Cover by invasives
Soil	Cores (for laboratory analysis) Refractometer Soil marker	Organic matter content Pore water salinity Sedimentation
Hydrology	Flood staff	Height of tide Water level fluctuations
Fish and Wildlife	Fish seines Small mammal traps Plots (within different zones) Visual and auditory observations	Species composition and density Species composition and density Counts of shellfish and other invertebrates Birds, large mammals (tracks/scat)

Notes: Vegetation should be evaluated during peak of the growing season (possibly twice yearly in Mediterranean climates), and wildlife observations should be made on at least two occasions (e.g., in June and July in Northeast and in winter if wetland is to be used by overwintering birds or spring and fall for migrating species). The regulatory agency or project sponsor will determine what features to monitor and the frequency of monitoring and reporting.

*Should be required as a minimum for proactive restoration sites; mitigation sites will require more work to compare functional equivalency with lost wetland.

Figure 11.7. Water level marker installed to monitor tidal flooding.

making observations during working hours. With a hammer or similar device, drive two long wooden stakes with water level marks in inches into the marsh substrate to a depth of about 1.5 to 2.0 feet (Figure 11.7). Water level marks, of course, should start at the marsh surface, so depth of flooding can be recorded. Depth intervals (e.g., 6-inch [10 cm] intervals) should be labeled so that they can be seen from the road using binoculars. One of the two stakes should be placed on the seaward side of the former restriction (to record tidal flooding levels in the unrestricted marsh), with the other stake placed upstream in the "restored" marsh (formerly restricted). Ideally the elevations at the stake locations should be recorded by a surveyor; at least relative elevations are needed for comparison. Monitoring periods may differ for other types of tidal wetlands. For tidal freshwater wetlands experiencing seasonal pulses of high water, hydrology observations should also be made at times of peak flooding.

On the day of observation, record the height of the tide at 15- or 30-minute intervals. After recording the depths at different times, a graph showing the two curves can be plotted: one for the restricted salt marsh above the culvert and the other for an unrestricted one below (Figure 11.8). This type of analysis will reveal the degree of the tidal restriction, which will aid in designing appropriate restoration (see Purinton and Mountain 1996 for other approaches). Hydrologic modeling programs are used to determine the culvert size necessary to "restore" tidal exchange, but where full tidal exchange and restoration of sheet flow is the objective, dike removal is necessary. This is the only way to truly restore the marsh to its original condition and functions. When used for assessing the success of restoration, the same steps should be followed to generate the needed data for preparing hydrographs at the downstream (unrestricted) site and upstream (restored) site. The hydrograph of the "restored" marsh should be similar to that of the "unrestricted" marsh for successful projects (as in Figure 11.8b). Hydrologic models must consider future changes in sea level. The effect of various sea-level rise scenarios over the next 50 years should be predicted for restoration and creation projects.

When evaluating the success of flooding of newly established pannes (e.g., created by ditch plugs), photographs taken at both high spring tide and low tide should show the success of these types of projects. For example, the presence of flooded pools at spring high tide and exposed surfaces at low tide provide some evidence of project success in many cases. Examining marsh processes beyond the pannes themselves may also be beneficial as outside influences can be significant.

HYDROPHYTIC VEGETATION

Permanent sampling stations should be established to analyze vegetation changes on

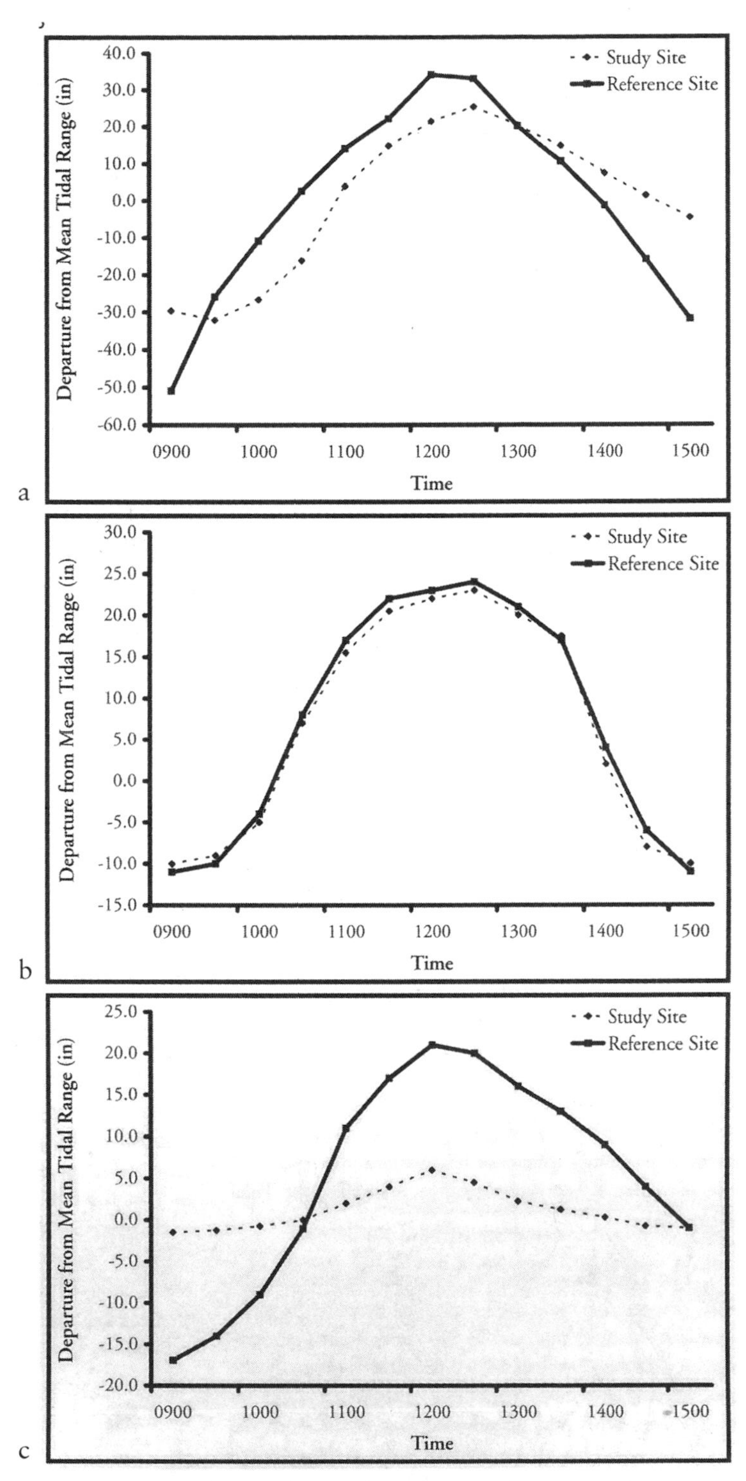

Figure 11.8. Hydrographs showing three situations comparing the marsh upstream of a structure with the downstream unrestricted marsh: (a) culvert too high preventing full range of flood tide and inadequate drainage from ebb tide; (b) bridge allows adequate tidal exchange, no hydrologic restriction; and (c) much lower tidal range indicates severe restriction attributed to undersized culvert or crushed culvert. (Carlisle et al. 2002)

a periodic basis. Vegetation sampling may involve two basic tasks: 1) making general observations in year 1; and 2) performing detailed assessment in future years. For sites in which vegetative succession is expected to occur after restoring tidal hydrology, sampling will often be a continuation of the baseline monitoring of permanent plots. For created sites, these plots will be established when site construction is completed.

At the end of the first full growing season, a cursory assessment of the vegetation should be conducted. This will involve making general observations of plant cover (e.g., detect patches of bare ground), species composition (i.e., a simple list of dominant and common species with estimated areal coverage), and for planted sites, a general assessment of survival and plant vigor. This step will help identify potential problems. More comprehensive vegetation analysis should be conducted beginning with the end of the second full growing season. A variety of vegetation sampling techniques may be employed including plot sampling, point intercept sampling, and line intercept (Figure 11.9; Mueller-Dombois and Ellenberg 1974; Tiner 1999; Kent and Coker 2002).

For planted sites, it is important to monitor the survival and growth of the plantings each year, since these species represent the desired plants for the restored or created wetland community. For all such sites, a general reconnaissance should be performed to identify areas of the project wetland where specimen plants are dead, dying, or showing signs of stress (no significant growth). These areas may then be studied in more detail to uncover the factors that may be limiting plant growth. These observations probably need only be done once a year, perhaps in mid- to late summer when comprehensive vegetation analysis is being performed. Lack of wetness in winter may be a factor leading to winter die-offs in some places, so early spring observations may also be worthwhile.

Survival of plantings is an important observation to record as is the coverage by seedlings. It is also necessary to assess the growth of the plantings and their production of flowers, fruits, and seeds to be certain that plants are actively colonizing

Figure 11.9. Sampling vegetation in plots along transects. This technique can be used to collect data on both reference and restored wetlands.

the site or, at least, growing vigorously rather than simply surviving. Measuring the annual growth of woody plants (e.g., height and diameter at breast height) and estimating the horizontal spread (areal cover) of herbaceous species at the end of the growing season are also recommended. It may be useful to record the number of living and dead woody plants and to show the location of the dead specimens on a sketch map if necessary. The same should be done for herbaceous plantings, although it may be more appropriate to count living and dead clumps rather than individual specimens. Height, density, and number of dead plants are also important metrics to be evaluated for species (e.g., common reed, purple loosestrife, cattails, and woody plants) targeted for elimination by projects restoring tidal flow to restricted areas (former salt marshes).

For restored or created wetlands, vegetation sampling should be done once a year at the peak of the growing season (e.g., mid-July to mid-August for the Northeast). Sampling of reference wetlands should be done simultaneously for comparative results. If serious problems are detected in years 1 or 2 (e.g., lack of significant cover by desired wetland species or equivalent types) that require a mid-course correction (adaptive management) in terms of project design/operation, then the sampling should start again after the correction is made (i.e., new year 1, etc.).

Aerial photos of the project area before project initiation and at periodic intervals after construction are useful, especially for mitigation and compensation projects. These aerial photographs (taken from low-flying aircraft; e.g., scale of 1:6,000 or larger) should be acquired in mid- to late summer to show vegetation at the peak of the growing season during low tide. This will give a good representation of the areal vegetative cover of the entire wetland in question.

In addition, on-site photos should be taken of the plots from permanent locations, so that the evolution of the vegetation pattern can be observed. Photo locations should be recorded on a map or large-scale aerial photograph (Hall 2001a), along with a brief description of the location noting any obvious landmarks (e.g., 10 feet east of end of stone wall). This will help others take photographs from these same locations to visually document changing vegetation patterns in the future. Photographs should be taken at least at the same frequency as the vegetation monitoring schedule. For restoration or creation sites one acre or smaller, it should be possible to replace the need for aerial photos with several well positioned on-the-ground shots taken from enough locations to provide a good perspective of the site. It may be necessary to take such photos from an aboveground location, such as a nearby tree, the top of a vehicle, or a 6- to 8-foot stepladder to provide the best overview.

When evaluating vegetation, it may be useful to prepare a sketch of the plant communities found in the wetland. For example, smooth cordgrass tall-form stands should be separated from stands of the short form of this species as well as from other species such as salt hay grass, salt grass, and high-tide bush (*Iva frutescens*). The distribution of these communities can then be compared with the "original plan" designed by project sponsors to see how well the project conforms to the original design over time. Keep in mind that departures from the original plan do not necessarily constitute a sign of failure provided the overall project objectives have been met. For example, the species may be different than intended but the wetland can have the appropriate hydrology and still serve the basic functions listed in the objectives.

SALINITY

Besides the typical vegetation and hydrology evaluations, assessment of the soil chemistry of restored salt or brackish marsh may be beneficial. These data are especially useful if there is a question as to whether the restored wetland is receiving too much freshwater inflow, such as surface water runoff from a storm drain. Salinity measurements should ideally be initiated before project construction as it represents a preexisting condition important in designing the restoration project. It is vital to assess post-construction results to see if soil-pore water salinity has increased to that of reference wetlands. Salinity of water extracted from peat cores can be determined and compared with the soil salinity of neighboring reference marshes. A refractometer can be used to determine the approximate salinity and to detect gross differences that may exist between the wetland in question and the unaffected salt marsh. The problem area of the marsh may also be a point of active groundwater discharge that lowers salinity naturally. If this situation has promoted the establishment of an undesirable plant such as common reed, restoration may not be warranted unless it shows signs of advancement or has already occupied a large portion of the marsh. If restoration is a viable solution, one might seek to divert some of this inflow to a marsh ditch or stream (by way of open marsh water management techniques) to reduce freshwater concentration of the marsh substrate. This may induce sufficient salt stress to reduce the vigor and areal cover of common reed to the benefit of native halophytic species.

Salinity should be monitored pre- and post-project in the subject wetland and the more seaward reference wetland on several occasions during the year in the Northeast: spring (March–April), summer (July–August), fall (October–November). Measurements should be made at high tide during two tidal cycles for each of these seasons: 1) predicted spring tide and 2) more typical tide. Measurements should be avoided after periods of heavy rainfall. Sampling should be conducted in years 1, 2, and 5 following salt marsh restoration or every year if necessary. Salinity measurements should include creek water salinity as well as interstitial soil salinity at several locations in the restored and reference wetlands. The locations of sampling sites in the restored wetland should include different elevations as well as some points near the marsh–upland border.

ACCUMULATION OF ORGANIC MATTER

For created tidal wetlands, the amount of organic matter is worth assessing on a periodic basis, since this is an important process that helps the marsh keep pace with rising sea level. At the end of the growing season (e.g., October), the upper surface layer can be examined to determine the buildup of organic matter. A simple measurement of the depth of the organic layer (O-horizon) may be sufficient for the wetter wetlands. Soil analysis (e.g., bulk density) can be used to determine the percentage of organic matter of restored or created wetlands compared with that of reference wetlands. Again, the percentages may not be the same initially, but over time, the percentages from soils at the restored and created wetlands should increase and approach that of the reference wetland. The time required for them to intersect is unknown, but based on work from North Carolina it will likely take longer than 25 years (Craft et al. 1999).

SEDIMENT ACCUMULATION AND CHANGES IN WETLAND SURFACE ELEVATION

Sedimentation is also vitally important to tidal marshes for the same reason that organic accumulation is—it helps marshes keep pace with rising sea level (see Chapter 12). These rates may be established by the horizon marker approach or by plastic disks attached to the soil surface. For the horizon marker approach, a layer of feldspar or colored sand, clay, or other material is initially deposited on the soil surface and after some time, a soil sample is taken to record the depth of sediment deposition and to calculate the rate of sedimentation. The amount of sediment and organic matter accumulating above these markers can be measured periodically. The colored sand approach usually works best for short-term studies. Marker studies are often done in conjunction with marsh elevation change studies. The latter studies require installation of surface elevation tables (SETs)—mechanical leveling devices for measuring relative changes in marsh elevations (see Chapter 12). Elevation change is caused by surface and subsurface processes. Surface processes are sedimentation and erosion, while subsurface processes involve root growth, decomposition, porewater flux, and compaction. As of 2010, these studies are being conducted in 22 states and 22 countries (Cahoon et al. 2002; Cahoon and Lynch 2010).

SOIL ANALYSIS FOR CREATED SITES

Created wetlands will require further evaluation of soil characteristics. If soils are imported from a destroyed wetland, soil examination will ensure that soils are maintaining hydric conditions. The hydrology measurements and assessment of organic matter accumulation should be sufficient for these sites. Other sites where hydric soils were not imported to the site require periodic examination to see if hydric soil properties are developing (e.g., look for oxidized rhizospheres, redox concentrations, and depletions).

At three or more locations in the created wetland, the soils should be described to a depth of 1 foot (30 cm) on an annual basis. Make observations at low tide, and describe soil texture and colors following standard protocols. Also, evidence of sulfidic odor should be recorded. Measuring bulk density, oxidation-reduction potential (Eh), sulfides in the soils, and hydrogen sulfide emissions may be done for more research-oriented projects (e.g., DeLaune et al. 2002; Rabenhorst et al. 2007).

WILDLIFE USE

Assessing wildlife use is especially important for projects seeking to restore, create, or enhance fish and wildlife habitat. Rather than simply assuming that if the plant community and hydrologic conditions are successfully established that the area is a viable wildlife habitat, verification may be required to ensure that fish and wildlife are really benefiting from the project. For mitigation projects, various fish and wildlife habitat restoration and similar projects, the project goals/objectives for fish and wildlife should have specified the target species and the intended usage.

To evaluate project success for fish and wildlife use, documentation of such use will require observations and collections during key periods, such as the breeding season and/or migration seasons (for migratory species), or perhaps during winter if such areas are intended to provide overwintering habitat. Follow standard fish and invertebrate sampling, small mammal trapping, and bird censusing techniques (e.g., Raposa and Roman 2001b; Erwin et al. 2003). Data collected needs to include date/time of sampling, methods employed, species recorded, and number of species observed/collected. Perform sampling at least three different times during a particular season (e.g., winter, spring, summer, and fall) relating to desired wildlife use (e.g., breeding, migration, or overwintering). Make observations during the peak of these periods and compare data with qualitative/quantitative measures for evaluating project success. Comparison with data from one or more neighboring reference wetlands is desirable; perform faunal surveys during similar periods and times of day to maximize correspondence. Document factors that may be adversely affecting wildlife use of the project wetland.

DETERMINING WETLAND EXTENT FOR MITIGATION PROJECTS

For mitigation projects, the area of the restored or created wetland is usually an important criterion for evaluating project compliance. Determining the limits of the subject wetland may be necessary for created wetlands following standard regulatory procedures for wetland delineation. The determination should include a narrative description of how the wetland boundary was established and an estimate of the acreage of the restored or created wetland.

Agency Guidance

Natural resource agencies have developed guidelines for restoration planning and/or monitoring to meet their program objectives. The National Oceanic and Atmospheric Administration (NOAA) has produced a comprehensive framework for designing monitoring plans for coastal habitat restoration projects in accordance with the Estuaries and Clean Water Act of 2000 (Thayer et al. 2003). This guidance emphasizes that habitat structure and function are key components for habitat recovery and that monitoring plans should include at least one structural parameter and one functional parameter in addition to assessing the area of habitat restored. The framework details numerous physical, chemical, and biological parameters that can be used to track and evaluate project success. Other agencies and organizations have developed guidance for tidal wetland restoration in particular geographic regions. For example, resource managers and scientists from the United States and Canada developed standard protocols for inventorying existing and potential salt marsh restoration sites and for evaluating project success for tidal wetland restoration in the Gulf of Maine (Neckles and Dionne 1999; Neckles et al. 2002; Table 11.7).

Table 11.7. Core variables for monitoring tidal wetland restoration in the Gulf of Maine region

Marsh Feature	Core Variable	Brief Definition
Hydrology	Tidal signal	Pattern of water level changes (no less than 15 minutes between measurements) with respect to a reference point; minimum of two weeks covering one lunar cycle of spring and neap tides (one month preferred); install recorders both upstream and downstream of restriction
	Surface elevations	Marsh elevations (contour intervals of 15 cm or less)
Soils and sediments	Pore water salinity	Parts dissolved salts per thousand (to nearest 1 ppt) referenced against a practical salinity scale; measured at low tide six times per year from April/May to July/August
Vegetation	Composition	Species in sample quadrats at peak biomass (midsummer: mid-July through August)
	Abundance	Cover class per m^2 for each species
	Height of species of concern	Mean height of three tallest individuals of each species per m^2
	Stem density of species of concern	Numbers of shoots per m^2
Nekton	Identity	Species of fish and crustaceans observed
	Density	Number of animals by species per unit sample
	Length	Total animal length/width by species to nearest 0.5mm
	Biomass	Wet weight of animals by species per sample
	Species richness	Number of species represented
Birds	Abundance	Number of birds by species per hectare
	Species richness	Number of species represented
	Feeding and breeding behavior	Type of behavior by species observed at 20-minute intervals

Source: Neckles and Dionne 1999.
Note: See source document for details and other variables.

Table 11.8. Core metrics for monitoring tidal wetland restoration projects in the Lower Columbia River and estuary

Indicator Category	Metric	Collection Method	Sampling Frequency
Hydrology	Surface water elevation	Data logger	Hourly
Water quality	Temperature, Salinity	Data logger	Hourly
Habitat	Landscape features Elevation	Photographs, GIS Ground survey	Annually Annually
Plants	Species composition Percent cover Elevation Planting success	Ground survey (for all)	Annually
Fish	Species composition Size/age structure Temporal presence	Ground survey (for all)	Monthly–seasonally

Source: Roegner et al. 2008.
Note: Most of the projects in this area involve tidal reconnection, so the focus of the metrics is on hydrology and physical and biological responses of floodplain wetlands.

Monitoring guidance has been prepared for other areas including New York (Niedowski 2000), California (Callaway et al. 2001), Pacific Northwest (Simenstad et al. 1991), and the Lower Columbia River and Estuary (Roegner et al. 2008; Table 11.8), while guidance for restoration site identification and prioritization as well as for project design have been drafted (e.g., North Carolina—Williams 2002, and San Francisco Bay—Williams & Associates Ltd and Faber 2004).

Restoration Prioritization

Like nontidal wetland restoration projects, most tidal wetland restoration is opportunistic—initiated because the wetlands have been identified as degraded and have generated interest by one or more parties either as compensation for altered wetlands (mitigation) or simply for recovering lost functions (proactive restoration). Despite this, some attention has been given to prioritizing sites for tidal wetland restoration and developing assessment procedures to evaluate and rank potential sites. Prioritization methods are either relatively simple ranking approaches relying on interpretation of existing maps, resource data, and aerial imagery, or more intensive methods employing geographic information system (GIS) technology and statistical analyses. The former methods are designed for use by resource planners or watershed associations, whereas the latter require technical analytical skills (e.g., integer programming) and can only be applied by technical experts. All these approaches are targeting tools that identify a subset of potential restoration sites for more time-consuming, expensive field assessments.

An example of a simple ranking method is one developed for prioritizing tidal wetland restoration and conservation sites in Oregon. It characterizes sites using existing maps and digital data, soil surveys, aerial photographs and Natural Heritage Program data; then ranks sites on six criteria to set priorities. Ranking criteria were chosen because they are likely to control or correlate closely to multiple wetland functions. Criteria include wetland size, tidal channel conditions, wetland connectivity, salmonid diversity, historic wetland type, and the diversity of current vegetation types (Brophy 2007). This method has been applied to several Oregon estuaries including the Siuslaw, Nehalem, Umpqua, Smith, Yaquina, and

Alsea (www.greenpointconsulting.com/reports.html). Examples of GIS-based prioritization methods have been developed for South Carolina, North Carolina, and the Pacific Northwest (Sutter 2001; Williams 2002; Kauffman-Axelrod and Steinberg 2010, respectively). These approaches consider both local site and landscape features such as area of adjacent wetlands within a certain distance, hydrologic connectivity, area of adjacent water within a certain distance, development along the perimeter, and the land cover, land use, or disturbances within the site's catchment (e.g., forest or impervious surface, number of tide gates, or road–stream intersections).

Another quantitative prioritization method has been constructed for South San Francisco Bay salt ponds to help identify whether salt ponds should be restored or managed to benefit birds (Stralberg et al. 2009). This method used extensive data on bird use of ponds and tidal marshes and existing habitat data to develop models and to test various restoration scenarios to evaluate their effect on bird use of the area. There are also attempts to add social values into the prioritization equation. This may require conducting public surveys and economic valuation (King et al. 2000; Johnston et al. 2002).

Further Readings

Creation and Restoration of Coastal Plant Communities (Lewis 1982)

"Tidal salt marsh restoration" (Broome et al. 1988)

Wetland Creation and Restoration: The State of the Science (Kusler and Kentula 1990)

Restoring the Nation's Marine Environment (Thayer 1992)

Restoring and Protecting Marine Habitat: The Role of Engineering and Technology (National Research Council 1994)

Handbook for Restoring Tidal Wetlands (Zedler 2000)

Tidal Wetland Restoration: Physical and Ecological Processes (Goodwin and Mehta 2001)

Restoration Ecology (Volume 10, no. 3, September 2002)

Delaware Bay Salt Marsh Restoration (Teal and Peterson 2005)

Science-based Restoration Monitoring of Coastal Habitats (Thayer et al. 2005)

Foundation of Restoration Ecology (Falk et al. 2006)

Tidal Marsh Restoration: A Synthesis of Science and Management (Roman and Burdick 2012)

12 The Future of Tidal Wetlands

Great progress has been made since the mid-1960s to control direct human impacts to coastal wetlands in the United States; and more recently, similar efforts by certain Canadian provinces have helped reduce salt marsh destruction from human activities. Various levels of government in the United States and Canada have enacted laws or policies to protect tidal wetlands to some degree, and the quality of many tidal wetlands has increased through restoration efforts by government agencies and private organizations. While these types of programs and policies have greatly improved the status of coastal wetlands over what they were prior to the 1960s, it must be recognized that laws can be changed by legislative bodies and the effectiveness of regulations at protecting wetlands can be weakened by the courts or by lack of enforcement. Nonetheless, even with existing regulations and policies in effect, serious threats to coastal wetlands remain. Population pressure on coastal resources continues as 53 percent of the U.S. population live in the nation's coastal counties (Crossett et al. 2004). This represents a 28 percent increase since 1980. Marsh degradation, loss of vegetative buffers, and estuarine enrichment (eutrophication) continue despite regulations that attempt to protect the physical integrity of tidal wetlands. Wetland wildlife is also suffering from these actions as habitat quality is diminished for fish and aquatic invertebrates with negative consequences felt by other organisms through the estuarine food web. Development in coastal watersheds adversely impacts wetland animals as well as terrestrial wildlife. It was disheartening to learn the conclusions of a recent study: when coastal watersheds in the Chesapeake Bay region experienced as little as 3.5 percent urban development, there was a significant negative impact on estuarine waterbirds—gulls, waders, raptors, kingfishers, and waterfowl (DeLuca et al. 2008). An earlier study by these researchers found that when 14 percent of the land within 1,640 feet (500 m) of a tidal marsh was developed and 25 percent of the land was developed within 3,280 feet (1,000 m), estuarine marsh bird communities were adversely impacted (DeLuca et al. 2004). If these statistics hold true for other estuaries, the future looks bleak for marsh birds and waterbirds as human population will continue to expand and convert upland forests to residential and commercial development in the coastal zone. Marsh view properties are high-valued real estate. To say that coastal resource planners have significant challenges before them is an understatement.

Habitat destruction and degradation by human actions are continuing problems for tidal wetlands, especially in the tropics. While laws and regulations have significantly reduced wetland losses in the United States, pollution (e.g., estuarine eutrophication) remains an issue for resource managers. The other main threat to tidal wetlands worldwide is not a point source that can be stopped by adopting a national wetland conservation policy or by enacting wetland or water quality laws, regulations, or other land use restrictions. Instead, it is a global force brought about by climate change. Increasing concentrations of carbon dioxide (CO_2), methane, chlorofluorocarbons, and other gases from the burning of fossil fuels, deforestation, and cement manufacture have released sufficient quantities of CO_2 to significantly raise the level of this gas in the atmosphere and contribute to global warming. Atmospheric CO_2 levels have increased 40 percent from pre-industrial levels (~280 ppm) to current levels (~388 ppm) (Langley et al. 2009). Annual emissions of CO_2 have increased by 80 percent from 1970 to 2004 (Intergovernmental Panel on Climate Change 2007). The International Panel on Climate Change (IPCC) predicts an increase of 2.0° to 11.5°F (1.1°–6.4°C) with a best estimate of 3.2° to 7.2°F (1.8°–4.0°C) by 2100 (IPCC 2007).

The effect of global warming on sea level has enormous consequences for coastal wetlands. When water is heated it expands (thermal expansion) and since the Earth is 75 percent water, this process in combination with accelerated melting polar and glacial ice (which adds more water to the oceans) cause a rise in sea level referred to as "eustatic sea-level rise." Local and regional conditions (e.g., subsidence and tectonic activity) may affect the position of the land relative to sea level. The combined effects of eustatic sea-level rise and local land elevation changes produce what is called "relative sea-level rise" (Figure 12.1). Relative sea-level rise is greater than eustatic sea-level rise where land subsides and is less where land rises (uplifts). The latter situation often results in "negative relative sea-level rise." Warming of the oceans also affects ocean currents, atmospheric currents, and other conditions that alter weather patterns contributing to what we now call "climate change." These changes are occurring whether or not one believes that industrialized societies are the main cause or a significant contributor to the situation. The U.S. Global Change Research Program (Karl et al. 2009) found that global warming is unequivocal and primarily human-induced, and that climate changes are already under way and projected to grow (e.g., sea-level rise, changing precipitation patterns, longer growing season, rise in temperature, longer ice-free season for oceans and fresh water bodies, earlier snow melt, and changes in river flows).

Changes in temperature and precipitation patterns will significantly affect life cycles of many plants and animals worldwide. Examples of these changes have already been observed in such events as earlier spring migrations and northward shifts in breeding grounds of some North American birds, and expansion of black mangroves in Gulf Coast wetlands. Lengthening the growing season in middle and northern latitudes will likely increase productivity of plants (Kirwan et al. 2009). Temperature changes will also affect plant community composition. While rising sea levels may create more waterlogged panne habitat in New England high marshes (Warren and Niering 1993), panne forbs may be replaced by salt hay grass (*Spartina patens*) prior to the formation of new pannes, thereby precluding forb colonization of the new pannes. A recent warming experiment demonstrated this change (Gedan and Bertness 2009). Exposing pannes to warmer temperatures of less than 39.2°F (4.0°C) forced salt panne species to expend more energy on transpiration, thereby reducing their cover and allowing salt hay grass colonization. After three growing seasons panne diversity was reduced by as much as 74 percent in study plots. At one site diversity dropped markedly after the first growing season of warmer temperatures.

While shifting temperatures will affect plant and animal distributions, rising sea level poses the greatest risks to the physical integrity of tidal wetlands as well as to coastal communities and Pacific island nations and territories. From 25 to 55 percent of the world's coastal wetlands could be lost from a 3.3 foot (1 m) rise in sea level, while about half of the world's coastal wetlands of international importance could be lost (Delft Hydraulics 1992; Nicholls et al. 1999). Its impact will undoubtedly be exacerbated by the likely societal response of shoreline armoring to protect private property (Figure 12.2). Along Great Bay in New Hampshire, the more populated areas have 43 percent of the marsh shoreline

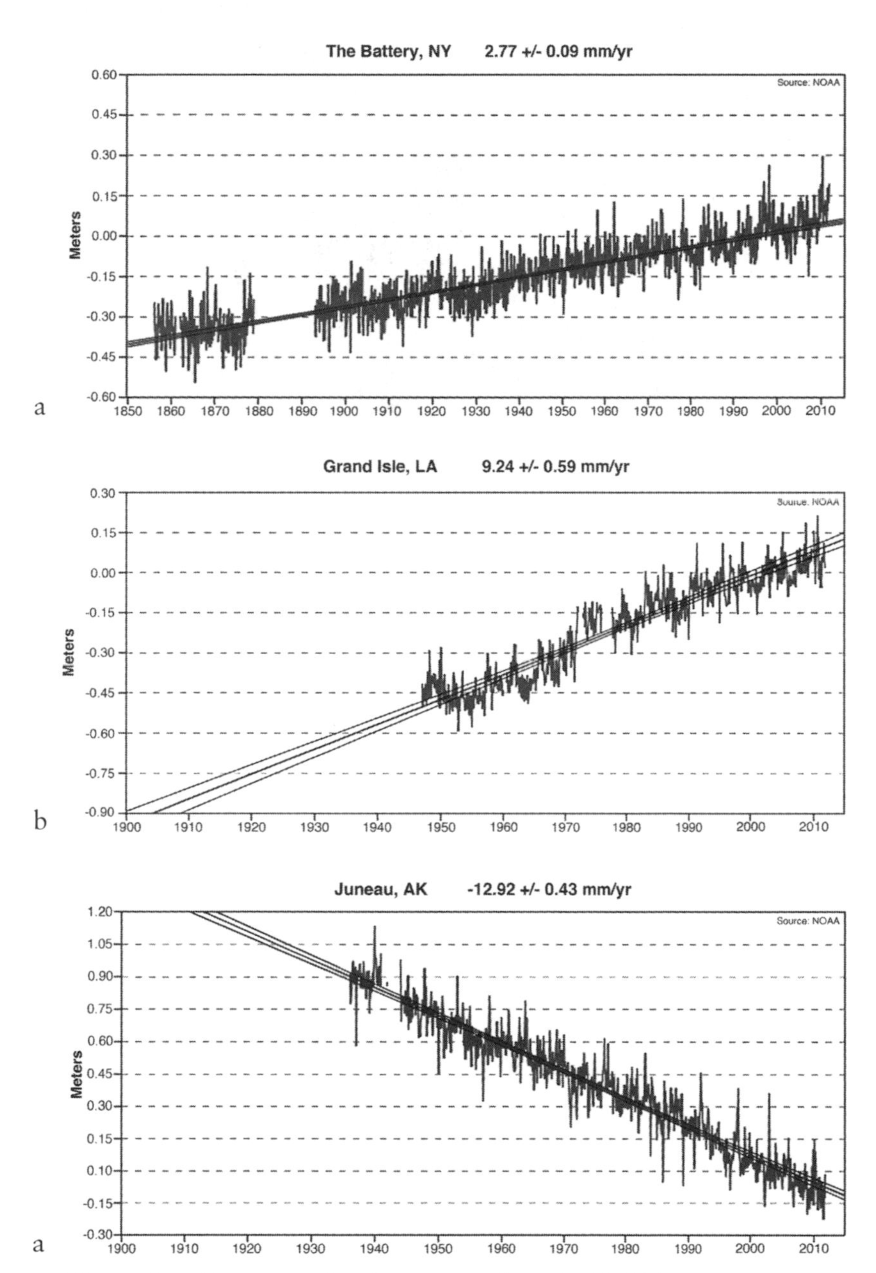

Figure 12.1. Three examples of relative sea-level rise (RSLR) for an area experiencing varying rates: (a) positive RSLR due to some subsidence (NY); (b) very high positive RSLR from extreme subsidence (Grand Isle, LA); and (c) negative RSLR due to extreme uplift from tectonic activity (Juneau, AK). (NOAA)

Figure 12.2. Bulkheading shores to protect private property prevent natural coastal processes such as barrier beach and salt marsh migration. Some bulkheading was done to fill tidal marshes and shallow waters.

armored while the bay as a whole has only 3.5 percent of its marsh shoreline protected (Bozak and Burdick 2005). Coastal wetlands are dynamic resources that move both landward and seaward in response to long-term changes in sea level. Over the Earth's history, sea level has been 9.8 to 65.6 feet (3–20 m) above current levels during past interglacial periods, and as much as 400 feet (125 m) below today's levels during the last glacial epoch—roughly 20,000 years ago (Fairbanks 1989; U.S. Geological Survey 2000; see Figure 2.4). Development of low-lying areas adjacent to tidal wetlands and construction of hardened shoreline (bulkheads, seawalls, and rip-rap) to protect those properties prevents the natural landward migration of salt marshes accompanying rising sea level. This has serious implications for the future of today's tidal wetlands.

In this chapter I intend to provide some insight into the problems that tidal wetlands face given accelerating rates of sea-level rise (SLR), current signs of tidal wetland response to changing sea levels, and predictions based on modeling studies. I also offer suggestions on what can be done by various organizations and individuals to help conserve tidal wetlands for future generations. More detailed treatments can be found in the publications listed under Further Readings.

Historic, Current, and Predicted Sea-Level Rise Rates

After 1,900 years of relatively little change in sea level, global sea levels began to gradually rise around the beginning of the 20th century (Intergovernmental Panel on Climate Change 2007). Recent studies have shown that somewhere between 1850 and 1920, annual rates of relative SLR increased to 0.06 inches (1.6 mm) for Nova Scotia, 0.06 to 0.07 inches (1.4–1.8 mm) for Connecticut, and 0.09 inches (2.2 mm) for North Carolina (Donnelly et al. 2004; Gehrels et al. 2005; Kemp et al. 2009).

Since 1993 an annual rise of 0.11 inch (2.8 mm) in the global rate has been reported (Gornitz 2007). Rates may increase further as satellite imagery shows thinning of the Greenland Ice Sheet and glaciers disgorging ice into the ocean more rapidly than in previous decades. Also the West Antarctic Ice Sheet is thinning, and if either ice sheet melts completely, sea level may rise 16.4 to 23.0 feet (5–7 m). The full effects of this meltdown on sea level, however, would not be realized for many centuries. The 2007 predictions by the IPCC estimate a global SLR of 7.4 to 22.8 inches (19 to 58 cm) by 2100, with possible additions of 2.0 to 4.3 inches (5 to 11 cm) from melting ice sheets (Parry et al. 2007). They mention that a more than 32.8-foot (10 m) rise is possible over many centuries if irreversible melting of the Greenland ice sheet occurs. The IPCC projections have been criticized by some researchers for assuming a near-zero net contribution from the Greenland and Antarctic ice sheets during the next century, in spite of observations of accelerating loss of their ice mass over the past two decades (Vermeer and Rahmstorf 2009; Grinsted et al. 2010; Rahmstorf 2010). A group of international climate scientists have claimed that the IPCC projections significantly underestimate the potential rise, and the calculations of this group indicate an expected rise well over 3.3 feet (1 m) by 2100 with an upper limit of nearly 6.6 feet (2 m)—at least twice the IPCC prediction (Allison et al. 2009). Granted, the IPCC report was based on data available prior to December 2005, so data from several key papers on potential ice sheet collapse were not included in the assessment (Don Cahoon, pers. comm. 2011; next set of IPCC reports are planned for publication in late 2013 and 2014). Regardless, all estimates project a significant increase in global sea levels due to warming that will put many tidal wetlands and coastal communities at risk. (*Note:* For interesting perspectives addressing the politics behind global warming and the scientific reticence on SLR that influence both the actual message delivered by the scientific community and the public's perception of the issue, see Hansen 2007 and Bradley 2011.)

While seasonal and decadal fluctuations in sea levels are the norm, global sea levels have risen 4 to 10 inches (10–25 cm) over the past 100 years (Figure 12.3; IPCC 1996, 2001, 2007; Gaffin 1997; Holgate 2007; Milne et al. 2009). Some areas have experienced higher century rates—in the 10 to 16-inch (25–40 cm) range—including Nova Scotia, Massachusetts, Chesapeake Bay, and the Mid-Atlantic region (Long Island, New York, to Cape May, New Jersey)

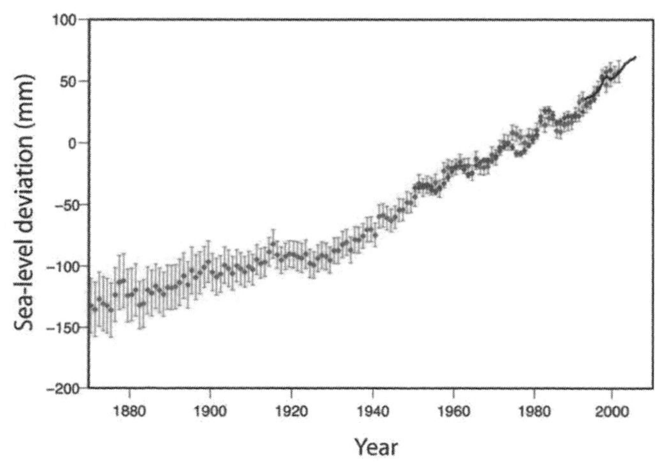

Figure 12.3. Global changes in sea level since 1880. (Williams et al. 2009)

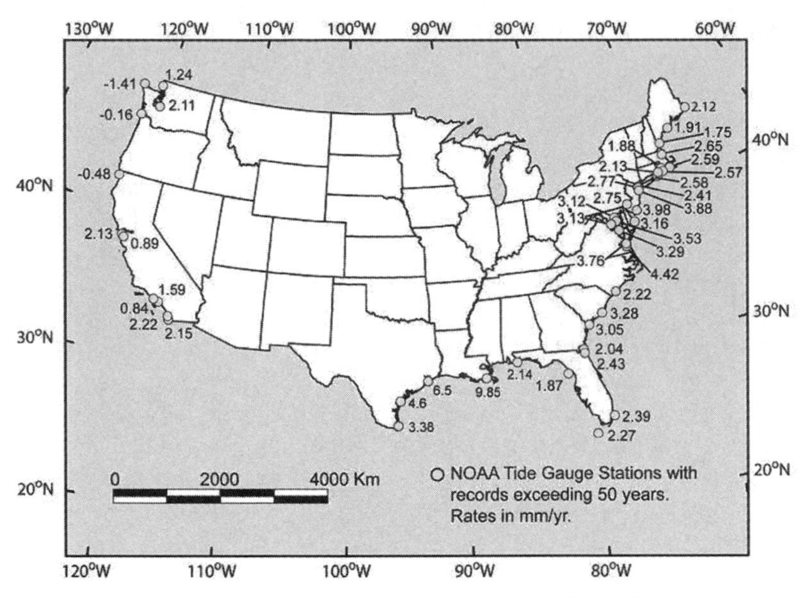

Figure 12.4. Annual relative sea-level rise rates averaged over the past 100 years for the coastal locations of the coterminous United States. Negative values occur in the Pacific Northwest, which is experiencing uplift due to tectonic forces; similar forces are uplifting much of the Alaskan coast. (Williams et al. 2009)

(Figure 12.4; Hicks et al. 1983; Emery and Aubrey 1991; Kearney and Stevenson 1991; Davis and Browne 1996; Peltier 2001; see examples of annual sea-level rise rates on the NOAA online sea-level trends website, http://tidesandcurrents.noaa.gov/sltrends/sltrends.shtml). Sections of Canada's Atlantic Coast including the Gulf of St. Lawrence have been listed among the nation's most threatened coastlines (Shaw et al. 1998). During the past 100 years, the rate of sea-level rise has been 2 to 4.5 times higher for Delaware Bay (0.10–0.16 inch/yr [2.5–4.0 mm/yr]) than rates were for the past 2,000 years (Valentine 2002). Similarly, in North Carolina, current rates of 0.12 to 0.13 inches per year (3.0–3.3 mm/yr) are three times that of the pre-Colonial rate (0.04 in/yr [1.0 mm/yr]) and about 50 percent higher than the rate (0.09 in/yr or 2.2 mm/yr) that accelerated between 1879 and 1915 (Kemp et al. 2009). In Louisiana's Barataria Bay, the rate of subsidence may be three to five times faster than the increase in eustatic sea level (DeLaune et al. 1990). Subsidence rates for Barataria Bay marshes are 0.24 to 0.31 inch/year (6–8 mm/yr) for brackish and tidal fresh marshes and 0.39 to 0.63 inch/year (10–16 mm/yr) for salt marshes; the latter amounts to a 3.3 to 5.3 foot (1.0–1.6 m) fall in 100 years.

The Probable Future of Tidal Wetlands on Submerging Coasts

While the oldest salt marshes have maintained themselves through accretion processes for a few thousand years, most coasts are sinking, with eustatic SLR and coastal plain subsidence having a tremendous impact on tidal wetlands and adjacent lowlands worldwide. Rising sea level may initiate a series of land transformations including: 1) lower-elevation salt marshes disappear, first transforming into tidal flats then into bay bottoms depending on tidal levels; 2) some portions of high salt marshes become subjected to daily tidal flooding and become low marshes; 3) unprotected lowlands (uplands and freshwater wetlands) adjacent to salt marshes become high salt marsh due to more frequent flooding by saltwater and colonization by halophytic plants; 4) tides move saltwater farther upstream, increasing the salinity of brackish marshes, possibly converting them to salt marshes, and increasing salinity of tidal freshwater wetlands transforming them into brackish marshes; and 5) tidal influence is extended farther upstream subjecting low-lying fields, forests, and freshwater wetlands to tidal flooding, thereby converting them to tidal freshwater wetlands. This process of "marine transgression" has continued for thousands of years at varying rates (Chapter 2). Such marine–land interactions have been described using examples along the Virginia coast, emphasizing the significance of slope and sediment supply (Brinson et al. 1995). Since many low-lying areas have been developed for cities and towns and are now bulkheaded or otherwise protected, the landward advance of the sea and coastal marshes has been prevented from occurring in those locales. This does not bode well for the future of tidal wetlands in developed areas. The ability of marshes to raise their elevations (e.g., accretion) in response to local SLR (e.g., eustatic SLR and subsidence) will largely determine their future at their present-day locations. Given that coastal wetlands in areas of low-tide range exist within a narrow elevation range, they have been identified as more vulnerable to rising sea level than similar wetlands in macrotidal areas (e.g., Stevenson et al. 1986; Day et al. 2000; Friedrichs and Perry 2001; Kirwan and Guntenspergen 2010; Kirwan et al. 2010).

Marsh Accretion Processes

Generally speaking, the buildup of marsh surfaces (marsh accretion) is dependent on two processes: 1) organic matter accumulation in the soil (autogenic processes); and 2) deposition of sediment from flood events (allogenic processes). These processes are influenced by a number of factors including tidal range, frequency and duration of flooding and soil saturation, storm activity, plant composition, organic productivity, autocompaction, and sediment supply (Figure 12.5; Roman et al. 1997; Nikitina et al. 2000; Friedrichs and Perry 2001; Culberson et al.

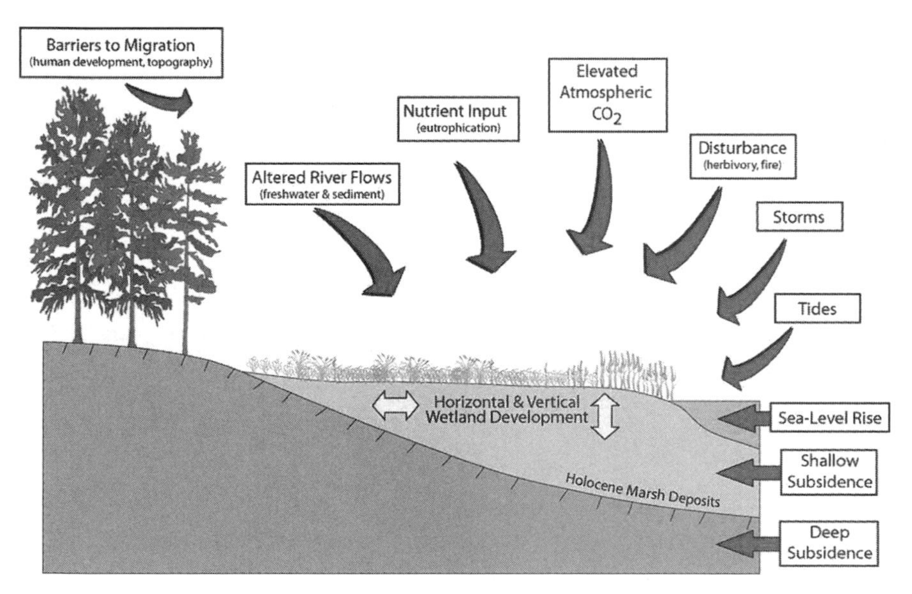

Figure 12.5. Many factors affect the ability of tidal marshes to adapt to changes in sea level. (Cahoon et al. 2009)

2004; Day et al. 2008; Cahoon et al. 2009; Mudd et al. 2009). An important distinction between tidal wetlands is whether the marsh is sediment-starved or not. If it is the former, the survival of the wetland in face of SLR will be dependent on organic accumulation rates (e.g., Nyman et al. 2006). If the wetland receives sediments, the combination of allogenic and autogenic processes working together will largely determine the ability of the marsh to maintain elevations suitable for sustaining plant growth in response to relative SLR.

ORGANIC ACCUMULATION (AUTOGENIC PROCESSES)

Producing more above- and belowground organic matter will help tidal marshes elevate their surfaces in an effort to maintain levels suitable for continued marsh growth and reproduction (Figure 12.6), which is particularly important for sediment-starved tidal wetlands. In these wetlands, belowground biomass dominates the vertical accretion (e.g., Kearney et al. 1988; Craft et al. 1993; Turner et al. 2000; Nyman et al. 2006). The importance of autogenic processes in helping plants convert tidal flats to marshes and then successfully respond to rising sea level has been attributed in large part to organic matter and its water-holding capacity. Organic matter and water/pore space accounted for an average of 91 percent of the vertical accretion in the regularly flooded zone and 96 percent in the irregularly flooded zone of Rhode Island salt marshes (Bricker-Urso et al. 1989). When water was discounted and only dry solids considered, equal amounts of organic and inorganic matter (9%) contributed to low marsh accretion, while for the high marsh, organic solid accumulation was more than twice (11%) that of inorganic solids (4%). This relationship is expected to be true for most tidal wetlands except low deltaic marshes of Louisiana where inorganic accretion is nearly twice that of organic accretion.

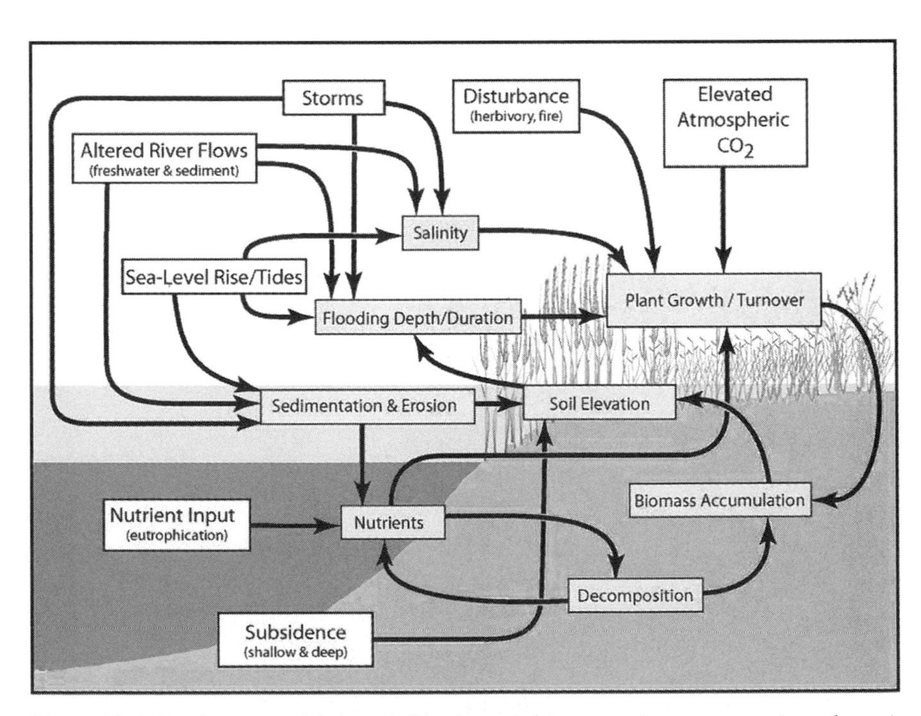

Figure 12.6. Environmental factors (white boxes) drive accretion processes (gray boxes) that determine a tidal wetland's ability to grow vertically and to keep pace with rising sea level. (Cahoon et al. 2009)

Studies have shown that elevated CO_2 levels stimulate plant productivity, shoot density, and fine root biomass (Drake et al. 1997; McKee and Rooth 2008; Langley et al. 2009) and reduce stress from salinity and flooding (Cherry et al. 2009). But, can marsh plants produce enough organic matter to survive rising seas? Plants differ in their response to elevated CO_2. Certain species called C3 species are more productive during times of high CO_2 than are C4 species, which have the advantage during periods of low CO_2 (Table 12.1). More than 95 percent of the Earth's plants (including trees) are C3 species. In salt marshes, C3 species are more characteristic of the upper high marsh, whereas C4 species are representative of the low and middle-high marsh. Examination of their stable carbon isotopes in soil cores documented shifts in vegetation (from C4 species to C3 species back to C4 species) over a 2,500-year period in a northern New England salt marsh (Johnson et al. 2007). Olney's three-square (*Schoenoplectus americanus*), a brackish marsh C3 species, was more productive than salt hay grass (*Spartina patens*) under laboratory-controlled conditions of elevated CO_2 (Cherry et al. 2009). Increased aboveground biomass production and shoot-base expansion resulted in a significant elevation change due to vertical soil displacement and organic matter buildup. Permanent flooding also expanded the soils and is likely to slow decomposition of organic matter by creating longer periods of anaerobic conditions. Enhanced tolerance of stressors (salinity and flooding) was cited as the reason that favored the C3 species over the C4 species in this experiment. Elevation change during this one-year mesocosm study ranged from a loss of 0.19 inch (4.7 mm) to a gain of 1.7 inches (44 mm), which was consistent with field observations (–0.6 to +1.4 inches per year [–15 mm to +36 mm/yr]) by one of the co-authors. The productivity of salt hay grass was significantly reduced under flooded conditions, falling from around 1,400 to 1,600 g/m^2 under unflooded and intermediate flooding (alternating wet and dry) conditions to about 400 g/m^2, while production of Olney's three-square increased by a factor of 2.5 from intermediate flooding conditions and fivefold over unflooded conditions. A similar study found that elevated CO_2 levels (current + 340 ppm) stimulated a 0.15-inch (3.9 mm) annual elevation gain in a brackish marsh (Langley et al. 2009). While this may be promising news, one must realize that the stimulatory effect of CO_2 on plant productivity and elevation gain may decline when atmospheric CO_2 rises above 720 ppm. Other factors such as increased nitrogen availability (e.g., estuarine eutrophication and runoff from farmland and fertilized

Table 12.1. Examples of C3 and C4 species found in North American salt marshes

C3 Species
Black grass (*Juncus gerardii*)
Salt marsh bulrush (*Schoenoplectus robustus, S. maritimus*)
Olney's three-square (*Schoenoplectus americanus*)
Glassworts (*Salicornia* spp.)
Northern seaside arrow-grass (*Triglochin maritima*)
Seaside goldenrod (*Solidago sempervirens*)
Common reed (*Phragmites australis*)
Northern sea lavender (*Limonium carolinianum*)
Seaside plantain (*Plantago maritima*)
Silverweed (*Argentina anserina*)
White sea blite (*Suaeda maritima*)
Marsh orach (*Atriplex patula*)
Black needlerush (*Juncus roemerianus*)
American alkali-grass (*Puccinellia americana*)
Sea ox-eye (*Borrichia frutescens*)
Annual salt marsh pink (*Sabatia stellaris*)
Water pimpernel (*Samolus valerandi* ssp. *parviflorus*)
Saltwort (*Batis maritima*)
C4 Species
Smooth cordgrass (*Spartina alterniflora*)
Salt hay grass (*Spartina patens*)
California cordgrass (*Spartina foliosa*)
Prairie cordgrass (*Spartina pectinata*)
Salt grass (*Distichlis spicata*)
Salt marsh fimbry (*Fimbristylis castanea*)

Sources: Haines and Montague 1979; Looney and Gibson 1995; Johnson et al. 2007; Tanner et al. 2007.

lawns) may, however, favor C4 species during times of enhanced CO_2 and reduce elevation gain by lowering root productivity and possibly stimulating organic matter decomposition (Langley et al. 2009; Langley and Megonigal 2010). Some researchers predict that rising temperatures and corresponding expansion of the growing season may increase plant productivity by 10 to 40 percent, which may help tidal wetlands, especially in middle and northern latitudes, mitigate some of the effects of marsh loss due to SLR (Kirwan et al. 2009). If precipitation increases simultaneously, decomposition of marsh litter in the high marsh may actually accelerate as microbial populations will likely increase with warming (Foote and Reynolds 1997; Charles and Dukes 2009). Regional variations may occur, but one study found that decomposition rates increase by about 20 percent per degree of warming (Kirwan and Blum 2011). This could significantly negate some of the gains in accretion from higher productivity and release more soil carbon into the environment, while making marshes more vulnerable to sea-level rise.

Each plant species has optimum elevations for shoot growth and for root production in tidal marshes and these elevations are not necessarily the same (Kirwan and Guntenspergen 2012). The effect of rising seas on a species' root growth may have more of an impact on marsh survival in some places. Where marsh elevations are above the optimal level for root production, increased duration of flooding from rising seas could stimulate more root growth, increase belowground biomass, and help maintain marsh elevations relative to sea level. Once this optimum elevation is exceeded by inundation, however, root production declines, less organic matter accumulates, and marshes begin to deteriorate, especially where sediment inputs are limited. The effect of rising sea level on a plant's ability to produce roots will likely determine the fate of many of today's marshes to survive into the next millennium (e.g., Blackwater National Wildlife Refuge in Maryland and Louisiana's coastal marshes; Kirwan and Guntenspergen 2012).

SEDIMENTATION (ALLOGENIC PROCESSES)

Sedimentation is crucial for salt marsh formation and for their survival under high rates of SLR. Smooth cordgrass (*Spartina alterniflora*) is the leading marsh builder on the East Coast, and its relative Pacific cordgrass (*S. foliosa*) plays a similar role in California, while salt grass (*Distichlis spicata*), pickleweed (*Salicornia depressa*), and seaside arrow-grass (*Triglochin maritima*) are pioneer species in the Pacific Northwest. These species become established on accreting tidal flats, colonizing the lowest elevations that emergent halophytes can tolerate (Figure 12.7). Once established these and other low marsh species slow tidal currents, change their direction within the marshes, trap sediments, stabilize the substrate, promote more rapid sedimentation, and add organic matter to the soil. These actions eventually raise the marsh surface to a level where other species can grow.

Since sediment is carried by tidal waters, the longer the duration of inundation, the greater the total amount of sediment deposited on the marsh (Friedrichs and Perry 2001). Therefore, sediment deposition tends to decrease with increasing marsh elevation. Regularly flooded marshes experience more sedimentation than irregularly flooded marshes, however, organic matter accumulates at higher rates in the latter due to infrequent tidal flushing (Craft et al. 1993). Accretion rates vary within marshes as creekside locations tend to receive more sediment than interior sites receive (Gammill and Hosier 1992; Chmura and Hung 2004). In a Delaware salt marsh, 80 percent of the suspended sediment was deposited within 40 feet (12 m) of the creekbank (Stumpf 1983). Some parts of marshes may keep pace with SLR while others may not. Wherever the buildup of marsh soils cannot keep pace with the rising sea level, marshes will become tidal flats and eventually open-water

Figure 12.7. Two patches of smooth cordgrass (*Spartina alterniflora*) colonized this Delaware Bay tidal flat.

habitats while other marshes face the problem of extreme waterlogging, which eventually leads to substrate instability and marsh deterioration. Evidence suggests that recent SLR is not constant and marsh accretion adjusts to the prevailing conditions (e.g., Kearney et al. 1994). Periodic changes in sea level (e.g., metonic cycle) may explain changes in vegetation patterns from salt marsh species to more brackish or freshwater species back to salt marsh species that have been observed in some marsh peat cores.

The tidal fresh to oligohaline marshes of the St. Lawrence River offer an interesting and perhaps unique example of the seasonal effect of vegetation on sedimentation (Serodes and Troude 1984). The combination of snow goose grazing and shearing by winter ice transforms the regularly flooded common three-square (*Schoenoplectus pungens*) marsh into a mudflat in winter. Plant growth resumes in late spring (June) and peaks in September. During this time, sediments build up in the marsh at an average rate of 0.08 inch (2 mm) per day due to the presence of vegetation. While the marsh plants did not appreciably slow the flow of flood currents, they did move flow direction landward. Vegetation did slow the speed of the ebb currents and alter their direction to flow perpendicular to the shore. This process enhanced sedimentation until migrating snow geese arrived in late September. The geese virtually eliminated the vegetation through grazing and walking through the marshes while probing the soil with their beaks further destabilized the soil. The activity of the geese destroyed the vegetative cover in a matter of days, thereby accelerating erosion until the winter freeze.

The morphology and aboveground biomass of plants may also affect sedimentation rates. These properties include surface roughness of the stems and leaves; the total surface area of vegetation per unit area; and the branching pattern of leaves, inflorescences, and fruiting structures (Yang 1998). Introduced smooth cordgrass, the species with the greatest biomass, was found to be more effective in trapping suspended sediments than an Asian bulrush (*Scirpus mariqueter*) and common reed (*Phragmites australis*) in the Yangtze Delta of China (Li and Yang 2009). On average, smooth cordgrass collected three times more sediment than did common reed and about seven times more than the bulrush. An analysis of sediment trapping per unit of biomass, however, showed that the shorter bulrush trapped more sediment per unit area: about three times that of common reed and 1.5 times that of smooth cordgrass. Individual plants of all species trapped more sediment at lower marsh elevations (creekside) than farther landward, which was expected due to higher suspended sediment concentrations at the marsh–water interface.

Accretion Rates

Annual marsh accretion rates vary due to several factors including tides, sediment sources, basin morphology, location within the marsh, and changes in elevations of land masses (e.g., submergence, emergence, or autocompaction of substrates). Given the concern about the impact of rising sea level on coastal marshes, numerous studies of marsh accretion have been, and continue to be, performed, with some referenced in this chapter. An examination of vertical accretion rates in 141 salt marshes from New England to the Gulf of Mexico found ranges from 0.04 to 0.25 inch per year (1.0–6.3 mm/yr) for the East Coast and 0.04 to 0.70 inch per year (1.0–17.8 mm/yr) for the Gulf Coast (Turner et al. 2000). Sediment rates may also vary considerably locally based on a study of Maine tidal marshes (Wood et al. 1989).

Higher rates of accretion occur in streamside marshes where suspended sediments are readily available. Regularly flooded freshwater tidal marshes in Virginia had annual accretion rates of 0.33 inches (8.4–8.5 mm), while high marsh stands dominated by big cordgrass (*Spartina cynosuroides*) and common reed had annual rates of 0.19 to 0.24 inches (4.9–6.0 mm) (Neubauer et al. 2002). In Louisiana, accretion rates at streamside marshes ranged from 0.08 to more than 0.63 inches per year (2.0–>15.9 mm/yr), whereas rates in the marsh interior (back marsh >49 feet [15 m] from streambank) varied from 0.17 to 2.04 inches (4.2–51.9 mm) annually (Jarvis 2010). Measuring accretion rates alone without considering subsidence can lead to substantial overestimates in increases in surface elevation (Cahoon et al. 1995). For example, after a couple of years, the marsh elevation change at the site with the highest accretion rate in the Louisiana example above was only 0.1 inch (2.9 mm) due to compaction (shallow subsidence). This difference underscores the problem of relying only on vertical accretion measurements to determine change in coastal elevation (Cahoon et al. 1995). Consequently, today many researchers monitoring coastal wetland elevations install surface elevation tables to account for subsidence in addition to laying down feldspar markers on the substrate (Figure 12.8; van Asselen et al. 2009).

Historically, marsh accretion rates in some regions could have been higher than current rates due to increased alluvial sediments from deforestation and agricultural expansion, and to other factors. Analysis of sediment cores from the province of New Brunswick revealed two different marsh accretion rates based on the presence of two soil horizons resulting from two massive land-clearing operations that occurred from 1790 to 1830. The marsh accretion rate averaged 0.08 inch per year (2.1 mm/yr) over a 200-year period but when calculated since 1830, the annual rate dropped to 0.05 inch (1.3 mm) (Chmura et al. 2001). Increased earthquake activity from 1790 to 1830 lowered land surfaces, thereby raising local sea level by as much as 0.14 inch per year (3.6 mm/yr) (Smith et al. 1989 a,b). This activity was the main factor responsible for the higher early accretion rate, with additional contributions from more ice-rafting of sediments to the high marsh during a colder era (Chmura et al. 2001).

Tidal wetlands undergoing restoration may experience higher sedimentation rates than natural wetlands because they may have lower elevations due to prior subsidence. The higher porosity of their sediments may also allow them to increase their elevation more rapidly (Anisfeld et al. 1999). As their elevation reaches that of unaltered wetlands, one would expect their accretion rates to decline and match those of neighboring natural wetlands. Where sediment loads are extremely high, accretion rates can be enormous, especially for wetland restoration sites. For example, one recently restored wetland in San Pablo Bay in California has an average sedimentation rate of 6.6 inches per year (168 mm/yr) over an 8-year period, while other sites had average annual rates ranging from 0.8 to 2.5 inches (20–63 mm) (Takekawa et al. 2010).

Prolonged droughts may affect the ability of salt marshes to maintain elevations necessary to support plant growth in light of SLR. Most of the volume of salt marsh soils is occupied by water; only 3.8 percent and 4.9 percent, on average, is composed of organic and inorganic materials, respectively (Turner et al. 2000). This makes them particularly susceptible to droughts. Studies of Texas and Louisiana marshes during droughts detected significant and rapid declines in marsh elevations related to lowered groundwater levels and reduced subsurface hydrology (Cahoon et al. 2010). These conditions may cause consolidation of shrink-and-swell soils that could affect the marsh's ability to return to its original elevation when the drought ends, although elevations rebounded some time after return to normal conditions. Where climate change causes more severe droughts, coastal wetlands may suffer from these effects and increased saltwater intrusion into tidal freshwater reaches putting additional stress on existing vegetation.

Several studies have identified wetlands where marsh accretion rates may be sufficient to help them keep pace with rising sea level (e.g., Chesapeake Bay—Ward et al. 1998, Hussein and Rabenhorst 2002, Griffin and Rabenhorst 1989, Neubauer et al. 2002; Connecticut—Anisfeld et al. 1999; Delaware—Nikitina et al. 2000; eastern Canada—Chmura and Hung 2004; Maine—Anderson et al. 1992; New Jersey—Orson et al. 1992; New York—Kolker et al. 2005; North Carolina—Craft

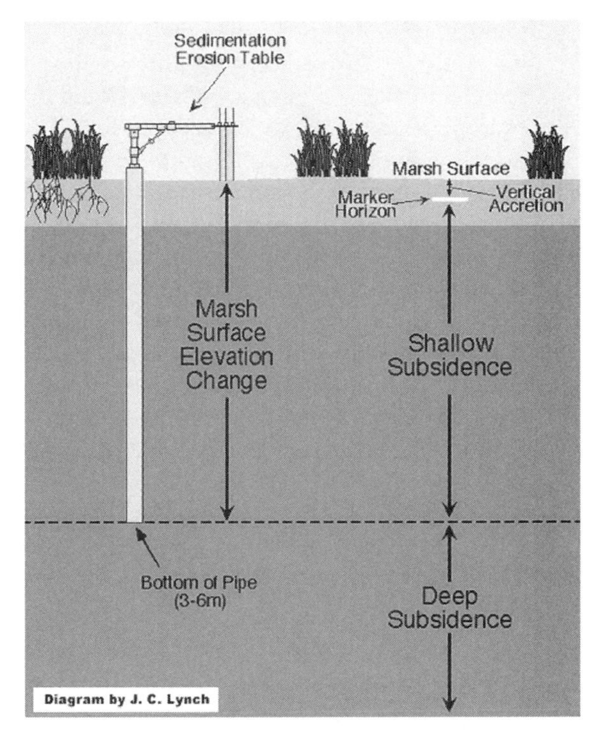

Figure 12.8. Surface elevation tables (SETs) are installed in marshes to determine elevation change in the marsh surface relative to sea-level rise. They are often used in combination with feldspar markers deposited on the soil. (U.S. Geological Survey)

et al. 1993; Oregon—Thom 1992; San Francisco Bay—Patrick and DeLaune 1990). The Chesapeake Bay examples are mostly brackish marshes that do not receive significant sediment input, yet appear to be successfully increasing their elevations by organic matter accumulation alone. One of these sites is a tidal freshwater marsh with accretion rates of 0.19 to 0.33 inches per year (4.9–8.5 mm/yr, with lower rates from high marsh communities) that exceed the local rate of relative SLR (0.18–0.24 inch/year [4.5–6.0 mm/yr]) (Neubauer et al. 2002). The New Jersey example, a tidal freshwater wetland on a tributary of the Delaware River, accumulated sediment at annual rates averaging 0.7 inches (17 mm) from 1954 to 1965 (a period of increased storm activity) and although the annual rate has since declined to 0.38 inch (9.7 mm), it is still collecting sediment at rates exceeding local SLR. Despite the devastating effects of hurricanes, they can provide sediments that support marsh accretion. For example, Hurricane Katrina deposited 1.2 to 3.2 inches (3 to 8 cm) of sediments in some areas on the Louisiana coast yielding net increases in elevation two years after the storm: a 0.7-inch (1.7 cm) increase in marsh elevation at a site with mostly sand deposition and a net gain of 0.3 inch (0.7 cm) where sediment with high organic matter content was deposited (McKee and Cherry 2009). It is important, however, to reiterate that accretion rates alone are not sufficient to assess a given marsh's ability to maintain its elevation relative to sea-level rise as they do not account for subsidence (Cahoon et al. 1995).

Salt Marsh Migration

While scientific research has focused on studying processes that affect the ability of existing marshes to survive, tidal wetland vegetation will, without question, relocate to suitable places for their establishment in the future. As coastal lowlands are inundated, forest vegetation will die and be replaced by halophytic species (Figure 12.9).

a

b

Figure 12.9. Examples of salt marsh migration: (a) Maine and (b) Maryland. (b: Drew Koslow, U.S. Fish and Wildlife Service)

Plenty of evidence of this process can be seen along the U.S. coastal zone (see Marine Transgression in Chapter 2, and the following section). A report on Maryland's tidal marsh soils listed the Elkton and Othello series as low soils that may convert to tidal marsh and identified over 50,000 acres (20,230 ha) of these soils for the state (Darmody and Foss 1978). The Northeast Region of the U.S. Fish and Wildlife Service working with other agencies and organizations has established permanent plots in the upper marsh and neighboring lowlands to monitor salt marsh migration at selected refuges and on other conservation lands. These sites will be evaluated periodically to track changes in vegetation with rising sea level. Undeveloped lowlands contiguous to coastal wetlands are vital for the future of salt marshes and other tidal wetlands given rising sea level. The challenge for coastal zone managers and planners is to maintain these areas in a natural state, so that natural marine transgression can occur.

Observed Effect of Recent Sea-Level Rise on Tidal Wetlands

Even where marshes appear to be keeping pace with sea-level changes, it is likely that many areas will at least experience an increase in low marsh at the expense of more diverse high marsh, and that some low marsh will become tidal flat or open water. The ability of marshes to raise their elevations to keep pace with SLR varies spatially within the marsh. With rising sea levels expected to continue, the future development of salt marshes will be influenced by two main factors besides organic accumulation, sedimentation, and subsidence: 1) the local topography and 2) society's attempts to protect low-lying areas from inundation (e.g., levees and hardening of shorelines). Where the topography landward of salt marshes is steep, little opportunity for salt marsh migration occurs, and the existing marsh will be drowned by higher water levels. These marshes will be virtually squeezed out of existence. The future of Delaware's salt hay grass community may depend on the degree of slope adjacent to existing marshes where it may be restricted to gently sloping former upland surfaces (Carey 1996). For example, a 76-foot-wide (23 m) salt hay zone that persisted on a steep slope for 190 years was reduced to just a 6.6-foot (2 m) belt in only 47 years. This Delaware wetland had only migrated 16.5 feet (5 m) landward over the steep slope and has only migrated landward at a similar rate today (0.23 feet/year [0.07 m/yr]). During the same time frame, a similar marsh bordered by lowlands migrated at rates four to five times that (i.e., 1.0–1.2 feet/year [0.30–0.36 m/yr]). Over the next 100 years, Delaware marshes adjacent to low-lying upland may migrate 83 to 248 feet (25–75 m) depending on projected water level increases, which are ranging from 14 to 33 inches (35–82 cm). Smooth cordgrass will likely become the sole species in areas with steeper slopes as the high marsh zone is expected to be eliminated from these sites. This process will likely occur in similar situations elsewhere along the Atlantic Coast.

In the Wequetequock-Pawcatuck marshes of Connecticut, areas dominated by salt hay grass in the 1940s are now colonized by the short form of smooth cordgrass, salt grass, and various flowering herbs with only patches of salt hay grass remaining (Warren and Niering 1993). During this time, the black grass (*Juncus gerardii*) belt was converted to a forb community dominated by seaside arrow-grass. Increased flood frequency and duration, prolonged soil saturation, increased salinities and sulfide concentrations, and accompanying anaerobic conditions may be largely responsible for this vegetation change. Today, vertical marsh accretion rates are 30 percent lower than the rate of relative SLR (Orson et al. 1998). Shifts in vegetation communities may be a normal occurrence as reflected in vegetation patterns interpreted from marsh peat cores (Figure 12.10). Similar changes have been reported in another salt

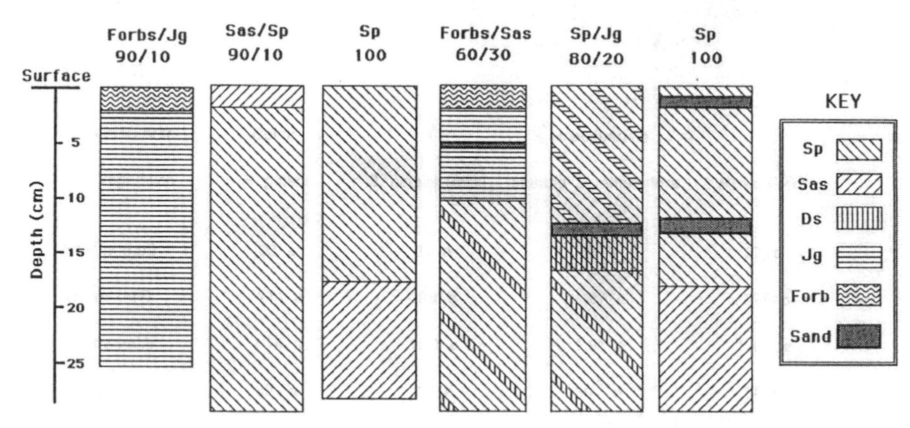

Figure 12.10. Core samples from Connecticut salt marshes showing changes in vegetation over time. Sp = *Spartina patens*; Sas = short form *Spartina alterniflora*; Ds = *Distichlis spicata*; Jg = *Juncus gerardii*; Forbs = mixture of *Limonium, Plantago, Triglochin,* and *Agalinis*. The numbers beneath the plant codes represent the percent cover by the species (Warren and Niering 1993; reprinted by permission from the Ecological Society of America, Inc.)

marsh in southwestern Connecticut (Biddle and Warren 1999) and elsewhere on the East Coast. Researchers studying Nauset Marsh on Cape Cod found that a portion of the marsh experienced a recent change in vegetation where salt grass replaced salt hay grass—evidence that the marsh is getting wetter, while marsh accretion rates suggested that the marsh is keeping pace with local SLR (Roman et al. 1997). Local subsidence may explain this contradiction. In Virginia, the same vegetation shift occurred due to increased flooding, which has caused more ponding and a change in the marsh substrate from turf to hummocks (Buck 2001). These observations suggest that these marshes, or at least portions of these marshes, may not be keeping up with rising sea level.

Rapid changes are occurring in some areas. In just four years (1995–1999), a significant increase in smooth cordgrass at the expense of salt hay grass marsh was observed in two Narragansett Bay (RI) marshes (Donnelly and Bertness 2001). Based on historic tide gauge readings and analysis of peat cores, this transgression has been occurring since the late 1800s. Comparison with findings from other studies suggests a regional trend for southern New England that high marsh will likely be replaced by smooth cordgrass marsh (e.g., Warren and Niering 1993; Roman et al. 1997; Biddle and Warren 1999). If sea level rises at annual rates of 0.24 inch (6 mm) or more, smooth cordgrass marshes will be drowned (Donnelly and Bertness 2001).

Where vacant low-lying lands border coastal wetlands, opportunities arise for salt marsh transgression and survival. For example, a Virginia salt marsh increased its area by 8.2 percent over a 50-year period due largely to migration over upland, while a nearby lagoonal (bayside) marsh lost 10.6 percent of its area (Kastler and Wiberg 1996). In northern Florida, salt marsh migrated landward 9.8 to 26.2 feet (3–8 m) into a neighboring forest from 1951 to 1997 (Wang et al. 2001). SLR elevates local water tables of lowland forests adjacent to estuarine marshes—a process that initiates changes in soil properties and vegetation patterns. In the short term, recruitment of tree seedlings may be impaired due to prolonged soil saturation, while mature trees may continue growing at normal rates as evidenced by loblolly pines (*Pinus taeda*) on the Delmarva Peninsula (Kirwan et al. 2007). In the long term, however, trees suffer storm-induced mortality and salt stress (chlorosis), and eventually the forest is overtaken by halophytic species—salt marsh transgression. An inventory of wetlands in Maryland reported the occurrence of "estuarine forests," former loblolly pine forests now subjected to tidal flooding and harboring an understory of halophytic (salt-tolerant) species (Tiner and Burke 1995). Examples of this phenomenon are common on the Delmarva Peninsula, especially on Maryland's Lower Eastern Shore along Chesapeake Bay (Dorchester and Somerset Counties). The presence of dead trees or stumps in salt and brackish marshes elsewhere provides ample evidence of this phenomenon along the Atlantic Coast (e.g., spruce in Maine and Atlantic Canada, white pine in New Hampshire, red cedar in Massachusetts, Atlantic white cedar in New Jersey, loblolly pine in Maryland and North Carolina, and cabbage palm in South Carolina and Florida; see Figure 12.9). Along the New Jersey coast, swamp and lowland forests along brackish marshes are changing—trees are dying and the canopy is opening, providing opportunities for light-loving herbs (Figure 12.11). Common reed, seaside goldenrod (*Solidago sempervirens*), annual salt marsh fleabane (*Pluchea odorata*), and groundsel-bush (*Baccharis halimifolia*) are among the first halophytic species to colonize these sites. In South Carolina, pine–oak–gum forest vegetation (*Pinus–Quercus–Nyssa*) was replaced by a succession of marsh species: high-tide bush (*Iva frutescens*), black needlerush (*Juncus roemerianus*), sea ox-eye (*Borrichia frutescens*), glasswort (*Salicornia*), short form of smooth cordgrass, and finally by medium-height smooth cordgrass (Gardner et al. 1992). In the Florida Keys, rising sea level has increased local groundwater levels and soil salinity resulting in the death of lowland slash pine (*Pinus elliottii*) forests (Ross et al. 1994). On Florida's west coast, cabbage palm forests (*Sabal palmetto*) are retreating as sea level rises and being replaced by salt marsh (Williams et al. 1999a). Exposure to salt was identified as the major cause preventing tree regeneration, which then allowed salt marsh plants

Figure 12.11. As lowland forest trees die from increased salinity and, in some cases, increased flooding and soil saturation, the canopy opens creating conditions for colonization by herbs such as common reed (*Phragmites australis*).

to successfully invade these sites. Increased height and frequency of storm surges may increase saltwater intrusion events into low-lying forests that may initiate tree mortality and create open canopies that promote colonization by herbaceous plants from neighboring marshes.

Rising sea levels will move saltwater farther upstream in estuaries causing the death of intolerant tidal freshwater vegetation and shifting vegetation patterns. Depending on the rate of colonization by halophytic species and the nature of the substrate, these wetlands may change to brackish marshes or be eroded and become either tidal flats or open-water habitat. Tidal fresh marshes may be pushed farther upstream depending on local topography. From 1954 to 1992, the shrub-marsh zone moved upstream from about 0.06 mile to almost 2 miles (100–3,400 m) in three tributaries to Delaware Bay (Valentine 2002). Since 1950, nine halophytic species migrated farther upstream in the Delaware River, while strictly freshwater species have either moved upriver or been extirpated (Schuyler et al. 1993). In the San Francisco Bay estuary, halophytic species are replacing more freshwater species along the channels, while in the interior where conditions are actually becoming less saline due to more frequent flooding, "salt-intolerant" species from the low marsh (e.g., *Schoenoplectus pungens*) are replacing the more "salt-tolerant" species (e.g., *Distichlis spicata*) on the high marsh (Watson and Byrne 2009). In Louisiana, black mangrove (*Avicennia germinans*) has expanded its range northward into smooth cordgrass marshes following two decades of warm weather (no killing freezes; Perry and Mendelssohn 2009). Rising sea level along the Gulf Coast have put Louisiana's coastal forests of bald cypress (*Taxodium distichum*) and water tupelo (*Nyssa aquatica*) and neighboring bottomland forests at risk in the Lake Verret watershed where the submergence rate is about four times the sedimentation rate (DeLaune et al. 1987a). Portions of these types of swamp elsewhere in Louisiana have become marshes and open water due to salt stress and increased flooding (Shaffer et al. 2009). Lowland coastal forests abutting tidal marshes have been experiencing vegetation changes due to rising sea levels for some time, as evidenced by stumps of freshwater or upland trees in salt marshes and mangrove swamps (Williams et al. 1999b). Failure of trees to regenerate (i.e., produce seedlings) usually provides the first sign of stress. Trees typically succumb to one or more of three factors that accompany rising sea level: increased salinity, increased hydroperiod (flooding and soil saturation), and erosion. The effect of these factors varies along the coast and within estuaries. Human activities (e.g., dredging, ditching, and reduced freshwater flows) may also contribute to these changes. For example, channel dredging of Cape Fear River in North Carolina caused an increase in tidal range and saltwater intrusion leading to the decline in upstream cypress swamps (Hackney and Yelverton 1990). More subtle changes may also be taking place. For example, a Virginia tidal fresh marsh experienced an increase in biomass of more salt-tolerant species (e.g., big cordgrass and shoreline sedge [*Carex hyalinolepis*]) at the expense of arrow arum (*Peltandra virginica*), the most abundant species, in almost two decades, presumably due to an increase in salinity caused by rising sea level (Perry and Herschner 1999). This vegetation pattern suggests a shift from a tidal fresh marsh to oligohaline conditions and movement of the tidal fresh marsh zone upstream. Saltwater intrusion into tidal freshwater marshes may also affect their ability to accumulate organic matter as studies have shown that changes in salinity change the rate and degree of organic matter mineralization by microbes (Neubauer 2011;Weston et al. 2011). Increased microbe-driven mineralization of organic matter will increase decomposition, thereby reducing the amount of organic matter accumulation. This is especially important in tidal freshwater marshes where organic matter is responsible for an average of 62 percent of their accretion (Neubauer 2008). Increased decomposition and reduced plant productivity accompanying saltwater intrusion will likely limit the ability of these marshes to adjust their elevations in response to rising sea levels.

Louisiana marshes are perhaps the best-known examples of tremendous marsh loss. In this oil-producing state, many factors besides SLR are contributing to this loss (e.g., oil and gas extraction, canal construction, coastal subsidence in the Mississippi delta due to consolidation of alluvium deposits, and reduced sediment loads from the Mississippi River; see Chapter 8). The rate of submergence (0.47 inch/year [12 mm/yr]) exceeds the vertical accretion rate (0.31 inch/year [8 mm/yr]) of a Louisiana salt marsh, and the marsh will likely become open water in 40 years (DeLaune et al. 1983b). Significant drowning of the Mississippi delta is expected even if sediment loads are restored because sea level is rising three times faster than when the delta was being formed (Blum and Roberts 2009). All Gulf Coast marshes may not be similarly impacted as four other wetlands have accretion rates that are compensating for SLR (Callaway et al. 1997). Freshwater floating marshes may be vulnerable to increased salinity because no floating salt marsh species are available to replace them, so they will likely become open water. Maidencane (*Panicum hemitomon*), a dominant flotant freshwater species, is already experiencing noticeable declines. It is particularly sensitive to salinity increases as laboratory experiments have shown significant reductions in productivity when salinity is increased to 1.5 and 3.0 ppt (Willis and Hester 2004).

Blackwater National Wildlife Refuge in Maryland may be the best example of marsh deterioration in the northeastern United States (Figure 12.12). Sea level in Chesapeake Bay has greatly increased since 1920 due to a combination of global effects as well as to increased precipitation and subsequent runoff from the bay's watershed (Kearney et al. 1988). Some of these brackish marshes have experienced accretion deficits of 3.1 inches (80 mm) since 1940, accounting for the loss of about half of the "submerged upland" marshes. About 0.5 percent of these marshes have been lost annually since 1938. With rising water

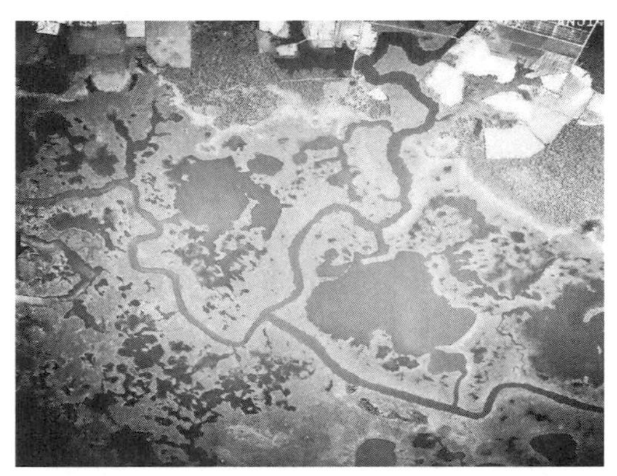

1938

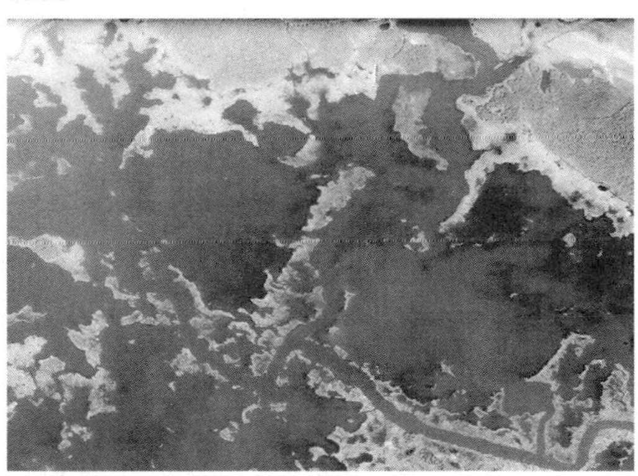

1978

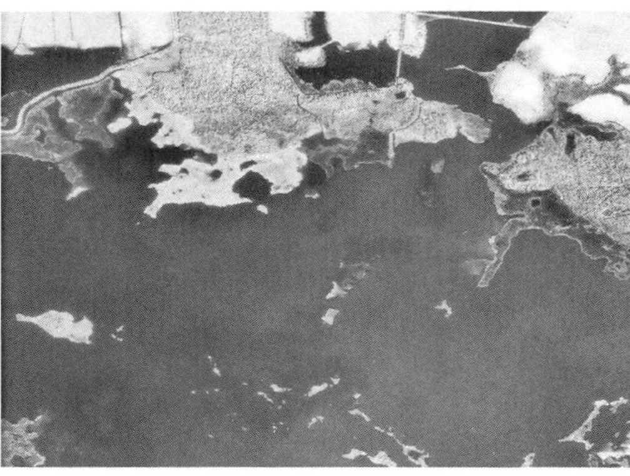

1989

Figure 12.12. Time series of aerial photos showing deterioration of marsh at Blackwater National Wildlife Refuge. (U.S. Fish and Wildlife Service)

levels, marsh soils have become excessively waterlogged and are prone to pond formation, development of new drainage channels, and subsequent erosion. Peat accumulation is the primary means of vertical accretion in these sediment-starved marshes. Upstream marshes lie in a zone of sediment trapping and there marsh losses are minimal.

Coastal marshes along Chesapeake and Delaware Bays are now recognized as extremely vulnerable to the effects of SLR (Kearney et al. 2002). About 63 percent of the marshes in these two estuaries show signs of deterioration like those observed at Blackwater. For Chesapeake Bay, about 20 percent of the marshes are severely or completely degraded with about 50 percent being in the slightly to moderately degraded condition. Only about 30 percent are not degraded. Marshes along Delaware Bay are in slightly better condition with about 44 percent not degraded, 16 percent severely or completely degraded, and the rest showing signs of slight to moderate degradation. Yet the areal coverage of deteriorated marshes more than doubled in just a decade (1984–1993) in Delaware Bay, with the marshes of the New Jersey shoreline being hardest hit. Massive export of this organic matter would increase turbidity with significant adverse impacts to estuarine and nearshore marine biota. The release of stored carbon would also eventually contribute more CO_2 to the atmosphere and add to global warming.

Changes are also occurring in tidal fresh marshes as sea level rises even where saltwater intrusion is not occurring. In the upper Delaware River estuary, tidal fresh marsh acreage did not change appreciably from the late 1970s to the late 1990s, but there was a marked increase in low marsh at the expense of high marsh—from 9 percent of the marshes to 34 percent (Field and Philipp 2000).

Predicted Changes in Coastal Wetlands

The U.S. government has identified the vulnerability of the nation's shorelines to change from future SLR (Thieler and Hammar-Klose 1999, 2000a,b; Cahoon et al. 2009). The most vulnerable shores are along the Atlantic Coast from Sandy Hook (NJ) to and including the Outer Banks (NC), the mid-coast of South Carolina, most of Florida's east coast, and along the Gulf Coast from the Alabama–Mississippi state line to Mexico (Table 12.2). Tidal wetlands in these areas are already showing signs of deterioration and change from rising seas (see above).

Coastal scientists have developed various models to predict the effect of projected rates of SLR on tidal wetlands (see Fagherazzi et al. 2012 for detailed review). Some models are relatively simple "static landscape" models where the expected rise is superimposed on the existing landscape (e.g., Titus 1988; Craft et al. 2009), while others are more complex including ecogeomorphic feedbacks between inundation and suspended sediments/sedimentation and elevated CO_2 levels and plant growth/organic accumulation (e.g., Morris et al. 2002; Temmerman et al. 2003; D'Alpaos et al. 2007; Kirwan and Murray 2007, 2008; Mudd et al. 2009). Models may include site variables such as different habitat types (e.g., salt marsh, mangrove, tidal swamp, or coastal forest), local accretion rates, local subsidence rates, elevation, slope, flooding duration, and overwash. Many scenarios for SLR (from IPCC reports and fixed rates such as 1 m, 1.5 m, or 2 m) are then applied to the model to predict wetland response to different rates of SLR.

For preliminary planning and resource management purposes, the most commonly used model is the Sea Level Affecting Marshes Model (SLAMM; Park et al. 1986, 1989, 1993; Lee et al. 1992; Galbraith et al. 2002, 2005; Craft et al. 2009; Warren Pinnacle Consulting, Inc. 2010). SLAMM simulates several dominant processes involved in marsh and shoreline dynamics related to SLR. It relies on existing data (e.g., National Wetlands Inventory maps, elevation data, and published rates of

Table 12.2. Range of wetland responses to three sea-level rise (SLR) scenarios

Geomorphic Setting	Region																							
	Long Island, NY			Raritan Bay, NY			New Jersey			Delaware Bay			Maryland–Virginia			Chesapeake Bay			Maryland Lower Eastern Shore			Virginia Beach–Currituck Sound		
	SLR	+2	+7	SLR	+2	+7	SLR	+2	+7	SLR	+2	+7	SLR	+2	+7	SLR	+2	+7	SLR	+2	+7	SLR	+2	+7
Back-barrier lagoon, other	K	K, M	K, L	—	—	—	K	M	L	—	—	—	K	M	L	—	—	—	—	—	—	M	M-L	L
Back-barrier lagoon, flood-tide delta	K	K	M	—	—	—	K	M	L	—	—	—	K	M	L	—	—	—	—	—	—	—	—	—
Back-barrier lagoon, lagoonal fill	K, L	K, M, L	K, L	—	—	—	K	M	L	—	—	—	K	M	L	—	—	—	—	—	—	—	—	—
Estuarine marsh	—	—	—	K	M	L	K	M	L	K, M	M, L	L	—	—	—	K, M, L	M-L	L	L, M	L	L	K	M	L
Estuarine fringe	—	—	—	K	M	L	K	M	L	—	—	—	—	—	—	—	—	—	—	—	—	M	M-L	L
Estuarine meander	—	—	—	K	M	L	K	M	L	—	—	—	—	—	—	—	—	—	—	—	—	—	—	—
Saline fringe	K	K, L	M	K	M	L	K, L	M, L	L	K	M	L	K, L	M, L	L	—	—	—	—	—	—	—	—	—
Tidal fresh forest	—	—	—	—	—	—	—	—	—	—	—	—	—	—	—	—	—	—	K	K	K	M	M-L	—
Tidal fresh marsh	—	—	—	K	K	K	K	M	L	K	K	K	—	—	—	K	K	K	K	K	K	K	K	K

Source: Cahoon et al. 2009.
Notes: Scenarios include 20th-century rate (SLR), 20th-century rate + 2 mm/yr, and 20th-century rate + 7 mm/yr, within and among geomorphic settings and subregions of the Mid-Atlantic Region from New York to Virginia. K = keeping pace, M = marginal, L = loss; multiple letters under single SLR scenario (e.g., K, M, or K, M, L) indicate more than one response for that geomorphic setting; M-L indicates that the wetland would be either marginal or lost.

accretion) and complex correlations to predict the effect of various SLR scenarios on coastal habitats. SLAMM has been applied to several large geographic areas including Delaware and Chesapeake Bays (Table 12.3) and to most coastal national wildlife refuges. For discussion of the implications of these changes on fish and wildlife resources consult the referenced reports. Delaware Bay may lose 57 percent of its intertidal habitats by 2100 under a 50 percent probability scenario for sea-level rise (i.e., 13.6 inches [34 cm]), with as much as 20 percent lost by 2050 (Galbraith et al. 2002). This raises serious concern for future welfare of shorebirds such as the already imperiled red knot that depend on these habitats for food during their migrations in the Western Hemisphere. Extensive losses of tidal flats have also been predicted for other areas, including southern San Francisco Bay and Humboldt Bay (Galbraith et al. 2005). Application of SLAMM5 to selected areas in coastal Georgia has predicted that 20 to 45 percent of the salt marshes will be submerged by 2100 using IPCC mean and maximum estimates of SLR (Craft et al. 2009). Low-lying coastal forests are also habitats predicted to undergo significant change by 2100.

Models incorporating ecogeomorphic feedbacks have been developed for research-driven assessments of the effect of SLR on salt marshes. Although useful for local assessments, these models may not be able to be scaled up for regional and national applications (Cahoon et al. 2009). One study applied five such models to three salt marsh ecosystems (North Inlet, SC; Venice Lagoon, Italy; Scheldt Estuary, Netherlands and Belgium) to examine the adaptability of coastal wetlands to rising seas (Kirwan et al. 2010) and reached the following conclusions. Marshes initially respond to rising rates of SLR by increasing productivity as the marsh platform becomes lower relative to sea level (i.e., more low marsh with its higher primary production). This also accelerates sedimentation. As rates of SLR increase, however, the marsh platform eventually falls below the level capable of supporting marsh vegetation. At this point, the plants die, accretion falls, and erosion increases, thereby creating an unvegetated subtidal surface (possibly intertidal depending on tidal range and erosion rates). Marshes in macrotidal regions are usually better able to adapt to higher SLR rates than are marshes in microtidal regions. Salt marshes are generally stable at more intermediate rates of SLR and suspended sediment concentrations. A critical SLR rate threshold of 0.4 inch per year (10 mm/year) was determined for areas where tidal range is greater than 3.3 feet (1 m) and suspended soil concentrations exceed 20 g/L. Also, if the SLR rate exceeds 0.8 inch per year (20 mm/year), the only marshes that will survive are likely to be those with tidal ranges greater than 9.9 feet (3 m) and suspended sediment concentrations greater than 30 g/L. Survival of marshes is largely dependent on sediment availability. Many of today's marshes that are deteriorating are located in areas where sediment supply has been reduced by dams, levees, reforestation, and soil conservation practices. Since many marshes formed during times of high sediment delivery (e.g., land-clearing for agriculture and other development), these marshes are experiencing a sediment deficit (Syvitski et al. 2009). When faced with accelerating rates of SLR, they are not likely to be capable of surviving the predicted higher rates of SLR. Finally, marsh submergence will depend on the magnitude of climate change and the fate of terrestrial ice sheets, which will determine the rate of eustatic SLR and eventual sea level (Kirwan et al. 2010).

If accretion rate alone was the determining factor in maintaining elevations suitable for plant survival, it would appear that many coastal wetlands can keep pace with the historic rate of eustatic SLR (global rate of 1.7 mm/yr; Bindoff et al. 2007), while marshes with lower accretion rates would be in jeopardy. With an annual rate of SLR in southern New England of nearly 0.12 inch (3.0 mm) and a marsh accretion rate of only

Table 12.3. Predicted changes in selected coastal habitats in the United States based on certain sea levels using the Sea Level Affecting Marshes Model (SLAMM)

Projected Sea-level Rise	Location	Habitat	Percentage of Gain or Loss (%)	Source
3.28 feet (1.0 m)	Chesapeake Bay Region (includes Delaware Bay and coastal bays on the Delmarva Peninsula)	Salt marsh	+100	Glick et al. 2008
		Tidal flat	+15	
		Estuarine water	+24	
		Brackish marsh	–88	
		Tidal fresh marsh	–36	
		Tidal swamp	–68	
		Riverine tidal	–25	
3.28 feet (1.0 m)	Delaware Bay	Salt marsh	+29	Glick et al. 2008
		Tidal flat	+2770	
		Estuarine water	+21	
		Brackish marsh	–96	
		Tidal fresh marsh	–32	
		Tidal swamp	–37	
		Riverine tidal	–60	
2.69 feet (0.82 m)	Georgia	Salt marsh	–45	Craft et al. 2009
		Tidal flat	+1,510	
		Brackish marsh	–1	
		Estuarine water	+85	
		Tidal fresh marsh	–2	
		Tidal swamp	+15	
2.26 feet (0.69 m)	Pacific Northwest	Salt marsh	+52	Glick et al. 2007
		Tidal flat	–44	
		Estuarine water	+11	
		Estuarine beach	–65	
		Brackish marsh	–52	
		Tidal fresh marsh	–25	
		Tidal swamp	–61	
		Riverine tidal	–43	
1.27 feet (0.39 m)	Pensacola Bay (FL)	Salt marsh	–73	Glick and Clough 2006
		Tidal flat	–4	
		Brackish marsh	+668	
		Estuarine water	+10	
		Tidal fresh marsh	0	
		Estuarine beach	–36	
		Ocean beach	–67	
1.27 feet (0.39 m)	Apalachicola Bay (FL)	Salt marsh	–61	Glick and Clough 2006
		Tidal flat	+613	
		Brackish marsh	+6,264	
		Estuarine water	+15	
		Tidal fresh marsh	–76	
		Estuarine beach	–87	
		Ocean beach	+8,978	
		Riverine tidal	–74	

Notes: The initial condition date varies depending on the available National Wetlands Inventory data (mostly 1980s–2000) and the end date is 2100. Brackish marsh includes the irregularly flooded marsh, whereas salt marsh is restricted to the regularly flooded marsh. The highest percentage gains are usually attributed to habitats that were uncommon at the initial time period.

0.08 to 0.10 inch (2.0–2.5 mm), scientists predict that these marshes will change significantly over the next 100 years (Donnelly and Bertness 2001). Coastal wetlands in other areas may be headed in the same direction. For example, salt marshes along barrier island lagoons in eastern North Carolina do not produce enough organic matter (0.05 inch per year [1.2 mm/yr]) to sustain themselves without sand inputs from marine sources (Hackney and Cleary 1987). Dredging marine sediments to maintain beaches in resort communities may be putting these marshes at risk of being drowned by rising sea levels. Marshes of Albemarle–Pamlico Sound are dependent on peat accumulation, and they are not expected to maintain elevations suitable to prevent drowning by rising sea levels (Moorhead and Brinson 1995). These and other microtidal wetlands are especially prone to pond formation and expansion—signs of marsh deterioration (Kirwan and Guntenspergen 2010).

Tidal marshes of the Mississippi River deltaic plain may last only another 100 years as these marshes appear unable to keep up with rising seas (DeLaune et al. 1990). They no longer receive sufficient alluvial sediments to maintain elevations that would support marsh vegetation over time. Before damming, the river carried 400 to 500 million tons of suspended sediment yearly to the delta (Blum and Roberts 2009). Today's sediment load is about 205 million tons, which is insufficient to maintain the existing delta. Construction of permanent levees now routes much sediment (about two-thirds of the river's discharge) directly into the deep waters of the Gulf of Mexico bypassing the delta's marshes. Coastal subsidence due to sediment loading, compaction, faulting, and human activities is causing a relative SLR of about 0.4 inch per year (10 mm/yr) versus the eustatic SLR of about 0.06 inch per year (1.5 mm/yr) (Day et al. 2007a). This process has enormous detrimental impacts on Louisiana coastal wetlands by 1) further lowering the marsh surface; 2) causing saltwater intrusion; 3) waterlogging soils; and 4) exposing vegetation to longer submergence periods. Plant productivity is reduced and soils destabilized, thereby limiting the rate at which marshes can build up their elevations. On the positive side, one-third of the Mississippi's discharge is directed to the Atchafalaya River where a new delta is forming.

What follows are some key summary points on the probable future of tidal wetlands on submerging coasts in the United States, compiled from several sources (Day et al. 1995; Moorhead and Brinson 1995; Howard and Mendelssohn 2000; Friedrichs and Perry 2001; Morris et al. 2002; Barras et al. 2003; Cahoon et al. 2009; Kirwan and Guntenspergen 2010; Kirwan and Murray 2008; Kirwan et al. 2008, 2010, 2012).

- Tidal marshes respond relatively slowly to changes in sea level and many wetlands are in disequilibrium because they were formed during periods of high sediment input (e.g., deforestation); current marsh platforms are deepening in spite of vertical accretion rates higher than relative SLR.
- Tidal wetlands now experiencing submergence and marsh surface deterioration will continue to decline in condition and area under accelerating rates of relative SLR.
- Tidal wetlands in microtidal ranges are more vulnerable to rising sea level than those in other regions for a number of possible reasons: 1) they may be more poorly drained; 2) they are generally ebb-dominated systems, which may promote export of sediment to the estuary; 3) their vegetation has a narrow growth range; and 4) deepening of the marsh platform, erosion, and expansion of the channel network by SLR should have a greater effect at low tidal ranges.
- Wetlands in areas of high tidal range and suspended sediment concentrations may persist even under SLR rates of more than 3.9 inches per year (>100 mm/yr) due to a plentiful supply of sediment (e.g., Yangtze River Delta is expanding seaward; Yang 1999).

- While stable at SLR rates of 0.08 to 0.12 inch per year (2–3 mm/yr), an SLR rate of 0.2 inch per year (5 mm/yr) for regions with low tidal ranges or sediment concentration will submerge most of the marshes in those regions (e.g., Plum Island Estuary, MA and Albemarle–Pamlico Sound, NC).
- The threshold SLR rate for most of the wetlands in the southeastern United States, where sediment concentrations are above 20 mg/L and tidal ranges are greater than 3.3 feet (1 m), is predicted at 0.4 inch per year (10 mm/yr).
- If the SLR rate exceeds 0.8 inch per year (20 mm/yr), marshes will survive only where tidal ranges are greater than 9.8 feet (3 m) and sediment concentrations are above 30 mg/L.
- For estuaries with high sediment loads (e.g., South Atlantic marshes), the limiting SLR would be at most 12 mm per yr which is 3.5 times the current rate.
- Tidal wetlands in sediment-poor estuaries (e.g., Northeast) have limited stability against a broad range of SLR rates compared with those in sediment-rich estuaries (e.g., Southeast).
- Where tidal range is low, higher suspended sediment concentrations will be needed for the marshes to survive.
- Relatively level marshes (marsh platforms) may change to more sloping marsh surfaces where sediment deposition increases due to longer periods of submergence.
- Marshes high in the intertidal zone may be more vulnerable to accelerated sea-level rise than lower-lying marshes as they appear unresponsive to interannual changes in sea level.
- The area of tidal wetlands in the United States is not likely to increase over the next 100 years, although some local areas may witness gains (e.g., active deltas such as Atchafalaya Delta in Louisiana).
- Louisiana, which contains the largest portion of U.S. tidal marshes, is predicted to lose 513 square miles (1,329 km^2) of land (mostly tidal wetlands) by 2050.
- Since many of today's coastal marshes were formed during periods of high sedimentation associated with major land-clearing operations, they are very likely to decrease in size due to a current lack of adequate sediments to maintain their elevations in the face of rising sea levels.
- Increasing salinity in brackish and tidal fresh marshes may enhance organic matter decomposition through sulfate reduction, making it more difficult for these organic-rich marshes to accumulate sufficient organic matter to keep pace with rising sea levels.
- With increasing saltwater intrusion, tidal freshwater plant communities will either change to more brackish types or to open water or nonvegetated areas in which vegetation is eliminated entirely, causing severe soil erosion.
- Temporary shifts in vegetation (e.g., due to heavy grazing by geese) can lower the marsh surface (platform) and promote pond formation on the platform and channel network expansion, which may ultimately lead to conversion of marshlands to mudflats; this will occur unless marsh vegetation can quickly recolonize the site.
- If the lower IPCC rates are realized and the ice sheets do not contribute much water to the ocean, many of the existing marshes will accrete sufficiently to maintain elevations that will sustain vegetation. If the rates are more rapid, however, most of the marshes will be submerged.

Recognize, however, that even where existing wetlands are lost, it may be possible for tidal wetlands to develop on low-lying uplands or contiguous nontidal wetlands where suitable lands are available for marsh migration. Fish and wildlife will also be affected by changes in the extent and quality of their habitats. These changes may include

- reduced habitat for wetland-specialist birds,
- loss of shorebird feeding habitat,
- increased turbidity of shallow-water habitat that may adversely affect submerged

aquatic vegetation and the species dependent on these resources,
- loss of diamondback terrapin habitat,
- loss of islands in the marsh that are important nesting habitat for species already displaced from barrier islands by development,
- loss of estuarine beaches that are important habitat for tiger beetles (some Federally Endangered), and
- loss of sea-level fens and their unique flora (Shellenbarger Jones et al. 2009).

The Probable Future of Tidal Wetlands along Emerging Coasts

Some coasts are emerging (increasing in elevation) due to postglacial rebound or tectonic uplifting. SLR is not a problem for coastal marshes in these regions as the coasts are uplifting at rates greater than predicted sea-level rise rates. Relative SLR in these areas is negative, which means local sea levels are "falling" relative to the land surface. Here the existing tidal marshes are threatened by tectonic or postglacial uplift processes that are raising them out of the intertidal zone and thereby converting them to nontidal wetlands or dry land. These wetlands should, however, be replaced by tidal wetlands that will form on newly exposed intertidal surfaces derived from raised sea floors. Several regions experiencing glacial rebound include the coasts of Labrador, northern Quebec (including the southwestern part of Hudson Bay), southeast Alaska, Russia's White Sea, and the Gulf of Bothnia in Sweden and Finland (Delft Hydraulics 1992; Svensson and Jeglum 2000; Larsen et al. 2005). Since the late 1700s, southeast Alaska's shorelines have been raised by 3.3 to 18.7 feet (1.0–5.7 m), with current uplift rates up to 1.26 inches per year (32 mm/yr)—the world's fastest measured rates of uplift (Larson et al. 2005). In this area, unloading (i.e., loss of glacial ice attributed to climate change) rather than tectonic processes are responsible for the extremely high rates of shoreline uplift, which produces a negative SLR (Motyka et al. 2007). Unloading plus tectonic processes may be expected to continue to uplift portions of the Pacific Coast on occasion as the Pacific plate slides under the North American plate ("subduction zone").

Other Climate Change Effects on Tidal Wetlands

While SLR is a major consequence of climate change, other factors will operate to affect tidal wetlands and their wildlife. Warmer ocean temperatures provide more energy for hurricane formation. Storm intensity is expected to increase in the North Atlantic, especially during the latter half of the 21st century (Karl et al. 2009). This will generate higher storm surges that will increase erosion of beaches, marshes, and other shorelines. Some scientists suggest that this impact poses a greater risk to marshlands than the SLR, especially in the short-term (Ray et al. 1992). Storm-induced erosion of marsh peat may be particularly troublesome for sediment-starved marshes in low-energy estuaries such as Albemarle and Pamlico Sounds (Moorhead and Brinson 1995). Wave action may increase along the coast to the point that inlets of estuaries in arid, semi-arid, and Mediterranean climates will experience more frequent and longer closures (Reddering and Rust 1990). In the summer of 2011, Hurricane Irene hammered the Outer Banks of North Carolina, breaking through the barrier island, creating new inlets, and temporarily cutting off Cape Hatteras and Ocracoke from the mainland (Figure 12.13).

Changes in regional precipitation patterns and timing of runoff (including snowmelt) can also have major impacts on tidal wetlands. Lack of precipitation will increase saltwater intrusion in estuaries, while more rainfall will have the opposite effect. Both conditions can, depending on their severity and persistence, have significant impacts on estuarine biota. Changes in freshwater flows and nutrient loadings

Figure 12.13. In the summer of 2011, Hurricane Irene pounded the coast of North Carolina destroying the bridge connection from the Outer Banks to the mainland, creating new inlets, and depositing overwash deposits on back-barrier marshlands. (Tom Mackenzie, U.S. Fish and Wildlife Service)

will strengthen or weaken the severity of hypoxia in the Gulf of Mexico depending on whether these inputs increase or decrease with climate change (Ning et al. 2003). Prolonged droughts may result in more incidences of marsh dieback. Increased evapotranspiration with warmer climates may also increase soil salinities in tidal marshes. Tidal freshwater wetlands may be more vulnerable to these changes than will salt and brackish wetlands, since the latter can tolerate freshwater inputs more than freshwater plants can tolerate salinity (Callaway et al. 2007). Changes in river discharge will also affect sediment loads in coastal rivers with obvious impacts on sedimentation rates that can help or hinder tidal marsh survival and development. Flow changes will also influence resource management decisions concerning river diversions with implications for tidal wetlands and their fish and wildlife. Marsh restorations may be at risk due to rising sea levels as well, and concerns abound regarding sufficient sediment sources for sustainability and plant recruitment with changing salinities. Timing of precipitation can negatively impact fish migration. Climate change will also influence bird migration patterns and southeastern marshes may not be as heavily used by some species as they find more northern marshes suitable for overwintering. Migratory shorebirds may be excellent indicators of climate change because they utilize a combination of habitats including tidal and nontidal wetlands and upland habitats on numerous continents (Piersma and Lindström 2004). Climate change will have additional impacts on wetland functions, estuaries, and associated wildlife as well as economic and social impacts (for an excellent summary see the report by the faculty of the University of North Carolina Wilmington; University of North Carolina Wilmington 2008).

Managing the Marsh Environment

With rising sea levels, tidal wetlands (e.g., beaches and marshes) naturally advance landward into neighboring low-lying areas (e.g., beaches over marshes and marshes into lowland forests). These lowlands are vital to the future of tidal wetlands. Where shorelines are bulkheaded or rip-rapped to protect development built on barrier islands and spits, this landward migration is impeded and beaches are simply eroded. Where development has been constructed along the edges of marshes, the landward movement of marshes is likewise stopped by these structures as well as by dikes, levees, roads, and railroad beds.

Sediment input into coastal waters is also of concern. With improved land management practices, less soil is eroding from the land. As learned in Chapter 2, sediments from land-clearing operations have contributed significantly to the formation of contemporary marshlands. Sediment delivery has been further reduced by damming of rivers. Sediment trapped behind dams is not available for maintaining coastal deltas and associated wetlands. Recent dam construction in Asia has had a major impact on delta integrity and coastal erosion (e.g., Chen et al. 2005; Chen and Saito 2011). Reduced sediment supplies pose serious problems for sustaining tidal wetlands.

Restoration and creation of tidal marshes can benefit both estuarine ecosystem functions and efforts to reduce civilization's contribution to global warming. Saltwater wetlands store more carbon per unit area than other ecosystems and will continue to do so with rising sea level. They also do not release as much methane as freshwater wetlands. Increasing the area of these wetlands through restoration and creation should play a significant role in any government's climate change initiative. Restoring tidal flow to diked agricultural lands (former marshes) is perhaps the easiest way to accomplish this. Estimates indicate that if the diked former tidelands of the Bay of Fundy are restored to their original salt marsh condition, the additional carbon stored would be equivalent to 4 to 6 percent of Canada's target reduction of 1990-level emissions of CO_2 under the Kyoto Protocol (Connor et al. 2001). Given the extent of coastal wetlands in the Mississippi delta, their economic importance, and their high rate of loss and degradation, restoration efforts are under way. Four general approaches to restoration are ongoing, planned, or being evaluated: 1) reconnecting the river to the deltaic plain (e.g., reopening distributaries and promoting crevasse-splay development; 2) using dredged material to create and restore wetlands; 3) restoring barrier islands by pumping sand from offshore, building various coastal structures to capture sediments, and planting dune vegetation to stabilize the islands; and 4) restoring hydrological processes by removing spoil banks, closing deep navigation channels, installing locks, trapping sediments, and protecting interior shorelines from erosion (Day et al. 2007a). Restoring tidal wetlands in the Mississippi delta is probably the most challenging of all wetland restoration projects in the United States as the causes of the problem are both natural (e.g., degradation of delta and rising sea level) and human-induced and will require enormous sums of money to implement.

The future of coastal wetlands in any locale will therefore depend on the relative rate of SLR, the ability of plant species to increase their above- and belowground biomass, sediment inputs, opportunities for the landward and upriver migration of tidal wetlands, as well as on society's actions to restore and create tidal wetlands, to allow marine transgression to take place, and to seek other measures to address climate change (Table 12.4).

Further Readings

"The impact of sea-level rise on coastal salt marshes" (Reed 1990)

"The response of coastal marshes to sea-level rise: survival or submergence?" (Reed 1995)

"Tidal salt marsh morphodynamics: a synthesis" (Friedrichs and Perry 2001)

Table 12.4. Possible actions that society can undertake to improve the future for tidal wetlands

Government Actions

1. Protect existing wetlands from human development, conserve wetland biodiversity, and reduce pollution in and around wetlands.
2. Acquire, protect and preserve low-lying uplands and nontidal wetlands adjacent to tidal wetlands as these are the places where tidal wetlands will migrate in the future. Prevent the armoring of undeveloped shorelines. Besides protecting existing wetlands from human development, this should be the next step in securing a promising future for tidal wetlands.
3. Rehabilitate damaged coastal wetlands by improving tidal flow and controlling invasive plant and animal species.
4. Restore lost tidal wetlands (e.g., through removal of fill and breaching of dikes).
5. Build tidal wetlands along eroding shorelines in areas of low to moderate wave action on public lands and encourage private landowners to do likewise on their property.
6. Initiate strategic planning to protect wetlands in the face of rising sea level and take necessary action including: a) Produce maps showing wetlands and other areas threatened by sea-level rise, b) incorporate wetland protection into infrastructure planning for transportation, utilities, and sewer systems, c) limit hardened shoreline construction, giving preference to construction of living shorelines, d) develop a strategy for assessing options for removing or breaching dikes and initiate removal or breaching when and where feasible (e.g., New Brunswick strategy [Marlin et al. 2007]), e) establish setbacks for buildings to allow for marsh and beach migration, f) adopt "rolling easement" policies that allow development but prohibit construction of bulkheads or other "protective" features with the understanding that the development will not be allowed to prevent the landward migration of wetlands, beaches, or dunes (e.g., Texas, South Carolina, Rhode Island, and Maine; www.grida.no/climate/ipcc/regional/224.htm), g) remove insurance subsidies for structures built in flood-prone areas, and h) provide support for increased monitoring and research related to climate change and fish and wildlife species and habitat impacts.
7. Explore ways of returning sediments captured by dams to estuaries to help improve accretion rates.
8. Continue to support research to improve our understanding of all tidal wetlands and their response to SLR.
9. Consider issuing special license plates to generate funds to support wetland restoration and rehabilitation and acquisition of low-lying uplands contiguous with tidal wetlands (e.g., Florida's Indian River Lagoon plate).
10. Improve public understanding of tidal wetlands, their values, and their vulnerability through various media.

Organization and Corporate Actions

1. Promote energy and habitat conservation measures in your business practices.
2. Restore and/or create tidal wetlands on your property where suitable conditions exist.
3. Encourage and support government initiatives to protect, restore, and conserve tidal wetlands and adjacent lowlands.
4. Help improve employees understanding of tidal wetlands, their values, and their vulnerability.
5. Initiate or encourage applicable actions listed under Individual Actions below.

Individual Actions

1. Improve your awareness of tidal wetlands and local wetland issues.
2. Reduce your carbon footprint and encourage family and friends to do likewise (e.g., replace standard light bulbs with compact fluorescent bulbs, use green energy, purchase energy-efficient appliances and automobiles, bike to work or use public transportation, and buy more locally produced items).
3. Restore and/or create tidal wetlands on your property where suitable conditions exist; government support may be available.
4. Support wetland conservation and restoration initiatives (e.g., buy duck stamps to support wetland acquisition or volunteer to work on local restoration and monitoring projects).
5. Plant native species and remove invasive species from your property; create a natural buffer between your lawn and coastal marshes to trap nutrients before they reach the wetlands.
6. Construct living shorelines instead of bulkheads to protect your shoreline from erosion.
7. Use phosphate-free laundry and dishwasher detergents; this will help reduce phosphate inputs into waterways.
8. Properly dispose of household waste products (e.g., motor oil, fertilizers, and pesticides).
9. Join a conservation organization.
10. Report filling and dredging activities to federal and state regulatory authorities to ensure that such work is permitted.

"Coastal wetland vulnerability to relative sea-level rise: wetland elevation trends and process controls" (Cahoon et al. 2006)

The Winds of Change: Climate, Weather, and the Destruction of Civilizations (Linden 2006)

Climate Change Synthesis Report (Intergovernmental Panel on Climate Change 2007)

Climate Change 2007: Impacts, Adaptation and Vulnerability (Parry et al. 2007)

The Potential Impacts of Climate Change on Coastal North Carolina (University of North Carolina Wilmington 2008)

Coastal Sensitivity to Sea-Level Rise: A Focus on the Mid-Atlantic Region (Climate Change Science Program 2009)

"Limits of the adaptability of coastal marshes to rising sea level" (Kirwan et al. 2010)

"Rapid wetland expansion during European settlement and its implication for marsh survival under modern sediment delivery rates" (Kirwan et al. 2011)

Appendix A

List of North American Wetlands of International Importance: Tidal Wetlands

The following is a list of internationally important wetlands that contain tidal wetlands and is a subset of a more complete listing designated in accordance with the Ramsar Convention (www.ramsar.org). The list is intended to establish a global network of significant wetlands for the conservation of biodiversity and for sustaining human life through their maintenance of ecosystem components, processes, and environmental services (benefits to humankind).

Canadian Sites

Alaksen (British Columbia; 1,448 acres [586 ha]). Part of the Fraser River delta; used by waterbirds migrating between Arctic breeding grounds and southern wintering grounds: up to 40,000 geese of the Wrangel Island breeding population and up to one million other shorebirds stage and winter here, with up to 25,000 ducks in fall and 10,000 surf scoters congregating to feed on the tidal flats in late summer.

Baie de l'Isle-Verte (Québec; 5,473 acres [2,215 ha]). Contains some of the last-remaining, unreclaimed salt marsh along the St. Lawrence River; important breeding habitat for black duck and serves as an important resting and feeding site for large numbers of migratory ducks, geese, and swans, particularly in spring; heavy shorebird use: 35,000 in spring and 10,000 in fall.

Cap Tourmente (Québec; 5,926 acres [2,398 ha]). Includes one-third of the remaining high-quality common three-square marshes of the St. Lawrence Estuary; vital staging grounds for greater snow geese (30% or more of world's population; to >1 million individuals) and thousands of ducks during fall migration to feed and pass through in spring; 700 plant species have been recorded (several rare), many reaching their northern limit.

Chignecto (Nova Scotia; 2,520 acres [1,020 ha]). Half the site consists of salt marsh; an important staging area for Canada geese (up to 6,000) and three species of ducks during spring migration; breeding by black ducks and greater yellowlegs.

Grand Codroy Estuary (Newfoundland; 2,286 acres [925 ha]). One of the most productive of Newfoundland's few estuarine wetland sites; mudflats support rich eelgrass beds—an important food source for up to 3,000 fall-staging Canada geese and >1,000 black ducks.

Malpeque Bay (Prince Edward Island; 60,390 acres [24,440 ha]). An estuarine embayment of extensive shallow open water, intertidal flats, and islands, fringed in places by salt marsh; intertidal flats support eelgrass beds; up to 20,000 Canada geese stage here in spring and fall; supports the largest great blue heron nesting colony in the province; also nesting by cormorants; seabird nesting on islands.

Mary's Point (New Brunswick; 2,965 acres [1,200 ha]). Contains gravel beaches, salt marshes, and extensive intertidal mudflats, supporting the world's highest known density of the crustaceans *Corophium volutator* (>60,000/m^2)—the principal food source for millions of migratory shorebirds staging here during the fall migration: up to 200,000 semipalmated sandpipers present in late summer and more than 2 million pass through in August.

McConnell River (Northwest Territories; 81,050 acres [32,800 ha]). A complex of coastal marshes and inland wet meadows around the mouth of the McConnell River; internationally important for breeding up to 200,000 pairs of snow geese as well as Canada geese and large numbers of nesting ducks and shorebirds.

Musquodoboit Harbour (Nova Scotia; 4,757 acres [1,925 ha]). An ice-free estuary in winter with a complex of intertidal sand and mudflats, scattered islands and fringing salt marsh; mudflats support extensive eelgrass beds and abundant invertebrates, providing food for large numbers of staging and wintering waterbirds; one of the most important coastal staging and wintering sites for Canada geese and black ducks supporting the largest wintering populations of these two species in eastern Canada.

Polar Bear Provincial Park (Ontario; 5,952,000 acres [2,408,700 ha]). Contains some marshes subject to saltwater inundation but includes the world's most southerly example of tundra ecosystem; supports hundreds of thousands of ducks, geese, and swans, a breeding colony of more than 50,000, and during migration more than one million snow geese, a substantial proportion of the central Arctic breeding population of red knot, and the entire breeding population of marbled godwit.

Queen Maud Gulf (Northwest Territories; 15,510,000 acres [6,278,200 ha]). Canada's largest Ramsar site encompasses a vast tundra plain including open sea, coastal bays, intertidal zones, tidal estuaries, deltas, lowland rivers, and freshwater lakes; internationally important for various species of nesting waterbirds, especially breeding geese that nest and molt here from May to August; the peregrine falcon (endangered in North America) is relatively common here; also important for seals and the large mammals including barren ground caribou (100,000 in 1988), and muskox.

Shepody Bay (New Brunswick; 30,150 acres [12,200 ha]). A Western Hemisphere Shorebird Reserve containing salt marshes, sand and gravel beaches, and extensive intertidal mudflats; the mudflats support internationally important numbers of the amphipod *Corophium volutator,* the principal food source for millions of fall migrating shorebirds: 400,000 semipalmated sandpipers may be present at one time in late summer while more than 2 million pass through in August, as do large numbers of other shorebirds; peregrine falcon nests on cliffs.

Southern Bight-Minas Basin (Nova Scotia; 66,220 acres [26,800 ha]). Another Western Hemisphere Shorebird Reserve containing salt marshes, extensive intertidal sand, and mudflats that attract large numbers of various species of staging waterbirds, including ducks, geese, swans, semipalmated sandpipers (over 100,000) and least sandpipers (up to 10,000).

Southern James Bay (Moose River and Hannah Bay) (Ontario; 62,470 acres [25,290 ha]). James Bay, an extension of Hudson Bay, is one of the most important staging areas in northern North America for migratory Arctic-breeding waterbirds; the area consists of mudflats, intertidal marsh, and freshwater wetlands and is a late fall staging ground for large numbers of geese (up to 75,000 at one time) and ducks.

Tabusintac Lagoon and River Estuary (New Brunswick; 12,350 acres [4,997 ha]). A coastal lagoon system that contains tidal flats, beaches, and salt marsh; 80% of the area supports eelgrass beds, while the beach supports the second largest nesting common tern colony in New Brunswick; also a major waterbird concentration area during spring and autumn migration for a high diversity of shore and waterbirds (>6,000 in spring and >5,000 in fall).

United States Sites

Bolinas Lagoon (California; 1,100 acres [445 ha]). Contains mudflats and tidal marshes located on the Pacific Flyway; the lagoon is a staging ground and stopover site for migratory birds and overwintering habitat for a wide array of ducks, geese, and shorebirds.

Chesapeake Bay Estuarine Complex (Virginia; 111,200 acres [45,000 ha]). A vast estuarine complex of multiple protected areas in the nation's largest estuary; contains sand beaches, dunes, mudflats, open water with submerged beds of aquatic vegetation, intertidal marshes, freshwater marshes, and lakes; an important staging and wintering for one million ducks, geese, and swans; loggerhead and Kemp's ridley turtles use the bay.

Connecticut River Estuary & Tidal Wetlands Complex (Connecticut; 16,020 acres [6,484 ha]). A complex of tidal and nontidal wetlands along largest river system in New England; use by piping plover, diamondback terrapin, >10,000 waterfowl (18 species of wintering ducks); migration route or spawning for Atlantic salmon, American shad, and shortnose and Atlantic sturgeons.

Delaware Bay Estuary (Delaware and New Jersey; 126,600 acres [51,252 ha]). A complex of protected areas containing tidal wetlands of varying salinities; an important staging area for more than 90% of the North American populations of several migratory shorebird species (including red knot); more than one million individuals use the region making it one of the two most important staging areas on the Atlantic Coast of North America.

Edwin B Forsythe National Wildlife Refuge (New Jersey; 32,320 acres [13,080 ha]). Contains beaches and salt marshes important for wintering habitat for >60,000 brant (⅓ of winter population) and 60,000 to 70,000 black ducks and greater snow geese; use by osprey and peregrine falcon; breeding by piping plover.

Everglades National Park (Florida; 1,399,000 acres [566,143 ha]). A World Heritage Site and Biosphere Reserve with one unit dominated by salt marshes, mangrove forests, beach and dune complexes, and brackish water estuaries; important for nesting, staging and wintering birds and supports a rich flora and several threatened and endangered species of flora and fauna; more than one thousand species of seed-bearing plants and 120 both tropical and temperate trees (60 of which are endemic) occur.

Izembek Lagoon National Wildlife Refuge (Alaska; 416,200 acres [168,433 ha]). Contains 96,370 acres (39,000 ha) of intertidal wetlands, with extensive coastal lagoons and freshwater marshes and meadows; critically important for migratory waterbirds, notably as spring and autumn staging sites for almost the entire world population of the black brant plus thousands of other geese and 300,000 ducks; nesting species include tundra swan, while moderate to large numbers of other ducks, geese, and swans occur in winter.

Pelican Island National Wildlife Refuge (Florida; 4,715 acres [1,908 ha]). The first National Wildlife Refuge and a National Historic Landmark contains estuarine wetlands important to thousands of migratory birds in winter as well as numerous nesting species and shorebirds.

Tijuana River National Estuarine Research Reserve (California; 2,523 acres [1,021 ha]). One of the few unfragmented estuaries and coastal lagoons in southern California; critical habitat for U.S. endangered species and subspecies such as the San Diego fairy shrimp, the light-footed clapper rail, and salt marsh bird's beak; nursery grounds for commercially important fish such as the diamond turbot and the California halibut.

Tomales Bay (California. 7,043 acres [2,850 ha]). An estuary with eelgrass beds, sand dune systems, and restored emergent tidal marshes; supports several endangered or threatened plant and animal species and is an important migratory waterbird stopover site and overwintering ground on the Pacific flyway with more than 20,000 individuals in winter, most notably surf scoter, bufflehead, and greater scaup.

Appendix B

Profiles of Some Tidal Wetland Restoration Projects on the Atlantic and Pacific Coasts of the United States

Project types (re-establishment, rehabilitation, and creation) are noted. Some projects involve more than one type of action, such as restoration of a diked tidally restricted marsh that involves dike removal (re-establishment) and restoration of tidal flow (rehabilitation). Also one of the case studies was restoration brought about by natural forces—storm breaching of dikes and a decision not to repair them. Results are based on published reports and reflect conditions for the reported period; for current findings, contact authors.

Case Study 1

Project type: Re-establishment and Rehabilitation

Location: Hatches Harbor, Cape Cod, Massachusetts

Initial condition: 200-acre (80.9 ha) former salt marsh; diked for mosquito control in 1930 with 2-foot (0.6 m) culvert and flap valve installed allowing outflow of water from restricted marsh. In 1987, the flap valve was removed providing limited tidal exchange.

Restoration objective: To restore 90 acres (36.4 ha) of salt marsh by increasing tidal exchange; also to provide additional flood storage and protection for adjacent airport.

Action taken: In 1997 installed four 7-foot by 3-foot (2.1 m x 0.9 m) box culverts with adjustable tide gates, progressively increased openings from 1999 to 2005 when maximum allowable tidal exchange was achieved. Monitoring commenced in 1997 including pre-restoration site assessment (vegetation, pore-water salinity, nekton, sedimentation, elevation, and fecal coliform; Farris et al. 1998).

Results: Tidal regime now equals 57% of neighboring unrestricted marsh; landward retreat of common reed, which is being replaced by smooth cordgrass and salt hay grass depending on elevation; landward expansion of halophytic vegetation and dieback of freshwater species; significant increase in abundance of smooth cordgrass at expense of both common reed and salt hay grass; increased pore-water salinities from fresh to >35 ppt in some places and from fresh to low salinities in uppermost marsh; and increased use by nekton for feeding, nursery grounds, and spawning (Portnoy et al. 2003; Smith et al. 2008, 2009).

Case Study 2

Project type: Rehabilitation

Location: Potter Pond, Prudence Island, Rhode Island

Initial condition: 5-acre (2.04 ha) tidally restricted estuarine pond with narrow band of salt marsh and 0.6 acre (0.24 ha) back marsh of common reed; tidally restricted since at least the 1930s by two roads with culverts that by 2002 were crushed, providing only a 1.6-inch (4 cm) tide range; pond had oxygen deficits and was often covered with macroalgae.

Restoration objective: To improve tidal exchange and restore salt marsh ecology; possibly to eliminate nuisance macroalgae.

Action taken: In 2003, replaced two 15-inch (38 cm) diameter culverts with a 50-foot (15.25 m) long, 4.9 foot x 3.9 foot (1.5 m x 1.2 m) aluminum arch culvert between the estuary and the pond, and the culvert to the back marsh with a 2-foot (61 cm) diameter culvert.

Results: Tidal range increased from 1.6 inches (4 cm) to 3.9 feet (120 cm), which converted pond to mostly intertidal mudflat with 93% reduction in open water at low tide; mudflat being colonized by smooth cordgrass, which increased its area by 67% and glasswort, which by 2004 occupied 0.7 acres (0.30 ha); 69% decline in common reed cover and the height of remaining reeds was reduced by 2.5 feet (76 cm); salinity of back marsh increased from 1.5 ppt to 26.8 ppt; bird use increased due to creating exposed mudflats, and nekton density decreased largely due to conversion of open water habitat to intertidal habitat (Raposa 2008).

Case Study 3

Project type: Re-establishment and Rehabilitation

Location: Galilee Salt Marsh, Point Judith, Rhode Island

Initial condition: *Phragmites*-dominated wetland where tidal flow was eliminated by road, and some portions of marsh filled with dredged spoil.

Restoration objective: Restore tidal flow and salt marsh community, while protecting adjacent low-lying development from flooding.

Action taken: Reconstructed channel, installed two box culverts with self-regulating tide gates, and removed dredge material from the west side of the marsh.

Results: Tidal flow restored; salt marsh vegetation recolonizing site; common reed reduced in stature and density; and improved aquatic productivity to about 100 acres (40.5 ha) of degraded marsh; won Coastal America Partnership award in 1999. See www.nae.usace.army.mil/projects/ri/GalileeSaltMarsh/InfoSheet.pdf and www.nae.usace.army.mil/projects/ri/GalileeSaltMarsh/galileeSaltMarsh.htm.

Case Study 4

Project type: Rehabilitation

Location: Barn Island Wildlife Management Area, Stonington, Connecticut

Initial condition: Former salt marsh ditched in the 1930s for mosquito control, then in 1940s diked to create a waterfowl impoundment resulting in a 52-acre (21 ha) oligohaline narrow-leaved cattail wetland.

Restoration objective: Restore tidal flow and salt marsh ecosystem.

Action taken: Installed 5-foot (1.5 m) diameter culvert in 1978 that restored partial tidal flow; in 1982, restored full tidal flow by installing a 7-foot (2.1 m) diameter culvert.

Results: After 10 years, smooth cordgrass increased cover along study transects from

<1% to 45%; high marsh plants had recolonized 20% of the transects; cattail cover was greatly reduced from 74% to 16% and where it survived was mostly reduced in stature; some brackish species were eliminated or substantially reduced from transects, and surprisingly common reed, although mostly stunted, increased from 6% to 17% along the transects (Sinicrope et al. 1990). Cattails remained mainly in the upper part of the wetland where fresher conditions persist (10 ppt or less), while common reed was typically found along the upland border and had best growth at moderate to low salinities (20 ppt or less) and stunted stature at salinities near sea strength (30 ppt). Marsh macroinvertebrates had recolonized marsh but the dominant salt marsh snail was not as abundant as in contiguous undiked marsh around that time (Fell et al. 1991). After 21 years, numbers of this snail were similar to the adjacent marsh, but numbers of other macroinvertebrates were either higher or lower than in adjacent marshes; killifish numbers were similar; and bird use was extensive. Restoration of animal populations may take two decades or more to fully recover (Swamy et al. 2002).

Case Study 5

Project type: Rehabilitation

Location: Arthur Kill, Staten Island, New York

Initial condition: On January 1, 1990, a ruptured underwater pipeline spilled 576,000 gallons (2.5 million liters) of No. 2 fuel oil into the Arthur Kill estuary impacting over 125 acres (50.6 ha) of low marsh and tidal creeks, eliminating approximately 11 acres (4.5 ha) of smooth cordgrass near the spill site, contaminating marsh soils with petroleum hydrocarbons, damaging local ribbed mussel beds (up to 100% mortality in places), killing about 700 aquatic birds, and disrupting the 1990 breeding season (Burger 1994; Bergen et al. 2000; Ofiara and Seneca 2001; Packer 2001).

Restoration objective: Remove as much oil as possible from soil and restore salt marsh vegetation and ecological functions to pre-damage conditions.

Action taken: Excavated trenches and vacuumed oil from soil; fertilized shoreline to accelerate microbial breakdown of oil; planted 9 acres (2.43 ha) of shoreline with more than 200,000 seedlings of nursery-grown smooth cordgrass from local and Maryland stock and mature transplants from non-oiled Staten Island marshes; and monitored vegetation, mussel beds, total petroleum hydrocarbons in soil, fish abundance and diversity, and wading bird forage success.

Results: Revegetation efforts were successful from a structural standpoint (aboveground biomass comparable with other marshes in region and 70% cover after three years) yet there was little or no colonization at unplanted oiled sites after 10 years; ribbed mussel beds at restored sites are less dense with smaller and slower growing mussels (could be due to other factors unrelated to spill); oil still in substrate (lower levels at planted oiled sites versus unplanted oiled sites but higher than reference non-oiled site); and questions remain as to long-term viability and functional equivalency to natural marshes. Assessment of recovery is complicated by nature of the ambient conditions—urbanized area subject to chronic pollution. Some problems noted regarding loss of plantings in some areas (erosion from waves generated by passing ships and predation by Canada geese, not from oil in substrate).

Case Study 6

Project type: Rehabilitation (naturally induced)

Location: Delaware Bay, New Jersey

Initial condition: 531-acre (215 ha) salt hay farm with dikes damaged by 1972 storm; dikes never repaired.

Restoration objective: Restoration of salt marsh.

Action taken: None required, tidal hydrology restored naturally through storm breach.

Results: Increased tidal flooding; converted salt hay meadow (372 acres [150.6 ha] to 46 acres [18.8 ha]) to open water and tall smooth cordgrass; 88% decline in salt hay grass marsh, with a 98% increase in smooth cordgrass marsh (7.2 acres [2.9 ha] to 257 acres [103.9 ha]) and 97% increase in open water (4 acres [1.6 ha] to 236 acres [95.5 ha]), and elimination of short-form smooth cordgrass (144 acres [58.4 ha] to zero); waterfowl use increased while use by passerines declined (Slavin and Shisler 1983).

Case Study 7

Project type: Rehabilitation

Location: Delaware Bay, New Jersey, and Delaware

Initial condition: Diked salt hay farms and *Phragmites*-dominated wetlands.

Restoration objective: Restore salt marsh and reduce cover of common reed.

Action taken: Restored tidal flow to dikelands by breaching dikes and constructed primary and secondary channels, allowing smaller creeks to develop naturally; controlled *Phragmites* through herbicide treatment (aerial spraying) and prescribed burns (to remove *Phragmites* mat). Project also included protecting buffers as well as installing fish ladders along tidal streams.

Results: Since 1996, restored tidal flow to 4,396 acres (1,779 ha) of former salt hay farms and treated 6,847 acres (2,771 ha) of *Phragmites* marsh allowing recolonization by native species (Weinstein et al. 2001; Balletto et al. 2005; Hinkle and Mitsch 2005; Teal and Weishar 2005).

Case Study 8

Project type: Re-establishment

Location: Pine Knoll Shores, Bogue Banks, North Carolina

Initial condition: Salt marsh filled with dredged material; shoreline began to erode, sheet-piling bulkhead was failing, homeowners association wanted to create living shoreline in front of bulkhead.

Restoration objective: Restore 39.4-foot (12 m) wide fringe marsh of smooth cordgrass.

Action taken: Obtained plantings from nearby dredged material disposal site and greenhouse-grown seedlings; planted at 17.7-inch and 23.6-inch (45 cm and 60 cm) spacings.

Results: After second growing season (18 months after planting), good recovery of vegetation, and by end of third growing season, a continuous stand of cordgrass covered most of the site. (For a full report see Broome et al. 1986.)

Case Study 9

Project type: Re-establishment and Rehabilitation

Location: Tomales Bay, Giacomini Ranch Restoration, Golden Gate National Recreation Area, California

Initial condition: 560-acre (227 ha) diked dairy farm (former tidal marsh)—wet pastures with some estuarine wetlands.

Restoration objective: Restore natural self-sustaining tidal and fluvial (stream) hydrology to improve water quality flowing into Tomales Bay and restore ecological functions and habitat for the bay area, while minimizing impacts to the federally endangered tidewater goby (*Eucyclogobius newberryi*) and federally threatened California red-legged frog (*Rana aurora draytonii*), to the extent practicable.

Action taken: Plans to remove levees, tide gates, culverts, agricultural infrastructure (barn, pipelines, and fences), and manure deposits; fill in ditches; recreate tidal sloughs and creeks; reorient creeks to historic alignment; and remove invasive species. Work was done in phases—first, removed agricultural infrastructure, manure, and one levee and then removed the main levees in the fall of 2008. Natural restoration allowed to occur thereafter.

Results: By 2009, significant gains in tidal wetland habitat have taken place at the restoration area: brackish marsh increased from 2% in 2003 to 12% of area, pannes and flats from 2% to 30%, and salt marsh from 9% to 15%; extent of grasslands decreased to 12% from 72% in 2003. Fish diversity increased with higher numbers of native top smelt (*Atherinops affinis*) and a decline in the nonnative mosquitofish (*Gambusia affinis*). More species of birds are using the area and water quality improvements exceeding original expectations have been realized after just one year of restoration (Parsons 2010). See www.nps.gov/pore/parkmgmt/planning_giacomini_wrp_1year_celebration.htm and www.nps.gov/pore/parkmgmt/planning_giacomini_wrp_restoration.htm.

Case Study 10

Project type: Re-establishment and Rehabilitation

Location: Salmon River estuary, Oregon

Initial condition: Two diked wetlands (pastures)—51.9-acre (21 ha) and 156-acre (63 ha) sites.

Restoration objective: Restore tidal flow regime and tidal wetland plant communities.

Action taken: Removed dike around 51.9-acre site in 1978 and remainder of dike in 1987 around the larger site.

Results: By 1998, marsh had become a low marsh dominated mostly by Lyngby's sedge, which was quite different from the reference wetland (desired outcome) (Janet Morlan, pers. comm. 2011); difference was due to fact that diked wetland had subsided 13.8 to 15.8 inches (35–40 cm) during the 17 years it was used as pasture. Current sedimentation rates averaged 2.0 to 2.4 inches per year (5–6 cm/yr) compared with a 1.6 inches per year (4 cm/yr) rate for the reference wetland. Net primary productivity was 2,300 g/m^2/yr in the restored marsh compared with 1,200 g/m^2/yr in the neighboring wetland. Other functions were not evaluated (Frenkel and Morlan 1990).

Case Study 11

Project type: Re-establishment and Rehabilitation

Location: Siuslaw River estuary, Oregon

Initial condition: 12-acre (4.8 ha) diked area containing 7 acres (2.8 ha) of nontidal forested wetland and 5 acres (2.0 ha) of reed canary grass marsh (former tidal swamp); diked for pasture prior to 1939, three culverts drained area allowing only inundation during very high river flows.

Restoration objective: Restore tidal swamp habitat.

Action taken: Breached dike, filled in ditches, removed culverts, created meandering tidal channel, and planted tidal shrubs and trees (saplings); completed in 2007.

Results: After one year, tidal regime was similar to nearby reference tidal swamp; cover by native species increased (excluding plantings); soil conditions were found to be optimal for growth of Sitka spruce with plantings well established (100% survival after driest part of summer) and site appears on the right trajectory for re-establishing tidal swamp. Most of site still contains reed canary grass but it is expected that eventually shading by woody plantings will reduce cover of this invasive grass (Brophy 2008, 2009). Site will be monitored for 15 years because project is mitigation for highway bridge work.

References

Able, K. W., and M. P. Fahey. 1998. *The First Year in the Life of Estuarine Fishes in the Middle Atlantic Bight*. Rutgers University Press, New Brunswick, NJ.

Able, K. W., and S. M. Hagan. 2003. Impact of common reed, *Phragmites australis,* on essential fish habitat: influence on reproduction, embryologic development, and larval abundance of mummichog (*Fundulus heteroclitus*). *Estuaries* 26:40–50.

Able, K. W., K. L. Heck, Jr., M. P. Fahay, and C. T. Roman. 1988. Use of salt-marsh peat reefs by small juvenile lobsters on Cape Cod, Massachusetts. *Estuaries* 11:83–86.

Able, K. W., D. M. Nemerson, R. Bush, and P. Light. 2001. Spatial variation in Delaware Bay (U.S.A.) marsh creek fish assemblages. *Estuaries* 24:441–52.

Adam, P. 1993. *Saltmarsh Ecology.* Cambridge University Press, Cambridge, England.

Adam, P. 2002. Saltmarshes in a time of change. *Environmental Conservation* 29:39–61.

Adamowicz, S. C., and C. T. Roman. 2005. New England salt marsh pools: a quantitative analysis of geomorphic and geographic features. *Wetlands* 25:279–88.

Adams, D. A. 1963. Factors influencing vascular plant zonation in North Carolina salt marshes. *Ecology* 44:445–56.

Adamus, P. R. 2005. Science Review and Data Analysis for Tidal Wetlands of the Oregon Coast. Part 2 of a Hydrogeomorphic Guidebook. Report to Coos Watershed Association, U.S. Environmental Protection Agency, and Oregon Department of State Lands, Salem, OR. www.oregon.gov/DSL/WETLAND/docs/tidal_HGM_pt2.pdf.

Adamus, P. R. 2006. Hydrogeomorphic (HGM) Assessment Guidebook for Tidal Wetlands of the Oregon Coast. Part 1: Rapid Assessment Method. Produced for the Coos Watershed Association, Oregon Department of State Lands, and U.S. Environmental Protection Agency, Region 10. Coos Watershed Association, Charleston, OR. www.oregon.gov/DSL/WETLAND/docs/tidal_HGM_pt1.pdf.

Adamus, P. R., J. Larsen, and R. Scranton. 2005. Watershed Profiles of Oregon's Coastal Watersheds and Estuaries. Part 3 of a Hydrogeomorphic Guidebook. Report to Coos Watershed Association, U.S. Environmental Protection Agency, and Oregon Department of State Lands, Salem, OR. www.oregon.gov/DSL/WETLAND/docs/tidal_HGM_pt3.pdf.

Adamus, P., J. Morlan, and K. Verble. 2009. Manual for the Oregon Rapid Wetland Assessment Protocol (ORWAP). Version 2.0. Oregon Department of State Lands, Wetlands Program, Salem, OR.

Adamus, P., J. Morlan, and K. Verble. 2010. Manual for the Oregon Rapid Wetland Assessment Protocol (ORWAP). Version 2.0.2. Oregon Department of State Lands, Wetlands Program, Salem, OR. www.oregon.gov/DSL/WETLAND/docs/orwap_manual_v2.pdf.

Alber, M., E. M. Swenson, S. C. Adamowicz, and I. A. Mendelssohn. 2008. Salt marsh dieback: an overview of recent events in the United States. *Estuarine, Coastal and Shelf Science* 80:1–11.

Alexander, H. D., and K. H. Dunton. 2002. Freshwater inundation effects on emergent vegetation of a hypersaline salt marsh. *Estuaries* 25:1426–35.

Alexander, H. D., and K. H. Dunton. 2006. Treated wastewater effluent as an alternative freshwater source in a hypersaline salt marsh: impacts on salinity, inorganic nitrogen, and emergent vegetation. *Journal of Coastal Research* 22:377–92.

Alexander, H. E. 1962. Changing concepts and needs in wildlife management. *Proceedings of the Annual Conference of South East Game and Fish Commissions* 16:161–67.

Allen, J. A. 1994. Intraspecific variation in the response of baldcypress (*Taxodium distichum*) seedlings to salinity. Louisiana State University, Baton Rouge, LA. Ph.D. dissertation.

Allen, J. A., J. L. Chambers, and M. Stine. 1994. Prospects for increasing the salt tolerance of forest trees: a review. *Tree Physiology* 14:843–53.

Allen, J. A., J. L. Chambers, and S. R. Pezeshki. 1997. Effects of salinity on baldcypress seedlings: physiological responses and their relation to salinity tolerance. *Wetlands* 17:310–20.

Allen, J. A., S. R. Pezeshki, and J. L. Chambers. 1996. Interaction of flooding and salinity stress on baldcypress (*Taxodium distichum*). *Tree Physiology* 16:307–13.

Allison, I., N. L. Bindoff, R. A. Bindschadler, P. M. Cox, N. de Noblet, M. H. England, J. E. Francis, N. Gruber, A. M. Haywood, D. J. Karoly, G. Kaser, C. Le Quéré, T. M. Lenton, M. E. Mann, B. I. McNeil, A. J. Pitman, S. Rahmstorf, E. Rignot, H. J. Schellnhuber, S. H. Schneider, S. C. Sherwood, R. C. J. Somerville, K. Steffen, E. J. Steig, M. Visbeck, and A. J. Weaver. 2009. *The Copenhagen Diagnosis, 2009: Updating the World on the Latest Climate Science.* The University of New South Wales Climate Change Research Centre, Sydney, Australia.

Alonzo-Pérez, F., A. Ruiz-Luna, J. Turner, C. A. Berlanga-Robles, and G. Mitchelson-Jacob. 2003. Land cover changes and impact of shrimp aquaculture on the landscape of the Ceuta coastal lagoon system, Sinaloa, Mexico. *Ocean & Coastal Management* 46:583–600.

Alpaugh, G. N., and F. Ferrigno. 1973. Ecology and management of New Jersey's wintering wetlands as related to the black duck and other wildlife. In: D. Thompson (tech. coord.). Waterfowl Management Symposium, Atlantic Flyway Council, Moncton, New Brunswick. 105–19.

Anderson, C. E. 1974. A review of structures of several North Carolina salt marsh plants. In: R. J. Reimold and W. H. Queen (eds.). *Ecology of Halophytes.* Academic Press, NY. 307–44.

Anderson, C. J., and B. G. Lockaby. 2007. Soils and biogeochemistry of tidal freshwater forested wetlands. In: W. H. Conner, T. W. Doyle, and K. W. Krauss (eds.). *Ecology of Tidal Freshwater Forested Wetlands of the Southeastern United States.* Springer, Dordrecht, The Netherlands. Chapter 3; 65–88.

Anderson, C. M., and M. Treshow. 1980. A review of environmental and genetic factors that affect height in *Spartina alterniflora* Loisel. (salt marsh cord grass). *Estuaries* 3:168–76.

Anderson, R. R., R. G. Brown, and R. D. Rappleye. 1968. Water quality and plant distribution along the upper Patuxent River, Maryland. *Chesapeake Science* 9:145–56.

Anderson, R. S., H. W. Borns, Jr., D. C. Smith, and C. Race. 1992. Implications of rapid sediment accumulation in a small New England salt marsh. *Canadian Journal of Earth Sciences* 29:2013–17.

Anderson, W. W. 1970. Contribution to the life histories of several penaeid shrimps (Penaeidae) along the south Atlantic coast of the United States. U.S. Fish and Wildlife Service Special Scientific Report no. 605.

Andresen, H., J. P. Bakker, M. Brongers, B. Heydemann, and U. Irmler. 1990. Long-term changes to salt marsh communities by cattle grazing. *Vegetatio* 89:137–48.

Aníbal, J., C. Rocha, and M. Sprung. 2007. Mudflat surface morphology as a structuring agent of algae and associated macroepifauna communities: a case study in the Ria Formosa. *Journal of Sea Research* 57:36–46.

Anisfeld, S. C., and G. Benoit. 1997. Impacts of flow restriction on salt marshes: an instance of acidification. *Environmental Science and Technology* 31:1650–57.

Anisfeld, S. C., M. J. Tobin, and G. Benoit. 1999. Sedimentation rates in flow-restricted and restored salt marshes in Long Island Sound. *Estuaries* 22(2A):231–44.

Antlfinger, A. E. 1976. Seasonal photosynthetic and water-use patterns in three salt marsh plants. University of Georgia, Athens. M.S. thesis.

Antlfinger, A. E. 1981. The genetic basis of micro differentiation in natural and experimental populations of *Borrichia frutescens* in relation to salinity. *Evolution* 35:1056–68.

Antlfinger, A. E., and E. L. Dunn. 1979. Seasonal patterns of CO_2 and water vapor exchange of three salt marsh succulents. *Oecologia* (Berlin) 43:249–60.

Antlfinger, A. E., and E. L. Dunn. 1983. Water use and salt balance in three salt marsh succulents. *American Journal of Botany* 70:561–67.

Archer, J. H., D. L. Connors, K. Laurence, S. C. Columbia, and R. Bowen. 1994. *The Public Trust Doctrine and the Management of America's Coasts.* University of Massachusetts Press, Amherst, MA.

Argow, B. A. 2004. Impact of winter processes on salt marsh accretion: variations in winter tidal deposition and ice-rafted sediment. *Geological Society of America Abstracts with Programs* Vol. 36:291.

Armstrong, W. 1982. Waterlogged soils. In: J. R. Etherington (ed.). *Environmental Plant Ecology.* John Wiley, Chichester, England. 290–330.

Arnold, A. F. 1901. *The Sea-beach at Ebb-tide.* The Century Company. Reprinted in 1968 by Dover Publications, New York.

Associated Press. 2010. Horseshoe crab populations rise, but not red knots. New Jersey 101.5 FM radio. Posted Monday June 14, 2010, 5:35AM. www.nj1015.com/pages/7459927.php.

Atwater, B. F., S. G. Conard, J. N. Dowden, C. W. Hedel, R. L. Macdonald, and W. Savage. 1979. History, landforms, and vegetation of the estuary's tidal marshes. In: *San Francisco Bay: the Urbanized Estuary. Investigations into the Natural History of San Francisco Bay and Delta with Reference to the Influence of Man.* 58th Annual Meeting of Pacific Division of the American Association for the Advancement of Science; 1977 June 12–16. San Francisco State University, San Francisco, CA. 347–86.

Austen, E., and A. Hanson. 2008. Identifying wetland compensation principles and mechanisms for Atlantic Canada using a Delphi approach. *Wetlands* 28:640–65.

Baca, B. J., T. E. Lankford, and E. R. Gundlach. 1987. Recovery of Brittany coastal marshes in the eight years following the Amoco Cadiz incident. *Proceedings of the 1987 Oil Spill Conference;* 1987 April 6–9; Baltimore, MD. American Petroleum Institute, Washington, DC. 459–64.

Baker, J. M. 1989. Oil in wetlands. In: B. Dicks (ed.). *Ecological Impacts of the Oil Industry.* John Wiley and Sons, New York. 287–316.

Baldwin, A. H. 2007 Vegetation and seed bank studies of salt-pulsed swamps of the Nanticoke River, Chesapeake Bay. In: W. H. Conner, T. W. Doyle, and K. W. Krause (eds.). *Ecology of Tidal Freshwater Forested Wetlands of the Southeastern United States.* Springer, Dordrecht, The Netherlands. Chapter 6; 139–60.

Baldwin, A. H., and I. A. Mendelssohn. 1998. Effects of salinity and water level on coastal marshes: an experimental test of disturbance as a catalyst for vegetation change. *Aquatic Botany* 61:255–68.

Baldwin, A. H., and F. N. Pendleton. 2003. Interactive effects of animal disturbance and elevation on vegetation of a tidal freshwater marsh. *Estuaries* 26:905–15.

Baldwin, A. H., K. L. McKee and I. A. Mendelssohn. 1996. The influence of vegetation, salinity, and inundation on seed banks of oligohaline coastal marshes. *American Journal of Botany* 83:470–79.

Baldwin, A. H., M. S. Egnotovich, and E. Clarke. 2001. Hydrologic change and vegetation of tidal freshwater marshes: field, greenhouse, and seed-bank experiments. *Wetlands* 21:519–31.

Baldwin, A. H., A. Barendregt, and D. F. Whigham. 2009. Tidal freshwater wetlands—an introduction to the ecosystem. In: A. Barendregt, D. F. Whigham, and A. H. Baldwin (eds.). *Tidal Freshwater Wetlands.* Backhuys Publishers, Leiden, The Netherlands. Chapter 1; 1–10.

Bales, J. D., C. J. Oblinger, and A. H. Sallenger, Jr. 2000. Two months of flooding in eastern North Carolina, September–October 1999: hydrologic water-quality, and geologic effects of Hurricanes Dennis, Floyd, and Irene. U.S. Geological Survey, Raleigh, NC. Water-Resources Investigations Report 00-4093.

Balletto, J. H., M. V. Heimbach, and H. J. Mahoney. 2005. Delaware Bay salt marsh restoration: mitigation for a power plant cooling water system in New Jersey, USA. *Ecological Engineering* 25:204–13.

Balling, S. S., and V. H. Resh. 1983. The influence of mosquito control recirculation ditches on plant biomass, production and composition in two San Francisco Bay salt marshes. *Estuarine, Coastal and Shelf Science* 16:151–61.

Baltzer, J. L., E. G. Reekie, H. L. Hewlin, P. D. Taylor, and J. S. Boates. 2002. Impact of flower harvesting on the salt marsh plant *Limonium carolinianum*. *Canadian Journal of Botany* 80:841–51.

Barbier, E. B. 2007. Valuing ecosystem services as productive inputs. *Economic Policy* 49:177–229.

Barbier, E. B., and M. Cox. 2004. An economic analysis of shrimp farm expansion and mangrove conversion in Thailand. *Land Economics* 80:389–407.

Barbier, E. B., and S. Sathirathai. 2004. *Shrimp Farming and Mangrove Loss in Thailand*. Edward Elgar Publishing, Northampton, MA.

Barbier, E. B., E. W. Koch, B. R. Silliman, S. D. Hacker, E. Wolanski, J. Primavera, E. F. Granek, S. Polasky, S. Aswani, L. A. Cramer, D. M. Storm, C. J. Kennedy, D. Bael, C. V. Kappel, G. M. E. Perillo, and D. J. Reed. 2008. Coastal ecosystem-based management with nonlinear ecological functions and values. *Science* 321:319–23.

Barbour, M. G. 1978. The effect of competition and salinity on the growth of a salt marsh plant species. *Oecologia* (Berlin) 37:93–99.

Barbour, M. G. 1970. Is any angiosperm an obligate hydrophyte? *American Midland Naturalist* 84:105–20.

Barendregt, A., D. F. Whigham, and A. H. Baldwin (eds.). 2009. *Tidal Freshwater Wetlands*. Backhuys Publishers, Leiden, The Netherlands.

Barnes, R. D. 1953. The ecological distribution of spiders in non-forest maritime communities at Beaufort, North Carolina. *Ecological Monographs* 23:315–22.

Barras, J. A. 2006. Land area changes in coastal Louisiana after the 2005 hurricanes: a series of three maps. U.S. Geological Survey, Reston, VA. Open-file Report 2006-1274. http://pubs.usgs.gov/of/2006/1274/.

Barras, J. A., S. Beville, D. Britsch, S. Hartley, S. Hawes, J. Johnston, P. Kemp, Q. Kinler, A. Martucci, J. Porthouse, D. Reed, K. Roy, S. Sapkota, and J. Suhayda. 2003. Historical and projected coastal Louisiana land changes: 1978–2050. U.S. Geological Survey, Reston, VA. Open-file Report 03-334. (Revised January 2004)

Barrett, N. E. 1989. Vegetation of the tidal wetlands of the lower Connecticut River: ecological relationships of the plant community types with respect to flooding and habitat. University of Connecticut, Biological Sciences, Storrs, CT. M.S. thesis.

Barske, P. 1961. Salt marshes and wildlife. In: Connecticut's Coastal Marshes—A Vanishing Resource. Connecticut Arboretum, Connecticut College, New London, CT. Bulletin no. 12. 13–15.

Bart, D., D. Burdick, R. Chambers, and J. M. Hartman. 2006. Human facilitation of *Phragmites australis* invasions in tidal marshes: a review and synthesis. *Wetlands Ecology and Management* 14:53–65.

Bartlett, H. H. 1909. The submarine Chamaecyparis bog at Woods Hole, Massachusetts. *Rhodora* 11:221–35.

Bartlett, H. H. 1911. Botanical evidence of coastal subsidence. *Science* 33:29–31.

Bartoldus, C. 1984. Relationship of salt marsh vegetation to tidal data at the Lloyd Point Marsh, Long Island, New York. Queens College, City University of New York, Flushing, NY. M.A. thesis.

Bartoldus, C. C. 1999. Comprehensive Review of Wetland Assessment Procedures: A Guide for Wetland Practitioners. Environmental Concern Inc., St. Michaels, MD.

Basquill, S. 2003. Marine and estuarine systems: herbaceous tidal wetlands. Taxonomy and occurrence data. Atlantic Canada Conservation Data Centre, Sackville, NB. Draft report.

Bauer, D. M., N. Cyr, and S. K. Swallow. 2004. Public preferences for compensatory mitigation of salt marsh losses: a contingent valuation choice of alternatives. *Conservation Biology* 18:401–11.

Baumann, R. H. 1987. Physical variables. In: W. H. Conner and J. W. Day, Jr. (eds.). *Ecology of the Barataria Basin, Louisiana: an Estuarine Profile*. U.S. Fish and Wildlife Service, Washington, DC. Biological Report 85(7.13):8–18.

Baye, P. R., P. M. Faber, and B. Grewell. 2000. Tidal marsh plants of the San Francisco

Estuary. In: P. R. Olofson (ed.). Baylands ecosystem species and community profiles: life histories and environmental requirements of key plants, fish and wildlife. Prepared by the San Francisco Bay Area Wetlands Ecosystem Goals Project. San Francisco Bay Regional Water Quality Control Board, Oakland, CA. http://66.147.242.191/~sfestuar/userfiles/ddocs/Species_and_Community_Profiles.pdf.

Bazely, D. R., and R. L. Jefferies. 1986. Changes in the composition and standing crop of salt-marsh plant communities in response to the removal of a grazer. *Journal of Ecology* 74:693–706.

Beck, M. W., M. Odaya, J. J. Bachant, J. Bergan, B. Keller, R. Martin, R. Mathews, C. Porter, and G. Ramseur. 2000. Identification of Priority Sites for Conservation in the Northern Gulf of Mexico: An Ecoregional Plan. The Nature Conservancy, Arlington, VA. www.masgc.org/gmrp/plans/TNC.pdf.

Bedford, B. L., M. R. Walbridge, and A. Aldous. 1999. Patterns in nutrient availability and plant diversity of temperate North American wetlands. *Ecology* 80:2151–69.

Beecher, C. B., and G. L. Chmura. 2004. Pollen-vegetation relationships in Bay of Fundy salt marshes. *Canadian Journal of Botany* 82:663–70.

Beeftink, W. G. 1977. Salt marshes. In: R. S. K. Barnes (ed.). *The Coastline.* John Wiley & Sons, London. 93–121.

Bélanger, L., and J. Bédard. 1994. Role of ice scouring and goose grubbing in marsh plant dynamics. *Journal of Ecology* 82:437–45.

Belknap, D. F., and Shipp, R. C. 1991. Seismic stratigraphy of glacial marine units, Maine inner shelf. In: J. B. Anderson and G. M. Ashley (eds.). Glacial Marine Sedimentation; Paleoclimatic Significance. Geological Society of America Special Paper 261, Boulder, Colorado. 137–57.

Benoit, L. K., and R. A. Askins. 1999. Impact of the spread of *Phragmites* on the distribution of birds in Connecticut tidal marshes. *Wetlands* 19:194–208.

Bergen, A., C. Alderson, B. Bergfors, C. Aquila, and M. A. Matsil. 2000. Restoration of a *Spartina alterniflora* salt marsh following a fuel oil spill, New York City, New York. *Wetlands Ecology and Management* 8:185–95.

Berggren, T. S., and J. T. Lieberman. 1977. Relative contribution of Hudson, Chesapeake, and Roanoke striped bass, *Morone saxatilis,* to the Atlantic coast fishery. *Fisheries Bulletin* 76:335–45.

Bertness, M. D. 1984a. Habitat and community modification by an introduced herbivorous snail. *Ecology* 65:370–81.

Bertness, M. D. 1984b. Ribbed mussels and *Spartina alterniflora* production in a New England salt marsh. *Ecology* 65:1784–1807.

Bertness, M. D. 1985. Fiddler crab regulation of *Spartina alterniflora* production on a New England salt marsh. *Ecology* 66:1043–55.

Bertness, M. D. 1988. Peat accumulation and the success of marsh plants. *Ecology* 69:703–13.

Bertness, M. D. 1991a. Interspecific interactions among high marsh perennials in a New England salt marsh. *Ecology* 72:125–37.

Bertness, M. D. 1991b. Zonation of *Spartina patens* and *Spartina alterniflora* in a New England salt marsh. *Ecology* 72:138–48.

Bertness, M. D. 1992. The ecology of a New England salt marsh. *American Scientist* 80:260–68.

Bertness, M. D., and A. M. Ellison. 1987. Determinants of pattern in a New England salt marsh plant community. *Ecological Monographs* 57:129–47.

Bertness, M. D., and E. Grosholz. 1985. Population dynamics of the ribbed mussel, *Geukensia demissa:* the costs and benefits of an aggregated distribution. *Oecologia* 67:192–204.

Bertness, M. D., and S. D. Hacker. 1994. Physical stress and positive associations among marsh plants. *The American Naturalist* 144:363–72.

Bertness, M. D., and G. H. Leonard. 1997. The role of positive interactions in communities: lessons from intertidal habitats. *Ecology* 78:1976–89.

Bertness, M. D., and T. Miller. 1984. The distribution and dynamics of *Uca pugnax* (Smith) burrows in a New England salt marsh. *Journal of Experimental Marine Biology and Ecology* 83:211–37.

Bertness, M. D., and S. W. Shumway. 1993. Competition and facilitation in marsh plants. *The American Naturalist* 142:718–24.

Bertness, M. D., K. Wikler, and T. Chatkupt. 1992a. Flood tolerance and the distribution of *Iva frutescens* across New England salt marshes. *Oecologia* 91:171–78.

Bertness, M. D., L. Gough, and S. W. Shumway. 1992b. Salt tolerances and the distribution of fugitive salt marsh plants. *Ecology* 73:1842–51.

Bertness, M. D., P. J. Ewanchuk, and B. R. Silliman. 2002. Anthropogenic modification of New England salt marsh landscapes. *Proceedings of the National Academy of Sciences* 99:1395–98.

Bertness, M. D., C. M Crain, C. Holdredge, and N. Sala. 2008. Eutrophication and consumer control of New England salt marsh primary productivity. *Conservation Biology* 22:131–39.

Biddle, A., and R. S. Warren. 1999. Sea-level, marsh surface microrelief, and vegetation: 25 years of change on a Long Island Sound tidal salt marsh. New England Estuarine Research Society, Spring 1999 Meeting, Oak Island, Nova Scotia.

Bigelow, H. B., and W. G. Schroeder. 1953. *Fishes of the Gulf of Maine*. Department of the Interior, Fish and Wildlife Service, Washington, DC. Fishery Bulletin Vol. 53.

Biggs, R. B. 1978. Coastal bays. In: R. A. Davis, Jr. (ed.). *Coastal Sedimentary Environments*. Springer-Verlag, New York. Chapter 2; 69–99.

Bindoff, N. L., J. Willebrand, V. Artale, A. Cazenave, J. M. Gregory, S. Gulev, K. Hanawa, C. Le Quere, S. Levitus, Y. Nojiri, C. K. Shum, L. D. Talley, and A. S. Unnikrishnan. 2007. Observations: oceanic climate change and sea level. In: S. Solomon, D. Qin, M. Manning, M. Marquis, K. B. Averyt, M. Tignor, H. L. Miller, and Z. Chen (eds.). *Climate Change 2007: The Physical Science Basis. Contribution of Working Group 1 to the Fourth Assessment Report of the Intergovernmental Panel on Climate Change*. Cambridge University Press, Cambridge, UK. 385–432.

Bishop, T. D., and C. T. Hackney. 1987. A comparative study of the mollusk communities of two oligohaline intertidal marshes: spatial and temporal distribution of abundance and biomass. *Estuaries* 10:141–52.

Blomberg, G., C. Simenstad, and P. Hickey. 1988. Changes in Duwamish estuary habitat over the past 125 years. In: *Proceedings First Annual Meeting on Puget Sound Research*, Vol. 2. Puget Sound Water Quality Authority, Seattle, WA. 437–54.

Bleakney, J. S., and K. B. Meyer. 1979. Observations on saltmarsh pools, Minas Basin, Nova Scotia, 1965–1977. *Proceedings of Nova Scotia Institute of Science* 29:353–71.

Bloom, A. L., and M. Stuiver. 1963. Submergence of the Connecticut coast. *Science* 139:332–34.

Blum, M. D., and H. H. Roberts. 2009. Drowning of the Mississippi delta due to insufficient sediment supply and global sea-level rise. *Nature Geoscience* 2:488–91.

Boesch, D. F., and R. E. Turner. 1984. Dependence of fishery species on salt marshes: the role of food and refuge. *Estuaries* 7:460–68.

Boesch, D. F., M. N. Josselyn, A. J. Mehta, J. T. Morris, W. K. Nuttle, C. A. Simenstad, and D. J. P. Swift. 1994. Scientific assessment of coastal wetland loss, restoration and management in Louisiana. *Journal of Coastal Research* Special Issue 20:1–103. www.eco-hydrology.com/boesch%20jcr20%20exsum.pdf.

Boggs, K. 2000. Classification of Community Types, Successional Sequences, and Landscapes of the Copper River Delta, Alaska. USDA Forest Service, Pacific Northwest Research Station, Portland, OR. General Technical Report PNW-GTR-469.

Bolen, E. G. 2000. Waterfowl management: yesterday and tomorrow. *Journal of Wildlife Management* 64:323–35.

Boorman, L. 2003. Saltmarsh Review. An overview of coastal saltmarshes, their dynamic and sensitivity characteristics for conservation and management. Joint Nature Conservation Committee, Peterborough, UK.

Boorman, L. A. 1996. Results for the Institute of Terrestrial Ecology, England. In: J. C. Lefeuvre (ed.). *The Effects of Environmental Changes on European Salt Marshes: Structure, Functioning and Exchange Potentialities with Marine Coastal Waters*. University of Rennes, France. Vol. 5.

Boothroyd, J. C. 1978. Mesotidal inlets and estuaries. In: R. A. Davis, Jr. (ed.). *Coastal Sedimentary Environments*. Springer-Verlag, New York. Chapter 6; 287–336.

Boshart, W. M. 1998. Three-year Comprehensive Monitoring Report: Naomi Freshwater Diversion BA-03. Louisiana Department of Natural Resources, Coastal Restoration Division, Baton Rouge, LA. www.lacoast.gov/reports/cmp/BA03cmp1.pdf.

Bouchard, V., M. Tessier, F. Digaire, J-P. Vivier, L. Valery, J-C. Gloaguen, and J-C. Lefeuvre. 2003. Sheep grazing as a management tool

in Western European saltmarshes. *Comptes Rendus Biologies* 326 (Suppl. 1): 148–57.
Boulé, M. E., N. Olmsted, and T. Miller. 1983. *Inventory of Wetland Resources and Evaluation of Wetland Management in Western Washington.* Report prepared for Washington State Department of Ecology. Shapiro Associates, Seattle, WA.
Bourn, W. S., and C. Cottam. 1950. Some Biological Effects of Ditching Tidewater Marshes. U.S. Fish and Wildlife Service, Washington, DC. Research Report 19.
Bourriaque, R. J. 2008. Spatial economics of the Louisiana wetland mitigation banking industry. Louisiana State University, Department of Agricultural Economics and Agribusiness, Baton Rouge, LA. M.S. thesis.
Bowditch, N. 1995. The American Practical Navigator. An Epitome of Navigation. National Imagery and Mapping Agency, Bethesda, MD. Publication no. 9.
Bowman, P. J. 2000. The Natural Communities of Delaware. Delaware Natural Heritage Program, Department of Natural Resources and Environmental Control, Dover, DE. Draft report.
Boyle, W. J., Jr. 1991. *A Guide to Bird Finding in New Jersey.* Rutgers University Press, New Brunswick, New Jersey.
Bozak, C. M., and D. M. Burdick. 2005. Impacts of seawalls on saltmarsh plant communities in the Great Bay Estuary, New Hampshire, USA. *Wetlands Ecology and Management* 13:553–68.
Bradley, P. M., and J. T. Morris. 1991. Relative importance of ion exclusion, secretion and accumulation in *Spartina alterniflora* Loisel. *Journal of Experimental Botany* 42:1525–32.
Bradley, P. M., B. Kjerfve, and J. T. Morris. 1990. Rediversion salinity change in the Cooper River, South Carolina: ecological implications. *Estuaries* 13:373–79.
Bradley, R. S. 2011. *Global Warming and Political Intimidation.* University of Massachusetts Press, Amherst, MA.
Bram, M. R., and J. A. Quinn. 2000. Sex expression, sex-specific traits, and the effects of salinity on growth and reproduction of *Amaranthus cannabinus* (Amaranthaceae), a dioecious annual. *American Journal of Botany* 87:1609–18.
Breitburg, D. L., J. K. Craig, R. S. Fulford, K. A. Rose, W. R. Boynton, D. Brady, B. J. Ciotti, R. J. Diaz, K. D. Friedland, J. D. Hagy, D. R. Hart, A. H. Hines, E. D. Houde, S. E. Kolesar, S. W. Nixon, J. A. Rice, D. H. Secor, T. E. Targett. 2009. Nutrient enrichment and fisheries exploitation: interactive effects on estuarine living resources and their management. *Hydrobiologia* 629:31–47.
Bretsch, K., and D. M. Allen. 2006. Tidal migration of nekton in salt marsh intertidal creeks. *Estuaries and Coasts* 29:474–86.
Brevik, E. C., and J. A. Homburg. 2004. A 500 year record of carbon sequestration from a coastal lagoon and wetland complex, Southern California, USA. *Catena* 57:221–32.
Brewer, J. E., G. P. Demas, and D. Holbrook. 1998a. Soil survey of Dorchester County, Maryland. USDA Natural Resources Conservation Service, in cooperation with Maryland Agricultural Experiment Station, Maryland Department of Agriculture, and Dorchester County Soil Conservation District. U.S. Government Printing Office, Washington, DC.
Brewer, J. S., J. M. Levine, and M. D. Bertness. 1998b. Interactive effects of elevation and burial with wrack on plant community structure in some Rhode Island salt marshes. *Journal of Ecology* 86:125–36.
Bricker, S. B., C. G. Clement, D. E. Pirhalla, S. P. Orlando, and D. R. G. Farrow. 1999. National Estuarine Eutrophication Assessment: Effects of Nutrient Enrichment in the Nation's Estuaries. NOAA, National Ocean Survey, Special Projects Office and the National Centers for Coastal Ocean Science, Silver Spring, MD.
Bricker-Urso, S., S. W. Nixon, J. K. Cochran, D. J. Hirschberg, and C. Hunt. 1989. Accretion rates and sediment accumulation in Rhode Island salt marshes. *Estuaries* 12:300–317.
Bridgham, S. D., J. P. Megonigal, J. K. Keller, N. B. Bliss, and C. Trettin. 2006. The carbon balance of North American wetlands. *Wetlands* 26:889–916.
Briggs, P. T., and J. S. O'Connor. 1971. Comparison of shore-zone fishes over naturally vegetated and sand-filled bottoms in Great South Bay. *New York Fish and Game Journal* 18:15–41.
Brinkhuis, B. H. 1976. The ecology of temperate salt-marsh fucoids. I. Occurrence and distribution of *Ascophyllum nodosum* ecads. *Marine Biology* 34:325–38.
Brinson, M. M. 1993. A Hydrogeomorphic Classification for Wetlands. U.S. Army Corps

of Engineers, Waterways Experiment Station, Wetlands Research Program, Vicksburg, MS. WRP-DE-4.
Brinson, M. M., and R. Rheinhardt. 1996. The role of reference wetlands in functional assessment and mitigation. *Ecological Applications* 6:69–76.
Brinson, M. M., R. R. Christian, and L. K. Blum. 1995. Multiple states in sea-level rise induced transition from terrestrial forest to estuary. *Estuaries* 18:648–59.
Bromberg, K. D., and M. D. Bertness. 2005. Reconstructing New England salt marsh losses using historical maps. *Estuaries* 28:823–32.
Broome, S. W., E. D. Seneca, and W. W. Woodhouse, Jr. 1986. Long-term growth and development of transplants of the salt-marsh grass *Spartina alterniflora*. *Estuaries* 9:63–74.
Broome, S. W., E. D. Seneca, and W. W. Woodhouse, Jr. 1988. Tidal salt marsh restoration. *Aquatic Botany* 32:1–22.
Brophy, L. 2005. Tidal wetland prioritization for the Siuslaw River estuary. Green Point Consulting, Corvallis, OR. Prepared for the Siuslaw Watershed Council, Mapleton, OR.
Brophy, L. 2007. Estuary Assessment: Component XII of the Oregon Watershed Assessment Manual. Green Point Consulting, Corvallis, OR. Prepared for the Oregon Department of Land Conservation and Development and Oregon Watershed Enhancement Board, Salem, OR.
Brophy, L. S. 2008. Annual compensatory wetland mitigation monitoring report, 2008: off-site mitigation, North Fork Siuslaw River Bridge Project. Green Point Consulting, Corvallis, OR. Prepared for the Oregon Department of Transportation, Salem, OR.
Brophy, L. 2009. Effectiveness monitoring of tidal wetland restoration and reference sites in the Siuslaw River Estuary: a tidal swamp focus. Green Point Consulting, Corvallis, OR. Prepared for Ecotrust, Portland, OR.
Brophy, L. S., C. E. Cornu, P. R. Adamus, J. A. Christy, A. Gray, M. A. MacClellan, J. A. Doumbia, and R. L. Tully. 2011. New tools for tidal wetland restoration: development of a reference conditions database and a temperature sensor method for detecting tidal inundation in least-disturbed tidal wetlands of Oregon, USA. Prepared for the Cooperative Institute for Coastal and Estuarine Environmental Technology (CICEET), University of New Hampshire, Durham, NH.
Brown, K. J., and G. B. Pasternack. 2005. A paleoenvironmental reconstruction to aid in the restoration of floodplain and wetland habitats on an upper deltaic plain, California, USA. *Environmental Conservation* 32:1–14.
Brown, P. H., and C. L. Lant. 1999. The effect of wetland mitigation banking on the achievement of no-net-loss. *Environmental Management* 23:333–45.
Brown, S., and P. Veneman. 1998. Compensatory wetland mitigation in Massachusetts. Massachusetts Agricultural Experiment Station, University of Massachusetts, Amherst, MA. Research Bulletin no. 746.
Brush, G. R. 1989. Rates and patterns of estuarine sediment accumulation. *Limnology and Oceanography* 34:1235–46.
Bryant, T. L., and J. R. Pennock (eds.). 1988. The Delaware Estuary: Rediscovering a Forgotten Resource. University of Delaware, Sea Grant College Program, Newark, DE.
Buchsbaum, R. N., L. A. Deegan, J. Horowitz, R. H. Garritt, A. E. Giblin, J. P. Ludlam, and D. H. Shull. 2009. Effects of regular salt marsh mowing on marsh plants, algae, invertebrates and birds at Plum Island Sound, Massachusetts. *Wetlands Ecology and Management* 17:469–87.
Buck, T. L. 2001. High marsh plant community response to sea-level induced high marsh subsidence and ecosystem state change. East Carolina University, Greenville, NC. M.S. thesis.
Buckley, F. G., and C. A. McCaffrey. 1978. Use of dredged material islands by colonial seabirds and wading birds in New Jersey. U.S. Army Engineer Waterways Experiment Station, Vicksburg, MS. Technical Report D-78-1.
Burdick, D. M. 1989. Root aerenchyma development in *Spartina patens* in response to flooding. *American Journal of Botany* 76:777–80.
Burdick, D. M., and I. A. Mendelssohn. 1987. Waterlogging responses in dune, swale and marsh populations of *Spartina patens* under field conditions. *Oecologia* (Berlin) 74:321–29.
Burdick, D. M., and I. A. Mendelssohn. 1990. Relationship between anatomical and

metabolic responses of waterlogging in the coastal grass *Spartina patens. Journal of Experimental Botany* 41:233–38.

Burdick, D. M., I. A. Mendelssohn, and K. L. McKee. 1989. Live standing crop and metabolism of the marsh grass *Spartina patens* as related to edaphic factors in a brackish, mixed marsh community in Louisiana. *Estuaries* 12:195–204.

Burdick, D. M., M. Dionne, R. M. Boumans, and F. T. Short. 1997. Ecological responses to tidal restorations of two northern New England salt marshes. *Wetlands Ecology and Management* 4:129–44.

Burger, J. (ed.). 1994. *Before and After An Oil Spill: The Arthur Kill.* Rutgers University Press, New Brunswick, NJ.

Burger, J., and J. Shisler. 1978. The effects of ditching a salt marsh on colony and nest site selection by herring gulls (*Larus argentatus*). *American Midland Naturalist* 100:54–63.

Burger, J., and J. Shisler. 1983. Succession and productivity on perturbed and natural *Spartina* salt-marsh areas in New Jersey. *Estuaries* 6:50–56.

Burger, J., J. Shisler, and F. H. Lesser. 1982. Avian utilization of six salt marshes in New Jersey. *Biological Conservation* 23:187–212.

Burke, D. G., and J. E. Dunn (eds.). 2010. *A Sustainable Chesapeake: Better Models for Conservation.* The Conservation Fund, Arlington, VA.

Butzer, K. W. 2002. French wetland agriculture in Atlantic Canada and its European roots: different avenues to historic diffusion. *Annals of the Association of American Geographers* 92:451–70.

Butzler, R. E., and S. E. Davis III. 2006. Growth patterns of Carolina wolfberry (*Lycium carolinianum* L.) in the salt marshes of Aransas National Wildlife Refuge, Texas, USA. *Wetlands* 26:845–53.

Byram, R. S., and R. Witter. 2000. Wetland landscapes and archaeological sites in the Coquille Estuary, Middle Holocene to recent times. R. Losey (ed.). *Proceedings of the Third Annual Coquille Cultural Preservation Conference,* 1999. Coquille Indian Tribe, North Bend, OR. 60–81.

Byrd, K. B., and M. Kelly. 2006. Salt marsh vegetation response to edaphic and topographic changes from upland sedimentation in a Pacific estuary. *Wetlands* 26:813–29.

Caffrey, J. M., M. C. Murrell, C. Wigand, and R. McKinney. 2007. Effect of nutrient loading on biogeochemical and microbial processes in a New England salt marsh. *Biogeochemistry* 82:251–64.

Cahoon, D. R. 2006. A review of major storm impacts on coastal wetland elevations. *Estuaries and Coasts* 29:889–98.

Cahoon, D. R., and J. Lynch. 2010. Surface Elevation Table (SET). U.S. Geological Survey, Patuxent Wildlife Research Center, MD. www.pwrc.usgs.gov/set/.

Cahoon, D. R., D. J. Reed, and J. W. Day, Jr. 1995. Estimating shallow subsidence in microtidal salt marshes of the southeastern United States: Kaye and Barghoorn revisited. *Marine Geology* 128:1–9.

Cahoon, D. R., J. C. Lynch, and A. N. Powell. 1996. Marsh vertical accretion in a southern California estuary, U.S.A. *Estuarine, Coastal and Shelf Science* 43:19–32.

Cahoon, D. R., J. Day, and D. Reed. 1999. The influence of surface and shallow subsurface soil processes on wetland elevation: a synthesis. *Current Topics in Wetland Biogeochemistry* 3:72–88.

Cahoon, D. R., J. C. Lynch, P. Hensel, R. Boumans, B. C. Perez, B. Segura, and J. W. Day, Jr. 2002. A device for high precision measurement of wetland sediment elevation: I. Recent improvements to the sedimentation-erosion table. *Journal of Sedimentary Research* 72:730–33.

Cahoon, D. R., P. R. Hensel, J. Rybczyk, K. L. McKee, E. E. Proffitt, and B. C. Perez. 2003. Mass tree mortality leads to mangrove peat collapse at Bay Islands, Honduras after Hurricane Mitch. *Journal of Ecology* 91:1093–1105.

Cahoon, D. R., P. F. Hensel, T. Spencer, D. J. Reed, K. L. McKee, and N. Saintilan. 2006. Coastal wetland vulnerability to relative sea-level rise: wetland elevation trends and process controls. In: J. T. A. Verhoeven, B. Beltman, R. Bobbink, and D. P. Whigham (eds.). *Wetlands and Natural Resource Management. Ecological Studies.* Springer-Verlag, Berlin/Heidelberg, Germany. Vol. 190. 271–92.

Cahoon, D. R., D. J. Reed, A. S. Kolker, and M. M. Brinson. 2009. Coastal sensitivity. In: Climate Change Science Program. *Coastal Sensitivity to Sea-level Rise: A Focus on the*

Mid-Atlantic Region. Synthesis and Assessment Product 4.1. Report prepared by the Climate Change Science Program and the Subcommittee on Global Change Research. U.S. Environmental Protection Agency, Washington, DC. Chapter 4; 57–72. http://downloads.climatescience.gov/sap/sap4-1/sap4-1-final-report-all.pdf.

Cahoon, D. R., B. C. Perez, B. D. Segura, and J. C. Lynch. 2010. Elevation trends and shrink-swell response of wetland soils to flooding and drying. *Estuarine, Coastal and Shelf Science.* doi: 10.1016/j.ecss.2010.03.022. Available online March 27, 2010.

Caldwell, F. A., and G. E. Crow. 1992. A floristic and vegetation analysis of a freshwater tidal marsh on the Merrimack River, West Newbury, Massachusetts. *Rhodora* 94:63–97.

California Invasive Plant Council (CIPC). 2006. California Invasive Plant Inventory. Berkeley, CA. CIPC Publication 2006-02. www.cal-ipc.org/ip/inventory/pdf/Inventory2006.pdf.

Callaway, J. C., and J. B. Zedler. 1998. Interactions between a salt marsh native perennial (*Salicornia virginica*) and an exotic annual (*Polypogon monspeliensis*) under varied salinity and hydroperiod. *Wetlands Ecology and Management* 5:179–94.

Callaway, J. C., R. D. DeLaune, and W. H. Patrick, Jr. 1997. Sediment accretion rates from four coastal wetlands along the Gulf of Mexico. *Journal of Coastal Research* 13:181–91.

Callaway, J. C., G. Sullivan, J. S. Desmond, G. D. Williams, and J. B. Zedler. 2001. Assessment and monitoring. In: J. B. Zedler (ed.). *Handbook for Restoring Tidal Wetlands.* CRC Press, Boca Raton, FL. 271–335.

Callaway, J. C., V. T. Parker, M. C. Vasey, and L. M. Schile. 2007. Emerging issues for the restoration of tidal marsh ecosystems in the context of predicted climate change. *Madroño* 54:234–48.

Callaway, R. M. 1994. Facilitation and interfering effects of *Arthrocnemum subterminalis* on winter annuals. *Ecology* 75:681–86.

Callaway, R. M., and S. C. Pennings. 1998. Impact of a parasitic plant on the zonation of two salt marsh perennials. *Oecologia* 114:100–105.

Callaway, R. M., and C. S. Sabraw. 1994. Effects of variable precipitation on the structure and diversity of a California salt marsh community. *Journal of Vegetation Science* 5:433–38.

Callaway, R. M., S. Jones, W. R. Ferren, and A. Parikh. 1990. Ecology of a Mediterranean-climate estuarine wetland at Carpinteria, California: plant distribution and soil salinities in the upper marsh. *Canadian Journal of Botany* 69:1139–46.

Canadian Wildlife Service. 1991. The Federal Policy on Wetland Conservation. Environment Canada, Ottawa, Ontario.

Candeletti, R. 2007. The history and application of open marsh water management in New Jersey. In: *Proceedings of the Ninety-Fourth Annual Meeting of the New Jersey Mosquito Control Association.* 2007 March 14–16; Atlantic City, NJ. 42–47.

Carey, W. L. 1996. Transgression of Delaware's fringing tidal salt marshes: surficial morphology, subsurface stratigraphy, vertical accretion rates, and geometry of adjacent and antecedent surfaces. University of Delaware, Newark, DE. Ph.D. dissertation.

Carletti, A., G. A. De Leo, and I. Ferrari. 2004. A critical review of representative wetland rapid assessment methods in North America. *Aquatic Conservation: Marine and Freshwater Ecosystems* 14 (Suppl. 1): S103–S113.

Carlisle, B. K., A. M. Donovan, A. L. Hicks, V. S. Kooken, J. P. Smith, and A. R. Wilbur. 2002. A Volunteer's Handbook for Monitoring New England Salt Marshes. Massachusetts Office of Coastal Zone Management, Boston, MA.

Carlisle, B. K., J. D. Baker, A. L. Hicks, J. P. Smith, and A. L. Wilbur. 2004a. Cape Cod Salt Marsh Assessment Project. Volume 1: Relationship of Salt Marsh Indices of Biotic Integrity to Surrounding Land Use, 1999. Massachusetts Office of Coastal Zone Management, Boston, MA.

Carlisle, B. K., J. D. Baker, A. L. Hicks, J. P. Smith, and A. L. Wilbur. 2004b. Cape Cod Salt Marsh Assessment Project. Volume 2: Response of Selected Salt Marsh Indicators to Tide Restriction, 2000–2003. Massachusetts Office of Coastal Zone Management, Boston, MA.

Carlisle, B. K., R. W. Tiner, M. Carullo, I. K. Huber, T. Nuerminger, C. Polzen, and M. Shaffer. 2005. 100 Years of Estuarine Marsh Trends in Massachusetts (1893–1995): Boston Harbor, Cape Cod, Nantucket,

Martha's Vineyard, and the Elizabeth Islands. Massachusetts Office of Coastal Zone Management, Boston, MA; U.S. Fish and Wildlife Service, Hadley, MA; and University of Massachusetts, Amherst, MA. Cooperative report. www.mass.gov/czm/estuarine_marsh_trend1.htm.

Carlson, R. R., Jr., and J. Forrest. 1982. Uptake of dissolved sulfide by *Spartina alterniflora:* evidence from natural sulfur isotope abundance ratios. *Science* 216:633–35.

Carney, J. A. 2001. *Black Rice: The African Origins of Rice Culture in the Americas.* Harvard University Press, Cambridge, MA.

Carson, R. 1947. Parker River: A National Wildlife Refuge. Conservation in Action no. 2. U.S. Government Printing Office, Washington, DC.

Carson, R. 1962. *Silent Spring.* Houghton Mifflin, Boston, MA.

Carson, R. 1998. *The Edge of the Sea.* Mariner Books, Houghton Mifflin Harcourt, New York. (Reprint of 1955 book by Houghton Mifflin)

Carter, V. 1996. Wetland hydrology, water quality, and associated functions. In: J. D. Fretwell, J. S. Williams, and P. J. Redman (comps.). *National Water Summary on Wetland Resources.* U.S. Geological Survey, Water-Supply Paper. 35–48.

Carullo, M., B. K. Carlisle, and J. P. Smith. 2007. A New England rapid assessment method for assessing condition of estuarine marshes: a Boston Harbor, Cape Cod and Island pilot study. Massachusetts Office of Coastal Zone Management, Boston, MA.

Casagrande, D. G. (ed.). 1997. Restoration of an Urban Salt Marsh: An Interdisciplinary Approach. Yale School of Forestry and Environmental Studies, New Haven, CT. Bulletin no. 100.

Casazza, M. 2010. Ecology of California Clapper Rails in the San Francisco Bay/Delta Region. U.S. Geological Survey, Western Ecological Research Center, Sacramento, CA. www.werc.usgs.gov/Project.aspx?ProjectID=152.

Caspers, H. 1967. Estuaries: analysis of definitions and biological considerations. In: G. H. Lauff (ed.). *Estuaries.* American Association for the Advancement of Science, Washington, DC. AAAS Publication 83. 6–8.

Cavalieri, A. J. 1983. Proline and glycinebetaine accumulation by *Spartina alterniflora* Loisel. in response to NaCl and nitrogen in a controlled environment. *Oecologia* 57:20–24.

Center for Science Advice. 2009. Use of the Lower Saint John River, New Brunswick, As Fish Habitat during the Spring Freshet. Fisheries and Oceans Canada, Maritimes Region, Dartmouth, NS. Canadian Science Advisory Secretariat Science Response 2009/014. www.dfo-mpo.gc.ca/CSAS/Csas/Publications/ScR-RS/2009/2009_014_E.pdf.

Chabreck, R. H. 1972. Vegetation, Water, and Soil Characteristics of the Louisiana Region. Louisiana State University, Agricultural Experiment Station, Baton Rouge, LA. Bulletin No. 664.

Chabreck, R. H. 1976. Management of wetlands for wildlife habitat improvement. In: M. Wiley (ed.). *Estuarine Processes.* Volume I. *Uses, Stresses, and Adaptation to the Estuary.* Academic Press, New York. 226–33.

Chabreck, R. H. 1981. Effect of burn date on regrowth rate of *Scirpus olneyi* and *Spartina patens. Proceedings of the Annual Conference of Southeastern Association of Fish and Wildlife Agencies* 35:201–10.

Chabreck, R. H. 1988. *Coastal Marshes: Ecology and Wildlife Management.* University of Minnesota Press, Minneapolis, MN.

Chabreck, R. H., and A. W. Palmisano. 1973. The effects of Hurricane Camille on the marshes of the Mississippi Delta Plain. *Ecology* 54:1118–23.

Chalmers, A. G., R. G. Wiegert, and P. L. Wolf. 1985. Carbon balance in a salt marsh: interaction of diffusive export, tidal deposition and rainfall-caused erosion. *Estuarine, Coastal and Shelf Science* 21:757–71.

Chambers, R. M., L. A. Meyerson, and K. Saltonstall. 1999. Expansion of *Phragmites australis* into tidal wetlands of North America. *Aquatic Botany* 64:261–73.

Chambers, R. M., D. T. Osgood, D. J. Bart, and F. Montalto. 2003. *Phragmites australis* invasion and expansion in tidal wetlands: interactions among salinity, sulfide, and hydrology. *Estuaries* 26:89–108.

Chapman, V. J. 1960. *Salt Marshes and Salt Deserts of the World.* Interscience Publishers, New York.

Chapman, V. J. (ed.). 1977. *Wet Coastal Ecosystems.* Elsevier Scientific Publishing Company, Amsterdam, The Netherlands.

Charles, H., and J. S. Dukes. 2009. Effects of warming and altered precipitation on plant and nutrient dynamics of a New England salt marsh. *Ecological Applications* 19:1758–73.
Chatry, M., R. J. Dugas, and K. A. Easley. 1983. Optimum salinity regime for oyster production on Louisiana's state seed grounds. *Contributions in Marine Science* 26:81–94.
Chen, Z., and Y. Saito. 2011. The mega-deltas of Asia: interlinkage of land and sea, and human development. *Earth Surface Processes and Landforms* 36:1703–4.
Chen, Z., Y. Saito, and S. L. Goodbred (eds.). 2005. *Mega-deltas of Asia—Geological Evolution and Human Impact.* China Ocean Press, Beijing, China.
Cherry, J. A., K. L. McKee, and J. B. Grace. 2009. Elevated CO_2 enhances biological contributions to elevation change in coastal wetlands by offsetting stressors associated with sea-level rise. *Journal of Ecology* 97:67–77.
Chesapeake Bay Executive Council. 1988. Chesapeake Bay Wetlands Policy. Chesapeake Bay Program, Annapolis, MD. Agreement commitment report.
Childers, D. L., J. W. Day, Jr., and R. A. Muller. 1990. Relating climatological forcing to coastal water levels in Louisiana estuaries and the potential importance of El Niño-Southern Oscillation events. *Climate Research* 1:31–42.
Childers, D. L., J. W. Day, Jr., and H. N. McKellar, Jr. 2000. Twenty more years of marsh and estuarine flux studies: revisiting Nixon (1980). In: M. P. Weinstein and D. A. Kreeger (eds.). *Concepts and Controversies in Tidal Marsh Ecology.* Kluwer Academic Publishers, Dordrecht, The Netherlands. 391–424.
Chmura, G. L. 2003. Salt marshes and mangrove swamps. In: F. I. Isla (ed.). *Coastal Zone and Estuaries, in Encyclopedia of Life Support Systems (EOLSS).* UNESCO, EOLSS Publishers, Oxford, UK. Chapter 3.5. www.eolss.net.
Chmura, G. L. 2009. Tidal marshes. In: D. d'A. Laffoley and G. Grimsditch (eds.). *The Management of Natural Coastal Carbon Sinks.* International Union for Conservation of Nature and Natural Resources (IUCN), Gland, Switzerland. 5–11.
Chmura, G. L., and G. A. Hung. 2004. Controls on salt marsh accretion: a test in salt marshes of eastern Canada. *Estuaries* 27:70–81.
Chmura, G. L., P. Chase, and J. Bercovitch. 1997. Climatic controls of the middle marsh zone in the Bay of Fundy. *Estuaries* 20:689–99.
Chmura, G. L., L. L. Helmer, C. B. Beecher, and E. M Sunderland. 2001. Historical rates of salt marsh accretion on the outer Bay of Fundy. *Canadian Journal of Earth Sciences* 38:1081–92.
Chmura, G. L., S. C. Anisfeld, D. R. Cahoon, and J. C. Lynch. 2003. Global carbon sequestration in tidal, saline soils. *Global Biogeochemical Cycles* 17:1–12.
Choi, Y., and Y. Wang. 2004. Dynamics of carbon sequestration in a coastal wetland using radiocarbon measurements. *Global Biogeochemical Cycles* 18. 12 pp. GB4016, doi: 10.1029/2004GB002261.
Christian, R. R., J. A. Hansen, R. E. Hodson, and W. J. Wiebe. 1983. Relationships of soil, plant, and microbial characteristics in silt-clay and sand, tall-form *Spartina alterniflora* marshes. *Estuaries* 6:43–49.
Christiansen, T. 1998. Sediment deposition on a tidal salt marsh. University of Virginia, Department of Environmental Sciences, Charlottesville, VA. Ph.D. dissertation. www.vcrlter.virginia.edu/thesis/Christiansen98.pdf.
Christy, J. A., and L. S. Brophy. 2007. Estuarine and freshwater tidal plant associations in Oregon. Oregon Natural Heritage Information Center, Oregon State University, Corvallis, OR.
Clark, R. C., and J. S. Finley. 1977. Effects of oil spills in arctic and subarctic environments. In: D. C. Malins (ed.). *Effects of Petroleum on Arctic and Subarctic Marine Organisms.* Vol. II. *Biological Effects.* Academic Press, New York. Chapter 9; 441–76.
Clark, J. S. 1986. Late-Holocene vegetation and coastal processes at a Long Island tidal marsh. *Journal of Ecology* 74:561–78.
Clark, J. S., and W. A. Patterson III. 1985. The development of a tidal marsh: upland and ocean influences. *Ecological Monographs* 55:189–217.
Clark, K. E., and L. J. Niles. 2000. North Atlantic Regional Shorebird Plan. Version 1.0. Prepared with the Members of the North Atlantic Shorebird Habitat Working Group. New Jersey Division of Fish and Wildlife, Endangered and Nongame Species

Program, Woodbine, NJ. www.fws.gov/shorebirdplan/RegionalShorebird/downloads/NATLAN4.pdf.

Clarke, J. A., B. A. Harrington, T. Hruby, and F. E. Wasserman. 1984. The effect of ditching for mosquito control on salt marsh use by birds in Rowley, Massachusetts. *Journal of Field Ornithology* 55:160–80.

Clemants, S. E., J. Marinellli, and G. Moore (eds.). 2004. History, Ecology, and Restoration of a Degraded Urban Wetland. Rutgers University, The Center for Urban Restoration Ecology. *Urban Habitats* 2(1). http://urbanhabitats.org/v02n01/urbanhabitats_v02n01_pdf.pdf.

Clements, B. W., and A. J. Rogers. 1964. Studies of impounding for the control of the salt marsh mosquitoes in Florida, 1958–1963. *Mosquito News* 24:265–76.

Climate Change Science Program. 2009. Coastal Sensitivity to Sea-level Rise: A Focus on the Mid-Atlantic Region. Synthesis and Assessment Product 4.1. Report prepared by the Climate Change Science Program and the Subcommittee on Global Change Research. U.S. Environmental Protection Agency, Washington, DC. http://downloads.climatescience.gov/sap/sap4-1/sap4-1-final-report-all.pdf.

Coker, R. E. 1962. *This Great and Wide Sea. An Introduction to Oceanography and Marine Biology*. Harper Torchbooks, Harper & Row Publishers, New York.

Collins, B. D., and A. J. Sheikh. 2005. Historical reconstruction, classification and change analysis of Puget Sound tidal marshes. University of Washington, Puget Sound River History Project, Department of Earth and Space Sciences, Seattle, WA.

Collins, J. N., and T. C. Foin. 1992. Evaluation of the impacts of aqueous salinity on the shoreline vegetation of tidal marshlands in the San Francisco Estuary. In: J. R. Schubel (ed.). *Managing Freshwater Discharge to the San Francisco Bay/San Joaquin Delta Estuary: The Scientific Basis for an Estuarine Standard*. San Francisco Estuary Project, U.S. Environmental Protection Agency, San Francisco CA. C1–C34.

Collins, J. N., and R. M. Grossinger. 2004. Synthesis of scientific knowledge concerning estuarine landscapes and related habitats of the South Bay ecosystem. Technical report of the South Bay Salt Pond Restoration Project. San Francisco Estuary Institute, Oakland, CA. www.southbayrestoration.org/pdf_files/Issues%201%20&%203%20Landscape%20&%20Marshes.pdf.

Collins, J. N., and V. H. Resh. 1985. Utilization of natural and man-made habitats by the salt marsh song sparrow *Melospiza melodia samuelis* (Baird). *California Fish and Game* 71:40–52.

Collins, J. N., L. M. Collins, L. B. Leopold, and V. H. Resh. 1986. The influence of mosquito control ditches on the geomorphology of tidal marshes in the San Francisco Bay Area: evolution of salt marsh mosquito habitat. *Proceedings and Papers of the California Mosquito and Vector Control Association* 54:91–95.

Collins, J. N., E. D. Stein, M. Sutula, R. Clark, A. E. Fetscher, L. Grenier, C. Grosso, and A. Wiskind. 2008. California Rapid Assessment Method (CRAM) for Wetlands. User's Manual. Version 5.0.2. San Francisco Estuary Institute, Oakland, CA; Southern California Coastal Water Research Project, Costa Mesa, CA; California Coastal Commission, Santa Cruz, CA; and Moss Landing Marine Laboratories, Moss Landing, CA. CRAM Field Books for Estuarine, Depressional and Riverine Wetlands. www.cramwetlands.org/.

Committee on the Alaska Earthquake. 1972. *The Great Alaska Earthquake of 1964*. Division of Earth Sciences, Oceanography and Coastal Engineering, Natural Research Council. National Academy of Sciences, Washington, DC.

Committee on Characterization of Wetlands. 1995. *Wetlands: Characteristics and Boundaries*. National Academy Press, Washington, DC.

Committee on Mitigating Wetland Losses. 2001. *Compensating for Wetland Losses Under the Clean Water Act*. National Academy Press, Washington, DC.

Committee on the Restoration and Protection of Coastal Louisiana. 2006. *Drawing Louisiana's New Map: Addressing Land Loss in Coastal Louisiana*. Ocean Studies Board, Division of Earth and Life Sciences, National Research Council of the National Academies. The National Academies Press, Washington, DC. www.nap.edu/openbook.php?record_id=11476&page=R1.

Commonwealth of Massachusetts. 1963. Report of the Department of Natural Resources relative to the coastal wetlands in the Commonwealth and certain shellfish grants (under Chapter 75 of the Resolves of 1962). Senate no. 635.

Conard, H. S. 1924. Second survey of the vegetation of a Long Island salt marsh. *Ecology* 5:379–88.

Conard, H. S. 1935. The plant associations of central Long Island. A study in descriptive plant sociology. *American Midland Naturalist* 16:433–516.

Conard, H. S., and G. S. Galligar. 1929. Third survey of a Long Island salt marsh. *Ecology* 10:326–36.

Conner, W. H. 1994. The effect of salinity and waterlogging on growth and survival of baldcypress and Chinese tallow seedlings. *Journal of Coastal Research* 10:1045–49.

Conner, W. H., and J. W. Day, Jr. 1987. The Ecology of Barataria Basin, Louisiana: An Estuarine Profile. U.S. Fish and Wildlife Service, Washington, DC. Biological Report 85(7.13).

Conner, W. H., and L. W. Inabinette. 2003. Tree growth in three South Carolina (USA) swamps after Hurricane Hugo. *Forest Ecology and Management* 182:371–80.

Conner, W. H., and L. W. Inabinette. 2005. Identification of salt tolerant baldcypress (*Taxodium distichum* (L.) Rich.) for planting in coastal areas. *New Forests* 29:305–12.

Conner, W. H., K. W. McLeod, and J. K. McCarron. 1997. Flooding and salinity effects on growth and survival of four common forested wetland species. *Wetlands Ecology and Management* 5:99–109.

Conner, W. H., T. W. Doyle, and K. W. Krauss (eds.). 2007a. *Ecology of Tidal Freshwater Forested Wetlands of the Southeastern United States.* Springer, Dordrecht, The Netherlands.

Conner, W. H., T. W. Doyle, and K. W. Krauss. 2007b. Ecology of tidal freshwater forests in coastal deltaic Louisiana and northeastern South Carolina. In: W. H. Conner, T. W. Doyle, and K. W. Krauss (eds.). *Ecology of Tidal Freshwater Forested Wetlands of the Southeastern United States.* Springer, Dordrecht, The Netherlands. Chapter 9; 223–53.

Connolly, K. D., S. M. Johnson, and D. R. Williams. 2005. *Wetlands Law and Policy: Understanding Section 404.* American Bar Association Publishing, Chicago, IL.

Connor, R., and G. L. Chmura. 2000. Dynamics of above- and belowground organic matter in a high latitude macrotidal saltmarsh. *Marine Ecology Progress Series* 204:101–10.

Connor, R. F., G. L. Chmura, and C. B. Beecher. 2001. Carbon accumulation in Bay of Fundy salt marshes: implications for restoration of reclaimed marshes. *Global Biogeochemical Cycles* 15:943–54.

The Conservation Foundation. 1988. Protecting America's Wetlands: An Action Agenda. The Final Report of the National Wetlands Policy Forum. Washington, DC.

Contra Costa County. 2010. Historical fresh water and salinity conditions in the Western Sacramento-San Joaquin Delta and Suisun Bay. Water Resources Department, Contra Costa Water District, Concord, CA. Technical Memorandum WR10-001. www.ccwater.com/salinity/HistoricalSalinityReport-2010Feb.pdf.

Cook, R. A., A. J. L. Stone, and A. P. Ammann. 1993. Method for the Evaluation and Inventory of Vegetated Tidal Marshes in New Hampshire (Coastal Method). Audubon Society of New Hampshire, Concord, NH.

Cooper, L. W., and C. P. McRoy. 1988. Anatomical adaptation to rocky substrates and surf exposure by the seagrass genus *Phyllospadix. Aquatic Botany* 32:365–81.

Copeland, B. J., R. G. Hodson, and S. R. Riggs. 1984. The Ecology of the Pamlico River, North Carolina: An Estuarine Profile. U.S. Fish and Wildlife Service, Washington, DC. FWS/OBS-82/06.

Costanza, R., M. Wilson, A. Troy, A. Voinov, S. Liu, and J. D'Agostino. 2006. The Value of New Jersey's Ecosystem Services and Natural Capital. Gund Institute for Ecological Economics, University of Vermont, Burlington, VT. www.state.nj.us/dep/dsr/naturalcap/nat-cap-2.pdf.

Costanza, R., O. Pérez-Maqueo, M. L. Martinez, P. Sutton, S. J. Anderson, and K. Mulder. 2008. The value of coastal wetlands for hurricane protection. *Ambio* 37:241–48.

Costanzo, G. R., and R. A. Malecki. 1989. Foods of black ducks wintering along coastal New Jersey. *Transactions of Northeast Section Wildlife Society* 46:7–16.

Couillard, L., and P. Grondin. 1986. *Végétation des milleux humides du Québec.* Les Publications du Québec, Québec, QC.

Coultas, C. L., and Y-P. Hsieh (eds.). 1997. *Ecology and Management of Tidal Marshes. A Model for the Gulf of Mexico*. St. Lucie Press, Delray Beach, FL.

Couvillion, B. R., J. A. Barras, G. D. Steyer, W. Sleavin, M. Fischer, H. Beck, N. Trahan, B. Griffin, and D. Heckman. 2011. Land area change in coastal Louisiana from 1932 to 2010. U.S. Geological Survey Scientific Investigations Map 3164, scale 1:265,000, pamphlet. http://pubs.usgs.gov/sim/3164/downloads/SIM3164_Pamphlet.pdf.

Cowardin, L. M., V. Carter, F. C. Golet, and E. T. LaRoe. 1979. Classification of Wetlands and Deepwater Habitats of the United States. U.S. Department of Interior, Fish and Wildlife Service, Washington, DC. FWS/OBS-79/31.

Craft, C., and J. Sacco. 2003. Long-term succession of benthic infauna communities on constructed *Spartina alterniflora* marshes. *Marine Ecology Progress Series* 257:45–58.

Craft, C. B., E. D. Seneca, and S. W. Broome. 1993. Vertical accretion in microtidal regularly and irregularly flooded estuarine marshes. *Estuarine, Coastal and Shelf Science* 37:371–86.

Craft, C., J. Reader, J. N. Sacco, and S. W. Broome. 1999. Twenty-five years of ecosystem development of *Spartina alterniflora* (Loiscl) marshes. *Ecological Applications* 9:1405–19.

Craft, C., S. Broome, and C. Campbell. 2002. Fifteen years of vegetation and soil development after brackish-water marsh creation. *Restoration Ecology* 10:248–58.

Craft, C., P. Megonigal, S. Broome, J. Stevenson, R. Freese, J. Cornell, L. Zheng, and J. Sacco. 2003. The pace of ecosystem development of constructed *Spartina alterniflora* marshes. *Ecological Applications* 13:1417–32.

Craft, C., J. Clough, J. Ehman, S. Joyce, R. Park, S. Pennings, H. Guo, and M. Machmuller. 2009. Forecasting the effects of accelerated sea-level rise on tidal marsh ecosystem services. *Frontiers in Ecology and the Environment* 7:73–78.

Craft, J. C. 1988. Geology. In: T. L. Bryant and J. R. Pennock (eds.). *The Delaware Estuary: Rediscovering a Forgotten Resource*. University of Delaware, Sea Grant Program, Newark, DE. 31–40.

Craig, N. J., L. M. Smith, N. M. Gilmore, G. D. Lester, and A. M. Williams. 1987. The Natural Communities of Coastal Louisiana: Classification and Description. Louisiana Department of Natural Resources, Coastal Management Division, Baton Rouge, LA.

Craig, R. J., and K. G. Beal. 1992. The influence of habitat variables on marsh bird communities of the Connecticut River estuary. *Wilson Bulletin* 104:295–311.

Crain, C. M. 2007. Shifting nutrient limitation and eutrophication effects in marsh vegetation across estuarine salinity gradients. *Estuaries and Coasts* 30:26–34.

Crain, C. M., B. R. Silliman, S. L. Bertness, and M. D. Bertness. 2004. Physical and biotic drivers of plant distribution across estuarine salinity gradients. *Ecology* 85:2539–49.

Crawford, D. W., N. L. Bonnevie, C. A. Gillis, and R. J. Wenning. 1994. Historical changes in the ecological health of the Newark Bay Estuary, New Jersey. *Ecotoxicology and Environmental Safety* 29:276–303.

Cronk, J. K., and M. S. Fennessy. 2001. *Wetland Plants: Biology and Ecology*. CRC Press, Boca Raton, FL.

Crooks, S., J. Schutten, G. D. Sheem, K. Pye, and A. J. Davy. 2002. Drainage and elevation as factors in the restoration of salt marsh in Britain. *Restoration Ecology* 10:591–602.

Crossett, K. M., T. J. Culliton, P. C. Wiley, and T. R. Goodspeed. 2004. *Population Trends Along the Coastal United States: 1980–2008*. National Oceanic and Atmospheric Administration, National Ocean Service, Management and Budget Office, Rockville, MD.

Culberson, S. D. 2001. The interaction of physical and biological determinants producing vegetation zonation in tidal marshes of the San Francisco Bay Estuary, California, USA. University of California at Davis. Ph.D. dissertation.

Culberson, S. D., T. C. Foin, and J. N. Collins. 2004. The role of sedimentation in estuarine marsh development within the San Francisco Estuary, California, USA. *Journal of Coastal Research* 20:970–79.

Daehler, C. C., and D. R. Strong. 1996. Status, prediction and prevention of introduced cordgrass *Spartina* spp. invasions in Pacific estuaries, USA. *Biological Conservation* 78:51–58.

Dahl, T. E. 2000. Status and trends of wetlands in the conterminous United States, 1986 to 1997. U.S. Department of the Interior, Fish and Wildlife Service, Washington, DC.

Dahl, T. E. 2011. Status and Trends of Wetlands in the Conterminous United States: 2004 to 2009. U.S. Department of the Interior, Fish and Wildlife Service, Washington, DC.
Dahl, T. E., and C. E. Johnson. 1991. Wetlands Status and Trends in the Conterminous United States Mid-1970's to Mid-1980's. U.S. Department of the Interior, Fish and Wildlife Service, Washington, DC.
Dai, T., and R. G. Wiegert. Ramet population dynamics and net aerial primary productivity of *Spartina alterniflora*. *Ecology* 77:276–88.
Daiber, F. C. 1982. *Animals of the Tidal Marsh*. Van Nostrand Reinhold, New York.
Daiber, F. C. 1986. *Conservation of Tidal Marshes*. Van Nostrand Reinhold, New York.
Daiber, F. C., L. L. Thornton, K. A. Bolster, T. G. Campbell, O. W. Crichton, G. L. Esposito, D. R. Jones, and J. M. Tyrawski. 1976. An Atlas of Delaware's Wetlands and Estuarine Resources. Delaware Coastal Management Program Technical Report No. 2.
D'Alpaos, A., S. Lanzoni, M. Marani, and A. Rinaldo. 2007. Landscape evolution in tidal embayments: modeling the interplay of erosion, sedimentation, and vegetation dynamics. *Journal of Geophysical Research* 112, FO1008. 17 pp.
Dame, R., M. Alber, D. Allen, M. Mallin, C. Montague, A. Lewitus, A. Chalmers, R. Gardner, C. Gilman, B. Kjerfve, J. Pinckney, and N. Smith. 2000. Estuaries of the South Atlantic Coast of North America: their geographical signatures. *Estuaries* 23:793–819.
Dames, R. F., J. D. Spurrier, T. M. Williams, B. Kjerfve, R. G. Zingmark, T. G. Wolaver, T. H. Chrzanowski, H. N. McKellar, and F. J. Vernberg. 1991. Annual material processing by a salt marsh-estuarine basin in South Carolina, USA. *Marine Ecology Progress Series* 72:153–66.
Danielsen, F., M. K. Sørensen, M. F. Olwig, V. Selvam, F. Parish, N. D. Burgess, T. Hiraishi, V. M. Karunagaran, M. S. Rasmussen, L. B. Hansen, A. Quarto, and N. Suryadiputra. 2005. The Asian tsunami: a protective role for coastal vegetation. *Science* 310:643.
Danielson, T. J., and D. G. Hoskins (eds.). 2003. Methods for evaluating wetland condition: wetland biological assessment case studies. U.S. Environmental Protection Agency, Office of Water, Washington, DC. EPA-822-R-03-013. www.epa.gov/owow/wetlands/bawwg.
Daoust, R. J., and D. L. Childers. 1998. Quantifying aboveground biomass and estimating net aboveground primary production for wetland macrophytes using a non-destructive phenometric technique. *Aquatic Botany* 62:115–33.
Darke, A. K., and J. P. Megonigal. 2003. Control of sediment deposition rates in two Mid-Atlantic coast tidal freshwater wetlands. *Estuarine, Coastal and Shelf Science* 57:255–68.
Darling, L. 1961. The death of a marsh: the story of Sherwood Island Marsh and its political consequences. In: Connecticut's Coastal Marshes—A Vanishing Resource. Connecticut Arboretum, Connecticut College, New London, CT. Bulletin no. 12. 21–27.
Darmody, R. G., and J. E. Foss. 1978. Tidal Marsh Soils of Maryland. University of Maryland, College Park, MD. Maryland Agricultural Experiment Station report MP 930.
Das, S., and J. R. Vincent. 2009. Mangroves protected villages and reduced death toll during Indian super cyclone. *Proceedings of the National Academy of Sciences of the United States* 106:7357–60.
Davies, J. L. 1973. *Geographic Variation in Coastal Development*. Hafner Press, New York.
Davies, J. L. 1974. A morphogenic approach to world shorelines. *Zeitschrift für Geomorphologie* 8:27–42.
Davis, C. A. 1910. Salt marsh formation near Boston and its geological significance. *Economic Geology* 5:623–39.
Davis, C. A. 1911. Salt marshes, a study in correlation. *Journal of the Association of American Geographers* 1:139–43.
Davis, D. S., and S. Browne (eds.). 1996. Natural History of Nova Scotia. Nova Scotia Museum of Natural History, Nova Scotia. http://museum.gov.ns.ca/mnh/nature/nhns.
Davis, G. M., M. Mazurkiewicz, and M. Mandracchia. 1982. *Spurwinkia:* morphology, systematics, and ecology of a new genus of North American marshland Hydrobiidae (Mollusca: Gastropoda). *Proceedings of the Academy of Natural Sciences of Philadelphia* 134:143–77.
Davis, J. L., B. Nowicki, and C. Wigand. 2004. Denitrification in fringing salt marshes

of Narragansett Bay, Rhode Island, USA. *Wetlands* 24:870–78.

Davis, L. V. 1978. Class Insecta. In: R. G. Zingmark (ed.). *An Annotated Checklist of Biota of the Coastal Zone of South Carolina*. Belle W. Baruch Institute for Marine Biology and Coastal Research. University of South Carolina Press, Columbia, SC.

Davis, L. V., and I. E. Gray. 1966. Zonal and seasonal distribution of insects in North Carolina salt marshes. *Ecological Monographs* 36:275–95.

Davis, R. A., Jr. 1978. Beach and nearshore zone. In: R. A. Davis, Jr. (ed.). *Coastal Sedimentary Environments*. Springer-Verlag, New York. Chapter 5; 237–85.

Davis, R. A., and G. A. Zarillo. 2003. Human-induced changes in back-barrier environment as factors in tidal inlet instability with emphasis on Florida. U.S. Army Engineer Research and Development Center, Vicksburg, MS. Coastal and Hydraulics Laboratory Engineering Technical Note ERDC/CHL CHETN-IV-57.

Dawicki, S. 2009. Surveys show increasing populations of gray and harbor seals in New England. Northeast Fisheries Science Center. *Science Spotlight* SS09.01.

Dawson, J. W. 1855. On a modern submerged forest at Fort Lawrence, Nova Scotia. *Quarterly Journal of Geological Society of London* 11:119–22. (Also in *American Journal of Science* II, 21:440–42 in 1856.)

Day, J. H. 1981. The nature, origin, and classification of estuaries. In: J. H. Day (ed.). *Estuarine Ecology with Particular Reference to Southern Africa*. A. A. Balkema, Cape Town, South Africa. 1–6.

Day, J. W., D. Pont, P. F. Hensel, and C. Ibanez. 1995. Impacts of sea-level rise on deltas in the Gulf of Mexico and Mediterranean: the importance of pulsing events to sustainability. *Estuaries* 18:636–47.

Day, J. W., Jr., L. D. Britsch, S. R. Hawes, G. P. Shaffer, D. J. Reed, and D. Cahoon. 2000. Pattern and process of land loss in the Mississippi Delta: a spatial and temporal analysis of wetland habitat change. *Estuaries and Coasts* 23:425–38.

Day, J. W., Jr., D. F. Boesch, E. J. Clairain, G. P. Kemp, S. B. Laska, W. J. Mitsch, K. Orth, H. Mashriqui, D. R. Reed, L. Shabman, C. A. Simenstad, B. J. Streever, R. R. Twilley, C. C. Watson, J. T. Wells, and D. F. Whigham. 2007a. Restoration of the Mississippi Delta: lessons from Hurricanes Katrina and Rita. *Science* 315:1679–84.

Day, R. H., T. M. Williams, and C. M. Swarzenski. 2007b. Hydrology of tidal freshwater forested wetlands of the southeastern United States. In: W. H. Conner, T. W. Doyle, and K. W. Krause (eds.). *Ecology of Tidal Freshwater Forested Wetlands of the Southeastern United States*. Springer, Dordrecht, The Netherlands. Chapter 2; 29–63.

Day, J. W., R. R. Christian, D. M. Boesch, A. Yáñez-Arancibia, J. Morris, R. R. Twilley, L. Naylor, L. Shaffner, and C. Stevenson. 2008. Consequences of climate change on the ecogeomorphology of coastal wetlands. *Estuaries and Coasts* 31:477–91.

Dedrick, K. G. 1979. Effects of levees on tidal currents in marshland creeks. In: Abstracts of the G. K. Gilbert Symposium: San Francisco Bay, Its Past, Present, and Future. Annual meeting of the Geological Society of America; 1979; San Jose, CA.

Deegan, L. A., and R. H. Garritt. 1997. Evidence for spatial variability in estuarine food webs. *Marine Ecology Progress Series* 147:31–47.

Deegan, L. A., B. J. Peterson, and R. Porter. 1990. Stable isotopes and cellulase activity as evidence for detritus as a food source for juvenile Gulf menhaden. *Estuaries* 13:14–19.

Deegan, L. A., J. E. Hughes, and R. A. Rountree. 2000. Salt marsh ecosystem support of marine transient species. In: M. P. Weinstein and D. A. Kreeger (eds.). *Concepts and Controversies in Tidal Marsh Ecology*. Kluwer Academic Publishers, Dordrecht, The Netherlands. 333–68.

Deegan, L. A., J. L. Bowen, D. Drake, J. W. Gleeger, C. T. Friedrichs, K. A. Galván, J. E. Hobbie, C. S. Hopkinson, D. S. Johnson, J. M. Johnson, L. E. LeMay, E. Miller, B. J. Peterson, C. Picard, S. Sheldon, M. Sutherland, J. Vallino, and R. S. Warren. 2007. Susceptibility of salt marshes to nutrient enrichment and predator removal. *Ecological Applications* 17 (Suppl.):S42–63.

DeLaune, R. D., R. J. Buresh, and W. H. Patrick, Jr. 1979. Relationship of soil properties to standing crop biomass of *Spartina alterniflora* in a Louisiana marsh. *Estuarine and Coastal Marine Science* 8:477–87.

DeLaune, R. D., C. N. Reddy, and W. H. Patrick, Jr. 1981. Accumulation of plant nutrients and heavy metals through sedimentation processes and accretion in a Louisiana salt marsh. *Estuaries* 4:328–34.

DeLaune, R. D., C. J. Smith, and W. H. Patrick, Jr. 1983a. Relationship of marsh elevation, redox potential, and sulfide to *Spartina alterniflora* productivity. *Soil Science Society of America Journal* 47:930–35.

DeLaune, R. D., R. H. Baumann, and J. G. Gosselink. 1983b. Relationships among vertical accretion, coastal submergence, and erosion in a Louisiana Gulf Coast marsh. *Journal of Sedimentary Petrology* 53:147–57.

DeLaune, R. D., C. J. Smith, W. H. Patrick, Jr., J. W. Fleeger, and M. D. Tolley. 1984. Effect on oil on salt marsh biota. Methods for restoration. *Environmental Pollution Series A, Ecological and Biological* 36:207–27.

DeLaune, R. D., W. H. Patrick, and S. R. Pezeshki. 1987a. Foreseeable flooding and death of coastal wetland forests. *Environmental Conservation* 14:129–33.

DeLaune, R. D., S. R. Pexeshki, and W. H. Patrick, Jr. 1987b. Response of coastal plants to increase in submergence and salinity. *Journal of Coastal Research* 3:533–48.

DeLaune, R. D., W. H. Patrick, Jr., and N. Van Breeman. 1990a. Processes governing marsh formation in a rapidly subsiding coastal environment. *Catena* 17:277–88.

DeLaune, R. D., S. R. Pezeshki, and J. H. Pardue. 1990b. An oxidation-reduction buffer for evaluating the physiological response of plants to root oxygen stress. *Environmental and Experimental Botany* 30:243–47.

DeLaune, R. D., I. Devai, C. R. Crozier, and P. Kelle. 2002. Sulfate reduction in Louisiana marsh soils of varying salinities. *Communications in Soil Science and Plant Analysis* 33:79–94.

DeLaune, R. D., A. Jugsujinda, G. W. Peterson, and W. H. Patrick, Jr. 2003. Impact of Mississippi River freshwater reintroduction on enhancing marsh accretionary processes in a Louisiana estuary. *Estuarine, Coastal and Shelf Science* 58:653–62.

Delft Hydraulics. 1992. *Sea Level Rise: A Global Vulnerability Assessment.* Tidal Waters Division, Rijkswaterstaat, Ministry of Transport, Public Works and Water Management, The Netherlands.

DeLuca, W. V., C. E. Studds, L. L. Rockwood, and P. P. Marra. 2004. Influence of land use on the integrity of marsh bird communities of Chesapeake Bay, USA. *Wetlands* 24:837–47.

DeLuca, W. V., C. E. Studds, R. S. King, and P. P. Marra. 2008. Coastal urbanization and the integrity of estuarine waterbird communities: threshold responses and the importance of scale. *Biological Conservation* 141:2669–78.

Demas, C. R., and D. K. Demcheck. 1996a. Louisiana wetland resources. In: J. D. Fretwell, J. S. Williams, and P. J. Redman (comps.). *National Water Summary of Wetland Resources.* U.S. Geological Survey, Reston, VA. Water-Supply Paper 2425. 207–12.

Demas, C. R., and D. K. Demcheck. 1996b. Mississippi wetland resources. In: J. D. Fretwell, J. S. Williams, and P. J. Redman (comps.). *National Water Summary of Wetland Resources.* U.S. Geological Survey, Reston, VA. Water-Supply Paper 2425. 243–48.

Demas, G. P., and M. C. Rabenhorst. 2001. Factors of subaqueous soil formation: a system of quantitative pedology for submersed environments. *Geoderma* 102:189–204.

de Mutsert, K. 2010. The effects of a freshwater diversion on nekton species biomass, distribution, food web pathways, and community structure in a Louisiana estuary. Louisiana State University, Department of Oceanography and Coastal Science, Baton Rouge, LA. Ph.D. dissertation.

DeRagon, W. R. 1988. Breeding ecology of seaside and sharp-tailed sparrows in Rhode Island salt marshes. University of Rhode Island, Kingston, RI. M.S. thesis.

DeSanto, T. L., J. W. Johnston, and K. L. Bildstein. 1997. Wetland feeding site use by white ibises (*Eudocimus albus*) breeding in coastal South Carolina. *Colonial Waterbirds* 20:167–76.

Deschenes, J., and J. B. Serodes. 1985. The influence of salinity on *Scirpus americanus* tidal marshes in the St. Lawrence River estuary, Quebec. *Canadian Journal of Botany* 63:920–27.

Desplanque, C., and D. J. Mossman. 1998. A review of ice and tide observations in the Bay of Fundy. *Atlantic Geology* 34:195–209.

DeVoe, M. R., and D. S. Baughman (eds.). 1987. Coastal Wetland Impoundments: Ecological Characterization, Management, Status, and Use. Vol. 1: Executive Summary.

South Carolina Sea Grant Consortium, Charleston, SC. Publication no. SC-SG-TR-86-1.

Digby, M. J., P. Saenger, M. B. Whelan, D. McConchie, B. Eyre, N. Holmes, and D. Bucher. 1998. A physical classification of Australian estuaries. Centre for Coastal Management, Southern Cross University, Lismore, NSW.

Dionne, J-C. 1969. Érosion glacielle littorale, estuaire de Saint-Laurent. *Revue de Géographie de Montréal* 23:5–20.

Division of Wetlands. 1978. A Guide to Coastal Wetlands Regulations. Massachusetts Department of Environmental Quality Engineering, Boston, MA.

Dolan, R., and E. R. Davis. 1992. An intensity scale for Atlantic coast northeast storms. *Journal of Coastal Research* 8:840–53.

Dolan, R., B. Hayden, G. Hornberger, J. Zieman, and M. Vincent. 1972. Classification of the Coastal Environments of the World. Part I, The Americas. University of Virginia, Charlottesville, VA. ONR Contract 389–158. Technical Report no. 1.

Dollopf, S. L., J-H. Hyun, A. C. Smith, H. J. Adams, S. O'Brien, and J. E. Kostka. 2005. Quantification of ammonia-oxidizing bacteria and factors controlling nitrification in salt marsh sediments. *Applied and Environmental Microbiology* 71:240–46.

Donnelly, J. P., and M. D. Bertness. 2001. Rapid shoreward encroachment of salt marsh cordgrass in response to accelerated sea-level rise. *Proceedings of the National Academy of Sciences of the United States* 98:14218–23.

Donnelly, J. P., P. Cleary, P. Newby, and R. Ettinger. 2004. Coupling instrumental and geological records of sea-level change evidence from southern New England of an increase in the rate of sea-level rise in the late 19th century. *Geophysical Research Letters* 31, L05203, doi: 10.1029/2003GL018933.

Doumlele, D. G. 1981. Primary production and seasonal aspects of emergent plants in a tidal freshwater marsh. *Estuaries* 4:139–42.

Dovel, W. L. 1971. Fish Eggs and Larvae of the Upper Chesapeake Bay. Natural Resources Institute, University of Maryland, College Park, MD. NRI Special Scientific Report no. 4.

Dovel, W. L. 1981. Ichthyoplankton of the lower Hudson Estuary, New York. *New York Fish and Game Journal* 28:21–39.

Doyle, T. W., W. H. Conner, M. Ratard, and L. W. Inabinette. 2007. Assessing the impact of tidal flooding and salinity on long-term growth of baldcypress under changing climate and riverflow. In: W. H. Conner, T. W. Doyle, and K. W. Krause (eds.). *Ecology of Tidal Freshwater Forested Wetlands of the Southeastern United States*. Springer, Dordrecht, The Netherlands. Chapter 15; 411–45.

Drake, B. G., M. A. Gonzalez-Meler, and S. P. Long. 1997. More efficient plants: a consequence of rising atmospheric CO_2. *Annual Review of Plant Physiology and Plant Molecular Biology* 48:609–39.

Drexler, J. Z., and K. C. Ewel. 2001. Effect of the 1997–1998 ENSO-related drought on hydrology and salinity in a Micronesian wetland complex. *Estuaries* 24:347–56.

Drexler, J. Z., C. S. de Fontaine, and S. J. Deverel. 2009. The legacy of wetland drainage on the remaining peat in the Sacramento-San Joaquin delta, California, USA. *Wetlands* 29:372–86.

Duarte, C. M., and J. Cebrián. 1996. The fate of marine autotrophic production. *Limnology and Oceanography* 41:1758–66.

Duarte, C. M., J. J. Middelburg, and N. Caraco. 2005. Major role of marine vegetation on the oceanic carbon cycle. *Biogeosciences* 2:1–8.

Duberstein, J., and W. Kitchens. 2007. Community composition of selected areas of tidal freshwater forest along the Savannah River. In: W. H. Conner, T. W. Doyle, and K. W. Krause (eds.). *Ecology of Tidal Freshwater Forested Wetlands of the Southeastern United States*. Springer, Dordrecht, The Netherlands. Chapter 12; 321–48.

Dubois, J-M. M. 1993. The Saint Lawrence River System, Atlantic Coast of Quebec. In: L. P. Hillebrand (ed.). *Coastlines of Canada*. American Society of Civil Engineers, New York. 159–69.

Dubois, J-M. M., and A. Grenier. 1993. The Magdalen Islands, Gulf of Saint Lawrence. In: L. P. Hillebrand (ed.). *Coastlines of Canada*. American Society of Civil Engineers, New York. 170–82.

Dufault, R. J., M. Jackson, and S. K. Salvo. 1993. Sweetgrass: history, basketry, and constraints to industry growth. In: J. Jancik and J. E. Simon (eds.). *New Crops*. Wiley, New York.

Dunne, P. 1984. 1983 Northern Harrier breeding survey in coastal New Jersey. NJ Audubon Society. *Records of New Jersey Birds* 10:3–5.
Dunton, K. H., B. Hardegree, and T. E. Whitledge. 2001. Response of estuarine marsh vegetation to interannual variations in precipitation. *Estuaries* 24:851–61.
Durako, M. J., M. Murphy, and K. Haddad. 1988. Assessment of Fisheries Habitat: Northeast Florida. Florida Marine Resources Publication no. 45.
Edelson, S. M. 2007. Clearing swamps, harvesting forests: trees and the making of a plantation landscape in the colonial South Carolina low country. *Agricultural History* 81:381–406.
Edgar, G. J., N. S. Barrett, and D. J. Graddon. 1999. A Classification of Tasmanian Estuaries and Assessment of Their Conservation Significance Using Ecological and Physical Attributes, Population, and Land Use. Tasmanian Aquaculture and Fisheries Institute, University of Tasmania, Tasmania. Report.
Edmonds, W. J., G. M. Silberhorn, P. R. Cobb, C. D. Peacock, Jr., N. A. McLoda, and D. W. Smith. 1985. Soil Classifications and Floral Relationships of Seaside Salt Marshes in Accomack and Northampton Counties, Virginia. Virginia Agricultural Experiment Station, Virginia Polytechnic Institute and State University, Blacksburg, VA. Bulletin 85-8.
Edwards, K. R., and C. E. Proffitt. 2003. Comparison of wetland structural characteristics between created and natural salt marshes in southwest Louisiana, USA. *Wetlands* 23:344–56.
Eilers, H. P. 1975. Plants, plant associations, net production, and tide levels: the ecological biogeography of the Nehalem salt marshes, Tillanook County, Oregon. Oregon State University, Corvallis, OR. Ph.D. dissertation.
Eisma, D., and K. S. Dijkema. 1997. The influence of salt marsh vegetation on sedimentation. In: D. Eisma (ed.). *Intertidal Deposits.* CRC Press, Boca Raton, FL. 403–14.
Eleuterius, L. N. 1989. Natural selection and genetic adaptation to hypersalinity in *Juncus roemerianus* Scheele. *Aquatic Botany* 36:465–53.
Eleuterius, L. N., and J. D. Caldwell. 1981. Growth kinetics and longevity of the salt marsh rush *Juncus roemerianus*. Gulf Research Reports 7:27–31.
Eleuterius, L. N., and C. K. Eleuterius. 1979. Tide levels and salt marsh zonation. *Bulletin of Marine Science* 29:394–400.
Eleuterius, L. N., and F. C. Lanning. 1987. Silica in relation to leaf decomposition of *Juncus roemerianus*. *Journal of Coastal Research* 3:531–34.
Eliot, J. 1748. Essays upon field husbandry in New England. Reprinted 1924. H. J. Carman and R. G. Tugwell (eds.). Columbia University Press.
Elliott, C. G. 1912. *Engineering for Land Drainage: A Manual for the Reclamation of Lands Injured by Water.* John Wiley & Sons, New York.
Elliott, M., and D. S. McLusky. 2002. The need for definitions in understanding estuaries. *Estuarine, Coastal and Shelf Science* 55:815–27.
Emanuel, K. A. 1987. The dependence of hurricane intensity on climate. *Nature* 326:483–85.
Emery, K. O., and D. G. Aubrey. 1991. *Sea Levels, Land Levels, and Tide Gauges.* Springer-Verlag, New York.
Emery, K. O., R. L. Wigley, and M. Rubin. 1965. A submerged peat deposit off the Atlantic Coast of the United States. *Limnology and Oceanography* 10 (Suppl.), R97–102.
Emery, N. C., P. J. Ewanchuk, and M. D. Bertness. 2001. Competition and salt-marsh plant zonation: stress-tolerators may be dominant competitors. *Ecology* 82:2471–85.
Emmett, R., R. Llansó, J. Newton, R. Thom, M. Hornberger, C. Morgan, C. Levings, A. Copping, and P. Fishman. 2000. Geographic signatures of North American west coast estuaries. *Estuaries* 23:765–92.
England, M. 1989. The breeding biology and status of the northern harrier (*Circus cyaneus*) on Long Island, New York. Long Island University, Greenvale, NY. M.S. thesis.
Environmental Laboratory. 1987. Corps of Engineers Wetlands Delineation Manual. U.S. Army Corps of Engineer Waterways Experiment Station, Vicksburg, MS. Technical Report Y-87-1.
Environmental Law Institute. 2006. 2005 Status Report on Compensatory Mitigation in the United States. Washington, DC.

Environmental Law Institute. 2008. State Wetland Protection: Status, Trends & Model Approaches. Washington, DC.

Environmental Services Office. 2000. Suisun Marsh Monitoring Program Reference Guide. Version 2. State of California, the Resource Agency, Department of Water Resources, Sacramento, CA. www.iep.ca.gov/suisun/dataReports/referenceGuide/SMSCGReferenceGuide_Version02.pdf.

Erwin, R. M., C. J. Conway, S. W. Hadden, J. S. Hatfield, and S. M. Melvin. 2003. Monitoring Protocol for Cape Cod National Seashore and Other Coastal Parks, Refuges, and Protected Areas. National Park Service, Cape Cod National Seashore, Long-term Coastal Ecosystem Monitoring Program, Wellfleet, MA. www.nature.nps.gov/im/monitor/protocoldb.cfm.

Erwin, R. M., G. M. Sanders, and D. J. Prosser. 2004. Changes in lagoonal marsh morphology at selected northeastern Atlantic coast sites of significance to migratory waterbirds. *Wetlands* 24:891–903.

Erwin, R. M., D. R. Cahoon, D. J. Prosser, G. M. Sanders, and P. Hensel. 2006. Surface elevation dynamics in vegetated *Spartina* marshes versus unvegetated tidal ponds along the Mid-Atlantic Coast, USA, with implications to waterbirds. *Estuaries and Coasts* 29:96–106.

Evanchuck, P. J., and M. D. Bertness. 2004. Structure and organization of a northern New England salt marsh community. *Journal of Ecology* 92:72–85.

Evers, D. E., C. E. Sasser, J. G. Gosselink, D. A. Fuller, and J. M. Visser. 1998. The impact of vertebrate herbivores on wetland vegetation in Atchafalaya Bay, Louisiana. *Estuaries* 21:1–13.

Fagherazzi, S., M. L. Kirwan, S. M. Mudd, G. R. Guntenspergen, S. Temmerman, A. D'Alpaos, J. van de Koppel, J. M. Rybczyk, E. Reyes, C. Craft, and J. Clough. 2012. Numerical models of salt marsh evolution: ecological, geomorphic, and climatic factors. *Reviews of Geophysics* 50, RG1002, doi:10.1029/2011RG000359.

Fahmy, S. H., S. W. R. Hann, and Y. Jiao. 2010. Soils of New Brunswick: The Second Approximation. Agriculture and Agri-Food Canada, Fredericton, New Brunswick, CN. www.ccse-swcc.nb.ca/publications/english/Soils%20of%20NB%202nd%20Approx.pdf.

Fahrenthold, D. A. 2004. Blackwater Refuge now nutria-free. *Washington Post,* November 17, 2004: page B01.

Fairbanks, R. G. 1989. A 17,000-year glacio-eustatic sea level record: influence of glacial melting rates on the Younger Dryas event and deep-ocean circulation. *Nature* 342:637–42.

Falk, D. A., M. A. Palmer, and J. B. Zedler (eds.). 2006. *Foundation of Restoration Ecology.* Island Press, Washington, DC.

Fanning, D. S. 2002. Acid sulfate soils. In: R Lal (ed.). *Encyclopedia of Soil Science.* Marcel Dekker, New York. 11–13.

Farnham, C. H. 1888. The lower St. Lawrence. *Harper's New Monthly Magazine* 77(462):814–25.

Farris, N., J. Portnoy, A. Bennett, C. Roman, N. Barrett, E. Gwilliam, E. Kinney, and K. Raposa. 1998. Hatches Harbor: Restoring a Salt Marsh. National Park Service, Cape Cod National Seashore, Wellfleet, MA. www.nps.gov/caco/naturescience/upload/Hatches_Harbor_Final_report_1998.pdf.

Fearn, M. L., D. W. Haywick, and J. M. Sanders. 2005. Changes in Water Conditions and Sedimentation Rates Associated with Construction of the Mobile Bay Causeway. www.southalabama.edu/geography/fearn/Causeway.htm.

Federal Interagency Committee for Wetland Delineation. 1989. Federal Manual for Identifying and Delineating Jurisdictional Wetlands. Cooperative technical publication. U.S. Army Corps of Engineers, U.S. Environmental Protection Agency, U.S. Fish and Wildlife Service, and USDA Soil Conservation Service, Washington, DC.

Fell, P. E., K. A. Murphy, M. A. Peck, and M. L. Recchia. 1991. Reestablishment of *Melampus bidentatus* (Say) and other macroinvertebrates in a restored impounded tidal marsh: comparison of populations above and below the impoundment. *Journal of Experimental Marine Biology and Ecology* 152:33–48.

Fennessy, M. S., A. D. Jacobs, and M. E. Kentula. 2004. Review of Rapid Methods for Assessing Wetland Condition. U.S. Environmental Protection Agency, National Health and Environmental Effects Laboratory, Corvallis, OR. EPA/620/R-04/009. www.epa.gov/owow/wetlands/monitor/RapidMethodReview.pdf.

Fennessy, M. S., A. D. Jacobs, and M. E. Kentula. 2007. An evaluation of rapid methods for assessing the ecological condition of wetlands. *Wetlands* 27:543–60.

Fensome, R. A., and G. L. Williams (eds.). 2001. *The Last Billion Years: A Geological History of the Maritime Provinces of Canada.* Atlantic Geoscience Society, Halifax, NS.

Fernald, M. L., and A. C. Kinsey. 1943. *Edible Wild Plants of Eastern North America.* Idlewild Press, Cornwall, NY.

Ferren, W. R., Jr., and A. E. Schuyler. 1980. Intertidal vascular plants of river systems near Philadelphia. *Proceedings of the Academy of Natural Sciences of Philadelphia* 132:86–120.

Ferren, W. R., Jr., R. E. Good, R. Walker, and J. Arsenault. 1981. Vegetation and flora of Hog Island, a brackish wetland in the Mullica River, New Jersey. *Bartonia* 48:1–10.

Ferrigno, F., and D. M. Jobbins. 1968. Open marsh water management. *Proceedings of the Fifty-fifth Annual Meeting of the New Jersey Mosquito Extermination Association.* 104–15.

Ferrigno, F., L. Widjeskog, and S. Toth. 1973. Marsh Destruction. New Jersey Department of Environmental Protection, Division of Fish, Game, and Wildlife, Trenton. Pittman-Robertson Report. Project W-53-R-1, Job I-G.

Fetscher, A. E., M. A. Sutula, J. C. Callaway, V. T. Parker, M. C. Vasey, J. N. Collins, and W. G. Nelson. 2010. Patterns in estuarine vegetation communities in two regions of California: insights from a probabilistic survey. *Wetlands* 30:833–46.

Field, D. W., A. J. Reyer, P. V. Genovese, and B. D. Shearer. 1991. *Coastal Wetlands of the United States. An Accounting of a Valuable Natural Resource.* Strategic Assessment Branch, National Ocean Service, National Oceanic and Atmospheric Administration, Rockville, MD.

Field, R. T., and K. R. Philipp. 2000. Vegetation changes in the freshwater tidal marsh of the Delaware estuary. *Wetlands Ecology and Management* 8:79–88.

Findlay, S. E. G., E. Kiviat, W. C. Nieder, and E. A. Blair. 2002. Functional assessment of a reference wetland set as a tool for science, management and restoration. *Aquatic Sciences* 64:107–17.

Findlay, S. E. G., W. C. Nieder, and S. Ciparis. 2009. Carbon flows, nutrient cycling, and food webs in tidal freshwater wetlands. In: A. Barendregt, D. F. Whigham, and A. H. Baldwin (eds.). *Tidal Freshwater Wetlands.* Backhuys Publishers, Leiden, The Netherlands. Chapter 12; 137–44.

Fitch, R., T. Theodose, and M. Dionne. 2009. Relationships among upland development, nitrogen, and plant community composition in a Maine salt marsh. *Wetlands* 29:1179–88.

Fleckenstein, H. A., Jr. 1983. *New Jersey Decoys.* Schiffer Publishing, Exton, PA.

Flowers, T. J., H. K. Galal, and L. Bromham. 2010. Evolution of halophytes: multiple origins of salt tolerance in land plants. *Functional Plant Biology* 37:604–12.

Foote, A. L., and K. A. Reynolds. 1997. Decomposition of saltmeadow cordgrass (*Spartina patens*) in Louisiana coastal marshes. *Estuaries* 20:579–88.

Forbes, M. G., H. D. Alexander, and K. H. Dunton. 2008. Effects of pulsed riverine versus non-pulsed wastewater inputs of freshwater on plant community structure in a semi-arid salt marsh. *Wetlands* 28:984–94.

Ford, M. A., and J. B. Grace. 1998. Effects of vertebrate herbivores on soil processes, plant biomass, litter accumulation and soil elevation change in a coastal marsh. *Journal of Ecology* 86:974–82.

Foulis, D. B., and R. W. Tiner. 1994a. Wetland status and trends in Calvert County, Maryland (1981 to 1988–89). U.S. Fish and Wildlife Service, National Wetlands Inventory, Region 5, Hadley, MA.

Foulis, D. B., and R. W. Tiner. 1994b. Wetland status and trends in Charles County, Maryland (1981–82 to 1988–89). U.S. Fish and Wildlife Service, National Wetlands Inventory, Region 5, Hadley, MA.

Foulis, D. B., and R. W. Tiner. 1994c. Wetland status and trends in St. Marys County, Maryland (1981–82 to 1988–89). U.S. Fish and Wildlife Service, National Wetlands Inventory, Region 5, Hadley, MA.

Foulis, D. B., and R. W. Tiner. 1994d. Wetland trends for selected areas of the Casco Bay Estuary of the Gulf of Maine (1974–77 to 1984–87). U.S. Fish and Wildlife Service, National Wetlands Inventory, Region 5, Hadley, MA.

Foulis, D. B., and R. W. Tiner. 1994e. Wetland trends for selected areas of the coast of

Massachusetts, from Plum Island to Scituate (1977 to 1985–86). U.S. Fish and Wildlife Service, National Wetlands Inventory, Region 5, Hadley, MA.

Foulis, D. B., and R. W. Tiner. 1994f. Wetland trends for selected areas of the Cobscook Bay/St. Croix River Estuary of the Gulf of Maine (1975/77 to 1983–85). U.S. Fish and Wildlife Service, National Wetlands Inventory, Region 5, Hadley, MA.

Foulis, D. B., J. A. Eaton, and R. W. Tiner. 1994. Wetland trends for selected areas of the Gulf of Maine, from York, Maine to Rowley, Massachusetts (1977 to 1985–86). U.S. Fish and Wildlife Service, National Wetlands Inventory, Region 5, Hadley, MA.

Foulis, D. B., T. W. Nuerminger, and R. W. Tiner. 1995. Wetland trends in Dorchester County, Maryland (1981–82 to 1988–89). U.S. Fish and Wildlife Service, National Wetlands Inventory, Region 5, Hadley, MA.

Fox, D. S., S. Bell, W. Nehlsen, and J. Damron. 1984. The Columbia River Estuary: Atlas of Physical and Biological Characteristics. Columbia River Estuary Data Development Program, Astoria, OR.

Frayer, W. E., and J. M. Hefner. 1991. Florida Wetlands Status and Trends, 1970's to 1980's. U.S. Fish and Wildlife Service, Southeast Regional Office, Atlanta, GA.

Frayer, W. E., T. J. Monahan, D. C. Bowden, and F. A. Graybill. 1983. Status and Trends of Wetlands and Deepwater Habitats in the Conterminous United States, 1950s to 1970s. Department of Forest and Wood Sciences, Colorado State University, Ft. Collins, CO.

Frenkel, R. E., and J. C. Morlan. 1990. Restoration of the Salmon River Salt Marshes: Retrospect and Prospect. Final Report to U.S. Environmental Protection Agency. Oregon State University, Department of Geosciences, Corvallis, OR.

Frey, R. W., and P. B. Basan. 1978. Coastal salt marshes. In: R. A. Davis, Jr. (ed.). *Coastal Sedimentary Environments.* Springer-Verlag, New York. Chapter 3; 101–69.

Friedrichs, C. T., and J. E. Perry. 2001. Tidal salt marsh morphodynamics: a synthesis. *Journal of Coastal Research* Special Issue 27:7–37.

Friends of the Bay. 2006. 2006 Annual Water Quality Report. Open Water Body Water Quality Monitoring Report. Oyster Bay, NY.

Fritz, H. M., C. Blount, R. Sokolosi, J. Singleton, A. Fuggle, B. G. McAdoo, A. Moore, C. Grass, and B. Tate. 2007. Hurricane Katrina storm surge distribution and field observations on the Mississippi barrier islands. *Estuarine, Coastal and Shelf Science* 74:12–20.

Frost, J. W., T. Schleicher, and C. Craft. 2009. Effects of nitrogen and phosphorus additions on primary production and invertebrate densities in a Georgia (USA) tidal freshwater marsh. *Wetlands* 29:196–203.

Frothingham, R. 1849. *The History of the Siege of Boston and of the Battles of Lexington, Concord, and Bunker Hill.* C. C. Little and J. Brown, Boston, MA.

Fukuyama, A. K., G. Shigenaka, and G. R. VanBlaricom. 1998. Oil Spill Impacts and the Biological Basis for Response Guidance: An Applied Synthesis of Research on Three Subarctic Intertidal Communities. National Oceanic and Atmospheric Administration, National Ocean Survey, Office of Ocean Resources Conservation and Assessment, Seattle, WA. NOAA Technical Memorandum NOS ORCA 125.

Furbish, C. E., and M. Albano. 1994. Selective herbivory and plant community in a Mid-Atlantic salt marsh. *Ecology* 75:1015–22.

Funk, D. W., L. E. Noel, and A. H. Freedman. 2004. Environmental gradients, plant distribution, and species richness in arctic salt marsh near Prudhoe Bay, Alaska. *Wetlands Ecology and Management* 12:215–33.

Gaffin, S. R. 1997. High Water Blues: Impacts of Sea Level Rise on Selected Coasts and Islands. Environmental Defense Fund, New York.

Galbraith, H., R. Jones, R. Park, J. Clough, S. Herrod-Julius, B. Harrington, and G. Page. 2002. Global climate change and sea level rise: potential losses of intertidal habitats for shorebirds. *Waterbirds* 25:173–83.

Galbraith, H., R. Jones, R. Park, J. Clough, S. Herrod-Julius, B. Harrington, and G. Page. 2005. Global Climate Change and Sea Level Rise: Potential Losses of Intertidal Habitats for Shorebirds. USDA Forest Service, Washington, DC. General Technical Report PSW-GTR-191. 1119–22. www.fs.fed.us/psw/publications/documents/psw_gtr191/Asilomar/pdfs/1119-1122.pdf.

Gallagher, J. L. 1974. Sampling macro-organic profiles in salt marsh plant root zones. *Proceedings Soil Science Society of America* 38:154–55.

Gallagher, J. L. 1998. Ecology and status of tidal wetlands in North America. In: S. K. Majumdar, E. W. Miller, and F. J. Brenner (eds.). *Ecology of Wetlands and Associated Systems.* The Pennsylvania Academy of Science, Lafayette College, Easton, PA. Chapter 4.

Gallagher, J. L., and F. C. Daiber. 1987. Primary production of edaphic algal communities in a Delaware salt marsh. *Limnology and Oceanography* 19:390–95.

Gallagher, J. L., R. J. Reimold, R. A. Linthurst, and W. J. Pfeiffer. 1980. Aerial production, mortality, and mineral accumulation-export dynamics in *Spartina alterniflora* and *Juncus roemerianus* plant stands in a Georgia salt marsh. *Ecology* 61:303–12.

Gallagher, J. L., G. F. Somers, D. M. Grant, and D. M. Seliskar. 1988. Persistent differences in two forms of *Spartina alterniflora:* a common garden experiment. *Ecology* 69:1005–8.

Gambrell, R. P., and W. H. Patrick, Jr. 1978. Chemical and microbial properties of anaerobic soils and sediments. In: D. D. Hook and R. M. M. Crawford (eds.). *Plant Life In Anaerobic Environments.* Ann Arbor Scientific Publications, Ann Arbor, MI. 375–423.

Gammill, S. P., and P. E. Hosier. 1992. Coastal saltmarsh development at southern Topsail Sound, North Carolina. *Estuaries* 15:122–29.

Ganju, N. K., D. H. Schoellhamer, and B. A. Bergamaschi. 2005. Suspended sediment fluxes in a tidal wetland: measurement, controlling factors, and error analysis. *Estuaries* 28:812–22.

Ganong, W. F. 1903. The vegetation of the Bay of Fundy salt and diked marshes: an ecological study. *Botanical Gazzette* 36:161–302, 349–67, 429–55.

Garbary, D. J., A. G. Miller, R. Scrosati, K-Y. Kim, and W. B. Schofield. 2008. Distribution and salinity tolerance of intertidal mosses from Nova Scotia salt marshes. *The Bryologist* 111:282–91.

Garbisch, E. W., Jr., P. B. Woller, and R. J. McCallum. 1975. Salt Marsh Establishment and Development. U.S. Army Corps of Engineers, Coastal Engineering Research Center, Ft. Belvoir, VA. Technical Memorandum no. 52.

Gardner, L. R., and D. E. Porter. 2001. Stratigraphy and geologic history of a southeastern salt marsh basin, North Inlet, South Carolina, USA. *Wetlands Ecology and Management* 9:371–85.

Gardner, L. R., B. R. Smith, and W. K. Michener. 1992. Soil evolution along a forest-salt marsh transect under a regime of slowly rising sea level, southeastern United States. *Geoderma* 55:141–57.

Gardner, L. R., H. W. Reeves, and P. M. Thibodeau. 2002. Groundwater dynamics along forest-marsh transects in a southeastern salt marsh, USA: Description, interpretation and challenges for numerical modeling. *Wetlands Ecology and Management* 10:145–59.

Gauthier, B., and M. Goudreau. 1983. Mares glacielles et non glacielles dans le marais salé de l'Isle-Verte, estuaire du Saint-Laurent, Québec. *Géographie physique et Quaternaire* 37:49–66.

Gedan, K. B., and M. D. Bertness. 2010. Human impacts, old and new, pushing the limits of northeastern high salt marshes. In: J. E. Potente (ed.). *Tidal Marshes of Long Island, New York.* Memoirs of the Torrey Botanical Society, Vol. 26:72–79.

Gedan, K. B., and M. D. Bertness. 2009. Experimental warming causes rapid loss of plant diversity in New England marshes. *Ecology Letters* 12:842–48.

Gedan, K. B., B. R. Silliman, and M. D. Bertness. 2009. Centuries of human-driven change in salt marsh ecosystems. *Annual Review of Marine Science* 1:117–41.

Gedan, K. B., M. L. Kirwan, E. Wolanski, E. B. Barbier, and B. R. Silliman. 2011. The present and future role of coastal wetland vegetation in protecting shorelines: answering recent challenges to the paradigm. *Climate Change* 106:7–29.

Gehrels, W. R., J. R. Kirby, A. Prokoph, R. M. Newnham, E. P. Achterberg, H. Evans, S. Black, and D. B. Scott. 2005. Onset of recent rapid sea-level rise in the western Atlantic Ocean. *Quaternary Science Review* 24:2083–2100.

Genesis Laboratories. 2002. Nutria (*Myocaster coypus*) in Louisiana. Wellington, CO. Prepared for the Louisiana Department of Wildlife and Fisheries. http://brownmarsh.com/data/IV-1/NutriaReport.pdf.

Gerard, V. A. 1999. Positive interactions between cordgrass, *Spartina alterniflora,* and the brown alga, *Ascophyllum nodosum* ecad *scorpioides,* in a Mid-Atlantic coast salt marsh. *Journal of Experiment Marine Biology and Ecology* 239:157–64.

Gilbert, G. K. 1917. Hydraulic-Mining Debris in the Sierra Nevada. U.S. Geological Survey Professional Paper 105.

Gilmore, R. G., D. W. Cooke, and C. J. Donohoe. 1982. A comparison of the fish population and habitat in open and closed salt marsh impoundments in east-central Florida. *Northeast Gulf Science* 5:25–37.

Giroux, J. F., and J. Bédard. 1987. The effects of grazing by greater snow geese on the vegetation of tidal marshes in the St. Lawrence Estuary. *Journal of Applied Ecology* 24:773–88.

Glibert, P. M. 2010. Long-term changes in nutrient loading and stoichiometry and their relationships with changes in the food web and dominant pelagic fish species in the San Francisco Estuary, California. *Reviews in Fisheries Science* 18:211–32.

Glick, P., and J. Clough. 2006. An Unfavorable Tide—Global Warming, Coastal Habitats and Sportfishing in Florida. National Wildlife Federation, Reston, VA and Florida Wildlife Federation, Tallahassee, FL.

Glick, P., J. Clough, and B. Nunley. 2008. Sea-level Rise and Coastal Habitats in the Chesapeake Bay Region. National Wildlife Federation, Reston, VA.

Glick, P., J. Clough, and B. Nunley. 2007. Sea-level Rise and Coastal Habitats in the Pacific Northwest. An Analysis for Puget Sound, Southwestern Washington, and Northwestern Oregon. National Wildlife Federation, Reston, VA.

Glooschenko, V., and P. Grondin. 1988. Wetlands of eastern temperate Canada. In: National Wetlands Working Group. *Wetlands of Canada.* Environment Canada, Sustainable Development Branch, Ottawa, and Polyscience Publications, Montreal, CN. Ecological Land Classification Series no. 24. Chapter 6; 199–248.

Glooschenko, W. A., and I. P. Martini. 1978. Hudson Bay lowlands baseline study. In: Coastal Zone '78. Symposium on Technical, Environmental, Sociometric and Regulatory Aspects of Coastal Zone Management. 1978 March 14–16; San Francisco. American Society of Civil Engineers, Waterway, Port, Coastal and Ocean Division and San Francisco Section, the Conservation Foundation, and U.S. Department of Commerce, Office of Coastal Zone Management. 663–79.

Glooschenko, W. A., I. P. Martini, and K. Clarke-Whistler. 1988. Salt marshes of Canada. In: National Wetlands Working Group. Wetlands of Canada. Environment Canada, Sustainable Development Branch, Ottawa, and Polyscience Publications, Montreal, CN. Ecological Land Classification Series no. 24. Chapter 9; 347–77.

Goals Project. 1999. Baylands Ecosystem Habitat Goals. A Report of Habitat Recommendations Prepared by the San Francisco Bay Area Wetlands Ecosystem Goals Project. U.S. Environmental Protection Agency, San Francisco, CA and San Francisco Bay Regional Water Quality Control Board, Oakland, CA.

Goals Project. 2000. Baylands Ecosystem Species and Community Profiles: Life Histories and Environmental Requirements of Key Plants, Fish and Wildlife. P. R. Olofson (ed.). Prepared by the San Francisco Bay Area Wetlands Ecosystem Goals Project. San Francisco Bay Regional Water Quality Control Board, Oakland, CA. http://66.147.242.191/~sfestuar/userfiles/ddocs/Species_and_Community_Profiles.pdf.

Godfrey, P. J. 1976. Barrier beaches of the East Coast. *Oceanus* 19:27–40.

Godfrey, P. J., and M. M. Godfrey. 1974. The role of overwash and inlet dynamics in the formation of salt marshes on North Carolina barrier islands. In: R. J. Reimold and W. H. Queen (eds.). *Ecology of Halophytes.* Academic Press, New York. 407–27.

Goldenberg, S. B., C. W. Landsea, A. M. Mestas-Nuñez, and W. M. Gray. 2001. The recent increase in Atlantic hurricane activity: causes and implications. *Science* 293:474–79.

Goldstein, W. A. 1971. Environmental Law—consideration must begin to ecological matters in federal agency decisions—Zabel v. Tabb. *Boston College Law Review* 12:674–85.

Good, J. W. 2000. Summary and current status of Oregon's estuarine systems. In: Oregon Progress Board. Oregon State of the Environment 2000 Report. Salem, OR. 33–44.

Good, J. W., J. W. Weber, J. W. Charland, J. V. Olson, and K. A. Chapin. 1998. National

Coastal Zone Management Effective Study: Protecting Estuaries and Coastal Wetlands. Oregon State University, Corvallis, OR. Oregon Sea Grant Special Report PI-98-001.
Good, J. W., J. W. Weber, and J. W. Charland. 1999. Protecting estuaries and coastal wetlands through state coastal zone management programs. *Coastal Management* 27:139–86.
Good, R. E. 1965. Salt marsh vegetation, Cape May, New Jersey. *Bulletin of the New Jersey Academy of Science* 10:1–11.
Good, R. E. 1972. Salt marsh production and salinity. *Bulletin of the Ecological Society of America* 53:22.
Good, R. E., R. W. Hastings, and R. E. Denmark. 1975. An Environmental Assessment of Wetlands: A Case Study of Woodbury Creek and Associated Marshes. Rutgers University, Marine Science Center, New Brunswick, NJ. Technical Report 75-2.
Good, R. E., N. F. Good, and B. R. Frasco. 1982. A review of primary production and decomposition dynamics of the belowground marsh component. In: V. S. Kennedy (ed.). *Estuarine Comparisons*. Academic Press, New York. 139–57.
Goodfriend, G. A., and H. B. Rollins. 1998. Recent barrier island retreat in Georgia: dating exhumed salt marshes by aspartic acid racemization and post-bomb radiocarbon. *Journal of Coastal Research* 14:960–69.
Goodman, P. J., and W. T. Williams. 1961. Investigations into "dieback" in *Spartina townsendii*. AggII. Physiological correlates of "dieback." *Journal of Ecology* 49:391–98.
Goodwin, P., and A. J. Mehta (eds.). 2001. Tidal Wetland Restoration: Physical and Ecological Processes. *Journal of Coastal Research* Special Issue 27.
Goodwin, R. H. 1961. The future: a call to action. In: Connecticut's Coastal Marshes—A Vanishing Resource. Connecticut Arboretum, Connecticut College, New London, CT. Bulletin no. 12. 31–35.
Gordon, D. C., Jr., and P. J. Cranford. 1994. Export of organic matter from macrotidal salt marshes in the upper Bay of Fundy, Canada. In: W. J. Mitsch (ed.). *Global Wetlands: Old World and New.* Elsevier Science B. V., Amsterdam, The Netherlands. 257–63.
Gordon, D. C., Jr., and C. Desplanque. 1983. Dynamics and environmental effects of ice in the Cumberland Basin of the Bay of Fundy (Canada). *Canadian Journal of Fisheries and Aquatic Sciences* 40:1331–42.
Gordon, D. C., Jr., P. J. Cranford, and C. Desplanque. 1985. Observations on the ecological importance of salt marshes in the Cumberland basin, a macrotidal estuary in the Bay of Fundy. *Estuarine, Coastal and Shelf Science* 20:205–27.
Gornitz, V. 2007. Sea Level Rises, after the Ice Melted and Today. Goddard Institute for Space Studies, New York. Science briefs. www.giss.nasa.gov/research/briefs/gornitz_09/.
Gosner, K. L. 1971. *Guide to Identification of Marine and Estuarine Invertebrates.* Cape Hatteras to the Bay of Fundy. Wiley-Interscience, John Wiley & Sons, New York.
Gosselink, J. G. 1980. *Tidal Marshes: The Boundary between Land and Ocean.* U.S. Fish and Wildlife Service, Washington, DC. FWS/OBS-80/15.
Gosselink, J. G. 1984. The Ecology of Delta Marshes of Coastal Louisiana: A Community Profile. U.S. Fish and Wildlife Service, Washington, DC. FWS/OBS-84/09.
Gosselink, J. G., and R. H. Baumann. 1980. Wetland inventories: wetland loss along the United States coast. *Zeitschrift für Geomorphologie N. F. Suppl. Bd.* 34:173–87.
Gosselink, J. G., E. P. Odum, and R. M. Pope. 1973. The Value of the Tidal Marsh. Urban and Regional Development Center, University of Florida, Gainesville, FL. Work Paper no. 3.
Gosselink, J. G., E. P. Odum, and R. M. Pope. 1974. The Value of the Tidal Marsh. Center for Wetland Resources, Louisiana State University, Baton Rouge, LA. Publication no. LSU-SG-74-03.
Gosselink, J. G., J. M. Coleman, and R. E. Stewart, Jr. 1998. Coastal Louisiana. In: M. J. Mac, P. A. Opler, C. E. Puckett Haecker, and P. D. Doran (eds.). *Status and Trends of the Nation's Biological Resources.* Vol. 1. U.S. Department of the Interior, Geological Survey, Reston, VA. 385–436.
Grant, R. R., Jr., and R. Patrick. 1970. Tinicum Marsh as a water purifier. In: J. McCormick, R. R. Grant, Jr., and R. Patrick. *Two Studies of Tinicum Marsh.* The Conservation Foundation, Washington, DC. 105–23.
Gratton, L., and C. Dubreuil. 1990. Portrait de la végétation et de la flore du Saint-Laurent. Ministère de l'Environnement, Direction de

la conservation et du patriomoine écologique, Québec, QC.

Green, J. 1971. *The Biology of Estuarine Animals.* University of Washington Press, Seattle, WA.

Greenberg, R., and J. E. Maldonado. 2006. Diversity and endemism in tidal-marsh vertebrates. *Studies in Avian Biology* 32:32–53.

Greenberg, R., J. E. Maldonado, S. Droege, and M. V. McDonald (eds.). 2006. Terrestrial Vertebrates of Tidal Marshes: Evolution, Ecology, and Conservation. Cooper Ornithological Society. Studies in Avian Biology no. 32. elibrary.unm.edu/sora/Condor/cooper/sab_032.pdf.

Greer, K., and D. Snow. 2003. Vegetation type conversion in Los Penasquitos Lagoon, California: an examination of the role of watershed urbanization. *Environmental Management* 31:489–503.

Gregory, C., W. Norden, E. Stancioff, and L. Watling. 1998. *A Guide to Common Marine Organisms along the Coast of Maine.* Maine/New Hampshire Sea Grant, University of Maine, Orono, ME.

Gregory, R. 2004. Eelgrass as nursery habitat for juvenile fish in the coastal marine environment. In: A. R. Hanson (ed.). Status and Conservation of Eelgrass (*Zostera marina*) in Eastern Canada. Canadian Wildlife Service, Atlantic Region, Sackville, NB. Technical Report no. 412. 18–19.

Greller, A. M. 2010. Salt marsh productivity. In: J. E. Potente (ed.). *Tidal Marshes of Long Island, New York.* Memoirs of the Torrey Botanical Society, Vol. 26:102–11.

Grenier, L., M. Marvin-DiPasquale, D. Drury, L. Kieu, E. Kakouros, J. Agee, A. Melwani, S. Bezalel, A. Robinson, L. Windham-Myers, J. Hunt, and J. Collins. 2010. South Baylands Mercury Project. Final report prepared for the California State Coastal Conservancy by the San Francisco Estuary Institute, U.S. Geological Survey, and Santa Clara Valley Water District. www.sfei.org/sites/default/files/SBMP_Final%20Report%2010FEB2010.pdf.

Griffin, T. M., and M. C. Rabenhorst. 1989. Processes and rates of pedogenesis in some Maryland tidal marsh soils. *Soil Science Society of America Journal* 53:862–70.

Griffith, R. E. 1940. Waterfowl management of Atlantic Coast refuges. *North American Wildlife Conference Transactions* 5:373–77.

Grinsted, A., J. C. Moore, and S. Jevrejeva. 2010. Reconstructing sea level from paleo and projected temperatures 200 to 2100 AD. *Climate Dynamics* 34:461–72.

Griswold, T. 1988. Physical factors and competitive interactions affecting salt marsh vegetation. San Diego State University, San Diego, CA. M.S. thesis.

Grossinger, R. M., E. D. Stein, K. N. Cayce, R. A. Askevold, S. Dark, and A. A. Whipple. 2011. Historical Wetlands of the Southern California Coast: An Atlas of US Coast Survey T-sheets, 1851–1889. San Francisco Estuary Institute Contribution no. 586 and Southern California Coastal Water Research Project Technical Report no. 589. www.sccwrp.org:8060/pub/download/DOCUMENTS/TechnicalReports/589_SoCalTsheetAtlas.pdf.

Guntenspergen, G. R., D. R. Cahoon, J. Grace, G. D. Steyer, S. Fournet, M. A. Townson, and A. L. Foote. 1995. Disturbance and recovery of the Louisiana coastal marsh landscape from the impacts of Hurricane Andrew. *Journal of Coastal Research* Special Issue 21:324–39.

Gunter, G. 1967. Some relationships of estuaries to the fisheries of the Gulf of Mexico. In: G. H. Lauff (ed.). *Estuaries.* American Association for the Advancement of Science Publication no. 83. 621–38.

Gustafson, D. J., J. Kilheffer and B. R. Silliman. 2006. Relative effects of *Littoraria irrorata* and *Prokelisia marginata* on *Spartina alterniflora. Estuaries and Coasts* 29:639–44.

Hacker, S. D., and M. D. Bertness. 1999. Experimental evidence for factors maintaining plant species diversity in a New England salt marsh. *Ecology* 80:2064–73.

Hackney, C. T., and T. D. Bishop. 1981. A note on the relocation of marsh debris during a storm surge. *Estuarine, Coastal and Shelf Science* 12:621–24.

Hackney, C. T., and W. J. Cleary. 1987. Saltmarsh loss in southeastern North Carolina lagoons: importance of sea level rise and inlet dredging. *Journal of Coastal Research* 3:93–97.

Hackney, C. T., and G. F. Yelverton. 1990. Effects of human activities and sea level rise on wetland ecosystems in the Cape Fear River estuary, North Carolina, USA. In: D. F. Whigham, R. E. Good, and J. Kvet (eds.). *Wetland Ecology and Management: Case Studies*. Kluwer Academic Publishers, Dordrecht, The Netherlands. 55–61.

Hackney, C. T., G. B. Avery, L. A. Leonard, M. Posey, and T. Alphin. 2007. Biological, chemical, and physical characteristics of tidal freshwater swamp forests of the lower Cape Fear River/Estuary, North Carolina. In: W. H. Conner, T. W. Doyle, and K. W. Krause (eds.). *Ecology of Tidal Freshwater Forested Wetlands of the Southeastern United States*. Springer, Dordrecht, The Netherlands. Chapter 8; 183–221.

Hagen, S. M., S. A. Brown, and K. W. Able. 2007. Production of mummichog (*Fundulus heteroclitus*): response in marshes treated fro common reed (*Phragmites australis*) removal. *Wetlands* 27:54–67.

Haines, E. B. 1976. Stable carbon isotope ratios in biota, soils and tidal water of a Georgia salt marsh. *Estuarine and Coastal Marine Science* 4:609–16.

Haines, E. B., and C. L. Montague. 1979. Food sources of estuarine invertebrates analyzed using $_{13}C/_{12}C$ ratios. *Ecology* 60:48–56.

Hall, F. C. 2001a. Ground-Based Photographic Monitoring. Portland, OR: U.S. Department of Agriculture, Forest Service, Pacific Northwest Research Station. General Technical Report PNW-GTR-503.

Hall, J. V. 2009. Tidal freshwater wetlands of Alaska. In: A. Barendregt, D. F. Whigham, and A. H. Baldwin (eds.). *Tidal Freshwater Wetlands*. Backhuys Publishers, Leiden, The Netherlands. Chapter 16; 179–84.

Hall, J. V., W. E. Frayer, and B. O. Wilen. 1994. Status of Alaska Wetlands. U.S. Fish and Wildlife Service, Alaska Region, Anchorage, AK.

Hall, J. W. 2001b. Wetlands Mitigation Program in South Carolina. http://fhwa.dot.gov/environment/greenerroadsides/sum01_3.htm.

Hall, S. L., and F. M. Fisher, Jr. 1985. Annual productivity and extracellular release of dissolved organic compounds by the epibenthic algal community of a brackish marsh. *Journal of Phycology* 21:277–81.

Halpin, P. M., and K. L. M. Martin. 1999. Aerial respiration in the salt marsh fish *Fundulus heteroclitus* (Fundulidae). *Copeia* 1999:743–48.

Hansen, D. J., P. Dayanandan, P. B. Kaufman, and J. D. Brotherson. 1976. Ecological adaptations of salt marsh grass, *Distichlis spicata* (Gramineae), and environmental factors affecting its growth and distribution. *American Journal of Botany* 63:635–50.

Hansen, J. E. 2007. Scientific reticence and sea level rise. *Environmental Research Letters* 2 (April–June 2007) 024002, doi: 10.1088/1748-9326/2/2/024002. http://iopscience.iop.org/1748-9326/2/2/024002/fulltext.

Hansknecht, K. A. 2008. Lingual luring by mangrove saltmarsh snakes (*Nerodia clarkii compressicauda*). *Journal of Herpetology* 42:9–15.

Hanson, A., L. Swanson, D. Ewing, G. Grabas, S. Meyer, L. Ross, M. Watmough, and J. Kirkby. 2008. Wetland ecological functions assessment: an overview of approaches. Canadian Wildlife Service, Atlantic Region. Technical Report Series no. 497. www.wetkit.net/docs/WA_TechReport497_en.pdf.

Hanson, A. R. 2004. Bird Use of Salt Marsh Habitat in the Maritime Provinces. Canadian Wildlife Service, Atlantic Region, Sackville, NB. Technical Report no. 414.

Hanson, A. R., and L. Calkins. 1996. Wetlands of the Maritime Provinces: Revised Documentation for the Wetlands Inventory. Canadian Wildlife Service, Atlantic Region, Sackville, NB. Technical Report no. 267.

Hanson, S. R., D. T. Osgood, and D. J. Yozzo. 2002. Nekton use of a *Phragmites australis* marsh on the Hudson River, New York, USA. *Wetlands* 22:326–37.

Hardaway, C. S., Jr., D. A. Milligan, K. P. O'Brien, C. A. Wilcox, J. Shen, and C. H. Hobbs, III. 2009. Encroachment of Sills onto State-owned Bottom: Design Guidelines for Chesapeake Bay. Virginia Institute of Marine Science, College of William & Mary, Gloucester Point, VA. www.jefpat.org/Documents/Sill_Encroachment.pdf.

Hardaway, C. S., Jr., D. A. Milligan, and K. Duhring. 2010. Living Shoreline Design Guidelines for Shore Protection in Virginia's Estuarine Environments, Version 1. Virginia Institute of Marine Science, College of William & Mary, Gloucester Point, VA. http://web.vims.edu/physical/research/shoreline/docs/LS_Design_final.pdf.

Hardisky, M. A., and V. Klemas. 1983. Tidal wetlands natural and human-made changes from 1973 to 1979 in Delaware: mapping techniques and results. *Environmental Management* 7:1–6.

Harper, R. M. 1918. Some dynamic studies of Long Island vegetation. *Plant World* 21:38–46.

Harrington, B. R. 2008. Coastal Inlets as Strategic Habitats for Shorebirds in the Southeastern United States. U.S. Army Engineer Research and Development Center, Vicksburg, MS. ERDC TN-DOER-E25.

Harris, V. T. 1953. Ecological relationships of meadow voles and rice rats in tidal marshes. *Journal of Mammalogy* 34:479–87.

Harrison, J. W., and P. Stango III. 2003. Shrubland Tidal Wetland Communities of Maryland's Eastern Shore: Identification, Assessment, and Monitoring. Maryland Natural Heritage Program, Maryland Department of Natural Resources, Annapolis, MD.

Harrison, R. W., and W. M. Kollmorgen. 1947. Past and prospective drainage reclamations in the coastal marshlands of the Mississippi River delta. *The Journal of Land & Public Utility Economics* 23:297–320.

Harshberger, J. W. 1909. The vegetation of salt marshes and of the salt and freshwater ponds of northern coastal New Jersey. *Proceedings Academy of Natural Sciences Philadelphia* 61:373–400.

Harshberger, J. W., and V. G. Burns. 1919. The vegetation of the Hackensack Marsh: a typical American fen. *Wagner Free Institute of Science Philadelphia Transactions* 9:1–35.

Hartig, E. K., V. Gornitz, A. Kolker, F. Mushacke, and D. Fallon. 2002. Anthropogenic and climate-change impacts on salt marshes of Jamaica Bay, New York City. *Wetlands* 22:71–89.

Hartig, E. K., F. Mushacke, D. Fallon, and A. Kolker. 2003. A wetlands climate impact assessment for the metropolitan East Coast region. Columbia University, Center for Climate Systems Research, New York.

Hartman, F. E. 1963. Estuarine wintering habitat for black ducks. *Journal of Wildlife Management* 27:339–47.

Hartman, J. M., H. Caswell, and I. Valiela. 1983. Effects of wrack accumulation on salt marsh vegetation. Proceedings of the 17th European Marine Biology Symposium, Brest, France. *Oceanologica Acta.* 99–102.

Hastings, R. W., and R. E. Good. 1977. Population analysis of the fishes of a freshwater tidal tributary of the lower Delaware River. *Bulletin of the New Jersey Academy of Science* 22:13–20.

Hatcher, A., and D. G. Patriquin. 1981. Salt Marshes of Nova Scotia: A Status Report of the Salt Marsh Working Group. Institute for Resource and Environmental Studies and Department of Biology, Dalhousie University, Halifax, NS.

Hatvany, M. G. 2001. "Wedded to the Marshes": Salt marshes and socio-economic differentiation in early Prince Edward Island. *Acadiensis* 30(2):40–55.

Hatvany, M. G. 2003. *Marshlands: Four Centuries of Environmental Change on the Shores of the St. Lawrence.* Les Presses de l'Université Laval, Sainte-Foy, Quebec, Canada.

Hatvany, M. G. 2009. Paysages de marais: quatre siècles de relations entre l'humain et les marais du Kamouraska. La Pocatière: Société historique de la Côte-du-Sud et Ruralys.

Havens, K. J., L. M. Varnell, and B. D. Watts. 2002. Maturation of a constructed tidal marsh relative to two natural reference marshes over 12 years. *Ecological Engineering* 18:305–15.

Havill, D. C., A. Ingold, and J. Pearson. 1985. Sulfide tolerance in coastal halophytes. *Vegetatio* 62:279–85.

Hawkins, A. S., R. C. Hanson, H. K. Nelson, and H. M. Reeves (eds.). 1984. *Flyways: Pioneering Waterfowl Management in North America.* U.S. Department of the Interior, Fish and Wildlife Service, Washington, DC.

Hayden, B. P., and N. R. Hayden. 2003. Decadal and century-long changes in storminess at long-term ecological research sites. In: D. Greenland, D. C. Goodin, and R. C. Smith (eds.). *Climate Variability and Ecosystem Response at Long-Term Ecological Research Sites.* Oxford University Press, New York. 262–85.

Heard, R. W. 1982. Guide to Common Tidal Marsh Invertebrates of the Northeastern Gulf of Mexico. Mississippi-Alabama Sea Grant Consortium. MASGP-79-004.

Heck, K. L., Jr., K. W. Able, M. P. Fahay, and C. T. Roman. 1989. Fishes and decapod crustaceans of Cape Cod eelgrass meadows: species composition, seasonal abundance patterns, and comparison with unvegetated substrates. *Estuaries* 12:59–65.

Heck, K. L., Jr., K. W. Able, C. T. Roman, and M. P. Fahay. 1995. Composition, abundance, biomass, and production of macrofauna in a New England estuary: comparisons among eelgrass meadows and other nursery habitats. *Estuaries* 18:379–89.

Helfield, J. M., and R. J. Naiman. 2006. Keystone interactions: salmon and bear in riparian forests of Alaska. *Ecosystems* 9:167–80.

Hellings, S. E., and J. L. Gallagher. 1992. The effects of salinity and flooding on *Phragmites australis*. *Journal of Applied Ecology* 29:41–49.

Herdendorf, C. E. 1990. Great Lakes estuaries. *Estuaries* 13:493–503.

Herke, W. H., and B. D. Rogers. 1984. Comprehensive estuarine nursery study completed. *Fisheries* (Bethesda) 9:12–16.

Herke, W. H., and B. D. Rogers. 1989. Threats to coastal fisheries. In: W. G. Duffy and D. Clark (eds.). Marsh Management in Coastal Louisiana: Effects and Issues. Proceedings of a symposium. U.S. Fish and Wildlife Service and Louisiana Department of Natural Resources. U.S. Fish and Wildlife Service Biological Report 89(22). 196–212.

Herke, W. H., E. E. Knudsen, P. A. Knudsen, and B. D. Rogers. 1992. Effects of semi-impoundment of Louisiana marsh on fish and crustacean nursery use and export. *North American Journal of Fisheries Management* 12:151–60.

Hester, M. V., I. A. Mendelssohn, and K. L. McKee. 1996. Intraspecific variation in salt tolerance and morphology in the coastal grass *Spartina patens*. *American Journal of Botany* 83:1521–27.

Hester, M. V., I. A. Mendelssohn, and K. L. McKee. 1998. Intraspecific variation in salt tolerance and morphology in *Panicum hemitomon* and *Spartina alterniflora* (Poaceae). *International Journal of Plant Science* 159:127–38.

Heusser, C. J. 1949. History of an estuarine bog at Secaucus, New Jersey. *Bulletin of the Torrey Botanical Club* 76:385–406.

Heusser, L. E., C. J. Heusser, and D. Weiss. 1975. Man's influence on the development of the estuarine marsh, Flax Pond, Long Island, New York. *Bulletin of the Torrey Botanical Club* 102:61–66.

Hewlett, R., and J. Bimie. 1996. Holocene environmental change in the inner Severn estuary, UK: an example of the response of estuarine sedimentation to relative sea-level change. *The Holocene* 6:49–61.

Hibbard, B. H. 1965. *A History of the Public Land Policies*. University of Wisconsin Press, Madison, WI.

Hicklin, P. W. 1987. The migration of shorebirds in the Bay of Fundy. *Wilson Bulletin* 99:540–70.

Hicks, S. D. 2006. Understanding Tides. U.S. Department of Commerce, National Oceanic and Atmospheric Administration, National Ocean Service. http://tidesandcurrents.noaa.gov/publications/Understanding_Tides_by_Steacy_finalFINAL11_30.pdf.

Hicks, S. D., H. A. Debaugh, and L. E. Hickman. 1983. Sea level variation for the United States, 1955–1980. U.S. Department of Commerce, National Oceanic Atmospheric Administration, Rockville, MD.

Hicks, S. D., R. L. Sillcox, C. R. Nichols, B. Via, and E. C. McCray. 2000. Tide and current glossary. National Oceanic and Atmospheric Administration, National Ocean Service, Silver Springs, MD.

Higgins, E. A. T., R. D. Rappleye, and R. G. Brown. 1971. The Flora and Ecology of Assateague Island. University of Maryland, Agricultural Experiment Station, College Park, MD. Bulletin A-172.

Hilgartner, W. B., and G. S. Brush. 2006. Prehistoric habitat stability and post-settlement habitat change in a Chesapeake Bay freshwater tidal wetland, USA. *The Holocene* 16:479–94.

Hilliard, S. B. 1975. The tidewater rice plantation: an ingenious adaptation to nature. *Geoscience and Man* 12:57–66.

Hilliard, S. B. 1978. Antebellum tidewater rice culture in South Carolina and Georgia. In: J. R. Gibson (ed.). *European Settlement and Development in North America*. University of Toronto Press, Toronto, Ontario. 91–115.

Hinkle, R., and W. J. Mitsch. 2005. Salt marsh vegetation recovery at salt hay farm wetland restoration sites on Delaware Bay. *Ecological Engineering* 25:240–51.

Hoff, R. Z. 1995. Responding to Oil Spills in Coastal Marshes: The Fine Line between Help and Hindrance. National Oceanic and Atmospheric Administration, Hazardous Materials Response and Assessment Division, Seattle, WA. HAZMAT Report 96-1.

Hoffman, J. A., J. Katz, and M. D. Bertness. 1984. Fiddler crab deposit-feeding and meiofaunal abundance in salt marsh habitats. *Journal of Experimental Marine Biology and Ecology* 82:161–74.
Hoffpauer, C. M. 1968. Burning for coastal marsh management. In: J. D. Newson (ed.). *Proceedings of the Marsh and Estuary Management Symposium.* 1967 July 19–20; Louisiana State University, Baton Rouge, LA. 134–39.
Hogarth, P. J. 2007. *The Biology of Mangroves and Seagrasses.* Oxford University Press, New York.
Holdredge, C., M. D. Bertness, and A. H. Altieri. 2008. Role of crab herbivory in die off of New England salt marshes. *Conservation Biology* 23:672–79.
Holgate, S. J. 2007. On the decadal rates of sea level change during the twentieth century. *Geophysical Research Letters* 34, L01602, 4 pp.
Homer, M. L. 1988. The impact of habitat loss on freshwater fish populations. University of South Carolina, Columbia, SC. Ph.D. dissertation.
Hood, W. G. 2007. Large woody debris influences vegetation zonation in an oligohaline tidal marsh. *Estuaries and Coasts* 30:441–50.
Hopkinson, C. S., J. G. Gosselink, and R. T. Parrondo. 1980. Productivity of coastal Louisiana marsh plants calculated from phenometric techniques. *Ecology* 61:1091–98.
Howard, R. 2010. Intraspecific variation in growth of marsh macrophytes in response to salinity and soil type: implications for wetland restoration. *Estuaries and Coasts* 33:127–38.
Howard, R. J., and I. A. Mendelssohn. 1999a. Salinity as a constraint on growth of oligohaline marsh macrophytes. I. Species variation in stress tolerance. *American Journal of Botany* 86:785–94.
Howard, R. J., and I. A. Mendelssohn. 1999b. Salinity as a constraint on growth of oligohaline marsh macrophytes. II. Salt pulses and recovery potential. *American Journal of Botany* 86:795–806.
Howard, R. J., and I. A. Mendelssohn. 2000. Structure and composition of oligohaline marsh plant communities exposed to salinity pulses. *Aquatic Botany* 68:143–64.
Howard, R. J., and P. S. Rafferty. 2006. Clonal variation in response to salinity and flooding stress in four marsh macrophytes of the Northern Gulf of Mexico, USA. *Environmental and Experimental Botany* 56:301–13.
Howes, B. L., and D. D. Goehringer. 1996. Ecology of Buzzards Bay: An Estuarine Profile. U.S. Department of the Interior, National Biological Service, Washington, DC. Biological Report 31.
Howes, B. L., R. W. Howarth, J. M. Teal, and I. Valiela. 1981. Oxidation-reduction potentials in salt marshes: spatial patterns and interactions with primary production. *Limnology and Oceanography* 26:350–60.
Howes, B. L., J. W. H. Dacey, and D. D. Goehringer. 1986. Factors controlling the growth form of *Spartina alterniflora:* feedbacks between above-ground production, sediment oxidation, nitrogen and salinity. *Journal of Ecology* 74:881–98.
Hsieh, Y. P. 1996. Assessing aboveground net primary production of vascular plants in marshes. *Estuaries* 19:82–85.
Hsieh, Y. P. 2001. Coastal salt barren as indicator to recent sea level change and wetland carbon dynamics. In: *Book of Abstracts,* Seventh International Symposium on the Biodiversity of Wetlands; 2001 June 17–20; Duke University Wetland Center, Duke University, Durham, NC. 49.
Hunter, W. C., W. Golder, S. Melvin, and J. Wheeler. 2006. Southeast United States Regional Waterbird Conservation Plan. U.S. Fish and Wildlife Service, Atlanta, GA. www.waterbirdconservation.org/pdfs/regional/seusplanfinal906.pdf.
Hurt, G. W., P. M. Whited, and R. F. Pringle (eds.). 2006. Field Indicators of Hydric Soils in the United States. Prepared in cooperation with the National Technical Committee for Hydric Soils. USDA Natural Resources Conservation Service, Fort Worth, TX.
Hussein, A. H., and M. C. Rabenhorst. 2001. Modeling the impact of tidal inundation on submerging coastal landscapes of the Chesapeake Bay. *Soil Science Society of America Journal* 65:932–41.
Hussein, A. H., and M. C. Rabenhorst. 2002. Modeling of nitrogen sequestration in coastal marsh soils. *Soil Science Society of America Journal* 66:324–30.
Hussein, A. H., M. C. Rabenhorst, and M. L. Tucker. 2004. Modeling of carbon

sequestration in coastal marsh soils. *Soil Science Society of America Journal* 68:1786–95.

Hutchinson, G. E. 1975. *A Treatise on Limnology*. Volume III. *Limnological Botany*. John Wiley & Sons, New York.

Hutchinson, I. 1988. Salinity Tolerance of Plants of Estuarine Wetlands and Associated Uplands. Washington Department of Ecology, Shorelands and Coastal Zone Management Program, Olympia, WA. www.ecy.wa.gov/biblio/0706018.html.

Ingold, A., and D. C. Havill. 1984. The influence of sulphide on the distribution of higher plants in salt marshes. *Journal of Ecology* 72:1043–54.

[IPCC] Intergovernmental Panel on Climate Change. 1996. *Climate Change 1995: The Science of Climate Change*. Cambridge University Press, New York.

[IPCC] Intergovernmental Panel on Climate Change. 2001. *Climate Change 2001: The Scientific Basis*. Cambridge University Press, New York.

[IPCC] Intergovernmental Panel on Climate Change. 2007. Climate Change Synthesis Report. Contribution of Working Groups I, II and III to the Fourth Assessment Report of the Intergovernmental Panel on Climate Change (Core Writing Team, R. K Pachauri and A. Reisinger [eds.]). IPCC, Geneva, Switzerland. www.ipcc.ch/publications_and_data/ar4/syr/en/contents.html.

International Marine. 2008. *Tide Current Tables 2007 East Coast of North and South America including Greenland*. Camden, ME.

Irlandi, E. A., and M. K. Crawford. 1997. Habitat linkages: the effect of intertidal saltmarshes and adjacent subtidal habitats on abundance, movement, and growth of an estuarine fish. *Oecologia* 110:222–30.

Jacobs, A., E. McLaughlin, and K. Havens. 2010. Mid-Atlantic Tidal Wetland Rapid Assessment Method. Version 3.0. Delaware Department of Natural Resources and Environmental Control, Dover, DE; Maryland Department of Natural Resources, Annapolis, MD; and Virginia Institute of Marine Science, Gloucester Point, VA. www.dnrec.delaware.gov/Admin/DelawareWetlands/Documents/Tidal%20Rapid_Protocol%203.0%20Jun10.pdf.

Jacobson, H. A., and G. L. Jacobson, Jr. 1989. Variability of vegetation in tidal marshes of Maine, U.S.A. *Canadian Journal of Botany* 67:230–38.

Jaffe, B. E., R. E. Smith, and L. Zink-Torresan. 1998. Sedimentation and Bathymetric Change in San Pablo Bay, 1856–1863. U.S. Geological Survey, Menlo Park, CA. U.S. Geological Survey Open File Report 98-759. http://geopubs.wr.usgs.gov/open-file/of98-759/of98-759.pdf.

James-Pirri, M-J., H. S. Ginsberg, R. M. Erwin, and J. Taylor. 2009. Effects of open marsh water management on numbers of larval salt marsh mosquitoes. *Journal of Medical Entomology* 46:1392–99.

Jarvis, J. C. 2010. Vertical Accretion Rates in Coastal Louisiana: A Review of the Scientific Literature. U.S. Army Engineer Research and Development Center, Vicksburg, MS. ERDC/EL TN-10-5. www.mvd.usace.army.mil/lcast/pdfs/Projects/33_Marsh%20Vertical%20Acc_sep10.pdf.

Jassby, A. D., W. J. Kimmerer, S. G. Monismith, C. Armor, J. E. Cloern, T. M. Powell, J. R. Schubel, and T. J. Vendlinski. 1995. Isohaline position as a habitat indicator for estuarine populations. *Ecological Applications* 5:272–89.

Jean, M., and G. Létourneau. 2011. Changes to the Wetlands of the St. Lawrence River from 1970 to 2002. Environment Canada, Science and Technology Branch, Québec Water Quality Monitoring and Surveillance Section, Montréal, QC. Technical Report No. 511. www.ec.gc.ca/Publications/E7C22846-04FE-4D8C-96BA-4E8DF853101D%5CChangesToTheWetlandsOfTheStLawrenceRiverFrom1970To2002.pdf.

Jean, M., G. Létourneau, C. Lavoie, and F. Delisle. 2002. Freshwater Wetlands and Exotic Species. Monitoring the State of the St. Lawrence River. St. Lawrence Vision 2000 Coordination Office, Québec, QC.

Jennings, A., T. Jennings, and B. Bailey. 2003. Estuary Management in the Pacific Northwest. An Overview of Programs and Activities in Washington, Oregon, and Northern California. Oregon Sea Grant, Oregon State University, Corvallis, OR.

Johnson, A. F. 1985. *A Guide to the Plant Communities of the Napeague Dunes*. Printed for the author by The Mad Printers of Mattituck, New York.

Johnson, B. J., K. A. Moore, C. Lehmann, C. Bohlen, and T. A. Brown. 2007. Middle to late Holocene fluctuations of C3 and C4 vegetation in a northern New England salt marsh, Sprague Marsh, Phippsburg, Maine. *Organic Geochemistry* 38:394–403.

Johnson, D. 1925. *The New England Acadian Shoreline*. Hafner Publications, New York.

Johnson, D. S., and H. H. York. 1915. *The Relation of Plants to Tide Levels*. Carnegie Inst. Wash. Publ. 206, New York.

Johnston, J. B., M. C. Watzin, J. A. Barras, and L. R. Handley. 1995. Gulf of Mexico coastal wetlands: case studies of loss trends. In: E. T. LaRoe, G. S. Farris, C. E. Puckett, P. D. Doran, and M. J. Mac (eds.). Our Living Resources: A Report to the Nation on the Distribution, Abundance, and Health of U.S. Plants, Animals, and Ecosystems. U.S. Department of the Interior, National Biological Service, Washington, DC. 269–77.

Johnston, R. J., G. Magnusson, M. J. Mazzotta, and J. J. Opaluch. 2002. Combining economic and ecological indicators to prioritize salt marsh restoration actions. *American Journal of Agricultural Economics* 84:1362–70.

Jorgenson, M. T., and D. Dissing. 2010. Landscape Changes in Coastal Ecosystems, Yukon-Kuskokwim Delta. ABR, Inc.-Environmental Services, Fairbanks, AK. Prepared for the U.S. Fish and Wildlife Service, National Wetlands Inventory Program, Region 7, Anchorage, AK.

Jorgenson, M. T., and J. E. Roth. 2010. Landscape Characterization and Mapping for the Yukon-Kuskokwim Delta, Alaska. ABR, Inc.-Environmental Services, Fairbanks, AK. Prepared for the U.S. Fish and Wildlife Service, National Wetlands Inventory Program, Region 7, Anchorage, AK.

Jorgenson, M. T., G. V. Frost, A. E. Miller, P. Spencer, M. Shepard, B. Mangipone, C. Moore, and C. Lindsay. 2010. Monitoring Coastal Salt Marshes in the Lake Clark and Katmai National Parklands of the Southwest Alaska Network. National Park Service, Ft. Collins, CO. Natural Resources Technical Report NPS/SWAN/NRTR-2010/338.

Joseph, V., A. Locke, and J-.G. Godin. 2004. Characterization and habitat use of eelgrass in Kouchibouguac Estuary, New Brunswick. In: A. R. Hanson (ed.). Status and Conservation of Eelgrass (*Zostera marina*) in Eastern Canada. Canadian Wildlife Service, Atlantic Region, Sackville, NB. Technical Report no. 412. 17.

Josselyn, Michael. 1983. The Ecology of San Francisco Bay Tidal Marshes: A Community Profile. US Fish and Wildlife Service, Washington, DC. Biological Services Program FWS/OBS-83/23.

Judd, F. W., and R. I. Lonard. 2002. Species richness and diversity of brackish and salt marshes in the Rio Grande Delta. *Journal of Coastal Research* 18:751–59.

Judd, F. W., and R. I. Lonard. 2004. Community ecology of freshwater, brackish and salt marshes of the Rio Grande delta. *Texas Journal of Science* 56:103–22.

Kane, R. 2001. *Phragmites* use by birds in New Jersey. *Phragmites:* a dissenting opinion. New Jersey Audubon Society Opinion: May 2001.

Kaplan, W., I. Valiela, and J. M. Teal. 1979. Denitrification in a salt marsh ecosystem. *Limnology and Oceanography* 24:726–34.

Karl, T. R., J. M. Melillo, T. C. Peterson, and S. J. Hassol (eds.). 2009. *Global Climate Change Impacts in the United States*. U.S. Global Climate Change Research Program. Cambridge University Press, New York.

Kastler, J. A., and P. L. Wiberg. 1996. Sedimentation and boundary changes of Virginia salt marshes. *Estuarine, Coastal and Shelf Science* 42:683–700.

Kathilankal, J. C., T. J. Mozdzer, J. D. Fuentes, P. D'Odorico, K. J. McGlathery and J. C. Zieman. 2008. Tidal influences on carbon assimilation by a salt marsh. Environmental Research Letters 3 (October–December 2008) 044010, doi: 10.1088/1748-9326/3/4/044010.

Kathiresan, K., and S. Z. Qasim. 2005. *Biodiversity of Mangrove Ecosystems*. Hindustan Publishing Corporation, New Delhi, India.

Katona, S. K., V. Rough, and D. T. Richardson. 1993. *A Field Guide to Whales, Porpoises, and Seals from Cape Cod to Newfoundland.* Smithsonian Institution Press, Washington, DC.

Kauffman-Axelrod, J. L., and S. J. Steinberg. 2010. Development and application of an automated GIS based evaluation to prioritize wetland restoration opportunities. *Wetlands* 30:437–48.

Kavenagh, W. K. 1980. Vanishing Tidelands: Land Use and Law in Suffolk County, N.Y.,

1650–1979. New York Sea Grant Institute Publication RS-80-28.
Kaye, C. A., and E. S. Barghoorn. 1964. Late Quaternary sea-level change and crustal rise at Boston, Massachusetts with notes on the autocompaction of peat. *Geological Society of America Bulletin* 75:63–80.
Kearney, M. S., and J. C. Stevenson. 1991. Island land loss and marsh vertical accretion rate evidence for historical sea-level changes in Chesapeake Bay. *Journal of Coastal Research* 7:403–15.
Kearney, M. S., R. E. Grace, and J. C. Stevenson. 1988. Marsh loss in Nanticoke Estuary, Chesapeake Bay. *The Geographical Review* 76:205–20.
Kearney, M. S., J. C. Stevenson, and L. G. Ward. 1994. Spatial and temporal changes in marsh vertical accretion rates at Monie Bay: implications for sea-level rise. *Journal of Coastal Research* 10:1010–20.
Kearney, M. S., A. S. Rogers, J. R. G. Townshend, E. Rizzo, D. Stutzer, J. C. Stevenson, and D. Sundborg. 2002. Landsat imagery shows decline of coastal marshes in Chesapeake and Delaware Bays. *Eos, Transactions, American Geophysical Union* 83:173, 177–78.
Kearney, T. H., Jr. 1900. The plant covering of Okracoke Island: A study in the ecology of the North Carolina strand vegetation. *U.S. Herbarium* 5:261–319.
Keeland, B. D., and J. M. McCoy. 2007. Plant community composition of a tidally influenced, remnant Atlantic white cedar stand in Mississippi. In: W. H. Conner, T. W. Doyle, and K. W. Krause (eds.). *Ecology of Tidal Freshwater Forested Wetlands of the Southeastern United States.* Springer, Dordrecht, The Netherlands. Chapter 4; 89–111.
Keene, H. W. 1971. Postglacial submergence and salt marsh evolution in New Hampshire. *Maritime Sediments* 7:64–68.
Keiper, R. R. 1990. Biology of large grazing mammals on the Virginia barrier islands. *Virginia Journal of Science* 41:353–63.
Kelley, J. T., D. F. Belknap, G. L. Jacobson, Jr., and H. A. Jacobson. 1988. The morphology and origins of salt marshes along the glaciated coastline of Maine, USA. *Journal of Coastal Research* 4:649–65.
Kelley, R. L. 1959. *Gold vs. Grain.* A. H. Clark Company, Glendale, CA.
Kelly, N. M. 2001. Changes to the landscape pattern of coastal North Carolina wetlands under the Clean Water Act, 1984–1992. *Landscape Ecology* 16:3–16.
Kemp, A. C., B. P. Horton, S. J. Culver, D. R. Corbett, O. van de Plassche, W. R. Gehrels, B. D. Douglas, and A. Parnell. 2009. The timing and magnitude of recent accelerated sea-level rise (North Carolina, USA). *Geology* 37:1035–38.
Kennish, M. J. 2001. Coastal salt marsh systems in the U.S.: a review of anthropogenic impacts. *Journal of Coastal Research* 17:731–48.
Kent, M., and P. Coker. 2002. *Vegetation Description and Analysis: A Practical Approach.* John Wiley & Sons, New York.
Kentula, M. E., R. P. Brooks, S. E. Gwin, C. C. Holland, A. D. Sherman, and J. C. Sifneos. 1992. An approach to improving decision making in wetland restoration and creation. U.S. Environmental Protection Agency, Washington, DC. EPA/600/R-92/150.
Khan, H., and G. S. Brush. 1994. Nutrient and metal accumulation in a freshwater tidal marsh. *Estuaries* 17:345–60.
Kiehl, K., P. Esselink, and J. P. Bakker. 1997. Nutrient limitation and plant species composition in temperate salt marshes. *Oecologia* 111:325–30.
King, D. M., and E. W. Price. 2004. Developing Defensible Wetland Mitigation Ratios. University of Maryland, Center for Environmental Science, Solomons, MD. Prepared for National Oceanic and Atmospheric Administration, Office of Habitat Conservation, Habitat Protection Division, Silver Spring, MD. www.nero.noaa.gov/hcd/socio/FinalNOAA%20Wetland%20mitigation%20ratio%20guidance.pdf.
King, D. M., L. A. Wainger, C. C. Bartoldus, and J. S. Wakeley. 2000. Expanding Wetland Assessment Procedures: Linking Indices of Wetland Functions with Services and Values. U.S. Army Corps of Engineers, Engineer Research and Development Center, Vicksburg, MS. Wetlands Research Program. ERDC?EL TR-00-17.
King, S. E., and J. N. Lester. 1995. The value of salt marsh as a sea defence. *Marine Pollution Bulletin* 30:180–89.
Kirby, C. J., and J. G. Gosselink. 1976. Primary production in a Louisiana Gulf

Coast *Spartina alterniflora* marsh. *Ecology* 57:1052–59.

Kirwan, M. L., and L. K. Blum. 2011. Enhanced decomposition offsets enhanced productivity and soil carbon accumulation in coastal wetlands responding to climate change. *Biogeosciences* 8:987–93.

Kirwan, M. L., and G. R. Guntenspergen. 2010. Influence of tidal range on the stability of coastal marshland. *Journal of Geophysical Research* 115, F02009, doi: 10.1029/2009JF001400.

Kirwan, M. L, and G. R. Guntenspergen. 2012. Feedbacks between inundation, root production, and shoot growth in a rapidly submerging brackish marsh. *Journal of Ecology* 100:764–70.

Kirwan, M. L., and A. B. Murray. 2007. A coupled geomorphic and ecological model of tidal marsh evolution. *Proceedings of the National Academy of Sciences of the United States, U.S.A.* 104:6118–22.

Kirwan, M. L., and A. B. Murray. 2008. Tidal Marshes As Disequilibrium Landscapes? Lags between Morphology and Holocene Sea Level Change. *Geophysical Research Letters* 35: L24401, 5 pp. doi: 10.1029/2008GL036050.

Kirwan, M. L., J. L. Kirwan, and C. A. Copenheaver. 2007. Dynamics of an estuarine forest and its response to rising sea level. *Journal of Coastal Research* 23:457–63.

Kirwan, M. L., A. B. Murray, and W. S. Boyd. 2008. Temporary vegetation disturbance as an explanation for permanent loss of tidal wetlands. *Geophysical Research Letters* 35: L05403, 5 pp. doi: 10.1029/2007GL032681.

Kirwan, M. L., G. R. Guntenspergen, and J. T. Morris. 2009. Latitudinal trends in *Spartina alterniflora* productivity and the response of coastal marshes to global change. *Global Change Biology* 15:1982–89.

Kirwan, M. L., G. R. Guntenspergen, A. D'Alpaos, J. T. Morris, S. M. Mudd, and S. Temmerman. 2010. Limits of the adaptability of coastal marshes to rising sea level. *Geophysical Research Letters* 37: L23401, doi: 10.1029/2010GL045489.

Kirwan, M. L., A. B. Murray, J. P. Donnelly, and D. R. Corbett. 2011. Rapid wetland expansion during European settlement and its implication for marsh survival under modern sediment delivery rates. *Geology* 39:507–10.

Kirwan, M. L., R. R. Christian, L. K. Blum, and M. M. Brinson. 2012. On the relationship between sea level and *Spartina alterniflora* production. *Ecosystems* 15:140–47.

Kistritz, R. U. 1978. An Ecological Evaluation of Fraser Estuary Tidal Marshes: The Role of Detritus and the Cycling of Elements. Westwater Research Centre, Vancouver, BC. Technical Report no. 15.

Kiviat, E. 1987. Common reed (*Phragmites australis*). In: D. Decker and J. Enck (eds.). *Exotic Plants with Identified Detrimental Impacts on Wildlife Habitats in New York.* New York Chapter, The Wildlife Society, Annandale, NY. 22–30.

Kjerfve, B., and K. E. Magill. 1990. Salinity changes in Charleston Harbor, 1922–1987. *Journal of Waterway, Port, Coastal, and Ocean Engineering* 116:153–68.

Kleindl, W., M. C. Rains, and F. R. Hauer. 2010. HGM is a rapid assessment: clearing the confusion. *Wetland Science and Practice* 27:17–22.

Kneib, R. T. 1984. Patterns of invertebrate distribution and abundance in the intertidal salt marsh: causes and questions. *Estuaries* 7:392–412.

Kneib, R. T. 1997. The role of tidal marshes in the ecology of estuarine nekton. *Oceanography, Marine Biology Annual Review* 35:163–220.

Kneib, R. T. 2000. Salt marsh ecoscapes and production by estuarine nekton in the southeastern United States. In: M. P. Weinstein and D. A. Kreeger (eds.). *Concepts and Controversies in Tidal Marsh Ecology.* Kluwer Academic Publishers, Dordrecht, The Netherlands. 267–91.

Knight, J. B. 1934. A salt-marsh study. *American Journal of Science* 28:161–81.

Knighton, A. D., K. Mills, and C. D. Woodroffe. 1991. Tidal-creek extension and saltwater intrusion in northern Australia. *Geology* 19:831–34.

Knutson, P. L., J. C. Ford, M. R. Inskeep, and J. Oyler. 1981. National survey of planted salt marshes (vegetative stabilization and wave stress). *Wetlands* 1:129–57.

Knutson, P. L., R. A. Brochu, W. N. Seelig, and M. Inskeep. 1982. Wave damping in *Spartina alterniflora* marshes. *Wetlands* 2:87–104.

Knutson, P. L., H. H. Allen, and J. W. Webb. 1990. Guidelines for Vegetative Erosion

Control on Wave-Impacted Coastal Dredged Material Sites. U.S. Army Engineer Waterways Experiment Station, Vicksburg, MS. Technical Report D-90-13.

Knutson, T. R., R. E. Tuleya, and Y. Kurihara. 1998. Simulated increase of hurricane intensities in a CO_2-warmed climate. *Science* 279:1018–20.

Ko, J-Y., and J. W. Day. 2004. Wetlands: impacts of energy development in the Mississippi Delta. In: C. J. Cleveland (ed.). *Encyclopedia of Energy.* Elsevier, New York. Vol. 6:397–408. www.lsu.edu/cei/research_projects/Wetlands_final.pdf.

Ko, J-Y., J. Day, J. Barras, R. Morton, J. Johnston, G. Steyer, G. P. Kemp, E. Clairain, and R. Theriot. 2004. Impacts of oil and gas activities on coastal wetland loss in the Mississippi Delta. In: K. Withers and M. Nipper (eds.). *Environmental Analysis of the Gulf of Mexico.* English translation of original Spanish version. Harte Research Institute for Gulf of Mexico Studies, Texas A&M University, Corpus Christi, TX. Special Publication Series no. 1. Chapter 33; 608–21. http://02a8a22.netsolhost.com/ebook/ch33-oil-gas-impacts-on-coastal-wetland-loss.pdf.

Koch, M. S., I. A. Mendelssohn, and K. L. McKee. 1990. Mechanism for the hydrogen sulfide-induced growth limitation in wetland macrophytes. *Limnology and Oceanography* 35:399–408.

Kolker, A., S. L. Goodbred, J. K. Cochran, and R. Aller. 2005. Marsh Loss on Long Island: Does Biochemistry Trump Climate Change? Marine Science Center, Stony Brook, NY.

Koneff, M. D., and J. A. Royle. 2004. Modeling wetland change along the United States Atlantic Coast. *Ecological Modelling* 177:41–59.

Konisky, R. A., D. M. Burdick, M. Dionne, and H. A. Neckles. 2006. A regional assessment of salt marsh restoration and monitoring in the Gulf of Maine. *Restoration Ecology* 14:516–25.

Koretsky, C. M., P. Van Cappellen, T. J. DiChristina, J. E. Kostka, K. L. Lowe, C. M. Moore, A. N. Roychoudhury, and E. Viollier. 2005. Salt marsh pore water geochemistry does not correlate with microbial community structure. *Estuarine, Coastal and Shelf Science* 62:233–51.

Kozlowski, T. T. 1984. *Flooding and Plant Growth.* Academic Press, Orlando, FL.

Kozlowski, T. T. 1997. Response of woody plants to flooding and salinity. *Tree Physiology Monograph* 1:1–29.

Kraeuter, J. N., and P. L. Wolf. 1974. The relationship of marine macroinvertebrates to salt marsh plants. In: R. J. Reimold and W. H. Queen (eds.). *Ecology of Halophytes.* Academic Press, New York. 449–62.

Kraft, H. C. 1986. *The Lenape: Archeology, History, and Ethnography.* The New Jersey Historical Society, Newark, NJ.

Kraft, J. C. 1971. Sedimentary facies patterns and geologic history of a Holocene marine transgression. *Geological Society of America Bulletin* 82:2131–58.

Kraft, J. C. 1988. Geology. In: T. L. Bryant and J. R. Pennock (eds.). *The Delaware Estuary: Rediscovering a Forgotten Resource.* University of Delaware Sea Grant Program, Newark, DE. 31–41.

Krauss, K. W., J. L. Chambers, and D. Creech. 2007. Selection for salt tolerance in tidal freshwater swamp species: advances using baldcypress as a model for restoration. In: W. H. Conner, T. W. Doyle, and K. W. Krauss (eds.). *Ecology of Tidal Freshwater Forested Wetlands of the Southeastern United States.* Springer, Dordrecht, The Netherlands. Chapter 14; 385–410.

Krauss, K. W., T. W. Doyle, T. J. Doyle, C. M. Swarzenski, A. S. From, R. H. Day, and W. H. Conner. 2009. Water level observations in mangrove swamps during two hurricanes in Florida. *Wetlands* 29:142–49.

Krebs, C. T., and K. A. Burns. 1977. Long-term effects of an oil spill on populations of the salt-marsh crab *Uca pugnax. Science* 197:484–87.

Krebs, C. T., and C. E. Tanner. 1981. Restoration of oiled salt marshes through sediment stripping and *Spartina* propagation. In: *Proceedings of the 1981 Oil Spill Conference.* American Petroleum Institute, Washington, DC. 375–85.

Kreeger, D. A., C. J. Langdon, and R. I. E. Newell. 1988. Utilization of refractory cellulosic carbon derived from *Spartina alterniflora* by the ribbed mussel *Geukensia demissa. Marine Ecology Progress Series* 42:171–79.

Kreeger, D. A., and R. L. E. Newell. 2000. Trophic complexity between produced and invertebrate consumers in salt marshes. In: M. P. Weinstein and D. A. Kreeger (eds.). *Concepts and Controversies in Tidal Marsh Ecology.* Kluwer Academic Publishers, Dordrecht, The Netherlands. 187–220.

Kritzer, J., and A. Hughes. 2009. The role of salt marshes in sustaining Long Island fisheries. In: *Tidal Marshes of Long Island.* Long Island Botanical Society. 40–47. www.edf.org/documents/9447_role-of-salt-marshes.pdf.

Kruczynski, W. L., C. B. Subrahmanyam, and S. H. Drake. 1978. Studies on the plant community of a north Florida salt marsh. Part I. Primary production. *Bulletin of Marine Science* 28:316–34.

Krull, K., and C. Craft. 2009. Ecosystem development of a sandbar emergent tidal marsh, Altamaha River Estuary, Georgia, USA. *Wetlands* 29:314–22.

Kudoh, H., and D. F. Whigham. 1997. Microgeographic genetic structure and gene flow in *Hibiscus moscheutos* (Malvaceae) populations. *American Journal of Botany* 84:1285–93.

Kuenzler, E. J. 1961. Structure and energy flow of a mussel population in a Georgia salt marsh. *Limnology and Oceanography* 6:191–204.

Kuenzler, E. J., and H. L. Marshall. 1973. Effects of Mosquito Control Ditching on Estuarine Ecosystems. Water Resources Research Institute, North Carolina State University, Raleigh, NC. Report No. 81.

Kuhn, N. L., and J. B. Zedler. 1997. Differential effects of salinity and soil saturation on native and exotic plants of a coastal salt marsh. *Estuaries* 20:391–403.

Kunza, A. E., and S. C. Pennings. 2008. Patterns of plant diversity in Georgia and Texas salt marshes. *Estuaries and Coasts* 31:673–81.

Kunze, L. M. 1994. Preliminary Classification of Native, Low Elevation, Freshwater Wetland Vegetation in Western Washington. Washington Department of Natural Resources, Natural Heritage Program, Olympia, WA.

Kushlan, J. A. 1986. The Everglades management of cumulative ecosystem degradation. In: E. D. Estevez, J. Miller, J. Morris, and R. Hamman (eds.). Proceedings of the conference: Managing Cumulative Effects in Florida Wetlands. Omnipress, Madison, WI. New College Environmental Studies Program Publication no. 37. 61–82.

Kusler, J. A., and M. E. Kentula (eds.). 1990. *Wetland Creation and Restoration: The State of the Science.* Island Press, Washington, DC.

Kusler, J., and T. Opheim. 1996. *Our National Wetland Heritage: A Protection Guide.* Environmental Law Institute, Washington, DC.

Lacerda, L. D. 2002. *Mangrove Ecosystems.* Springer, Berlin, Germany.

Laessle, A. M. 1942. The Plant Communities of the Welaka Area with Special Reference to Correlations between Soils and Vegetational Succession. University of Florida, Gainesville, FL. Biological Science Series Vol. 4, no. 1.

Laffoley, D. d'A., and G. Grimsditch (eds.). 2009. The Management of Natural Coastal Carbon Sinks. International Union for Conservation of Nature and Natural Resources, Gland, Switzerland. http://data.iucn.org/dbtw-wpd/edocs/2009-038.pdf.

Lagna, L. 1975. The relationship of *Spartina alterniflora* to mean high water. Marine Sciences Research Center, State University of New York, Stony Brook, NY.

Landin, M. C., M. R. Palermo, L. J. Hunt, R. T. Huffman, C. V. Klimas, M. K. Vincent, and J. S. Wilson. 1978. Wetland Habitat Development with Dredged Material: Engineering and Plant Propagation. U.S. Army Engineer Waterways Experiment Station, Vicksburg, MS. Technical Report DS-78-16.

Lane, R. R., J. W. Day, Jr., and B. Thibodeaux. 1999. Water quality analysis of a freshwater diversion at Caernarvon, Louisiana. *Estuaries* 22:327–36.

Lang, J. W., and H. V. Andrews. 1994. Temperate-dependent sex determination in crocodilians. *Journal of Experimental Zoology* 270:28–44.

Langdon, C. J., and R. I. E. Newell. 1990. Utilization of detritus and bacteria as food sources by two bivalve suspension-feeders, the oyster *Crassostrea virginica* and the mussel *Genkensia demissa. Marine Ecology Progress Series* 58:299–310.

Langley, J. A., and J. P. Megonigal. 2010. Ecosystem response to elevated CO_2 levels limited by nitrogen-induced plant species shift. *Nature* 466:96–99.

Langley, J. A., K. L. McKee, D. R. Cahoon, J. A. Cherry, and J. P. Megonigal. 2009. Elevated CO_2 stimulates marsh elevation gain, counterbalancing sea-level rise. *Proceedings National Academy of Sciences* 106:6182–86.

Langlois, E., A. Bonis, and J. B. Bouzillé. 2003. Sediment and plant dynamics in salt marshes pioneer zone: *Puccinellia maritima* as a key species? *Estuarine, Coastal and Shelf Science* 56:239–49.

Larsen, C. F., R. J. Motyka, J. T. Freymueller, K. A. Echelmeyer, and E. R. Ivins. 2005. Rapid viscoelastic uplift in southeast Alaska caused by post-Little Ice Age glacial retreat. *Earth and Planetary Science Letters* 237:548–60.

LaSalle, M. W., and A. A. de la Cruz. 1985. Seasonal abundance and diversity of spiders in two intertidal marsh plant communities. *Estuaries and Coasts* 8:381–93.

Lasalle, P., and R. J. Rogerson. 2007. The Champlain Sea. In: *The Canadian Encyclopedia.* http://thecanadianencyclopedia.com/index.cfm?PgNm=TCE&Params=A1ARTA0001507.

Leatherman, S. P. 1979. *Barrier Island Handbook.* National Park Service, Cooperative Research Unit, University of Massachusetts, Amherst, MA.

Leatherman, S. P., and R. E. Zaremba. 1987. Overwash and aeolian processes on a U.S. northeast coast barrier. *Sedimentary Geology* 52:183–206.

Leck, M. A. 2003. Seed-bank and vegetation development in a created tidal freshwater wetland on the Delaware River, Trenton, New Jersey, USA. *Wetlands* 23:310–43.

Leck, M. A., and C. M. Crain. 2009. Northeastern North American case studies—New Jersey and New England. In: A. Barendregt, D. F. Whigham, and A. H. Baldwin (eds.). *Tidal Freshwater Wetlands.* Backhuys Publishers, Leiden, The Netherlands. Chapter 13; 145–56.

Leck, M. A., and R. L. Simpson. 1987. Seed bank of a freshwater tidal wetland: turnover and relationship to vegetation change. *American Journal of Botany* 74:360–70.

Leck, M. A., and R. L. Simpson. 1994. Tidal freshwater wetland zonation: seed and seedling dynamics. *Aquatic Botany* 47:61–75.

Leck, M. A., and R. L. Simpson. 1995. Ten-year seed bank and vegetation dynamics of a tidal freshwater wetland. *American Journal of Botany* 82:1547–57.

Leck, M. A., A. H. Baldwin, V. T. Parker, L. Schile, and D. F. Whigham. 2009. Plant communities of tidal freshwater wetlands of the continental USA and southeastern Canada. In: A. Barendregt, D. F. Whigham, and A. H. Baldwin (eds.). *Tidal Freshwater Wetlands.* Backhuys Publishers, Leiden, The Netherlands. Chapter 5; 41–58.

Lee, J. K., R. A. Park, and P. W. Mausel. 1992. Application of Geoprocessing and Simulation Modeling to Estimate Impacts of Sea Level Rise on the Northeast Coast of Florida. *Photogrammetric Engineering and Remote Sensing* 58:1579–86.

Lee, V. 1980. An Elusive Comprise: Rhode Island Coastal Ponds and Their People. University of Rhode Island, Coastal Resources Center, Narragansett, RI. Marine Technical Report 73.

Leeworthy, V. R., and P. C. Wiley. 2001. Current Participation Patterns in Marine Recreation. U.S. Department of Commerce, National Oceanic and Atmospheric Administration, National Ocean Service, Silver Spring, MD. Special Projects.

Lefeuvre, J-C., V. Bouchard, E. Feunteun, S. Grare, P. Laffaille, and A. Radureau. 2000. European salt marshes diversity and functioning: the case study of the Mont Saint-Michel Bay, France. *Wetlands Ecology and Management* 8:147–61.

LeMay, L. E. 2007. The impact of drainage ditches on salt marsh flow patterns, sedimentation and morphology: Rowley River, Massachusetts. College of William and Mary, School of Marine Science, Williamsburg, VA. M.S. thesis. http://web.vims.edu/physical/projects/CHSD/publications/reports/LeMay2007_MS.pdf.

Lent, R. A., T. Hruby, D. P. Cowan, and T. S. Litwin. 1990. Open Marsh Water Management on Great South Bay, Islip, NY: Final Report. Seatuck Research Program, Cornell University, Islip, NY.

Leonard, L. A., and M. E. Luther. 1995. Flow hydrodynamics in tidal marsh canopies. *Limnology and Oceanography* 40:1474–84.

Les, D. H., M. A. Cleland, and M. Waycott. 1997. Phylogenetic studies in Alismatidae, II. Evolution of marine angiosperms (seagrasses) and hydrophily. *Systematic Botany* 22:443–63.

Lesser, C. A. 1971. Memorandum to Secretary Austin N. Heller re: 1971 wetland inventory (corrected). Delaware Department of Natural Resources and Environmental Control, Dover, DE.

Létourneau, G., and M. Jean. 1996. Cartographie des marais, marécages et herbiers aquatiques le long du Saint-Laurent par telédétection aéroportée. Rapport scientifique et technique ST-61. Environnement Canada-Région du Québec, Conservation de l'environnement, Centre Saint-Laurent, Montréal, QC.

Levin, L. A., and T. S. Talley. 2000. Influence of vegetation and abiotic environmental factors in salt marsh invertebrates. In: M. P. Weinstein and D. A. Kreeger (eds.). *Concepts and Controversies in Tidal Marsh Ecology.* Kluwer Academic Publishers, Dordrecht, The Netherlands. 661–707.

Levin, L. A., C. Neira, and E. D. Grosholz. 2006. Invasive cordgrass modifies wetland trophic function. *Ecology* 87:419–32.

Levin, P. S., J. Ellis, R. Petrik, and M. E. Hay. 2002. Indirect effects of feral horses on estuarine communities. *Conservation Biology* 16:1364–71.

Levine, J. M., J. S. Brewer, and M. D. Bertness. 1998. Nutrients, competition, and plant zonation in a New England salt marsh. *Journal of Ecology* 86:285–92.

Levings, C. D., and R. M. Thom. 1994. Habitat changes in Georgia Basin: Implications for resource management and restoration. In: R. C. H. Wilson, R. J. Bemish, F. Aitkens, and J. Bell (eds.). Review of the Marine Environment and Biota of Strait of Georgia, Puget Sound, and Juan de Fuca Strait. *Proceedings of British Columbia/Washington Symposium on the Marine Environment;* 1994 January 13–14. Canadian Technical Report of Fisheries and Aquatic Sciences 1948. 330–51.

Lewis, J. C., and R. L. Garrison. 1984. Habitat Suitability Index Models: American Black Duck (Wintering). U.S. Fish and Wildlife Service, Washington, DC. FWS/OBS-82/10.68.

Lewis, R. 1995. Geologic history of Long Island Sound. In: G. D. Dreyer and W. A. Niering (eds.). Tidal Marshes of Long Island Sound: Ecology, History and Restoration. The Connecticut College Arboretum, New London, CT. Bulletin no. 34. 12–16.

Lewis, R. R. (ed.). 1982. *Creation and Restoration of Coastal Plant Communities.* CRC Press, Boca Raton, FL.

Li, H., and S. L. Yang. 2009. Trapping effect of tidal marsh vegetation on suspended sediment, Yangtze Delta. *Journal of Coastal Research* 25:915–24, 930.

Li, H., L. Li, and D. Lockington. 2005. Aeration for plant root respiration. *Water Resources Research* 41. W06023, doi: 10.1029/2004WR003759.

Libes, S. M. 2009. *Introduction to Marine Biochemistry.* 2nd edition. Academic Press, Burlington, MA.

Light, H. M., M. R. Darst, and R. A. Mattson. 2007. Ecological characteristics of tidal freshwater forests along the lower Suwanne River, Florida. In: W. H. Conner, T. W. Doyle, and K. W. Krause (eds.). *Ecology of Tidal Freshwater Forested Wetlands of the Southeastern United States.* Springer, Dordrecht, The Netherlands. Chapter 11; 291–320.

Lin, Q., and I. A. Mendelssohn. 1996. A comparative investigation of the effects of south Louisiana crude oil on the vegetation of fresh, brackish and salt marshes. *Marine Pollution Bulletin* 32:202–9.

Lin, Q., I. A. Mendelssohn, N. P. Bryner, and W. D. Walton. 2005. In-situ burning of oil in coastal marshes. I. Vegetation recovery and soil temperature as a function of water depth, oil type, and marsh type. *Environmental Science & Technology* 39:1848–54.

Linden, E. 2006. *The Winds of Change: Climate, Weather, and the Destruction of Civilizations.* Simon & Schuster Paperbacks, New York.

Linhart, Y. B., and M. C. Grant. 1996. Evolutionary significance of local genetic differentiation in plants. *Annual Review of Ecology and Systematics* 27:237–77.

Linthurst, R. A., and R. J. Reimold. 1978. Estimated net aerial primary productivity for selected estuarine angiosperms in Maine, Delaware, and Georgia. *Ecology* 59:945–55.

Lippson, A. J. 1973. *The Chesapeake Bay in Maryland: An Atlas of Natural Resources.* The John Hopkins University Press, Baltimore, MD.

Little, C. 2000. *The Biology of Soft Shores and Estuaries.* Oxford University Press, Oxford, England.

Little, C., and J. A. Kitching. 1996. *The Biology of Rocky Shores.* Oxford University Press, Oxford, England.

Little, C., G. Williams, and C. Trowbridge. 2009. *The Biology of Rocky Shores.* Second edition, Oxford University Press, Oxford, UK.

Livingston, D. C., and D. G. Patriquin. 1981. Belowground growth of *Spartina alterniflora* Loisel: habit, functional biomass and non-structural carbohydrates. *Estuarine, Coastal and Shelf Science* 12:579–87.

Loch, D. S., E. Barrett-Lennard, and P. Truong. 2003. Role of salt tolerant plants for production, prevention of salinity and amenity values. *Proceedings of 9th National Conference on Productive Use of Saline Lands (PURSL);* 2003 September 21–October 2. Rockhampton, Queensland, Australia.

Long, S. P., and C. F. Mason. 1983. *Saltmarsh Ecology.* Blackie, Glasgow, Scotland.

Loomis, M. A., and C. Craft. 2010. Carbon sequestration and nutrient (nitrogen, phosphorus) accumulation in river-dominated tidal marshes, Georgia, USA. *Soil Science Society of America Journal* 74:1028–36.

Looney, P. B., and D. J. Gibson. 1995. The relationship between the soil seed bank and above-ground vegetation of a coastal barrier island. *Journal of Vegetation Science* 6:825–36.

Lopez, J. A. 2009. The environmental history of human-induced impacts to the Lake Pontchartrain basin in southeastern Louisiana since European settlement—1718 to 2002. *Journal of Coastal Research* 54:1–11.

The Louis Berger Group. 2004. Regional Guidebook for Hydrogeomorphic Assessment of Tidal Fringe Wetlands in the Hackensack Meadowlands. East Orange, NJ.

Louisiana Coastal Wetlands Conservation and Restoration Task Force and the Wetlands Conservation and Restoration Authority. 1998. Coast 2050: Toward a Sustainable Coastal Louisiana. Louisiana Department of Natural Resources, Baton Rouge, LA. www.lacoast.gov/programs/2050/MainReport/report1.pdf.

Love, J. W., J. Gill, and J. J. Newhard. 2008. Saltwater intrusion impacts fish diversity and distribution in the Blackwater River drainage (Chesapeake Bay watershed). *Wetlands* 28:967–74.

Lovelace, J. K. 1994. Storm-tide Elevations Produced by Hurricane Andrew along the Louisiana coast, August 25–27, 1992. U.S. Geological Survey, Baton Rouge, LA. Open-file Report 94-371.

Lovelace, J. K., and B. F. McPherson. 1996. Effect of Hurricane Andrew (1992) on Wetlands in Southern Florida and Louisiana. In: J. D. Fretwell, J. S. Williams, and P. J. Redman (comps.). National Water Summary of Wetland Resources. U.S. Geological Survey, Reston, VA. Water-Supply Paper 2425. 93–96.

Lubchenco, J. 1978. Plant species diversity in a marine intertidal community: importance of herbivore food preference and algal competitive abilities. *American Naturalist* 112:23–39.

Lugo, A. E., and S. C. Snedaker. 1974. The ecology of mangroves. *Annual Review of Ecology and Systematics* 5:39–64.

Lunz, J. D., T. W. Zeigler, R. T. Huffman, R. J. Diaz, E. J. Clairain, and L. J. Hunt. 1978. Habitat Development Field Investigations, Windmill Poin Marsh Development Site, James River, Virginia, Summary Report. U.S. Army Engineer Waterways Experiment Station, Vicksburg, MS. Technical Report D-77-23.

Lyles, L. D., and F. Namwamba. 2005. Louisiana Coastal Zone Erosion: 100+ Years of Landuse and Land Loss Using GIS and Remote Sensing. *Proceedings of ESRI User Conference.* Paper 1222. http://proceedings.esri.com/library/userconf/educ05/papers/pap1222.pdf.

Lynch-Stewart, P. 1983. Land Use Change on Wetlands in Southern Ontario: Review and Bibliography. Lands Directorate, Environment Canada, Ottawa, Ontario. Working Paper no. 26.

Lynch-Stewart, P. 1992. No Net Loss: Implementing "No Net Loss" Goals to Conserve Wetlands in Canada. North American Wetlands Conservation Council (Canada), Ottawa, Ontario. Sustaining Wetlands Issues Paper no. 1992-2. www.wetlandscanada.org/No%20Net%20Loss%201992-2.pdf.

Lynch-Stewart, P., P. Neice, C. Rubec, and I. Kessel-Taylor. 1996. The Federal Policy on Wetland Conservation Implementation Guide for Federal Land Managers. Wildlife Conservation Branch, Canadian Wildlife Service, Environment Canada, Ottawa, Ontario. http://dsp-psd.pwgsc.gc.ca/Collection/CW66-145-1996E.pdf.

Lynch-Stewart, P., I. Kessel-Taylor, and C. Rubec. 1999. Wetlands and Government: Policy and Legislation for Wetland Conservation in Canada. North

American Wetlands Conservation Council (Canada), Ottawa, Ontario. Sustaining Wetlands Issues Paper no. 1999-1. www.wetlandscanada.org/1999-1%20Wetlands%20and%20Government.pdf.

MacDonald, G. K., P. E. Noel, D. van Proosdij, and G. L. Chmura. 2010. The legacy of agricultural reclamation on channel and pool networks of Bay of Fundy salt marshes. *Estuaries and Coasts* 33:151–60.

Macdonald, K. B. 1977. Plant and animal communities of Pacific North American salt marshes. In: V. J. Chapman (ed.). *Wet Coastal Ecosystems.* Elsevier Scientific Publishing, Amsterdam, The Netherlands. 167–91.

Macdonald, K. B., and M. G. Barbour. 1974. Beach and salt marsh vegetation of the North American Pacific Coast. In: R. J. Reimold and W. H. Queen (eds.). *Ecology of Halophytes.* Academic Press, New York. 175–233.

MacKenzie, W. H., and J. R. Moran. 2004. Wetlands of British Columbia: A Guide to Identification. British Columbia Ministry of Forests, Research Branch, Victoria, BC. Land Management Handbook 52.

MacKinnon, K., and D. B. Scott. 1984. An Evaluation of Salt Marshes in Atlantic Canada. Centre for Marine Geology, Dalhousie University, Halifax, Nova Scotia and Lands Directorate, Environment Canada, Dartmouth, Nova Scotia. Technical Report no. 1.

MacLennan, A. 2005. An analysis of large woody debris in two Puget Sound salt marshes: Elger Bay, Camano Island and Sullivan Minor Marsh, Padilla Bay. Western Washington University, Bellingham, WA. M.S. thesis.

Madden, C. J., J. W. Day, Jr., and J. M. Randall. 1988. Freshwater and marine coupling in estuaries of the Mississippi River deltaic plain. *Limnology and Oceanography* 33:982–1004.

Madrone Associates, Philip William and Associates, J. R. Cherniss, and N. Wakeman. 1983. Ecological Values of Diked Historic Baylands. A Technical Report Prepared for the San Francisco Bay Conservation and Development Commission, San Francisco, CA.

Magenheimer, J. F., T. R. Moore, G. L. Chmura, and R. J. Daoust. 1996. Methane and carbon dioxide flux from a macrotidal salt marsh, Bay of Fundy, New Brunswick. *Estuaries* 19:139–45.

Mahall, B. E., and R. B. Park. 1976. The ecotone between *Spartina foliosa* Trin., and *Salicornia virginica* L. in salt marshes of northern San Francisco Bay: III. Soil aeration and tidal immersion. *Journal of Ecology* 64:811–19.

Maia, L. P., L. H. U. Monteiro, G. M. Sousa, and L. D. Lacerda. 2006. Changes in mangrove extension along the northeastern Brazilian coat (1978–2003). International Society for Mangrove Ecosystems, Okinawa, Japan. *ISME/GLOMIS* 5(1):1–5.

Major, C., and K. M. Lusht. 2004. Beach proximity and the distribution of property values in shore communities. *The Appraisal Journal* (Fall):4433–38.

Marcus, L. 2000. Restoring tidal wetlands at Sonoma Baylands, San Francisco Bay, California. *Ecological Engineering* 15:373–83.

Maricle, B. R., and R. W. Lee. 2002. Aerenchyma development and oxygen transport in the estuarine cordgrasses, *Spartina alterniflora* and *S. anglica. Aquatic Botany* 74:109–20.

Marie-Victorin. Frère. 1964. *Flore Laurentienne.* Les Presses de l'Université de Montréal.

Marinucci, A. C. 1982. Trophic importance of *Spartina alterniflora* production and decomposition to the marsh-estuarine ecosystem. *Biological Conservation* 22:35–58.

Marks, M., B. Lapin, and J. Randall. 1994. *Phragmites australis* (*P. communis*): threats, management, and monitoring. *Natural Areas Journal* 14:285–94.

Marples, T. G. 1966. A radionuclide tracer study of arthropod food chains in a *Spartina* salt marsh ecosystem. *Ecology* 47:270–77.

Marquardt, W. H. 1996. Four discoveries: environmental archaeology in Southwest Florida. In: E. J. Reitz, L. A. Newsom, and S. J. Scudder (eds.). *Case Studies in Environmental Archaeology.* Plenum Press, New York. 17–32.

Marshall, D. E. 1970. Characteristics of *Spartina* marsh which is receiving treated municipal sewage wastes. In: H. T. Odum and A. F. Chestnut (eds.). *Studies of Marine Estuarine Ecosystems Developing With Treated Sewage Wastes.* Institute of Marine Science, University of North Carolina, Chapel Hill, NC. 317–58.

Martin, A. C., H. S. Zim, and A. L. Nelson. 1961. *American Wildlife and Plants.* Dover Publications, New York.

Massachusetts Office of Coastal Zone Management. 2002. Massachusetts Aquatic Invasive Species Management Plan. Boston, MA. www.anstaskforce.gov/Mass_AIS_Plan.pdf.

Mattheus, C. R., A. B. Rodriguez, and B. A. McKee. 2009. Direct connectivity between upstream and downstream promotes rapid response of lower coastal-plain rivers to land-use change. *Geophysical Research Letters* 36, L20401, doi: 10.1029/2009GL039995. 6 pp.

Mattila, J., G. Chaplin, M. R. Eilers, K. L. Heck, Jr., J. P. O'Neal, and J. F. Valentine. 1999. Spatial and diurnal distribution of invertebrate and fish fauna of a *Zostera marina* bed and nearby unvegetated sediments in Damariscotta River, Maine (USA). *Journal of Sea Research* 41:321–32.

Mattoon, W. R. 1915. The Southern Cypress. U.S. Department of Agriculture, Washington, DC. Bulletin no. 272.

Mawhinney, K., P. W. Hicklin, and J. S. Boates. 1993. A re-evaluation of the numbers of migrant semipalmated sandpipers, *Calidris pusila,* in the Bay of Fundy during fall migration. *Canadian Field-Naturalist* 107:19–23.

McAloney, K. 1981. Waterfowl use of Nova Scotian salt marshes. In: A. Hatcher and D. G. Patriquin (eds.). Salt Marshes in Nova Scotia. A Status Report of the Salt Marsh Working Group. Institute of Resource and Environmental Studies and Department of Biology, Dalhousie University, Halifax, Nova Scotia. 60–66.

McCormick, J., and T. Ashbaugh. 1972. Vegetation of a section of Oldmans Creek tidal marsh and related areas in Salem and Gloucester Counties, New Jersey. *Bulletin of the New Jersey Academy of Science* 17:31–37.

McCormick, M. K., K. M Kettenring, H. M. Baron, and D. F. Whigham. 2010. Extent and reproduction mechanisms of *Phragmites australis* spread in brackish wetlands in Chesapeake Bay, Maryland (USA). *Wetlands* 30:67–74.

McCraith, B. J., L. R. Gardner, D. S. Wethey, and W. S. Moore. 2003. The effect of fiddler crab burrowing on sediment mixing and radionuclide profiles along a topographic gradient in a southeastern salt marsh. *Journal of Marine Research* 61:359–90.

McFalls, T. B., P. A. Keddy, D. Campbell, and G. Shaffer. 2010. Hurricanes, floods, levees, and nutria: vegetation response to interacting disturbance and fertility regimes with implications for coastal wetland restoration. *Journal of Coastal Research* 26:901–11.

McHugh, J. L. 1975. Estuarine fisheries: are they doomed. In: M. Wiley (ed.). *Estuarine Processes.* Volume I: *Uses, Stresses, and Adaptation to the Estuary.* Academic Press, New York. 15–23.

McHugh, L. L. 1966. Management of Estuarine Fishes. American Fisheries Society Special Publication no. 3. 133–54.

McKee, K. 2004. Global change impacts on mangrove ecosystems. U.S. Geological Survey, National Wetlands Research Center, Lafayette, LA. Fact sheet 2004-3125.

McKee, K. L., and J. A. Cherry. 2009. Hurricane Katrina sediment slowed elevation loss in subsiding brackish marshes of the Mississippi River delta. *Wetlands* 29:2–15.

McKee, K. L., and I. A. Mendelssohn. 1989. Response of a freshwater marsh plant community to increased salinity and increased water level. *Aquatic Botany* 34:301–16.

McKee, K. L., and W. H. Patrick, Jr. 1988. The relationship of smooth cordgrass (*Spartina alterniflora*) to tidal datums: a review. *Estuaries* 11:143–51.

McKee, K. L., and J. E. Rooth. 2008. Where temperate meets tropical: multi-factorial effects of elevated CO_2, nitrogen enrichment, and competition on a mangrove-salt marsh community. *Global Change Biology* 14:971–84.

McKee, K. L., I. A. Mendelssohn, and M. D. Materne. 2004. Acute salt marsh dieback in the Mississippi River deltaic plain: a drought-induced phenomena. *Global Ecology and Biogeography* 13:65–73.

McKee, K. L., I. A. Mendelssohn, and M. D. Materne. 2006. Salt Marsh Dieback in Coastal Louisiana: Survey of Plant and Soil Conditions in Barataria and Terrebone Basins, June 2000–September 2001. U.S. Geological Survey Open-file Report 2006-1167.

McLaren, J. R., and R. L. Jefferies. 2004. Initiation and maintenance of vegetation mosaics in an Arctic salt marsh. *Journal of Ecology* 92:648–60.

McMahon, R. F., and W. D. Russell-Hunter. 1981. The effects of physical variables and acclimation on survival and oxygen consumption in the high littoral salt-marsh snail, *Melampus bidentatus* Say. *Biological Bulletin* 161:246–69.

McNutt, M. 2010. Summary Preliminary Report from the Flow Rate Technical Group. U.S. Geological Survey. www.doi.gov/deepwaterhorizon/loader.cfm?csModule=security/getfile&PageID=33972.

McPherson, B. F. 1996. Alabama wetland resources. In: J. D. Fretwell, J. S. Williams, and P. J. Redman (comps.). National Water Summary of Wetland Resources. U.S. Geological Survey, Reston, VA. Water-Supply Paper 2425. 101–6.

Meade, R. H. 1982. Sources, sinks, and storage of river sediment in the Atlantic drainage of the United States. *Journal of Geology* 90:235–52.

Meadows, A., P. S. Meadows, D. M. Wood, and J. M. H. Murray. 1994. Microbiological effects on slope stability: an experimental analysis. *Sedimentology* 41:423–35.

Melville, M. D., and I. White. 2002. Acid sulfate soils, management. In: R Lal (ed.). *Encyclopedia of Soil Science.* Marcel Dekker, New York. 19–22.

Mendelssohn, I. A., and D. P. Batzer. 2006. Abiotic constraints for wetland plants and animals. In: D. P. Batzer and R. R. Sharitz (eds.). *Ecology of Freshwater and Estuarine Wetlands.* University of California Press, Berkeley, CA. Chapter 4; 82–114.

Mendelssohn, I. A., and N. L. Kuhn. 2003. Sediment subsidy: effects on soil-plant responses in a rapidly submerging coastal salt marsh. *Ecological Engineering* 21:115–28.

Mendelssohn, I. A., and J. T. Morris. 2000. Eco-physiological controls on the productivity of *Spartina alterniflora* Loisel. In: M. P. Weinstein and D. A. Kreeger (eds.). *Concepts and Controversies in Tidal Marsh Ecology.* Kluwer Academic Publishers, Dordrecht, The Netherlands. 59–80.

Mendelssohn, I. A., and M. T. Postek. 1982. Elemental analysis of deposits on the roots of *Spartina alterniflora* Loisel. *American Journal of Botany* 69:902–12.

Mendelssohn, I. A., M. W. Hester, C. Sasser, and M. Fischel. 1990. The effect of a Louisiana crude oil discharge from a pipeline break on the vegetation of a southeast Louisiana brackish marsh. *Oil & Chemical Pollution* 7:1–15.

Mendelssohn, I. A., G. L. Andersen, D. M. Baltz, R. H. Caffey, K. R. Carman, J. W. Fleeger, S. B. Joye, Q. Lin, E. Maltby, E. B. Overton, and L. R. Rozas. 2012. Oil impacts on coastal wetlands: implications for the Mississippi River Delta Ecosystem after the Deepwater Horizon Oil Spill. *BioScience* 62:562–74.

Meredith, W. H., and C. R. Lesser. 2007. Open marsh water management in Delaware: 1979–2007. *Proceedings of New Jersey Mosquito Control Association* 94:56–69.

Merino, J. H., D. Huval, and A. J. Nyman. 2010. Implications of nutrient and salinity interaction on the productivity of *Spartina patens. Wetlands Ecology and Management* 18:111–17.

Meyerson, L. A., K. Saltonstall, L. Windham, E. Kiviat, and S. Findlay. 2000. A comparison of *Phragmites australis* in freshwater and brackish marsh environments in North America. *Wetlands Ecology and Management* 8:89–103.

Micheli, F., and C. H. Peterson. 1999. Estuarine vegetated habitats as corridors for predator movements. *Conservation Biology* 13:869–81.

Middleton, B. A. 2009. Effects of Hurricane Katrina on the forest structure of *Taxodium distichum* swamps of the Gulf Coast, USA. *Wetlands* 29:80–87.

Milankovitch, M. 1941. Kanon der Erdbestrahlungen und seine Anwendung auf das Eiszeitenproblem. Koniglich Serbische Akademie, Belgrade, Serbia. (New English Translation: M. Milankovic. 1998. Canon of Insolation and the Ice-Age Problem. Alven Global.)

Miller, C. D. 1988. Predicting the impact of vegetation on storm surges. *Proceedings of National Wetland Symposium on Wetland Hydrology;* 1987 September 16–18. Association of State Wetland Managers, Berne, NY. 113–21.

Miller, D. L., F. E. Smeins, J. W. Webb, and L. Yager. 2005. Mid-Texas, USA coastal marsh vegetation patterns and dynamics as influenced by environmental stress and snow goose herbivory. *Wetlands* 25:648–58.

Miller, W. B., and F. E. Egler. 1950. Vegetation of the Wequetequock-Pawcatuck tidal marshes, Connecticut. *Ecological Monographs* 20:143–72.

Millero, F. J. 1974. The physical chemistry of seawater. *Annual Review of Earth and Planetary Sciences* 2:101–50.

Milligan, D. C. 1987. *Maritime Dykelands: The 350 Year Struggle*. Nova Scotia Department of Government Services, Publishing Division, Halifax, NS.

Milne, G. A., W. R. Gehrels, C. W. Hughes, and M. E. Tamisiea. 2009. Identifying the causes of sea-level change. *Nature Geoscience* 2:471–78.

Milner, C., and R. E. Hughes. 1968. Methods for the Measurement of the Primary Production of Grassland. Blackwell Scientific Publications, Oxford, UK. I. B. P. Handbook no. 6.

Minello, T. J., and J. W. Webb, Jr. 1997. Use of natural and created *Spartina alterniflora* salt marshes by fishery species and other aquatic fauna in Galveston Bay, Texas, USA. *Marine Ecology Progress Series* 151:165–79.

Minnesota Board of Water and Soil Resources. 2008. MnRam 3.2. St. Paul, MN.

Mitchell, L. R., S. Gabrey, P. P. Marra, and R. M. Erwin. 2006. Impacts of marsh management on coastal-marsh bird habitats. *Studies in Avian Biology* 32:155–75.

Mitsch, W. J. (ed.). 2006. *Wetland Creation, Restoration, and Conservation: The State of the Science*. Elsevier, Amsterdam.

Mitsch, W. J., and J. G. Gosselink. 2007. *Wetlands*. 4th edition. John Wiley & Sons, Hoboken, NJ.

Mogensen, R., and P. McBrien. 2010. Richard P. Kane wetland mitigation bank and restoration in the Hackensack Meadowlands. 5th National Conference on Coastal and Estuarine Habitat Restoration. Preparing for Climate Change: Science, Practice, and Policy; 2010 November 13–17; Galveston, TX. Presentation. Restore America's Estuaries, Arlington, VA. www.estuaries.org.

Möller, I., T. Spencer, J. R. French, D. J. Leggett, and M. Dixon. 1999. Wave transformation over salt marshes: a field and numerical modeling study from North Norfolk, England. *Estuarine, Coastal and Shelf Science* 49:411–26.

Montague, C. L. 1982. The influence of fiddler crab burrows and burrowing on metabolic processes in salt marsh sediments. In: V. S. Kennedy (ed.). *Estuarine Comparisons*. Academic Press, New York. 283–301.

Montague, C. L., and R. G. Wiegert. 1991. Salt marshes. In: R. L. Myers and J. J. Ewel (eds.). *Ecosystems of Florida*. University of Central Florida Press, Orlando, FL. Chapter 14; 481–516.

Montague, C. L., A. V. Zale, and H. F. Percival. 1987. Ecological effects of coastal marsh impoundments: a review. *Ecological Management* 11:743–56.

Montalto, F. A., T. S. Steenhuis, and J-Y. Parlange. 2006. The hydrology of Piermont Marsh, a reference for tidal marsh restoration in the Hudson river estuary, New York. *Journal of Hydrology* 316:108–28.

Moonsammy, R. Z., D. S. Cohen, and L. E. Williams (eds.). 1987. *Pinelands Folklife*. Rutgers University Press, New Brunswick, NJ.

Moore, K. 2009. NERRS SWMP Biomonitoring Protocol. Long-term Monitoring of Estuarine Submersed and Emergent Vegetation Communities. National Estuarine Research Reserve System (NERRS) Technical Report. Chesapeake Bay NERRS-Virginia, Gloucester Point, VA. www.nerrs.noaa.gov/Doc/PDF/Research/TechReportSWMPBio-MonitoringProtocol.pdf.

Moore, R. H. 1992. Low-salinity backbays and lagoons. In: C. T. Hackney, S. M. Adams, and W. H. Martin (eds.). *Biodiversity of the Southeastern United States: Aquatic Communities*. John Wiley & Sons, New York. Chapter 13; 541–614.

Moorhead, K. K., and M. M. Brinson. 1995. Response of wetlands to rising sea level in the lower coastal plain of North Carolina. *Ecological Applications* 5:261–71.

Mooring, M. T., A. W. Cooper, and E. D. Seneca. 1971. Seed germination response and evidence for height ecophenes in *Spartina alterniflora* from North Carolina. *American Journal of Botany* 58:48–55.

Morgan, P. A., and F. T. Short. 2002. Using functional trajectories to track constructed salt marsh development in the Great Bay Estuary, Maine/New Hampshire, U.S.A. *Restoration Ecology* 10:461–73.

Morgan, P. A., D. M. Burdick, and F. T. Short. 2009. The functions and values of fringing salt marshes in northern New England, USA. *Estuaries and Coasts* 32:483–95.

Morris, J. T. 2007. Estimating net primary production of salt-marsh macrophytes. In:

T. J. Fahey and A. K. Knapp (eds.). *Principles and Standards for Measuring Primary Production*. Oxford University Press, Cary, NC. 106–19.

Morris, J. T., and J. W. H. Dacey. 1984. Effects of O_2 on ammonium uptake and root respiration by *Spartina alterniflora*. *American Journal of Botany* 71:979–85.

Morris, J. T., and B. Haskin. 1990. A 5-yr record of aerial primary production and stand characteristics of *Spartina alterniflora*. *Ecology* 71:2209–17.

Morris, J. T., P. V. Sundareshwar, C. T. Nietch, B. Kjerfve, and D. R. Cahoon. 2002. Response of coastal wetlands to rising sea level. *Ecology* 83:2869–72.

Morton, R. A., and J. C. Bernier. 2010. Recent subsidence-rate reductions in the Mississippi Delta and their geological implications. *Journal of Coastal Research* 26:555–61. doi: 10.2112/JCOASTRES-D-09-00014R1.1.

Morton, R. A., N. A. Buster, and M. D. Krohn. 2002. Subsurface controls on historical subsidence rates and associated wetland loss in south-central Louisiana. *Transactions, Gulf Coast Association of Geological Societies* 52:767–78.

Morton, R. A., G. Tiling, and N. F. Ferina. 2003. Causes of hot-spot wetland loss in the Mississippi Delta plain. *Environmental Geosciences* 10:71–80.

Motyka, R. J., C. F. Larsen, J. T. Freymueller, and K. A. Echelmeyer. 2007. Post Little Ice Age glacial rebound in Glacier Bay National Park and surrounding areas. *Alaska Park Science Journal* 6:37–41.

Moulton, D. W., T. E. Dahl, and D. M. Dall. 1997. Texas Coastal Wetlands: Status and Trends, Mid-1950s to Early 1990s. U.S. Fish and Wildlife Service, Southwestern Region, Albuquerque, NM.

Moy, L. D., and L. A. Levin. 1991. Are *Spartina* marshes a replaceable resource? A functional approach to evaluation of marsh creation efforts. *Estuaries* 14:1–16.

Mudd, S. M., S. M. Howell, and J. T. Morris. 2009. Impact of dynamic feedbacks between sedimentation, sea-level rise, and biomass production on near-surface marsh stratigraphy and carbon accumulation. *Estuarine, Coastal and Shelf Science* 82:377–89.

Mudge, B. F. 1862. The salt marsh formation of Lynn. *Essex Institute Proceedings* 2:117–19.

Mueller-Dombois, D., and H. Ellenberg. 1974. *Aims and Methods of Vegetation Ecology.* John Wiley & Sons, New York.

Nahkeeta Northwest Wildlife Services. 2008. Puget Sound Marine Invasive Species Identification Guide. Bow, WA. http://vmp.bioe.orst.edu/Documents/mism_ID_Cards5print.pdf.

Naidoo, G., K. L. McKee, and I. A. Mendelssohn. 1992. Anatomical and metabolic responses to waterlogging and salinity in *Spartina alterniflora* and *S. patens* (Poaceae). *American Journal of Botany* 79:765–70.

National Invasive Species Council. 2008. 2008–2012 National Invasive Species Management Plan. Department of the Interior, Office of the Secretary, Washington, DC. www.invasivespeciesinfo.gov/council/mp2008.pdf.

National Marine Fisheries Service. 1983. Unpublished Coastal Wetland (National and Southeast) estimates. St. Petersburg, FL.

National Oceanic and Atmospheric Administration [NOAA]. 2010. Beach Nourishment: A Guide for Local Officials. www.csc.noaa.gov/beachnourishment/html/geo/barrier.htm.

National Park Service and California State Lands Commission. 2007. Giacomini Wetland Restoration Project: Final Environmental Impact Statement/Environmental Impact Report. Point Reyes National Seashore, Point Reyes, CA. www.nps.gov/pore/parkmgmt/planning_giacomini_wrp_eiseir_final_2007.htm.

National Research Council. 1992. *Restoration of Aquatic Ecosystems: Science, Technology, and Public Policy.* Washington, DC. www.nap.edu/catalog.php?record_id=1807.

National Research Council. 1994. *Restoring and Protecting Marine Habitat: The Role of Engineering and Technology.* Washington, DC. www.nap.edu/catalog.php?record_id=2213.

National Research Council. 1995. *Wetland Characteristics and Boundaries.* National Academy Press, Washington, DC. www.nap.edu/catalog.php?record_id=4766.

National Research Council. 2001. *Compensating for Wetland Losses under the Clean Water Act.* National Academy Press, Washington, DC. www.nap.edu/catalog.php?record_id=10134.

National Wetlands Working Group. 1997. The Canadian Wetland Classification System.

2nd edition. The Wetland Research Centre, University of Waterloo, Waterloo, ON.

Neal, W. J., O. H. Pilkey, and J. T. Kelley. 2007. *Atlantic Coast Beaches: A Guide to Ripples, Dunes, and Other Natural Features of the Seashore.* Mountain Press Publishing, Missoula, MT.

Neckles, H., and M. Dionne (eds.). 1999. Regional Standards to Identify and Evaluate Tidal Wetland Restoration in the Gulf of Maine. Wells National Estuarine Research Reserve, Wells, ME. www.pwrc.usgs.gov/resshow/neckles/Gpac.pdf.

Neckles, H. A., M. Dionne, D. M. Burdick, C. T. Roman, R. Buchsbaum, and E. Hutchins. 2002. A monitoring protocol to assess tidal restoration of salt marshes on local and regional scales. *Restoration Ecology* 10:556–63.

Neff, K. P., and A. H. Baldwin. 2005. Seed dispersal into wetlands: techniques and results for a restored tidal freshwater marsh. *Wetlands* 25:392–404.

Nepf, H. M. 1999. Drag, turbulences and diffusion in flow through emergent vegetation. *Water Resources Research* 35:479–89.

Nesbit, D. M. 1885. Tide Marshes of the United States. U.S. Department of Agriculture, Washington, DC. Special Report no. 7.

Nestler, J. 1977. Interstitial salinity as a cause of ecophenic variation in *Spartina alterniflora. Estuarine and Coastal Marine Science* 5:707–14.

Neubauer, S. C. 2008. Contributions of mineral and organic components to tidal freshwater marsh accretion. *Estuarine, Coastal and Shelf Science* 78:78–88.

Neubauer, S. C. 2011. Ecosystem responses of a tidal freshwater marsh experiencing saltwater intrusion and altered hydrology. *Estuaries and Coasts.* Oneline First article. DOI 10.1007/s12237-011-9455-x.

Neubauer, S. C., I. C. Anderson, J. A. Constantine, and S. A. Kuehl. 2002. Sediment deposition and accretion in a mid-Atlantic (U.S.A.) tidal freshwater marsh. *Estuarine, Coastal and Shelf Science* 54:713–27.

Neuman, M. J., G. Ruess, and W. Able. 2004. Species composition and food habits of dominant fish predators in salt marshes of an urbanized estuary, the Hackensack Meadowlands, New Jersey. *Urban Habitats* 2:62–82.

Neumeier, U., and P. Ciavola. 2004. Flow resistances and associated sedimentary processes in a *Spartina maritima* salt marsh. *Journal of Coastal Research* 20:435–47.

New Brunswick Department of the Environment and Local Government. 2002. A Coastal Areas Protection Policy for New Brunswick. The Sustainable Planning Branch, Fredericton, NB.

New Brunswick Department of Natural Resources and Energy and Department of Environment and Local Government. 2002. New Brunswick Wetlands Conservation Policy. Fredericton, NB.

New Brunswick Maritime Ringlet Recovery Team. 2005. Recovery Strategy and Action Plan for the Maritime Ringlet (*Coenonympha nipisiquit*) in New Brunswick. New Brunswick Department of Natural Resources, Fredericton, NB.

New Hampshire Department of Environmental Services. 2009. Aquatic Resource Mitigation. Concord, NH. Fact sheet WD-WB-17. http://des.nh.gov/organization/commissioner/pip/factsheets/wet/documents/wb-17.pdf.

Newell, S. Y., and D. Porter. 2000. Microbial secondary production from saltmarsh-grass shoots, and its known and potential fates. In: M. P. Weinstein and D. A. Kreeger (eds.). *Concepts and Controversies in Tidal Marsh Ecology.* Kluwer Academic Publishers, Dordrecht, The Netherlands. 159–85.

Newson, J. D. (ed.). 1968. *Proceedings of the Marsh and Estuary Management Symposium.* 1967 July 19–20; Louisiana State University, Baton Rouge, LA.

Nicholls, R. J., F. M. J. Hoozemans, and M. Marchand. 1999. Increasing flood risk and wetland losses due to global sea-level rise: regional and global analyses. *Global Environmental Change* 9:S69-S87.

Nichols, G. E. 1920. The vegetation of Connecticut. VII. The associations of depositing areas along the seacoast. *Bulletin of the Torrey Botanical Club* 47:511–48.

Niedowski, N. L. 2000. New York State Salt Marsh Restoration and Monitoring Guidelines. New York State Department of State, Albany, NY and New York Department of Environmental Conservation, East Setauket, NY. www.dec.ny.gov/docs/wildlife_pdf/saltmarsh.pdf.

Niering, W. A. 1961. Tidal marshes: their use in scientific research. In: Connecticut's Coastal

Marshes: A Vanishing Resource. Connecticut Arboretum, Connecticut College, New London, CT. Bulletin 12. 2–7.

Niering, W. A. 1966. *The Life of the Marsh.* McGraw-Hill, New York.

Niering, W. A., and R. S. Warren. 1980. Vegetation patterns and processes in New England salt marshes. *BioScience* 30:301–7.

Nikitina, D. L., J. E. Pizzuto, R. A. Schwimmer, and K. W. Ramsey. 2000. An updated Holocene sea-level curve for the Delaware coast. *Marine Geology* 171:7–20.

Ning, Z. H., R. E. Turner, T. Doyle, and K. Abdollahi. 2003. Preparing for a Changing Climate: The Potential Consequences of Climate Variability and Change. Gulf Coast Region. Findings of the Gulf Coastal Regional Assessment. Gulf Coast Regional Climate Change, Baton Rouge, LA. www.usgcrp.gov/usgcrp/Library/nationalassessment/gulfcoast/gulfcoast-brief.pdf.

Nixon, S. W. 1980. Between coastal marshes and coastal water—a review of twenty years of speculation and research in the role of salt marshes in estuarine productivity and water chemistry. In: P. Hamilton and K. B. MacDonald (eds.). *Wetland Processes with Emphasis on Modelling.* Plenum Press, New York. 437–525.

Nixon, S. W. 1982. The Ecology of New England High Salt Marshes: A Community Profile. U.S. Fish and Wildlife Service, Washington, DC. FWS/OBS-81/55.

Nixon, S. W., and C. A. Oviatt. 1973. Analysis of local variation in the standing crop of *Spartina alterniflora. Botanica Marina* 16:103–9.

Nixon, S. W., C. A. Oviatt, J. Frithsen, and B. Sullivan. 1986. Nutrients and the productivity of estuarine and coastal marine ecosystems. *Journal of the Limnology Society of South Africa* 12:43–71.

Noe, G. B., and J. B. Zedler. 2000. Differential effects of four abiotic factors on the germination of salt marsh annuals. *American Journal of Botany* 87:1679–92.

Nordlie, F. G. 2003. Fish communities of estuarine salt marshes of eastern North America, and comparisons with temperate estuaries of other continents. *Reviews in Fish Biology and Fisheries* 13:281–325.

North Carolina Department of Environment and Natural Resources. 2009. North Carolina Wetland Assessment Method. Raleigh, NC.

Nova Scotia Department of Agriculture and Fisheries. 2003. Dykeland history archive. Accessed at www.gov.ns.ca/nsaf/rs/marsh/history.htm. Excerpt from a book entitled *Maritime Dykelands—The 350 Year Struggle* published by the Department in 1987.

Nova Scotia Department of Environment and Labour. 2006. Wetland Designation Policy. February 14, 2006. Halifax, NS.

Nova Scotia Environment. 2011. NovaWET Version 3.0. Halifax, NS. www.gov.ns.ca/nse/wetland/assessing.wetland.function.asp.

Nuttle, W. K. 1988. The extent of lateral water movement in the sediments of a New England salt marsh. *Water Resources Research* 24:2077–85.

Nyman, J. A., and R. H. Chabreck. 1995. Fire in coastal marshes: history and recent concerns. In: S. I. Cerulean and R. T. Engstrom (eds.). Fire in Wetlands: A Management Perspective. *Proceedings of the Tall Timbers Fire Ecology Conference.* Number 19. Tall Timbers Research Station, Tallahassee, FL. 134–41. www.rnr.lsu.edu/nyman/Fire%20in%20Coastal%20Marshes.pdf.

Nyman, J. A., C. R. Crozier, and R. D. DeLaune. 1995. Roles and patterns of hurricane sedimentation in an estuarine marsh landscape. *Estuarine, Coastal and Shelf Science* 40:665–79.

Nyman, J. A., R. J. Walters, R. D. DeLaune, and W. H. Patrick, Jr. 2006. Marsh vertical accretion via vegetative growth. *Estuarine, Coastal and Shelf Science* 69:370–80.

Oberrecht, K. 1997. Classification of Estuaries. Accessed at www.harborside.com/~ssnerr/class1.htm.

O'Connell, J. L., and J. A. Nyman. 2010. Marsh terraces in coastal Louisiana increase marsh edge and densities of waterbirds. *Wetlands* 30:125–35.

Odum, E. P. 1961. The role of tidal marshes in estuarine production. *New York State Conservationist* 15:12–15.

Odum, E. P. 1980. The status of three ecosystem-level hypotheses regarding salt marsh estuaries: tidal subsidy, outwelling, and detritus-based food chains. In. V. S. Kennedy (ed.). *Estuarine Perspectives.* Academic Press, New York. 485–95.

Odum, E. P. 2000. Tidal marshes as outwelling/pulsing systems. In: M. P. Weinstein and D. A. Kreeger (eds.). *Concepts and Controversies in Tidal Marsh Ecology.* Kluwer Academic Publishers, Dordrecht, The Netherlands. 3–7.

Odum, E. P., and A. A. de la Cruz. 1967. Particulate organic detritus in a Georgia salt marsh-estuarine ecosystem. In: G. H. Lauff (ed.). *Estuaries.* American Association for the Advancement of Science, no. 83. 383–88.

Odum, W. E. 1988. Comparative ecology of tidal freshwater and salt marshes. *Annual Review of Ecology and Systematics* 19:147–76.

Odum, W. E., J. S. Fisher, and J. C. Pickral. 1979. Factors controlling the flux of particulate organic carbon from estuarine wetlands. In: R. J. Livingston (ed.). *Ecological Processes in Coastal and Marine Systems.* Plenum Press, New York. 69–79.

Odum, W. E., C. C. McIvor, and T. J. Smith III. 1982. The Ecology of the Mangroves of South Florida: A Community Profile. U.S. Fish and Wildlife Service, Washington, DC. FWS/OBS-81/24.

Odum, W. E., T. J. Smith III, J. K. Hoover, and C. C. McIvor. 1984. The Ecology of Tidal Freshwater Marshes of the United States East Coast: A Community Profile. U.S. Fish and Wildlife Service, Washington, DC. FWS/OBS-83/17.

Odum, W. E., E. P. Odum, and H. T. Odum. 1995. Nature's pulsing paradigm. *Estuaries* 18:547–55.

Oertel, G. F., and H. J. Woo. 1994. Landscape classification and terminology for marsh in deficit coastal lagoons. *Journal of Coastal Research* 10:919–32.

Officer, C. B. 1976. Physical oceanography of estuaries. *Oceanus* 19:3–9.

Ofiara, D. D., and J. J. Seneca. 2001. *Economic Losses from Marine Pollution: A Handbook for Assessment.* Island Press, Washington, DC.

Ohimain, E. I. 2003. Available options for the bioremediation and restoration of abandoned pyretic dredge spoils causing the death of fringing mangroves in the Niger Delta. Presented at the International Biohydrometallurgy Symposium; 2003 September 14–19; Athens, Greece.

Olsen, S., and V. Lee. 1993. Rhode Island lagoons. In: O. T. Magoon. *The Management of Coastal Lagoons and Enclosed Bays.* American Society of Civil Engineers, New York. 221–33.

O'Meara, G. F. 1992. The eastern saltmarsh mosquito *Aedes sollicitans. Wing Beats* 3:5. www.rci.rutgers.edu/~insects/sp7.html.

Opler, P. A., and V. Malikul. 1992. *Peterson Field Guides: Eastern Butterflies.* Houghton Mifflin, New York.

Ornes, W. H., and D. I. Kaplan. 1989. Macronutrient status of tall and short forms of *Spartina alterniflora* in a South Carolina salt marsh. *Marine Ecology Progress Series* 55:63–72.

Orson, R. A., R. S. Warren, and W. A. Niering. 1987. Development of a tidal marsh in a New England river valley. *Estuaries* 10:20–27.

Orson, R. A., R. S. Warren, and W. A. Niering. 1998. Interpreting sea level rise and rates of vertical marsh accretion in a southern New England tidal salt marsh. *Estuarine, Coastal and Shelf Science* 47:419–29.

Orson, R. A., R. L. Simpson, and R. E. Good. 1990. Rates of sediment accumulation in a tidal freshwater marsh. *Journal of Sedimentary Petrology* 60:859–69.

Orson, R. A., R. L. Simpson, and R. E. Good. 1992. The paleoecological development of a late Holocene, tidal freshwater marsh of the Upper Delaware River Estuary. *Estuaries* 15:130–46.

Orth, R. J., and K. L. Heck. 1980. Structural components of eelgrass (*Zostera marina*) meadows in the Lower Chesapeake Bay—fishes. *Estuarine Research* 3:278–88.

Osgood, D. T., M. C. F. V. Santos, and J. Zieman. 1995. Sediment physio-chemistry associated with natural marsh development on storm-deposited sandbar. *Marine Ecology* 120:271–83.

Osteen, D. V., A. G. Eversole, and R. W. Christie. 1989. Spawning utilization of an abandoned ricefield by Blueback Herring. In: R. R. Sharitz and J. W. Gibbons (eds.). *Freshwater Wetlands and Wildlife.* Proceedings of a symposium held 1989 March 24–27; Charleston, South Carolina. CONF-8603101. U.S. Department of Energy, Oak Ridge, TN. DOE Symposium Series no. 61. 552–65.

Ovenshine, A. T., and S. Bartsch-Winkler. 1978. Portage, Alaska: Case history of an earthquake's impact on an estuarine system. In: M. L. Wiley (ed.). *Estuarine Interactions.* Academic Press, New York. 275–84.

Pacific Flyway Council. 2002. Pacific Flyway Management Plan for Pacific Brant. Pacific Flyway Study Committee, U.S. Fish and Wildlife Service, Portland, OR.

Packer, D. B. (ed.). 2001. Assessment and Characterization of Salt Marshes in the Arthur Kill (New York and New Jersey) Replanted after a Severe Oil Spill. U.S. Department of Commerce, National Oceanic and Atmospheric Administration, National Marine Fisheries Service, Northeast Fisheries Science Center, Woods Hole, MA. National Oceanic and Atmospheric Administration Technical Memorandum NMFS-NE-167. www.nefsc.noaa.gov/publications/tm/tm167/tm167.pdf.

Packham, J. R., and A. J. Willis. 1997. *Ecology of Dunes, Salt Marshes and Shingle*. Chapman and Hall, London.

Padgett, D. E., and J. L. Brown. 1999. Effects of drainage and soil organic content on growth of *Spartina alterniflora* (Poaceae) in an artificial salt marsh mesocosm. *American Journal of Botany* 86:697–702.

Page, G. W., W. D. Shuford, J. E. Kjelmyr, and L. E. Stenzel. 1992. Shorebird Numbers in Wetlands of the Pacific Flyway: A Summary of Counts from April 1988 to January 1992. Point Reyes Bird Observatory, Stinson Beach, CA.

Page, G. W., L. E. Stenzel, and J. E. Kjelmyr. 1999. Overview of shorebird abundance and distribution in wetlands of the Pacific Coast of the contiguous United States. *Condor* 101:461–71.

Park, R. A., T. V. Armentano, and C. L. Cloonan. 1986. Predicting the effects of sea level rise on coastal wetlands. In: J. G. Titus (ed.). *Effects of Changes in Stratospheric Ozone and Global Climate,* Vol. 4: *Sea Level Rise.* U.S. Environmental Protection Agency, Washington, DC. 129–52.

Park, R. A., M. S. Trehan, P. W. Mausel, and R. C. Howe. 1989. The Effects of Sea Level Rise on U.S. Coastal Wetlands and Lowlands. Final Report for Cooperative Agreement CR814578-01, U.S. Environmental Protection Agency. Holcomb Research Institute (HRI), Butler University, Indianapolis, IN. HRI Report 164.

Park, R. A., J. K. Lee, and D. Canning. 1993. Potential effects of sea level rise on Puget Sound wetlands. *Geocarto International* 8:99–110.

Parkes, G. S., L. A. Ketch, C. T. O'Reilly, J. Shaw, and A. Ruffman. 1999. The Saxby Gale of 1869 in the Canadian Maritimes. A Case Study of Flooding Potential in the Bay of Fundy. Environment Canada, Canadian Hurricane Centre, Dartmouth, Nova Scotia, Canada.

Parnell, J. F., D. M. DuMond, and R. N. Needham. 1978. A Comparison of Plant Succession and Bird Utilization on Diked and Undiked Dredged Material Islands in North Carolina Estuaries. U.S. Army Engineer Waterways Experiment Station, Vicksburg, MS. Technical Report D-78-9.

Parrondo, R. T., J. G. Gosselink, and C. S. Hopkinson. 1978. Effects of salinity and drainage on the growth of three salt marsh grasses. *Botanical Gazette* 139:102–7.

Parry, M. L., O. F. Canziani, J. P. Palutikof, P. J. van der Linden, and C. E. Hanson (eds.). 2007. *Climate Change 2007—Impacts, Adaptation and Vulnerability.* Contribution of Working Group II to the Fourth Assessment Report of the Intergovernmental Panel on Climate Change. Cambridge University Press, Cambridge, UK.

Parsons, K. A., and A. A. de la Cruz. 1980. Energy flow and grazing behavior of conocephaline grasshoppers in a *Juncus roemerianus* marsh. *Ecology* 61:1045–50.

Parsons, L. 2010. Year one of the Giacomini Wetland Restoration Project: Analysis of Changes in Water Quality Conditions in the Project Area and Downstream. Report to the State Water Control Board [CA]. www.nps.gov/pore/parkmgmt/upload/planning_giacomini_wrp_restoration_report_water_quality_1001223.pdf.

Patrick, W. H., Jr., and R. D. DeLaune. 1990. Subsidence, accretion and sea level rise in south San Francisco Bay marshes. *Limnology and Oceanography* 35:1389–95.

Partridge, T. R., and J. B. Wilson. 1987. Salt tolerance of salt marsh plants of Otago, New Zealand. *New Zealand Journal of Botany* 25:559–66.

Patriquin, D. G. 1981. The general biology of salt marshes. In: A. Hatcher and D. G. Patriquin (eds.). Salt Marshes in Nova Scotia. A Status Report of the Salt Marsh Working Group. Institute of Resource and Environmental Studies and Department of Biology, Dalhousie University, Halifax, Nova Scotia. 4–27.

Patterson, S. G. 1986. Mangrove Community Boundary Interpretation and Detection of Areal Changes in Marco Island, Florida: Application of Digital Image Processing and Remote Sensing Techniques. U.S. Fish and Wildlife Service, Washington, DC. Biological Report 86(10).

Pattullo, J., W. Munk, R. Revelle, and E. Strong 1955. The seasonal oscillation in sea level. *Journal of Marine Research* 14:88–156.

Pearcy, R. W., D. E. Bayer, and S. L. Ustin. 1981. Salinity-Productivity relationships of selected plant species from Suisan Marsh, California. California Water Resources Center, University of California, Davis, CA.

Pearlstine, L. G., W. M. Kitchens, P. J. Latham, and R. D. Bartleson. 1993. Tide gate influences on a tidal marsh. *Water Resources Bulletin* 29:1009–19.

Pearse, A. S. 1936. Estuarine animals at Beaufort, North Carolina. *Journal of the Elisha Mitchell Scientific Society* 32:174–222. (Now the *Journal of the North Carolina Academy of Science*)

Pearse, A. S., H. J. Humm, and G. W. Wharton. 1942. Ecology of sand beaches at Beaufort, North Carolina. *Ecological Monographs* 12:135–90.

Pederson, D. C., D. M. Peteet, K. Kurdyla, and T. Guilderson. 2005. Medieval warming, Little Ice Age, and European impact on the environment during the last millennium in the lower Hudson River Valley, New York, USA. *Quaternary Research* 63:238–49.

Peinado, M., F. Alcaraz, J. Delgadillo, M. De La Cruz, J. Alvarez, and J. L. Aguirre. 1994. The coastal marshes of California and Baja California. *Vegetatio* 110:55–66.

Peltier, W. R. 2001. Global isostatic adjustment and modern instrumental records of sea level history. In: B. C. Douglas, M. S. Kearney, and S. P. Leatherman (eds.). *Sea Level Rise: History and Consequences.* Academic Press, San Diego, CA.

Pendleton, L. 2008. The economic and market value of coasts and estuaries: What's at stake? *National Wetlands Newsletter* 30:11–13.

Penfound, W. T., and E. S. Hathaway. 1938. Plant communities in the marshlands of southeastern Louisiana. *Ecological Monographs* 8:1–56.

Penhallow, D. P. 1907. A contribution to our knowledge of the origin and development of certain marsh lands on the coast of New England. *Transactions of the Royal Society of Canada* 1:13–45.

Pennings, S. C., and M. D. Bertness. 2001. Salt marsh communities. In: M. E. Hay (ed.). *Marine Community Ecology.* Sinauer Associates, Sunderland, MA. Chapter 11; 289–316.

Pennings, S. C., and R. M. Callaway. 1992. Salt marsh plant zonation: the relative importance of competition and physical factors. *Ecology* 73:681–90.

Pennings, S. C., and B. R. Silliman. 2005. Linking biogeography and community ecology: latitudinal variation in plant-herbivore interaction strength. *Ecology* 86:2310–19.

Pennings, S. C., E. L. Siska, and M. D. Bertness. 2001. Latitudinal differences in plant palatability in Atlantic coast salt marshes. *Ecology* 82:1344–59.

Pennings, S. C., L. E. Stanton, and J. S. Brewer. 2002. Nutrient effects on the composition of salt marsh plant communities along the southern Atlantic and Gulf coasts of the United States. *Estuaries* 25:1164–73.

Pennings, S. C., E. R. Selig, L. T. Houser, and M. D. Bertness. 2003. Geographic variation in positive and negative interactions among salt marsh plants. *Ecology* 84:1527–38.

Pennings, S. C., M. B. Grant, and M. D. Bertness. 2005. Plant zonation in low-latitude salt marshes: disentangling the role of flooding, salinity, and competition. *Journal of Ecology* 93:159–67.

Pepper, M. A. 2008. Salt marsh bird community responses to open marsh water management. University of Delaware, Newark, DE. M.S. thesis.

Pepper, M. A., and W. G. Shriver. 2010. Effects of open marsh water management on the reproductive success and nesting ecology of seaside sparrows in tidal marshes. *Waterbirds* 33:381–88.

Percy, J. A. 1996a. Tides of change: natural processes in the Bay of Fundy. Bay of Fundy Estuary Program. The Clean Annapolis River Project, Annapolis Royal, Nova Scotia. Fundy Issues no. 2.

Percy, J. A. 1996b. Dykes, dams, and dynamics: the impacts of coastal structures. Bay of Fundy Estuary Program. The Clean Annapolis River Project, Annapolis Royal, Nova Scotia. Fundy Issues no. 9.

Perillo, G. M. E., E. Wolanski, D. R. Cahoon, and M. M. Brinson (eds.). 2009. *Coastal Wetlands: An Integrated Ecosystem Approach.* Elsevier, Amsterdam, The Netherlands.

Perret, W. S. 1971. Phase IV, Biology. In: Louisiana Wild Life and Fisheries Commission. Cooperative Gulf of Mexico Estuarine Inventory and Study, Louisiana. New Orleans, LA.

Perry, C. L., and I. A. Mendelssohn. 2009. Ecosystem effects of expanding populations of Avicennia germinans in a Louisiana salt marsh. *Wetlands* 29:396–406.

Perry, J. E., and R. B. Atkinson. 1997. Plant diversity along a salinity gradient of four marshes on the York and Pamunkey Rivers in Virginia. *Castanea* 62:112–18.

Perry, J. E., and C. H. Herschner. 1999. Temporal changes in the vegetation pattern in a tidal freshwater marsh. *Wetlands* 19:90–99.

Perry, J. E., D. M. Bilkovic, K. J. Havens, and C. H. Hershner. 2009. Tidal freshwater wetlands of the Mid-Atlantic and southeastern United States. In: A. Barendregt, D. F. Whigham, and A. H. Baldwin (eds.). *Tidal Freshwater Wetlands.* Backhuys Publishers, Leiden, The Netherlands. Chapter 14; 157–66.

Perry, L., and K. Williams. 1996. Effect of salinity and flooding on seedlings of cabbage palm (*Sabal palmetto*). *Oecologia* 105:428–34.

Perry, M. C. 2010. The creation of forested wetlands for wildlife. *Jalaplavit* 2:16–18.

Perry, M. C., P. C. Osenton, and C. S. Stoll. 1998. Biological diversity of created forested wetlands in comparison to reference forested wetlands in the Bay watershed. In: G. D. Therres (ed.). *Conservation of Biological Diversity: A Key to the Restoration of the Chesapeake Bay Ecosystem and Beyond.* Maryland Department of Natural Resources, Annapolis, MD. 261–68.

Perry, M. C., A. S. Jacobs, S. B. Pugh, and P. C. Osenton. 2000. Evaluation of Forested Wetlands Constructed for Mitigation in Comparison to Natural Systems. U.S. Geological Survey, Patuxent Wildlife Research Center, Laurel, MD. www.pwrc.usgs.gov/resshow/perry/forwetlands.htm.

Peterson, B. J., and R. W. Howarth. 1987. Sulfur, carbon and nitrogen isotopes used to trace organic matter flow in the salt-marsh estuaries of Sapelo Island, Georgia. *Limnology and Oceanography* 32:1195–1213.

Peterson, B. J., R. W. Howarth, and R. H. Garritt. 1985. Multiple stable isotopes used to trace the flow of organic matter in estuarine food webs. *Science* 227:1361–63.

Peterson, C. H., and N. H. Peterson. 1979. The Ecology of Intertidal Flats of North Carolina: A Community Profile. U.S. Fish and Wildlife Service, Washington, DC. FWS/OBS-79/39.

Peterson, G. W., and R. E. Turner. 1994. The value of salt marsh edge vs interior as a habitat for fish and decapod crustaceans in a Louisiana tidal marsh. *Estuaries* 17:235–62.

Peterson, M. S., B. H. Comyns, J. R. Hendon, P. J. Bond, and G. A. Duff. 2000. Habitat use by early life-history stages of fishes and crustaceans along a changing estuarine landscape: differences between natural and altered shoreline sites. *Wetlands Ecology and Management* 8:209–19.

Petit, D. R., and K. L. Bildstein. 1986. Development of formation flying in juvenile White Ibises (*Eudocimus albus*). *Auk* 103:244–46.

Pezeshki, S. R. 2001. Wetland plant responses to soil flooding. *Environmental and Experimental Botany* 46:299–312.

Pezeshki, S. R., and J. L. Chambers. 1986. Effect of soil salinity on stomatal conductance and photosynthesis of green ash (*Fraxinus pennsylvanica*). *Canadian Journal of Forest Research* 16:569–73.

Pezeshki, S. R., and R. D. DeLaune. 1996. Responses of *Spartina alterniflora* and *Spartina patens* to rhizosphere oxygen deficiency. *Acta Oecologica* 17:365–78.

Pezeshki, S. R., R. D. DeLaune, and W. H. Patrick, Jr. 1989. Assessment of saltwater intrusion impact on gas exchange behavior of Louisiana Gulf Coast wetland species. *Wetlands Ecology and Management* 1:21–30.

Pezeshki, S. R., R. D. DeLaune, and W. H. Patrick, Jr. 1990. Flooding and saltwater intrusion: potential effects on survival and productivity of wetland forest along the U.S. Gulf Coast. *Forest Ecology and Management* 33/34:287–301.

Pezeshki, S. R., M. W. Hester, Q. Lin, and J. A. Nyman. 2000. The effects of oil spill and clean-up on dominant US Gulf coast marsh macrophytes: a review. *Environmental Pollution* 108:129–39.

Pfeiffer, W. J., and R. G. Wiegert. 1981. Grazers on Spartina and their predators. In: L. R. Pomeroy and R. G. Wiegert (eds.). *The*

Ecology of a Salt Marsh. Springer-Verlag, New York. 87–112.

Philip Williams & Associates, EDAW, H. T. Harvey & Associates, and Brown & Caldwell. 2006. South Bay Salt Pond Restoration Project. Final Alternatives Report. Prepared for the California State Coastal Conservancy, U.S. Fish and Wildlife Service, and California Department of Fish and Game. www.southbayrestoration.org/pdf_files/alternatives/final/Final_Alternatives_Rpt-TEXT.pdf.

Phleger, C. F. 1971. Effect of salinity on growth of a salt marsh grass. *Ecology* 52:908–11.

Piazza, B. P., P. D. Banks, and M. K. La Peyre. 2005. The potential for created oyster shell reefs as a sustainable shoreline protection strategy in Louisiana. *Restoration Ecology* 13:499–506.

Pickral, J. C., and W. E. Odum. 1977. Benthic detritus in a salt marsh creek. In: M. Wiley (ed.). *Estuarine Processes*. Volume II. *Circulation, Sediments, and Transfer of Material in the Estuary*. Academic Press, New York. 280–92.

Pidgeon, E. 2009. Carbon sequestration by coastal marine habitats: important missing links. In: D. d'A. Laffoley and G. Grimsditch (eds.). *The Management of Natural Coastal Carbon Sinks*. International Union for Conservation of Nature and Natural Resources (IUCN), Gland, Switzerland. 47–51.

Pielou, E. C. 1991. *After the Ice Age. The Return of Life to Glaciated North America*. University of Chicago Press, Chicago, IL.

Pielou, E. C., and R. D. Routledge. 1976. Salt marsh vegetation: latitudinal gradients in the zonation patterns. *Oecologia* 24:311–21.

Piersma, T., and Å. Lindström. 2004. Migrating shorebirds as integrative sentinels of global environmental change. *Ibis* 146 (Suppl. 1):61–69.

Pilon, P., and M. A. Kerr. 1984. Land Use Monitoring on Wetlands in the Southwestern Fraser Lowland, British Columbia. Lands Directorate, Environment Canada, Vancouver, British Columbia. Working Paper no. 34.

Pinckney, J. L., and R. G. Zingmark. 1993. Modeling the annual production of intertidal benthic microalgae in estuarine ecosystems. *Journal of Phycology* 29:396–407.

Platts, N. G., S. E. Shields, and J. B. Hull. 1943. Diking and pumping for control of sand flies and mosquitoes in Florida salt marshes. *Journal of Economic Entomology* 36:409–12.

Polidoro, B. A., K. E. Carpenter, L. Collins, N. C. Duke, A. M Ellison, J. C. Ellison, E. J. Farnsworth, E. S. Fernando, K. Kathiresan, N. E. Koedam, S. R. Livingstone, T. Miyagi, G. E. Moore, V. N. Nam, J. E. Ong, J. H. Primerva, S. G. Salmo, III, J. C. Sanciangco, S. Sukardjo, Y. Wang, and J. W. H. Yong. 2010. The loss of species: mangrove extinction risk and geographic areas of global concern. *PLoS ONE* 5: e10095, doi: 10.1371/journal.pone.0010095.

Poljakoff-Mayber, A., and J. Gale. 1975. *Plants in Saline Environments*. Springer-Verlag, New York.

Pomeroy, L. R. 1959. Algal productivity in salt marshes of Georgia. *Limnology and Oceanography* 4:386–97.

Pomeroy, L. R., and R. G. Wiegert (eds.). 1981. *The Ecology of a Salt Marsh*. Springer-Verlag, New York.

Pomeroy, L. R., W. M. Darley, E. L. Dunn, J. L. Gallagher, E. B. Haines, and D. M. Whitney. 1981. Primary production. In: L. R. Pomeroy and R. G. Wiegert (eds.). *The Ecology of a Salt Marsh*. Springer-Verlag, New York. 39–67.

Pore, N. A., and C. S. Barrientos. 1976. Storm Surge. New York Sea Grant Institute, Albany, NY. MESA New York Bight Atlas Monograph 6.

Portnoy, J., C. Roman, S. Smith, and E. Gwilliam. 2003. Estuarine habitat restoration at Cape Cod National Seashore—the Hatches Harbor prototype. *Park Science* 22:51–58. www.nps.gov/caco/naturescience/upload/HatchesHarborParkScience.pdf.

Portnoy, J. W. 1999. Salt marsh diking and restoration: biogeochemical implications of altered wetland hydrology. *Environmental Management* 24:111–20.

Portnoy, J. W., and I. Valiela. 1997. Short-term effects of salinity reduction and drainage on salt-marsh biogeochemical cycling and *Spartina* (cordgrass) production. *Estuaries* 20:569–78.

Posey, M. H., T. D. Alphin, H. Harwell, and B. Allen. 2005. Importance of low salinity areas for juvenile blue crabs, *Callinectes sapidus* Rathburn, in river-dominated estuaries of southeastern United States. *Journal of Experimental Marine Biology and Ecology* 319:81–100.

Potente, J. E. (ed.). 2010a. *Tidal Marshes of Long Island, New York*. Memoirs of the Torrey Botanical Society, Vol. 26.
Potente, J. E. 2010b. Fundamental flaws of open marsh water management. In: J. E. Potente (ed.). *Tidal Marshes of Long Island, New York*. Memoirs of the Torrey Botanical Society, Vol. 26:122–53.
Pough, R. H. 1961. Valuable vistas: a way to protect them. In: Connecticut's Coastal Marshes—A Vanishing Resource. Connecticut Arboretum, Connecticut College, New London, CT. Bulletin no. 12. 28–30.
Poulakis, G. R., J. M. Shenker, and D. S. Taylor. 2002. Habitat use by fishes after tidal reconnection of an impounded estuarine wetland in the Indian River Lagoon, Florida (USA). *Wetlands Ecology and Management* 10:51–69.
Poulin, P., É. Pelletier, V. G. Koutitonski, and U. Newmeier. 2009. Seasonal nutrient fluxes variability of northern salt marshes in the lower St. Lawrence Estuary. *Wetlands Ecology and Management* 17:655–73.
Price, C. H. 1984. Tidal migrations of the littoral salt marsh snail *Melampus bidentatus* Say. *Journal of Experimental Marine Biology and Ecology* 78:111–26.
Prince Edward Island (PEI), Department of Environment, Energy, and Forestry. 2003. A Wetland Conservation Policy for Prince Edward Island. Charlottetown, PEI, Canada. www.gov.pe.ca/photos/original/2007wetlands-po.pdf.
Pritchard, D. W. 1952. Salinity distribution and circulation in the Chesapeake Bay estuarine system. *Journal of Marine Research* 11:106–23.
Pritchard, D. W. 1967. What is an estuary? In: G. H. Lauff (ed.). *Estuaries*. American Association for the Advancement of Science, Washington, DC. AAAS Publication 83. 3–5.
Pritchard, E. S. (ed.). 2008. Fisheries of the United States 2008. National Marine Fisheries Service, Office of Science and Technology, Silver Spring, MD. Current Fisheries Statistics no. 2008.
Proffitt, C. E., S. E. Travis, and K. R. Edwards. 2003. Genotype and elevation influence *Spartina alterniflora* colonization and growth in a created salt marsh. *Ecological Applications* 13:180–92.
Proffitt, C. E., R. L. Chiasson, A. B. Owens, K. R. Edwards, and S. E. Travis. 2005. *Spartina alterniflora* genotype influences facilitation and suppression of high marsh species colonizing an early successional salt marsh. *Journal of Ecology* 93:404–16.
Province of Nova Scotia. 2011. Nova Scotia Wetland Conservation Policy. www.gov.ns.ca/nse/wetland/docs/Nova.Scotia.Wetland.Conservation.Policy.pdf.
Provost, M. W. 1973. Mean high water mark and use of tidelands in Florida. *Florida Scientist* 36:50–66.
Provost, M. W. 1976. Tidal datum planes circumscribing salt marshes. *Bulletin of Marine Science* 26:558–63.
Public Broadcasting Service. 2008. *Crash: A Tale of Two Species*. A film of the series—Nature. Posted online in February 2008. www.pbs.org/wnet/nature/episodes/crash-a-tale-of-two-species/video-full-episode/4772/.
Pung, O. J., C. B. Grinstead, K. Kersten, and C. L. Edenfield. 2008. Spatial distribution of hydrobiid snails in salt marsh along the Skidaway River in southeastern Georgia with notes on their larval trematodes. *Southeastern Naturalist* 7:717–28.
Pupedis, R. J. 1997. Aquatic insects of the West River and salt marshes of Connecticut. In: D. G. Casagrande (ed.). Restoration of an Urban Salt Marsh: An Interdisciplinary Approach. Yale School of Forestry and Environmental Studies, New Haven, CT. Bulletin no. 100. 162–77.
Purer, E. A. 1942. Plant ecology of the coastal salt marshes of San Diego County, California. *Ecological Monographs* 12:83–111.
Purinton, T. A., and D. C. Mountain. 1996. Tidal Crossing Handbook. A Volunteer Guide to Assessing Tidal Restriction. Parker River Clean Water Association, Byfield, MA. www.businessevision.info/parker_river/TCHandbook.html.
Pyle, R. M. 1995. *National Audubon Society Field Guide to North American Butterflies*. Alfred A. Knopf, New York.
Raabe, E. A., L. C. Roy, and C. C. McIvor. 2012. Tampa Bay coastal wetlands: nineteenth to twentieth century tidal marsh-to-mangrove conversion. *Estuaries and Coasts* 35:1145–62.
Rabenhorst, M. C. 2001. Soils of tidal and fringing wetlands. In: J. L. Richardson and M. J. Vepraskas (eds.). *Wetland Soils: Genesis, Hydrology, Landscapes, and*

Classification. Lewis Publishers, CRC Press, Boca Raton, FL. Chapter 13; 301–15.

Rabenhorst, M. C., D. S. Fanning, and S. N. Burch. 2002. Acid sulfate soils, formation. In: R. Lal (ed.). *Encyclopedia of Soil Science.* Marcel Dekker, New York. 14–18.

Rabenhorst, M. C., W. D. Hively, and B. R. James. 2007. Measurements of soil redox potential. *Soil Science Society of America Journal* 73:668–74.

Ragotzkie, R. A., L. R. Pomeroy, J. M. Teal, and D. C. Scott (eds.). 1959. *Proceedings of the Salt Marsh Conference;* 1958 March 25–28; Marine Institute of the University of Georgia, Sapelo Island, Georgia. www.biodiversitylibrary.org/bibliography/10197.

Rahmstorf, S. 2010. A new view on sea level rise. Nature Reports Climate Change. www.nature.com/climate/2010/1004/full/climate.2010.29.html.

Rai, P. K. 2008. Heavy metal pollution in aquatic ecosystems and its phytoremediation using wetland plants: an ecosustainable approach. *International Journal of Phytoremediation* 10:133–60.

Rampino, M. R., and J. E. Sanders. 1981. Episodic growth of Holocene tidal marshes in the northeastern United States: a possible indicator of eustatic sea-level fluctuations. *Geology* 9:63–67.

Rangeley, R., and D. Kramer. 1995. Use of rocky intertidal habitats by juvenile pollock *Pollachius virens. Marine Ecology Progress Series* 126:9–17.

Rankin, J. S. 1961. Salt marshes as a source of food. In: Connecticut's Coastal Marshes—A Vanishing Resource. Connecticut Arboretum, Connecticut College, New London, CT. Bulletin no. 12. 8–12.

Ranwell, D. S. 1972. *Ecology of Salt Marshes and Sand Dunes.* Chapman and Hall, London, England.

Raposa, K. 2003. Overwintering habitat selection by the mummichog, *Fundulus heteroclitus,* in a Cape Cod (USA) salt marsh. *Wetlands Ecology and Management* 11:175–82.

Raposa, K. 2008. Early ecological responses to hydrologic restoration of a tidal pond and salt marsh complex in Narragansett Bay, Rhode Island. *Journal of Coastal Research* 55 (Special Issue): 180–92.

Raposa, K. B., and C. A. Oviatt. 2000. The influence of contiguous shoreline type, distance from shore, and vegetation biomass on nekton community structure in eelgrass beds. *Estuaries* 23:46–55.

Raposa, K. B., and C. T. Roman. 2001a. Seasonal habitat-use patterns of nekton in a tide-restricted and unrestricted New England salt marsh. *Wetlands* 21:451–61.

Raposa, K. B., and C. T. Roman. 2001b. Monitoring Nekton in Shallow Estuarine Habitats. National Park Service, Cape Cod National Seashore, Long-term Coastal Ecosystem Monitoring Program, Wellfleet, MA. http://science.nature.nps.gov/im/monitor/protocols/caco_nekton.pdf.

Rathbun, G. B., J. P. Reid, and G. Carowan. 1990. Distribution and Movement Patterns of Manatees (*Trichechus manatus*) in Northwestern Peninsular Florida. Florida Marine Research Institute, St. Petersburg, FL. Florida Marine Research Publication no. 48.

Raup, H. M. 1959. Archaeology and salt marsh problems in the Taunton River valley, Massachusetts. In: *Proceedings of the Salt Marsh Conference;* 1958 March 25–28; Marine Institute of the University of Georgia, Sapelo Island, GA. 116–18.

Ray, G. C., B. P. Hayden, A. J. Bulger, Jr., and M. G. McCormick-Ray. 1992. Effects of global warming on the biodiversity of coastal-marine zones. In: R. L. Peters and T. E. Lovejoy (eds.). *Global Warming and Biological Diversity.* Yale University Press, New Haven, CT. 91–104.

Ray, G. L. 2005. Invasive Marine and Estuarine Animals of the Gulf of Mexico. U.S. Army Engineer Research and Development Center, Vicksburg, MS. ERDC/TN ANSRP-05-4.

Reddering, J. S. V., and I. C. Rust. 1990. Historical changes and sedimentary characteristics of Southern African estuaries. *South African Journal of Science* 86:425–28.

Reddy, C. M., T. I. Eglinton, A. Hounshell, H. K. White, L. Xu, R. B. Gaines, and G. S. Frysinger. 2002. The West Falmouth oil spill after thirty years: the persistence of petroleum hydrocarbons in marsh sediments. *Environmental Science and Technology* 36:4754–60.

Reddy, K. R., and R. D. DeLaune. 2008. *Biogeochemistry of Wetlands: Science and Applications.* CRC Press, Taylor & Francis Group, Boca Raton, FL.

Redfield, A. C. 1965. The ontogeny of a salt marsh estuary. *Science* 147:50–55.
Redfield, A. C. 1972. Development of a New England salt marsh. *Ecological Monographs* 42:201–37.
Redfield, A. C., and M. Rubin. 1962. The age of salt marsh peat and its relation to recent changes in sea level at Barnstable, Massachusetts. *Proceedings of the National Academy of Sciences of the United States* 48:1728–35.
Reed, A., and G. Moisan. 1971. The *Spartina* tidal marshes of the St. Lawrence Estuary and their importance to aquatic birds. *Naturaliste Canada* 98:905–22.
Reed, D. J. 1990. The impact of sea-level rise on coastal salt marshes. *Progress in Physical Geography* 14:465–81.
Reed, D. J. 1995. The response of coastal marshes to sea-level rise: survival or submergence? *Earth Surface Processes and Landforms* 20:39–48.
Reed, D. J., A. M. Commagere, and M. W. Hester. 2009. Marsh elevation response to Hurricanes Katrina and Rita and the effect of altered nutrient regimes. *Journal of Coastal Research* 54:166–73.
Reed, J. F. 1947. The relation of the *Spartinetum glabrae* near Beaufort, North Carolina, to certain edaphic factors. *American Midland Naturalist* 38:605–14.
Rees, H. W., J. P. Duff, S. Colville, and T. L. Chow. 1996. Soils of Selected Agricultural Areas of Moncton Parish, Westmorland County, New Brunswick. Centre for Land and Biological Resources Branch, Agriculture and Agri-Food Canada, Ottawa, Ontario. New Brunswick Soil Survey Report no. 15.
Reid, G. K. 1961. *Ecology of Inland Waters and Estuaries.* Van Nostrand Reinhold, New York.
Reidenbaugh, T. G., and W. C. Banta. 1980. Origin and effects of tidal wrack in a Virginia salt marsh. *Gulf Research Reports* 6:393–401.
Reimold, R. J. 1976. Grazing on wetland meadows. In: M. Wiley (ed.). *Estuarine Processes.* Volume I. *Uses, Stresses, and Adaptation to the Estuary.* Academic Press, New York. 219–25.
Reimold, R. J., and R. A. Linthurst. 1977. Primary Productivity of Minor Marsh Plants in Delaware, Georgia, and Maine. U.S. Army Engineer Waterways Experiment Station, Vicksburg, MS. Technical Report D-77-36.
Reimold, R. J., and W. H. Queen (eds.). 1974. *Ecology of Halophytes.* Academic Press, New York.
Reimold, R. J., R. A. Linthurst, and P. L. Wolf. 1975. Effects of grazing on a salt marsh. *Biological Conservation* 8:105–25.
Reinert, S. E., and M. J. Mello. 1995. Avian community structure and habitat use in a southern New England estuary. *Wetlands* 15:9–19.
Reinert, S. E., F. C. Golet, and W. R. DeRagon. 1981. Avian use of ditched and unditched salt marshes in southeastern New England: a preliminary report. *Proceedings from the 27th Annual Meeting of the Northeastern Mosquito Control Association;* 1981 November 2–4; Newport, Rhode Island. 24 pp.
Resh, V. H., S. S. Balling, M. A. Barnby, and J. N. Collins. 1980. Ecological impact of marshland recirculation ditches. *California Agriculture* 34:38–39.
Rey, J. R., and E. D. McCoy. 1986. Terrestrial arthropods of Northwest Florida salt marshes: Diptera (Insecta). *Florida Entomologist* 69:197–205.
Rey, J. R., and E. D. McCoy. 1997. Terrestrial arthropods. In: C. L. Coultas and Y. Hsieh (eds.). *Ecology and Management of Tidal Marshes: A Model from the Gulf of Mexico.* St. Lucie Press, Delray Beach, FL. 175–208.
Rey, J. R., J. Shaffer, D. Tremain, R. A. Crossman, and T. Kain. 1990. Effects of re-establishing tidal connections in two impounded subtropical marshes in fishes and physical conditions. *Wetlands* 10:27–47.
Rheinhardt, R. D. 2007. Tidal freshwater swamps of a lower Chesapeake Bay subestuary. In: W. H. Conner, T. W. Doyle, and K. W. Krause (eds.). *Ecology of Tidal Freshwater Forested Wetlands of the Southeastern United States.* Springer, Dordrecht, The Netherlands. Chapter 7; 161–81.
Rice, D., J. Rooth, and J. C. Stevenson. 2000. Colonization and expansion of *Phragmites australis* in upper Chesapeake Bay tidal marshes. *Wetlands* 20:280–99.
Richard, G. A. 1978. Seasonal and environmental variations in sediment accretion in a Long Island salt marsh. *Estuaries* 1:29–35.
Riepe, D. 2010. Open marsh water management: impacts on tidal wetlands. In: J. E. Potente (ed.). *Tidal Marshes of Long Island,*

New York. Memoirs of the Torrey Botanical Society, Vol. 26:80–101.

Roach, E. R., M. C. Watzin, J. D. Scurry, and J. B. Johnston. 1987. Wetland changes in coastal Alabama. In: T. A. Lowry (ed.). *Proceedings of Symposium on the Natural Resources of Mobile Bay Estuary;* 1987 February 10–12; Mobile, Alabama. Auburn University, Alabama Sea Grant Extension Service and Alabama Cooperative Extension Service. 92–101.

Rochlin, I., T. Iwanejko, M. E. Dempsey, and D. V. Ninivaggi. 2009. Geostatistical evaluation of integrated marsh management impact on mosquito vectors using before-after-control impact (BACI) design. *International Journal of Health Geographics* 8:35–54.

Roe, G. 2006. In defense of Milankovitch. *Geophysical Research Letters* 33: L24703.

Roegner, G. C., H. L. Diefenderfer, A. B. Borde, R. M. Thom, E. M. Dawley, A. H. Whiting, S. A. Zimmerman, and G. E. Johnson. 2008. Protocols for Monitoring Habitat Restoration Projects in the Lower Columbia River and Estuary. Pacific Northwest National Laboratory, Richland, WA. PNNL-15793. www.pnl.gov/main/publications/external/technical_reports/PNNL-15793.pdf.

Rogers, B. D., and W. H. Herke. 1985. Temporal Patterns and Size Characteristics of Migrating Juvenile Fishes and Crustaceans in a Louisiana Marsh. Louisiana Cooperative Fishery Research Unit, Louisiana Agricultural Experiment Station, Louisiana State University, Baton Rouge, LA. Research Report no. 5.

Rogers, D. R., B. D. Rogers, and W. H. Herke. 1994. Structural marsh management effects on coastal fishes and crustaceans. *Environmental Management* 18:351–69.

Rogers, S. G., T. E. Targett, and S. B. Van Sant. 1984. Fish-nursery use in Georgia salt-marsh estuaries: the influence of springtime freshwater conditions. *Transactions of the American Fisheries Society* 113:596–606.

Rogerson, A., A. Jacobs, and A. Howard. 2010. Condition of Wetlands in the St. Jones River Watershed. Delaware Department of Natural Resources and Environmental Control, Water Resources Division, Watershed Assessment Section, Dover, DE. www.dnrec.delaware.gov/Admin/DelawareWetlands/Documents/St.%20Jones%20Watershed%20Wetland%20Condition%20Report%20Final.pdf.

Roland, A. E. 1982. Geological Background and Physiography of Nova Scotia. The Nova Scotian Institute of Science, Halifax, NS.

Roland, R. M., and S. L. Douglass. 2005. Estimating wave tolerance of *Spartina alterniflora* in coastal Alabama. *Journal of Coastal Research* 21:453–63.

Roman, C. T., and D. M. Burdick (eds.). 2012. *Tidal Marsh Restoration: A Synthesis of Science and Management*. Island Press, Washington, DC.

Roman, C. T., and F. C. Daiber. 1984. Aboveground and belowground primary production dynamics of two Delaware Bay tidal marshes. *Bulletin of the Torrey Botanical Club* 111:34–41.

Roman, C. T., and F. C. Daiber. 1989. Organic carbon flux through a Delaware salt marsh: tidal exchange, particle size distribution, and storms. *Marine Ecology Progress Series* 54:149–56.

Roman, C. T., W. A. Niering, and R. S. Warren. 1984. Salt marsh vegetation change in response to tidal restriction. *Environmental Management* 8:141–50.

Roman, C. T., K. W. Able, M. A. Lazzari, and K. L. Heck. 1990. Primary production of angiosperm and macroalgae dominated habitats in a New England salt marsh: a comparative analysis. *Estuarine, Coastal and Shelf Science* 30:35–45.

Roman, C. T., J. A. Peck, J. R. Allen, J. W. King, and P. G. Appleby. 1997. Accretion of a New England (U.S.A.) salt marsh in response to inlet migration, storms, and sea-level rise. *Estuarine, Coastal and Shelf Science* 45:717–27.

Roman, C. T., N. Jaworski, F. T. Short, S. Findlay, and R. S. Warren. 2000. Estuaries of the northeastern United States: habitat and land use signatures. *Estuaries* 23:743–64.

Roman, C. T., M-J. James-Pirri, and J. F. Heltsche. 2001. Monitoring Salt Marsh Vegetation. A Protocol for the Long-term Coastal Ecosystem Monitoring Program at Cape Cod National Seashore. National Park Service, Cape Cod National Seashore, Wellfleet, MA. http://science.nature.nps.gov/im/monitor/protocols/caco_marshveg.pdf.

Roman, C. T., K. B. Raposa, S. C. Adamowicz, M-J. James-Pirri, and J. G. Catena. 2002.

Quantifying vegetation and nekton response to tidal restoration of a New England salt marsh. *Restoration Ecology* 10:450–60.

Ross, M. S., J. J. O'Brien, and L. de Silveira Lobo Sternberg. 1994. Sea-level rise and the reduction of pine forests in the Florida Keys. *Ecological Applications* 4:144–56.

Rountree, R. A. 1992. Fish and macroinvertebrate community structure and habitat use patterns in salt marsh creeks of southern New Jersey, with a discussion of marsh carbon export. Rutgers University, New Brunswick, NJ. Ph.D. dissertation.

Rousseau, J. 1967. Pour une esquisse biogéographique du Saint-Laurent. *Cahiers de Géographie du Québec* 11:181–241.

Rozas, L. P. 1995. Hydroperiod and its influence on nekton use of the salt marsh: a pulsing ecosystem. *Estuaries* 18:579–90.

Rozas, L. P., and T. J. Minello. 2001. Marsh terracing as a wetland restoration tool for creating fishery habitat. *Wetlands* 21:327–41.

Rozas, L. P., and T. J. Minello. 2010. Nekton density patterns in tidal ponds and adjacent wetlands related to pond size and salinity. *Estuaries and Coasts* 33:652–67.

Rozsa, R. 1995. Human impacts on tidal wetlands: history and regulations. In: G. D. Dreyer and W. A. Niering (eds.). Tidal Marshes of Long Island Sound: Ecology, History and Restoration. The Connecticut College Arboretum, New London, CT. Bulletin no. 34. 42–50.

Rozsa, R. 2005. Twenty-five years of tidal wetland restoration in Connecticut. *Proceedings of the 14th Biennial Coastal Zone Conference;* 2005 July 17–21; New Orleans, LA. www.csc.noaa.gov/cz/CZ05_Proceedings/pdf%20files/Rozsa.pdf.

Rubec, C. D. A. 1994. Wetland Policy Implementation in Canada. Proceedings of a national workshop; 1994 June 12–14; Winnipeg and Stonewall, Manitoba. North America Wetlands Conservation Council, Ottawa, Ontario. Report no. 94-1.

Rubec, C. D. A., and A. R. Hanson. 2009. Wetland mitigation and compensation: Canadian experience. *Wetlands Ecology and Management* 17:3–14.

Russell, H. S. 1976. *A Long, Deep Furrow. Three Centuries of Farming in New England.* University of New England Press, Hanover, NH.

Russell-Hunter, W. D., M. L. Apley, and R. D. Hunter. 1972. Early life-history of Melampus and the significance of semilunar synchrony. *Biological Bulletin* 143:623–56.

Russo, M. (ed.). 1991. Final Report on Horr's Island: The Archeology of Archaic and Glades Settlement and Subsistence Patterns. University of Florida, Institute of Archeology and Paleoenvironmental Studies, Gainesville, FL.

Rutherford, S., and S. D'Hondt. 2000. Early onset and tropical forcing of 100,000-year Pleistocene glacial cycles. *Nature* 408:72–75.

Rybczyk, J. M., X. W. Zhang, J. W. Day, Jr., I. Hesse, and S. Feagley. 1995. The impact of Hurricane Andrew on tree mortality, litterfall, nutrient flux, and water quality in a Louisiana coastal swamp forest. *Journal of Coastal Research* 21:340–53.

Ryer, C. H., J. van Montfrans, and R. J. Orth. 1990. Utilization of a seagrass meadow and tidal marsh creek by blue crabs *Callinectes sapidus*. II. Spatial and temporal patterns of molting. *Bulletin of Marine Science* 46:95–104.

Saenger, P. 2002. *Mangrove Ecology, Silviculture and Conservation.* Kluwer Academic Publishers, Dordrecht, The Netherlands.

Salinas, L. M., R. D. DeLaune, and W. H. Patrick, Jr. 1986. Changes occurring along a rapidly submerging coastal area: Louisiana, USA. *Journal of Coastal Research* 2:269–84.

Saltonstall, K. 2002. Cryptic invasion by a non-native genotype of the common reed, *Phragmites australis,* into North America. *Proceedings of the National Academy of Sciences of the United States* 99:2445–49.

Salzman, J. E., and J. B. Ruhl. 2007. "No Net Loss": instrument choice in wetlands protection. In: J. Freeman and C. Kolstad (eds.). *Moving to Markets in Environmental Regulations: Twenty Years of Experience.* Oxford University Press. New York. 323–52.

Sánchez-Núñez, D. A., and J. E. Mancera-Pineda. 2011. Flowering pattern in three neotropical mangrove species: evidence from a Caribbean island. *Aquatic Botany.* doi: 10.1016/j.aquabot.2011.02.005.

Sandifer, P. A., J. V. Miglarese, D. R. Calder, J. J. Manzi, and L. S. Barclay. 1980. Ecological Characterization of the Sea Island Coastal Region of South Carolina and Georgia. Volume III. Biological Features of the

Characterization Area. U.S. Fish and Wildlife Service, Office of Biological Sciences, Washington, DC. FWS/OBS-79/42.

Sasser, C. E., J. G. Gosselink, E. M. Swenson, C. M. Swarzenski, and N. C. Leibowitz. 1996. Vegetation, substrate and hydrology in floating marshes in the Mississippi river delta plain wetlands, USA. *Vegetatio* 122:129–42.

Sasser, C. E., J. G. Gosselink, G. O. Holm, Jr., and J. M. Visser. 2009. Tidal freshwater wetlands of the Mississippi River delta. In: A. Barendregt, D. F. Whigham, and A. H. Baldwin (eds.). *Tidal Freshwater Wetlands*. Backhuys Publishers, Leiden, The Netherlands. Chapter 15; 167–78.

Saunders, R. 2007. Feasting and foodways at Fig Island 38CH42. In: We Are What They Ate: A History of Food in South Carolina. South Carolina Archaeology Month, October 2007. Poster. The South Carolina Institute of Archeology & Anthropology, University of South Carolina, Columbia, SC.

Schelske, C. L., and E. P. Odum. 1961. Mechanisms maintaining high productivity in Georgia estuaries. *Proceedings of Gulf Caribbean Fisheries Institute* 14:75–80.

Schneider, R. W., and S. Useman. 2005. The possible role of plant pathogens in Louisiana's brown marsh syndrome. In: *Proceedings of the 14th Biennial Coastal Zone Conference;* 2005 July 17–21; New Orleans, LA.

Schroeder, W. W. 1978. Riverine influence on estuaries: a case study. In: M. L. Wiley (ed.). *Estuarine Interactions*. Academic Press, New York. 347–64.

Schubel, J. R. 1974. Effects of tropical storm Agnes on the suspended solids of the northern Chesapeake Bay. In: R. J. Gibbs (ed.). *Suspended Solids in Water.* Plenum Press, New York. 113–32.

Schubel, J. R. 1986. *The Life & Death of the Chesapeake Bay.* Maryland Sea Grant Publication, University of Maryland, College Park, MD.

Schubel, J. R. (ed.). 1992. Managing Freshwater Discharge to the San Francisco Bay/San Joaquin Delta Estuary: The Scientific Basis for an Estuarine Standard. San Francisco Estuary Project, U.S. Environmental Protection Agency, San Francisco CA.

Schuyler, A. E., S. B. Andersen, and V. J. Kolaga. 1993. Plant zonation changes in the tidal portion of the Delaware River. *Proceedings of the Academy of Natural Sciences of Philadelphia* 144:263–66.

Schuyt, K., and L. Brander. 2004. *The Economic Value of the World's Wetlands. Living Waters Conserving the Source of Life.* World Wildlife Fund, Zeist, The Netherlands.

Schwartz, A. M. 2003. Spreading mangroves: a New Zealand phenomena or global trend? *Water & Atmosphere* 11:8–10.

Sculthorpe, C. D. 1967. *The Biology of Aquatic Vascular Plants.* Koeltz Scientific Books, Köningstein, Germany. (Reprinted in 1985).

Seasholes, N. 2003. *Gaining Ground: A History of Landmaking in Boston, Massachusetts.* Massachusetts Institute of Technology Press, Cambridge, MA.

Sebold, K. R. 1992. From Marsh to Farm: The Landscape Transformation of Coastal New Jersey. U.S. Department of the Interior, National Park Service, Washington, DC. www.nps.gov/history/history/online_books/nj3/index.htm.

Seitz, R. D., R. N. Lipcius, and M. S. Seebo. 2005. Food availability and growth of the blue crab in seagrass and unvegetated nurseries of Chesapeake Bay. *Journal of Experimental Marine Biology and Ecology* 319:57–68.

Seliskar, D. M. 1983. Root and rhizome distribution as an indicator of upper salt marsh wetland limits. *Hydrobiologia* 107:231–36.

Seliskar, D. M. 1985. Morphometric variations of five tidal marsh hydrophytes along environmental gradients. *American Journal of Botany* 72:1340–52.

Seliskar, D. M., and J. L. Gallagher. 1983. The Ecology of Tidal Marshes of the Pacific Northwest Coast: A Community Profile. U.S. Fish and Wildlife Service, Washington, DC. FWS/OBS-82/32.

Seliskar, D. M., J. L. Gallagher, D. M. Burdick, and L. A. Mutz. 2002. The regulation of ecosystem function by ecotypic variation in the dominant plant: a *Spartina alterniflora* salt-marsh case study. *Journal of Ecology* 90:1–11.

Seneca, E. D., and S. W. Broome. 1992. Restoring tidal marshes in North Carolina and France. In: G. W. Thayer (ed.). *Restoring the Nation's Marine Environment.* Maryland Sea Grant Program, University of Maryland, College Park, MD. Chapter 2; 53–78.

Serodes, J-P., and J-P. Troude. 1984. Sedimentation cycle of a freshwater tidal flat in the St. Lawrence Estuary. *Estuaries* 7:119–27.

Sewall, J. O. 2008. Cenozoic Era. In: S. G. Philander (ed.). *Encyclopedia of Global Warming and Climate Change*. Vol. 1. Sage Publications, Thousand Oaks, CA. 171–73.

Shafer, D. J., B. Herczeg, D. W. Moulton, A. Sipocz, K. Jaynes, L. R. Rozas, C. P. Onuf, and W. Miller. 2002. Regional Guidebook for Applying the Hydrogeomorphic Approach to Assessing Wetland Functions of Northwest Gulf of Mexico Tidal Fringe Wetlands. U.S. Army Engineer Research and Development Center, Vicksburg, MS. ERDC/EL TR-02-5.

Shafer, D. J., R. Roland, and S. L. Douglass. 2003. Preliminary Evaluation of Critical Wave Energy Thresholds at Natural and Created Coastal Wetlands. U.S. Army Engineer Research and Development Center, Vicksburg, MS. ERDC TN-WRP-HS-CP-2.2.

Shafer, D. J., T. H. Roberts, M. S. Peterson, and K. Schmid. 2007. A Regional Guidebook for Applying the Hydrogeomorphic Approach to Assessing Wetland Functions of Tidal Fringe Wetlands Along the Mississippi and Alabama Coast. U.S. Army Engineer Research and Development Center, Vicksburg, MS. ERDC/EL TR-07-02.

Shaffer, G. P., C. E. Sasser, J. G. Gosselink, and M. Rejmánek. 1992. Vegetation dynamics in the emerging Atchafalaya Delta, Louisiana, USA. *Journal of Ecology* 80:677–87.

Shaffer, G. P., W. B. Wood, S. S. Hoeppner, T. E. Perkins, J. Zoller, and D. Kandalepas. 2009. Degradation of baldcypress–water tupelo swamp to marsh and open water in southeastern Louisiana, U.S.A.: an irreversible trajectory? *Journal of Coastal Research: Special Issue 54—Geologic and Environmental Dynamics of the Pontchartrain Basin* (FitzGerald & Reed, eds.): 152–65.

Shaler, N. S. 1886. Sea-coast swamps of the eastern United States. *U.S. Geological Survey, 6th Annual Report*. 359–68.

Sharp, G., and R. Semple. 2004. Status of eelgrass bed in south-western Nova Scotia. In: A. R. Hanson (ed.). Status and Conservation of Eelgrass (*Zostera marina*) in Eastern Canada. Canadian Wildlife Service, Atlantic Region, Sackville, NB. Technical Report no. 412. 8.

Sharpe, P. J., and A. H. Baldwin. 2009. Patterns of wetland plant species richness across estuarine gradients of Chesapeake Bay. *Wetlands* 29:225–35.

Shaw, J., R. B. Taylor, D. L. Forbes, M.-H. Ruz, and S. Solomon. 1998. Sensitivity of the Coasts of Canada to Sea-level Rise. Geological Survey of Canada. Bulletin 505.

Shea, M. L., R. S. Warren, and W. A. Niering. 1975. Biochemical and transplantation studies of the growth form of *Spartina alterniflora* in Connecticut salt marshes. *Ecology* 56:461–66.

Shellenbarger Jones, A., C. Bosch, and E. Strange. 2009. Vulnerable species: the effects of sea-level rise on coastal habitats. In: Climate Change Science Program. *Coastal Sensitivity to Sea-level Rise: A Focus on the Mid-Atlantic Region. Synthesis and Assessment Product 4.1*. Report prepared by the Climate Change Science Program and the Subcommittee on Global Change Research. U.S. Environmental Protection Agency, Washington, DC. Chapter 5; 73–83. http://downloads.climatescience.gov/sap/sap4-1/sap4-1-final-report-all.pdf.

Shelton, A. O. 2010. Temperature and community consequences of the loss of foundation species: surfgrasses (*Phyllospadix* spp. Hooker) in tidepools. *Journal of Experimental Marine Biology and Ecology* 391:35–42.

Shennan, I., and S. Hamilton. 2006. Coseismic and pre-seismic subsidence associated with great earthquakes in Alaska. *Quarternary Science Reviews* 25:1–8.

Shew, D. M., R. A. Linthurst, and E. D. Seneca. 1981. Comparison of production computation methods in a southeastern North Carolina *Spartina alterniflora* salt marsh. *Estuaries* 4:97–109.

Shirer, B., T. Tear, K. Dolan, N. Salafsky, and C. Stem. 2005. Advancing Biodiversity Conservation in the Hudson River Estuary Watershed: A Report on the Products of Multi-Stakeholder Workshop Series. The Nature Conservancy, Eastern New York Chapter, Troy, NY.

Short, F. T., D. M. Burdick, C. A. Short, R. C. Davis, and P. A. Morgan. 2000. Developing success criteria for restored eelgrass, salt marsh and mud flat habitats. *Ecological Engineering* 15:239–52.

Shreve, F., M. A. Chrysler, F. H. Bodgett, and F. W. Besley. 1910. *The Plant Life of Maryland.* John Hopkins Press, Baltimore, MD. Special Publications Vol. III.

Silander, J. A. 1979. Microevolution and clone structure in *Spartina patens. Science* 203:658–60.

Silliman, B. R., and M. D. Bertness. 2002. A trophic cascade regulates salt marsh primary production. *Proceedings of the National Academy of Sciences of the United States* 99:10500–505.

Silliman, B. R., and M. D. Bertness. 2004. Shoreline development drives invasion of *Phragmites australis* and the loss of plant diversity on New England marshes. *Conservation Biology* 18:1424–34.

Silliman, B. R., and J. C. Zieman. 2001. Top-down control of *Spartina alterniflora* production by periwinkle grazing in a Virginia salt marsh. *Ecology* 82:2830–45.

Silliman, B. R., J. van de Koppel, M. D. Bertness, L. E. Stanton, and I. A. Mendelssohn. 2005. Drought, snails, and large-scale die-off of southern U.S. salt marshes. *Science* 310:1803–6.

Silliman, B. R., E. D. Grosholtz, and M. D. Bertness (eds.). 2009. *Human Impacts in Salt Marshes: A Global Perspective.* University of California Press, Berkeley, CA.

Simenstad, C. A., and R. M. Thom. 1992. Restoring wetland habitats in urbanized Pacific Northwest estuaries. In: G. W. Thayer (ed.). *Restoring the Nation's Marine Environment.* Maryland Sea Grant College, University of Maryland, College Park, MD. Chapter 10; 423–72.

Simenstad, C. A., K. L. Fresh, and E. O. Salo. 1982. The role of Puget Sound and Washington coastal estuaries in the life of the Pacific salmon: an unappreciated function. In: V. S. Kennedy (ed.). *Estuarine Comparisons.* Academic Press, New York. 343–64.

Simenstad, C. A., C. D. Tanner, R. M. Thom, and L. Conquest. 1991. Estuarine Habitat Assessment Protocol. Wetland Ecosystem Team, Fisheries Research Institute, University of Washington, Seattle, WA. UW-FRI-8918/-8919.

Simpson, R. L., and R. E. Good. 1985. The role of tidal wetlands in the retention of heavy metals. In: H. A. Groman, T. R. Henderson, E. J. Meyers, D. G. Burke, and J. A. Kusler (eds.). Wetlands of the Chesapeake: Protecting the Future of the Bay. Proceedings of the conference held 1985 April 9–11; Easton, MD. Environmental Law Institute, Washington, DC. 164–68.

Simpson, R. L., R. E. Good, M. A. Leck, and D. F. Whigham. 1983a. The ecology of freshwater tidal wetlands. *BioScience* 33:255–59.

Simpson, R. L., R. E. Good, R. Walker, and B. R. Frasco. 1983b. The role of Delaware River freshwater tidal wetlands in the retention of nutrients and heavy metals. *Journal of Environmental Quality* 12:41–48.

Sinicrope, T. L., P. G. Hine, R. S. Warren, and W. A. Niering. 1990. Restoration of an impounded salt marsh in New England. *Estuaries* 13:25–30.

Sipple, W. S. 1971. The past and present flora and vegetation of the Hackensack Meadowlands. *Bartonia* 41:4–56.

Sipple, W. S. 1988. Wetland Identification and Delineation Manual. Volume I. Rationale, Wetland Parameters, and Overview of Jurisdictional Approach. Volume II. Field Methodology. U.S. Environmental Protection Agency, Washington, DC.

Slade, D. C., R. K. Kehoe, and J. K. Stahl. 1997. Putting the Public Trust Doctrine to Work. The Application of the PTD to the Management of Lands, Waters and Living Resources of the Coastal States. 2nd edition. Coastal States Organization, Washington, DC. http://media.coastalstates.org/Public%20Trust%20Doctrine%202nd%20Ed%20%201997%20CSO.pdf.

Slaughter, T. H. 1967. Shore erosion control in tidewater Maryland. *Journal of the Washington Academy of Sciences* 57:117–29.

Slavin, P., and J. K. Shisler. 1983. Avian utilisation of a tidally restored salt hay farm. *Biological Conservation* 26:271–85.

Slocum, M. G., I. A. Mendelssohn, and N. L. Kuhn. 2005. Effects of sediment slurry enrichment on salt marsh rehabilitation: plant and soil responses over seven years. *Estuaries* 28:519–28.

Smalley, A. E. 1958. The role of two invertebrate populations, *Littorina irrorata* and *Orchelimum fidicinium,* in the energy flow of a salt marsh ecosystem. University of Georgia, Athens, GA. Ph.D. dissertation.

Smalley, A. E. 1960. Energy flow of a salt marsh grasshopper population. *Ecology* 41:672–77.

Smart, R. M., and J. W. Barko. 1978. Influence of sediment salinity and nutrients on the physiological ecology of selected salt marsh plants. *Estuarine and Coastal Marine Science* 7:487–95.
Smart, R. M., and J. W. Barko. 1980. Nitrogen nutrition and salinity tolerance of *Distichlis spicata* and *Spartina alterniflora*. *Ecology* 61:630–38.
Smith, D. C., C. Fox, B. Craig, and A. E. Bridges. 1989a. A contribution to the earthquake history of Maine. In: W. A. Anderson and H. W. Borns, Jr. (eds.). *Neotectonics of Maine*. Maine Geological Survey, Department of Conservation, Augusta, ME. Bulletin 40. 139–48.
Smith, D. C., H. W. Borns, Jr., and R. S. Anderson. 1989b. Relative sea-level changes measured by historic records and structures in coastal Maine. In: W. A. Anderson and H. W. Borns, Jr. (eds.). *Neotectonics of Maine*. Maine Geological Survey, Department of Conservation, Augusta, ME. Bulletin 40. 127–37.
Smith, D. D. 1977. Dredging and spoil disposal—major geological processes in San Diego Bay, California. In: M. Wiley (ed.). *Estuarine Processes*. Volume II. *Circulation, Sediments, and Transfer of Material in the Estuary*. Academic Press, New York. 150–66.
Smith, G. S., and R. W. Tiner. 1993. Status and Trends of Wetlands in Cape May County, New Jersey and Vicinity (1977 to 1991). U.S. EPA, Region II, Marine and Wetlands Protection Branch, New York, NY and U.S. Fish and Wildlife Service, Region 5, Ecological Services, Hadley, MA. Cooperative Report R-93/13.
Smith, J. B. 1907. The New Jersey Salt Marsh and Its Improvement. N.J. Agricultural Experiment Station Bulletin 207.
Smith, J. K., G. L. Taghon, and K. W. Able. 2000. Trophic linkages in marshes: ontogenetic changes in diet for young-of-the-year mummichog, *Fundulus heteroclitus*. In: M. P. Weinstein and D. A. Kreeger (eds.). *Concepts and Controversies in Tidal Marsh Ecology*. Kluwer Academic Publishers, Dordrecht, The Netherlands. 221–37.
Smith, J. P., and M. Carullo. 2007. Survey of Potential Marsh Dieback Sites in Coastal Massachusetts. Massachusetts Bays National Estuary Program and Massachusetts Office of Coastal Zone Management, Boston, MA. http://ellisvillemarsh.org/FEMcontent/marshdieback.pdf.
Smith, L. M. 1993. Estimated Presettlement and Current Acres of Natural Plant Communities in Louisiana. Louisiana Natural Heritage Program, Louisiana Department of Wildlife and Fisheries, Baton Rouge, LA.
Smith, L. M., R. L. Pederson, and R. M. Kaminski (eds.). 1989c. *Habitat Management for Migrating and Wintering Waterfowl in North America*. Texas Tech University Press, Lubbock, TX.
Smith, S. M. 2006. Report on Salt Marsh Dieback on Cape Cod (2006). National Park Service, Cape Cod National Seashore, Wellfleet, MA. Draft report.
Smith, S. M., K. Lellis-Dibble, K. Chapman, H. Bayley, and L. Curtis. 2008. Hatches Harbor Monitoring Report 2008. National Park Service, Cape Cod National Seashore, Wellfleet, MA. www.nps.gov/caco/naturescience/upload/Hatches_Harbor_Final_report_2008.pdf.
Smith, S. M., C. T. Roman, M-J. James-Pirri, K. Chapman, J. Portnoy, and E. Gwilliam. 2009. Responses of plant communities to incremental hydrologic restoration of a tide-restricted salt marsh in southern New England (Massachusetts, U.S.A.). *Restoration Ecology* 17:606–18.
Smith, T. J., III, and W. E. Odum. 1981. The effects of grazing by snow geese on coastal salt marshes. *Ecology* 62:98–106.
Smith, T. S., and S. T. Partridge. 2004. Dynamics of intertidal foraging by coastal brown bears in southwestern Alaska. *Journal of Wildlife Management* 68:233–40.
Soil Survey Division Staff. 1993. Soil Survey Manual. U.S. Department of Agriculture, Washington, DC. Handbook no. 18. http://soils.usda.gov/technical/manual/.
Soil Survey Staff. 1998. *Keys to Soil Taxonomy*, 8th edition. USDA Natural Resources Conservation Service, Washington, DC.
Soil Survey Staff. 2010. *Keys to Soil Taxonomy*, 11th edition. USDA Natural Resources Conservation Service, Washington, DC.
Soots, R. F., Jr., and M. C. Landin. 1978. Development and Management of Avian Habitat on Dredged Material Islands. U.S. Army Engineer Waterways Experiment Station, Vicksburg, MS. Technical Report DS-78-18.

Sorenson, R. M. 1973. Water waves produced by ships. *Journal of the Waterways and Harbors Division* 99 (WW2):245–56.

Southwick Associates. 2008. The Economic Benefits of Fisheries, Wildlife and Boating Resources in the State of Louisiana—2006. Prepared for the Louisiana Department of Wildlife and Fisheries.

Squiers, E. R., and R. E. Good. 1974. Seasonal changes in the productivity, caloric content, and chemical composition of a population of salt-marsh cord-grass (*Spartina alterniflora*). *Chesapeake Science* 15:63–71.

St. Lawrence Centre. 1996. State of the Environment on the St. Lawrence River. Volume 1: The St. Lawrence Ecosystem. Environment Canada, Conservation and Éditions MultiMondes, Quebec Region, Montreal. St. Lawrence Update Collection.

St. Lawrence Centre and Université Laval. 1991. The Environmental Atlas of the St. Lawrence—Wetlands: Habitats on the Edge of Land and Water. Environment Canada, Conservation and Protection, Quebec Region, Montreal. St. Lawrence Update Collection.

Stagg, C. L. 2009. Remediating impacts of global climate change-induced submergence on salt marsh ecosystem functions. Louisiana State University and Agricultural and Mechanical College, Department of Oceanography and Coastal Sciences, Baton Rouge, LA. Ph.D. dissertation.

Stagg, C. L., and I. A. Mendelssohn. 2010. Restoring ecological function to a submerged salt marsh. *Restoration Ecology* 18:10–17.

Stalter, R., and E. E. Lamont. 2002. Vascular flora of Jamaica Bay Wildlife Refuge, Long Island, New York. *Journal of the Torrey Botanical Club* 129:346–58.

Standards Working Group. 2010. Coastal and Marine Ecological Classification Standard, Version 3.1 (Working Draft). Federal Geographic Data Committee, U.S. Geological Survey, Reston, VA.

State Mosquito Control Commission. 2008. Open Marsh Water Management Standards for Salt Marsh Mosquito Control. New Jersey Department of Environmental Protection, Office of Mosquito Control Coordination, Trenton, NJ. www.nj.gov/dep/mosquito/docs/omwm_full.pdf.

Stedman, S., and T. E. Dahl. 2008a. Coastal wetlands of the eastern United States: 1998–2004 status and trends. *National Wetlands Newsletter* 30:18–20.

Stedman, S., and T. E. Dahl. 2008b. Status and Trends of Wetlands in the Coastal Watersheds of the Eastern United States 1998 to 2004. National Oceanic and Atmospheric Administration, National Marine Fisheries Service and Department of the Interior, Fish and Wildlife Service, Washington, DC.

Stedman, S-M, J. Linn, and K. Kutschenreuter. 2010. Celebrate coastal wetlands connecting us all. *National Wetlands Newsletter* 32:20–23.

Steenis, J. H., N. G. Wilder, H. P. Cofer, and R. A. Beck. 1954. The marshes of Delaware, their improvement and preservation. Delaware Board of Game and Fish Commissions. *Pittman-Robertson Bulletin* 2:1–42.

Steever, E. Z. 1972. Productivity and vegetation studies of a tidal salt marsh in Stonington, Connecticut: Cottrell marsh. Connecticut College, New London, CT. M.A. thesis.

Steever, E. Z., R. S. Warren, and W. A. Niering. 1976. Tidal energy subsidy and standing crop production of *Spartina alterniflora*. *Estuarine and Coastal Marine Science* 4:473–78.

Stevenson, J. C., L. G. Ward, and M. S. Kearney. 1986. Vertical accretion in marshes with varying rates of sea level rise. In: D. A. Wolfe (ed.). *Estuarine Variability.* Academic Press, New York. 241–59.

Stevenson, J. C., J. E. Root, M. S. Kearney, and K. L. Sundberg. 2000. The health and long term stability of natural and restored marshes in Chesapeake Bay. In: M. P. Weinstein and D. A. Kreeger (eds.). *Concepts and Controversies in Tidal Marsh Ecology.* Kluwer Academic Publishers, Dordrecht, The Netherlands. 709–35.

Steyer, G. D., B. C. Perez, S. Piazza, and G. Suir. 2007. Potential consequences of saltwater intrusion associated with hurricanes Katrina and Rita. In: G. S. Farris, G. J. Smith, M. P. Crane, C. R. Demas, L. L. Robbins, and D. L. Lavoie (eds.). Science and the Storms: The USGS Response to the Hurricanes of 2005. U.S. Geological Survey Circular 1306. 138–47. http://pubs.usgs.gov/circ/1306/.

Stiven, A. E., and J. T. Hunter. 1976. Growth and mortality of *Littorina irrorata* Say in three North Carolina marshes. *Chesapeake Science* 17:168–76.

Stockton, M. B., and C. J. Richardson. 1987. Wetland development trends in coastal North Carolina, USA, from 1970 to 1984. *Environmental Management* 11:649–57.
Stone, G. W., J. M. Grymes, J. R. Dingler, and D. A. Pepper. 1997. Overview and significance of hurricanes on the Louisiana coast, USA. *Journal of Coastal Research* 13:656–69.
Stout, J. P. 1984. The Ecology of Irregularly Flooded Salt Marshes of the Northeastern Gulf of Mexico: A Community Profile. U.S. Fish and Wildlife Service, Washington, DC. Biological Report 85 (7.1).
Stralberg, D., D. L. Applegate, S. J. Phillips, M. P. Herzog, N. Nur, and N. Warnock. 2009. Optimizing wetland restoration and management for avian communities using a mixed integer programming approach. *Biological Conservation* 142:94–109.
Strand, M. N., and L. M. Rothschild. 2010. *Wetlands Deskbook*. 3rd edition. Island Press, Washington, DC.
Streever, B. 2001. *Saving Louisiana? The Battle for Coastal Wetlands.* University Press of Mississippi, Jackson, MS.
Stribling, J. M., and J. C. Cornwell. 1997. Identification of important primary producers in a Chesapeake Bay tidal creek system using stable isotopes of carbon and sulfur. *Estuaries* 20:77–85.
Stroud, L. M. 1976. Net primary production of belowground material and carbohydrate patterns in two height forms of *Spartina alterniflora* Loisel in two North Carolina marshes. North Carolina State University, Raleigh, NC. Ph.D. dissertation.
Stumpf, R. P. 1983. The process of sedimentation on the surface of a salt marsh. *Estuarine and Coastal Shelf Science* 17:495–508.
Subrahmanyam, C. B., and C. L. Coultas. 1980. Studies on the animal communities in two North Florida salt marshes. Part III. Seasonal fluctuations of fish and macroinvertebrates. *Bulletin of Marine Science* 30:790–818.
Subrahmanyam, C. B., and S. H. Drake. 1975. Studies on the animal communities in two North Florida salt marshes. Part I. Fish communities. *Bulletin of Marine Science* 25:445–65.
Subrahmanyam, C. B., W. L. Kruczynski, and S. H. Drake. 1976. Studies on the animal communities in two north Florida salt marshes. Part II. Macroinvertebrate communities. *Bulletin of Marine Science* 26:172–95.
Sugihara, T., C. Yearsley, J. B. Durand, and N. P. Psuty. 1979. Comparison of Natural and Altered Estuarine Systems: Analysis. Rutgers University, Center for Coastal and Environmental Studies, New Brunswick, NJ. CCES Publication no. NJ/RU-DEP-11-9-79.
Sullivan, M. J., and C. A. Currin. 2000. Community structure and functional dynamics of benthic microalgae in salt marshes. In: M. P. Weinstein and D. A. Kreeger (eds.). *Concepts and Controversies in Tidal Marsh Ecology.* Kluwer Academic Publishers, Dordrecht, The Netherlands. 81–106.
Sullivan, M. S., and C. A. Moncreiff. 1988. Primary production of edaphic algal communities in a Mississippi salt marsh. *Journal of Phycology* 24:49–58.
Sullivan, M. S., and C. A. Moncreiff. 1990. Edaphic algae are an important component of salt marsh food-webs: evidence from multiple stable isotope analyses. *Marine Ecology Progress Series* 62:149–59.
Sullivan, R. 1998. *The Meadowlands—Wilderness Adventures on the Edge of a City.* Doubleday Press, New York.
Sutter, L. 2001. Spatial wetland assessment for management and planning (SWAMP): technical discussions. National Oceanic and Atmospheric Administration Coastal Services Center, Charleston, SC. Publication no. 20129-CD.
Svensson, J. S., and J. K. Jeglum. 2000. Primary Succession and Dynamics of Norway Spruce Coastal Forests on Land-Uplift Ground Moraine. Swedish University of Agricultural Sciences, Faculty of Forestry, Uppsala, Sweden. Studia Forestalia Suecica no. 209.
Svensson, J. S., and J. K. Jeglum. 2001. Structure and dynamics of an undisturbed old-growth Norway spruce forest on the rising Bothnian coastline. *Forest Ecology and Management* 151:67–79.
Swain, P. C., and J. B. Kearsley. 2001. Classification of the Natural Communities of Massachusetts. Massachusetts Division of Fisheries and Wildlife, Natural Heritage and Endangered Species Program, Westborough, MA.
Swamy, V., P. E. Fell, M. Body, M. B. Keaney, M. K. Nyaku, E. C. McIlvain, and A. L. Keen. 2002. Macroinvertebrate and fish populations in a restored impounded salt marsh 21 years after the reestablishment of

tidal flooding. *Environmental Management* 29:516–30.
Swann, L. 2008. The use of living shorelines to mitigate the effects of storm events on Dauphin Island, Alabama, USA. *American Fish Society Symposium* 64:47–57.
Swarth, C. W., and E. Kiviat. 2009. Animal communities in North American tidal freshwater wetlands. In: A. Barendregt, D. F. Whigham, and A. H. Baldwin (eds.). *Tidal Freshwater Wetlands.* Backhuys Publishers, Leiden, The Netherlands. Chapter 7; 71–88.
SWS Wetland Concerns Committee. 2009. Current practices in wetland management for mosquito control. Society of Wetland Scientists. White paper. www.sws.org/wetland_concerns/docs/SWS-MosquitoWhitePaperFinal.pdf.
Syvitski, J. P. M., A. J. Kettner, I. Overeem, E. W. H. Hutton, M. T. Hannon, G. R. Brakenridge, J. Day, C. Vörösmarty, Y. Saito, L. Giosan, and R. J. Nicholls. 2009. Sinking deltas due to human activities. *Nature Geosciences* 2:681–86.
Taggart, J. B. 2008. Management of feral horses at North Carolina National Estuarine Research Reserve. *Natural Areas Journal* 28:187–95.
Takekawa, J. Y., I. Woo, H. Spautz, N. Nur, J. L. Grenier, K. Malamud-Roam, J. C. Nordby, A. N. Cohen, F. Malamud-Roam, and S. E. Wainwright-De La Cruz. 2006. Environmental threats to tidal-marsh vertebrates of the San Francisco Bay Estuary. *Studies in Avian Biology* 32:176–97.
Takekawa, J. Y., I. Woo, N. D. Ahearn, S. Demers, R. J. Gardiner, W. M. Perry, N. K. Ganju, G. G. Shellenbarger, and D. H. Schoellhamer. 2010. Erratum to: Measuring sediment accretion in early tidal marsh restoration. *Wetlands Ecology and Management* 18:755.
Talbot, C. W., and K. W. Able. 1984. Composition and distribution of larval fishes in New Jersey high marshes. *Estuaries* 7:433–43.
Talley, T. S., and L. A. Levin. 1999. Macrofaunal succession and community structure in *Salicornia* marshes of southern California. *Estuarine, Coastal and Shelf Science* 49:713–31.
Tan, S., and H. Sung. 2008. Carlos Juan Finlay (1833–1915): of mosquitoes and yellow fever. *Singapore Medical Journal* 49:370–71.
Tande, G. F. 1996. Mapping and Classification of Coastal Marshes: Lake Clark National Park and Preserve, Alaska. Prepared for Lake Clark National Park and Preserve, Kenai, AK.
Tanner, B. R., M. E. Uhle, J. T. Kelley, and C. I. Mora. 2007. C3/C4 variations in salt-marsh sediments: an application of compound specific isotopic analysis of lipid biomarkers to late Holocene paleoenvironmental research. *Organic Geochemistry* 38:474–84.
Taylor, J. 1998. Guidance for Meeting U.S. Fish and Wildlife Service Trust Resource Needs When Conducting Coastal Marsh Management for Mosquito Control on Region 5 National Wildlife Refuges. U.S. Fish and Wildlife Service, Region 5, Great Bay National Wildlife Refuge, Newington, NH.
Taylor, K. L., J. B. Grace, and B. D. Marx. 1997. The effects of herbivory on neighbor interactions along a coastal marsh gradient. *American Journal of Botany* 84:709–15.
Taylor, N. 1938. A preliminary report on the salt marsh vegetation of Long Island, New York. *Bulletin of the New York State Museum* 316:21–84.
Taylor, N. 1939. Salt Tolerance of Long Island Salt Marsh Plants. New York State Museum Circular 23.
Teal, J., and M. Teal. 1969. *Life and Death of the Salt Marsh.* Audubon/Ballantine Books, New York.
Teal, J. M. 1958. Distribution of fiddler crabs in Georgia salt marshes. *Ecology* 39:185–93.
Teal, J. M. 1962. Energy flow in the salt marsh ecosystem of Georgia. *Ecology* 43:614–24.
Teal, J. M. 1986. The Ecology of Regularly Flooded Salt Marshes of New England: A Community Profile. U.S. Fish and Wildlife Service, Washington, DC. Biological Report 95 (7.4).
Teal, J. M., and B. L. Howes. 1996. Interannual variability of a salt-marsh ecosystem. *Limnology and Oceanography* 41:802–6.
Teal, J. M., and B. L. Howes. 2000. Salt marsh values: retrospection from the end of the century. In: M. Weinstein and D. Kreeger (eds.). *Concepts and Controversies in Tidal Marsh Ecology.* Kluwer Academic Publishers, Dordrecht, The Netherlands. 9–22.
Teal, J. M., and J. Kanwisher. 1961. Gas exchange in a Georgia salt marsh. *Limnology and Oceanography* 6:388–99.
Teal, J. M., and J. W. Kanwisher. 1966. Gas transport in the marsh grass, *Spartina alterniflora. Journal of Experimental Botany* 17:355–61.

Teal, J. M., and J. B. Peterson (eds.). 2005. Delaware Bay salt marsh restoration. *Ecological Engineering* 25:199–314.
Teal, J. M., and L. Weishar. 2005. Ecological engineering, adaptive management, and restoration management in Delaware Bay salt marsh restoration. *Ecological Engineering* 25:304–15.
Temmerman, S., G. Govers, P. Meire, and S. Wartel. 2003. Modeling long-term tidal marsh growth under changing tidal conditions and suspended sediment concentrations, Scheldt Estuary, Belgium. *Marine Geology* 193:151–69.
Thayer, G. W. (ed.). 1992. *Restoring the Nation's Marine Environment.* Maryland Sea Grant College, University of Maryland, College Park, MD.
Thayer, G. W., T. A. McTigue, R. J. Bellmer, F. M. Burrows, D. H. Merkey, A. D. Nickens, S. J. Lozano, P. F. Gayaldo, P. J. Polmateer, and P. T. Pinit. 2003. Science-based Restoration Monitoring of Coastal Habitats. Volume 1: A Framework for Monitoring Plans Under the Estuaries and Clean Waters Act of 2000 (Public Law 160–457). NOAA National Centers for Coastal Ocean Science, Silver Spring, MD. NOAA Coastal Ocean Program Decision Analysis Series no. 23. Volume 1. www.era.noaa.gov/pdfs/Sci-based%20Restoration%20Monitoring-%20Vol%201.pdf.
Thayer, G. W., T. A. McTigue, R. J. Salz, D. H. Merkey, F. M. Burrows, and P. F. Gayaldo (eds.). 2005. Science-based Restoration Monitoring of Coastal Habitats. Volume 2: Tools for Monitoring Coastal Habitats. NOAA National Centers for Coastal Ocean Science, Silver Spring, MD. NOAA Coastal Ocean Program Decision Analysis Series no. 23. Vol. 2.
Thibodeau, P. M. 1997. Groundwater flow dynamics across the forest-salt marsh interface: North Inlet, South Carolina. University of South Carolina, Department of Geological Sciences, Columbia, SC. Ph.D. dissertation.
Thibodeau, P. M., L. R. Gardner, and H. W. Reeves. 1998. The role of groundwater flow in controlling the spatial distribution of soil salinity and rooted macrophytes in a southeastern salt marsh, USA. *Mangroves and Salt Marshes* 2:1–13.
Thieler, E. R., and E. S. Hammar-Klose. 1999. National Assessment of Coastal Vulnerability to Future Sea-Level Rise: Preliminary Results for the U.S. Atlantic Coast. U.S. Geological Survey, Open-File Report 99-593. http://pubs.usgs.gov/dds/dds68/htmldocs/ofreport.htm.
Thieler, E. R., and E. S. Hammar-Klose. 2000a. National Assessment of Coastal Vulnerability to Future Sea-Level Rise: Preliminary Results for the U.S. Gulf of Mexico Coast. U.S. Geological Survey, Open-File Report 00-179. http://pubs.usgs.gov/dds/dds68/htmldocs/ofreport.htm.
Thieler, E. R., and E. S. Hammar-Klose. 2000b. National Assessment of Coastal Vulnerability to Future Sea-Level Rise: Preliminary Results for the U.S. Pacific Coast. U.S. Geological Survey, Open-File Report 00-178. http://pubs.usgs.gov/dds/dds68/htmldocs/ofreport.htm.
Thilenius, J. F. 1990. Plant succession on earthquake uplifted coastal wetlands, Copper River Delta, Alaska. *Northwest Science* 64:259–62.
Thilenius, J. F. 1995. Phytosociology and Succession on Earthquake-uplifted Coastal Wetlands, Copper River Delta, Alaska. U.S.D.A. Forest Service, Pacific Northwest Research Station, Juneau, AK. General Technical Report PNW-GTR-346. www.fs.fed.us/pnw/pubs/pnw_gtr346.pdf.
Thom, R. M. 1992. Accretion rates of low intertidal salt marshes in the Pacific Northwest. *Wetlands* 12:147–56.
Thom, R. M., R. Zeigler, and A. B. Borde. 2002. Floristic development patterns in a restored Elk River estuarine marsh, Grays Harbor, Washington. *Restoration Ecology* 10:487–96.
Thomas, S., R. Milner, and J. Buchanan (eds.). 2004. Summaries of Current Projects that Benefit Shorebirds in the Coastal Region of Oregon and Washington. Northern Pacific Coast Shorebird Working Group. www.fws.gov/shorebirdplan/RegionalShorebird/downloads/OregonWashingtonSummary2004.pdf.
Thompson, J. D. 1991. The biology of an invasive plant. What makes *Spartina anglica* so successful? *BioScience* 41:393–401.
Tiner, R. W. 1974. The ecological distribution of the invertebrate macrofauna in the Cottrell Salt Marsh, Stonington, Connecticut. University of Connecticut, Storrs, CT. M.S. thesis.
Tiner, R. W. 1985a. *Wetlands of Delaware.* U.S. Fish and Wildlife Service, Region 5, Newton Corner, MA and Delaware Department of Natural Resources and Environmental

Control, Dover, DE. Cooperative NWI publication.

Tiner, R. W. 1985b. *Wetlands of New Jersey.* U.S. Fish and Wildlife Service, Region 5, Newton Corner, MA. NWI publication.

Tiner, R. W. 1987. *A Field Guide to Coastal Wetland Plants of the Northeastern United States.* University of Massachusetts Press, Amherst, MA.

Tiner, R. W. 1991. Recent changes in estuarine wetlands of the coterminous United States. In: Coastal Wetlands. Coastal Zone '91 Conference-ASCE, Long Beach, CA. 100–109.

Tiner, R. W. 1993a. *Field Guide to Coastal Wetland Plants of the Southeastern United States.* University of Massachusetts Press, Amherst, MA.

Tiner, R. W. 1993b. The primary indicators method—a practical approach to wetland recognition and delineation in the United States. *Wetlands* 13:50–64.

Tiner, R. W. 1998. Managing Common Reed (*Phragmites australis*) in Massachusetts: An Introduction to the Species and Control Techniques. U.S. Fish and Wildlife Service, Ecological Services, Northeast Region, Hadley, MA.

Tiner, R. W. 1999. *Wetland Indicators: A Guide to Wetland Identification, Delineation, Classification, and Mapping.* Lewis Publishers, CRC Press, Boca Raton, FL.

Tiner, R. W. 2003a. Dichotomous Keys and Mapping Codes for Wetland Landscape Position, Landform, Water Flow Path, and Waterbody Type Descriptors. U.S. Fish and Wildlife Service, National Wetlands Inventory Program, Northeast Region, Hadley, MA. http://library.fws.gov/wetlands/dichotomouskeys0903.pdf.

Tiner, R. W. 2003b. Correlating Enhanced National Wetlands Inventory Data with Wetland Functions for Watershed Assessments: A Rationale for Northeastern U.S. Wetlands. U.S. Fish and Wildlife Service, National Wetlands Inventory Program, Northeast Region, Hadley, MA. www.fws.gov/northeast/wetlands/pdf/CorrelatingEnhancedNWIDataWetlandFunctionsWatershedAssessments%5B1%5D.pdf.

Tiner, R. W. 2005a. *In Search of Swampland: A Wetland Sourcebook and Field Guide.* Revised and Expanded Version. Rutgers University Press, New Brunswick, NJ.

Tiner, R. W. 2005b. Assessing cumulative loss of wetland functions in the Nanticoke River watershed using enhanced National Wetlands Inventory data. *Wetlands* 25:405–19.

Tiner, R. W. 2009. *Field Guide to Tidal Wetland Plants of the Northeastern United States and Neighboring Canada. Vegetation of Beaches, Tidal Flats, Rocky Shores, Marshes, Swamps, and Coastal Ponds.* University of Massachusetts Press, Amherst, MA.

Tiner, R. W. 2010. NWIPlus: Geospatial database for watershed-level functional assessment. *National Wetlands Newsletter* 32:4–7, 23. www.fws.gov/northeast/wetlands/Publications%20PDFs%20as%20of%20March_2008/Mapping/NWIPlus_NWN.pdf.

Tiner, R. W. 2011a. Predicting Wetland Functions at the Landscape Level for Coastal Georgia using NWIPlus Data. U.S. Fish and Wildlife Service, National Wetlands Inventory Program, Northeast Region, Hadley, MA. www.fws.gov/northeast/wetlands/publications/CORRELATIONREPORT_GeorgiaFINAL092011.pdf.

Tiner, R. W. 2011b. Dichotomous Keys and Mapping Codes for Wetland Landscape Position, Landform, Water Flow Path, and Waterbody Type Descriptors. Version 2.0. U.S. Fish and Wildlife Service, National Wetlands Inventory Program, Northeast Region, Hadley, MA. www.fws.gov/northeast/wetlands/publications/DichotomousKeys_090611wcover.pdf.

Tiner, R. W., and J. T. Finn. 1986. Status and Recent Trends of Wetlands in Five Mid-Atlantic States: Delaware, Maryland, Pennsylvania, Virginia, and West Virginia. U.S. Fish and Wildlife Service, National Wetlands Inventory Project, Newton Corner, MA.

Tiner, R. W., and D. G. Burke. 1995. *Wetlands of Maryland.* U.S. Fish and Wildlife Service, Ecological Services, Region 5, Hadley, MA and Maryland Department of Natural Resources, Annapolis, MD. Cooperative NWI publication.

Tiner, R. W., I. Kenenski, T. Nuerminger, J. Eaton, D. B. Foulis, G. S. Smith, and W. E. Frayer. 1994. Recent Wetland Status and Trends in the Chesapeake Watershed (1982 to 1989): Technical Report. U.S. Fish and Wildlife Service, Region 5, Ecological Services, Hadley, MA.

Tiner, R. W., D. B. Foulis, C. Nichols, S. Schaller, D. Petersen, K. Andersen, and J. Swords. 1998. Wetland Status and Recent Trends for the Neponset Watershed, Massachusetts (1977–1991). U.S. Fish and Wildlife Service, National Wetlands Inventory Program, Region 5, Hadley, MA and University of Massachusetts, Department of Plant and Soil Sciences, Natural Resources Assessment Group, Amherst, MA. Cooperative Report.

Tiner, R. W., J. Q. Swords, and B. J. McClain. 2002. Wetland Status and Trends for the Hackensack Meadowlands. U.S. Fish and Wildlife Service, Northeast Region, Hadley, MA. http://library.fws.gov/wetlands/hackensack.pdf.

Tiner, R. W., K. McGuckin, and M. Fields. 2012. Wetland Changes for Long Island, New York: circa 1900 to 2004. U.S. Fish and Wildlife Service, Northeast Region, Hadley, MA.

Titus, J. G. (ed.). 1988. Greenhouse Effect, Sea Level Rise and Coastal Wetlands. U.S. Environmental Protection Agency, Office of Policy, Planning and Evaluation, Washington, DC. EPA-230-05-86-013.

Tomczak, M. 1996. Definition of estuaries; empirical estuary classification. Accessed at www.es.flinders.edu.au/!mattom/ShelfCoast/notes/chapter11.html.

Tomlinson, P. B. 1986. *The Botany of Mangroves*. Cambridge University Press, London, UK.

Törnqvist, T. E., D. J. Wallace, J. E. A. Storms, J. Wallinga, R. L. van Dam, M. Blaauw, M. S. Derksen, C. J. W. Klerks, C. Meijneken, and E. M. A. Snijders. 2008. Mississippi Delta subsidence primarily caused by compaction of Holocene strata. *Nature Geoscience* 1:173–76.

Townshend, C. W. 1913. *Sand Dunes and Salt Marshes*. I. C. Page and Company, Boston, MA.

Transeau, E. N. 1909. Successional relations of the vegetation about Yarmouth, Nova Scotia. *The Plant World* 12:271–81.

Transeau, E. N. 1913. The vegetation of Cold Spring Harbor, Long Island. I. The littoral successions. *The Plant World* 16:189–209.

Travis, S. E., C. E. Proffitt, and K. Ritland. 2004. Population structure and inbreeding vary with successional stage in created *Spartina alterniflora* marshes. *Ecological Applications* 14:1189–1202.

Trenhaile, A. S. 1990. *The Geomorphology of Canada*. Oxford University Press, Toronto, ON.

Trites, M., I. Kaczmarska, J. M. Ehrman, P. W. Hicklin, and J. Ollerhead. 2005. Diatoms from two macro-tidal mudflats in Chignecto Bay, Upper Bay of Fundy, New Brunswick, Canada. *Hydrobiologia* 544:299–319.

Troyer, W. 2005. *Into Brown Bear Country*. University of Alaska Press, Fairbanks, AK.

Tufford, D. L. 2005. State of Knowledge Report South Carolina Coastal Wetland Impoundments. South Carolina Sea Grant Consortium.

Tunnell, J. W., Jr., N. L. Hilbun, and K. Withers. 2002. Comprehensive Bibliography of the Laguna Madre of Texas & Tamaulipas. Center for Coastal Studies, Texas A&M University, Corpus Christi, TX. TAMU-CC-0202-CCS.

Tupper, M., and K. W. Able. 2000. Movements and food habits of striped bass (*Morone saxatilis*) in Delaware Bay (USA) salt marshes: comparison of a restored and a reference marsh. *Marine Biology* 137:1049–58.

Turner, M. G. 1987. Effects of grazing by feral horses, clipping, trampling, and burning on a Georgia salt marsh. *Estuaries* 10:54–60.

Turner, R. E. 1976. Geographic variations in salt marsh macrophyte production: a review. *Contribution in Marine Science* 20:47–68.

Turner, R. E. 1987. Relationship Between Canal and Levee Density and Coastal Land Loss in Louisiana. U.S. Fish and Wildlife Service, Washington, DC.

Turner, R. E. 1992. Coastal wetlands and penaeid shrimp habitat. In: R. H. Stroud (ed.). Stemming the Tide of Coastal Fish Habitat Loss. National Coalition for Marine Conservation, Savannah, GA. 97–104.

Turner, R. E. 1997. Wetland loss in the northern Gulf of Mexico: multiple working hypotheses. *Estuaries* 20:1–13.

Turner, R. E., and B. Streever. 2002. *Approaches to Coastal Wetland Restoration: Northern Gulf of Mexico*. SPB Academic Publishing bv, the Hague, The Netherlands.

Turner, R. E., J. M. Lee, and C. Neill. 1993. Backfilling Canals as a Wetland Restoration Technique in Coastal Louisiana. U.S. Department of the Interior, Minerals Management Service, Gulf of Mexico OCS Region, New

Orleans, LA. www.gomr.boemre.gov/PI/PDFImages/ESPIS/1/1219.pdf,

Turner, R. E., J. M. Lee, and C. Neill. 1994. Backfilling canals to restore wetlands: empirical results in coastal Louisiana. *Wetlands Ecology and Management* 3:63–78.

Turner, R. E., E. M. Swenson and C. S. Milan. 2000. Organic and inorganic contributions to vertical accretion in salt marsh sediments. In: M. P. Weinstein and D. A. Kreeger (eds.). *Concepts and Controversies in Tidal Marsh Ecology.* Kluwer Academic Publishers, Dordrecht, The Netherlands. 583–95.

Turner, R. E., A. M. Redmond, and J. B. Zedler. 2001. Count it by acre or function—mitigation adds up to net loss of wetlands. *National Wetlands Newsletter* 23:5–6, 14–15.

Turner, R. E., B. L. Howes, J. M. Teal, C. S. Milan, E. M. Swenson, and D. D. Goehringer-Toner. 2009. Salt marshes and eutrophication: an unsustainable outcome. *Limnology and Oceanography* 54:1634–42.

Tyrrell, M. C. 2004. Gulf of Maine Marine Habitat Primer. Gulf of Maine Council on the Marine Environment. www.gulfofmaine.org. Report.

Udell, H. F., J. Zarudsky, T. E. Doheny, and P. R. Burkholder. 1969. Productivity and nutrient values of plants growing in the salt marshes of the town of Hempstead, Long Island. *Bulletin of the Torrey Botanical Club* 96:42–51.

Underwood, G. J. C., and D. M. Patterson. 1993. Seasonal changes in diatom biomass, sediment stability and biogenic stabilization in the Severn Estuary. *Journal of Marine Biological Association of the United Kingdom* 76:431–50.

Ungar, I. A. 1978. Halophyte seed germination. *The Botanical Review* 44:233–64.

Ungar, I. A. 1991. *Ecophysiology of Vascular Plants.* CRC Press, Boca Raton, FL.

Ungar, I. A. 1998. Are biotic factors significant in influencing the distribution of halophytes in saline habitats? *The Botanical Review* 64:176–99.

University of North Carolina Wilmington. 2008. The Potential Impacts of Climate Change on Coastal North Carolina. A Report by the Faculty of the University of North Carolina Wilmington. Prepared for the President of the University of North Carolina-Chapel Hill. http://uncw.edu/aa/documents/UNCW%20Global%20Climate%20Change%20Final%20Report.pdf.

U.S. Army Corps of Engineers. 1991. Engineering and Design—Tidal Hydraulics. Washington, DC. Engineer Manual EM 1110-2-1607.

U.S. Army Corps of Engineers. 2004. Louisiana Coastal Area (LCA), Louisiana—Ecosystem Restoration Study. New Orleans District, LA.

U.S. Army Corps of Engineers. 2010. Regional Supplement to the Corps of Engineers Wetland Delineation Manual: Atlantic and Gulf Coastal Plain Region (Version 2.0). U.S. Army Engineer Research and Development Center, Vicksburg, MS. ERDC/EL TR-10-20.

U.S. Army Corps of Engineers and Washington State Department of Ecology. 2010. Joint public notice—proposal for a wetland mitigation bank: Blue Heron Slough Conservation and Bank. February 3, 2010. Regulatory Branch, Seattle District, WA.

U.S. Department of the Interior. 2006. 2006 National Survey of Fishing, Hunting, and Wildlife-Associated Recreation. U.S. Fish and Wildlife Service, U.S. Department of Commerce, Bureau of the Census, Washington, DC. http://library.fws.gov/pubs/nat_survey2006_final.pdf.

U.S. Environmental Protection Agency. 1998. Wetland Bioassessment Fact Sheet 8. Office of Wetlands, Oceans and Watersheds, Washington, DC. EPA843-F-98-001h. http://water.epa.gov/type/wetlands/assessment/fact8.cfm.

U.S. Environmental Protection Agency. 2011. USA RAM manual. Version 11. Corvallis, OR.

U.S. Fish and Wildlife Service. 1959. Wetlands Inventory of Connecticut. Office of River Basin Studies, Boston, MA.

U.S. Fish and Wildlife Service. 1965. A Supplementary Report on the Coastal Wetlands Inventory of Connecticut. Division of River Basin Studies, Boston, MA.

U.S. Fish and Wildlife Service. 1993. Anchorage Wetlands Trend Study (1950 to 1990). Ecological Services, Region 7, Anchorage, AK.

U.S. Fish and Wildlife Service. 1997. Regionally Significant Habitats and Habitat Complexes of the New York Bight Watershed. Southern New England-New York Bight, Coastal Ecosystems Program, Charlestown, RI. http://library.fws.gov/pubs5/begin.htm.

U.S. Fish and Wildlife Service. 2002. The Federal Duck Stamp Story. Federal Duck Stamp

Office, Arlington, VA. Accessed at http://duckstamps.fws.gov.
U.S. Fish and Wildlife Service. 2007. The Hackensack Meadowlands Initiative. Preliminary Conservation Planning for the Hackensack Meadowlands Hudson and Bergen Counties, New Jersey. New Jersey Field Office, Pleasantville, NJ. www.fws.gov/northeast/njfieldoffice/PCP_2007/HMI_PCP.html.
U.S. Fish and Wildlife Service and U.S. Census Bureau. 2006. 2006 National Survey of Fishing, Hunting, and Wildlife-Associated Recreation. Washington, DC.
U.S. Geological Survey. 2000. Sea Level and Climate. http://pubs.usgs.gov/fs/fs2-00/.
U.S. Geological Survey. 2006. A Century of Wetland Exploitation. www.npwrc.usgs.gov/resource/wetlands/uswetlan/century.htm.
U.S. Government Accounting Office. 2005. Wetlands Protection: Corps of Engineers Does Not Have an Effective Oversight Approach to Ensure That Compensatory Mitigation Is Occurring. Washington, DC. GAO-05-898.
U.S. Government Accounting Office. 2007. Coastal Wetlands: Lessons Learned from Past Efforts in Louisiana Could Help Guide Future Restoration and Protection. Washington, DC. GAO-08-130.
Valentin, H. 1954. *Die Küsten der Erde*. VEB Geographisch-Kartographische Anstalt Gotha, Berlin.
Valentine, J. M., Jr. 1984. Cajun country marshes. In: A. S. Hawkins, R. C. Hanson, H. K. Nelson, and H. M. Reeves (eds.). Flyways: Pioneering Waterfowl Management in North America. U.S. Department of the Interior, Fish and Wildlife Service, Washington, DC. 439–51.
Valentine, V. J. 2002. Scrub-shrub/emergent wetland ecotone migration along Delaware tidal rivers in response to sea-level change, natural impacts, and human modifications. University of Delaware, College of Marine Studies, Newark, DE. Ph.D. dissertation.
Valiela, I., and J. M. Teal. 1974. Nutrient limitation of salt marsh vegetation. In: R. J. Reimold and W. H. Queen (eds.). *Ecology of Halophytes*. Academic Press, New York. 547–63.
Valiela, I., J. M. Teal, and N. Y. Persson. 1976. Production and dynamics of experimentally enriched salt marsh vegetation: belowground biomass. *Limnology and Oceanography* 21:245–52.
Valiela, I., J. M. Teal, and W. J. Sass. 1975. Production and dynamics of salt marsh vegetation and the effects of experimental treatment with sewage sludge. *Journal of Applied Ecology* 12:973–82.
Valiela, I., J. Costa, K. Foreman, J. M. Teal, B. Howes, and D. Aubrey. 1990. Transport of groundwater-borne nutrients from watersheds and their effects on coastal waters. *Biogeochemistry* 10:177–97.
Valiela, I., M. L. Cole, J. McClelland, J. Hauxwell, J. Cebrian, and S. B. Joye. 2000. Role of salt marshes as part of coastal landscapes. In: M. P. Weinstein and D. A. Kreeger (eds.). *Concepts and Controversies in Tidal Marsh Ecology*. Kluwer Academic Publishers, Dordrecht, The Netherlands. 23–38.
Valiela, I., J. L. Bowen, and J. K. York. 2001. Mangrove forests: one of the world's threatened major tropical environments. *BioScience* 51:807–15.
Valiela, I., D. Rutecki, and S. Fox. 2004. Salt marshes: biological controls of food webs in a diminishing environment. *Journal of Experimental Marine Biology and Ecology* 300:131–59.
Van Andel, T., E. Zangger, and A. Demitrack. 1990. Land use and soil erosion in prehistoric and historical Greece. *Journal of Field Archeology* 17:379–96.
Van Asselen, E., E. Stouthamer, and T. W. J. van Asch. 2009. Effects of peat compaction on delta evolution: a review of processes, responses, measuring, and modeling. *Earth-Science Reviews* 92:35–51.
Van de Plassche, O., W. G. Mook, and A. L. Bloom. 1989. Submergence of coastal Connecticut 6000–3000 years B.P. *Marine Geology* 86:349–54.
Van der Brink, F. W. B., G. der Velde, W. W. Bosman, and H. Coops. 1995. Effects of substrate parameters on growth of eight halophyte species in relation to flooding. *Aquatic Botany* 50:79–97.
Vanderhoof, M., B. A. Holzman, and C. Rogers. 2009. Predicting the distribution of perennial pepperweed (*Lepidium latifolium*), San Francisco Bay Area, California. *Invasive Plant Science and Management* 2:260–69.
Van Doren, M. (ed.). 1928. *Travels of William Bartram*. Dover Publications, New York.

Van Dyke, J. C. 1898. *Nature for Its Own Sake.* Charles Scribner's Sons, New York. (5th edition, 1908, accessed online at www.archive.org/details/natureforitsowns00vandrich).

Van Dyke, J. C. 1901. *The Desert. Further Studies in Natural Appearances.* Charles Scribner's Sons, New York.

Van Proosdij, D., J. Ollerhead, and R. G. D. Davidson-Arnott. 2000. Controls on suspended sediment deposition over single tidal cycles in a macrotidal saltmarsh, Bay of Fundy, Canada. In: K. Pye and J. R. L. Allen (eds.). *Coastal and Estuarine Environments: Sedimentology, Geomorphology, and Geoarcheology.* Geological Society of London, London, England. 43–57.

Van Raalte, D. D., I. Valiela, and J. M. Teal. 1976. Production of epibenthic salt marsh algae: light and nutrient limitation. *Limnology and Oceanography* 21:862–72.

Van Zandt, P. A., and S. Mopper. 2002. Delayed and carryover effects of salinity on flowering in *Iris hexagona* (Iridaceae). *American Journal of Botany* 89:1847–51.

Vasilas, L. M., G. W. Hurt, and C. V. Noble (eds.). 2010. Field Indicators of Hydric Soils in the United States. A Guide for Identifying and Delineating Hydric Soils, Version 7.0. USDA Natural Resources Conservation Service in cooperation with the National Technical Committee for Hydric Soils. ftp://ftp-fc.sc.egov.usda.gov/NSSC/Hydric_Soils/FieldIndicators_v7.pdf.

Vermeer, M., and S. Rahmstorf. 2009. Global sea level linked to global temperature. *Proceedings of the National Academy of Sciences of the United States* 106:21527–32.

Vernberg, W. B., and F. J. Vernberg. 1972. *Environmental Physiology of Marine Animals.* Springer-Verlag, New York.

Viereck, L. A., C. T. Dyrness, A. R. Batten, and K. J. Wenzlik. 1992. The Alaska Vegetation Classification System. U.S.D.A. Forest Service, Pacific Northwest Research Station, Portland, OR. General Technical Report PNW-GTR-286. www.fs.fed.us/pnw/publications/pnw_gtr286/pnw_gtr286a.pdf.

Vileisis, A. 1997. *Discovering the Unknown Landscape: A History of America's Wetlands.* Island Press, Washington, DC.

Viosca, P. 1928. Louisiana wet lands and the value of their wild life and fishery resources. *Ecology* 9:216–29.

Virginia Marine Resources Commission and Virginia Institute of Marine Science. 1997. Guidelines for the Establishment, Use and Operation of Tidal Wetland Mitigation Banks in Virginia. www.mrc.virginia.gov/regulations/bankguide.shtm.

Virginia Natural Heritage Program. 2003. Virginia's rare natural environments: sea-level fens. Natural Heritage Resources Fact Sheet. www.dcr.state.va.us/dnh/wfens.htm.

Visser, J. M., C. E. Sasser, R. H. Chabreck, and R. G. Linscombe. 2002. The impact of a severe drought on the vegetation of a subtropical estuary. *Estuaries* 25:1184–95.

Visser, J. M., C. E. Sasser, R. H. Chabreck, and R. G. Linscombe. 1998. Marsh vegetation types of the Mississippi River deltaic plain. *Estuaries* 21:818–28.

Wainwright, S. J. 1980. Plants in relation to salinity. *Advances in Botanical Research* 8:221–61.

Wainwright, S. J. 1984. Adaptations of plants to flooding with salt water. In: T. T. Kozlowski (ed.). *Flooding and Plant Growth.* Academic Press, Orlando, FL. 295–343.

Waisel, Y. 1972. *Biology of Halophytes.* Academic Press, New York.

Wakeley, J. S. 2002. Developing a "Regionalized" Version of the Corps Wetlands Delineation Manual: Issues and Recommendations. U.S. Army Engineer Research and Development Center, Vicksburg, MS. ERDC/EL TR-02-20. http://el.erdc.usace.army.mil/elpubs/pdf/trel02-20.pdf.

Walkup, C. J. 2007. *Spartina alterniflora.* In: Fire Effects Information System. U.S. Department of Agriculture, Forest Service, Rocky Mountain Research Station, Fire Sciences Laboratory, Ogden, UT (Producer). Accessed at www.fs.fed.us/database/feis/.

Wallace, A. L., A. S. Klein, and A. C. Mathieson. 2004. Determining the affinities of salt marsh fucoids using microsatellite markers: evidence of hybridization and introgression between two species of *Fucus* (*Phaeophyta*) in a Maine estuary. *Journal of Phycology* 40:1013–27.

Walls, E. A., J. Berkson, and S. A. Smith. 2002. The horseshoe crab, *Limulus polyphemus:* 200 million years of existence, 100 years of study. *Reviews in Fisheries Science* 10:39–73.

Walsh, D. C., and R. G. LaFleur. 1995. Landfills in New York City: 1844–1994. *Ground Water* 33:556–60.

Walter, H. 1977. Climate. In: V. J. Chapman (ed.). *Wet Coastal Ecosystems*. Elsevier Scientific Publishing, Amsterdam, The Netherlands. 61–67.

Wang, Y., Y. Choi, Y-P. Hsieh, P. Gong, and L. Robinson. 2001. Sea level rise and carbon sequestration in coastal wetlands. In: *Book of Abstracts*. Seventh International Symposium on the Biogeochemistry of Wetlands. Duke University Wetland Center, Duke University, Durham, NC. 51.

Ward, K. M., J. C. Callaway, and J. B. Zedler. 2003. Episodic colonization of an intertidal mudflat by native cordgrass (*Spartina foliosa*) at Tijuana Estuary. *Estuaries* 26:116–30.

Ward, L. G., M. S. Kearney, and J. C. Stevenson. 1998. Variations in sedimentary environments and accretionary patterns in estuarine marshes undergoing rapid submergence, Chesapeake Bay. *Marine Geology* 151:111–34.

Waring, G. E. 1867. *Draining for Profit, and Draining for Health*. Orange Judd and Company, New York. Accessed online through Project Gutenberg at www.gutenberg.org/files/19465/19465-h/19465-h.html#toc61.

Warming, E. 1909. *Oecology of Plants: An Introduction to the Study of Plant-communities*. (Updated English version of a 1896 text.) Clarendon Press, Oxford, England.

Warren, R. S. 1995. Evolution and development of tidal marshes. In: G. D. Dreyer and W. A. Niering (eds.). Tidal Marshes of Long Island Sound: Ecology, History and Restoration. The Connecticut College Arboretum, New London, CT. Bulletin no. 34:17–21.

Warren, R. S., and P. E. Fell. 1995. Tidal wetland ecology of Long Island Sound. In: G. D. Dreyer and W. A. Niering (eds.). Tidal Marshes of Long Island Sound: Ecology, History and Restoration. The Connecticut College Arboretum, New London, CT. Bulletin no. 34. 22–41.

Warren, R. S., and W. A. Niering. 1993. Vegetation change on a Northeast tidal marsh: interaction of sea-level rise and marsh accretion. *Ecology* 74:96–103.

Warren, R., P. Fell, J. Grimsby, E. Buck, C. Rilling, and R. Fertik. 2001. Rates, patterns, and impacts of *Phragmites* expansion and effects of experimental *Phragmites* control on vegetation, macroinvertebrates, and fish within tidelands of the lower Connecticut River. *Estuaries* 24:90–107.

Warren, R. S., P. E. Fell, R. Rozsa, A. H. Brawley, A. C. Orsted, E. T. Olson, V. Swamy, and W. A. Niering. 2002. Salt marsh restoration in Connecticut: 20 years of science and management. *Restoration Ecology* 10:497–513.

Warren Pinnacle Consulting. 2010. SLAMM: Sea Level Affecting Marshes Model. http://warrenpinnacle.com/prof/SLAMM/.

Wass, M. S., and T. D. Wright. 1969. Coastal Wetlands of Virginia: Interim Report to the Governor and General Assembly. Virginia Institute of Marine Sciences, Gloucester Point, VA. VIMS Special Report in Applied Marine Science and Oceanographic Engineering 10.

Watson, E. B., and R. Byrne. 2009. Abundance and diversity of tidal marsh plants along the salinity gradient of the San Francisco Estuary: implications for global change ecology. *Plant Ecology* 205:113–28.

Wayne, C. J. 1976. The effects of sea and marsh grass on wave energy. *Coastal Research Notes* 4(7):6–8.

Webb, J. W., G. T. Tanner, and B. H. Koerth. 1981. Oil spill effects on smooth cordgrass in Galveston Bay, Texas. *Contribution in Marine Science* 24:107–14.

Webster, P. J., C. J. Holland, J. A. Curry, and H. R. Chang. 2005. Changes in tropical cyclone number, duration, and intensity in a warming environment. *Science* 309:1844–46.

Weinstein, M. P., and J. H. Balletto. 1999. Does the common reed, *Phragmites australis*, affect essential fish habitat? *Estuaries* 22:793–802.

Weinstein, M. P., and D. A. Kreeger (eds.). 2000. *Concepts and Controversies in Tidal Marsh Ecology*. Kluwer Academic Publishers, Dordrecht, The Netherlands.

Weinstein, M. P., K. R. Philipp, and P. Goodwin. 2000. Catastrophe, near-catastrophe and the bounds of expectations for macroscale marsh restoration. In: M. P. Weinstein and D. A. Kreeger (eds.). *Concepts and Controversies in Tidal Marsh Ecology*. Kluwer Academic Publishers, Dordrecht, The Netherlands. 777–804.

Weinstein, M. P., J. M. Teal, J. H. Balleto, and K. A. Strait. 2001. Restoration principles emerging from one of the world's largest tidal marsh restoration projects. *Wetlands Ecology and Management* 9:387–407.

Weis, J. S., and C. A. Butler. 2009. *Salt Marshes: A Natural and Unnatural History.* Rutgers University Press, New Brunswick, NJ.

Weldon, R. J., R. J. Burgette, and D. Schmidt. 2006. Along-Strike Variation in Locking on the Cascadian Subduction Zone, Oregon and Northern California. American Geophysical Union, Fall Meeting. Abstract no. T41A-1556.

Wellner, R. W., G. M. Ashley, and R. E. Sheridan. 1993. Seismic stratigraphic evidence for a submerged middle Wisconsin barrier: implication for sea-level history. *Geology* 21:109–12.

Wells, B. W. 1928. Plant communities of the coastal plain of North Carolina and their successional relations. *Ecology* 9:230–42.

Wells, E. D., and H. E. Hirvonen. 1988. Wetlands of Atlantic Canada. In: National Wetlands Working Group. *Wetlands of Canada.* Environment Canada, Sustainable Development Branch, Ottawa, and Polyscience Publications, Montreal, CN. Ecological Land Classification Series no. 24. Chapter 7; 249–303.

Wells, J. T., S. J. Chinburg, and J. M. Coleman. 1982. Development of the Atchafalaya River Deltas: Generic Analysis. Coastal Studies Institute, Center for Wetland Resources, Louisiana State University, Baton Rouge, LA. Prepared for the U.S. Army Engineers Waterways Experiment Station, Vicksburg, MS.

Wells, P. G. 1999. Environmental Impacts of Barriers on Rivers Entering the Bay of Fundy: Report of an Ad Hoc Environment Canada Working Group. Canada Wildlife Service, Dartmouth, NS. Technical Report Series 334.

Welsh, B. L., J. P. Herring, and L. M. Read. 1978. The effects of reduced wetlands and storage basins on the stability of a small Connecticut estuary. In: M. L. Wiley (ed.). *Estuarine Interactions.* Academic Press, New York. 381–401.

Werme, C. E. 1981. Resource partitioning in a salt marsh fish community. Boston University, Boston, MA. Ph.D. dissertation.

Westad, K. E., and E. Kiviat. 1986. Flora of freshwater tidal swamps at Tivoli Bays, Hudson River National Estuarine Sanctuary. In: J. C. Cooper (ed.). Polgar Fellowship Reports on the Hudson River National Estuarine Sanctuary Program, 1985. Hudson River Foundation, New York. III-1–20.

Weston, N. B., M. A. Vile, S. C. Neubauer, and D. J. Velinsky. 2011. Accelerated microbial organic matter mineralization following saltwater intrusion into tidal freshwater marsh soils. *Biogeochemistry* 102:135–51.

Wetland Studies and Solutions. 2010. Julie J. Metz Wetlands Bank, Woodbridge, Virginia (Prince William County). Gainesville, VA. www.mitigationbanking.org/PDFs/VA-WSSIJullieMetz.pdf.

Whelan, K. R. T., T. J. Smith, III, G. H. Anderson, and M. L. Ouellette. 2009. Hurricane Wilma's impact on overall soil elevation and zones within the soil profile in a mangrove forest. *Wetlands* 29:16–23.

Wherry, E. T. 1920. Plant distribution around salt marshes in relation to soil acidity. *Ecology* 1:42–48.

Whigham, D. F. 2009. Primary production in tidal freshwater wetlands. In: A. Barendregt, A., D. F. Whigham, and A. H. Baldwin (eds.). *Tidal Freshwater Wetlands.* Backhuys Publishers, Leiden, The Netherlands. Chapter 10; 115–22.

Whigham, D. P., and S. M. Nusser. 1990. The response of *Distichlis spicata* (L.) Greene and *Spartina patens* (Ait.) Muhl. to nitrogen fertilization in hydrologically altered wetlands. In: D. F. Whigham, R. E. Good, and J. Kvet (eds.). *Wetland Ecology and Management: Case Studies.* Kluwer Academic Publishers, Dordrecht, The Netherlands. 31–38.

Whigham, D. P., and R. L. Simpson. 1975. Ecological Studies of the Hamilton Marshes. Rider College, Biology Department, Trenton, NJ.

Whigham, D. P., and R. L. Simpson. 1976. The potential use of freshwater tidal marshes in the management of water quality in the Delaware River. In: J. Tourbier and R. W. Pierson, Jr. (eds.). *Biological Control of Water Pollution.* University of Pennsylvania Press, Philadelphia, PA. 173–86.

Whigham, D. P., T. E. Jordan, and J. Miklas. 1989. Biomass and resource allocation of *Typha angustifolia* L. (Typhaceae): the effect of within and between year variations in salinity. *Bulletin of the Torrey Botanical Club* 116:364–70.

Whitcraft, C. R., D. M. Talley, J. A. Crooks, J. Boland, and J. F. Gaskins. 2007. Invasion of tamarisk (*Tamarix* spp.) in a southern California salt marsh. *Biological Invasions*

9:875–79. www.csulb.edu/~cwhitcra/PDF/Biology.pdf.

White, D. A. 1993. Vascular plant development on mudflats in the Mississippi River Delta, Louisiana, USA. *Aquatic Botany* 45:171–94.

White, J. (ed.). 1913. *Handbook of Indians of Canada.* Tenth Report of the Geographic Board of Canada, Ottawa, CN. Appendix. http://faculty.marianopolis.edu/c.belanger/quebechistory/encyclopedia/barkuse.htm.

White, S. N., and M. Alber. 2009. Drought-associated shifts in *Spartina alterniflora* and *S. cynosuroides* in the Altamaha River estuary. *Wetlands* 29:215–24.

White, W. A., T. A. Tremblay, E. G. Wermund, Jr., and L. R. Handley. 1993. Trends and Status of Wetland and Aquatic Habitats in the Galveston Bay System, Texas. The Galveston Bay National Estuary Program. GBNEP-31.

White House Office of Environmental Policy. 1993. Protecting America's Wetlands—A Fair, Flexible, and Effective Approach. Washington, DC.

Whitlatch, R. B. 1982. The Ecology of New England Tidal Flats: A Community Profile. U.S. Fish and Wildlife Service, Washington, DC. FWS/OBS-81/01.

Wicklund, R. E., and K. K. Langmaid. 1953. Soil Survey of Southwestern New Brunswick. Canada Department of Agriculture and New Brunswick Department of Agriculture, Fredericton, NB.

Wiegert, R. G., and F. C. Evans. 1964. Primary production and the disappearance of dead vegetation on an old field in southeastern Michigan. *Ecology* 35:49–63.

Wiegert, R. G., and B. J. Freeman. 1990. Tidal Salt Marshes of the South Atlantic Coast: A Community Profile. U.S. Fish and Wildlife Service, Washington, DC. Biological Report 85 (7.29).

Wiegert, R. G., and L. R. Pomeroy. 1981. The salt-marsh ecosystem: a synthesis. In: L. R. Pomeroy and R. G. Wiegert (eds.). *The Ecology of a Salt Marsh.* Springer-Verlag, New York. 219–30.

Wiegert, R. G., L. R. Pomeroy, and W. J. Wiebe. 1981. Ecology of salt marshes: an introduction. In: L. R. Pomeroy and R. G. Wiegert (eds.). *The Ecology of a Salt Marsh.* Springer-Verlag, New York. 3–19.

Wieski, K., H. Guo, C. B. Craft, and S. C. Pennings. 2010. Ecosystem functions of tidal fresh, brackish, and salt marshes on the Georgia Coast. *Estuaries and Coasts* 33:161–69.

Wigand, C., R. A. McKinney, M. A. Charpentier, M. Chintala, and G. Thursby. 2003. Relationships of nitrogen loads, residential development, and physical characteristics with plant structures in New England salt marshes. *Estuaries* 26:1494–1504.

Wigand, C., R. A. McKinney, M. Chintala, M. A. Charpentier, and P. M. Groffman. 2004. Denitrification enzyme activity in fringe salt marshes in New England (USA). *Journal of Environmental Quality* 33:1144–51.

Wigand, C., R. McKinney, M. Chintala, S. Lussier, and J. Heltshe. 2010. Development of a reference coastal wetland set in southern New England (USA). *Ecological Monitoring and Assessment* 161:583–98.

Wigand, C., B. Carlisle, J. Smith, M. Carullo, D. Fillis, M. Charpentier, R. McKinney, R. Johnson, and J. Heltshe. 2011. Development and Validation of Rapid Assessment Indices of Condition for Coastal Tidal Wetlands in Southern New England, USA. *Environmental Monitoring and Assessment* 182:31–46.

Williams, A. B. 1965. Marine decapod crustaceans of the Carolinas. *Fishery Bulletin* 65:1–298.

Williams, G. D., and J. B. Zedler. 1999. Fish assemblage composition in constructed and natural tidal wetlands of San Diego Bay: relative influence of channel morphology and restoration history. *Estuaries* 22:702–16.

Williams, K. B. 2002. The Potential Wetland Restoration and Enhancement Site Identification Procedure: A Geographic Information System for Target Wetland Restoration and Enhancement. North Carolina Division of Coastal Management, Department of Environment and Natural Resources, Morehead City, NC. http://dcm2.enr.state.nc.us/Wetlands/RESTORATIONDOC.pdf.

Williams, K., M. V. Meads, and D. A. Sauerbrey. 1998. The roles of seedling salt tolerance and resprouting in forest zonation on the west coast of Florida, USA. *American Journal of Botany* 85:1745–52.

Williams, K., K. C. Ewel, R. P. Stumpf, F. E. Putz, and T. W. Workman. 1999a. Sea-level rise and coastal forest retreat on the West Coast of Florida, USA. *Ecology* 80:2045–63.

Williams, K., Z. S. Pinzon, R. P. Stumpf, and E. A. Raabe. 1999b. Sea-level Rise and Coastal Forests of the Gulf of Mexico. U.S. Geological Survey, Center for Coastal Geology, St. Petersburg, FL. Open-file Report 99-441. http://coastal.er.usgs.gov/wetlands/ofr99-441/OFR99-441.pdf.

Williams, P. & Associates, Ltd., and P. M. Faber. 2004. Design Guidelines for Tidal Wetland Restoration in San Francisco Bay. The Bay Institute and California State Coastal Conservancy, Oakland, CA. www.wrmp.org/design/Guidelines_Report-Final.pdf.

Williams, J. S., B. T. Gutierrez, J. G. Titus, S. K. Gill, D. R. Cahoon, E. R. Thieler, and K. E. Anderson. 2009. Sea-level rise and its effects on the coast. Chapter 1. In: Climate Change Science Program. *Coastal Sensitivity to Sea-level Rise: A Focus on the Mid-Atlantic Region. Synthesis and Assessment Product 4.1.* Report prepared by the Climate Change Science Program and the Subcommittee on Global Change Research. U.S. Environmental Protection Agency, Washington, DC. 11–24. http://downloads.climatescience.gov/sap/sap4-1/sap4-1-final-report-all.pdf.

Williams, P. B., and M. K. Orr. 2002. Physical evolution of restored breached levee salt marshes in the San Francisco Bay estuary. *Restoration Ecology* 10:527–42.

Willis, J. M., and M. W. Hester. 2004. Interactive effects of salinity, flooding, and soil type on *Panicum hemitomon. Wetlands* 24:43–50.

Willis, P. L., L. K. Blum, and P. L. Wiberg. 2005. Impact of Hurricane Isabel on Elevation and Vertical Accretion of a Virginia Salt Marsh. Poster abstract, Eastern Virginia State College. www.evsc.virginia.edu/~evscgsa/2005/abstracts/Willis.htm.

Wilson, B. C., C. A. Manlove, and C. G. Esslinger. 2002. North American Waterfowl Management Plan, Gulf Coast Joint Venture: Mississippi River Coastal Wetlands Initiative. North American Waterfowl Management Plan, Albuquerque, NM. www.gcjv.org/docs/MSRcoastpub.pdf.

Wilson, K. A. 1968. Fur production on southeastern coastal marshes. In: J. D. Newson (ed.). *Proceedings of the Marsh and Estuary Management Symposium;* 1967 July 19–20; Louisiana State University, Baton Rouge, LA. 149–62.

Windham, L., and R. G. Lathrop, Jr. 1999. Effects of *Phragmites australis* (common reed) invasion on aboveground biomass and soil properties in brackish tidal marsh of the Mullica River, New Jersey. *Estuaries* 22:927–35.

Winogrond, H. G., and E. Kiviat. 1997. Invasion of *Phragmites australis* in the tidal marshes of the Hudson River. In: W. C. Nieder and J. R. Waldman (eds.). Final Reports of the Tibor T. Polgar Fellowship Program 1996. Hudson River Foundation and New York State Department of Environmental Conservation, Hudson River National Estuarine Research Reserve, New York. 1–29.

Wipfli, M. S., J. P. Hudson, J. P. Caquette, and D. T. Chaloner. 2003. Marine subsidies in freshwater ecosystems: salmon carcasses increase the growth rate of stream-resident salmonids. *Transactions of the American Fisheries Society* 132:371–81.

Woerner, L. S., and C. T. Hackney. 1997. Distribution of *Juncus roemerianus* in North Carolina tidal marshes: the importance of physical and biotic variables. *Wetlands* 17:284–91.

Wolanski, E. 2007. *Estuarine Ecohydrology.* Elsevier, Amsterdam, The Netherlands.

Wolfe, P. E. 1977. *The Geology and Landscapes of New Jersey.* Crane, Russak, and Co., Division of Taylor and Francis, Bristol, PA.

Wood, F. J. 2001. The role of the lunar nodical cycle and heightened declination of the moon in prolonging periods of exceptionally high tides. In: F. J. Wood. Tidal Dynamics. Volume II: Extreme Tidal Peaks and Coastal Flooding. *Journal of Coastal Research* Special Issue 31. Chapter 5; 65–76.

Wood, G. W., M. T. Mengak, and M. Murphy. 1987. Ecological importance of feral ungulates at Shackleford Banks, North Carolina. *American Midland Naturalist* 118:236–44.

Wood, M. E., J. T. Kelley, and D. F. Belknap. 1989. Patterns of sediment accumulation in the tidal marshes of Maine. *Estuaries* 12:237–46.

Woodhouse, W. W., Jr., E. D. Seneca, and S. W. Broome. 1972. Marsh Building with Dredge Spoil in North Carolina. North Carolina Agricultural Experiment Station, Raleigh, NC. Bulletin 445.

Woodhouse, W. W., Jr., E. D. Seneca, and S. W. Broome. 1974. Propagation of *Spartina alterniflora* for Substrate Stabilization and Salt Marsh

Development. U.S. Army Corps of Engineers, Coastal Engineering Research Center, Ft. Belvoir, VA. Technical Memorandum no. 46.

Woodwell, G. M., D. E. Whitney, C. A. S. Hall, and R. A. Houghton. 1977. The Flax Pond ecosystem study: exchange of carbon in water between a salt marsh and Long Island Sound. *Limnology and Oceanography* 22:823–28.

Woodwell, S. R. J. 1985. Salinity and seed germination patterns in coastal plants. *Vegetatio* 61:223–29.

World Wildlife Fund. 1992. *Statewide Wetlands Strategies: A Guide to Protecting and Managing the Resources*. Island Press, Washington, DC.

Wright, L. D. 1978. River delta. In: R. A. Davis, Jr. (ed.). *Coastal Sedimentary Environments*. Springer-Verlag, New York. Chapter 1; 5–68.

Wu, J., D. M. Seliskar, and J. L. Gallagher. 1998. Stress tolerance in the marsh plant *Spartina patens:* impact of NaCl on growth and root plasma membrane lipid compositions. *Physiologia Plantarum* 102:307–17.

Yancey, R. K. 1964. Matches and marshes. In: J. P. Linduska (ed.). *Waterfowl Tomorrow*. U.S. Department of the Interior, Bureau of Sport Fisheries and Wildlife, Fish and Wildlife Service, Washington, DC. 619–26.

Yang, S. L. 1998. The role of *Scirpus* marsh in attenuation of hydrodynamics and retention of fine-grained sediment in the Yangtze Estuary. *Estuarine, Coastal and Shelf Science* 47:227–33.

Yang, S. L. 1999. Sedimentation on a growing intertidal island in the Yangtze River mouth. *Estuarine, Coastal and Shelf Science* 49:401–10.

Yanosky, T. M., C. R. Hupp, and C. T. Hackney. 1995. Chloride concentrations in growth rings of *Taxodium distichum* in a saltwater intruded estuary. *Ecological Applications* 5:785–82.

Yearicks, E. F., R. C. Wood, and W. S. Johnson. 1981. Hibernation of the northern diamondback terrapin, *Malaclemys terrapin terrapin*. *Estuaries* 4:78–80.

Young, R. F., and H. D. Phillips. 2002. Primary production required to support bottlenose dolphins in a salt marsh creek system. *Marine Mammal Science* 18:358–73.

Yozzo, D. J., and D. E. Smith. 1998. Competition and abundance of resident marsh-surface nekton: comparison between tidal freshwater and salt marshes in Virginia, USA. *Hydrobiologia* 362:9–19.

Yozza, D. J., and J. P. Titre. 1997. Coastal Wetland Restoration Bibliography. U.S. Army Corps of Engineers, Waterways Experiment Station, Vicksburg, MS. Wetland Research Program Technical Report. WRP-RE-20.

Zangger, E. 1991. Prehistoric coastal environments in Greece: the vanished landscapes of Dimini Bay and Lake Lerna. *Journal of Field Archeology* 18:1–15.

Zapryanova, N., and B. Atanassova. 2009. Effects of salt stress on growth and flowering of ornamental annual species. *Biotechnology and Biotechnological Equipment* 23:177–79.

Zedler, J. B. 1980. Algal mat productivity: comparisons in a salt marsh. *Estuaries and Coasts* 3:122–31.

Zedler, J. B. 1982. The Ecology of Southern California Coastal Salt Marshes: A Community Profile. U.S. Fish and Wildlife Service, Washington, DC. FWS/OBS-81/54.

Zedler, J. B. 1992. Restoring cordgrass marshes in southern California. In: G. W. Thayer (ed.). *Restoring the Nation's Marine Environment*. Maryland Sea Grant College, University of Maryland, College Park, MD. Chapter 1; 7–51.

Zedler, J. B. 1993. Canopy architecture of natural and planted cordgrass marshes: selecting habitat evaluation criteria. *Ecological Applications* 3:123–38.

Zedler, J. B. 1997. Adaptive management of coastal ecosystems to support endangered species. *Ecology Law Quarterly* 24:735–43.

Zedler, J. B. (ed.) 2000. *Handbook for Restoring Tidal Wetlands*. CRC Press, Boca Raton, FL. CRC Marine Science Series 25.

Zedler, J. B. 2006. Wetland restoration. In: D. P. Batzer and R. R. Sharitz (eds.). *Ecology of Freshwater and Estuarine Wetlands*. University of California Press, Berkeley, CA. Chapter 10; 348–406.

Zedler, J. B., and J. C. Callaway. 1999. Tracking wetland restoration: do mitigation sites follow desired trajectories? *Restoration Ecology* 7:69–73.

Zedler, J. B., and J. C. Callaway. 2000. Evaluating the progress of engineered tidal wetlands. *Ecological Engineering* 15:211–25.

Zedler, J. B., and J. C. Callaway. 2003. Adaptive restoration: A strategic approach for integrating research into restoration projects.

In: D. J. Rapport, W. L. Lasley, D. E. Rolston, N. O. Nielsen, C. O. Qualset, and A. B. Damania (eds.). *Managing for Healthy Ecosystems.* Lewis Publishers, Boca Raton, FL. 16–174.

Zedler, J. B., and S. Kercher. 2004. Causes and consequences of invasive plants in wetlands: opportunities, opportunists, and outcomes. *Critical Reviews in Plant Sciences* 23:431–52.

Zedler, J. B., and R. Lindig-Cisneros. 2000. Functional equivalency of restored and natural salt marshes. In: M. D. Weinstein and D. A. Kreeger (eds.). *Concepts and Controversies in Tidal Marsh Ecology.* Kluwer Academic Publishers, Dordrecht, The Netherlands. 565–82.

Zedler, J. B., and J. S. Nordby. 1986. *The Ecology of the Tijuana Estuary: An Estuarine Profile.* U.S. Fish and Wildlife Service, Washington, DC. Biological Report 85(7.5).

Zedler, J. B., and J. M. West. 2008. Declining diversity in natural and restored salt marshes: a 30-year study of Tijuana Estuary. *Restoration Ecology* 16:249–62.

Zedler, J. B., J. Covin, C. Nordby, P. Williams, and J. Boland. 1986. Catastrophic events reveal the dynamic nature of salt-marsh vegetation in southern California. *Estuaries* 9:75–80.

Zhang, K., B. C. Douglas, and S. P. Leatherman. 1997. East coast storm surges provide unique climate record. *Eos* 78:389, 396–97.

Zimmerman, R. J., T. J. Minello, and L. P. Rozas. 2000. Salt marsh linkages to productivity of penaeaid shrimps and blue crabs in the northern Gulf of Mexico. In: M. D. Weinstein and D. A. Kreeger (eds.). *Concepts and Controversies in Tidal Marsh Ecology.* Kluwer Academic Publishers, Dordrecht, The Netherlands. 293–314.

Index

Note: Although many plants and animals are referenced in this book, the index focuses on those where significant information is presented; plants are referenced by both common name and scientific name, while animals are listed by common name.

Ralph W. Tiner was born in Stuttgart, Germany, and grew up in Somerville, New Jersey. He received degrees from the University of Connecticut and Harvard University. He has directed the National Wetlands Inventory (NWI) Program in the Northeast for the U.S. Fish and Wildlife Service for over thirty-five years and has been an adjunct professor at the University of Massachusetts Amherst for more than two decades. Tiner has written over 250 publications on a variety of wetland topics including two field guides published by the University of Massachusetts Press. His *Field Guide to Nontidal Wetland Identification* (1988) won the Blue Pencil Award from the National Association of Government Communicators for outstanding government publication. *In Search of Swampland: A Wetland Sourcebook and Field Guide* (1998) was named one of the Best Science Books for Junior High and High School Readers in Energy, Environment, and Natural Resources by the American Association for the Advancement of Science. His "Geographically Isolated Wetlands of the United States" (2003) was recognized as one of the thirty most important papers for furthering the field of wetland science in the past thirty years by the Society of Wetland Scientists. Tiner lives with his wife, Barbara, and their children, Andrew, Avery, and Dillon, and Simon, his playful Westie, in Leverett, Massachusetts.